INBRED AND GENETICALLY DEFINED STRAINS OF LABORATORY ANIMALS

Part 1

Mouse and Rat

III

INBRED AND GENETICALLY DEFINED STRAINS OF LABORATORY ANIMALS

Part 1
Mouse and Rat

COMPILED AND EDITED BY

Philip L. Altman and Dorothy Dittmer Katz

Federation of American Societies for Experimental Biology

BETHESDA, MARYLAND

PRINTED IN THE UNITED STATES OF AMERICA

Library of Congress Catalog Card Number: 78-73555

International Standard Book Number: 0-913822-12-4

FOREWORD

In approaching the many complex problems of biology, the ability to restrict the biological variables affecting an experiment has led to major discoveries about fundamental mechanisms and to rapid developments in the biomedical sciences. Paramount among these experimental tools are the inbred and genetically defined animals. Their use has led to profound insights into the mechanisms of gene expression in mammals, tumor growth, tissue transplantation, host defense, responsiveness to drugs, and many other biological processes of basic scientific and medical importance.

The tremendous power that genetic insight gives to biological experiments has led to the rapidly increasing use of inbred and genetically defined laboratory animals. The difficulties in organizing the genetic information about the different strains and species, and in developing a perspective on the number and availability of such animals, have presented a formidable barrier to those who are not primarily geneticists. The objective in preparing *Part 1* and *Part 2* of this volume was to organize and present the information on inbred and genetically defined laboratory animals in such a way that the range and nature of the data would be immediately accessible to those who wish to use the information for experimental purposes. It is expected also that the volume will be a useful reference work for geneticists.

The spectrum of resources among inbred and genetically defined animals is quite broad and probably not adequately appreciated. The mouse is the prototype of such animals, and it has been the best studied and most extensively developed. The information on the rat is also fairly well developed, although it is not yet near the level attained by the mouse. It has some interesting differences from the mouse, and further studies on these differences may increase our understanding of the organization and expression of genetic information in mammals. The data for these species are summarized in *Part 1* of the volume. The hamster, guinea pig, rabbit, and chicken have been studied in varying, but lesser, detail, and the information about them is tabulated in *Part 2* of the volume.

The Editorial Board wishes to acknowledge a grant from the American Cancer Society which provided partial support for the preparation of this book.

6 November 1978
Pittsburgh, Pennsylvania

Thomas J. Gill III, M.D., *Chairman*
Biological Handbooks Editorial Board

BIOLOGICAL HANDBOOKS EDITORIAL BOARD

ADVISORY COMMITTEE ON INBRED AND GENETICALLY DEFINED STRAINS OF LABORATORY ANIMALS

FASEB PUBLICATIONS COMMITTEE

OFFICE OF BIOLOGICAL HANDBOOKS STAFF

CONTRIBUTORS AND REVIEWERS

ADAMS, RICHARD A.
Bio-Research Institute, Inc.
Cambridge, Massachusetts 02141

AIZAWA, M.
Hokkaido University School of Medicine
060 Sapporo, Japan

ALTMAN, NORMAN H.
Papanicolaou Cancer Research Institute
Miami, Florida 33136

ARTZT, KAREN
Memorial Sloan-Kettering Cancer Center
New York, New York 10021

BACON, LARRY D.
USDA, Regional Poultry Research Laboratory
East Lansing, Michigan 48823

BAILEY, DONALD W.
Jackson Laboratory
Bar Harbor, Maine 04609

BALDWIN, R. W.
University of Nottingham
Nottingham, NG7 2RD, England

BAZIN, HERVÉ
University of Louvain
1200 Brussels, Belgium

BEAMER, WESLEY G.
Jackson Laboratory
Bar Harbor, Maine 04609

BECKERS, ANDRÉE
University of Louvain
1200 Brussels, Belgium

BEECHEY, C. V.
MRC Radiobiology Unit
Harwell, Didcot, Oxon, OX11 0RD, England

BERNSTEIN, SELDON E.
Jackson Laboratory
Bar Harbor, Maine 04609

BOYSE, EDWARD A.
Memorial Sloan-Kettering Cancer Center
New York, New York 10021

BRDIČKA, RADIM
Charles University
Prague 2, Czechoslovakia

BURTONBOY, GUY
University of Louvain
1200 Brussels, Belgium

CARPENTER, CHARLES B.
Peter Bent Brigham Hospital
Boston, Massachusetts 02115

CHAPMAN, VERNE M.
Roswell Park Memorial Institute
Buffalo, New York 14263

CHUSED, THOMAS M.
NIH, National Institute of Dental Research
Bethesda, Maryland 20014

COHEN, BENNETT J.
University of Michigan Medical School
Ann Arbor, Michigan 48109

COHEN, CARL
University of Illinois Medical Center
Chicago, Illinois 60612

CRISPENS, CHARLES G., JR.
University of Alabama
Birmingham, Alabama 35294

CROW, JAMES F.
University of Wisconsin
Madison, Wisconsin 53706

CULLEN, SUSAN E.
Washington University School of Medicine
St. Louis, Missouri 63110

DAVID, CHELLA S.
Mayo Medical School
Rochester, Minnesota 55901

DE MAEYER, EDWARD
Institut du Radium
91405 Orsay, France

DÉMANT, PETER
Antoni van Leeuwenhoekhuis
108 Amsterdam-C, Netherlands

DERINGER, MARGARET K.
NCI, Carcinogenesis Branch
Bethesda, Maryland 20014

DeWITT, C. W.
University of Utah College of Medicine
Salt Lake City, Utah 84132

FESTING, MICHAEL F. W.
MRC Laboratory Animals Centre
Carshalton, Surrey, SM5 4EF, England

FINLAYSON, J. S.
FDA, Bureau of Biologics
Bethesda, Maryland 20014

FOREJT, JIŘI
Czechoslovak Academy of Sciences
Prague 4, Czechoslovakia

GARDNER, MURRAY B.
University of Southern California School of Medicine
Los Angeles, California 90033

GAY, VERNON L.
University of Pittsburgh School of Medicine
Pittsburgh, Pennsylvania 15261

GILL, THOMAS J., III
University of Pittsburgh School of Medicine
Pittsburgh, Pennsylvania 15261

GOODMAN, DAWN G.
NCI, Experimental Pathology Branch
Bethesda, Maryland 20014

GÖTZE, DIETRICH
Wistar Institute of Anatomy and Biology
Philadelphia, Pennsylvania 19104

GRABIN, MYRA E.
NCI, Tumor Pathology Branch
Bethesda, Maryland 20014

GRAFF, RALPH J.
Jewish Hospital of St. Louis
St. Louis, Missouri 63110

GREEN, MARGARET C.
Seely Road
Bar Harbor, Maine 04609

GÜNTHER, EBERHARD
Max Planck Institute of Immunobiology
Freiburg, German Federal Republic

GUTTMANN, RONALD D.
Royal Victoria Hospital
Montreal, Quebec, H3A 1A1, Canada

HANSEN, CARL T.
NIH, Division of Research Services
Bethesda, Maryland 20014

HARRINGTON, GORDON M.
University of Northern Iowa
Cedar Falls, Iowa 50613

HESTON, WALTER E.
1380 Burgundy Drive, S. W.
Fort Myers, Florida 33907

HILGERS, JO
Antoni van Leeuwenhoekhuis
108 Amsterdam-C, Netherlands
HOMBURGER, FREDDY
Bio-Research Institute, Inc.
Cambridge, Massachusetts 02141
HULETT, NANCY
University of Michigan Medical School
Ann Arbor, Michigan 48109

IVANYI, PAVOL
Central Laboratory of the Netherlands Red Cross Transfusion Service
Amsterdam, Netherlands

KALTER, HAROLD
Children's Hospital Medical Center
Cincinnati, Ohio 45229
KLEIN, JAN
Max Planck Institute for Biology
7400 Tubingen 1, Federal Republic of Germany
KUNZ, H. W.
University of Pittsburgh School of Medicine
Pittsburgh, Pennsylvania 15261

LANE, PRISCILLA W.
Jackson Laboratory
Bar Harbor, Maine 04609
LEHMAN, JOHN M.
University of Colorado Medical Center
Denver, Colorado 80262
LIEBELT, ANNABEL G.
Northeastern Ohio Universities College of Medicine
Kent, Ohio 44240
LIEBERMAN, ROSE
NIH, National Institute of Allergy and Infectious Diseases
Bethesda, Maryland 20014
LYON, M. F.
MRC Radiobiology Unit
Harwell, Didcot, Oxon, OX11 0RD, England

MARSH, R. F.
University of Wisconsin
Madison, Wisconsin 53706
MARSHALL, JOE T., JR.
National Museum of Natural History
Washington, D.C. 20560
MAYER, THOMAS C.
Rider College
Lawrenceville, New Jersey 08648
McKENZIE, IAN F. C.
University of Melbourne
Heidelberg, Victoria, Australia 3084
MELVOLD, ROGER W.
Shields Warren Radiation Laboratory
Boston, Massachusetts 02115
MICKOVÁ, MILADA
Institute of Molecular Genetics
Prague 4, Czechoslovakia
MORSE, HERBERT C.
NIH, National Institute of Allergy and Infectious Diseases
Bethesda, Maryland 20014
MOUTIER, R.
Centre de Sélection et d'élevage des Animaux de Laboratoire
45045 Orleans Cedex, France
MURPHY, DONAL B.
Stanford Medical Center
Palo Alto, California 94304
MURPHY, EDWIN D.
Jackson Laboratory
Bar Harbor, Maine 04609
MURPHY, GWENDOLYN
University of Pennsylvania School of Medicine
Philadelphia, Pennsylvania 19174
MURPHY, MICHAEL R.
NIH, National Institute of Mental Health
Bethesda, Maryland 20014
MYEROWITZ, RICHARD L.
Presbyterian-University Hospital
Pittsburgh, Pennsylvania 15213
MYERS, DAVID D.
Jackson Laboratory
Bar Harbor, Maine 04609

NESBITT, MURIEL N.
University of California at San Diego
La Jolla, California 92093
NEZLIN, ROALD
Institute of Molecular Biology
Moscow B-312, USSR

O'CONNELL, ROBERT J.
Worcester Foundation for Experimental Biology
Shrewsbury, Massachusetts 01545

PAIGEN, KENNETH
Roswell Park Memorial Institute
Buffalo, New York 14263
PASSMORE, HOWARD C.
Rutgers University
New Brunswick, New Jersey 08903
POOLE, TIMOTHY W.
University of Pennsylvania School of Medicine
Philadelphia, Pennsylvania 19174
POPP, RAYMOND A.
Oak Ridge National Laboratory
Oak Ridge, Tennessee 37830
POTTER, MICHAEL
NCI, Laboratory of Cell Biology
Bethesda, Maryland 20014
POTTER, TERRY
University of Melbourne
Heidelberg, Victoria, Australia 3084
PRESSMAN, DAVID
Roswell Park Memorial Institute
Buffalo, New York 14263

RISSER, REX
McArdle Laboratory for Cancer Research
Madison, Wisconsin 53706
ROBINSON, PETER J.
Northwestern University
Chicago, Illinois 60611
ROBINSON, ROY
St. Stephens Road Nursery, Ealing
London, W13, 8HB, England
ROSE, NOEL R.
Wayne State University School of Medicine
Detroit, Michigan 48201

SASAKI, MOTOMICHI
Hokkaido University
Sapporo, Japan
SEARLE, A. G.
MRC Radiobiology Unit
Harwell, Didcot, Oxon, OX11 0RD, England
SELANDER, ROBERT K.
University of Rochester
Rochester, New York 14627
SHEN, FUNG-WIN
Memorial Sloan-Kettering Cancer Center
New York, New York 10021
SHIMKIN, MICHAEL B.
University of California School of Medicine
La Jolla, California 92037
SHREFFLER, DONALD C.
Washington University School of Medicine
St. Louis, Missouri 63110
SHULTZ, LEONARD D.
Jackson Laboratory
Bar Harbor, Maine 04609

SILVERS, WILLYS K.
University of Pennsylvania School of Medicine
Philadelphia, Pennsylvania 19174

SIRACUSA, LINDA
Roswell Park Memorial Institute
Buffalo, New York 14263

SPROTT, RICHARD L.
Jackson Laboratory
Bar Harbor, Maine 04609

STAATS, JOAN
Jackson Laboratory
Bar Harbor, Maine 04609

ŠTARK, OTAKAR
Charles University
Prague 2, Czechoslovakia

STEVENS, LEROY C.
Jackson Laboratory
Bar Harbor, Maine 04609

TANIGAKI, NOBUYUKI
Roswell Park Memorial Institute
Buffalo, New York 14263

TAYLOR, BENJAMIN A.
Jackson Laboratory
Bar Harbor, Maine 04609

THEILER, K.
University of Zurich
CH-8006 Zurich, Switzerland

THOENES, GUNTHER H.
University of Munich
8000 Munich 2, German Federal Republic

TURTON, JON A.
Middlesex Hospital Medical School
London, W1P 7LD, England

WARD, JERROLD M.
NCI, Tumor Pathology Branch
Bethesda, Maryland 20014

WOMACK, JAMES E.
Texas A & M University
College Station, Texas 77843

YAMORI, YUKIO
Kyoto University
Sakyoku, Kyoto, Japan 606

YOSIDA, TOSHIDE H.
National Institute of Genetics
Misima, Sizuoka-ken, 411 Japan

CONTENTS

INTRODUCTION xiii

HOLDERS xv

INBRED ANIMALS

History, Uses, and Classification 1
Definition of Inbred Strains 4

I. MOUSE

Introductory Statements
Nomenclature and Rules for Mouse Genetics 9
Genealogy of the More Commonly Used Inbred Mouse Strains [*chart*] 16
Genealogies of Long-separated Sublines in Six Major Inbred Mouse Strains [*charts*] 18

Inbred and Genetically Defined Strains
1. Inbred Strains: Mouse 21
2. Congenic Lines: Mouse 29
3. Mutations on Inbred Backgrounds: Mouse 29
Part I. Maintained by The Jackson Laboratory 29
Part II. Maintained in the United Kingdom 36
4. Recombinant Inbred Lines: Mouse 37

Karyology
5. Linkage Map: Mouse 38
6. Chromosome Banding Patterns: Mouse [*figures*] 40
7. Chromosomal Aberrations: Mouse 41
Part I. Robertsonian Translocations Giving Metacentrics 41
Part II. Reciprocal Translocations 42
Part III. Inversions 44
8. Centromeric Heterochromatin Variations: Mouse 44

Development
9. Reproduction and Growth Characteristics: Mouse 45
Part I. Basic Data 45
Part II. Supplemental Information 47
10. Normal Embryo Characteristics: Mouse 50
11. Comparative Ages: Mouse and Other Mammals 52
12. Existing *T/t* Mutant Stocks: Mouse 53
Part I. Dominant Mutations 53
Part II. Recessive Mutations 54
13. Mutant Gene Effects: Mouse 55
Part I. Congenital Malformations 55
Part II. Lethal and Semilethal Genes 63
Part III. Lethal *t* Haplotypes 64
14. Mutant Genes Affecting Development of Anemia and Erythrocyte Production: Mouse 65
15. Mutant Genes Affecting Muscle Development: Mouse 66
16. Mutant Genes Affecting Development of the Immune System: Mouse 67

Epidermis
17. Mutant Genes Affecting Color: Mouse 70
18. Mutant Genes Affecting Skin and Hair: Mouse 74

Biochemical and Endocrine Mutants
19. Biochemical Variation: Mouse 77
Part I. Proteins Controlled and Tissue Distribution 77
Part II. Phenotypes and Strain Distribution 80

20. Biochemical Loci: Mouse 96
Part I. Map 96
Part II. References 98
21. Mutant Genes with Endocrine Effects: Mouse 101
22. Variants of Complement and Other Proteins: Mouse 103
Part I. Complement 103
Part II. Other Proteins 104
23. Hemoglobin Genes: Mouse 105
24. Urinary Proteins: Mouse 105
Immunology
25. Immunoglobulin Allotypes: Mouse 107
Part I. Immunoglobulin Classes and Corresponding Loci 107
Part II. Alleles of IgC_H Genes 108
Part III. Determinants Assigned to Specific IgC_H Loci 110
Part IV. Heavy Chain Linkage Groups in Inbred and Wild Mice 110
Part V. Origin and Characteristics of Allotype Congenic Strains 111
26. Strain Distribution of *If-1* Alleles: Mouse 112
27. Immune Response Genes, except *Ir-1*: Mouse 112
28. Cellular Alloantigens: Mouse 113
29. Cellular Alloantigens in Congenic Lines: Mouse 115
30. Lymphocyte Alloantigens: Mouse 116
31. Histocompatibility Systems, except *H-2*: Mouse 118
Part I. Standard Loci 118
Part II. Temporary Designations for New Loci 121
Part III. Histocompatibility Typing Results 121
32. Histocompatibility Linkage Map: Mouse 122
33. Standard Independent *H-2* Haplotypes and Markers of the *H-2* Gene Complex: Mouse 123
34. *H-2* Congenic Lines Carrying Standard Haplotypes: Mouse 124
35. Composition of Recombinant Haplotypes of the *H-2* Gene Complex: Mouse 125
36. *H-2* Congenic Lines Carrying Recombinant Haplotypes: Mouse 127
37. Mutant *H-2* Haplotypes of the *H-2* Gene Complex: Mouse 129
38. Wild-derived Haplotypes, Congenic Strains, and Specificities of the *H-2* Gene Complex: Mouse 130
Part I. B10.W Congenic Line Origins 130
Part II. *H-2* Chart of B10.W and *T/t* Strains of Dallas Group 130
Part III. *H-2* Chart of B10.W Strains of Prague Group 132
39. Maps of the *H-2* Gene Complex: Mouse 134
Part I. Genetic Fine Structure Map of *H-2* Complex 134
Part II. Recombinant Haplotypes Defining Discrete Regions and Subregions of *H-2* Gene Complex 135
Part III. Broad Classes of Functions Associated with Various Regions of *H-2* Gene Complex 135
40. Principal Genetic Traits Controlled by the *H-2-Tla* Gene Complex: Mouse 136
41. *H-2K, H-2D, H-2L,* and *H-2G* Specificities by Haplotype of the *H-2* Gene Complex: Mouse 137
Part I. *H-2* Chart 138
Part II. *H-2L* Chart 138
Part III. *H-2G* Chart 139
Part IV. Specificities of the *K* and *D* Regions 140
42. *Ss* Alleles and Properties in the *H-2* Gene Complex: Mouse 140
43. Traits of the *I* Region of the *H-2* Gene Complex: Mouse 141
44. *H-2* Gene Complex-associated Immune Responses: Mouse 142
45. Ia Specificities by Haplotypes of the *H-2* Gene Complex: Mouse 145
Part I. Definition of Ia Specificities 145
Part II. Distribution of Ia Specificities by Haplotypes 146
46. Summary of Molecular Properties of Gene Products of the *H-2* Complex: Mouse 147
Neurology
47. Neurologic Mutants: Mouse 149
48. Behavioral Gene Effects: Mouse 154
Part I. Single Locus Behavior 154
Part II. Mutations with Pleiotropic Behavior Effects 155

Tumors
49. Viruses and Incidence of Mammary Tumors: Mouse 157
Part I. Genes Controlling Induction of Endogenous Mammary Tumor Viruses 157
Part II. Genetic Resistance to Exogenous Mammary Tumor Viruses 158
Part III. Specific Genes Influencing Mammary Tumorigenesis 161
50. Pulmonary Adenomas: Mouse 161
Part I. Incidence of Spontaneous Tumors 161
Part II. Effect of Specific Genes or Linkage Groups on Tumor Incidence 162
Part III. Effect of Diet on Tumor Incidence 162
51. Hepatomas: Mouse 163
Part I. Incidence of Spontaneous Tumors 163
Part II. Effect of Genetic Factors on Tumor Incidence 164
Part III. Effect of Diet and Bedding on Tumor Incidence 164
52. Type C Leukemia Viruses: Mouse 165
Part I. Non-defective Murine Leukemia Viruses 165
Part II. Defective Murine Leukemia Viruses 166
Part III. Genetically Transmitted Type C Viruses or Viral Antigens 167
Part IV. Genes Affecting Murine Leukemogenesis or Leukemia Antigens 167
53. Strains with a High Spontaneous Incidence of Lymphoma and Leukemia: Mouse 168
54. Reticulum Cell Sarcomas: Mouse 169
55. Occurrence of Hemangiomas and Hemangioendotheliomas: Mouse 173
56. Occurrence of Adrenal Cortical Tumors: Mouse 178
57. Pituitary Tumors: Mouse 181
Part I. Spontaneous 181
Part II. Induced 181
58. Ovarian Tumors: Mouse 193
Part I. Spontaneous 194
Part II. Hormonal Effects 197
Part III. Other Induction Methods 199
59. Interstitial Cell Testicular Tumors: Mouse 206
60. Spontaneous Testicular Teratomas: Mouse 207
61. Intestinal Tumors: Mouse 208
Part I. Naturally Occurring Tumors 208
Part II. Experimentally Induced Tumors 209
62. Strain Incidence of Spontaneous Fibrosarcoma 210
Wild Mice
63. Classification of Genus *Mus* 212
Part I. List of Better-known Species and Principle Subspecies of the House Mouse and Allies . . . 212
Part II. Wild Mice of Eurasia [*figures*] 212
64. Polymorphisms: Wild Mouse 220
Part I. Electrophoretic Polymorphisms in Feral Populations of *Mus musculus* 220
Part II. Loci with Unusual or Rare Electrophoretic Alleles in Subspecies and Feral *Mus musculus* 225
Part III. Electrophoretic Differences Between *Mus musculus* and Other *Mus* Species 226
65. Type C RNA Viruses: Wild Mouse 227
Part I. Biological Characteristics 227
Part II. Relationship to Other Retroviruses 229

II. RAT

Introductory Statements
Rat Taxonomy 233
Origin of the Laboratory Rat 234
Suitability of the Rat for Different Investigations 236
Inbred and Genetically Defined Strains
66. Inbred Strains: Rat 238
67. Congenic Strains: Rat 255
68. Mutant Genes: Rat 257

69. Linkage Groups: Rat . . . 259
70. Recombinant Strains: Rat . . . 260
71. Chromosome Polymorphisms: Rat . . . 260
72. Karyology: Rat . . . 261
Part I. Modern Cytogenetic Techniques (Conventional Staining) . . . 261
Part II. Differential Staining Techniques [*figures*] . . . 266
Part III. Chromosome Numbering Systems: Conventional vs. Differential Staining Techniques . . 271
Morphological Traits
73. Coat and Eye Color Genes: Rat . . . 272
74. Length and Weight of Body Structures: Rat . . . 273
75. Aging: Rat . . . 277
Part I. Longevity . . . 277
Part II. Physiological and Biochemical Alterations . . . 279
Part III. Lesions . . . 289
76. Tumors: Rat . . . 298
Part I. Spontaneous . . . 298
Part II. Transplantable . . . 301
Part III. Radiation-Induced . . . 302
Part IV. Virally Induced . . . 302
Part V. Chemically Induced . . . 303
Biochemical Polymorphisms
77. Genetically Controlled Biochemical Variants: Rat . . . 305
78. Immunoglobulin Polymorphisms: Rat . . . 305
Part I. Nomenclature of Kappa-Chain Allotypic Alleles . . . 306
Part II. Allotype Distribution . . . 306
79. Blood Protein Polymorphisms: Rat . . . 308
80. Enzyme Polymorphisms: Rat . . . 309
Immunogenetics
81. Immunogenetic Characterization: Rat . . . 313
82. Alloantigenic Systems: Rat . . . 315
Part I. Tissue Distribution of Alloantigenic Alleles . . . 315
Part II. Strain Distribution of *Ag-B* 〈*H-1*〉 Haplotypes . . . 316
Part III. Strain Distribution of Other Alloantigenic Alleles . . . 317
83. Immunological Responsiveness: Rat . . . 317
84. Chemistry of Alloantigens: Rat . . . 319
85. Transplant Survival Times: Rat . . . 321
86. Allograft Survival: Rat . . . 324
87. Tissue and Organ Transplantation: Rat . . . 327
Part I. Graft-versus-Host Reaction . . . 327
Part II. Graft Survival Time . . . 328
Part III. Enhancement [*graph*] . . . 332
88. Immunologically Mediated Diseases: Rat . . . 334
89. Susceptibility to Infectious Diseases: Rat . . . 337
Regulatory Systems
90. Hormone Polymorphisms: Rat . . . 339
91. Reproductive Endocrinology: Rat . . . 340
92. Hypertension: Rat . . . 348
Part I. Strains and Origins of Hypertensive Rats . . . 348
Part II. Blood Pressure in Stroke-Prone and Stroke-Resistant Spontaneously Hypertensive Rats [*graph*] . . . 349
Part III. Incidence of Complications in Autopsied Spontaneously Hypertensive Rats [*graph*] . . . 350
93. Behavior Genetics: Rat . . . 350
Part I. Strains and Other Stocks Used in Behavioral Research . . . 351
Part II. Characteristics of Genetically Defined Strains . . . 354

INDEX . . . 363

INTRODUCTION

INBRED AND GENETICALLY DEFINED STRAINS OF LABORATORY ANIMALS is volume III in the new series of Biological Handbooks. This volume is divided into two parts with *Part 1* covering Mouse and Rat, and *Part 2* covering Hamster, Guinea Pig, Rabbit, and Chicken.

Contents and Review

Part 1 of this volume contains a short chapter on history, uses, classification, and definition of inbred animal strains; three introductory statements and 65 tables on the mouse; and three introductory statements and 28 tables on the rat. The contents of *Part 1* were authenticated by 106 outstanding experts in animal genetics. The review process to which the data were subjected was designed to eliminate, insofar as possible, material of questionable validity and errors of transcription.

Headnote

An explanatory headnote, serving as an introduction to the subject matter, may precede a table. More frequently, tables are prefaced by a short headnote containing such important information as units of measurement, abbreviations, definitions, and estimate of the range of variation. To interpret the data, it is essential to read the related headnote.

Exceptions

Occasionally, differences in values for the same specifications, certain inconsistencies in nomenclature, and some overlapping of coverage may occur among tables. These result, not from oversight or failure to choose between alternatives, but from a deliberate intent to respect the judgment and preferences of the individual contributors.

Conventions and Terminology

The main conventions used throughout this volume were adapted from the fourth edition of the *Council of Biology Editors Style Manual,* published in 1978 for the Council by the American Institute of Biological Sciences. Terminology was checked against *Webster's Third New International Dictionary,* published in 1961 by G. & C. Merriam Company.

Contributors and References

Appended to the tables are the names of the contributors, and a list of the literature citations arranged in alphabetical sequence. The reference abbreviations conform to those in the *Bibliographic Guide for Editors and Authors,* published by The American Chemical Society in 1974. References in some tables are to review articles rather than to the original papers from which the data were obtained. The objective was to conserve space while providing the user with the latest citation from which earlier references could be identified and retrieved.

Enzyme Nomenclature

Enzyme names and Enzyme Commission numbers were verified in *Enzyme Nomenclature,* the 1972 recommendations of the Commission on Biochemical Nomenclature, published by Elsevier Scientific Publishing Company for the International Union of Pure and Applied Chemistry and the International Union of Biochemistry.

Range of Variation

Values are generally presented as either the mean plus and minus the standard deviation, or the mean and the lower and upper limit of the range of individual values about the mean (either observed or statistical). Usually, it is of greater importance that the range be given rather than the mean. The several methods used to estimate the range—depending on the information available—are designated by the letters "a, b, c, or d" to identify the type of range in descending order of accuracy.

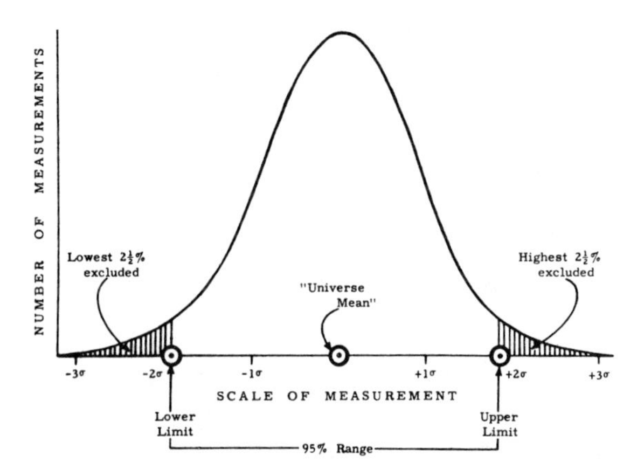

"a"—When the group of values is relatively large, a 95% range is derived by curve fitting. A recognized type of normal frequency curve is fitted to a group of measured values, and the extreme 2.5% of the area under the curve at each end is excluded (*see* illustration).

"b"—When the group of values is too small for curve fitting, as is usually the case, a 95% range is estimated by a simple statistical calculation. Assuming a normal symmetrical distribution, the standard deviation is multiplied by a factor of 2, then subtracted from and added to the mean to give the lower and upper range limits.

"c"—A less dependable, but commonly applied, procedure takes as range limits the lowest value and the highest value of the reported sample group of measurements. It underestimates the 95% range for small samples and overestimates for larger sample sizes, but where there is marked asymmetry in the position of the mean within the sample range, this method may be used in preference to the preceding one.

"d"—Another estimate of the lower and upper limits of the range of variation is based on the judgment of an individual experienced in measuring the quantity in question.

HOLDERS

With few exceptions, the abbreviations for HOLDERS of animal strains that appear in the tables that follow are from the three references cited at the end of this list.

Abbreviation	Holder
A	Antoni van Leeuwenhoekhuis The Netherlands Cancer Institute Sarphatistraat 108 Amsterdam-C, The Netherlands (O. Mühlbock)
ALB	Albert Einstein College of Medicine Department of Genetics Bronx, New York 10461
Ald	A. Léonard Laboratoire de Génétique Department de Radiobiologie Centre d'Etude de l'Energie Nucleaire Mol, Belgium
An	H. B. Andervont (retired) National Cancer Institute Bethesda, Maryland 20014
ANN	J. Klein Department of Oral Biology University of Michigan Ann Arbor, Michigan 48109
Ao AO	D. B. Amos Duke University School of Medicine Durham, North Carolina 27706
ARG	Argonne National Laboratory Division of Biological and Medical Research 9700 Cass Avenue Argonne, Illinois 60439
Bcr	University of Birmingham Cancer Research Laboratories The Medical School Birmingham B15 2TJ, England (June Marchant)
BELL	D. Bellamy Department of Zoology University College Cardiff, South Wales
Bg	Behavior Genetics Laboratory University of Connecticut Storrs, Connecticut 06268 (B. E. Ginsburg)
Bi	J. J. Bittner (deceased)
Bl	M. Bielschowsky Department of Tumor Biology University of Otago Medical School Great King Street Dunedin C.1, New Zealand
Bln	Deutsche Akademie der Wissenschaften zu Berlin Abteilung Versuchstiere Lindenberger Weg 70 1115 Berlin-Buch Democratic Republic of Germany
Blu BLU	Blue Spruce Farms, Inc. Gardner Road Altamont, New York 12009 (Hans K. Kappel)
Bn	S. E. Bernstein The Jackson Laboratory Bar Harbor, Maine 04609
BNL	L. K. Dahl (deceased)
Bnr	Department of Pathology University of Bonn/Rhein Federal Republic of Germany (A. Gropp)
Bom	Laboratory Animals Breeding and Research Center 8680 Ry Gl. Bomholtgård, Denmark (C. W. Friis)
Bon	Bonn University Zoolgisches Forschungsinstitut und Museum Alexander Koenig 53 Bonn 1 Federal Republic of Germany (E. von Lehmann)
Boy BOY	E. A. Boyse Sloan-Kettering Institute Division of Immunology New York, New York 10021
Br	G. M. Bonser (retired) Department of Experimental Pathology and Cancer Research School of Medicine Leeds 2, England

BRC	University of British Columbia Department of Medical Genetics 2075 Westbrook Place Vancouver, British Columbia V6T 1W5, Canada
Brh BRH	P. L. Broadhurst Department of Psychology University of Birmingham P.O. Box 363 Birmingham B15 2TT, England
BRN	Brown University Division of Biological and Medical Sciences Providence, Rhode Island 02912 (Herman B. Chase)
BU	Walter J. Burdette University of Texas M. D. Anderson Hospital and Tumor Institute Houston, Texas 77025
BUA	University of Birmingham Department of Anatomy The Medical School Birmingham B15 2TJ, England
By BY	Donald W. Bailey The Jackson Laboratory Bar Harbor, Maine 04609
Ca	T. C. Carter University of Edinburgh Edinburgh, Scotland
Cal CAL	California State University Biology Department Long Beach, California 90840
Cam CAM	University of Cambridge Genetics Department Milton Road Cambridge, England (M. E. Wallace)
Caw CAW	Carworth-Division of Charles River Breeding Labs, Inc. 251 Ballardvale Street Wilmington, Massachusetts 01887 (Henry L. Foster)
Cbi CBI	Chester Beatty Research Institute Institute of Cancer Research Fulham Road London S.W. 3, England (P. C. Koller)
Cd	Albert Claude Université Libre de Bruxelles Laboratoire de Cytologie et de Cancérologie Expérimentale Rue Héger-Bordet 1 1000 Bruxelles, Belgium
Ch	Herman B. Chase Department of Biology Brown University Providence, Rhode Island 02912
Chbb	Karl Thomae GmbH Department for Laboratory Animal Science 795 Biberach/Riss Federal Republic of Germany
CIN	H. Kalter Children's Hospital Research Foundation Cincinnati, Ohio 45229
Ckc	C. K. Chai The Jackson Laboratory Bar Harbor, Maine 04609
Cl	Ruth Clayton Institute of Animal Genetics West Mains Road Edinburgh 9, Scotland
CMML	Istituto di Clinica Medica I dell' Università di Milano Via Francesca Sforza 35 Milano, Italy (G. Bianchi)
CON	University of Connecticut Behavior Genetics Laboratory Storrs, Connecticut 06268 (John S. Cowen)
COR	Cornell University Medical College Department of Anatomy New York, New York 10021
CP	A. B. Chapman Department of Genetics University of Wisconsin Madison, Wisconsin 53706
Cpb CPB	Centraal Proefdierenbedrijf TNO P.O. B.167 Woudenbergseweg 25 NL-Zeist, The Netherlands (J. C. J. van Vliet)

Cr
CR
Division of Cancer Treatment
National Cancer Institute
Bethesda, Maryland 20014
(Samuel M. Poiley)

Crc
CRC
Clinical Research Centre
Northwick Park Hospital
Watford Road
Harrow, Middlesex HA1 3UJ, England
(C. Hetherington)

Crgl
Cancer Research Genetics Laboratory
University of California
Berkeley, California 94720

Crl
CRL
The Charles River Breeding Laboratories, Inc.
251 Ballardvale Street
Wilmington, Massachusetts 01887
(Henry Foster)

CSR
Charles Salt Research Centre
Orthopaedic Hospital
Oswestry, Shropshire SY10 7AG, England
(N. Nisbett)

Ct
Bruce M. Cattanach
Medical Research Council
Radiobiology Unit, Harwell
Didcot, Oxfordshire OX11 0RD, England

Cub
CUB
Charles University
Faculty of General Medicine
Department of Biology
Albertov 4
Prague 2, Czechoslovakia

Cum
CUM
Cumberland View Farms
Route 3
Clinton, Tennessee 37716
(James C. Kile, Jr.)

Cv
Verne M. Chapman
Roswell Park Memorial Institute
666 Elm Street
Buffalo, New York 14203

CWRP
Case Western Reserve University
Medical School
Department of Pathology
Cleveland, Ohio 44106
(S. Koletsky)

Cz
Gustavo Cudkowicz
Roswell Park Memorial Institute
666 Elm Street
Buffalo, New York 14203

Da
D. A. Davies
Searle Research Laboratories
High Wycombe
Buckinghamshire HP12 4HL, United Kingdom

DBW
Darcy B. Wilson
Department of Pathology
University of Pennsylvania
School of Medicine
Philadelphia, Pennsylvania 19174

De
Margaret K. Deringer (retired)
5528 Johnson Avenue
Bethesda, Maryland 20034

DET
University of Detroit
4001 W. McNichols Road
Detroit, Michigan 48221

Dg
C. P. Dagg
University of Alabama
College of General Studies
Department of Biology
Birmingham, Alabama 35233

Di
Margaret M. Dickie (deceased)

Dp
Giuseppe Della Porta
Section of Experimental Carcinogenesis
Istituto Nazionale per lo
Studio e la Cura dei Tumori
Via G. Venezian 1-20133
Milano, Italy

DR
Herman Druckrey
Laboratorium D. Chirurg. Univ. Klinik
Hugstetterstr. 55
Freiburg im Breisgau
Federal Republic of Germany

DU
Wilhelmina F. Dunning
Papanicolaou Cancer Research Institute
1445 N.W. 14th Street
Miami, Florida 33136

DUB
An Foras Taluntaes
(The Agricultural Institute)
Dunsinea, Castleknock
Co. Dublin, Ireland

EDI
Institute of Animal Genetics
West Mains Road
Edinburgh EH9 3JN, Scottland

Eg
EG
I. Egorov
Institute of General Genetics
USSR Academy of Sciences
Moscow B-133, USSR

Eh	J. Engelbreth-Holm (deceased)
Ei	Eva M. Ficher The Jackson Laboratory Bar Harbor, Maine 04609
Ep	Carlos Epper Radiobiology Laboratory Atomic Energy National Commission Avda. del Libertador Gen. San Martin 8250 Buenos Aires, Argentina
Fa	D. S. Falconer Institute of Animal Genetics West Mains Road Edinburgh EH9 3JN, Scotland (R. C. Roberts)
Fe	Elizabeth Fekete (retired) The Jackson Laboratory Bar Harbor, Maine 04609
Fg	Frank H. J. Figge University of Maryland School of Medicine Department of Anatomy Baltimore, Maryland 21201
Fn	Paul F. Fenton Brown University Biology Department Providence, Rhode Island 02912
Fo	P. Forsthoefel University of Detroit Biology Department 4001 W. McNichols Road Detroit, Michigan 48221
Fr	F. Clarke Fraser Department of Biology McGill University P.O. Box 6070 Montreal, Quebec, Canada
FRA	K. H. Degenhardt Institut für Humangenetik J. W. Goethe Universität Frankfurt/Main Federal Republic of Germany
Fs	Morris Foster University of Michigan Department of Zoology Ann Arbor, Michigan 48104
Fu	J. Furth Department of Pathology Francis Delafield Hospital 99 Fort Washington Avenue New York, New York 10032
G	Glaxo Laboratories, Ltd. Greenford, Middlesex, England
Gd GD	C. M. Goodall National Cancer Research Laboratories University of Otago Dunedin, New Zealand
Ge	H. N. Green (deceased)
Gf	Anna Goldfeder Cancer and Radiobiological Research Laboratory City of New York Department of Hospitals 99 Fort Washington Avenue New York, New York 10032
Gh	Douglas Grahn Argonne National Laboratory Division of Biological and Medical Research Argonne, Illinois 60439
Gif	Centre de Sélection et d'Elevage d'Animaux de Laboratoire C.N.B.S. 45 Orleans-la-Source, France
GLA	University of Glasgow Institute of Genetics Glasgow G11 5JS, Scotland (J. G. M. Shire)
Gn	E. L. and M. C. Green (retired) The Jackson Laboratory Bar Harbor, Maine 04609
Go	Peter Gorer (deceased)
Gr	H. Grüneberg (retired), Department of Zoology or G. Truslove, Department of Genetics University College London Gower Street London WC1, England
Grf	Ralph J. Graff Jewish Hospital of St. Louis Department of Surgery 216 South Kingshighway St. Louis, Missouri 63110
Gs	Ludwik Gross Cancer Research Laboratory Veterans Administration Hospital 130 W. Kingsbridge Road Bronx, New York 10468
GUE	University of Guelph Department of Biomedical Sciences Ontario Veterinary College Guelph, Ontario, Canada

Gw John W. Gowen (deceased)

H Medical Research Council
Radiobiology Unit, Harwell
Didcot, Oxfordshire OX11 0RD, England
(M. F. Lyon)

Ha T. S. Hauschka
Roswell Park Memorial Institute
Buffalo, New York 14203

HAL Martin-Luther-Universität Halle
Biologisches Institut des Bereiches Medizin
402 Halle/Saale
Universitätsplatz 7
Democratic Republic of Germany

HALLE Martin-Luther-Universität Halle-Wittenberg
Lehrstuhl für Industrietoxikologie
402 Halle (Saale) Leninallee 4
Democratic Republic of Germany
(E. Matthies or R. Schmidt)

Han
HAN Zentralinstitut für Versuchstierzucht
Lettrow-Vorbeck-Allee 57
D-3000 Hannover 91
Federal Republic of Germany
(W. Heine)

Har G. M. Harrington
University of Northern Iowa
Department of Psychology
Cedar Falls, Iowa 50613

HAR Medical Research Council
Radiobiology Unit, Harwell
Didcot, Oxfordshire OX11 0RD, England
(M. F. Lyon)

Hb H. Everett Hrubant
California State College at Long Beach
Department of Biology
Long Beach, California 90804

HBS Philip Harris Biological Ltd.
Oldmixon, Weston-Super-Mare
Avon BS24 9BJ, England

He W. E. Heston (retired)
National Cancer Institute
Bethesda, Maryland 20014

Hf Harold A. Hoffman
Laboratory of Biology
37/2D25 National Cancer Institute
Bethesda, Maryland 20014

HKM Hokkaido University School of Medicine
Institute for Animal Experiments
Sapporo, Japan

HLA Hilltop Lab Animals, Inc.
P.O. Box 195
Scottdale, Pennsylvania 15683
(James F. Miedel)

Hn F. K. Hoornbeek
University of New Hampshire
Zoology Department
Spaulding Building
Durham, New Hampshire 03824

HOK Hokkaido University
Chromosome Research Unit
Faculty of Science
North 10 West 8, Sapporo, Japan
(M. Sasaki)

How Alma Howard
Paterson Laboratories
Christie Hospital and Holt Radium Institute
Manchester 20, England

Hu Katharine P. Hummel (retired)
The Jackson Laboratory
Bar Harbor, Maine 04609

HV Children's Hospital Medical Center
Department of Neuroscience
Boston, Massachusetts 02115
(R. Sidman)

HVD Harvard Medical School
Department of Neuroscience
Children's Hospital Medical Center
Boston, Massachusetts 02115
(Richard J. Mullen)

Hz L. A. Herzenberg
Department of Genetics
Stanford University School of Medicine
Palo Alto, California 94304

IAP Institute of Animal Physiology
Agricultural Research Council
Department of Immunology
Babraham, Cambridge CB2 4AT, England

Icgn Institute of Cytology and Genetics
Siberian Branch of the Academy of Sciences
Novosibirsk 630090, USSR

ICO IFFA CREDO
Domaine des Oncins
Saint Germain sur l'Arbresle 69210, France
(A. Perrot)

Icr Institute for Cancer Research
7701 Burholme Avenue
Fox Chase, Philadelphia, Pennsylvania 19111

Icrc	Indian Cancer Research Centre Parel, Bombay 12, India (B. K. Batra)
Icrf ICRF	Imperial Cancer Research Fund Burtonhole Lane, Mill Hill London NW7 1AD, England (J. Craigie)
Iem	Institute for Experimental Medicine Department of Embryology USSR Academy of Medical Sciences Pavlov str. 12 Leningrad 197022, USSR (A. P. Dyban)
ILL	University of Illinois Department of Genetics and Development 515 Morrill Hall Urbana, Illinois 61801
IS	M. Ishibashi Laboratory of Animal Sciences Azabu Veterinary College Sagamichara, Kanagawa, Japan
J Jax JAX	The Jackson Laboratory Bar Harbor, Maine 04609
Jcl	Central Laboratory for Experimental Animals of Japan, Inc. Aobadai 2-20-14 Meguro-ku, Tokyo, Japan
Jd	A. Jurand Institute of Animal Genetics West Mains Road Edinburgh 9, Scotland
JSP	Japan Stroke Prevention Center Izumo, Japan 693 (Y. Yamori)
Ka	H. S. Kaplan Department of Radiology Stanford University School of Medicine Palo Alto, California 94304
KAN	University of Kansas Department of Physiology and Cell Biology Lawrence, Kansas 66045 (Mary Makepeace)
Kh	Henry I. Kohn Shields Warren Radiation Laboratory New England Deaconess Hospital 50 Binney Street Boston, Massachusetts 02115
Ki	Kirschbaum Memorial Laboratory Northeastern Ohio Universities College of Medicine 275 Martinel Drive Kent, Ohio 44240 (A. G. Liebelt)
KIEL	W. Müller-Ruchholtz Division of Immunology Hygiene-Institut der Universität Brunswikerstr. 2-6 D-2300 Kiel, Federal Republic of Germany
Kl KL	G. and E. Klein Department of Cell Research Karolinska Institutet Stockholm 60, Sweden
Klj KLJ	Jan Klein Department of Microbiology University of Texas Southwestern Medical School Dallas, Texas 75235
KNOX	W. E. Knox New England Deaconess Hospital 194 Pilgrim Road Boston, Massachusetts 02215
Ks	N. Kaliss The Jackson Laboratory Bar Harbor, Maine 04609
Kw	Department of Animal Genetics and Organic Evolution Jagiellonian University Krupnieza 50 Kraków 2, Poland (Halina Krzanowska)
L	Lilly Research Laboratories Eli Lilly & Co. Indianapolis, Indiana 46206 (Terence T. T. Yen)
Lac LAC	Laboratory Animals Centre Medical Research Council Woodmansterne Road Carshalton, Surrey SM5 4EF, England
Ld	J. Russell Lindsey University of Alabama in Birmingham Department of Comparative Medicine 1919 Seventh Avenue South Birmingham, Alabama 35233
Lds	University of Leeds School of Medicine Department of Pure and Applied Zoology Leeds LS2 9JT, England

Le Priscilla Lane
The Jackson Laboratory
Bar Harbor, Maine 04609

Lee Leeds University
School of Medicine
Laboratory Animals Department
Thoresby Place
Leeds LS2 9NL, England

LHR London Hospital Research Laboratories
Animal House
Ashfield Street
London E1 2AD, England

Li
LI Frank Lilly
Department of Genetics
Albert Einstein College of Medicine
Bronx, New York 10461

LIL Lilly Research Laboratories
Eli Lilly & Co.
Indianapolis, Indiana 46206

LMX University of London
The Middlesex Hospital Medical School
Department of Biology as
Applied to Medicine
Cleveland Street
London W1P 7PN, England

Ln J. B. Lyon
Department of Biochemistry
Emory University
Atlanta, Georgia 30322

LN R. Leyton
Profdierencentrum KUL
de Croylaan 34
B-3030 Heverlee-Leuven, Belgium

LUB Klinikum der Medizinische
Hochschule Lübeck
Abteilung für Pathologie
D-24 Lübeck, Federal Republic of Germany
(A. Gropp)

LUC Medical Research Council
Mammalian Development Unit
Wolfson House
University College London
4 Stephenson Way
London NW1 2HE, England
(G. M. Truslove)

Lw L. W. Law
National Cancer Institute
Bethesda, Maryland 20014

M Memorial Sloan-Kettering Cancer Center
New York, New York 10021

Mai
MAI Microbiological Associates
Building #1, Biggs Ford Road
Walkersville, Maryland 21793
(Wilbur L. Athey)

Man Stanley J. Mann
Temple University Health Sciences Center
Skin and Cancer Hospital of Philadelphia
3322 North Broad Street
Philadelphia, Pennsylvania 19140

MAS University of Massachusetts
Department of Zoology
Morrill Science Center
Amherst, Massachusetts 01002
(H. Rauch)

MAX Max-Planck-Institut für Immunobiologie
Stübeweg 51
D-78 Freiburg im Breisgau
Federal Republic of Germany

MCD Hugh O. McDevitt
Stanford University School of Medicine
Division of Immunology
Stanford, California 94305

Mcl Anne McLaren
Medical Research Council
Mammalian Development Unit
Wolfson House
University College London
4 Stephenson Way
London NW1 2HE, England

MCM McMaster University Medical Centre
Department of Neuroscience
1200 Main Street, W.
Hamilton, Ontario L8S 4J9, Canada
(Bruce Douglas)

MGM McGill University
Medical Clinics
Montreal, Quebec H3A 2B4, Canada
(J. B. Martin)

MIC University of Michigan
Mammalian Genetics Center
Department of Zoology
1250 N. Hospital Drive
Ann Arbor, Michigan 48104

MIS University of Missouri
Sinclair Comparative Medicine Research Farm
Route 3
Columbia, Missouri 65201

Mk
MK
Sajiro Makino
Zoological Instutute
Hokkaido University
Sapporo, Japan

Ml
James R. Miller
The University of British Columbia
Department of Medical Genetics
2075 Westbrook Place
Vancouver, British Columbia V6T 1W5
Canada

MOB
Larry Mobraaten
The Jackson Laboratory
Bar Harbor, Maine 04609

MOS
Histocompatibility Genetics Laboratory
Institute of General Genetics
USSR Academy of Sciences
Moscow B-133, USSR
(J. K. Egorov)

Ms
MS
National Institute of Genetics
Yata 1111
Misima, Sizuoka-ken, Japan
(Tosihide H. Yosida)

MSN
R. M. Mason
Charing Cross Hospital Medical School
Department of Biochemistry
Fulham Palace Road
London W6, England

Mtz
Carlos Martinez
Department of Physiology and Pediatrics
University of Minnesota Medical School
Rochester, Minnesota 55901

Mv
N. N. Medvedev
Gamaleya Institute of Epidemiology and
Microbiology
Schukinskaya 33
Moscow D-98, USSR

My
W. S. Murray (deceased)

N
National Institutes of Health
Small Animal Resources Section
Building 14A
Bethesda, Maryland 20014
(Carl Hansen)

NAG
Nagoya University
Research Institute of Environmental Medicine
Furo-cho, Chikusa-ku
Nagoya, Japan 464
(Sen-ichi Oda)

NCI
National Cancer Institute
Mammalian Genetics and Animal
Production Section
Drug Research and
Development, Chemotherapy
Bethesda, Maryland 20014

NCT
National Center for Toxicological Research
Jefferson, Arkansas 72079

Nga
Nagoya University School of Agriculture
Department of Animal Genetics
Furo-cho, Chikusa-ku
Nagoya-shi, Japan
(Kyoji Kondo)

NHA
University of New Hampshire
Zoology Department
Spaulding Building
Durham, New Hampshire 03824

NICHD
National Institute of Child
Health and Human Development
Bethesda, Maryland 20014
(D. W. Nebert)

NIDR
National Institute of Dental Research
Human Genetics Branch
Bethesda, Maryland 20014
(K. S. Brown)

NIH
National Institutes of Health
Veterinary Resources Branch
Division of Research Services
Bethesda, Maryland 20014

NIJ
University of Nijmegen
Department of Zoology
Toernooiveld, Nijmegen, The Netherlands

Nimr
National Institute for Medical Research
Mill Hill
London NW7 1AA, England

Nmg
University of Nijmegen
Department of Zoology
Toernooiveld, Nijmegen, The Netherlands
(J. H. F. van Abeelen)
or
Mouse Research Group
Genetics Laboratory
Driehuzerweg 200
Nijmegen, The Netherlands
(P. H. W. van der Kroon)

NMI
Northern Michigan University
Department of Biology
Marquette, Michigan 49855
(Frank A. Verley)

Not	Cancer Research Laboratory University Park Nottingham NG7 2RD, England (R. W. Baldwin)
NOVO	P. M. Borodin Institute of Cytology and Genetics Siberian Branch of the Academy of Sciences Novosibirsk 630090, USSR
NYB	State University of New York at Buffalo Medical Genetics Unit Department of Medicine Buffalo General Hospital Buffalo, New York 14203
NYH	New York State Department of Health Division of Laboratories and Research New Scotland Avenue Albany, New York 12201 (Lorraine Flaherty)
OAK	Oak Ridge National Laboratory Mammalian Genetics Section Biology Division P.O. Box Y Oak Ridge, Tennessee 37830 (W. L. and L. B. Russell)
OKJ	K. Okamoto Kinki University School of Medicine Osaka, Japan 589
OMR	University of Otago Medical School Department of Medicine Wellcome Medical Research Institute P.O. Box 913 Dunedin, New Zealand
Orl ORL	Centre de Sélection et d'Elevage d'Animaux de Laboratoire C.N.B.S. 45 Orleans-la-Source, France
Os	Osaka University Institute for Cancer Research Dojima-Hamadori, Fukushima-ku Osaka, Japan (Yasuyuki Akamatsu)
OSU	University of Otago Medical School Department of Surgery P.O. Box 913 Dunedin, New Zealand (Barbara F. Heslop)
OX	Oxford University Oxford, England (James Gowans)
OXF	Oxford University Cellular Immunology Unit Sir William Dunn School of Pathology South Parks Road Oxford OX1 3RE, England
Pas PAS	Institute Pasteur de Paris Unité de Génétique Cellulaire Département de Biologie Moléculaire 25 Rue du Docteur Roux 75015 Paris, France (Jean-Louis Guénet)
PAT	Paterson Laboratories Christie Hospital and Holt Radium Institute Manchester M20 9BX, England (A. Howard (and D. D. Porteous))
PFD	Proefdierencentrum de Croylaan 34 Katholieke Universiteit Leuven B-3030 Heverlee-Leuven, Belgium
Ph PH	Czechoslovak Academy of Sciences Institute of Experimental Biology and Genetics Budejovicka 1083 Prague 4, Czechoslovakia (Milan Hašek)
Pi	H. I. Pilgrim University of Utah College of Medicine Department of Surgery Salt Lake City, Utah 84112
PIT	University of Pittsburgh School of Medicine Department of Pathology, 716A Scaife Hall Pittsburgh, Pennsylvania 15261 (T. J. Gill III)
Pl PL	Samuel M. Poiley (retired) Mammalian Genetics and Animal Production Section, CCNSC National Cancer Institute Bethesda, Maryland 20014
PM	Joy Palm (deceased)
PP	Raymond and Diana Popp Biology Division Oak Ridge National Laboratory Oak Ridge, Tennessee 37830
PRA	Institute of Experimental Biology and Genetics Czechoslovak Academy of Sciences Budejovická 1083 124 20 Praha 4, Czechoslovakia

PSC	Skin and Cancer Hospital Temple University Health Sciences Center Philadelphia, Pennsylvania 19140 (Donald Forbes)
Psy	Institute of Psychiatry Animal Psychology Laboratory Bethlem Royal Hospital Monks Orchard Road Beckenham, Kent, England
QEH	The Queen Elizabeth Hospital Woodville, South Australia 5011
Rd	G. Rudali Foundation Curie 26 rue d'Ulm Paris 75005, France
Re	Elizabeth S. Russell The Jackson Laboratory Bar Harbor, Maine 04609
Rij RIJ	Radiobiological Institute TNO 151 Lange Kleiweg Rijswijk (Z. H.), The Netherlands (D. W. van Bekkum)
Rk	Thomas Roderick The Jackson Laboratory Bar Harbor, Maine 04609
Rl RL	W. L. and L. B. Russell Biology Division Oak Ridge National Laboratory P.O. Box Y Oak Ridge, Tennessee 37830
Rma	Università di Roma Istituto di Anatomia Comparata Via Borelli 50 I-00161 Roma, Italy (E. Capanna)
RMD	University of Rochester Medical Center Division of Genetics and Department of Radiation Biology and Biophysics Rochester, New York 14642
ROC	University of Rochester Biology Department River Campus Station Rochester, New York 14627
Rr	M. N. Runner University of Colorado Department of Biology Boulder, Colorado 80304
SATO	Sadako Sato Nippon Rat Co. Ltd. 608-3, Negishi Urawa Saitama 336, Japan
Sc	J. P. Scott Bowling Green State University Department of Psychology Bowling Green, Ohio 43402
Se	Lucio Severi Istituto di Anatomia e Istologia Patologica dell' Università degli Studi di Perugia Perugia, Italy
Sf SF	Donald C. Shreffler Department of Genetics Washington University School of Medicine St. Louis, Missouri 63110
Sg SG	Jack Stimpfling McLaughlin Research Institute Columbus Hospital Great Falls, Montana 59401
SHA	Mount Holyoke College South Hadley, Massachusetts 01075 (K. F. Stein)
Sim	Harry C. Simonsen Simonsen Laboratories, Inc. 1180C Day Road Gilroy, California 95020
SKI	Sloan-Kettering Institute for Cancer Research Laboratory of Developmental Genetics 1275 York Avenue New York, New York 10021
Sm	David Steinmuller Department of Pathology and Surgery University of Utah Salt Lake City, Utah 84112
Sn SN	G. D. Snell (retired) The Jackson Laboratory Bar Harbor, Maine 04609
Sp	William L. Simpson Detroit Institute of Cancer Research 4811 John R Street Detroit, Michigan 48201
Sr	Howard A. Schneider Institute of Nutrition University of North Carolina School of Medicine Chapel Hill, North Carolina 27514

Ss
SS
Willys K. Silvers
Department of Medical Genetics
University of Pennsylvania School of Medicine
Philadelphia, Pennsylvania 19104

St
L. C. Strong
Leonell C. Strong Research Foundation, Inc.
Del Mar, California 92014

STK
Stockholm University
Laboratory of Radiation Genetics
Institute of Genetics
Box 6801, S-113
86 Stockholm Va, Sweden
(W. Sheridan)

Sto
Breeding Farm Stolbovaya
Laboratory of Inbred Animals
Academy of Medical Science
Solyanka 14, Chehov Town
Moscow, USSR

Sv
L. C. Stevens
The Jackson Laboratory
Bar Harbor, Maine 04609

SYD
CSIRO, Division of Animal Genetics
Sydney P.O. Box 90
Epping 2121, New South Wales, Australia
(D. Sleeman)

T
University of Toronto
Department of Zoology
Toronto 5, Canada
(L. Butler)

TBI
Tokyo Biochemical Research Institute
41-8 Takada 3
Toshima, Tokyo, Japan

TOR
S. Torres
Instituto Nacional do Cancer
Secao de Pesquisas
Pca, da Druz Vermelha 23
Rio de Janeiro, Guanabara, Brazil

Tw
Noboru Takasugi
Zoological Institute Faculty of Science
University of Tokyo
Hongo, Bunkyo-ku
Tokyo, Japan

TXA
University of Texas Medical School
Anatomy Department
San Antonio, Texas 78284

Umc
University of Minnesota Medical School
Department of Physiology
Minneapolis, Minnesota 55455
(June Smith)

Un
A. C. Upton
Biology Division
Oak Ridge National Laboratory
P.O. Box Y
Oak Ridge, Tennessee 37831

Up
Delta E. Uphoff
National Cancer Institute
Bethesda, Maryland 20014

W
Instytut Onkologii im.
Marii Sklowdowskiej-Curie
Zaklad Biologii Nowotworow Warszawa
ul. Wawelska 15, Poland
(Kazimierz Dux and Alina Czarnomska)

WAG
Wageningen Agricultural University
Department of Genetics
Gen. Foulkesweg 53
Wageninen, The Netherlands

We
J. A. Weir
University of Kansas
Department of Zoology
Lawrence, Kansas 66044

Wf
George L. Wolff
National Center for Toxicological Research
Division of Special Studies
Jefferson, Arkansas 72079

Wh
Beverly J. White
Laboratory of Experimental Pathology
National Institute of Arthritis,
Metabolism and Digestive Diseases
National Institutes of Health
Bethesda, Maryland 20014

Wi
J. W. Wilson (deceased)

WIS
Wistar Institute of Anatomy and Biology
36th St. at Spruce
Philadelphia, Pennsylvania 19104

Wl
H. G. Wolfe
Department of Physiology and Cell Biology
University of Kansas
Lawrence, Kansas 66045

Wn
Weizmann Institute of Science
Laboratory Animals Breeding Center
Rehovot, Israel
(A. Meshorer)

Wsl
WSL
Universite Catholique de Louvain
Ecole de Sante Publique
Av. Chapelle aux Champs 4
1200 Bruxelles, Belgium
(H. Bazin)

Wt	Wesley K. Whitten The Jackson Laboratory Bar Harbor, Maine 04609	Y YUR	Yurlovo Department of Genetics Laboratory of Experimental Animals Yurlovo Post Office, Khimki Town Moscow Region, USSR (A. M. Madashenko, Z. K. Blandova, or V. P. Krishkina)
Ww	E. F. Woodworth The Jackson Laboratory Bar Harbor, Maine 04609	ZBZ	Universität Zürich Biologisches Zentrallaboratorium Kantonsspital Ramistr. 100 Zürich, Switzerland
Wy	G. W. Woolley Division of General Medical Sciences National Institutes of Health Bethesda, Maryland 20014	ZUR	Anatomisches Institut der Universität Gloriastrasse 19 8006 Zürich, Switzerland

References

1. Festing, M. F. W., ed. 1975. International Index of Laboratory Animals. Ed. 3. Medical Research Council Laboratory Animals Centre, Carshalton, Surrey SM5 4EF, United Kingdom.
2. National Research Council, Institute of Laboratory Animal Resources. 1975. Animals for Research. Ed. 9 (Rev.). National Academy of Sciences, Washington, DC 20418.
3. Staats, J. 1976. Standardized Nomenclature for Inbred Strains of Mice: Sixth Listing. Cancer Res. 36(Dec.):4333-4377.

INBRED AND GENETICALLY DEFINED STRAINS OF LABORATORY ANIMALS

Part 1

Mouse and Rat

INBRED ANIMALS

HISTORY, USES, AND CLASSIFICATION

The investigator requiring animals for research can choose his subjects from a large number of species. These include a variety of domestic animals (ranging from cattle, swine, and chickens, to dogs and cats), non-human primates (such as monkeys, chimpanzees, baboons, and marmosets), and the small laboratory mammals. The latter are by far the most commonly used.

In addition to selecting the ideal organism for the experiment, the investigator must also choose the genetic state that makes the model a valid representation of the target population. In experiments using mice, rats, hamsters, guinea pigs, rabbits, dogs, or chickens, the research worker's options are random-bred, inbred, F_1 hybrids, and haphazardly bred animals. In this volume, the inbred or genetically characterized animal is presented as the research tool of choice.

Historical Development of Inbred Lines

The great variability of living organisms accounts in part for the fact that biology has lagged far behind physics and chemistry in precision. The ideal scientific method enables the investigator to distinguish the variables that play roles in a particular phenomenon, and to reduce to one the number of variables being tested. With the development of measurement standards and the preparation of reagent and analytical grade chemicals, repetition of experiments in various laboratories, under essentially defined conditions, became possible. Such experiments were the foundations on which physics and chemistry have been built.

The development of lines of inbred animals can be considered an equivalent achievement in the biological and biomedical sciences. The "reagent grade" animal, the result of carefully controlled inbreeding, now permits repetition of experiments requiring fine discrimination within an animal species.

The inbred line is a population of animals that has attained homozygosis at nearly every locus through the use of a mating system that reduces the number of genetically unlike ancestors. The most common practice is rigid brother by sister mating over many generations, although other systems may be used (*see* Definition of Inbred Strains, p. 4). Thus, all members of an inbred line have essentially the same genetic constitution.

The history of the inbred laboratory mammal is actually the history of the use of the mouse in cancer studies and the guinea pig in studies of multifactorial inheritance. The rediscovery of Mendel's work in 1900 led to the reexamination of cancer as an inherited disease. Studies of human families over successive generations disclosed that the occurrence of cancer was more frequent in some families than in others. Although common environmental conditions might well have been implicated, it apparently was intellectually more satisfying at the turn of the century to seek possible genetic factors as an explanation for such incidences. Unfortunately, there was little evidence that the mode of cancer inheritance conformed to any of the Mendelian patterns.

The observed inconsistencies stimulated the use of animal systems in which uniform or controlled conditions could be applied. The initial experiments involved spontaneous cancer in breeds of mice raised by fanciers. Attempts by Jensen, Ehrlich, Tyzzer, Loeb, and others to transplant these tumors to other mice produced some successes and many failures. The variance in results was explained when Little and Tyzzer found that several loci were involved in transplanting tumors between two mouse populations. Partly in response to this finding, Wright, who had a particular interest in the mathematics of genetics, developed the theoretical expression of the manner in which genes segregate during the inbreeding process.

In order to attain repeatable and controlled systems for studying factors affecting tumor transplantation and for providing insight into why cancer develops, Little began to inbreed mice to obtain the necessary genetic homogeneity. His inbreeding originated in a line that had been maintained for coat color studies since 1909. By the early 1920's, a number of other lines were developed by Little, Murray, Strong, Bagg, and Lynch.

Many of the early inbred strains of mice originated from a small number of stocks (*see* Genealogy of the More Commonly Used Inbred Mouse Strains, p. 16). This relatively restricted gene pool accounts for the similarities and differences in the classic inbred lines. The history of each inbred mouse line is included in the listing of Standardized Nomenclature of Inbred Strains of Mice published regularly in *Cancer Research.*

The original lines of inbred guinea pigs were started by Wright who published an important series of papers dealing with the effect of inbreeding and crossbreeding on general characteristics such as vigor and reproductive effectiveness (*see* Origin of the Domestic Guinea Pig and of Inbred Strains. Part 2 of this volume). More importantly, his work on the inbreeding of the guinea pig led to an understanding of some of the complexities of multiple factor inheritance. Wright's scientific studies also gave rise to the formalization of a method of analysis of breeding systems, called "path analysis," which has been most useful in estimating levels of in-

continued

breeding. The coefficient of inbreeding is the calculated probability that two allelic genes united in a zygote are both descended from the same gene found in an ancestor common to both parents (*see* Definition of Inbred Strains, p. 4).

Philosophy Underlying Use of Inbred Animals

An inbred line consists of a population of great genetic homogeneity, and experiments performed within such a population can test the treatment variability which in the usual, haphazardly bred animal might be confused with genetic variability. Thus, homogeneity of animal stock allows for more sophisticated experiments.

Although the inbred mouse was not originally developed for the purpose of improving experimental design, the use of specifically named and defined animals has served this purpose extremely well. New inbred lines permitted comparison of the characteristics of individual inbred strains within a species. Characterization of each of the several inbred lines also demonstrated that each line might have unique properties because of the random fixation of the genome. The fixed genome of one strain did not resemble that of another strain, and gene action during development and during the life-span, after endogenous or exogenous stimulation, resulted in differences in phenotypic expression.

The use of inbred animals permits (i) the testing of the influences of a specific treatment or manipulation on the response of the animal, with the potential for repetition of the experiments at another time or in another laboratory; and (ii) the exploitation of the unique properties of a particular strain obtained by inbreeding or by selection during the inbreeding process. The concept of uniqueness also includes the introduction of specific alleles into an inbred background to provide a line specifically created for particular studies.

The number of inbred lines and sublines available for experimental use is constantly increasing as new lines are created by inbreeding, by genetic manipulations to create a specifically characterized, genetically homogeneous animal, or by variations discovered within sublines derived from the original parental stock. The qualities of the inbred animal—namely, genetic homogeneity, defined responsiveness to certain stimuli, and the possession of unique behavioral, biochemical, or developmental attributes—allow the investigator to carry out experiments with a much-reduced number of variables. Great precision with small numbers of animals thus becomes possible. One attribute that is lost, of course, is genetic variability, and the investigator who fails to realize that inbred animals may be missing some of the characteristics essential to his research could be in difficulty. He may select a line popular for cancer research but one which is totally unresponsive to the biochemical or behavioral treatment under investigation.

Experiments Requiring Use of Genetically Defined Stocks

Although the well-known inbred lines of mice and the fewer lines of inbred rats, hamsters, guinea pigs, rabbits, and chickens have been available for a long time, it is only in the last 15-20 years that investigators without extensive genetic training have used inbred strains rather than face possible criticism for not having done so. In many instances, both the investigator and critic fail to recognize that the purpose of the experiment may not require the more expensive, inbred animal. Some investigations may even lose their meaningfulness if an inbred animal is used. For example, such an animal is obviously inappropriate for experiments designed to exploit the variability present in a population.

Research requiring laboratory animals falls into three major categories: (i) bioassay, (ii) studies of the attributes of the specific animal or species, and (iii) studies in which the animal serves as the model for another species.

For the bioassay, the ideal animal is one in which the genetic variance is reduced so that the effect of the test substance on the animal can be attributed to the dose, route, or exposure time. The number of animals used can be restricted to the number necessary to give statistically meaningful results, especially if a pilot experiment was performed to establish the variance for the test. The investigator should not assume, however, that the marked reduction in genotypic variance achieved by inbreeding will necessarily eliminate all phenotypic variance. Some experimental evidence suggests that homozygous individuals are less well buffered against minor environmental agents, and that inbred animals may be no more uniform in response than random-bred animals. Use of the F_1 hybrid from two inbred strains, however, retains the advantage of genetic uniformity while adding superior development and physiological homeostasis. The cost of the hybrid also is generally lower than that of pure strains.

When the animal species itself is the target of investigation—either for the development of concepts or for investigation of the animal's attributes—the selection of the strain, strain combination, or substrain should be based on the characteristics necessary to answer the scientific question under consideration. The well-defined inbred strain is probably the most suitable for this kind of study.

The object of the third type of investigation is to discover models for the exploration of human problems, particularly those in which the nature of the knowledge sought involves a treatment or manipulation that for ethical or legal reasons cannot be carried out in man. Since the genetic variability and resultant physiologic differences in man are the products of a haphazard breeding system based on geographic, cultural, socio-economic and other influences, and since an important consideration is that the study be based on population characteristics, the non-inbred is the most desirable animal for such research.

An exact animal model which retains the characteristic of human genetic variability cannot be achieved, but there are several types of heterozygous populations available from which to select the one most closely approximating the extent of variability for the specified line of investigation. The population most commonly used, but in many ways the least suitable, is the so-called random-bred animal, such as the Swiss mouse or the commercial rabbit, rat, or

continued

guinea pig. These populations have an unknown amount of genetic variability, and frequently have much more restricted variability than can be assumed from the term "random-bred."

A wild population or a mixed multi-strain population is useful when a truly heterogenous population is desired. These animals are suitable for the study of population dynamics, but they are not convenient for the separation of genetic from environmental factors.

An interesting population, not commonly available, is one made up of F_1 offspring from many distantly related, highly inbred strains in many combinations of crossbreeding. Although this population is quite different from the human population, it allows the investigator to retain genetic control, as well as to reproduce the identical, defined population. It has the further great advantage of mimicking the genetic diversity of man's genotype, including heterozygosis and its concomitant variability and epistatically acting genotypes.

Since a great many different kinds of experimental animals are now available, the experimental design must include consideration of the best choice of animal. The value of the research results ultimately may be dependent on the selection of the experimental animal. It is essential, therefore, that the investigator be deliberate in his choice of animal and be cognizant of the reputation of the supplier. Fortunately, many sources of laboratory animals use breeding systems that maintain either the homogeneity of the inbred animal or the heterogeneity of the random-bred animal.

Nomenclature

More than 300 strains of inbred lines of mammals are available today; over 250 of these are mouse strains. The nomenclature for all inbred strains reflects both the independent development of the strains and the rugged individualism of the investigators who attach a designation to their particular inbred creation.

Efforts, started in 1932, to achieve a standardized nomenclature for the inbred mouse, have resulted in some degree of uniformity and orderliness (*see* Nomenclature and Rules for Mouse Genetics, p. 9). Wide variety in names and symbols still persists, however, and it is essential that an investigator recognize and understand the symbols designating the specific strain or substrain with which he is working.

There is a standing committee for the nomenclature of the mouse, and regular publication of a standardized nomenclature. The most recent compendium appears in *Cancer Research* 36:4333, 1976, and includes the rules governing the committee's judgments. No other species except the rat has had the benefit of committee action and symbol designation.

A key rule set forth by the committee on mouse nomenclature pertains to the definition of an inbred strain: For a strain to be considered inbred, it must have had the equivalent of 20 generations of brother by sister mating. This rule is rigidly applied in the designation of the mouse, but, because of operational factors, may need to be modified for other species.

In some instances, the designation "inbred" indicates only reduced genetic variability within the strain. Such use of the term generally is in connection with (i) species having lines which represent fewer than 20 inbred generations, but in which skin grafts may be exchanged between all members of the designated line; and (ii) species with long intervals between generations, but to which the term "inbred" may be applied after only a few generations of brother by sister mating.

Other important rules which apply to mouse nomenclature include the designation for substrains and the factors which define a substrain, and the designation for animals having an unusual derivation or experimental background (congenic stocks, foster nursing, ova or ovary transplantation, complex crosses, or embryos preserved by freezing).

It is strongly recommended that the most recent publication of the list of inbred strains be consulted so that an investigator using an inbred mouse fully understands the development and the qualities of the animal used. Although the inbred mouse is referred to as a "reagent," each strain has a long and interesting history which merits attention.

Contributor: Carl Cohen

Inbred strains of experimental animals have come into widespread use in research because of their increased genetic uniformity and constancy. These highly valued attributes for experimental strategy in the biomedical and agricultural sciences are gained through the genetic effects of inbreeding.

Inbreeding is the mating of individuals that have at least one ancestor in common. Any one gene from the common ancestor has a chance of being transmitted through both parents, so the resulting offspring will have two homologous alleles identical by descent. In this way the genetic homogeneity in inbred individuals will tend to increase, and progressively so if inbreeding is continuously applied generation after generation.

The *inbreeding coefficient*, F, is a useful theoretical measure of the progress of inbreeding. It is defined as the probability that both alleles at a locus are identical by descent. It therefore indicates the proportionate decrease of heterozygous loci in the inbred individual relative to those in a representative individual of the starting population.

F increases at different rates, depending on the amount of ancestry shared by the systematically mated individuals. Repeated backcrossing to an already existing, highly inbred strain, as used in constructing the congenic lines discussed later, is the most intensive inbreeding attainable in vertebrates. If there is no existing inbred strain, then brother by sister, or younger parent by offspring, is the most intense type of inbreeding. Brother by sister mating is the regimen conveniently applied to laboratory mammals; regimens of more distant relationships have been found more practicable for experimental farm animals.

The changing values of F in three different mating regimens are presented in Table 1. In all three, F theoretically approaches but never reaches unity. In practice, however, inbreeding gains will tend to level off even before unity, due to the countering effect of new mutations. This reservoir of mutations, although very small, prevents the highly inbred strain from ever being genetically constant through time, and it permits separately maintained sublines of the strain to become genetically divergent because of the genetic fixation of mutant alleles at different loci. Long-separated sublines of inbred strains therefore should always be appropriately identified by the conventional symbols to avoid confusion in comparing research results that would otherwise be taken without question as applying to the strain in general.

F is a theoretical value calculated from a pedigree, and not only does it ignore mutational effects, it also ignores effects of selection favoring heterozygotes. For a species in which inbreeding is difficult, i.e., when many matings fail to produce viable offspring, one can assume that selection has favored heterozygotes. The calculated F in such cases is an over-estimate.

Table 1. **Inbreeding Coefficients Under Three Systems of Close Inbreeding**

Data were modified from reference 2.

Generation	F		
	Repeated Backcrosses to Highly Inbred Line 1/	Full Brother x Sister, or Offspring x Younger Parent	Half Sib 2/
0	0	0	0
1	0.500	0.250	0.125
2	0.750	0.375	0.219
3	0.875	0.500	0.305
4	0.938	0.594	0.381
5	0.969	0.672	0.449
6	0.984	0.734	0.509
7	0.992	0.785	0.563
8	0.996	0.826	0.611
9	0.998	0.859	0.654
10	0.999	0.886	0.691
11	1	0.908	0.725
12	1	0.926	0.755
13	1	0.940	0.782
14	1	0.951	0.806
15	1	0.961	0.827
16	1	0.968	0.846
17	1	0.974	0.863
18	1	0.979	0.878
19	1	0.983	0.891
20	1	0.986	0.903

1/ Generation 0 is equivalent to generation N_1 in Figure 1. 2/ Females are half sisters to males.

continued

The ultimate objective of inbreeding is to attain uniformity and constancy of genotype. Theoretical probabilities of attaining three different aspects of this objective by brother by sister mating are presented in Table 2. One aspect is *homozygosity* [ref. 4]—the homozygous condition *an individual* attains at a locus that was heterozygous at the outset of inbreeding. Another is *genetic fixation*—the homozygous condition *both individuals* of a generation attain for the same allele at a locus that was heterozygous at the outset of inbreeding [ref. 3]. Such a genotypic mating combination perpetuates itself in succeeding generations; it is irreversible, barring mutation.

A third aspect is *genome purity*—the condition in which the entire genome is free of *heterogenic tracts* (chromosomal segments of heterogeneous origin) in both individuals of a generation. As inbreeding progresses, the number and mean length of heterogenic tracts decrease [ref. 3]. The proportion of the genome involved in these tracts is that proportion not yet genetically fixed (i.e., the difference between unity and the value in the genetic-fixation column of Table 2). From this it is clear that relatively few genes would be involved in these tracts in later generations. Nevertheless, there is a sufficient number of such tracts present so that it is not until generation 60 that one has a 99 percent chance of being completely rid of them.

Table 2. **Effects of Brother by Sister Inbreeding on the Genotype**

Calculations are based on the methods of Fisher [ref. 3]. For **Probability of Homozygosity** and for **Probability of Genetic Fixation**, a locus is selected at which the two parents in generation 0 are both heterozygous with four alleles, i.e., a double cross (ab x cd). **Probability of Purity**: The probability that no heterogenic tract (chromosomal segment of heterogeneous origin) remains in the genome; it has been assumed that the genome length is 1500 centi-Morgans.

Generation	Probability of Homozygosity 1/	Probability of Genetic Fixation	Probability of Purity
0	0	0	0
5	0.594	0.409	0
10	0.859	0.785	0
15	0.951	0.925	0
20	0.983	0.974	0
25	0.994	0.991	0.009
30	0.998	0.997	0.140
35	0.999	0.999	0.500
40	1	1	0.728
45	1	1	0.883
50	1	1	0.953
55	1	1	0.982
60	1	1	0.993

1/ Homozygosity is equivalent to F, but, unlike F in Table 1, it is delayed one generation due to starting with a double instead of a single cross.

For those species easily inbred, such as the laboratory mouse, the conventional criterion for declaring a strain "inbred" is that it be continuously maintained by matings of brother by sister for 20 generations (when F = 0.986). Before then, it is an "incipient" inbred strain. For other species, such as poultry, which are difficult if not impossible to inbreed at the intensity of the brother by sister regimen, the term "inbred" is commonly used to describe any line which is more inbred than the population from which it arose.

Once inbred strains are established, they can be genetically manipulated to establish yet other strains with special attributes for genetic analysis and control, namely the *congenic* lines, the *coisogenic* lines and the *recombinant-inbred* strains.

A *congenic line* is an inbred strain genetically identical to an already established inbred strain except for a short chromosomal segment that bears a distinctive gene of interest. The congenic line is created by crossing the established inbred strain with the individual bearing the distinctive gene. This is followed by repeatedly crossing selected carriers of the distinctive gene back to the established inbred strain (Figure 1). In time, all introduced genes except the distinctive gene and closely linked genes will have been purged from the inbred strain by the backcrossing procedure. Sometimes other mating systems, such as the cross-intercross and the cross-backcross-intercross, are employed instead of the backcross. The choice of system depends on properties of the selected trait.

continued

The locus at which the distinctive gene resides is known as the *differential* locus, the linked genes carried along on the introduced segment are called *passenger* genes, and the original established inbred strain is termed the *partner* or *background* strain.

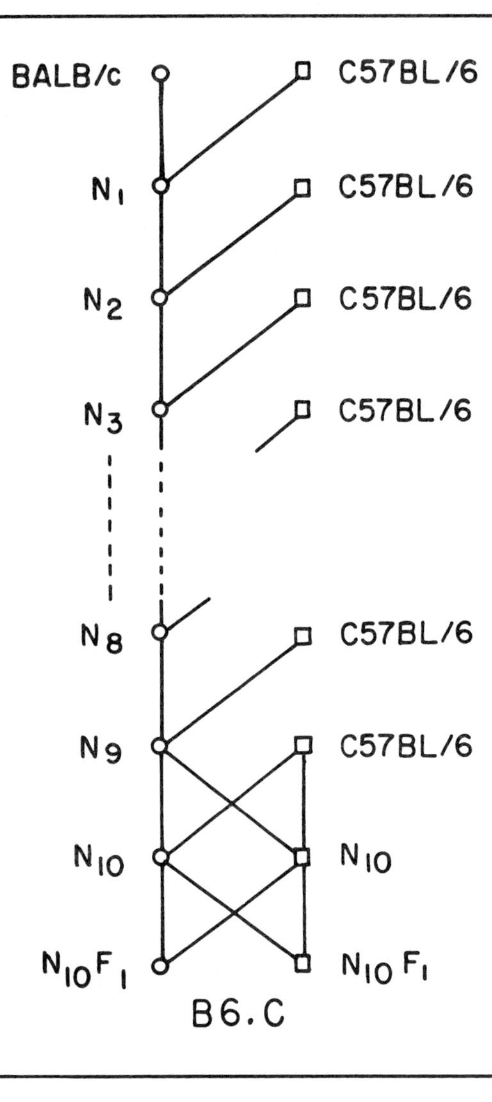

Figure 1

A congenic line is developed by a series of matings to a background strain (C57BL/6 in this example). In each generation (N), a carrier of the differential gene is chosen as a parent of the next generation. The line is usually made homozygous for the differential gene by reverting to full-sib mating, as indicated in generation N_{10}. Circles represent females, squares represent males, and line segments represent gametic pathways.

Congenic lines are used (i) to compare effects of genes without the interference of genes in the background, (ii) to identify easily, by the congenic line in which they are carried, individual genes that have similar phenotypic effects, such as histocompatibility genes, and (iii) to assist in linkage studies—for example, discovery of a passenger gene in a congenic line is, in itself, evidence of linkage to the differential locus.

In contrast to the congenic line, a *coisogenic* line is one that differs from its partner strain at a single locus. This situation obtains from a mutation occurring in an established inbred strain.

A *recombinant-inbred (RI) strain* is derived from a cross of two already established highly inbred strains (the *progenitor* strains), followed by systematic inbreeding as for any other inbred strain [ref. 1]. This procedure, with no conscious selection applied, allows the reassortment and fixation of genes from the two progenitor strains.

A series of such RI strains, each of the strains independently derived from the same two progenitor strains as in Figure 2, provides a useful device for linkage analysis. To this end, one can type each RI strain to find which progenitor-strain allele is genetically fixed at a specified locus. Because linked genes from the same progenitor strain will tend to become established together in any one RI strain, the strain distribution patterns (SDPs) of alleles at loci that are linked will tend to be alike. Linkage is suggested by concordant, or nearly concordant, SDPs for two loci. However, concordance can occur by chance, and the probability is a function of the number of RI strains and the number of SDPs one attempts to match. Therefore, if the number of

continued

RI strains in a series is not large, an SDP match is only a guide for making conventional linkage tests. If the match is not perfect, then at least that is good evidence that the two genes being compared are not alleles at the same locus.

A series of recombinant-inbred strains is derived from a cross of two highly inbred progenitor strains (BALB/c and C57BL/6 strains in the Figure below). After the cross, inbreeding continues, although independently, for each new strain in the series. Circles represent females, squares represent males, and line segments represent gametic pathways.

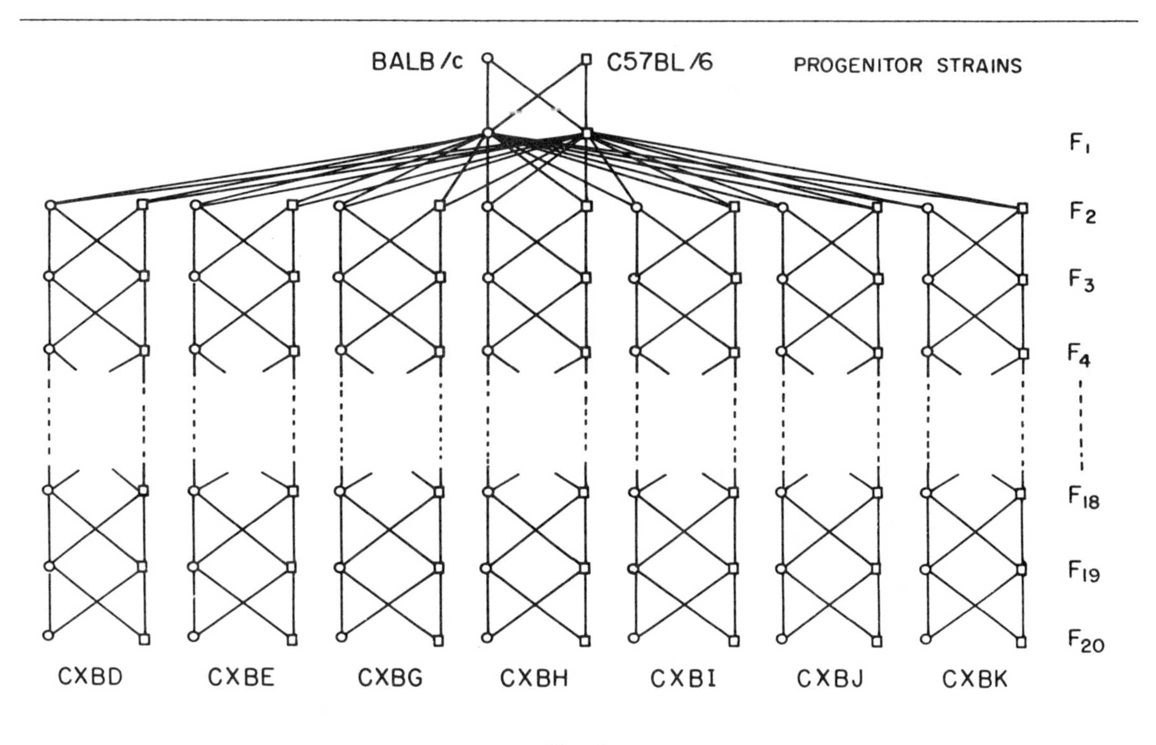

Figure 2

Each recombinant genotype of an RI strain is replicable in an indefinite number of individuals within that strain. This property allows linkage analyses of traits that could not be made if there were but one individual of a given recombinant genotype, as is the case in a segregating test cross generation of a conventional linkage test. For example, linkage of a trait that depends on the proportion of individuals showing the trait (i.e., incidence), such as "susceptibility to tumor induction," is analyzable with RI strains. Also, two or more traits that are not measurable on the same individual, such as a fetal trait and an adult trait, or a female trait and a male trait, are analyzable by this method.

Once the RI strains are typed, the data are available for comparison with those of any new trait for which the strains are typed in the future. Thus, linkage information is cumulative. This is not true for the recombinant genotype in a conventional test cross where the recombinant individual has a limited life-span.

Contributor: Donald W. Bailey

References

1. Bailey, D. W. 1971. Transplantation 11:325-327.
2. Falconer, D. S. 1960. Introduction to Quantitative Genetics. Oliver and Boyd, Edinburgh.
3. Fisher, R. A. 1949. Theory of Inbreeding. Oliver and Boyd, London.
4. Wright, S. 1921. Genetics 6:167-178.

I. MOUSE

NOMENCLATURE AND RULES FOR MOUSE GENETICS

As the contents of this volume indicate, the genetics of the mouse has become a highly complex and intricate field, embracing a wide body of knowledge, and fundamental to those aspects of biology in which the greatest progress is being made at present.

Of overriding importance to the development of this knowledge has been the availability of numerous inbred strains, and congenic strains differing from a standard strain at one or a few known loci. Present-day geneticists owe a great debt to those scientists who foresaw the need for such strains. Among the early investigators were C. C. Little, who was the first to develop inbred strains of mice, and G. D. Snell who was the pioneer of congenic strains. It is perhaps less appreciated that the foresight of the individuals who developed the present system of nomenclature for inbred strains and for genetic variants was equally remarkable and of comparable importance.

The first moves to organize a Committee on Mouse Genetics Nomenclature were made as long ago as 1939, at a time when there were only 31 known gene loci and 7 linkage groups. Again, G. D. Snell was in the forefront. A significant achievement of the first Committee was the inauguration of the *Mouse News Letter* [ref. 5] which is today an important organ for the dissemination of information on mouse genetics nomenclature.

At first there were separate committees for the nomenclature of genetic variants and for inbred strains. In 1958, however, these two committees were merged into the Committee on Standardized Genetic Nomenclature for Mice which is the body now responsible for coordinating and regulating all problems concerned with the nomenclature of strains, genes, and chromosomal variants of the mouse. (For further information on the history of mouse genetics nomenclature, consult references 7 and 9.)

The Committee on Standardized Genetic Nomenclature for Mice comes under the aegis of the International Committee on Laboratory Animals, and hence under UNESCO. It thus has official international status. However, a feature of perhaps more importance to scientists is that all its decisions are reached by democratic processes. After consultation, scientists active in the field vote on the issue. Problems that arise in particular branches of mouse genetics are handled by ad hoc subcommittees of individuals working in the involved areas. Recent examples of such subcommittees include those on Chromosomal Changes [ref. 11] and on nomenclature for inbred strains preserved by freezing [ref. 6]. Views put forward by subcommittees are circulated to interested scientists, and are made available for comment through the *Mouse News Letter*, before any decisions are taken by the main Committee.

The chairmen of the main Committee since 1958 have been George D. Snell, Margaret C. Green, and now Mary F. Lyon. The following were members of the Committee in 1977:

- P. Démant, Netherlands Cancer Institute, Sarphatistraat 108, Amsterdam-C, The Netherlands
- I. K. Egorov, Institute of General Genetics, USSR Academy of Sciences, Moscow B-133, USSR
- J. J. Hutton, University of Kentucky, Lexington, Kentucky 40506
- K. Kondo, Department of Animal Genetics, Nagoya University, Nagoya-shi, Japan
- M. F. Lyon, MRC Radiobiology Unit, Harwell, England
- T. H. Roderick, The Jackson Laboratory, Bar Harbor, Maine 04609
- M. Sabourdy, C.S.E.A.L., C.N.R.S., 45-Orleans-la-Source, France (ICLA representative)
- R. Schmidt, Biologisches Institut, Halle-Saale, Germany
- A. G. Searle, MRC Radiobiology Unit, Harwell, England
- J. Staats, The Jackson Laboratory, Bar Harbor, Maine 04609

In making its decisions on nomenclature, the Committee in general follows the rules put forward by the International Committee on Genetic Symbols and Nomenclature [ref. 3]. Moreover, the Committee bears in mind the possible use of computer storage of information, and therefore keeps to a minimum the use of sub- and super-scripts (such forms as C_3H and $C_{57}Bl$ are *incorrect*) and of differing type fonts (such as italics and bold face). Gene symbols are always italicized, however, in accordance with the usage adopted for many other organisms.

The symbols must be appropriate for manipulation in pedigrees and genetic formulae of animals used in experimental breeding. Therefore, the rules follow, where possible, those for *Drosophila melanogaster*, this being the most widely used experimental animal. The rules used in human genetics ordinarily are *not* suitable since, for obvious reasons, they were not designed for use in selective breeding. For instance, the mouse rules for the designation of chromosomal changes follow fairly closely those for *Drosophila*, whereas those for man are entirely different.

continued

In addition to the present volume, symbols currently in use are listed in the *Mouse News Letter* [ref. 5] (genetic variants), and in *Inbred Strains of Mice* [ref. 8] and *Standardized Nomenclature for Inbred Strains of Mice* [ref. 10].

The use of correct nomenclature is extremely important in avoiding confusion in the literature. Scientists are therefore exhorted to follow closely the rules of nomenclature.

In naming new strains or variants, (i) consult the rules, (ii) avoid duplication of symbols by perusing the most recent lists in *Mouse News Letter, Inbred Strains of Mice,* or *Standardized Nomenclature,* and (iii) inform the editor or compiler of the list concerned, or the chairman of the Nomenclature Committee.

When using existing strains or variants, the *full nomenclature* should be given at least once in any publication, perhaps in the "Materials and Methods" section. Thereafter, in cases where the full designation is cumbersome, it may be appropriate to use an abbreviated symbol.

Rules for Symbols to Designate Inbred Strains of Mice (reprinted from ref. 10).

1. Definition of Inbred Strain. A strain shall be regarded as inbred when it has been mated brother x sister (hereafter called b x s) for 20 or more consecutive generations. Parent x offspring matings may be substituted for b x s matings provided that, in the case of consecutive parent x offspring matings, the mating in each case is to the younger of the 2 parents.

2. Symbols for Inbred Strains. Inbred strains shall be designated by a capital letter or letters in roman type. It is urged that anyone naming a new stock consult reference 10 of this article or *Inbred Strains of Mice* to avoid duplication. Brief symbols are preferred.

An exception is allowed in the case of stocks already widely used and known by a designation that does not conform.

3. Definition of Substrain. The definition of substrain presents some of the same problems as the definition of species. In practice the determination of whether 2 related strains should be treated as substrains and of whether, in published articles, substrain symbols should be added to the strain symbol must rest with the investigators using them. The following rules, however, may be of help.

Any strains separated after 8 to 19 generations of b x s inbreeding and maintained thereafter in the same laboratory without intercrossing for a further 12 or more generations shall be regarded as substrains. It shall also be considered that substrains have been constituted (a) if pairs from the parent strain (or substrain) are transferred to another investigator or (b) if detectable genetic differences become established.

4. Designation of Substrains. A substrain shall be known by the name of the parent strain followed by a slant line and an appropriate substrain symbol. Substrain symbols may be of 2 types.

(a) Abbreviated name as substrain symbol: The symbol for substrains should usually consist of an abbreviation of the name of the person or laboratory maintaining it. The initial letter of this symbol should be set in roman capitals; all other letters should be lower case. Abbreviations should be brief, should as far as possible be standardized, and should be checked with published lists to avoid duplication. Examples: A/He (Heston substrain of strain A), A/Icrc (Indian Cancer Research Centre substrain of strain A).

When a new substrain is created by transfer, the old symbol may be retained and a new one added. Example: YBR/He, on transfer from Heston to Wilson, becomes YBR/HeWi. The accumulation of substrain symbols in this fashion provides a history of the strain. If the substrain symbols are not accumulated, the history of transfers should be recorded in *Inbred Strains of Mice*.

(b) Numbers or lowercase letters: numbers or lowercase letters may be used as substrain symbols in certain circumstances. The position of these relative to other parts, if any, of the substrain symbol should be suggestive of a historic or time sequence. Thus, two substrain branches, separated in and maintained by 1 laboratory, may be designated by terminal numbers, with or without a preceding slant line. Example: two sublines of A/HeCrgl, separated and maintained by Crgl, become A/HeCrgl/1 (or A/Crgl/1) and A/HeCrgl/2 (or A/Crgl/2). Lowercase letters immediately following the strain symbol, with a slant line only intervening, may be used when 2 substrains are separated from a common strain prior to complete inbreeding. Example: C57BR/a and C57BR/cd. (These were separated after 9 generations of b x s). The use of numbers or lowercase letters immediately after the slant line, to designate lines separated after 20 or more generations b x s, is ordinarily not recommended but may occasionally be justifiable for sublines widely recognized as different. Example: DBA/1 and DBA/2. Appropriate checks to avoid duplication should be made before this type of symbol is adopted.

5. Congenic Stocks. Congenic stocks produced by the occurrence of a single major mutation within an inbred

continued

strain, or by the introduction of a gene into an inbred background by a series of crosses, shall be designated by the strain symbol and, where appropriate (*see* Rule 7), substrain symbol, followed by a hyphen and the gene symbol (in italics in printed articles)[1/]. Example: DBA/Ha-*D*. When the mutant or introduced gene is maintained in the heterozygous condition, this may be indicated by including a + in the symbol. Examples: A/Fa-+*c*, C3H/N-+W^{j}. (The term congenic is used to designate stocks that approximate the coisogenic state but, because they are derived through a limited series of crosses rather than by mutation, may be presumed to carry, in addition to the distinguishing foreign allele, some other contaminant genes).

When a coisogenic strain is produced by inbreeding with forced heterozygosis, indication of the segregating locus is strictly optional. Examples: 129 or 129-$c^{ch}c$ (129 is customary); SEAC-*d*+/+*se* or SEAC/Gn.

In the case of congenic stocks produced by repeated crosses of a dominant gene into a standard inbred strain, it may be desirable to indicate the number of backcross generations. Strains shall be regarded as congenic when at least 7 such crosses have been made[2/]. Example: C57BL/6-+W^{v}(N8). The first hybrid or F1 generation should be counted as Generation 1, the first backcross generation as Generation 2 etc.

6. Substrains Developed through Foster Nursing, Ova Transfer, or Ovary Transplant. Substrains developed by foster nursing shall be indicated by appending an "f" to the strain symbol. Example: C3Hf. The strain used as foster parent may be indicated, if desired, by the addition of its symbol or an abbreviation for the same. Example: C3HfC57BL or C3HfB (C3H fostered on C57BL). In like manner strains developed through egg transfer or ovary transplant shall be indicated by adding an "e" or "o", respectively. Example: AeB (A ova transferred to C57BL). When the symbol for fostering or transfer might be confused with an adjoining substrain symbol, it may be used in a subscript position. Example: A/He_fB (Heston substrain of strain A fostered on C57BL).

7. Compound Substrain Symbols for Stocks of Complex Origin. When a stock has been produced by manipulation of a standard inbred strain, as, for example, by fostering or by introduction of a foreign gene, compound substrain symbols may be necessary. In general, the elements of such a compound symbol should be arranged in an order indicating a historic or time sequence. Specifically, different positions should be interpreted as follows:

(a) Substrain symbol that immediately follows strain symbol: Examples: BALB/cf, DBA/2eB, C3H/Ha-*p*. In this position the substrain symbol (c, 2, or Ha in examples given) designates the substrain that was fostered or otherwise manipulated, or in which a mutation occurred.

(b) Substrain symbols following symbol for manipulative process or introduced gene: Substrain symbols in this position refer to the person performing the fostering or other manipulation, to the person or laboratory currently maintaining the strain, or to both. The symbol or symbols may or may not be immediately preceded by a slant line. Examples: DBA/2eB/De or DBA/2eBDe (strain derived from ova of DBA/2 transferred by Deringer to C57BL, maintained by Deringer); C3H/He_f/Ha (C3H/He fostered by Heston, currently maintained by Hauschka); CBA/Ca-*se*/Gn (Carter's substrain of CBA with mutation to *se*, maintained by Green). Since a single symbol in this position (e.g. the De in DBA/2eBDe) May refer either to the person producing or to the person maintaining the strain, the intended meaning should be clearly recorded.

8. Indication of Inbreeding. When it is desired to indicate the number of generations of b x s inbreeding, this shall be done by appending, in parentheses, an F followed by the number of inbred generations. Example: A(F87). If, because of incomplete information, the number given represents only part of the total inbreeding, this should be indicated by preceding it with a question mark and plus sign. Example: YBL(F?+10).

9. Priority in Strain Symbols. If two inbred strains are assigned the same symbol, the symbol to be retained shall be determined by priority in publication. For this purpose, listing in *Mouse News Letter* or *Inbred Strains of Mice* shall be regarded as publication.

1/ Since this rule was formulated (in 1952) other acceptable types of designation have come into use and are given by Klein [ref. 4]. Thus, the three now acceptable types of symbol for congenic lines each consists of two parts separated by a period or dash. The first part is the full or abbreviated symbol of the background strain, followed by

(i) a period and an abbreviated symbol of the donor strain e.g. B10.129 (a strain with the genetic background of C57BL/10Sn (=B10) but which differs in a differential allele derived from strain 129/J). If several lines are derived from the same donor, the individual lines are distinguished by a number and/or letter in parentheses e.g. B10.129(5M), B10.129(21M)

or

(ii) a hyphen and the gene symbol of the differential allele (as described above)

or

(iii) a combination of both types of symbol, e.g. B10.129-$H\text{-}1^{b}$, B10.129-$H\text{-}4^{b}$.

Method (iii) is the most informative and is to be preferred.

2/ However it is preferable to make at least 10 such backcrosses or backcross equivalents (*see* ref. 4).

continued

Rules for Symbols to Designate Inbred Strains of Mice Preserved by Freezing (reprinted from ref. 6).

1. The strain name, as normally used in accordance with Standardized Nomenclature, should be followed by the letter p to indicate preservation by freezing.

2. If the uterine foster mother is of the same strain as the thawed embryos transferred to her, no further designation is necessary.

If the uterine foster mother is of a different strain from the embryos, the letter p should be followed by the usual nomenclature for egg transfer, i.e. the letter e followed by the strain designation of the foster mother; e.g. C57BL/6p means strain C57BL/6 preserved by freezing and transferred to a C57BL/6 foster mother; C57BL/6peCBA/H means strain C57BL/6 preserved by freezing and transferred to a CBA/H foster mother.

In accordance with Standardized Nomenclature, abbreviations may be used for strain names in these complex symbols.

3. The substrain symbol of the person or laboratory carrying out the preservation should follow; e.g. C57BL/6pe-CBA/HBy, denoting that the freezing has been performed by D. W. Bailey.

4. In denoting the generation of inbreeding, the years of freezing and of thawing should be indicated in parentheses, together with the letter P; e.g. F37 + (P1975-1977) + F16 means frozen at generation 37 in 1975, thawed in 1977, with 16 generations of inbreeding since thawing.

If the freezing and thawing were carried out by different people or laboratories, this might be indicated in the parentheses; e.g. (P1975By-1977H), frozen in 1975 by D. W. Bailey and thawed in 1977 at Harwell.

Gene Nomenclature Rules for Mice (Revision of 1963) (reprinted from ref. 1).

1. The names of genes, as distinguished from the symbols, shall be written with a lowercase initial letter, regardless of whether the mutant is dominant or recessive, except at the beginning of a sentence or other place in which capitalization would normally be used, or when the word is a proper noun. Examples: albinism, rex, Jay's dominant spotting.

2. The symbols for genes should typically be abbreviations of the name, e.g. *d* for dilution, *dw* for dwarf, *ac* for absence of corpus callosum. For convenience in alphabetical listings, the initial letters of names and symbols should be the same. Certain exceptions are indicated in subsequent rules.

3. Recessive mutations shall be indicated by the use of a small initial letter for the symbol of the mutant gene, e.g. *a* for non-agouti.

4. Dominant mutations (mutations with consistent heterozygous expression) shall be indicated by the use of a capital initial letter for the symbol of the mutant gene, e.g., *Re* for rex.

5. The locus symbol shall be the symbol of the first named mutant gene or allelic pair, except that any superscript indicative of a specific allele shall be omitted. Examples: *d* for the dilution locus, *Re* for the rex locus, *H-2* for the histocompatibility-2 locus.

6. The wild-type may be designated by:

(a) The locus symbol with a small initial letter and a plus superscript, e.g. t^{+}, re^{+}.

(b) A + sign only, when the context leaves no doubt as to the locus represented.

(c) The same symbol as the mutant gene but with a capital initial letter for recessive mutants, a small initial letter for dominant mutants. Examples: *D* for the wild-type allele of *d; re* for the wild-type allele of *Re*.

Alternatives (a) and (b) are recommended, except for teaching purposes or when there is doubt as to which allele is wild-type, e.g. biochemical polymorphisms, the agouti locus.

7. Mutants of similar phenotype but different location (mimics) shall be indicated either by entirely different names and symbols (e.g., *ln* for leaden and *d* for dilution) or by the same name and symbol with the addition of a hyphen and a distinguishing number. Examples: *H-1, H-2, H-3* etc. for different histocompatibility loci; *wa-1* (waved-1) and *wa-2* (waved-2). The preferred system is to give mimics distinctive names.

8. Multiple alleles determining visible or other clearly characterized distinctions shall be represented by the locus symbol with an added superscript, typically a small letter or letters suggestive of the name. The initial letter of the symbol shall be capitalized according as the allele behaves as a dominant or a recessive. Examples: c^{ch}, (chinchilla allele of *c* or albino), A^{y} (the yellow allele of *A* or agouti), $H\text{-}2^{k}$, (the *k* allele of histocompatibility-2).

9. Variants which are members of a series (e.g. reoccurrences, reversions to wild-type, translocations, some series of multiple alleles) should usually be distinguished one from another by the use of a series symbol. For this purpose an Arabic numeral corresponding to the serial number of the variant in any given laboratory, plus an abbreviation indicative of the name of the laboratory or discoverer, shall be used. Where the laboratory or discoverer has already been assigned an abbreviation for the designation of inbred substrains, this shall be the abbreviation used. Where there is no pre-assigned abbreviation, any appropriate abbreviation may be used, except that it should not duplicate an existing symbol in the standard list of abbreviations. In every case the first letter of the abbreviation should be cap-

continued

italized. Example: 7Rl (the 7th reoccurrence of any particular type of variant found by Russell). To avoid the confusion of the numeral one and the letter *l*, any first-discovered variant may be left unnumbered. The second variant is then numbered 2.

10. Indistinguishable alleles of independent origin (supposed reoccurrences) shall be designated by the existing gene symbol with the series symbol (rule 9) appended as a superscript. If the gene symbol already has a superscript, this shall be separated from the appended superscript by a hyphen. Examples: c^{4Rl} (the fourth reoccurrence of *c* found by Russell); $a^{t\text{-}7J}$ (the seventh reoccurrence of a^t found at The Jackson Laboratory).

11. Reversions to wild-type shall be designated by the symbol for the wild-type allele (rule 6a) with the series symbol (rule 9) appended as a superscript. Examples: d^{+J}, d^{+2J} (the first and second reversions from *d* to d^+ found at The Jackson Laboratory).

12. Translocations, when initially discovered, shall be designated by a T followed by the series symbol (rule 9). Example: T138Ca (the 138th translocation found by Carter). When the chromosomes[3] involved in the translocation have been identified, these shall be indicated by adding the appropriate numbers. These shall be inserted between the T and the series symbol, shall be Arabic rather than Roman numerals (except that the symbol for the X chromosome may be either X or 20, and that the Y chromosome shall be indicated by Y), shall be enclosed in parentheses, and shall be separated by a semicolon. Example: after it is found that T138Ca involves chromosomes 9 and 17 the symbol becomes T(9;17)138Ca. If one of the chromosomes involved in the translocation has been identified but the other has not, the unknown chromosome shall be indicated by a question mark. Example: T(6;?)7Ca.

In general, the translocation symbol should be reserved for reciprocal translocations. If a translocation is shown to be something other than reciprocal, e.g. a transposition, some other appropriate initial letter or letters should be used.

13. In published articles in which symbols are used, the symbols should be set in italics.

Rules for Designation of Chromosome Anomalies of the Mouse (reprinted from ref. 11)

1. Symbols for chromosome anomalies

(a) The initial letter symbols used in terminology for *Drosophila* should be adopted as follows:

T	Translocation	Dp	Duplication
R	Ring	Df	Deficiency
In	Inversion	C	Compound
Tp	Transposition		

In addition the following symbols are recommended for the mouse:

Rb	Robertsonian translocation
Ts	Trisomy
Ms	Monosomy
I	Isochromosome

The order of precedence of the symbols should be
T > R > In > Ts > Ms > Tp > Dp > Df > Rb

(b) In Robertsonian translocations the centromere should be indicated by a point rather than a semi-colon, e.g. T(9;19)163H becomes Rb(9.19)163H.

The chromosome numbers only (i.e. 9.19) may be used to denote the Robertsonian when it is involved in translocations with other chromosomes e.g. T(1;9.19) (preferably with a half-space after the semi-colon).

(c) The symbols L and S should be used to denote the long and short arms of mouse chromosomes. In translocations, breaks in the short arm should be so designated, but the L for long arm may be omitted if the meaning is clear e.g. T(4S; 5)99H, a translocation involving a break in the short arm of chromosome 4 and the long arm of 5.

(d) When Robertsonian translocations are involved in other aberrations the arm involved should be indicated by L or S.

e.g. T(1; 9.19L) - a translocation involving chromosome 1 and chromosome 9 of a Robertsonian.

In(9.19S) - inversion in chromosome 19
In(9.19LS) - pericentric inversion.

(e) Trisomies and monosomies should be denoted by the appropriate initial symbol, followed by the chromosome(s) concerned in parentheses, and by the translocation number if the trisomy etc. is derived from a translocation, e.g. $Ts(1^{13})89H$ - trisomy for the proximal end of 1 and the distal end of 13, derived from T(1; 13)89H.

2. Nomenclature for variations in heterochromatin

(a) The symbol NO should be reserved for nucleolus organizers. Different organizers could be distinguished by numbers.

(b) Centric heterochromatin.

The symbol H should be used for heterochromatin, followed by a symbol indicating the chromosome region involved, in this case c for centromeric, and a number indicating the chromosome on which it lies, e.g. Hc14 - centric heterochromatin on chromosome 14. Variations in size etc., of any block should be indicated

[3] The rules as published in 1963 gave "linkage groups" throughout in place of "chromosomes." This has been amended here in the light of current knowledge and practice.

continued

by superscripts, e.g. $Hc14^s$ - standard centric heterochromatin.

3. The systems proposed here do not preclude the use of Denver-type nomenclature, as for human chromosomes, for cytogenetic papers dealing with whole arm changes or mosaics e.g. 41, XY + 13, or 39, X/41, XYY.

Guidelines for Nomenclature of BioChemical Variants in the Mouse (reprinted from ref. 2)

1. Biochemical nomenclature should be in accord with the rules of the International Union of Biochemistry, Commission on Biochemical Nomenclature. The nomenclature recommended by the Commission is published periodically in major international biochemical journals such as the *Journal of Biological Chemistry* and the *Biochemical Journal.* Enzymes and other biochemicals have both trivial and formal names. The correct formal name should be given the first time a substance is mentioned in a publication; trivial or abbreviated names can be used subsequently. For example: Glucose-6-phosphate dehydrogenase (E.C. 1.1.1.49; D-glucose-6-phosphate:NADP 1-oxidoreductase) could be abbreviated G6PD, ĜPD, or Gd. This nomenclature is used in periodicals, reference works, and textbooks of biochemistry.

2. Symbols of structural loci should typically be two- or three-letter abbreviations (italic type) of the official Commission name of the enzyme, protein, or other substance affected. The initial letter of the symbol should be capitalized. Example: The symbol *Gpi-1* is used to designate the first identified structural locus of glucosephosphate isomerase (E.C. 5.3.1.9; D-glucose-6-phosphate ketol-isomerase). A proposed new symbol must not be the same as one previously used to designate another locus. Lists of existing symbols are published in *Mouse News Letter*. In the case of biochemical variants, the use of the locus symbol with a small initial letter to indicate recessive mutant genes and with a capital initial letter to indicate dominant mutant genes should generally be avoided. Such nomenclature is not suited to polymorphic systems of alleles, and the dominance-recessive relationship usually varies and depends on the method used to assess it.

3. Alleles should be designated by the locus symbol with an added superscript and should be italicized. The superscript typically should be one or two small letters. Example: $Pgm\text{-}1^a$, the allele at the *Pgm-1* locus in the C57BL/6J strain; $Pgm\text{-}1^b$, the allele in the DBA/2J strain. The superscript can be used to convey additional information about the allele. Example: Hbb^d, the hemoglobin β-chain allele that gives a *diffuse* band after electrophoresis. In describing alleles, whether found in inbred strains or in the wild it is desirable that the phenotypes of a number of widely used inbred strains be reported. One strain should arbitrarily be designated the prototype strain for each allele, since variation that has not been detected by the methods used may be present within each allelic class. If an apparently identical allele in other strains is found by new methods to be different from that in the prototype strain, it should be assigned a new alphabetical symbol as a superscript and a prototype strain designated. This system permits the orderly assignment of symbols to newly identified alleles and allows ready comparisons of new variants with previously reported variants.

Locus and allele symbols are necessarily brief and cannot contain more than a small fraction of the known information. Additional information may be contained in gene descriptions which in some cases can be collected in catalogues or tables. The hemoglobin α-chain locus *Hba*, for example, specifies at least four different polypeptides, with the additional complication that in some strains two different polypeptides are both produced. The alleles can be assigned letter designations, and information about the amino acid composition of the chains produced by the alleles can be shown in tables. This is somewhat similar to the methods already in use to record the specificities determined by alleles at loci for antigenic variants.

4. A series of loci specifying the structure of isoenzymes that catalyze the same or similar reactions but are structurally different can be designated by the same letter symbol for the structural locus with the addition of a hyphen and a distinguishing number. Example: Phosphoglucomutase (E.C. 2.7.5.1; α-D-glucose-1,6-bisphosphate:α-D-glucose-1-phosphate phosphotransferase); loci of structurally different isoenzymes are *Pgm-1* and *Pgm-2*.

5. Identification of a locus should not be assumed from the discovery of phenotypic structural variation; crosses should be made to show mendelian segregation of the alleles. Official gene symbols should not be assigned to variants found in wild mice unless appropriate genetic tests for allelism with known similar variants are carried out. In the absence of genetic tests phenotypic symbols (*see* par. 6) should be used, together with a description of the criteria used to establish identity with phenotypes of inbred strains.

6. Phenotype symbols should be the same as genotype symbols except that symbols for the phenotypes should be in capitals, not italicized, and the superscript should be lowered to the line. The phenotypes of heterozygotes should be written as in the following example: Phenotypes associated with the *Gpi-1* locus are GPI-1A, GPI-1B, GPI-1AB.

7. Genetic variants affecting enzyme activity may do so for reasons other than a direct change in the catalytic activity per molecule of the enzyme under study. Presumptive mutations in this group include those producing activity differences with no discernible alteration in physical or chemical properties of the enzyme and those producing tissue-specific differences in activity. Mutations producing this type of quantitative variation may or may not prove to be allelic with the structural locus of the enzyme in ques-

continued

tion. When allelic with the structural locus, they should be so designated following the above rules. Even when not allelic, or when the structural locus has not been identified, the new locus should be named on the basis of its discernible phenotype, following the above rules. Examples: *Lv*, levulinate dehydratase, a locus affecting the amount of enzyme present; *Ah*, aromatic hydrocarbon responsiveness, a locus affecting the level of hydroxylase induced by aromatic hydrocarbons.

Contributor: M. F. Lyon

References

1. Committee on Standardized Genetic Nomenclature for Mice. 1963. J. Hered. 54:159-162.
2. Committee on Standardized Genetic Nomenclature for Mice. 1973. Biochem. Genet. 9:369-374.
3. International Union of Biological Sciences. 1957. Int. Union Biol. Sci., Ser. D, 30:1-6.
4. Klein, J. 1973. Transplantation 15:137-153.
5. Searle, A. G., ed. 1949-. Mouse News Letter. Laboratory Animals Centre, Carshalton, Surrey, U.K.
6. Searle, A. G., ed. 1976. Ibid. 54:2-3.
7. Snell, G. D. 1974. Ibid. 50:7-8.
8. Staats, J., ed. 1959-. Inbred Strains of Mice. The Jackson Laboratory, Bar Harbor, ME.
9. Staats, J. 1966. In E. L. Green, ed. Biology of the Laboratory Mouse. McGraw-Hill, New York. pp. 1-9.
10. Staats, J. 1976. Cancer Res. 36:4333-4377.
11. Subcommittee on Cytogenetics. 1974. Mouse News Lett. 51:2-3.

GENEALOGY OF THE MORE COMMONLY USED INBRED MOUSE STRAINS

Chart is based, in part, on data provided by Michael Potter and Rose Lieberman in 1967; it was extended by Jan Klein in 1975; and it was revised by Potter in 1978. *H-2* haplotypes are shown in parentheses.

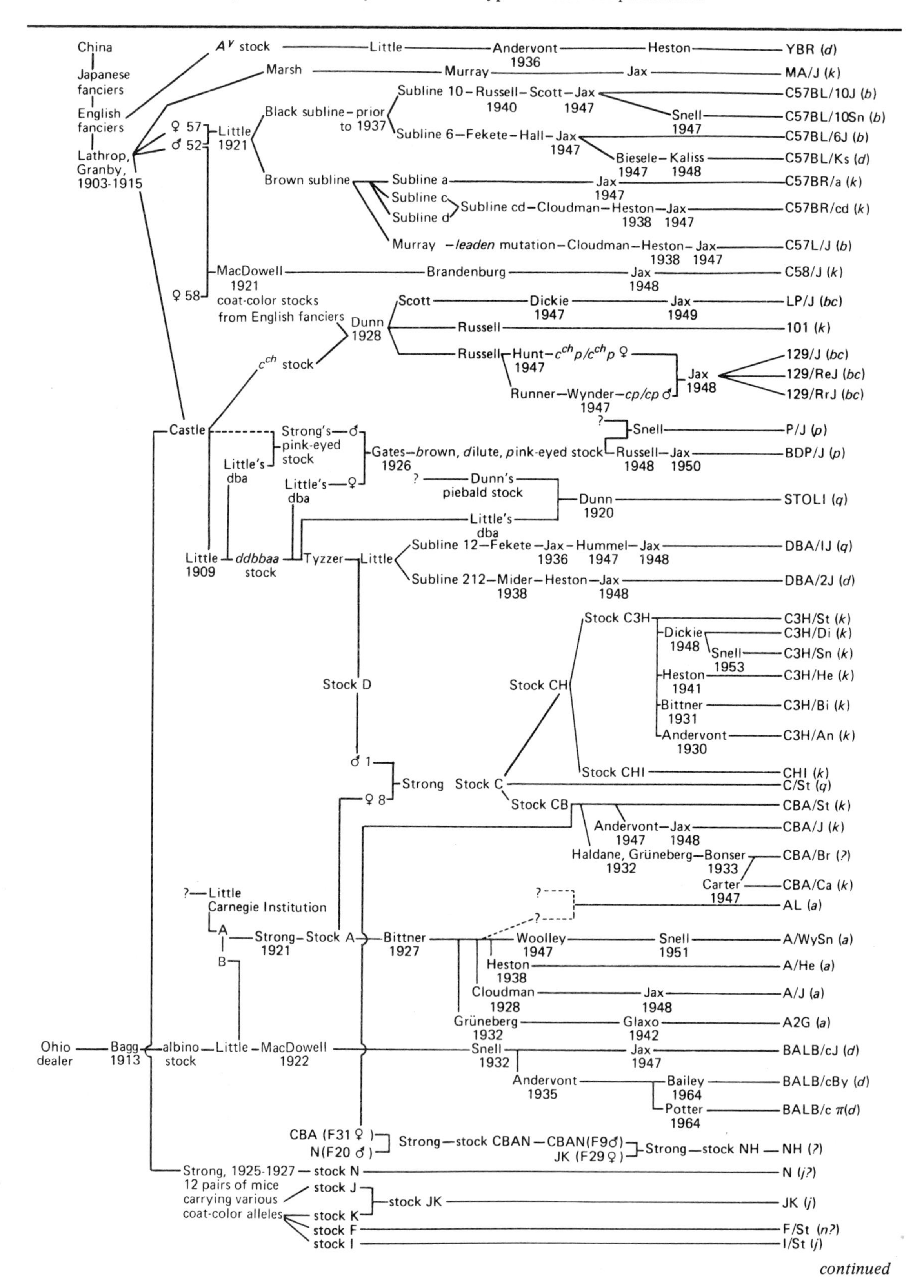

continued

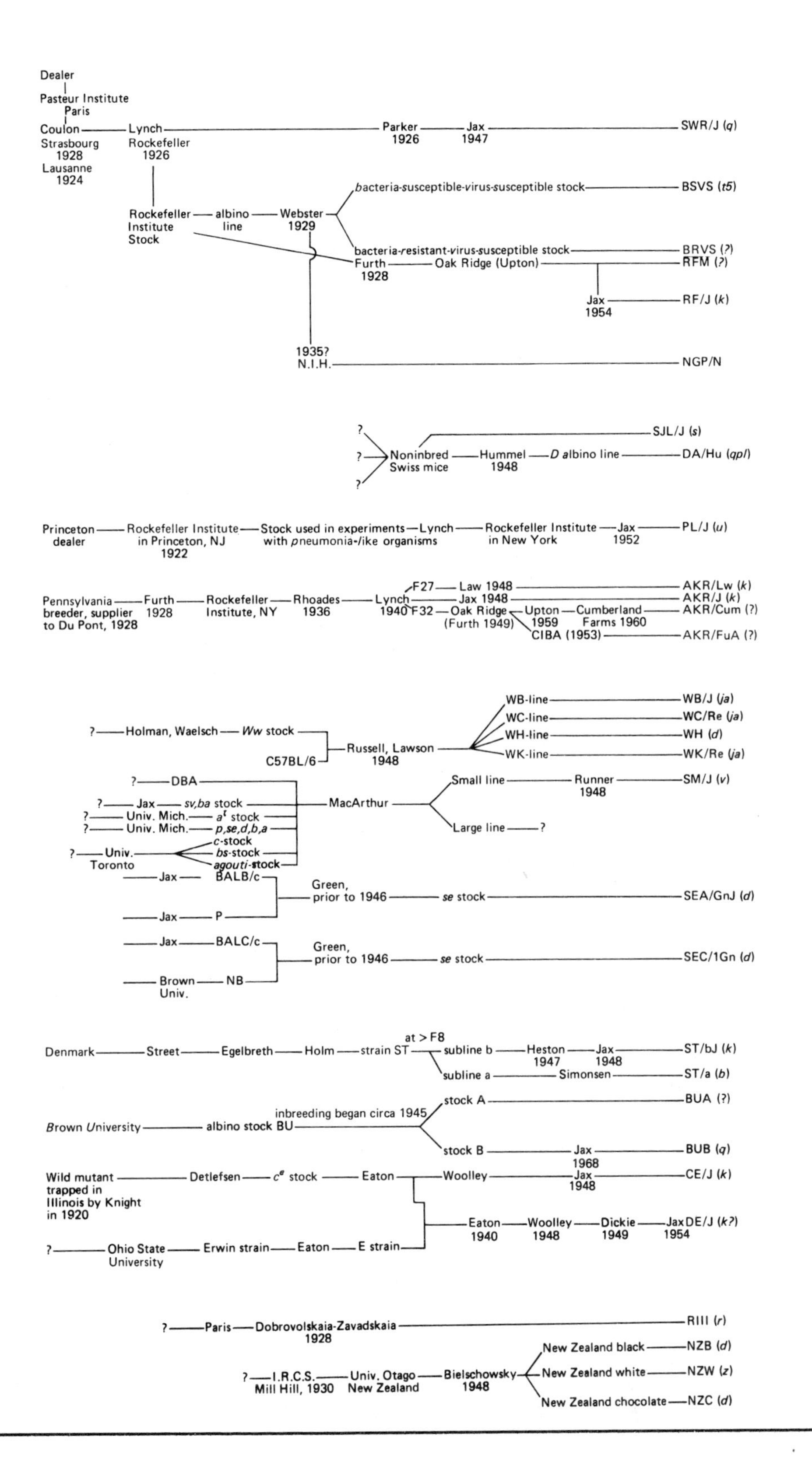

Contributors: Michael Potter; Jan Klein

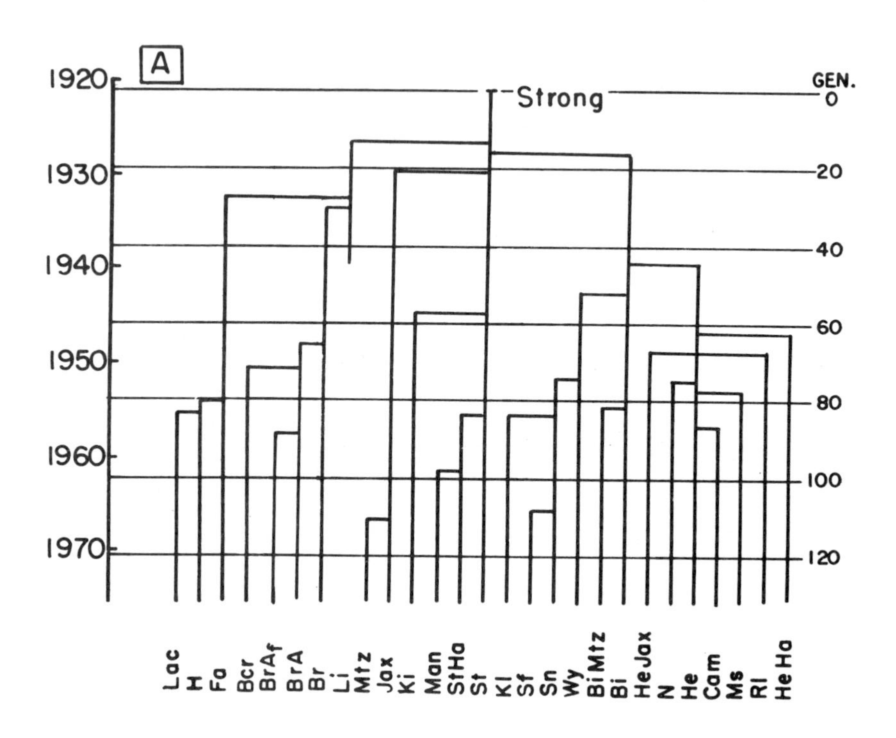
A
Strong
GEN.
1920
1930
1940
1950
1960
1970
0
20
40
60
80
100
120
Lac
H
Fa
Bcr
BrAf
BrA
Br
Li
Mtz
Jax
Ki
Man
StHa
St
Kl
Sf
Sn
Wy
BiMtz
Bi
HeJax
N
He
Cam
Ms
Rl
HeHa

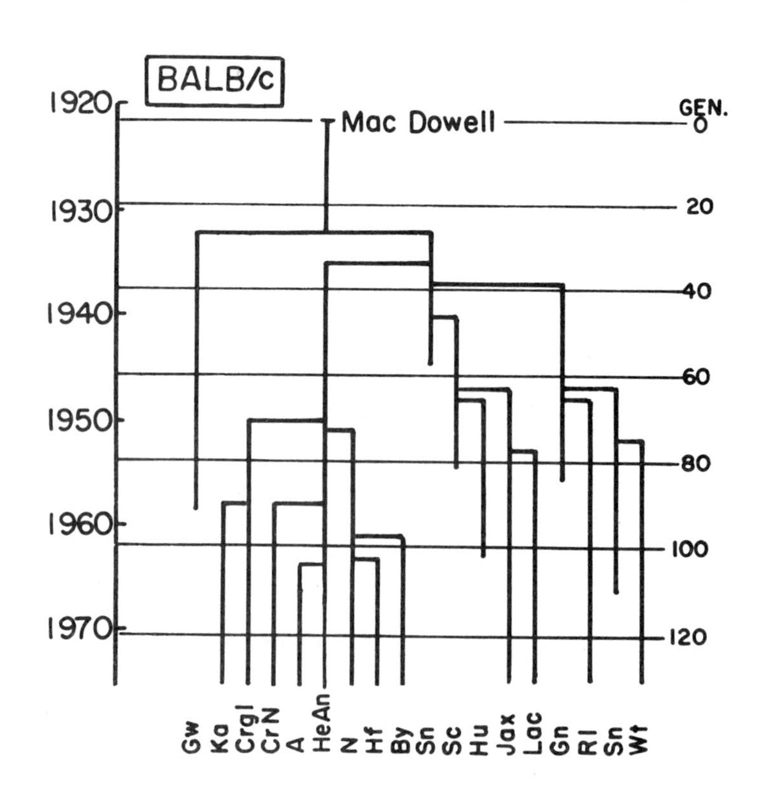
BALB/c
Mac Dowell
GEN.
1920
1930
1940
1950
1960
1970
0
20
40
60
80
100
120
Gw
Ka
Crgl
CrN
A
HeAn
N
Hf
By
Sn
Sc
Hu
Jax
Lac
Gn
Rl
Sn
Wt

continued

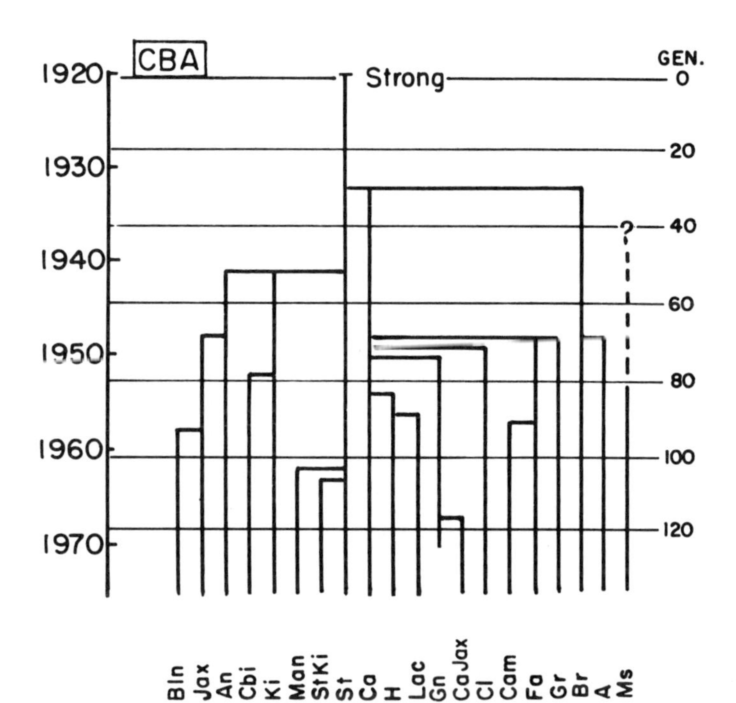

CBA
Strong
GEN.
1920
1930
1940
1950
1960
1970
0
20
40
60
80
100
120
?
Bln
Jax
An
Cbi
Ki
Man
StKi
St
Ca
H
Lac
Gn
CaJax
Cl
Cam
Fa
Gr
Br
A
Ms

C3H
Strong
GEN.
1920
1930
1940
1950
1960
1970
0
20
40
60
80
100
120
Bcr
Lee
BiIcrf
Ki
Mai
Gs
Bi
Ha
StKi
St
An
fB/HeHa
fB/HeDe
fB/HeN
fB/He
DiSn
Jax
Fs
Sf
eB/FeJax
eB/Fe
eB/De
De
Ms
Ep
He
Gif
HeA
HeAf
N
fC/Crgl
Crgl
Rl
Ca
H
Lac
HeIcrf

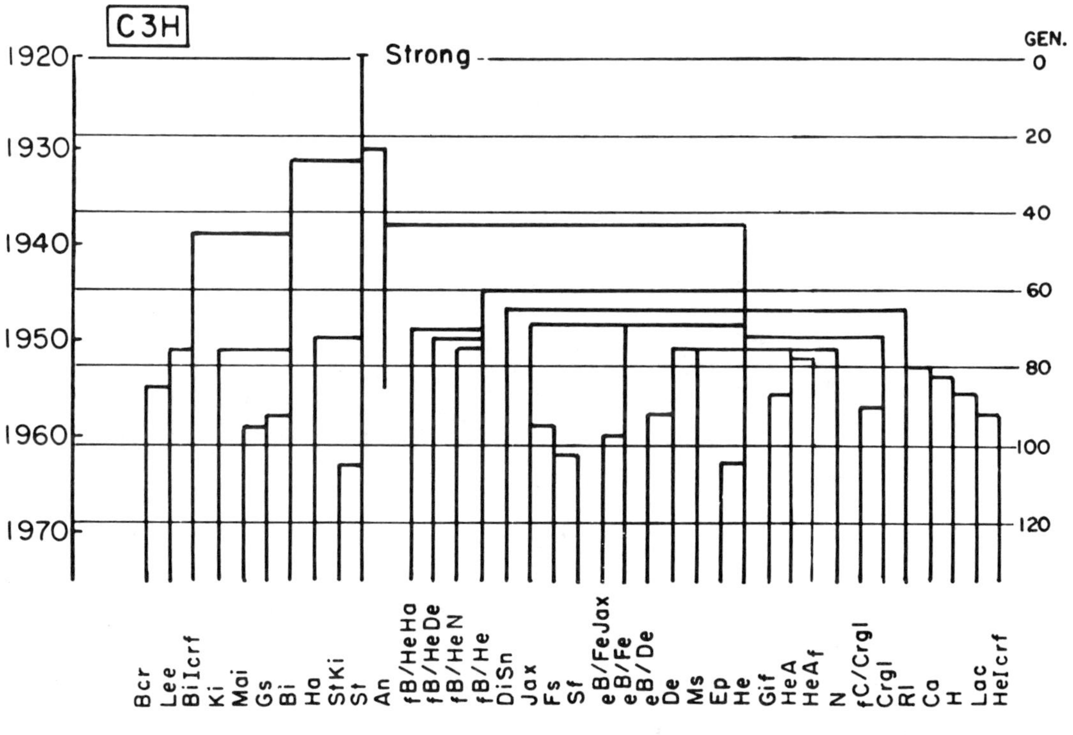

continued

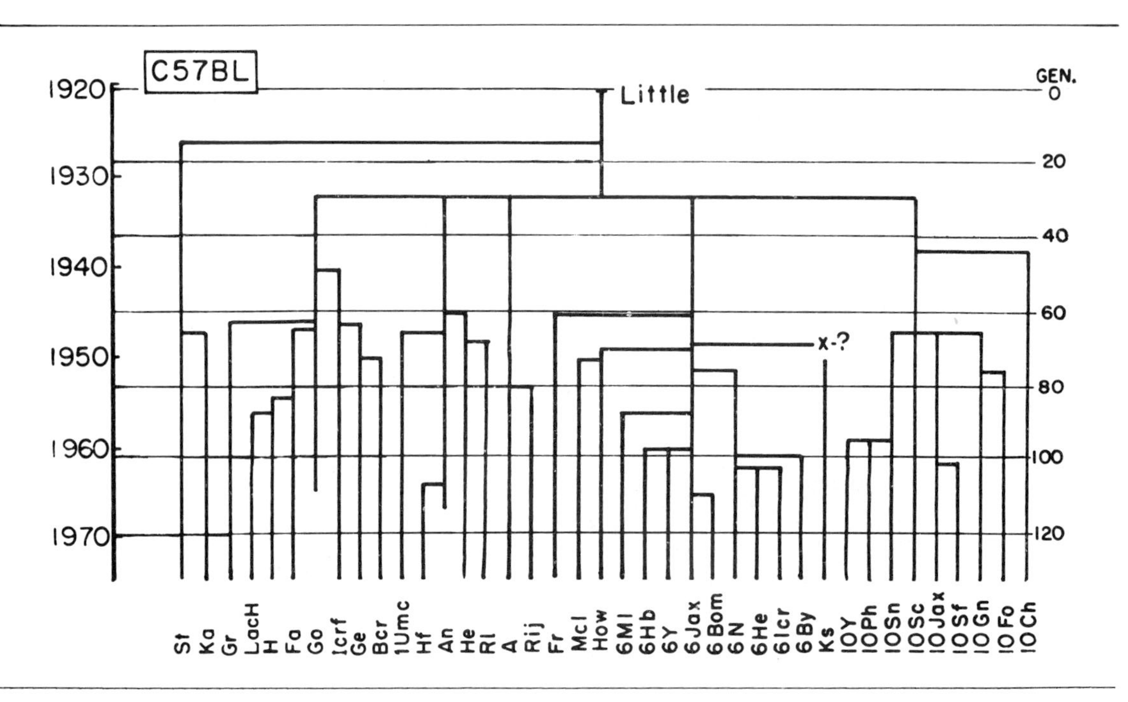

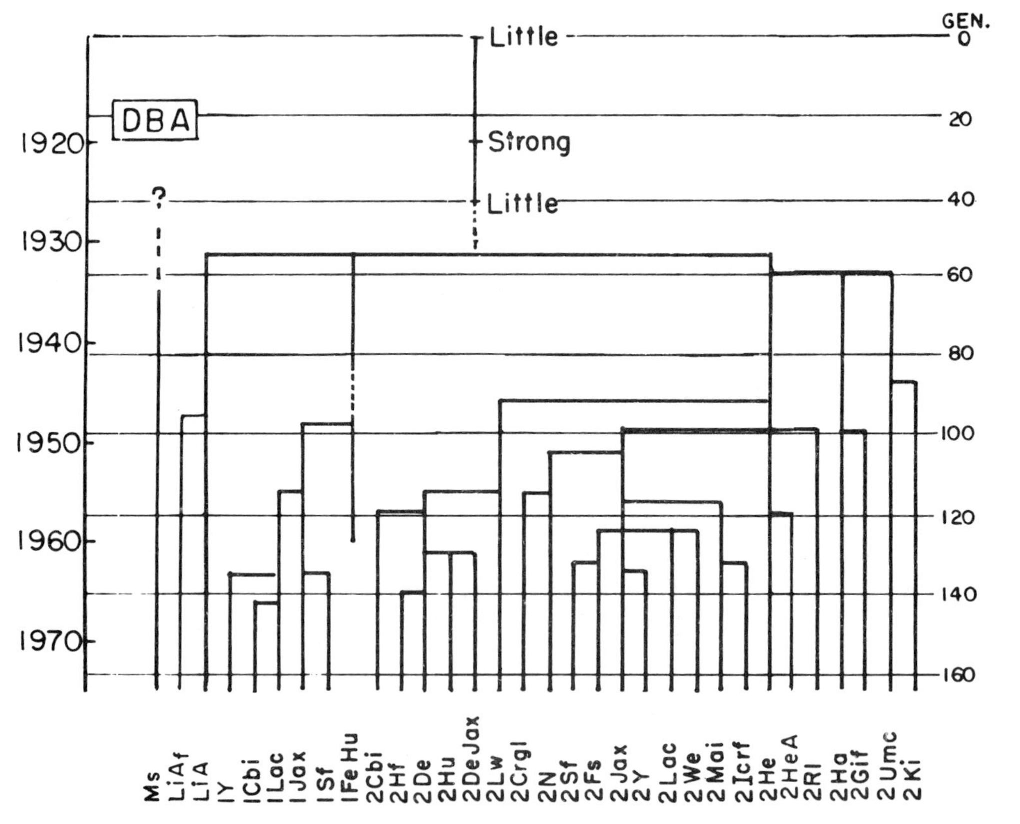

Contributor: Donald W. Bailey

Reference: Bailey, D. W. 1978. In H. C. Morse III, ed. Origins of Inbred Mice. Academic, New York.

1. INBRED STRAINS: MOUSE

Mus musculus is the animal of choice in experimental cancer research, but wide use has also been made of inbred strains in almost every field of biomedical research. A strain is considered inbred after 20 or more generations of brother-to-sister (b x s) matings. The total list of inbred strains and clearly defined substrains now exceeds 300; only the older, more widely used, or more unusual strains are given below. For more information about these and other strains, and for nomenclatural conventions, consult reference 5. **Genes Carried:** Most of the genes listed are mutations affecting coat color, biochemical polymorphisms, or histocompatibility factors. Data in brackets refer to the column heading in brackets.

Strain	Inbred Generations	Genes Carried [Excluded Alleles]	Characteristics	Reference
A	130	*a, b, c, Ea-4*^{a}, *Ea-5*^{a}, *Ea-6*^{a}, *Ea-7*^{b}, *Fv-1*^{b}, *Fv-2*^{s}, *H-2*^{a}, *H-31*^{a}, *H-32*^{a}, *Ly-1*^{b}, *Ly-2*^{b}, *Ly-3*^{b}, *Ly-5*^{a}, *Pca-1*^{a}, *Thy-1*^{b}, *Tla*^{a}	Mammary cancer high in breeders, low in virgins. Cleft lip & cleft palate occur sporadically; easily induced with variety of agents. Highly susceptible to measles virus.	5
A/He	172	*a, Acf-1*^{a}, *Ah*^{b}, *As-1*^{b}, *b, Bgs*^{h}, *c, Car-1*^{a}, *Car-2*^{b}, *Dip-1*^{b}, *Es-1*^{b}, *Es-2*^{b}, *Es-3*^{c}, *Es-6*^{a}, *Es-8*^{a}, *Es-10*^{a}, *Gdr-2*^{h}, *Got-2*^{b}, *Gpd-1*^{b}, *Gpi-1*^{a}, *Gpt-1*^{a}, *Gr-1*^{a}, *Hba*^{a}, *Hbb*^{d}, *Hc*^{0}, [*H-1 a* or *c*], *H-2*^{a}, [*H-3 a* or *b*], [*H-4 a* or *b*], *H-7*^{b}, [*H-8 a* or *b*], *H-9*^{a}, [*H-12 a* or *b*], *H-13*^{a}, *Id-1*^{a}, *If-1*^{l}, *Ig-1*^{e}, *Ig-3*$^{H-}$, *Lap-1*^{b}, *Ldr-1*^{a}, *Lv*^{a}, *Mod-1*^{a}, *Mod-2*^{b}, *Mpi-1*^{b}, *Mud*^{2}, *Mup-1*^{a}, *Pgm-1*^{a}, *Pgm-2*^{a}, *Pre-1*^{o}, *Pre-4*^{b}, *Prt-1*^{a}, *Prt-2*^{b}, *Ss*^{h}, *Trf*^{b}	Same as for strain A. Renal disease in older animals. Most susceptible to ectromelia virus of 8 strains tested.	1,2, 5
A/J	168	*a, Acf-1*^{a}, *Ah*^{b}, *b, Bgs*^{h}, *c, Car-1*^{a}, *Car-2*^{b}, *Ema*^{h}, *Es-10*^{a}, *Gdc-1*^{b}, *Gdr-2*^{h}, *Gur*^{a}, *Gus*^{a}, *Hbb*^{d}, *Hc*^{0}, *If-1*^{l}, *Ig-1*^{e}, *Lap-1*^{b}, *Mls*^{c}, *Mph-1*^{b}, *Mud*^{2}, *Pre-1*^{o}, *Pre-4*^{b}, *Ss*^{h}, *Trf*^{b}	Same as for strain A. Lower percentage of granulocytes than in other strain A sublines. Blood pressure lowest of 8 strains tested; some heart disease in old breeders.	1,5
A/Sn	165	*a, b, c, Ea-2*^{2}, *Es-5*^{b}, [*H-1 a* or *c*], *H-2*^{a}, [*H-3 a* or *b*], [*H-4 a* or *b*], *H-7*^{b}, [*H-8 a* or *b*], *H-12*^{a}, *H-13*^{a}, *Ly-4*^{a}	25 presacral vertebrae (7 cervical, 13 thoracic, 5 lumbar). High level of lipid in adrenal glands. Used in transplantation studies.	5
AKR	131	*a, Acf-1*^{b}, *Ah*^{d}, *Amy-1*^{a}, *Bgs*^{d}, *c, Car-2*^{a}, *Dip-1*^{b}, *Ea-2*^{2}, *Ea-4*^{a}, *Ea-6*^{a}, *Ea-7*^{b}, *Ema*^{h}, *Es-1*^{b}, *Es-2*^{b}, *Es-3*^{c}, *Es-5*^{b}, *Es-6*^{a}, *Es-8*^{a}, *Es-10*^{b}, *Fv-1*^{n}, *Fv-2*^{s}, *Gdc-1*^{b}, *Gdr-2*^{l}, *Got-2*^{b}, *Gpd-1*^{b}, *Gpi-1*^{a}, *Gpt-1*^{a}, *Gr-1*^{a}, *Gus*^{h}, *Hba*^{a}, *Hbb*^{d}, *Hc*^{0}, [*H-1 a* or *c*], *H-2*^{k}, [*H-3 a* or *b*], *H-4*^{a}, [*H-7 a* or *b*], [*H-8 a* or *b*], [*H-12 a* or *b*], *H-13*^{a}, *Id-1*^{b}, *If-1*^{h}, *Ig-1*^{d}, *Ig-3*$^{H-}$, *Lap-1*^{b}, *Ldr-1*^{a}, *Lv*^{a}, *Ly-1*^{b}, *Ly-2*^{a}, *Ly-3*^{a}, *Ly-4*^{a}, *Ly-5*^{a}, *Mls*^{a}, *Mod-1*^{b}, *Mph-1*^{b}, *Mpi-1*^{b}, *Mud*^{1}, *Mup-1*^{a}, *Pca-1*^{a}, *Pgm-1*^{a}, *Pgm-2*^{a}, *Pre-1*^{o}, *Pre-4*^{a}, *Sas-1*^{a}, *Slp*^{o}, *Ss*^{l}, *Svp-1*^{a}, *Svp-2*^{b}, *Thy-1*^{a}, *Tla*^{b}, *Trf*^{b}	Blood catalase activity high; low concentration of lipids in adrenals. Subline AKR/FuA differs at some polymorphic loci. Lymphatic leukemia in different sublines, 70-100%.	5
AL/N	166	*a, b, Bgs*^{h}, *c, H-2*^{a}, *Ig-3*$^{H-}$, *Mup-1*^{a}, *Svp-1*^{b}	Very low incidence of mammary tumors; good fertility. Believed to have originated from strain A outcrossed to unknown strain, followed by b x s matings. Should not be considered a strain A subline.	1,5
AU	>35	*a, Acf-1*^{a}, *Ah*^{b}, *Bgs*^{d}, *Car-1*^{a}, *Car-2*^{b}, *Dip-1*^{b}, *Es-1*^{b}, *Es-2*^{b}, *Es-3*^{c}, *Es-6*^{a}, *Es-8*^{a}, *Es-10*^{b}, *Got-2*^{b}, *Gpd-1*^{b}, *Gpi-1*^{b}, *Gr-1*^{a}, *Gur*^{a}, *Gus*^{b}, [*Hba*^{c}], *Hbb*^{p}, *Hc*^{0}, *H-2*^{q}, *Id-1*^{b}, *Lap-1*^{b}, *Ldr-1*^{a}, *Mod-1*^{b}, *Mod-2*^{b}, *Mpi-1*^{b}, *Pgm-1*^{b}, *Pgm-2*^{a}, *Pre-1*^{o}, *Trf*^{a}	♀ do not reject ♂ isografts. Unusual induction of kidney β-glucuronidase by testosterone.	5

continued

Strain	Inbred Generations	Genes Carried [Excluded Alleles]	Characteristics	Reference
BALB/c	121	Aal^{h}, Acf-1^{b}, Ah^{b}, As-1^{a}, b, Bgs^{h}, c, Car-1^{a}, Car-2^{b}, Ce-1^{b}, Cpl-1^{l}, Cpl-2^{h}, Cs-1^{a}, Dao^{a}, Dip-1^{a}, Ea-2^{2}, Ea-3^{b}, Ea-4^{a}, Ea-6^{b}, Ea-7^{b}, Ema^{l}, Erp-1^{a}, Es-1^{b}, Es-2^{b}, Es-3^{a}, Es-5^{b}, Es-6^{a}, Es-10^{a}, Exa^{l}, Fv-1^{b}, Fv-2^{s}, Gdc-1^{c}, Gdr-2^{l}, Got-2^{b}, Gpd-1^{b}, Gpi-1^{a}, Gpt-1^{a}, Gr-1^{a}, Gur^{a}, Gus^{a}, Hao-1^{a}, Hba^{b}, Hbb^{d}, Hc^{1}, Hnl^{h}, H-2^{d}, [H-3 a or b], [H-4 a or b], H-7^{b}, H-8^{c}, H-9^{b}, H-12^{a}, [H-13 a or b], H-15 through H-30^{c}, H-33^{b}, H-34 through H-38^{c}, Id-1^{a}, If-1^{l}, If-2^{l}, Ig-1^{a}, Ig-3^{H9}, Ig-4^{a}, Lap-1^{b}, Ldr-1^{a}, Lv^{a}, Ly-1^{b}, Ly-2^{b}, Ly-3^{b}, Ly-4^{a}, Ly-5^{a}, Mls^{b}, Mod-1^{a}, Mod-2^{b}, Mph-1^{b}, Mpi-1^{b}, Mud^{1}, Mup-1^{a}, Pca-1^{a}, Pd^{a}, Pgm-1^{a}, Pgm-2^{a}, Pre-1^{a}, Pre-4^{a}, Rv-1^{s}, Rv-2^{S}, Sas-1^{a}, Sco^{a}, Slp^{a}, Ss^{h}, Svp-1^{b}, Svp-2^{c}, Thy-1^{b}, Tla^{c}, Tol-1^{b}, Trf^{b}	Low incidence of mammary tumors (susceptible to mammary tumor agent) & of leukemia; moderate incidence of lung tumors; extremely sensitive to radiation. Blood pressure highest of 8 strains tested; some heart lesions in old breeders; arteriosclerosis common in ♂♀. Relatively non-aggressive. Very large reticuloendothelial organs relative to body weight; spleen amyloidosis in ♂ by 20 mo.	2,3, 5
BALB/cCd	48	....	60-70% of ♂♀ develop spontaneous bilateral adenocarcinoma of the kidney at 9-15 mo.	5
BDP	94	a, b, Bgs^{d}, Car-1^{a}, Car-2^{a}, d, Dip-1^{b}, Ea-2^{2}, Ea-4^{a}, Ea-7^{b}, Ema^{h}, Es-1^{b}, Es-2^{b}, Es-3^{a}, Es-6^{a}, Es-8^{a}, Es-10^{a}, Gpd-1^{a}, Gpi-1^{a}, Gr-1^{a}, Gur^{b}, Hba^{c}, Hbb^{d}, Hc^{1}, H-1^{a}, H-2^{p}, [H-3 a or b], H-4^{b}, H-7^{a}, [H-8 a or b], [H-9 a or b], H-12^{a}, [H-13 a or b], Id-1^{b}, Ig-1^{h}, Ig-2^{a}, Ig-3^{H9}, Ig-4^{a}, Lap-1^{b}, Ldr-1^{a}, Ly-1^{b}, Ly-2^{a}, Mod-1^{a}, Mpi-1^{b}, Mud^{1}, p, Pd^{a}, Pgm-1^{b}, Pgm-2^{a}, Pre-1^{a}, rd, Sas-1^{a}, se, Slp^{a}, Thy-1^{a}, Tla^{a}	Frequent mammary tumors; kidneys polycystic or granular; hemorrhagic ovaries	5
BRVR	>83	Bgs^{d}, c, Hc^{1}, H-2^{k}	Resistant to *Salmonella,* some encephalitic viruses, & experimental allergic encephalomyelitis. Have larger thymus than BSVS; thymus increases more under antigenic stimulation.	1,5
BSVS	>99	Bgs^{d}, c, Hc^{1}	Susceptible to *Salmonella,* some encephalitic viruses, & experimental allergic encephalitis	5
CBA/Ca	90	Bgs^{d}, Car-1^{a}, Car-2^{a}, Es-10^{b}, Gur^{b}, Hba^{d}, H-1^{a}, H-2^{k}, [H-3 a or b], H-4^{a}, [H-7 a or b], [H-8 a or b], H-9^{b}, [H-12 a or b], H-13^{a}, Lap-1^{b}, Ly-3^{b}, Ly-5^{a}, Mls^{b}, Mod-2^{b}, Mph-1^{b}, Pgm-1^{b}, Pre-1^{a}, Ss^{h}, Trf^{a}	Some mammary tumors in breeders; ∿18% lack lower third molars; few skeletal variants. CBA/N has an X-linked defect, with very low serum IgM levels and failure to respond to type III pneumococcal polysaccharide.	5
CBA/J	166	Acf-1^{b}, Ah^{b}, Bgs^{d}, Car-1^{a}, Car-2^{b}, Dip-1^{b}, Ea-1^{c}, Ea-2^{2}, Ea-4^{a}, Ea-6^{a}, Ema^{h}, Es-1^{b}, Es-2^{b}, Es-3^{c}, Es-5^{b}, Es-6^{a}, Es-8^{a}, Es-10^{b}, Fv-1^{n}, Fv-2^{s}, Gdc-1^{c}, Gdr-2^{l}, Got-2^{b}, Gpd-1^{b}, Gpi-1^{b}, Gpt-1^{a}, Gr-1^{a}, Gus^{h}, [Hba^{c}], Hbb^{d}, Hc^{1}, H-1^{a}, [H-3 a or b], H-4^{a}, [H-7 a or b], [H-8 a or b], H-9^{b}, [H-12 a or b], H-13^{a}, Id-1^{b}, If-1^{l}, Ig-1^{a}, Ig-2^{a}, Ig-3^{H9}, Ig-4^{a}, Ir-5^{h}, Lap-1^{b}, Ldr-1^{a}, Lv^{a}, Ly-1^{a}, Ly-2^{a}, Mls^{d}, Mod-1^{b}, Mod-2^{b}, Mpi-1^{b}, Mud^{2}, Pgm-1^{a}, Pgm-2^{a}, Phk^{a}, Pre-1°, Pre-4^{a}, Prt-1^{a}, Prt-2^{b}, Sas-1°, Slp^{o}, Ss^{l}, Svp-1^{b}, Svp-2^{b}, Thy-1^{b}, Tla^{b}, Trf^{b}	Mammary tumor incidence, 35-65%; hepatomas in ♂, 25-65%. In mid-range of radiation resistance; highly susceptible to measles virus.	5
CBA/H-T6	55	Car-1^{a}, Car-2^{a}, Fv-1^{n}, Fv-2^{s}, Ly-2^{a}, Ly-3^{b}, Ly-5^{a}, Mls^{b}	Homozygous for cytologic marker translocation T(14;15)6Ca	5

continued

Strain	Inbred Generations	Genes Carried [Excluded Alleles]	Characteristics	Reference
CE	>68	A^{w}, Acf-1^{a}, Bgs^{h}, c^{e}, Car-1^{a}, Car-2^{b}, Dip-1^{b}, Ea-2^{2}, Ea-4^{a}, Ea-7^{b}, Es-1^{b}, Es-2^{b}, Es-3^{c}, Es-6^{a}, Es-8^{a}, Es-10^{b}, Fv-1^{n}, Fv-2^{s}, Got-2^{b}, Gpd-1^{a}, Gpi-1^{a}, Gr-1^{a}, Gur^{b}, Gus^{b}, Hba^{a}, Hbb^{s}, Hc^{0}, [H-1 a or c], H-2^{k}, [H-4 a or b], [H-7 a or b], H-9^{a}, H-12^{a}, H-13^{b}, Id-1^{a}, If-1^{l}, Ig-1^{f}, Ig-3^{H9}, Ig-4^{a}, Lap-1^{b}, Ldr-1^{a}, Lv^{a}, Ly-1^{b}, Ly-2^{a}, Ly-3^{b}, Mod-1^{a}, Mpi-1^{b}, Mud^{2}, Mup-1^{a}, Pca-1^{a}, Pgm-1^{a}, Pgm-2^{a}, Pre-1^{o}, Pre-4^{b}, Sas-1^{o}, Thy-1^{b}, Tla^{b}	Low incidence of mammary tumors; ovarian tumors, 10-33%; high incidence of adrenocortical carcinoma in ♂♀ after neonatal gonadectomy. Large amount of lipids in adrenals.	5
C3H/He	147	Acf-1^{a}, Ah^{b}, Amy-1^{a}, Asr-1^{b}, Bgs^{h}, Car-1^{a}, Car-2^{b}, Ce-1^{b}, Dao^{a}, Dip-1^{b}, Ea-1^{c}, Ea-2^{2}, Ea-3^{b}, Ea-4^{a}, Ea-5^{a}, Ea-6^{a}, Ea-7^{a}, Ema^{h}, Erp-1^{a}, Es-1^{b}, Es-2^{b}, Es-3^{c}, Es-5^{b}, Es-8^{a}, Es-10^{b}, Fv-1^{n}, Fv-2^{s}, Gdc-1^{b}, Gdr-2^{l}, Gpd-1^{b}, Gpi-1^{b}, Gr-1^{a}, Gus^{h}, Hao-1^{a}, Hba^{c}, Hbb^{d}, Hc^{1}, H-1^{a}, H-2^{k}, H-3^{b}, H-4^{a}, H-7^{b}, [H-8 a or b], H-9^{a}, H-12^{a}, [H-13 a or b], Ig-4^{a}, Ir-5^{h}, Lap-1^{b}, Lv^{a}, Ly-1^{a}, Ly-2^{a}, Ly-3^{b}, Ly-4^{a}, Ly-5^{a}, Mls^{c}, Mod-1^{a}, Mod-2^{b}, Mph-1^{b}, Mup-1^{a}, Pca-1^{a}, Pgm-1^{b}, Pre-1^{o}, Prt-1^{a}, Prt-2^{b}, Sas-1^{o}, Slp^{o}, Ss^{l}, Svp-1^{a}, Svp-2^{c}, Thy-1^{b}, Tla^{b}, Trf^{b}	Mammary tumor incidence, 85% in breeders; 28% in virgins. High mortality in ♂ exposed to chloroform & turpentine fumes; hepatomas in ♂. Low erythrocyte & leukocyte counts. Fails to respond strongly to bacterial lipopolysaccharides. Other important sublines are An, Bi, & St, which may differ in skeletal and biochemical factors.	2,3, 5
C57BL/Ks	95	a, Bgs^{d}, Car-1^{a}, Car-2^{a}, Ea-2^{2}, Ea-4^{b}, Ema^{l}, Es-10^{a}, Gdc-1^{b}, Hbb^{s}, H-2^{d}, H-3^{a}, H-4^{a}, H-7^{a}, [H-8 a or b], [H-12 a or b], H-13^{a}, If-1^{h}, Lap-1^{a}, Ly-1^{b}, Ly-2^{b}, Pre-1^{o}	High incidence of toe malformations; microphthalmia; uniform high fertility. The mutation "diabetes" arose in this strain which is used as a control in diabetes studies.	5
C57BL/6	120	a, Aal^{l}, Acf-1^{b}, Ah^{b}, As-1^{a}, Asr-1^{a}, Bgs^{h}, Car-1^{a}, Car-2^{a}, Ce-1^{c}, Cpl-1^{h}, Cpl-2^{l}, Cs-1^{g}, Dip-1^{a}, Ea-2^{2}, Ea-6^{a}, Ea-7^{b}, Ema^{l}, Es-1^{a}, Es-2^{b}, Es-3^{a}, Es-5^{b}, Es-6^{a}, Es-8^{a}, Es-10^{a}, Exa^{h}, Fv-1^{b}, Fv-2^{r}, Gdc-1^{b}, Gdr-2^{l}, Got-2^{b}, Gpd-1^{a}, Gpi-1^{b}, Gpt-1^{a}, Gr-1^{a}, Gur^{b}, Gus^{b}, Hba^{a}, Hbb^{s}, Hc^{1}, Hnl^{l}, H-1^{c}, H-2^{b}, H-3^{a}, H-4^{a}, H-7^{a}, H-8^{a}, [H-9 a or b], H-12^{a}, H-13^{a}, H-15 through 30^{b}, H-31^{b}, H-32^{b}, H-34 through 38^{b}, Id-1^{a}, If-1^{h}, If-2^{h}, Ig-1^{b}, Ig-3^{H9}, Ig-4^{b}, Ir-5^{l}, Lap-1^{a}, Ldr-1^{a}, Lv^{b}, Ly-1^{b}, Ly-2^{b}, Ly-3^{b}, Ly-4^{b}, Ly-5^{a}, Mls^{b}, Mod-1^{b}, Mod-2^{b}, Mph-1^{b}, Mpi-1^{b}, Mud^{l}, Mup-1^{b}, Ox-1^{b}, Ox-2^{b}, Pca-1^{b}, Pd^{a}, Pgm-1^{a}, Pgm-2^{a}, Phk^{a}, Phr^{l}, Pp^{b}, Pre-1°, Pre-4^{b}, Pro-1^{a}, Prt-1^{a}, Prt-2^{b}, Sas-1^{a}, Sco^{p}, Sip^{b}, Slp^{o}, Spl^{h}, Ss^{h}, Svp-1^{b}, Svp-2^{a}, Thy-1^{b}, Tla^{b}, Trf^{b}	Mammary tumor incidence low; high incidence of hepatomas after irradiation; high ethanol preference; low concentration of lipids in adrenals; high complement activity; blood pressure in mid-range. Least susceptible to ectromelia virus of 8 strains tested.	2,4, 5, 6
C57BL/10	126	a, Ah^{b}, Bgs^{h}, Car-1^{a}, Car-2^{a}, Cd-1^{c}, Dip-1^{a}, Ea-2^{2}, Ea-3^{b}, Ema^{l}, Es-1^{a}, Es-3^{a}, Es-5^{b}, Es-10^{a}, Gr-1^{a}, Gur^{b}, Gus^{b}, Hba^{a}, Hbb^{s}, Hc^{1}, H-1^{c}, H-2^{b}, H-3^{a}, H-4^{a}, H-7^{a}, H-8^{a}, H-9^{a}, H-12^{a}, H-13^{a}, Id-1^{a}, If-1^{h}, Ig-1^{b}, Ig-3^{H9}, Ig-4^{b}, Ir-5^{l}, Lap-1^{a}, Lv^{c}, Ly-1^{b}, Ly-2^{b}, Ly-3^{b}, Ly-4^{b}, Ly-5^{a}, Mls^{b}, Mph-1^{b}, Mud^{l}, Mup-1^{b}, Pd^{a}, Pre-1^{o}, Rv-1^{R}, Rv-2^{r}, Sas-1^{a}, Slp^{o}, Ss^{h}, Svp-1^{b}, Thy-1^{b}, Tla^{b}, Trf^{b}	Occasional vaginal septa; skeletal abnormalities	5
C57BR/cd	154	a, b, Bgs^{h}, Car-1^{a}, Car-2^{b}, Ce-1^{c}, Dip-1^{a}, Ea-1^{c}, Ea-2^{2}, Ea-4^{a}, Ea-7^{b}, Ema^{l}, Es-1^{a}, Es-2^{b}, Es-3^{a}, Es-6^{a}, Es-8^{a}, Es-10^{a}, Fv-1^{n}, Fv-2^{r}, Gdc-1^{b}, Got-2^{b}, Gpd-1^{a}, Gpi-1^{a}, Gr-1^{a}, Gur^{b}, Gus^{b}, Hba^{a}, Hbb^{s}, Hc^{1}, H-1^{c}, H-2^{k}, H-3^{a}, H-4^{a}, H-7^{a}, H-8^{a}, H-9^{a}, H-12^{a}, H-13^{a}, Id-1^{b}, If-1^{h}, Ig-1^{a}, Ig-2^{a}, Ig-3^{H9}, Ir-5^{h}, Lap-1^{a}, Ldr-1^{a}, Lv^{a}, Ly-1^{b}, Ly-2^{b}, Ly-3^{b}, Ly-4^{b}, Ly-5^{a}, Mod-1^{b}, Mod-2^{b}, Mpi-1^{b}, Mud^{l}, Mup-1^{a}, Pd^{a}, Pgm-1^{a}, Pgm-2^{a}, Pre-1°, Pre-4^{a}, Sas-1^{a}, Slp^{o}, Ss^{l}, Svp-1^{b}, Svp-2^{b}, Tla^{a}, Trf^{b}	Low incidence of mammary tumors; some hepatomas in ♂; very sensitive to insulin; very resistant to X-radiation	5

continued

Strain	Inbred Generations	Genes Carried [Excluded Alleles]	Characteristics	Reference
C57L	130	*a*, *Ah*^{b}, *Bgs*^{h}, *Car-1*^{a}, *Car-2*^{b}, *Ce-1*^{c}, *Ea-2*^{2}, *Ea-4*^{a}, *Ea-7*^{b}, *Ema*^{l}, *Es-1*^{a}, *Es-8*^{a}, *Es-10*^{a}, *Fv-1*^{n}, *Fv-2*^{r}, *Gpd-1*^{a}, *Gus*^{b}, *Hba*^{a}, *Hbb*^{s}, *Hc*^{1}, *H-2*^{b}, *H-3*^{a}, *H-4*^{a}, *H-7*^{a}, *H-8*^{a}, *H-9*^{a}, *H-12*^{a}, *Ig-1*^{a}, *Ig-2*^{a}, *Ig-3*^{H9}, *Ig-4*^{a}, *Lap-1*^{a}, *Ldr-1*^{a}, *Lv*^{a}, *Ly-1*^{b}, *Ly-2*^{b}, *Ly-3*^{b}, *Ly-4*^{b}, *Ly-5*^{a}, *Mls*^{b}, *Mpi-1*^{b}, *Mud*^{1}, *Mup-1*^{b}, *Pgm-1*^{a}, *Pre-1*^{o}, *Prt-1*^{a}, *Prt-2*^{b}, *Sas-1*^{a}, *Slp*^{o}, *Ss*^{h}, *Thy-1*^{b}, *Tla*^{b}	Low mammary tumor incidence; some pituitary tumors in old animals; type B reticular cell tumors in 50% of breeders. Hematocrit extremely high; congenital cystic ovaries frequent.	5
C58	166	*a*, *Acf-1*^{a}, *Ah*^{b}, *Bgs*^{h}, *Car-1*^{a}, *Car-2*^{a}, *Ce-1*^{c}, *Dip-1*^{a}, *Ea-2*^{2}, *Ea-4*^{a}, *Ea-5*^{b}, *Ea-6*^{a}, *Ea-7*^{a}, *Ema*^{l}, *Es-1*^{b}, *Es-2*^{b}, *Es-3*^{c}, *Es-6*^{a}, *Es-8*^{a}, *Es-10*^{a}, *Fv-1*^{n}, *Fv-2*^{r}, *Gdr-2*^{l}, *Got-2*^{b}, *Gpd-1*^{a}, *Gpi-1*^{a}, *Gpt-1*^{a}, *Gus*^{b}, *Hba*^{b}, *Hbb*^{s}, *Hc*^{1}, *H-1*^{c}, *H-2*^{k}, *H-3*^{b}, *H-4*^{a}, *H-7*^{a}, *H-8*^{a}, *H-12*^{a}, *H-13*^{b}, *Id-1*^{a}, *If-1*^{h}, *Ig-1*^{a}, *Ig-2*^{a}, *Ig-3*^{H9}, *Lap-1*^{a}, *Ldr-1*^{a}, *Ly-1*^{b}, *Ly-2*^{a}, *Ly-3*^{a}, *Ly-4*^{b}, *Ly-5*^{a}, *Mod-1*^{b}, *Mud*^{1}, *Mup-1*^{b}, *Pca-1*^{b}, *Pd*^{a}, *Pgm-1*^{a}, *Pgm-2*^{a}, *Pre-1*^{o}, *Sas-1*^{a}, *Slp*^{o}, *Thy-1*^{b}, *Tla*^{a}	Incidence of lymphatic leukemia in ♂♀ before 1 yr, 75-100%; frequent polyovular follicles; aplasia of kidney, ∿10%	5
DBA/1	>88	*a*, *Ah*^{b}, *b*, *Bgs*^{d}, *Car-1*^{a}, *Car-2*^{a}, *d*, *Ea-2*^{2}, *Ea-4*^{a}, *Ea-5*^{b}, *Ea-6*^{b}, *Ea-7*^{b}, *Ema*^{h}, *Es-3*^{c}, *Es-10*^{b}, *Fv-1*^{n}, *Fv-2*^{s}, *Gur*^{b}, [*Hba*^{c}], *Hbb*^{d}, *Hc*^{1}, *H-1*^{a}, *H-3*^{b}, *H-7*^{a}, *H-12*^{a}, *H-13*^{b}, *Id-1*^{b}, *If-1*^{l}, *Ig-1*^{c}, *Ig-2*^{c}, *Ig-3*^{H9}, *Lap-1*^{b}, *Ldr-1*^{a}, *Lv*^{a}, *Ly-1*^{a}, *Ly-2*^{a}, *Ly-3*^{b}, *Ly-4*^{a}, *Mls*^{a}, *Mph-1*^{b}, *Mud*^{2}, *Mup-1*^{a}, *Pgm-1*^{b}, *Pre-1*^{a}, *Rv-1*^{R}, *Rv-2*^{S}, *Sas-1*^{o}, *Slp*^{o}, *Ss*^{h}, *Tla*^{b}, *Trf*^{b}	Resistant to most DBA/2 tumors (melanoma S91 grows in both sublines); mammary tumors high in some sublines. High erythrocyte count; calcareous heart deposits in almost all retired breeders.	5
DBA/2	118	*a*, *Acf-1*^{b}, *Ah*^{d}, *As-1*^{a}, *b*, *Bgs*^{d}, *Car-1*^{a}, *Car-2*^{b}, *d*, *Dip-1*^{b}, *Ea-1*^{c}, *Ea-2*^{2}, *Ea-5*^{b}, *Ea-6*^{b}, *Ea-7*^{b}, *Ema*^{h}, *Es-1*^{b}, *Es-2*^{b}, *Es-3*^{c}, *Es-6*^{a}, *Es-8*^{a}, *Es-10*^{b}, *Fv-1*^{n}, *Fv-2*^{s}, *Gdc-1*^{b}, *Gdr-2*^{l}, *Got-2*^{b}, *Gpi-1*^{a}, *Gpt-1*^{a}, *Gr-1*^{a}, *Gur*^{b}, *Gus*^{b}, *Hba*^{a}, *Hbb*^{d}, *Hc*^{0}, *H-1*^{a}, *H-2*^{d}, *H-3*^{b}, [*H-4 a* or *b*], *H-7*^{a}, *H-8*^{b}, *H-12*^{a}, *H-13*^{b}, *Id-1*^{b}, *If-1*^{h}, *Ig-1*^{c}, *Ig-2*^{c}, *Ig-3*^{H9}, *Ig-4*^{a}, *Lap-1*^{b}, *Ldr-1*^{a}, *Lv*^{a}, *Ly-1*^{a}, *Ly-2*^{a}, *Ly-3*^{b}, *Ly-4*^{a}, *Ly-5*^{a}, *Mls*^{a}, *Mod-1*^{a}, *Mod-2*^{b}, *Mph-1*^{b}, *Mpi-1*^{b}, *Mud*^{2}, *Ox-1*^{d}, *Ox-2*^{d}, *Pgm-1*^{b}, *Pgm-2*^{a}, *Pp*^{d}, *Pre-1*^{a}, *Pre-4*^{a}, *Pro-1*^{a}, *Prt-1*^{a}, *Prt-2*^{b}, *Sas-1*^{o}, *Sip*^{d}, *Slp*^{a}, *Ss*^{h}, *Svp-1*^{a}, *Tol-1*^{a}, *Trf*^{b}	Resistant to most DBA/1 tumors (melanoma S91 grows in both sublines; leukemia P1534 grows in 50%); widely variable incidence of mammary tumors among sublines. Audiogenic seizures, 100% at 35 d; 5% after 55 d. High mortality in ♂ exposed to chloroform fumes & to vitamin K deficiency. Low ethanol preference; high erythrocyte count; calcareous heart deposits; low concentration of lipids in adrenals.	2,4,5,6
DE	70	*Acf-1*^{a}, *Bgs*^{h}, *c*^{e}, *Dip-1*^{b}, *Es-1*^{b}, *Es-2*^{b}, *Es-3*^{c}, *Es-6*^{a}, *Es-8*^{a}, *Got-2*^{b}, *Gpd-1*^{b}, *Gpi-1*^{a}, *Hba*^{c}, *Hbb*^{s}, *Hc*^{0}, *H-2*^{k}(?), *Id-1*^{a}, *Ig-1*^{f}, *Ig-3*^{H9}, *Ig-4*^{a}, *Ldr-1*^{b}, *Mod-1*^{a}, *Mod-2*^{b}, *Mpi-1*^{b}, *Mud*^{2}, *Pgm-1*^{a}, *Pgm-2*^{a}, *Sas-1*^{o}	High incidence of amyloidosis; polydipsia (30-50 ml/d) & polyuria (25-45 ml/d) more noticeable in ♀; specific gravity of urine, 1.005	5
DW	60	*Bgs*^{h}, *Car-1*^{a}, *Car-2*^{b}, *Dip-1*^{b}, *dw*, *Es-1*^{b}, *Es-2*^{b}, *Es-3*^{a}, *Es-6*^{a}, *Es-8*^{a}, *Es-10*^{a}, *Got-2*^{b}, *Gpd-1*^{a}, *Gpi-1*^{a}, *Gr-1*^{a}, *Hba*^{b}, *Hbb*^{s}, *H-2*^{b}, *Id-1*^{a}, *Ldr-1*^{b}, *Mod-1*^{a}, *Mpi-1*^{b}, *Pgm-1*^{a}, *Pgm-2*^{a}, *Svp-2*^{c}	Dwarf; useful for endocrine studies	5
F	95	*a*, *b*, *Bgs*^{d}, *c*^{ch}, *d*, *Hc*^{1}, *s*	High leukemia in older animals regardless of breeding conditions; white head markings	1,5
FL/1Re	65	*a*, *Bgs*^{h}, *Car-1*^{a}, *Car-2*^{b}, *Es-10*^{a}, *f*, *Hbb*^{d}, *H-2*^{k}, *Mup-1*^{a}, *rd*	Diffuse hemoglobin; both α-chain & β-chain characterized. Fetuses & newborn show microcytic siderocytic anemia; completely cured by 14 d.	5
GRS	76	*Ah*^{b}, *Bgs*^{d}, *Dip-1*^{b}, *Es-1*^{b}, *Es-2*^{b}, *Es-3*^{b}, *Es-5*^{b}, *Fv-1*^{n}, *Fv-2*^{s}, *Gpd-1*^{b}, [*Hba*^{c}], *Hbb*^{s}, *Hc*^{0}, *H-2*^{dx}, *H-7*^{a}, *Id-1*^{a}, *If-1*^{h}, *Ldr-1*^{a}, *Ly-1*^{b}, *Ly-2*^{a}, *Ly-3*^{b}, *Ly-5*^{a}, *Mod-1*^{a}, *Mod-2*^{a}, *Mup-1*^{b}, *Pgm-1*^{b}, *Pgm-2*^{a}, *Ss*^{h}, *Thy-1*^{b}, *Tla*^{c}, *Trf*^{b}	Not related to American mice. Nearly 100% mammary tumors; latent period, 3 mo. Carries a mammary tumor agent different from mammary tumor inciter ⟨MTI⟩ of American strains; not dependent on milk transmission.	1,5

Strain	Inbred Generations	Genes Carried [Excluded Alleles]	Characteristics	Reference
HRS/J	54	Bgs^h, Car-1^a, Car-2^b, Dip-1^a, Ea-2^2, Ea-4^a, Ea-7^b, Es-1^b, Es-3^a, Es-6^a, Es-8^a, Es-10^a, Gpd-1^a, Gpi-1^a, Gur^a, Hba^b, Hbb^s, hr, H-2^k, Id-1^a, Lap-1^b, Ldr-1^a, Lv^a, Mod-1^b, Mud^l, p, Pgm-1^a, Pgm-2^a, Pre-1^o, Tla^b	Permanent near-hairlessness after first molt; tylotrichs may persist up to 1.5 yr; leukemia in hr/hr at 18 mo, 72%	5
I	117	a, Acf-1^b, Ah^d, b, Bgs^d, Car-1^a, Car-2^a, d, Dip-1^b, Ea-2^2, Ea-4^a, Ea-5^a, Ea-6^a, Ea-7^b, Es-1^b, Es-2^b, Es-3^c, Es-10^b, Fv-1^b, Fv-2^s, Gpd-1^b, [Hba^c], Hbb^d, Hc^0, H-2^j, Id-1^a, Ig-1^c, Ig 3^{H9}, Ldr-1^a, Ly-1^b, Ly-2^a, Ly-3^b, Mod-1^b, Mph-1^a, Mup-1^a, p, Pca-1^b, Pgm-1^b, Pgm-2^a, Phk^b, Pre-1^a, s, Thy-1^b, Tla^b	Some lines carry ln and/or c^{ch}, and some do not. Resistant to chemical induction of tumors.	5
IF	44	a^t, Es-5^b, Fv-2^r, Hbb^d, Hc^0, Ly-1^b, Ly-2^a, Ly-3^b, Ly-5^a, Ss^h, Thy-1^b, Tla^c, Trf^b	No spontaneous mammary tumors, but high sensitivity to agent or to chemical induction; susceptible to ovarian tumor induction by chemicals; high incidence of spontaneous pseudopregnancy	5
IS[1/]	55	Amy-1^a, Amy-2^b, Car-1^b, Car-2^a, Es-1^b, Es-2^a, Es-3^c, Es-10^a, Got-2^a, Gpd-1^b, Gpi-1^a, Hbb^d, Ldr-1^b, Mod-1^a, Pgm-1^b, Pgm-2^a	....	5
JU	69	a, c, Gpi-1^b, Hbb^d, Hc^0, Mod-2^b, Trf^b	Average first litter size, 9; 50% prenatal mortality in second litter when gestation is concurrent with suckling first litter	5
KE	69	a, b, Bgs^d, c	Mean litter size, 6.1; low percentage of fertilized ova; 17.6% of spermatozoa have abnormal heads	1,5
KK	65	a, A^y, c, Es-1^b, Es-2^a, Gpd-1^a, Hbb^s, Hc^0, Id-1^a, Mod-1^a, Trf^b	Small litters; frequent diabetes mellitus; occasional obesity in old animals, especially in presence of yellow gene	5
KP	69	a, b, p	Mean litter size, 5; high embryonal & postnatal mortality; frequent sterile matings; low sperm production; testis abnormalities	5
LG	67	a, Acf-1^b, Bgs^d, c, Car-1^a, Car-2^a, Dip-1^b, Es-1^b, Es-2^b, Es-3^c, Es-6^a, Es-8^a, Es-10^b, Gpd-1^b, Gpi-1^a, Gr-1^a, Gus^a, [Hba^c], Hbb^d, H-2^d, Id-1^a, Lap-1^b, Ldr-1^a, Ly-4^b, Mod-1^a, Mod-2^b, Mpi-1^b, Pgm-1^a, Pgm-2^a, Trf^b	Birth wt, 1.8 g; 28-d wt, 20.1 g; 60-d wt, 37.4 g; average litter size, 6; sex ratio, 56% ♂. Docile; thyroid activity low.	5
LP	97	A^w, Acf-1^a, Ah^d, Bgs^d, Car-1^a, Car-2^a, Dip-1^a, Ea-2^2, Ea-4^a, Ea-7^a, Es-1^b, Es-2^b, Es-3^c, Es-6^a, Es-8^a, Es-10^b, Gdc-1^b, Got-2^b, Gpd-1^b, Gpi-1^a, Gpt-1^a, Gur^b, Gus^b, [Hba^c], Hbb^s, Hc^1, [H-1 a or c], H-2^{bc}, H-3^b, [H-4 a or b], [H-7 a or b], H-8^b, [H-12 a or b], H-13^b, Id-1^a, Ig-1^b, Ig-3^{H9}, Ig-4^b, Lap-1^b, Ldr-1^b, Lv^c, Ly-1^b, Ly-2^b, Ly-3^b, Mod-1^a, Mpi-1^b, Mud^l, Mup-1^a, Pd^a, Pgm-1^a, Pgm-2^a, Pre-1°, Pre-4^a, s, Sas-1^a, Thy-1^b	Late-occurring tumors of various types; large amount of lipids in adrenals of both sexes	5
LS/Le	50	a^t, Hba^c, Hbb^s, ls	Homozygous ls/ls develop megacolon, but some live and breed	5
LT	94	a, B^{lt}, Car-1^a, Car-2^b, Hba^b, Hbb^s, Lap-1^a, Ly-4^a, Pre-1^a	Hair almost white except for blackish tip	5
MA	97	Acf-1^b, Bgs^h, c, Car-1^a, Car-2^a, Dip-1^b, Ea-2^2, Ea-4^a, Ea-7^b, Es-1^b, Es-2^b, Es-3^c, Es-6^a, Es-8^a, Es-10^a, Got-2^b, Gpd-1^b, Gpi-1^a, Gr-1^a, Gus^b, Hba^a, Hbb^s, Hc^1, H-2^k, Id-1^a, Ig-1^a, Ig-2^a, Ig-3^{H9}, Lap-1^b, Ldr-1^a, Lv^a, Ly-1^a, Ly-2^a, Ly-4^b, Mod-1^a, Mod-2^b, Mpi-1^a, Mud^l, Mup-1^a, Pca-1^a, Pgm-1^b, Pgm-2^a, Pre-1^o, Slp^o, Thy-1^a, Tla^b	Tumor incidence, 0% in ♂; 5% in ♀. Incidence of polydipsia-polyuria, 12% in ♂; 79% in virgin ♀. Gross kidney disease, 25% in ♂. Pituitary cysts in >50%. High resistance to chronic, whole-body X-irradiation. Reduction in size of lateral septal nucleus.	5

1/ *Mus musculus praetextus* x *M. musculus musculus.*

continued

Strain	Inbred Generations	Genes Carried [Excluded Alleles]	Characteristics	Reference
MK	40	Bgs^{h}, Car-1^{a}, Car-2^{a}, Es-10^{a}, Hba^{c}, Hbb^{d}	All *mk/mk* born alive; 1/3 to 1/2 die before 3 wk, developing skin lesions & tail amputation; surviving *mk/mk* are fertile and appear normal except for supernormal numbers of very small erythrocytes	5
NH	82	a, Bgs^{d}, d, Ig-1^{f}, Ig-3^{H9}, Mup-1^{a}, p, s	High in renal amyloidosis; nodular hyperplasia & adrenal adenomas in at least 10% of old ♂♀; incidence increased with gonadectomy. Lung tumors in very old animals, 30-40%.	1,5
NZB	96	a, Acf-1^{a}, Bgs^{d}, Car-1^{a}, Car-2^{a}, Ea-2^{2}, Ea-4^{a}, Es-1^{b}, Es-2^{b}, Es-3^{a}, Es-6^{a}, Es-10^{b}, Fv-1^{n}, Fv-2^{s}, Gpd-1^{b}, Gpi-1^{a}, Gpt-1^{b}, Gr-1^{a}, Gur^{b}, [Hba^{c}], Hbb^{d}, Hc^{0}, H-2^{d}, Id-1^{a}, Ig-1^{e}, Lap-1^{b}, Ldr-1^{a}, Ly-1^{b}, Ly-2^{b}, Ly-3^{b}, Ly-5^{a}, Mls^{a}, Mod-1^{b}, Mod-2^{b}, Mph-1^{b}, Pca-1^{a}, Pgm-2^{a}, Pre-1^{a}, Tla^{a}	Autoimmune hemolytic anemia; extramedullary erythropoiesis; glomerular & tubular renal lesions; lupus-like nephritis	5
NZC	109	a, b, Bgs^{d}, Dao^{b}, Hao-1^{b}	Low mammary cancer; congenital cystic kidney; increasing incidence of adenoacanthoma	1,3,5
NZO	100	Bgs^{d}, Hc^{1}, Ig-1^{e}, Mud^{1}	Low mammary tumor; some lung & ovarian tumors; obese	1,5
NZW	70	b, Bgs^{d}, c, H-2^{z}, Ig-1^{e}, Mph-1^{b}	Hybrids between NZB & NZW develop lupus nephritis, with positive LE cell tests & antinuclear antibodies	1,5
NZX	62	....	Some ♀ have imperforate vagina; low incidence of megacolon in ♂♀	5
NZY	93	b, Bgs^{d}, Ig-1^{a}, s	High mammary cancer incidence associated with pituitary enlargement in breeders; heritable megacolon; highly susceptible to ovarian tumor induction by chemicals	1,5
O20	164	a, Ah^{b}, Bgs^{d}, c, Dip-1^{b}, Es-1^{b}, Es-3^{c}, Es-5^{b}, Fv-1^{n}, Fv-2^{s}, Gpd-1^{a}, Hbb^{d}, Hc^{0}, Id-1^{a}, Ldr-1^{a}, Ly-1^{b}, Ly-2^{a}, Ly-3^{b}, Mod-1^{a}, Mup-1^{b}, Ss^{h}, Thy-1^{b}, Trf^{b}	Very low mammary tumor incidence; susceptible to mammary tumor agent and can transfer it to offspring	1,5
P/J	124	a, Acf-1^{b}, b, Bgs^{d}, Car-1^{a}, Car-2^{a}, d, Dip-1^{b}, Es-1^{b}, Es-2^{b}, Es-3^{a}, Es-6^{a}, Es-8^{a}, Es-10^{a}, Got-2^{a}, Gpd-1^{a}, Gpi-1^{a}, Gus^{b}, Hbb^{d}, Hc^{1}, H-2^{p}, Id-1^{b}, Ig-1^{h}, Ig-2^{a}, Ig-4^{a}, Lap-1^{b}, Ldr-1^{a}, Ly-4^{a}, Mod-1^{b}, Mpi-1^{b}, Mud^{1}, Mup-1^{b}, p, Pgm-1^{b}, Pgm-2^{a}, Pre-1^{a}, rd, se, Slp^{a}	Fairly low tumor incidence; high resistance to chronic, whole-body X-irradiation; some polycystic or granular kidneys in ♂♀	5
PL	100	Acf-1^{b}, Ah^{b}, Bgs^{d}, c, Car-1^{a}, Car-2^{b}, Dip-1^{b}, Ea-2^{2}, Ea-4^{a}, Ea-7^{b}, Ema^{h}, Es-1^{b}, Es-2^{c}, Es-3^{c}, Es-6^{a}, Es-8^{a}, Es-10^{b}, Got-2^{b}, Gpd-1^{a}, Gpi-1^{a}, Gpt-1^{a}, Gr-1^{a}, Gur^{b}, Gus^{b}, [Hba^{c}], Hbb^{d}, Hc^{1}, [H-1 *a* or *b*], H-2^{u}, [H-3 *a* or *b*], [H-4 *a* or *b*], [H-7 *a* or *b*], [H-8 *a* or *b*], H-9^{a}, H-12^{a}, [H-13 *a* or *b*], Id-1^{b}, Ig-1^{a}, Ig-3^{H9}, Lap-1^{b}, Ldr-1^{a}, Lv^{c}, Ly-1^{b}, Ly-2^{a}, Ly-3^{a}, Ly-4^{a}, Mod-1^{a}, Mod-2^{b}, Mpi-1^{b}, Mud^{2}, Pca-1^{a}, Pd^{a}, Pgm-1^{b}, Pgm-2^{a}, Sas-1°, Slp^{o}, Thy-1^{a}, Tla^{a}	High leukemia incidence; low mammary tumor incidence	5
RF	80	a, Acf-1^{b}, Ah^{d}, Bgs^{d}, c, Car-1^{a}, Car-2^{a}, Dip-1^{b}, Ea-2^{1}, Ea-4^{a}, Ea-5^{b}, Ea-6^{b}, Ea-7^{b}, Ema^{l}, Es-1^{b}, Es-2^{b}, Es-3^{b}, Es-6^{a}, Es-8^{a}, Es-10^{b}, Fv-1^{n}, Fv-2^{s}, Gdc-1^{b}, Gdr-2^{l}, Got-2^{b}, Gpd-1^{a}, Gpi-1^{a}, Gpt-1^{a}, Gr-1^{a}, Gur^{b}, Gus^{b}, Hba^{a}, Hbb^{d}, Hc^{0}, [H-1 *a* or *c*], H-2^{k}, [H-3 *a* or *b*], [H-4 *a* or *b*], [H-7 *a* or *b*], [H-8 *a* or *b*], [H-9 *a* or *b*], [H-12 *a* or *b*], H-13^{a}, Id-1^{a}, If-1^{h}, Ig-1^{c}, Ig-3^{H9}, Lap-1^{b}, Ldr-1^{a}, Lv^{b}, Ly-2^{a}, Ly-4^{a}, Ly-5^{a}, Mod-1^{a}, Mpi-1^{b}, Mud^{2}, Mup-1^{a}, Pca-1^{a}, Pgm-1^{a}, Pgm-2^{a}, Pre-1^{a}, Slp^{o}, Ss^{l}, Thy-1^{a}, Tla^{b}	Leukemia incidence, up to 50%; radiation resistance & blood pressure in mid-range; glomerulosclerosis progressive with age. Differs from RFM at Es-2 & H-2 loci.	5

continued

Strain	Inbred Generations	Genes Carried [Excluded Alleles]	Characteristics	Reference
RIII	78	Ah^{b}, Bgs^{d}, Car-1^{a}, Car-2^{b}, Dip-1^{b}, Ea-2^{l}, Ea-4^{a}, Ea-5^{b}, Ea-7^{b}, Ema^{h}, Es-1^{b}, Es-2^{b}, Es-3^{c}, Es-5^{b}, Es-6^{a}, Es-8^{a}, Es-10^{b}, Fv-2^{s}, Got-2^{b}, Gpd-1^{b}, Gpi-1^{a}, Gr-1^{a}, Hbb^{s}, Hc^{l}, H-2^{r}, Id-1^{a}, Ig-1^{g}, Ig-3^{H9}, Lap-1^{a}, Ldr-1^{a}, Ly-2^{b}, Ly-3^{b}, Ly-5^{b}, Mod-1^{b}, Mod-2^{b}, Mud^{2}, Mup-1^{b}, Pgm-1^{a}, Pgm-2^{a}, Pre-1^{a}, Ss^{h}, Thy-1^{b}, Tla^{b}, Trf^{b}	Mammary tumor incidence varies between sublines, 45-96%; carries mammary tumor agent; low leukemia incidence; audiogenic seizures at sufficiently loud sounds. RIII/SeA subline differs at several polymorphic loci.	5
RNC	>20	nu, H-2^{k}, Mls^{a}, Re	Nude; athymic	5
SEC/1Gn	112	a, b, c^{ch} +/+ se, H-2^{d}, Ig-1^{h}	Multiple lung cysts, especially in se/se	5
SF/Cam	50	Amy-1^{a}, Amy-2^{a}, Car-1^{a}, Car-2^{a}, Es-1^{b}, Es-2^{b}, Es-3^{c}, Es-10^{b}, Gpd-1^{a}, Gpi-1^{b}, Hbb^{d}, Id-1^{a}, Lap-1^{b}, Ldr-1^{a}, Mod-1^{a}	Adrenal X zone relatively & absolutely large in ♀; very poorly developed zona glomerulosa	5
SJL	73	Acf-1^{b}, Ah^{d}, Bgs^{d}, Car-1^{a}, Car-2^{b}, Dip-1^{b}, Ea-2^{2}, Ea-4^{a}, Ea-7^{b}, Ema^{h}, Es-1^{b}, Es-2^{b}, Es-3^{c}, Es-6^{a}, Es-8^{a}, Es-10^{b}, Gdc-1^{b}, Gdr-2^{l}, Got-2^{b}, Gpd-1^{b}, Gpi-1^{a}, Gpt-1^{a}, Gr-1^{b}, Gur^{b}, Gus^{b}, Hba^{c}, Hbb^{s}, Hc^{l}, [H-1 a or c], H-2^{s}, [H-3 a or b], H-4^{a}, [H-7 a or b], [H-8 a or b], [H-9 a or b], [H-12 a or b], [H-13 a or b], Id-1^{b}, Ig-1^{b}, Ig-3^{H9}, Ig-4^{b}, Lap-1^{b}, Ldr-1^{a}, Lv^{a}, Ly-2^{b}, Ly-3^{b}, Ly-4^{a}, Ly-5^{b}, Mod-2^{b}, Mpi-1^{b}, Mud^{2}, Pca-1^{a}, Pd^{b}, Pgm-1^{b}, Pgm-2^{a}, Pre-1^{a}, Sas-1°, Tla^{a}	Multicellular reticulum cell sarcomas, 88-95%; high resistance to chronic whole-body X-irradiation; γ_1- & γ_2-paraproteinemia; ♂ have low erythrocyte count; low incidence of audiogenic seizures; hepatosplenic amyloidosis after 6 mo; blood pressure in mid-range; fastest heart rate of 7 strains tested	5
SK	24	Amy-1^{a}, Amy-2^{b}, Es-1^{b}, Es-2^{a}, Es-3^{c}, Es-6^{b}, Es-9^{b}, Es-10^{b}, Got-2^{a}, Gpd-1^{a}, Gpi-1^{a}, Hbb^{d}, Ldr-1^{b}, Mod-1^{b}, Pgm-1^{a}, Pgm-2^{a}	Fairly large litters	5
SM	80	Acf-1^{b}, Ah^{b}, Bgs^{d}, Car-1^{b}, Car-2^{b}, Dip-1^{b}, Ea-2^{2}, Ea-4^{a}, Ea-7^{b}, Es-1^{b}, Es-2^{b}, Es-3^{c}, Es-6^{a}, Es-10^{b}, Got-2^{b}, Gpd-1^{b}, Gpi-1^{a}, Gpt-1^{a}, Gr-1^{a}, Gur^{a}, Gus^{a}, Hba^{b}, Hbb^{s}, Hc^{l}, H-2^{v}, Id-1^{b}, Ig-1^{b}, Ig-3^{H9}, Ig-4^{b}, Lap-1^{b}, Ldr-1^{a}, Lv^{b}, Ly-2^{a}, Ly-3^{b}, Mod-1^{a}, Mod-2^{a}, Mpi-1^{b}, Mud^{2}, Pgm-2^{b}, Pre-1^{o}, Sas-1^{o}, Thy-1^{b}, Tla^{b}	Small body size at birth & at weaning, but relatively small size tends to disappear at maturity; very low tumor incidence; amyloidosis at 1 yr, 100%	5
ST/b	112	Acf-1^{b}, Ah^{d}, Bgs^{d}, Car-1^{a}, Car-2^{b}, Dip-1^{b}, Ea-2^{2}, Ea-4^{a}, Ea-7^{b}, Ema^{l}, Es-1^{b}, Es-2^{b}, Es-3^{b}, Es-6^{a}, Es-10^{a}, Fv-1^{n}, Fv-2^{s}, Got-2^{b}, Gpd-1^{b}, Gpi-1^{a}, Gpt-1^{a}, Gr-1^{a}, Gus^{b}, Hba^{c}, Hbb^{d}, Hc^{0}, H-2^{k}, Id-1^{a}, If-1^{l}, Ig-1^{a}, Ig-2^{a}, Ig-3^{H9}, Lap-1^{b}, Ldr-1^{a}, Lv^{a}, Ly-2^{b}, Ly-3^{b}, Ly-5^{a}, Mod-1^{b}, Mpi-1^{b}, Mud^{l}, Mup-1^{a}, Pgm-1^{a}, Pgm-2^{a}, Phr^{h}, Pre-1^{a}, Sas-1°, Ss^{l}, Thy-1^{b}, Tla^{b}, Trf^{b}	1-2% leukemia, including some plasma cell type; other tumors, including pulmonary adenomas & mammary carcinomas, ∿3%	5
SWR	117	Acf-1^{b}, Ah^{d}, As-1^{a}, Bgs^{d}, Car-1^{a}, Car-2^{b}, Dip-1^{b}, Ea-2^{2}, Ea-4^{a}, Ea-7^{b}, Ema^{l}, Erp-1^{a}, Es-1^{b}, Es-2^{b}, Es-3^{c}, Es-6^{a}, Es-8^{a}, Es-10^{a}, Gdc-1^{b}, Gdr-2^{l}, Got-2^{a}, Gpd-1^{b}, Gpi-1^{b}, Gr-1^{b}, Gur^{b}, Gus^{b}, Hba^{c}, Hbb^{s}, Hc^{0}, [H-1 a or b], H-2^{q}, H-3^{b}, [H-4 a or b], [H-7 a or b], [H-8 a or b], H-12^{b}, H-13^{a}, Id-1^{a}, Ig-1^{c}, Ig-2^{c}, Ig-3^{H9}, Ig-4^{a}, Lap-1^{b}, Ldr-1^{b}, Lv^{a}, Ly-1^{b}, Ly-2^{b}, Ly-3^{b}, Ly-4^{a}, Ly-5^{a}, Mls^{c}, Mod-1^{a}, Mod-2^{b}, Mph-1^{b}, Mpi-1^{b}, Mud^{l}, Mup-1^{a}, Pd^{a}, Pgm-1^{a}, Pgm-2^{a}, Pre-1^{o}, Sas-1^{a}, Thy-1^{b}, Tla^{a}, Trf^{b}	Mammary tumors, up to 30% in some sublines; primary lung tumors, ∿50%; some leukemia; ♀ develop polydipsia at 6-10 mo; high mortality in ♂ exposed to ethylene oxide products & to vitamin K deficiency; in mid-range of radiation resistance; arteriosclerosis common in ♂♀. Resting O_2 consumption highest of 7 strains tested; highest blood pressure of 19 strains.	2,5
SWV	46	....	♀ over 8 mo have nephrogenic diabetes insipidus; vasopressin resistant; ♀ have unique progressive kidney defect resembling nephronophthisis in some aspects & hypokalemia in others	5

continued

Strain	Inbred Generations	Genes Carried [Excluded Alleles]	Characteristics	Reference
TM	65	....	No mammary or liver tumors; some adrenocortical tumors. Unusually large postcastrational accumulation of body fat.	5
WB	90	a, Bgs^h, Car-1^b, Car-2^a, Ea-2^2, Ea-4^a, Ea-7^b, Es-10^a, Gus^b, Hba^c, Hbb^d, H-1^a, H-2^{ja}, H-3^a, [H-4 a or b], [H-7 a or b], [H-8 a or b], H-9^a, [H-12 a or b], [H-13 a or b], Ig-1^b, Ly-1^b, Ly-2^b, Ly-4^b, Mup-1^a, Pre-1^a, rd, Ss^h, Thy-1^b, Tla^b, Trf^b, W	Homozygotes (W/W) are anemic, sterile, and lack coat pigmentation. Heterozygotes have normal blood picture & fertility, but white ventral spotting. Anemics die at ∿11 d.	5
WC	89	a, Bgs^d, Car-1^b, Car-2^a, Es-10^a, Hba^c, Hbb^d, H-1^a, H-2^{ja}, H-3^a, [H-4 a or b], [H-7 a or b], [H-8 a or b], [H-9 a or b], [H-12 a or b], H-13^a, Ig-1^b, Ly-4^b, Mup-1^a, rd, Ss^h, Trf^b, W	Like WB except that anemics die at 14 d. Some WC survive to adulthood.	5
WH	90	a, Bgs^d, Hba^c, Hbb^d, H-2^d, Ig-1^b, rd, W	Like WB, except that anemics die at ∿5 d	5
WHT	80	c, Hc^1	Hardy. Carries spontaneous squamous carcinoma which metastasizes to the lymph nodes.	5
WK	74	a, Bgs^d, Car-2^a, Es-10^b, Hba^a, Hbb^s, [H-1 a or c], H-2^{ja}, [H-3 a or b], H-4^a, H-7^a, [H-8 a or b], [H-9 a or b], [H-12 a or b], H-13^a, Ly-4^b, Pre-1°, rd, W	Like WB except that anemics die at ∿10 d	5
X/Gf	58	a, B/b, c	No tumors of any kind; does not carry milk agent but produces antibodies against it; resistant to many viruses & to many carcinogens; radiation not carcinogenic	5
XLII	45	b, Es-1^b, Hbb^d, Id-1^b, Mod-1^b, Trf^a	Very low tumor incidence; hepatomas induced by radiation	5
XVII	80	Bgs^h, c	Low tumor incidence; very susceptible to chemical induction of lung tumors	1,5
YBR	104	A^y/a, b, Bgs^h, Ce-1^b, Ea-1^c, Ea-2^2, Hc^0, H-2^d, Ig-3^{H9}, Mup-1^b, Slp^a, Ss^h	Amyloid in ♂♀ with A^y/a or a/a, at least 50%; obesity in hybrids between A^y/a & other strains; low tumor incidence. Lacks microsomal β-glucuronidase.	5
101	66	A^w, Bgs^d, Gpi-1^a, Hc^1, H-2^k, Ig-1^b, Mod-2^b	Low mammary tumor incidence; susceptible to skin & lung tumor induction	1,5
129	80	A^w, Acf-1^b, Ah^d, Bgs^d, c^{ch}, Car-1^a, Car-2^a, Dip-1^b, Ea-2^2, Ea-4^a, Ea-5^a, Ea-6^a, Ea-7^a, Ema^h, Es-1^b, Es-2^b, Es-3^c, Es-6^a, Es-8^a, Es-10^b, Fv-1^n, Fv-2^s, Gdr-2^l, Got-2^b, Gpd-1^a, Gpi-1^a, Gpt-1^a, Gr-1^a, Gus^b, Hba^a, Hbb^d, Hc^1, H-1^b, H-2^{bc}, H-3^b, H-4^b, H-7^a, H-8^b, H-12^b, H-13^b, Id-1^a, Ig-1^a, Ig-2^a, Ig-3^{H9}, Ig-4^a, Lap-1^b, Ldr-1^a, Lv^b, Ly-1^b, Ly-2^b, Ly-3^a, Ly-4^a, Ly-5^a, Mod-1^a, Mod-2^b, Mph-1^b, Mud^l, Mup-1^a, p, Pca-1^b, Pd^a, Pgm-1^a, Pgm-2^a, Pre-1^a, Pro-1^a, Sas-1^a Svp-1^a, Thy-1^b	Incidence of spontaneous testicular teratoma, 5%. Not susceptible to mammary tumor agent; highly sensitive to estrogen at all ages; blood pressure relatively low; very resistant to X-radiation; high incidence of urinary calculi. Slowest heart rate of 7 strains tested.	5
129/Sv-*ter*	27	....	Incidence of spontaneous congenital testicular teratomas, ∿30%, appearing at 12-13 d of gestation. Tumors may be uni- or bilateral, and may contain tissues derived from all germ layers; rarely metastasize.	5

continued

1. INBRED STRAINS: MOUSE

Contributor: Joan Staats

References

1. Breen, G. A. M., et al. 1977. Genetics 85:73-84.
2. Daniel, W. L. 1976. Biochem. Genet. 14:1003-1018.
3. Holmes, R. S. 1976. Ibid. 14:981-987.
4. Kumaki, K., et al. 1977. J. Biol. Chem. 252:157-165.
5. Staats, J. 1976. Cancer Res. 36(12):4333-4377.
6. Symons, J. P., and R. L. Sprott. 1976. Physiol. Behav. 17:837-839.

2. CONGENIC LINES: MOUSE

Congenic lines, by specific genes or traits, can be found on the indicated pages of the Tables listed below. Also some congenic-strain information can be found on pages 44, 81-83, 86-89, 92, 103-107, 112, 114, 124, 126-127, 140, 145-147, 159-160, 163, 167, 169-172, 194, 201, and 211.

Table No.	Table Title	Pages
3	Mutations on Inbred Backgrounds	29-36
14	Mutant Genes Affecting Development of Anemia and Erythrocyte Production	65-66
25	Immunoglobulin Allotypes	111
29	Cellular Alloantigens in Congenic Lines	115
31	Histocompatibility Systems, except *H-2*	119-122
34	*H-2* Congenic Lines Carrying Standard Haplotypes	124-125
36	*H-2* Congenic Lines Carrying Recombinant Haplotypes	127-128
37	Mutant *H-2* Haplotypes of the *H-2* Gene Complex	129-130
38	Wild-derived Haplotypes, Congenic Strains, and Specificities of the *H-2* Gene Complex	130-133
60	Spontaneous Testicular Teratomas	207-208

Contributor: Donald C. Shreffler

3. MUTATIONS ON INBRED BACKGROUNDS: MOUSE

Generations Inbred refers to the number of generations since the mutation occurred or was placed on the specific background; N = number of successive backcrosses to inbred strain; F = number of generations of brother x sister matings. Data in brackets or parentheses refer to the column heading in brackets or parentheses, respectively.

Part I. Maintained by The Jackson Laboratory

Strain: Data are for three kinds of stocks: (i) those with mutations that occurred on other backgrounds and have had seven or more generations of crossing into a standard inbred strain; (ii) those with genes that occurred as mutations in an already inbred strain and have been maintained within that strain either by repeated backcrossing or by brother x sister matings; (iii) those stocks inbred with forced heterozygosis for a mutation for at least 20 generations. For the first two kinds, the strain name is separated from the mutation by a hyphen. For the third, if there is a strain designation, the **Genotype** follows the strain name and is enclosed in parentheses; when there is no strain name, the genotype of the stock is listed. **Origin:** M indicates mutation occurred in inbred strain. **Reference** in each case refers to the gene, not to the inbred strain. *Abbreviation:* Superscript "J" attached to a gene symbol, e.g. "bg^J," indicates that the mutation occurred at The Jackson Laboratory.

Gene Symbol	Gene Name	Strain (Genotype) [Origin]	Generations Inbred	Reference
A^{vy}	Viable yellow	C57BL/6J-A^{vy}	N18	15
		$C3H_eB$/FeJ-A^{vy}	N21	

continued

3. MUTATIONS ON INBRED BACKGROUNDS: MOUSE

Part I. Maintained by The Jackson Laboratory

Gene Symbol	Gene Name	Strain (Genotype) [Origin]	Generations Inbred	Reference
A^{y}	Yellow	C57BL/6J-A^{y}	N95	15
		SEC/1Re-A^{y}	N67	
plus A^{w}	White-bellied agouti	129/Sv-c^{+} p^{+} A^{y}/A^{w}	N9, F32	15
plus *a*	Non-agouti	(A^{y}/a c^{ch}/c^{ch})	F55	15
$A^{w\text{-}J}$	White-bellied agouti-J	C57BL/6J-$A^{w\text{-}J}/A^{w\text{-}J}$ [M]	F55	15
plus a^{e}	Extreme non-agouti	C57BL/6J-$A^{w\text{-}J}/a^{e}$	N10, F14	15
		AEJ/Gn-$A^{w\text{-}J}/a^{e}$ *bd*/+ [M]	F42	
A^{i}	Intermediate agouti	C57BL/6J-A^{i}/a *jv*/+ [M]	F9	15
A^{iy}	Intermediate yellow	C57BL/6J-A^{iy}/a *pi*/+	N12, F3	5
A^{sy}	Sienna yellow	C57BL/6J-A^{sy}/a *jc*/+ [M]	N13, F2	9
A	Agouti	C57BL/6J-*A md*	N10	15
a^{t}	Black and tan	LS/Le (a^{t} *ls*/+ +)	F58	15
		MWT/Le (a^{t}/a^{t} Mi^{wh}/+ W^{v}/+ *T*/+)	F43	
		(*we un* a^{t}/+ + a^{t})	F40	
$a^{t\text{-}33J}$	Black and tan-33J	C57BL/6J-$a^{t\text{-}33J}$ [M]	N22	15
a^{td}	Tanoid	C57BL/6J-a^{td} [M]	N84	15
a	Non-agouti	$C3H_{e}B$/FeJ-*a*/*a*	N7, F7	15
ab^{J}	Asebia-J	(ab^{J}/+ *b*/*b*)	F32	15
ak	Aphakia	C57BL/Ks-*ak*	N5, F22	46
an	Anemia	C57BL/6J-*an* B^{lt}	N49	15
		WB/Re-*an* B^{lt}	N22	
ax^{J}	Ataxia-J	C57BL/6J-ax^{J}	N10	15
b	Brown	C57BL/6J-*b m*	N39	15
		C57BL/6J-*wi b*	N14	
b^{J}	Brown-J	C57BL/6J-b^{J} [M]	N32	15
$b^{c\text{-}J}$	Cordovan-J	C57BL/6J-$b^{c\text{-}J}$	N17	15
B^{lt}	Light	C57BL/6J-*an* B^{lt}	N49	15
		WB/Re-*an* B^{lt}	N22	
bd	Bradypneic	AEJ/Gn-$A^{w\text{-}J}/a^{e}$ *bd*/+ [M]	F42	19
bf	Buff	C57BL/6J-*bf* [M]	N9	15
bg	Beige	C57BL/6J-*bg*/+	N11, F7	15
bg^{J}	Beige-J	C57BL/6J-bg^{J} [M]	N13	15
bg^{2J}	Beige-2J	C3H/HeJ-bg^{2J}/+ [M]	F18	15
bm	Brachymorphic	(*a*/*a bm*/+)	F40	30,37
Bn	Bent-tail	(*a*/*a Bn*/+)	F44	15
bp^{J}	Brachypodism-J	A/J-bp^{J} [M]	N2	15
bt^{2J}	Belted-2J	C57BL/6J-*uw* bt^{2J}/+ + [M]	N4, F4	15
c^{J}	Albino-J	C57BL/6J-c^{J} [M]	N37	15
c^{2J}	Albino-2J	C57BL/6J-c^{2J} [M]	N5	15
c^{4J}	Albino-4J	B10.D2/nSn-c^{4J}/+ [M]	F16	15
c^{ch}	Chinchilla	SEC/1GnLe (*a*/*a b*/*b* c^{ch}/c^{ch} *se*/+)	F120	15
		(A^{y}/a c^{ch}/c^{ch})	F55	
		(c^{ch}/c^{ch} *se* +/+ *sv*)	F45	
		(c^{ch} *sh-1*/c^{ch} +)	F50	
c^{h}	Himalayan	C57BL/6J-c^{h}	N34	15
c^{p}	Platinum	C57BL/6J-c^{p}	N19	5
Ca	Caracul	C57BL/6By-*Ca Sl*	N24	15
Ca^{J}	Caracul-J	C57BL/6J-Ca^{J} *Hm Sl* [Ca^{J} M]	N15	15
Cm	Coloboma	C57BL/6By-*Cm*	N23	41
cn	Achondroplasia	(*a*/*a c*/*c cn*/+)	F28	4,30
cri	Cribriform degeneration	DBA/2J-*cri*/+ [M]	N2, F18	20

continued

3. MUTATIONS ON INBRED BACKGROUNDS: MOUSE

Part I. Maintained by The Jackson Laboratory

Gene Symbol	Gene Name	Strain (Genotype) [Origin]	Generations Inbred	Reference
d	Dilute	C57BL-*sg* + +/+ *d se*	B6 N4 x B10 N4, F50	15
		(*a*/*a d fd*/+ +)	F30	
d^{l}	Dilute-lethal	DLS/Le (*a*/*a* d^{l} +/+ *se*)	F66	15
d^{+2J}	Dense-2J	DBA/2J-d^{+2J} [M]	N73	15
db	Diabetes	C57BL/Ks-*db* +/+ *m* [*db* M]	N5, F21	25
Dc	Dancer	(*Dc*/+)	F73	15
Den	Denuded	C57BL/6By-*Den*	N11	42
dm	Diminutive	C57BL/6J-*dm*	N12	15
		WB/Re-*dm*	N11	
dt^{J}	Dystonia musculorum-J	C57BL/6J-dt^{J} [M]	F33, N13	15
du	Ducky	(*a*/*a d* + *du*/*d tk* +)	F41	15
dw	Dwarf	C57BL/6J-*dw*	N9	15
		DW/J(*dw*/+ *ln*/*ln*)	F72	
dw^{J}	Dwarf-J	C3H/HeJ-dw^{J}/+ [M]	F12	15
dy	Dystrophia muscularis	C57BL/6J-*dy*	N28	15
		129/ReJ-*dy*/+ [M]	F70	
dy^{2J}	Dystrophia muscularis-2J	C57BL/6J-dy^{2J}	N11	35
e	Recessive yellow	C57BL/6J-*e* [M 1/]	J N15 2/	23
		C57BL/6J-*e*/*e*-Mi^{wh}	N15	
E^{so}	Sombre	(E^{so}/E^{so} *Tw*/+ Xt^{J}/+)	F34	15
Eh	Hairy ears	C57BL/6By-*Eh Sig*	N20	1
ep	Pale ear	C57BL/6J-*ep*/+	N20, F25	31
Er	Repeated epilation	C57BL/6By-*Er*	N21	15
f	Flexed-tail	FL/1Re (*a*/*a f*/+ *W*/+)	N69	15
		(*a*/*a f*/*f je*/+ *ru*/*ru*)	F22	
fd	Fur deficient	(*a*/*a d fd*/+ +)	F30	32
fi	Fidget	(*stb* + *a*/+ *fi a*)	F42	15
fs	Furless	FSB/GnEi (*fs*/+)	F66	15
Fu	Fused	129/Rr-*Fu*/+	N31, F34	15
fz	Fuzzy	C57BL/6J-*fz ln*/+ +	N9, F64	15
		(*a*/*a fz ln*/*fz ln s*/*s v*/+)	F28	
gm	Gunmetal	C57BL/6J-*gm*/+ [M]	F40	15
go	Angora	C57BL/6J-*go*/+	N12, F6	15
gr	Grizzled	(*a*/*a gr* +/+ *ji*)	F32	15
ha	Hemolytic anemia	C57BL/6J-*ha*	N36	15
		WB/Re-*ha*	N22	
Hba	Hemoglobin α-chain	C57BL/6J-*wa-2* Hba^{c} *sh-2*	N9	15
		SEC/1Re-*wa-2* Hba^{c} *sh-2*	N7	
Hbb	Hemoglobin β-chain	C57BL/6J-Hbb^{p}/Hbb^{p}	N30, F4	15
Hk	Hook	(*c*/*c Hk*/+)	F59	15
Hm	Hammer-toe	(57BL/6J-Ca^{J} *Hm Sl*	N15	15
ho^{2J}	Hotfoot-2J	AKR/J-ho^{2J}/+ [M]	F19	5
ho^{3J}	Hotfoot-3J	C57BL/6J-ho^{3J}	N9	5
hr	Hairless	HRS/J (*b*/*b c*/*c d*/*d hr*/+)	F63	15
		WLHR/Le (*a*/*a b*/*b wl* +/+ *hr*)	F70	
$hr^{rh\text{-}J}$	Rhino-J	(*a*/*a b*/*b c*/*c* $hr^{rh\text{-}J}$/+)	F44	15
Hx	Hemimelic extra toes	B10.D2/nSn-*Hx*/+ [M]	F26	8
hy-3	Hydrocephalus-3	(*hy-3*/+)	F56	15
Hyp	Hypophosphatemia	C57BL/6J-*Hyp*	N14	13

1/ Mutation occurred in C57BL/6Ha. 2/ Backcrossed to C57BL/6J 15 times.

continued

3. MUTATIONS ON INBRED BACKGROUNDS: MOUSE

Part I. Maintained by The Jackson Laboratory

Gene Symbol	Gene Name	Strain (Genotype) [Origin]	Generations Inbred	Reference
ic	Ichthyosis	IC/Le (*ic*/+)	F36	21
ic^J	Ichthyosis-J	C57BL/6J-ic^J/+ [M]	F3	21
ja	Jaundiced	C57BL/6J-*ja* WB/Re-*ja*	N39 N43	15
jc	Jackson circler	C57BL/6J-A^{sy}/a *jc*/+ [M]	N13, F2	43
je	Jerker	(*a*/*a* *f*/*f* *je*/+ *ru*/*ru*)	F22	15
ji	Jittery	(*a*/*a* *gr* +/+ *ji*)	F32	15
js	Jackson shaker	C57BL/6J-*js*/+ Sp^d/+ [Sp^d M]	N14, F3	7
jv	Jackson waltzer	C57BL/6J-A^i/a *jv*/+ [M]	F9	6
ld^J	Limb-deformity-J	(ld^J + *a*/+ *mg a*)	F44	15
le	Light ear	C57BL/6J-*rd le*	N39	31
lit	Little	C57BL/6J-*lit*/+ [M]	F12, N2, F3	11
lm	Lethal milk	C57BL/6J-*lm*/*lm* [M]	F36	9,18
ln	Leaden	C57BL/6J-*fz ln*/+ + DW/J (*dw*/+ *ln*/*ln*) (*a*/*a fz ln*/*fz ln s*/*s v*/+)	N9, F64 F72 F28	15
Lp	Loop-tail	(*Lp*/+)	F63	15
ls	Lethal spotting	LS/Le (a^t *ls*/+ +)	F58	15
lt	Lustrous	C57BL/6J-*lt*/+	N12, F8	15
lu	Luxoid	C57BL/6J-*lu*	N23	15
lx	Luxate	C57BL/6J-*lx* W^v WB/Re-*lx* W^v	N95 N82	15
m	Misty	C57BL/Ks-*db* +/+ *m* C57BL/6J-*b m*	N5, F21 N39	15
ma	Matted	C57BL/6J-*ma*	N20	15
md	Mahoganoid	C3H/HeJ-*md*/+ [M] C57BL/6J-*A md*	F72 N10	15
mdf	Muscle deficient	C57BL/6J-*mdf*	N7	47
me	Motheaten	$C3H_eB$/FeJ-*a*/*a*-*me*	N7	17
mg	Mahogany	(ld^J + *a*/+ *mg a*)	F44	15
mg^{2J}	Mahogany-2J	$C3H_eB$/FeJ-mg^{2J} [M]	N5	15
Mi^{wh}	White	C57BL/6By-Mi^{wh} W^v *T* C57BL/6J-Mi^{wh} C57BL/6J-*e*/*e*-Mi^{wh} MWT/Le (a^t/a^t Mi^{wh}/+ W^v/+ *T*/+)	N25 N122 N15 F43	15
plus *mi*	Microphthalmia	C57BL/6J-Mi^{wh}/mi	N64	15
plus mi^{rw}	Red-eyed white	C57BL/6J-Mi^{wh}/mi^{rw}	N13	44
plus mi^{sp}	Microphthalmia-spotted	C57BL/6J-Mi^{wh}/mi^{sp} [M]	N59	15
mk	Microcytic anemia	C57BL/6J-*mk* MK/Re (*mk*/+) SEC/1 Re-*mk* WB/Re-*mk*	N17 F43 N20 N18	15
Mo^{blo}	Blotchy	C57BL/6J-Mo^{blo}	N10	15
$Mo^{br\text{-}J}$	Brindled-J	C3H/HeJ-$Mo^{br\text{-}J}$	N12	15
Mp	Micropinna-microphthalmia	C57BL/6By-*Mp Sk*	N27	39
myd	Myodystrophy	MYD/Le (*myd*/+)	F45	33
N	Naked	(*a*/*a N*/+ *Sp*/+)	F41	15
nb	Normoblastic anemia	C57BL/6J-*nb* WB/Re-*nb*	N21 N21	2
nr	Nervous	BALB/cGr-*nr*/+ [M] $C3H_eB$/FeJ-*nr*	F30 N8	27

continued

3. MUTATIONS ON INBRED BACKGROUNDS: MOUSE

Part I. Maintained by The Jackson Laboratory

Gene Symbol	Gene Name	Strain (Genotype) [Origin]	Generations Inbred	Reference
nu^{str}	Nude-streaker	AKR/J-nu^{str} [M]	N3	38
ob	Obese	C57BL/6J-*ob*	N36	15
or^J	Ocular retardation-J	129/Sv-$c^+ p^+ Sl^J$/+ - or^J/+ [or^J M]	F30	15
Os	Oligosyndactylism	C57BL/6By-*Ra Os Pt* C57BL/6J-*Os* +/+ tg^{la} ROP/GnLe (*Ra*/+ *Os*/+ *Pt*/+)	N18 N11, F3 F46	15
p^J	Pink-eyed dilution-J	C3H/HeJ-p^J/+ [M]	F73	15
p^{un}	Pink-eyed unstable	C57BL/6J-p^{un} [M]	N43	36
pa	Pallid	C57BL/6J-*pa* C57BL/6-*pa* +/+ *Tsk*	N42 By N13 x J N35, F10	15
pe	Pearl	C57BL/6J-*pe*	N19	15
Ph	Patch	C57BL/6J-*Ph* +/+ W^v PH/Re (*Ph*/+)	N54 F82	15
pi	Pirouette	C57BL/6J-A^{iy}/*a pi*/+	N12, F3	15
pk	Plucked	C57BL/6J-*pk*/+	N12, F6	15
Pro-1	Proline oxidase-1	PRO/Re-$Pro\text{-}1^a$	N14	3
Ps	Polysyndactyly	C57BL/6By-*Ps Ve*	N22	26
Pt	Pintail	C57BL/6By-*Ra Os Pt* ROP/GnLe (*Ra*/+ *Os*/+ *Pt*/+)	N18 F46	15
qk	Quaking	C57BL/6J-T^{2J} +/+ *qk*	N14	15
Ra	Ragged	C57BL/6By-*Ra Os Pt* C57BL/6J-*Ra* ROP/GnLe (*Ra*/+ *Os*/+ *Pt*/+)	N18 N26 F46	15
rc	Rough coat	C57BL/6J-*rc*/+ [M]	F29	5
rd	Retinal degeneration	C57BL/6J-*rd le*	N39	15
Re	Rex	C57BL/6By-*Re Sd* Va^J RSV/Le (*Re*/*Re Sd*/+ *Va*/+) SHR/GnEi (*Re* +/+ *shm*)	N8 F38 F33	15
Rf	Rib fusions	(*Rf*/+)	F60	15
Rn	Roan	C57BL/10Gn-*Rn*/+	N10, F28	34
rs	Recessive spotting	C57BL/6J-*rs*	N13	6
ru	Ruby-eye	C57BL/6J-*ru* (*a*/*a f*/*f je*/+ *ru*/*ru*)	N39 F22	15
$ru\text{-}2^J$	Ruby eye-2-J	C57BL/6J-$ru\text{-}2^J$/+ [M]	F17	10
$ru\text{-}2^{2J}$	Ruby eye-2-2J	C3H/HeJ-$ru\text{-}2^{2J}$/+ [M]	F20	10
$ru\text{-}2^{hz}$	Haze	C57BL/6J-$ru\text{-}2^{hz}$	N14	12
s	Piebald spotting	(*a*/*a s*/+) (*a*/*a fz ln*/*fz ln s*/*s v*/+)	F28 F28	15
plus s^l	Piebald lethal	(*a*/*a s*/s^l)	F28	15
sch	Scant hair	C57BL/Ks-*sch*/+ [M]	F19	24
Sd	Danforth's short tail	C57BL/6By-*Re Sd* Va^J RSV/Le (*Re*/*Re Sd*/+ *Va*/+)	N8 F38	15
se	Short ear	C57BL/6J-*se* C57BL-*sg* + +/+ *d se* DLS/Le (*a*/*a* d^l +/+ *se*) SEA/GnJ (*b*/*b d se*/*d* +) SEC/1GnLe (*a*/*a b*/*b* c^{ch}/c^{ch} *se*/+) (c^{ch}/c^{ch} *se* +/+ *sv*)	N28 B6 N4 x B10 N4, F50 F66 F120 F120 F45	15
sg	Staggerer	C57BL-*sg* + +/+ *d se*	B6 N4 x B10 N4, F50	15
sh-1	Shaker-1	(c^{ch} *sh-1*/c^{ch} +)	F50	15

continued

3. MUTATIONS ON INBRED BACKGROUNDS: MOUSE

Part I. Maintained by The Jackson Laboratory

Gene Symbol	Gene Name	Strain (Genotype) [Origin]	Generations Inbred	Reference
sh-2	Shaker-2	C57BL/6J-*wa-2* Hba^c *sh-2*	N9	15
		SEC/1Re-*wa-2* Hba^c *sh-2*	N7	
shm	Shambling	SHR/GnEi (*Re* +/+ *shm*)	F33	14
Sig	Sightless	C57BL/6By-*Eh Sig*	N20	40
Sk	Scaly	C57BL/6By-*Mp Sk*	N27	15
Sl	Steel	C57BL/6By-*Ca Sl*	N24	15
		C57BL/6J-Ca^J *Hm Sl*	N15	
		WC/Re-*Sl*	N89	
Sl^J	Steel-J	129/Sv-c^+ p^+ Sl^J/+	N11, F15	15
Sl^d	Steel-Dickie	C57BL/6J-Sl^d	N59	15
		WB/Re-Sl^d	N32	
slt	Slaty	C57BL/6J-*slt*/+	N10, F16	16
sm	Syndactylism	(*sm*/+)	F35	15
Sp	Splotch	(*a*/*a* *N*/+ *Sp*/+)	F41	15
Sp^d	Delayed-splotch	C57BL/6J-*js*/+ Sp^d/+ [Sp^d M]	N14, F3	15
spa	Spastic	C57BL/6J-*spa*	N12	15
sph	Spherocytic anemia	C57BL/6J-*sph*	N9	15
		WB/Re-*sph*	N10	
stb	Stubby	(*stb* + *a*/+ *fi a*)	F42	30
sv	Snell's waltzer	(c^{ch}/c^{ch} *se* +/+ *sv*)	F45	15
sy^{fp}	Fused phalanges	C3H$_e$B/FeJ-sy^{fp}	N7	24,29
T	Brachyury	C57BL/6By-Mi^{wh} W^v *T*	N25	15
		C57BL/10ScSn-*T*	N33	
		MWT/Le (a^t/a^t Mi^{wh}/+ W^v/+ *T*/+)	F43	
		TF/GnLe (*T tf*/+ *tf*)	F50	
plus t^6	t^6	(*a*/*a* *T tf*/t^6 +)	F25	15
T^{2J}	Brachyury-2J	C57BL/6J-T^{2J} +/+ *qk*	N14	15
tf	Tufted	TF/GnLe (*T tf*/+ *tf*)	F50	15
		(*a*/*a* *T tf*/t^6 +)	F25	
tg	Tottering	C57BL/6J-*tg*	B10 N13, B6 N11	15
tg^{la}	Leaner	C57BL/6J-*Os* +/+ tg^{la}	N11, F3	45
tk	Tail kinks	(*a*/*a* *d* + *du*/*d* *tk* +)	F41	15
tn	Teetering	C57BL/6J-*tn*	N5, F56	15
To	Tortoiseshell	C57BL/6J-*To*	N75	15
		TR/DiEi (*a*/*a* *To*/+)	F55	
tp	Taupe	(*a*/*a* *tp*/+)	F42	15
Ts	Tail short	TSJ/Le (*b*/*b* *Ts*/+)	F41	15
Tsk	Tight skin	C57BL/6-*pa* +/+ *Tsk*	By N13 x J N35, F10	22
Tw	Twirler	(E^{so}/E^{so} *Tw*/+ Xt^J/+)	F34	15
un	Undulated	(*we un* a^t/+ + a^t)	F40	15
uw	Underwhite	C57BL/6J-*uw* bt^{2J}/+ + [M]	N4, F4	15
v	Waltzer	(*a*/*a* *fz ln*/*fz ln* *s*/*s* *v*/+)	F28	15
Va	Varitint-waddler	C57BL/6J-*Va*	N61	15
		RSV/Le (*Re*/*Re* *Sd*/+ *Va*/+)	F38	
Va^J	Varitint-waddler-J	C57BL/6By-*Re Sd* Va^J	N8	28
vb	Vibrator	(*a*/*a* *vb*/+)	F42	15
Ve	Velvet coat	C57BL/6By-*Ps Ve*	N22	15
W	Dominant spotting	C57BL/6J-*W*	N121	15
		FL/1Re (*a*/*a* *f*/+ *W*/+)	N69	
		WB/Re (*a*/*a* *W*/+)	F101	
		WC/Re (*a*/*a* *W*/+)	F96	
		WH/Re (*a*/*a* *W*/+)	N92	
		WK/Re (*a*/*a* *b*/*b* *W*/+)	F72	

continued

3. MUTATIONS ON INBRED BACKGROUNDS: MOUSE

Part I. Maintained by The Jackson Laboratory

Gene Symbol	Gene Name	Strain (Genotype) [Origin]	Generations Inbred	Reference
W^{2J}	Dominant spotting-2J	C57BL/6J-W^{2J}	N67	15
W^v	Viable dominant spotting	C57BL/6By-Mi^{wh} W^v T	N25	15
		C57BL/6J-*lx* W^v	N95	
		C57BL/6J-*Ph* +/+ W^v	N54	
		C57BL/6J-W^v	N125	
		MWT/Le (a^t/a^t Mi^{wh}/+ W^v/+ T/+)	F43	
		WB/Ro *lx* W^v	N82	
W^x	Dominant spotting-x	C3H/HeJ-W^x/+	F75	15
wa-2	Waved-2	C57BL/6J-*wa-2* Hba^c *sh-2*	N9	15
		SEC/1Re-*wa-2* Hba^c *sh-2*	N7	
		(*a/a wa-2*/+)	F27	
we	Wellhaarig	(*we un* a^t/+ + a^t)	F40	15
wi	Whirler	C57BL/6J-*wi b*	N14	15
wl	Wabbler-lethal	WLHR/Le (*a/a b/b wl* +/+ *hr*)	F70	15
Xt^J	Extra-toes-J	C57BL/6J-Xt^J	N26	15
		(E^{so}/E^{so} Tw/+ Xt^J/+)	F34	

Contributor: Priscilla W. Lane

References

1. Bangham, J. W. 1965. Mouse News Lett. 33:68.
2. Bernstein, S. E. 1969. Ibid. 40:28.
3. Blake, R. L., and E. S. Russell. 1973. Genetics 74:S26.
4. Bonucci, E., et al. 1976. Growth 40:241-251.
5. Dickie, M. M. 1966. Mouse News Lett. 34:30.
6. Dickie, M. M. 1966. Ibid. 35:31.
7. Dickie, M. M. 1967. Ibid. 36:39.
8. Dickie, M. M. 1968. Ibid. 38:24.
9. Dickie, M. M. 1969. Ibid. 41:31.
10. Eicher, E. M. 1970. Genetics 64:495-510.
11. Eicher, E. M., and M. G. Beamer. 1976. J. Hered. 67: 87:91.
12. Eicher, E. M., and S. Fox. 1977. Mouse News Lett. 56:42.
13. Eicher, E. M., et al. 1976. Proc. Natl. Acad. Sci. USA 73:4667-4671.
14. Green, E. L. 1968. J. Hered. 59:59.
15. Green, M. C. 1966. In E. L. Green, ed. Biology of the Laboratory Mouse. Ed. 2. McGraw-Hill, New York. pp. 87-150.
16. Green, M. C. 1972. Mouse News Lett. 47:46.
17. Green, M. C., and L. D. Shultz. 1975. J. Hered. 66: 250-258.
18. Green, M. C., and H. O. Sweet. 1973. Mouse News Lett. 48:35.
19. Green, M. C., and H. O. Sweet. 1976. Ibid. 56:40.
20. Green, M. C., et al. 1972. Science 176:800-803.
21. Green, M. C., et al. 1975. Transplantation 20:172-175.
22. Green, M. C., et al. 1976. Am. J. Pathol. 82:493-507.
23. Hauschka, T. S., et al. 1968. J. Hered. 59:339-341.
24. Hummel, K. P. 1971. Mouse News Lett. 45:28.
25. Hummel, K. P., et al. 1972. Biochem. Genet. 7: 1-13.
26. Johnson, D. R. 1969. J. Embryol. Exp. Morphol. 21: 285-294.
27. Landis, S. C. 1972. J. Cell Biol. 57:782-797.
28. Lane, P. W. 1972. J. Hered. 63:135-140.
29. Lane, P. W. 1973. Mouse News Lett. 49:32.
30. Lane, P. W., and M. M. Dickie. 1968. J. Hered. 59:300-308.
31. Lane, P. W., and E. L. Green. 1967. Ibid. 58:17-20.
32. Lane, P. W., and H. O. Sweet. 1973. Mouse News Lett. 48:34.
33. Lane, P. W., et al. 1976. J. Hered. 67:135-138.
34. Lyon, M. 1973. Mouse News Lett. 48:31.
35. Meier, H., and J. L. Southard. 1970. Life Sci. 9:137-144.
36. Melvold, R. W. 1971. Mutat. Res. 12:171-174.
37. Orkin, R. W., et al. 1976. Dev. Biol. 50:82-94.
38. Pantelouris, E. M. 1971. Immunology 20:247-252.
39. Phipps, E. L. 1965. Mouse News Lett. 33:68.
40. Searle, A. G. 1965. Ibid. 33:29.
41. Searle, A. G. 1966. Ibid. 35:27.
42. Snell, G. D. 1968. Ibid. 39:28.
43. Southard, J. L. 1970. Ibid. 42:30.
44. Southard, J. L. 1973. Ibid. 51:23.
45. Tsuji, S., and H. Meier. 1971. Genet. Res. 17:83-88.
46. Varnum, D., and L. C. Stevens. 1968. J. Hered. 59: 147-150.
47. Womack, J., et al. 1976. Mouse News Lett. 56:41.

continued

3. MUTATIONS ON INBRED BACKGROUNDS: MOUSE

Part II. Maintained in the United Kingdom

Generations Inbred: M indicates mutation occurred in inbred strain; Fx indicates unknown number of generations of brother x sister mating. **Holder**: For full name, *see* list of HOLDERS at front of book.

Gene Symbol	Gene Name	Strain	Generations Inbred	Holder
A^{vy}	Viable yellow	C3H/He-A^{vy}/Icrf	F30, F36	ICRF
A^{y}	Yellow	AG/Cam	F>20, N5, F36	CAM
A^{w}	White-bellied agouti	AG/Cam	F>20, N5, F36	CAM
A^{s}	Agouti-suppressor	AG/Cam	F>20, N5, F36	CAM
A	Agouti	AG/Cam	F>20, N5, F36	CAM
		C3H/He-A^{vy}/Icrf	F30, F36	ICRF
		JU/FaCt-*A C*	N5, F14	H
a^{t}	Black and tan	AG/Cam	F>20, N5, F36	CAM
		C57BL/GoH-a^{t}	F6, M, F55	H
		C57BL/Go-a^{t}/Icrf	F50, M, F60	ICRF
a	Non-agouti	AG/Cam	F>20, N5, F36	CAM
		C57BL/GoH-a^{t}	F6, M, F55	H
		C57BL/Go-a^{t}/Icrf	F50, M, F60	ICRF
a^{e}	Extreme non-agouti	AG/Cam	F>20, N5, F36	CAM
bt^{Crc}	Belted-Crc	CBA/H-T6-bt^{Crc}	M, Fx	CRC
C	Colored	JU/FaCt-*A C*	N5, F14	H
		JU/FaCt-*C*	N5, F23	H
c	Albino	C3H/He$_f$CBA/CaLac-*c*	N8, F2	LAC
		C57BL/10J$_f$C3H/He$_f$Lac-*c*	N8, F2	LAC
c^{e}	Extreme dilution	C3H/He$_f$CBA/CaLac-c^{e}	N8, F2	LAC
		C57BL/10J$_f$C3H/He$_f$Lac-c^{e}	N8, F2	LAC
dw	Dwarf	CBA/Cam-*dw*/+	N5, F26	HBS-CAM [1]
dy^{2J}	Dystrophia muscularis-2J	C57BL/6J-$dy^{2J}$$_f$C3H/He$_f$Lac	N11, F5	LAC
hr	Hairless	A2G$_f$C3H/He$_f$Lac-*hr*/+	N8, F4	LAC
kd	Kidney disease	CBA/CaH-*kd*	F35, M, F47	H
ky	Kyphoscoliosis	BDL	F30	MSN
lx	Luxate	CBA/Cl-*lx*/+	?	CL
mg	Mahogany	C3H/He$_f$C3H$_f$/Lac-*mg*	F27, M, F11	LAC, CRC
mi^{Crc}	Microphthalmia-Crc	CBA/Ca-mi^{Crc}	M, Fx	CSR
nu	Nude	AKR$_f$C3H/He$_f$Lac-*nu*/+	N7	LAC
		BALB/c-*nu*/+	N>10	CRC
		BALB/c-*nu*/+	N9, Fx	NIMR
		BALB/c$_f$C3H/He$_f$Lac-*nu*/+	N8	LAC
		CBA/Ca-*nu*/+	N15	CRC
		CBA/Ca-*nu*/+	N14	NIMR
		CBA/Ca$_f$NMRI/Lac-*nu*/+	N8	LAC
		C57BL/10J$_f$CBA/CaLac-*nu*/+	N7	LAC
		NZB/Lac-*nu*/+	N6	LAC
		NZW$_f$C3H/He$_f$-*nu*/+	N3	LAC
ob	Obese	C57BL/6J-*ob*$_f$C3H/He$_f$Lac	Fx, F4	LAC
p	Pink-eyed dilution	CBA/Cam-*p*	M, F58	CAM
		CBA/Cam-*p*/H	M, F56, F5	H
		C3H/He$_f$CBA/CaLac-*p*	N8, F2	LAC
		C57BL/10J$_f$C3H/He$_f$Lac-*p*	N8, F2	LAC
rl	Reeler	C57BL/6J-*rl*/+	Fx	NIMR
se	Short ear	CBA/Cam-*se*	M, F58	CAM
Sey	Small eye	C57BL/Go-*Sey*	?	CL
		JU/Fa-*Sey*	?	CL
tk	Tail-kinks	BALB/c-*tk*/+	F51	GR

[1] HBS, supervised by CAM.

Contributor: M. F. Lyon

4. RECOMBINANT INBRED LINES: MOUSE

Holder: For full name, *see* list of HOLDERS at front of book.

Progenitors		Designation	No. of Lines	Generations of Inbreeding	Holder	Reference
AKR/J	C57L/J	AKXL	20	10-33	JAX	1,2,5,9,13,18,20,25,27,28,31
AKR/J	DBA/2J	AKXD	31	6-7	JAX	11
BALB/cBy	C57BL/6By	CXB	7	56-65	JAX	3,4,6-8,15,16,19,22
BALB/cJPas	DBA/2JPas	CXD	11	10-15	PAS	10
C57BL/6J	C3H/HeJ	BXH	13	25-32	JAX	13,20,23,29,32
C57BL/6J	DBA/2J	BXD	24	24-35	JAX	5,9,20,26,27,29,30,33
C57BL/6N	AKR/N	BNXAKN	12	6-18	NICHD	12
C57BL/6N	C3H/HeN	BXHN	5	8-12	NICHD	14
C57L/J	C57BL/6J	LXB	5	12-19	JAX	24
LT/Sv	C57BL/6J	LTXB	4	11	JAX	21
NZB/Icr	C58/J	NX8	14	12-15	ICR	17
SWR/J	C57L/J	SWXL	8	12-32	JAX	23
129/SvPas-*CPSI*/+	C57BL/6JPas	129XB	15	12-13	PAS	10

Contributor: Benjamin A. Taylor

References

1. Abramson, R. K., et al. 1977. Biochem. Genet. 15: 723-740.
2. Atlas, S. A., et al. 1976. Genetics 83:537-550.
3. Bailey, D. W. 1971. Transplantation 11:325-327.
4. Bailey, D. W. 1975. Immunogenetics 2:249-256.
5. Berek, C., et al. 1976. J. Exp. Med. 144:1164-1174.
6. DeMaeyer, E., et al. 1974. J. Gen. Virol. 23:209-211.
7. DeMaeyer, E., et al. 1975. Immunogenetics 1:438-443.
8. Eleftheriou, B. E., et al. 1972. J. Endocrinol. 55:225-226.
9. Festenstein, H., et al. 1977. Immunogenetics 5:357-361.
10. Guénet, J. L. 1976. Mouse News Lett. 55:22.
11. Heiniger, H. J. Unpublished. Jackson Laboratory, Bar Harbor, ME, 1978.
12. Kouri, R. E. 1976. In R. I. Freudenthal and P. W. Jones, ed. Polynuclear Aromatic Hydrocarbons. Raven, New York. pp. 139-151.
13. Mishkin, J. D., et al. 1976. Biochem. Genet. 14:635-640.
14. Nebert, D. W. Unpublished. N.I.C.H.H.D., National Institutes of Health, Bethesda, MD, 1978.
15. Oliverio, A., et al. 1973. Physiol. Behav. 10:893-899.
16. Potter, M., et al. 1975. J. Natl. Cancer Inst. 54:1413-1417.
17. Riblet, R. Unpublished. Institute of Cancer Research. Philadelphia, PA, 1978.
18. Riblet, R., et al. 1975. Eur. J. Immunol. 5:775-777.
19. Shultz, L. D., et al. 1975. Immunogenetics 1:570-583.
20. Stern, R. H., et al. 1976. Biochem. Genet. 14:373-381.
21. Stevens, L. C. Unpublished. Jackson Laboratory, Bar Harbor, ME, 1978.
22. Swank, R. T., et al. 1973. Science 181:1249-1252.
23. Taylor, B. A. 1976. Genetics 83:373-377.
24. Taylor, B. A. Unpublished. Jackson Laboratory, Bar Harbor, ME, 1978.
25. Taylor, B. A., et al. 1971. Proc. Natl. Acad. Sci. USA 68:3190-3194.
26. Taylor, B. A., et al. 1973. Proc. Soc. Exp. Biol. Med. 143:629-633.
27. Taylor, B. A., et al. 1975. Nature (London) 256:644-646.
28. Taylor, B. A., et al. 1976. Genet. Res. 26:307-312.
29. Taylor, B. A., et al. 1977. J. Virol. 23:106-109.
30. Taylor, B. A., et al. 1977. Immunogenetics 4:597-599.
31. Thomas, P. E., et al. 1973. Genetics 74:655-659.
32. Watson, J., et al. 1977. J. Immunol. 118:2088-2093.
33. Womack, J. E., et al. 1975. Biochem. Genet. 13:511-518.

Chromosome numbers are represented by Arabic numerals at one end of the chromosome, while linkage group numbers are represented by Roman numerals at the opposite end. The smaller Arabic numerals to the left of the chromosome/linkage group symbol represent map distances, in centimorgans, between gene loci. Loci having an uncertain

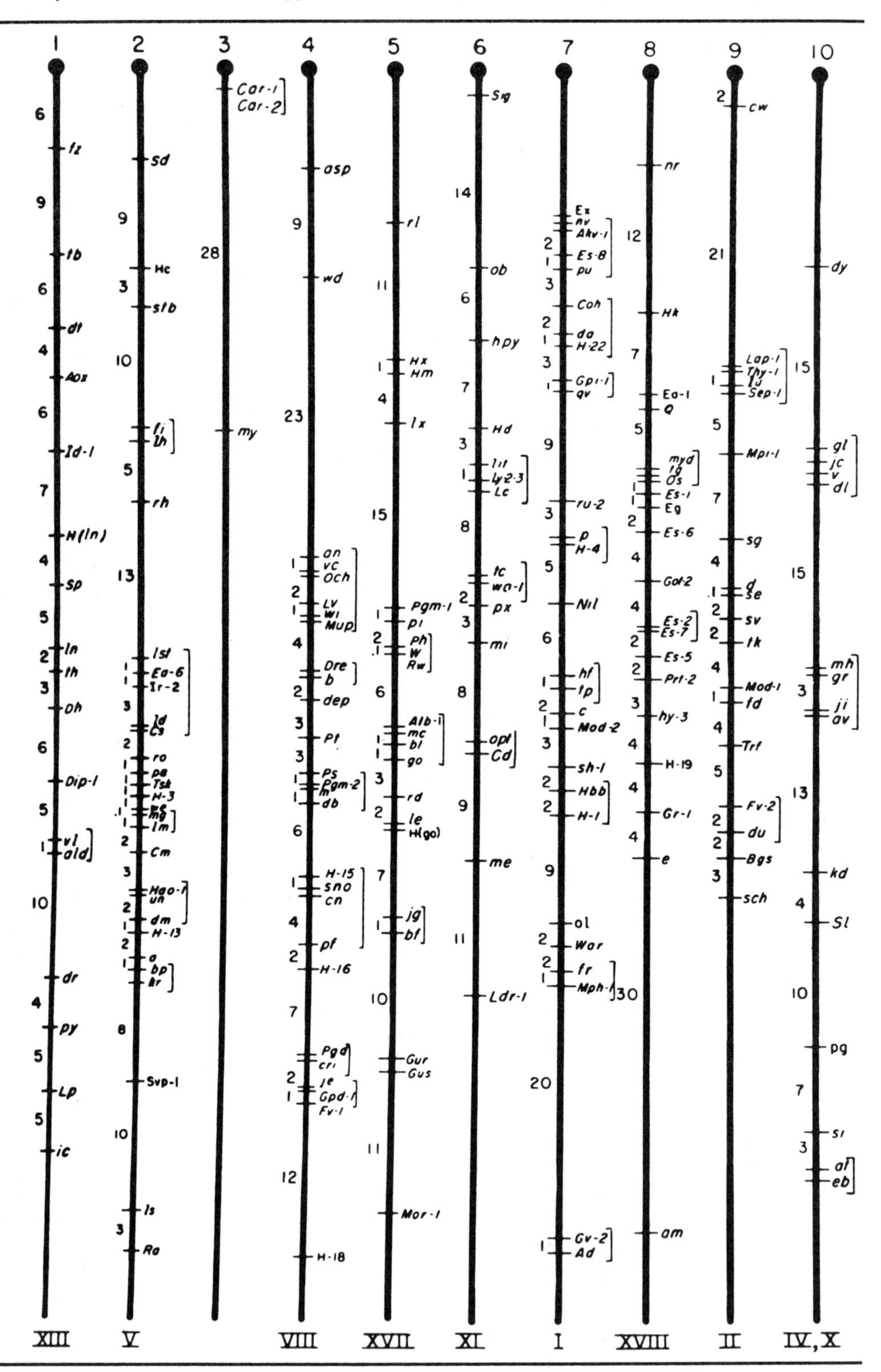

Contributor: James E. Womack

position are not italicized. Brackets indicate that the order within the bracketed group is unknown. Knobs indicate the location of those centromeres that are known. Dr. Margaret C. Green compiled most of the data represented in this map prior to her retirement from The Jackson Laboratory in 1975.

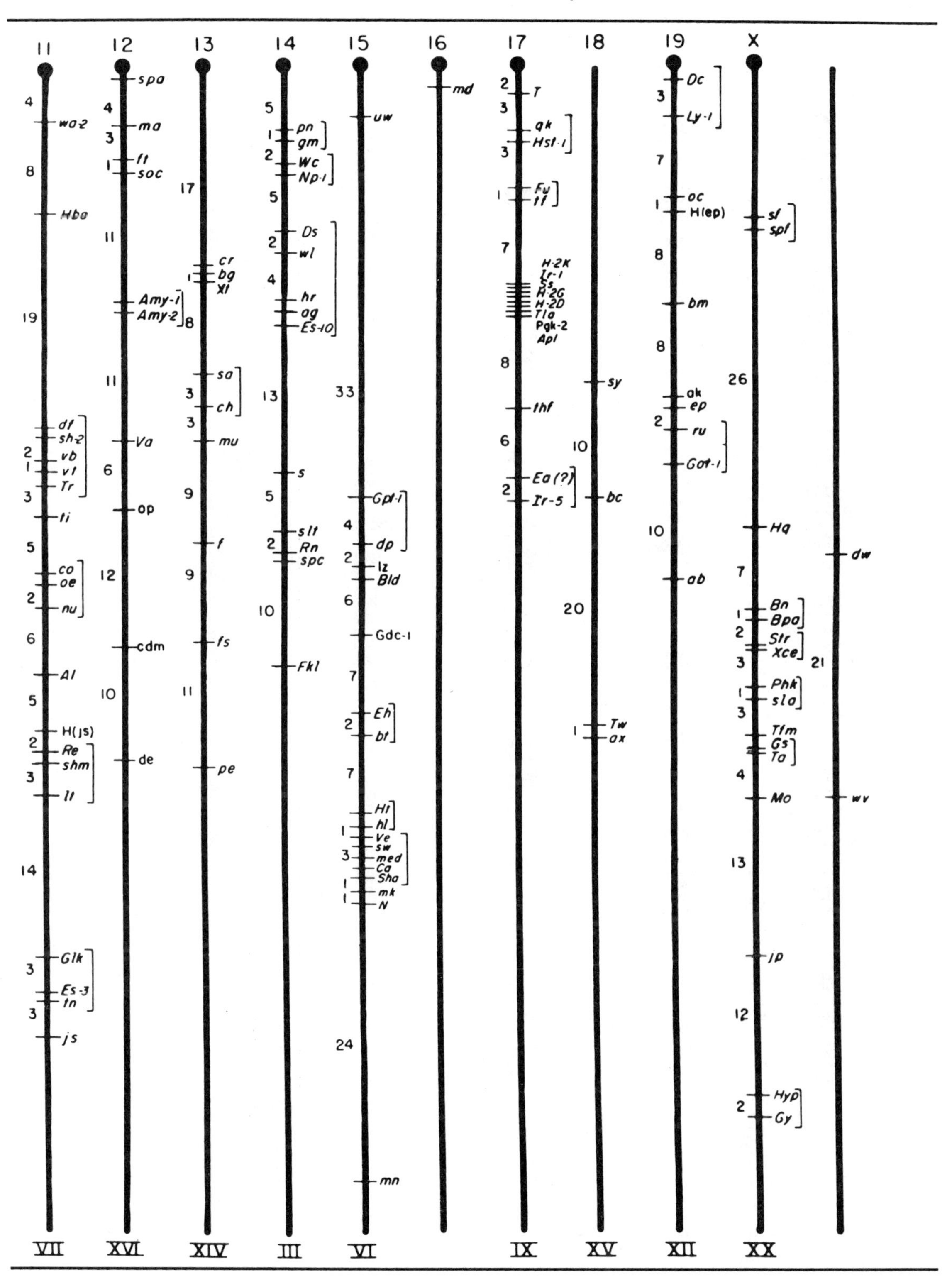

6. CHROMOSOME BANDING PATTERNS: MOUSE

A standard nomenclature for the band patterns in mouse chromosomes has been adopted (*see* reference). The autosomes are numbered in descending order of length. Major bands, which are so prominent that they can be seen even in relatively poor preparations, are designated with the number of their chromosome and a capital letter, in alphabetical order from the centromere to the distal end of the chromosome. In good preparations, the major bands are often resolved into several smaller bands. The smaller bands are then designated with the symbol for the major band of which they are a part and, in addition, are numbered from the most proximal to the most distal. Thus, band 7D2 is the second most proximal of the minor bands into which the fourth major band from the centromere of the seventh longest autosome can be resolved.

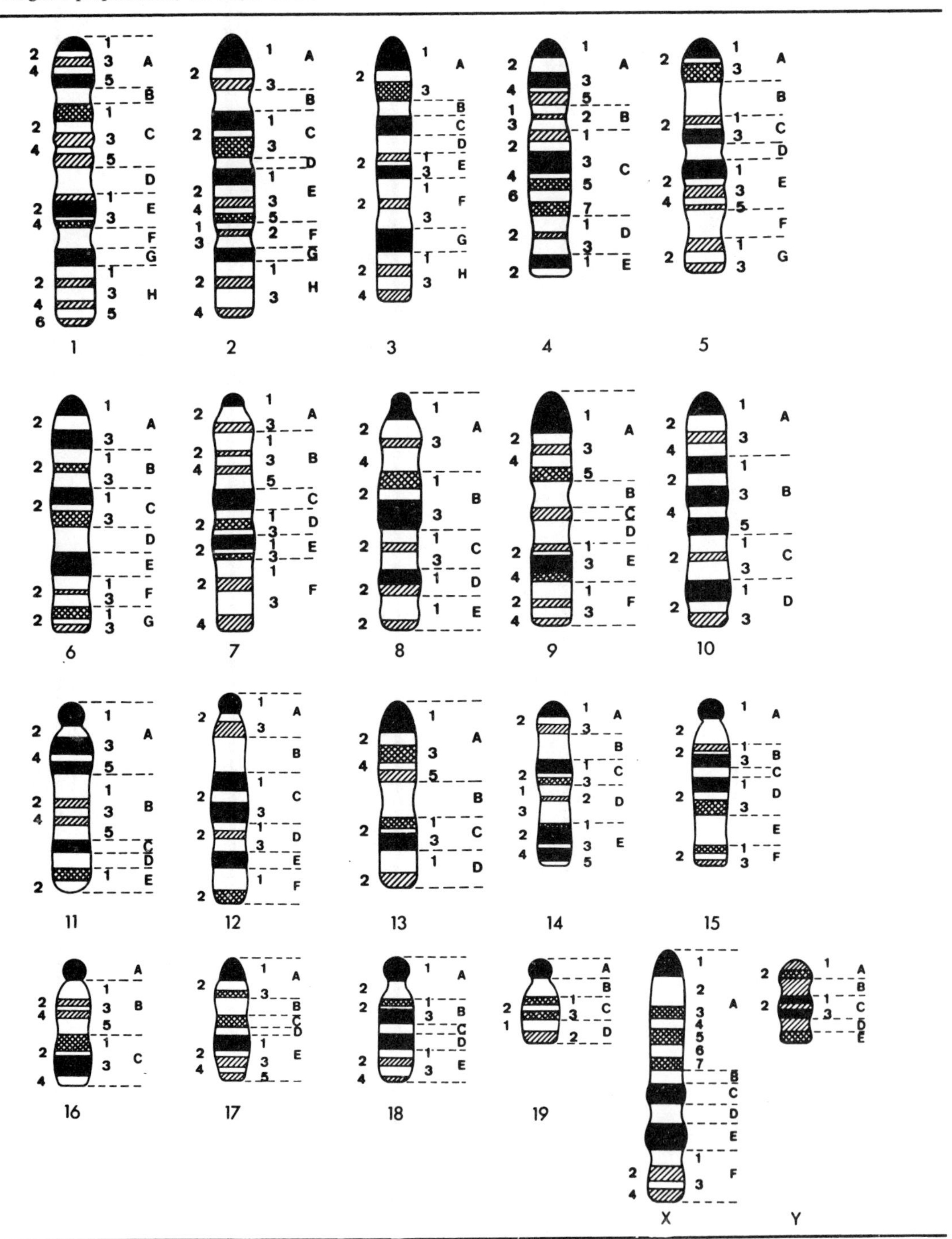

Contributor: Muriel N. Nesbitt

Reference: Nesbitt, M. N. and U. Francke. 1973. Chromosoma 41:145-158.

7. CHROMOSOMAL ABERRATIONS: MOUSE

Part I. Robertsonian Translocations Giving Metacentrics

Name: Chromosomes involved in translocations are shown in parentheses. **Origin:** Alp = Rhaetian Alps; CD = Cittaducale, Appenines; Pos = Poschiavo Valley (*Mus poschiavinus*). Many other wild-derived Robertsonians not listed here are included in references 3 and 9. Note that reference 9 gives a revision of Table 1 from reference 3, with corrected designations for a number of translocations as shown below. **Holder:** For full name, *see* list of HOLDERS at front of book.

Name	Origin	Holder	Reference
Rb(1.3)1Bnr	Wild: "Pos"	HAR, JAX, LUB	7,8
Rb(1.7)1Rma	Wild: "CD"	LUB	2,3,9
Rb(1.10)10Bnr	Wild: "Alp"	HAR, LUB	3
Rb(1.15)2Ct	Laboratory	HAR	5
Rb(2.14)11Bnr	Wild: "Alp"	LUB	3
Rb(2.18)6Rma	Wild: "CD"	LUB	2,3,9
Rb(3.8)2Rma	Wild: "CD"	LUB	2,3
Rb(4.6)2Bnr	Wild: "Pos"	HAR, JAX, LUB	7,8
Rb(4.12)9Bnr	Wild: "Alp"	LUB	2,3
Rb(4.15)4Rma	Wild: "CD"	LUB	2,3
Rb(5.15)3Bnr	Wild: "Pos"	HAR, JAX, LUB	7,8
Rb(5.17)7Rma	Wild: "CD"	LUB	2,3,9
Rb(5.19)1Wh	Laboratory	HAR	14
Rb(6.13)1H	Laboratory	HAR	13
Rb(6.13)3Rma	Wild: "CD"	LUB	2,3,9
Rb(6.15)1Ald	Laboratory	HAR, JAX, KAN, LUB, PAS	10,12
Rb(7.8)12Bnr	Wild: "Alp"	LUB	3
Rb(8.12)5Bnr	Wild: "Pos"	HAR, JAX, LUB	7,8
Rb(8.17)1Iem	Laboratory	LUB, PAS, PRA	1
Rb(8.19)1Ct	Laboratory	HAR, LUB	4
Rb(9.14)6Bnr	Wild: "Pos"	HAR, JAX, LUB	7,8
Rb(9.16)9Rma	Wild: "CD"	LUB	2,3,9
Rb(9.19)163H	Laboratory	HAR, JAX, KAN, LUB, OAK, PAS	6
Rb(10.11)8Bnr	Wild: "Alp"	HAR, LUB	2,3
Rb(10.11)5Rma	Wild: "CD"	LUB	7,8
Rb(11.13)4Bnr	Wild: "Pos"	HAR, LUB	7,8
Rb(11.16)2H	Laboratory	HAR	11
Rb(12.14)8Rma	Wild: "CD"	LUB	2,3,9
Rb(13.16)13Bnr	Wild: "Alp"	LUB	3
Rb(16.17)7Bnr	Wild: "Pos"	HAR, JAX, LUB	7,8

Contributors: A. G. Searle and C. V. Beechey

References

1. Baranov, V. S., and E. Dyban. 1971. Ontogenez 2: 164-176.
2. Capanna, E., et al. 1975. Experientia 31:294-296.
3. Capanna, E., et al. 1976. Chromosoma 58:341-353.
4. Cattanach, B. M., and J. R. K. Savage. 1976. Mouse News Lett. 54:38.
5. Cattanach, B. M., et al. 1977. Ibid. 56:37.
6. Evans, E. P., et al. 1967. Cytogenetics 6:105-119.
7. Gropp, A., et al. 1970. Ibid. 9:9-23.
8. Gropp, A., et al. 1972. Chromosoma 39:265-288.
9. Gropp, A., et al. 1977. Mouse News Lett. 57:26-28.
10. Léonard, A., and G. H. Deknudt. 1967. Nature (London) 214:504-505.
11. Lyon, M. F., et al. 1977. Mouse News Lett. 56: 37.
12. Miller, O. J., et al. 1971. Proc. Natl. Acad. Sci. USA 68:1530-1533.

continued

7. CHROMOSOMAL ABERRATIONS: MOUSE

Part I. Robertsonian Translocations Giving Metacentrics

13. Phillips, R. J. S., and J. R. K. Savage. 1976. Mouse News Lett. 55:14.

14. White, B. J., et al. 1972. Cytogenetics 11:363-378.

Part II. Reciprocal Translocations

Name and **Genetic Tag**: Chromosomes involved in translocations are shown in parentheses. **Breakpoint**: Consult reference 40 for definition by band nomenclature. **Genetic Tag**: Location is within approximately 8 units of the breakpoint. **Special Property**: CB = C-band involved; LM = long somatic marker chromosome; MS = causes sterility in male; P = phenotypic effect with genetic tag; SM = short marker (same size as chromosome 19 or shorter); V = may give coat-color variegation in heterozygous females. **Holder**: For full name, *see* list of HOLDERS at front of book. For a general review of mouse cytogenetics, consult reference 33.

Name	Breakpoint	Genetic Tag		Special Property	Holder	Reference
T(X;4)1Rl	XF2, 4A5	*Ta*(X)	....	MS, LM, V	OAK	19,43,45,48
T(X;4)7Rl	XA2, 4D1	....	....	MS, LM, V	OAK	19,44,48
T(X;4)8Rl	....	....	....	MS, V	OAK	48
T(X;4)37H	....	....	....	MS, LM, SM, V	HAR, JAX	6,7
T(X;7)2Rl	XA4, 7D3	....	*c*(7)	MS, LM, V	OAK	43,46-48
T(X;7)3Rl	XA2, 7F1	....	*c*(7)	MS, LM, V	OAK	43,46-48
T(X;7)5Rl	XF1, 7A3	*Ta*(X)	*p*(7)	MS, LM, SM, V	OAK	43,46-48
T(X;7)6Rl	XF1, 7B3	*Ta*(X)	*p*(7)	MS, LM, SM, V	OAK	43,46-48
T(X;7)1Ct[1/]	....	*Ta*(X)	*p*(7)	LM, V	HAR, JAX, OAK	15,16,18,19,21,24,41
T(X;11)38H	....	....	Re^{wc}(11)	MS, LM, SM	HAR, JAX	4,7
T(X;12)13Rl	....	....	....	MS	OAK	49
T(X;16)16H	XD, 16B5	*Ta*(X)	....	MS	HAR, JAX, OAK	19,21,22,29
T(1;2)5Ca	1H3, 2H1	....	*a*(2)	....	HAR, JAX	13,14,21
T(1;7)40H	....	....	....	MS, LM, SM	HAR	4,5,7
T(1;9)27H	....	....	*se*(9)	P	HAR	1,3,36
T(1;13)70H[2/]	1A4, 13D1	....	....	LM, SM	HAR, JAX, PAS, WAG	10,11,21,35,54,56
T(1;17)190Ca	1H5, 17B	....	*T*(17)	LM, SM	HAR, JAX, PAS, PRA	9,13,14,21,30,34
T(2;3)24H	....	*we*(2)	....	LM	HAR	6,35,40
T(2;4)1Sn	2H3, 4E2	*bp*(2)	....	....	JAX, OXF	21,32,57
T(2;4)1Ca	....	*Sd*(2)	*b*(4)	....	HAR, JAX	2,13,14
T(2;4)13H	....	*fi*(2)	*m*(4)	....	HAR	2,5,35
T(2;6)7Ca	2D, 6C2	*pa*(2)	*Lc*(6)	....	HAR	13,14,31,38
T(2;8)26H	2H1, 8A4	*a*(2)	....	P, LM	HAR, JAX, WAG	1,10,11,31,53,56
T(2;9)11H	2E4, 9A4	*pa*(2)	*se*(9)	....	HAR	37,52
T(2;?)28H	....	*ls*(2)	....	....	HAR	23,52
T(4;8)36H	....	....	....	MS, sub-metacentric	HAR	7
T(5;10)9Rl	5E, 10D	W^v(5)	*Sl*(10)	P	OAK	12
T(5;12)31H	5B, 12F1	....	....	MS, LM, SM	HAR	7,8
T(5;13)264Ca	....	W^v(5)	*f*(13)	....	HAR, JAX	13,14,31,39,42,56
T(7;15)9H	....	*p*(7)	....	....	HAR	4,55
T(7;18)50H	....	*sh-1*(7)	....	LM	HAR	2,35,55
T(7;19)145H	7B3, 19D1	*p*(7)	*ru*(19)	MS, SM	HAR, JAX, KAN, PRA	21,26,28,35
T(8;16)17H	....	E^{so}(8)	....	....	HAR	2,35
T(9;17)138Ca	9B, 17D	*H-2*(17)	....	....	COR, HAR, JAX, PRA	9,13,14,21,25,30,34

[1/] Actually non-reciprocal; an inverted insertion of part of chromosome 7 into the X = Cattanach's translocation.

[2/] Also available as tertiary trisomic for small marker (WAG).

continued

7. CHROMOSOMAL ABERRATIONS: MOUSE

Part II. Reciprocal Translocations

Name	Breakpoint	Genetic Tag	Special Property	Holder	Reference
T(10;12)10Rl	10D, 12A	*Sl*(10)	P	OAK	12
T(10;13)199H	10C, 13A1	*gr*(10)	MS, CB	HAR, JAX	17,27,28,35
T(10;17)11Rl	10D, 17C	*Sl*(10)	P	OAK	12
T(10;18)12Rl	10D, 18D	*Sl*(10)	P	OAK	12
T(10;18)18H	....	*v*(10)	....	HAR, LUB	3,35,51,55
T(11;19)42H	....	Re^{wc}(11)	MS, SM	HAR	4,6,7
T(14;15)6Ca[3/]	14E5, 15B3	*s*(14) *uw*(15)	LM, SM	COMMON	13,14,20,21,34,39
T(16;17)43H	....	....	MS, SM, CB	HAR	7,50

[3/] Gives T6 marker chromosome.

Contributors: A. G. Searle and C. V. Beechey

References

1. Batchelor, A. L., et al. 1966. Mutat. Res. 3:218-229.
2. Beechey, C. V., and A. G. Searle. 1972. Mouse News Lett. 46:28.
3. Beechey, C. V., and A. G. Searle. 1975. Ibid. 53:31.
4. Beechey, C. V., and A. G. Searle. 1976. Ibid. 54:38-39.
5. Beechey, C. V., and A. G. Searle. 1976. Ibid. 55:14.
6. Beechey, C. V., and A. G. Searle. 1977. Ibid. 56:39.
7. Beechey, C. V., et al. 1975. Ibid. 53:30.
8. Beechey, C. V., et al. 1976. Ibid. 54:38.
9. Bennett, D. 1965. Proc. Natl. Acad. Sci. USA 53:730-737.
10. Boer, P. de. 1976. Genet. Res. 27:369-388.
11. Boer, P. de, and M. van Gijsen. 1974. Can. J. Genet. Cytol. 16:783-788.
12. Cacheiro, N. L. A., and L. B. Russell. 1975. Genet. Res. 25:193-195.
13. Carter, T. C., et al. 1955. J. Genet. 53:154-166.
14. Carter, T. C., et al. 1956. Ibid. 54:462-473.
15. Cattanach, B. M. 1961. Z. Vererbungsl. 92:165-182.
16. Cattanach, B. M. 1966. Genet. Res. 8:253-256.
17. Dev, V. G., et al. 1974. Cytogenet. Cell Genet. 13:256-267.
18. Eicher, E. M. 1970. Genetics 64:495-510.
19. Eicher, E. M. 1970. Adv. Genet. 15:175-259.
20. Eicher, E. M., and M. C. Green. 1972. Genetics 71:621-632.
21. Eicher, E. M., and L. Washburn. 1977. Mouse News Lett. 56:43.
22. Eicher, E. M., et al. 1972. Genetics. 71:643-648.
23. Ford, C. E., et al. 1969. Cytogenetics 8:447-470.
24. Francke, U., and M. Nesbitt. 1971. Proc. Natl. Acad. Sci. USA 68:2918-2920.
25. Klein, J., and D. Klein. 1972. Genet. Res. 19:177-179.
26. Lyon, M. F., and P. Glenister. 1971. Mouse News Lett. 45:24.
27. Lyon, M. F., and P. Glenister. 1972. Ibid. 46:29.
28. Lyon, M. F., and R. Meredith. 1966. Cytogenetics 5:335-354.
29. Lyon, M. F., et al. 1964. Ibid. 3:306-323.
30. Miller, D. A., and O. J. Miller. 1972. Science 178:949-955.
31. Miller, D. A., et al. 1971. Proc. Natl. Acad. Sci. USA 68:2699-2702.
32. Miller, D. A., et al. 1972. Genetics 71:633-637.
33. Miller, O. J., and D. A. Miller. 1975. Annu. Rev. Genet. 9:285-303.
34. Miller, O. J., et al. 1971. Proc. Natl. Acad. Sci. USA 68:1530-1533.
35. Miller, O. J., et al. 1971. Cytogenetics 10:452-464.
36. Nesbitt, M. N. 1973. Mouse News Lett. 49:23.
37. Nesbitt, M. N. 1975. Ibid. 52:31.
38. Nesbitt, M. N. 1975. Ibid. 53:68.
39. Nesbitt, M. N., and U. Francke. 1971. Genetics 69:517-522.
40. Nesbitt, M. N., and U. Francke. 1973. Chromosoma 41:145-158.
41. Ohno, S., and B. M. Cattanach. 1972. Cytogenetics 11:129-140.
42. Phillips, R. J. S. 1961. Mouse News Lett. 24:34.
43. Russell, L. B. 1963. Science 140:976-978.
44. Russell, L. B. 1972. Genetics 71:s53-s54.
45. Russell, L. B., and J. W. Bangham. 1961. Ibid. 46:509-525.
46. Russell, L. B., and C. S. Montgomery. 1969. Ibid. 63:103-120.
47. Russell, L. B., and C. S. Montgomery. 1970. Ibid. 64:281-312.
48. Russell, L. B., et al. 1974. Oak Ridge Natl. Lab. Biol. Div. Annu. Prog. Rep. ORNL-4993:121-122.
49. Russell, L. B., et al. 1975. Mouse News Lett. 53:51-52.
50. Searle, A. G. 1974. Genetics 78:173-186.
51. Searle, A. G., and C. V. Beechey. 1970. Mouse News Lett. 42:27-28.
52. Searle, A. G., and C. V. Beechey. 1970. Ibid. 43:28-29.
53. Searle, A. G., and C. V. Beechey. 1971. Ibid. 44:28.
54. Searle, A. G., and C. V. Beechey. 1971. Ibid. 48:32.
55. Searle, A. G., and C. V. Beechey. 1974. Ibid. 50:40.
56. Searle, A. G., et al. 1971. Genet. Res. 18:215-235.
57. Snell, G. D. 1946. Genetics 31:157-180.

continued

7. CHROMOSOMAL ABERRATIONS: MOUSE

Part III. Inversions

Name: Chromosomes involved in inversions are shown in parentheses. **Breakpoint**: Consult reference 3 for definition by band nomenclature. **Relative Length** is given as percent of whole chromosome. **Holder**: For full name, *see* list of HOLDERS at front of book.

Name	Breakpoint	Relative Length	Genetic Tag	Holder	Reference
In(X)1H	XA1, XF	85%	*Bpa*	HAR	2
In(1)1Rk	1B, 1F	42%	*Sp*	JAX	1,4-6
In(2)5Rk	2C, 2G	37%	*Sd*	JAX	1,6
In(5)2Rk	....	...	W^v	JAX	4-6
In(5)9Rk	....	...	....	HAR, JAX	6
In(10)6Rk	....	...	*gr*	JAX	6

Contributors: A. G. Searle and C. V. Beechey

References

1. Davisson, M. T., and T. H. Roderick. 1973. Cytogenet. Cell Genet. 12:398-403.
2. Evans, E. P., and R. J. S. Phillips. 1975. Nature (London) 256:40-41.
3. Nesbitt, M. N., and U. Francke. 1973. Chromosoma 41:145-158.
4. Roderick, T. H. 1971. Mutat. Res. 11:59-69.
5. Roderick, T. H., and N. L. Hawes. 1970. Proc. Natl. Acad. Sci. USA 67:961-967.
6. Roderick, T. H., and N. L. Hawes. 1974. Genetics 76: 109-117.

8. CENTROMERIC HETEROCHROMATIN VARIATIONS: MOUSE

Nomenclature of centromeric heterochromatin is based on rules recommended by the Nomenclature Committee [ref. 1]. The symbols Hc^h, Hc^s, and Hc^o—for large, standard, or low quantity of centromeric heterochromatin—are represented in the table by h, s, or o, respectively. Symbols in parentheses indicate either slight deviation from standard or disagreement in reported data. In general, only prominent Hc variants have been included. In wild *Mus musculus musculus* and *M. musculus molossinus,* all animals studied displayed Hc^h, Hc^s, and Hc^o variants in homozygous and heterozygous form in autosomes, but only Hc^o on X and Y chromosomes [ref. 2, 3, 5].

Strain	Hc Region																					Reference
	1	2	3	4	5	6	7	8	9	10	11	12	13	14	15	16	17	18	19	X	Y	
A/Ph	s	s	s	s	s	s	s	s	s	s	s	s	s	s	s	s	s	s	s	s	o	2,4
AKR	s	s	o	s	s	s	s	s	s	s	s	s	s	s	s	s	s	s	o	s	o	2
BALB/cJ	o	s	s	s	s	s	o	s	s	s	s	s	s	s	s	s	s	o	s	s	o	6
CBA/H, CBA/J	(o)	s	s	s	s	s	s	s	s	s	s	s	s	o	s	s	s	(o)	s	s	o	2,4-6
C3H/Di, C3H/HeJ	(o)	s	s	s	s	s	s	s	s	s	s	s	s	o	s	s	s	(o)	s	s	o	2,4-6
C57BL/10J, C57BL/10ScSnPh, B10.A, B10.D2	s	s	s	s	s	s	s	s	s	s	o	s	s	s	s	s	s	s	s	s	o	4-6
C57BR/cdJ	s	s	s	s	s	s	s	s	s	o	s	h	o	s	s	s	s	s	s	s	o	2,6
C57L/J	s	s	s	s	s	s	s	s	s	o	s	h	o	s	s	s	s	s	s	s	o	2,6
DBA/1J, DBA/2J	(h)	s	s	s	s	s	s	s	s	s	s	s	s	o	s	s	s	s	s	s	o	2,6
129	s	s	s	s	s	s	s	o	s	s	s	s	s	o	s	s	s	s	s	s	o	2

Contributor: Jiři Forejt

References

1. Committee on Standardized Genetic Nomenclature for Mice. 1974. Mouse News Lett. 51:2-3
2. Dev, V. G., et al. 1973. Genetics 75:663-670.
3. Dev, V. G., et al. 1975. Chromosoma 53:335-344.
4. Forejt, J. 1972. Folia Biol. (Prague) 18:213-215.
5. Forejt, J. 1973. Chromosoma 43:187-201.
6. Miller, D. A., et al. 1976. Genetics 84:67-75.

9. REPRODUCTION AND GROWTH CHARACTERISTICS: MOUSE

Part I. Basic Data

For additional information on body weights, and for information on organ weights, consult reference 4. Plus/minus (±) values are standard error, unless otherwise indicated. **Sex Ratio**: (B) = at birth; (W) = at weaning.

Strain	Litter Size	Sex Ratio % ♂	Body Weight		Life-Span, d		Reference
			Age, d	g	Virgin	Breeder	
A/HeJ	4.0	47.9 (B)	120	♂, 33.9 ± 1.5 ♀, 32.1 ± 1.2	♀, 520 ± 9	♂, 478 ± 7 ♀, 512 ± 11	6,14,15,18
A/J	5.0	48.6 (B)	120	♂, 33.1 ± 0.7 ♀, 29.7 ± 0.9	♂, 662 ± 20 ♀, 688 ± 22	♂, 503 ± 11 ♀, 512 ± 9	5,6,14,15,18
AKR/J	5.3	51.4 (B)	120	♂, 40.5 ± 1.1 ♀, 32.4 ± 1.3	♂, 326 ± 16 ♀, 276 ± 9	♂, 272 ± 5 ♀, 269 ± 3	6,14,15,18,21
A2G	5.4	48.0 (W)	112	♂, 28.7 ± 0.7 ♀, 24.4 ± 0.2	....	....	1,2
BALB/c	5.2	52.5 (B) 49.2 (W)	60	♂, 24.0 ± 1.8 1/ ♀, 20.4 ± 1.7 1/	....	....	2,9,24
BALB/cJ	5.3	49.2 (B)	120	♂, 28.3 ± 0.5 ♀, 27.3 ± 0.4	♂, 648 ± 21 ♀, 816 ± 32	♂, 485 ± 9 ♀, 462 ± 7	5,6,10,14,15
BDP/J	4.9	47.9 (B)	120	♂, 26.1 ± 1.0 ♀, 23.1 ± 0.7	♂, 421 ± 12 ♀, 468 ± 17	....	6,18,21
BRVR/SrCr	....	....	56	♂, 25.9 ♀, 20.4	....	....	13
BSVS/SrCr	....	....	70	♂, 29.2 ♀, 24.8	....	....	13
CBA	6.3	49.9 (W)	182	♂, 28.4 ± 0.6 ♀, 22.7 ± 1.8	....	....	2,17
CBA/J	5.3	52.2 (B)	120	♂, 37.3 ± 0.7 ♀, 33.4 ± 0.9	♂, 527 ± 17 ♀, 527 ± 15	♀, 523 ± 8	6,15,18,21
C3H	6.0	52.2 (B)	60	♂, 21.7 ± 2.2 1/ ♀, 18.1 ± 1.6 1/	♂, 791	....	9,20,24
C3H/He	5.6	50.3 (W)	168	♂, 34.0 ♀, 32.0	....	....	2,13
C3H/HeJ	6.2	51.9 (B)	120	♂, 35.1 ± 0.7 ♀, 29.6 ± 0.9	....	....	6,18
C3H/J	....	....	...	....	♀, 398 ± 6	♂, 407 ± 7 ♀, 280 ± 1	15
$C3H_eB$/FeJ	5.4	51.2 (B)	120	♂, 38.9 ± 1.0 ♀, 33.7 ± 0.8	♂, 652 ± 17 ♀, 657 ± 15	....	6,14,18,21
$C3H_f$/He	5.5	50.9 (B)	168	♂, 34.4 ♀, 31.3	....	....	13,23
C57BL	7.0	52.9 (B) 51.9 (W)	60	♂, 20.3 ± 2.2 1/ ♀, 17.9 ± 1.8 1/	....	....	2,8,9
C57BL/6	6.3	52.3 (B)	140	♂, 27.7 ♀, 26.3	....	....	13,24
C57BL/6J	6.0	51.6 (B)	120	♂, 30.1 ± 0.6 ♀, 24.7 ± 0.4	♂, 827 ± 34 ♀, 818 ± 21	♂, 539 ± 7 ♀, 695 ± 9	5,6,14,15,18
C57BL/10J	6.8	49.1 (B)	120	♂, 33.0 ± 0.5 ♀, 26.7 ± 0.6	....	....	6,18
C57BL/10ScSn	....	....	168	♂, 37.0 ♀, 36.0	♂, 826 ♀, 693	....	13,20
C57BR/cdJ	6.4	48.8 (B)	120	♂, 33.4 ± 0.7 ♀, 26.8 ± 0.1	♂, 703 ± 17 ♀, 694 ± 20	♂, 475 ± 13 ♀, 512 ± 10	6,14,15,18,21
C57L/J	5.4	51.1 (B)	120	♂, 33.6 ± 0.5 ♀, 25.6 ± 0.4	♀, 577 ± 19	♂, 532 ± 13 ♀, 492 ± 13	6,14,15,18
C58/J	4.7	55.2 (B)	120	♂, 37.7 ± 1.1 ♀, 26.9 ± 1.6	♂, 373 ± 19 ♀, 351 ± 13	....	6,18,21

1/ Plus/minus (±) value is standard deviation.

continued

Part I. Basic Data

Strain	Litter Size	Sex Ratio % ♂	Body Weight		Life-Span, d		Reference
			Age, d	g	Virgin	Breeder	
DBA/1J	4.5	51.8 (B)	120	♂, 29.9 ± 1.0 ♀, 21.7 ± 0.7	♂, 433 ± 31 ♀, 750 ± 20	♂, 438 ± 9 ♀, 602 ± 13	6,15,18,21
DBA/2	4.9	51.8 (B)	140	♂, 27.9 ♀, 25.7	....	....	13,24
DBA/2J	4.9	51.3 (B)	120	♂, 30.8 ± 1.0 ♀, 23.4 ± 0.6	♂, 722 ± 30 ♀, 683 ± 26	♂, 415 ± 9 ♀, 661 ± 11	5,6,15,18
IVCE	9.0	....	70	♂, 29.9 ± 0.9[1] ♀, 25.1 ± 0.6[1]	....	....	12
IVCS	10.1	....	70	♂, 32.6 ♀, 26.2 ± 0.3[1]	....	....	11,12
LP/J	5.1	50 (B)	120	♂, 31.4 ± 0.6 ♀, 23.5 ± 0.4	♂, 748 ± 19 ♀, 799 ± 20	♀, 517 ± 7	6,15,18,21
MA/J	6.0	49.5 (B)	120	♂, 27.1 ± 1.1 ♀, 26.2 ± 0.7	♂, 459 ± 19 ♀, 585 ± 14	....	6,18,21
MK/Re-*mk*/+	....	....	60	♂, 26.9 ± 0.6 ♀, 20.8 ± 0.6	....	....	16
MK/Re-*mk*/*mk*	....	....	60	♂, 23.0 ± 0.6 ♀, 18.7 ± 0.7	....	....	16
NZW/Cr	....	....	112	♂, 35.0 ♀, 30.0	....	....	13
OUCW	6.0	....	...	....	♂, 640 ♀, 540	....	7
P/J	4.8	47.4 (B)	120	♂, 28.0 ± 0.8 ♀, 23.2 ± 0.6	♂, 384 ± 14 ♀, 510 ± 15	♀, 369 ± 9	6,15,18,21
PHH	6.4	47.3 (B) 52.8 (W)	...	....	....	....	25,26
PHL	5.7	41.0 (B) 41.8 (W)	...	....	....	....	25,26
PL/J	5.2	48.5 (B)	...	....	♂, 517 ± 17 ♀, 448 ± 14	....	18,21
RF/J	5.5	51.2 (B)	120	♂, 41.5 ± 1.3 ♀, 38.5 ± 0.8	♂, 651 ± 25 ♀, 452 ± 13	....	6,18,21
RIII/An	6.2	49.7 (B)	120	♂, 29.8 ± 0.9 ♀, 26.1 ± 1.1	♂, 685 ± 21 ♀, 655 ± 23	....	6,18,21
SJL/J	7.9	51.3 (B)	120	♂, 29.5 ± 0.7 ♀, 23.1 ± 0.4	♂, 472 ± 25 ♀, 395 ± 11	....	6,18,21
SJL/JDg	6.8	45.9 (W)	170	♂, 28.4 ± 0.6 ♀, 23.2 ± 0.4	....	....	3
SM/J	5.3	49.6 (B)	120	♂, 29.9 ± 0.6 ♀, 24.3 ± 1.0	♂, 572 ± 17 ♀, 591 ± 15	....	6,18,21
ST/bJ	5.6	50.5 (B)	120	♂, 40.2 ♀, 30.9	♂, 433 ± 23 ♀, 511 ± 15	....	18,21,22
STR/Cr	....	....	84	♂, 29.3 ♀, 27.8	....	....	13
STR/1N	...	....	480	♂, 46.4 ♀, 40.8	....	....	19
SWR/J	6.7	50.7 (B)	120	♂, 28.4 ± 0.7 ♀, 22.7 ± 0.4	♂, 616 ± 29 ♀, 496 ± 21	....	6,18,21
129/J	5.8	49.5 (B)	120	♂, 27.0 ± 0.4 ♀, 22.1 ± 0.1	♂, 679 ± 19 ♀, 648 ± 24	♀, 504 ± 11	6,14,15,18,21

[1] Plus/minus (±) value is standard deviation.

continued

9. REPRODUCTION AND GROWTH CHARACTERISTICS: MOUSE

Part I. Basic Data

Contributor: Charles G. Crispens, Jr.

References

1. Barnett, S. A., and E. M. Widdowson. 1965. Proc. R. Soc. London B162:502-516.
2. Cook, M. J., and A. Vlcek. 1961. Nature (London) 191:89.
3. Crispens, C. G., Jr. 1973. Lab. Anim. Sci. 23:408-413.
4. Crispens, C. G., Jr. 1975. Handbook on the Laboratory Mouse. Thomas, Springfield.
5. Goodrick, C. L. 1975. J. Gerontol. 30:257-263.
6. Green, E. L., ed. 1962. Handbook on Genetically Standardized JAX Mice. Jackson Laboratory, Bar Harbor, ME.
7. Hall, W. O., and L. O. Simpson. 1975. Lab. Anim. 9: 139-142.
8. Howard, A., et al. 1955. J. Genet. 53:200-214.
9. Howe, W. L., and P. A. Parsons. 1967. J. Embryol. Exp. Morphol. 17:283-292.
10. Maeda, Y. Y., and G. Chihara. 1973. Int. J. Cancer 11:153-161.
11. Nagai, J., et al. 1971. Can. J. Genet. Cytol. 13:20-28.
12. Nobunaga, T. 1973. Lab. Anim. Sci. 23:803-811.
13. Poiley, S. M. 1972. Ibid. 22:759-779.
14. Roderick, T. H., and J. B. Storer. 1961. Science 134: 48-49.
15. Russell, E. S. 1966. In E. L. Green, ed. Biology of the Laboratory Mouse. Ed. 2. McGraw-Hill, New York. pp. 511-519.
16. Russell, E. S., et al. 1970. Transplant. Proc. 2:144-151.
17. Santisteban, G. A. 1960. Anat. Rec. 136:117-126.
18. Schlager, G., and T. H. Roderick. 1968. J. Hered. 59: 363-365.
19. Silverstein, E., et al. 1960. Am. J. Physiol. 199:203-208.
20. Smith, G. S., et al. 1973. J. Natl. Cancer Inst. 50: 1195-1213.
21. Storer, J. B. 1966. J. Gerontol. 21:404-409.
22. Storer, J. B. 1967. Exp. Gerontol. 2:173-182.
23. Verley, F. A., et al. 1967. J. Hered. 58:285-290.
24. Weir, J. A. 1960. Genetics 45:1539-1552.
25. Weir, J. A. 1962. Ibid. 47:881-897.
26. Weir, J. A., et al. 1958. J. Hered. 49:217-222.

Part II. Supplemental Information

Strain	Remarks	Reference
A/HeJ	Dorsoventral vaginal septa in 8% of weaned ♀	7
AKR/J	Dorsoventral vaginal septa in 4% of weaned ♀	7
AL	Good fertility. Body weight at 140 d: ♂ = 34.0 g; ♀ = 33.6 g.	33,39
BA	Good reproductive performance	39
BALB/cJ	Spermatozoa exhibit abnormally low levels of β-glucuronidase. Dorsoventral vaginal septa in 38% of weaned ♀.	7,10
BALB/cWt	♂ induces both estrus synchrony & pregnancy block; ♀ exhibits high incidence of pregnancy on 1st night after mating. Low ♂ sex ratio (41%). High incidence of XO ♀ & hermaphrodites.	5,6,44
BLN/Nmg	Excellent reproductive performance	5
BUA	Selected for good growth & reproductive performance	39
BUB	Selected for good growth & reproductive performance	39
CBA	5.5% of spermatozoa have morphologically deformed heads	1
CBA/H	50% survival: 940 d for ♂, 960 d for ♀. 10% survival: 1030 d for ♂, 1140 d for ♀.	40
CBA/J	Dorsoventral vaginal septa in 3% of weaned ♀	7
CE/J	Poor breeders; avg litter size = 6.8. Sex ratio at birth = 53.7% ♂. Body weight at 390-480 d for ♂ breeders = 38.5 g.	36,37
CPB-H	Most ♀ have pair of supernumerary non-lactating nipples lateral to pair IV	39
CS	Good reproductive performance	3
C3H	Spermatozoa exhibit high in vitro fertilization rate (72-84%)	32
$C3H_eB$/FeJ	Dorsoventral vaginal septa in 5% of weaned ♀	7
$C3H_f$/Umc	50% survival: 610 d for ♂, 657 d for ♀. 10% survival: 705 d for ♂, 720 d for ♀.	40
C57BL	26.4% of spermatozoa have morphologically deformed heads	27
C57BL/6	Spermatozoa exhibit low in vitro fertilization rate (14-30%). 50% survival: 910 d for ♂, 900 d for ♀. 10% survival: 1000 d for ♂, 1080 d for ♀.	32,40

continued

9. REPRODUCTION AND GROWTH CHARACTERISTICS: MOUSE

Part II. Supplemental Information

Strain	Remarks	Reference
C57BL/6J	Dorsoventral vaginal septa in 4% of weaned ♀	7
C57L/J	Frequent occurrence of polycystic ovaries	12
C58/J	Frequent occurrence of polyovular follicles	11
DBA/2J	Dorsoventral vaginal septa in 6% of weaned ♀	7
DDI	Selected for good growth & reproductive performance. Pregnancy rate = 82.7%; avg litter size = 8.2; weaning rate = 97.8%	4,5
DDK	Low fertility of ♀ in intra-strain matings due to embryonic death at morula-blastocyst stage	43
DLS	Homozygous d^l/d^l mice die before weaning	5
DMC	Good reproductive performance. Avg litter size = >7.	39
FL/1Re	Initially good breeding; subsequently impaired by obesity	39
IF	High incidence of spontaneous pseudopregnancy	39
IVCE	Regular 4-d estrous cycle	31
IVCS	Regular 4-d estrous cycle; when ♂ brought into proximity, 3-d cycle appears; estrus appears in suckling ♀ after mother's postpartum estrus	31,41,42
IXBL	Good reproductive performance	39
JU/Fa	Avg litter size = 9.5. 50% prenatal mortality in 2nd litter when gestation concurrent with suckling 1st litter.	29,39
KE	>20% of ova remain unfertilized in fertile matings. High incidence of sperm head abnormalities (22.1%) shown to be determined largely by Y chromosome. Avg litter size = 6.8. Body weight of ♂ at 90 d = 27.4 ± 0.9 g.	1,24,25,26,27
KI	Homozygous *Ki/Ki* mice die before birth	39
KK	Avg litter size = 4.7. Occasional obesity in old mice.	21,22
KP	In ♂, low libido, multiple testicular abnormalities (including necrosis of seminiferous tubules, increased amounts of interstitial tissue, & pyknosis of spermatogonia & primary spermatocytes); low sperm production (7.7% of spermatozoa have morphologically deformed heads). Frequent sterile matings. High embryonal & post-embryonal mortality occur in fertile matings. Avg litter size = 5. Body weight of ♂ at 90 d = 26.4 ± 0.7 g.	1,15,27
KR	Good reproductive performance. Avg litter size = 7.4. Imperforate vagina in 2% of ♀.	23,39
KSB	Small litters. Occasional obesity in old mice.	3
LG	Selected for large body size. Avg litter size = 6. Sex ratio at weaning = 56% ♂. Life-span = 491 d in both sexes.	17,39
LS/Le	Many homozygous *ls/ls* mice die before weaning, but some survive & breed	5
LT/ChRe	Eggs tend to undergo cleavage without fertilization	5
MK/Re	30-50% of homozygous *mk/mk* mice die before weaning, but those surviving to adulthood are fertile	30,35
MYD/Le	Homozygous *myd/myd* mice die between 5 wk & 5 mo	39
NBR	Poor reproductive performance	3
NC	Good reproductive performance	3
NH	Breeding ceases at ∿9 mo	39
NZB	Body weight at 112 d: ♂ = 33.5 g, ♀ = 31.2 g. 50% survival: 280 d for ♂, 270 d for ♀. 10% survival: 450 d for ♂, 330 d for ♀.	33,40
NZO	Body weight of ♂ at 120 d = 42.7 g. Life-span of ♀ = 800 d; slightly shorter in ♂.	38,39
NZX	Some ♀ have imperforate vagina	39
PBB/Ld	Good breeders. Body weight = ∿60 g in both sexes, but some mice exceed 90 g. Life-span = >20 mo.	20
PH	Homozygous *Ph/Ph* mice die before birth	19
QF-5604	Selected for early sexual maturity of ♂; mean age of first mating = 52 d. Body weight of ♂ at 100 d = 39.6 g.	2,39
QF-5612	Selected for early sexual maturity of ♂; mean age of first mating = 50 d. Body weight of ♂ at 100 d = 35.9 g.	2,39
SD	Homozygous *Sd/Sd* mice die within 24 h after birth	14
SEC/1Re	Excellent breeders. Sex ratio at birth = 50.4% ♂.	37,39

continued

9. REPRODUCTION AND GROWTH CHARACTERISTICS: MOUSE

Part II. Supplemental Information

Strain	Remarks	Reference
SF-5613	Selected for late sexual maturity of ♂; mean age of first fertile mating initially 59 d, but now 52 d. Body weight of ♂ at 100 d = 28.9 g.	2,39
SF-5621	Selected for late sexual maturity of ♂; mean age of first fertile mating = 58 d. Body weight of ♂ at 100 d = 32.3 g.	2,39
SHM/2Gn	Homozygous *shm/shm* mice exhibit low fertility, small body size, & short life-span	18
SHN	Avg litter size = 7.6; weaning rate = 89%	4,20
SJL/J	Dorsoventral vaginal septa in 4% of weaned ♀	7
SJL/Wt	♂ induces & ♀ exhibits both estrus synchrony & pregnancy block	5,44
SK/Cam	Fairly large litter sizes which exhibit little decline on inbreeding	39
SLN	Avg litter size = 5.8; weaning rate = 60%	5
SM	Selected for small body size	28
SWV	Selected for good reproductive performance	5
TP	Homozygous *tp/tp* ♀ unable to nurse young because of abnormal development of nipples	13
TR	Hemizygous *To* ♂ die before birth	8
WB	Homozygous *W/W* mice die at ∿11 d	34
WC	Most homozygous *W/W* mice die at mean age of 14 d, but a few survive to adulthood	34
WH	Homozygous *W/W* mice die at ∿5 d	34
WK	Homozygous *W/W* mice die at ∿10 d	34
WLHR/Le	Homozygous *wl/wl* mice die before weaning	9
WN	Most homozygous W^n/W^n mice die at 16-18 d gestation; newborns survive only a few days	5
X/Gf	Avg litter size = 8-12. Life-span = 728 d in both sexes.	16
101/H	Slow to breed; moderate fertility	5

Contributor: Charles G. Crispens, Jr.

References

1. Bartke, A., and H. Krzanowska. 1972. J. Hered. 63: 172-174.
2. Bartke, A., and J. A. Weir. 1973. Genetics 74:s17.
3. Committee for Laboratory Animal Strains, JEARA. 1966. Gann Monogr. 5:123-128.
4. Committee for Laboratory Animal Strains, JEARA. 1973. Exp. Anim. (Tokyo) 22:271-273.
5. Committee on Standardized Genetic Nomenclature for Mice, Jackson Laboratory. 1975. Inbred Strains Mice 9.
6. Crispens, C. G., Jr., and W. K. Whitten. 1971. Experientia 27:41-42.
7. Cunliffe-Beamer, T. L., and D. B. Feldman. 1976. Lab. Anim. Sci. 26:895-898.
8. Dickie, M. M. 1954. J. Hered. 45:158, 190.
9. Dickie, M. M., et al. 1952. Ibid. 43:283-286.
10. Erickson, R. P. 1976. Genet. Res. 28:139-145.
11. Fekete, E. 1950. Anat. Rec. 108:699-708.
12. Fekete, E. 1953. Ibid. 117:93-114.
13. Fielder, J. H. 1952. J. Hered. 43:75-76.
14. Gluecksohn-Schoenheimer, S. 1943. Genetics 28: 341-348.
15. Godowicz, B. 1965. Folia Biol. (Warsaw) 13:297-309.
16. Goldfeder, A., et al. 1966. Br. J. Cancer 20:361-374.
17. Goodale, H. D. 1941. Science 94:442-443.
18. Green, E. L. 1967. J. Hered. 58:65-68.
19. Gruneberg, H., and G. M. Truslove. 1960. Genet. Res. 1:69-90.
20. Hunt, C. E., et al. 1976. Fed. Proc. Fed. Am. Soc. Exp. Biol. 35:1206-1217.
21. Ino, T., and S. Yoshikawa. 1966. Bull. Exp. Anim. 15:115-116.
22. Iwatsuka, H., et al. 1970. Endocrinol. Jpn. 17:23-35.
23. Kondo, K. 1965. Bull. Exp. Anim. 14:53-70.
24. Krzanowska, H. 1960. Folia Biol. (Warsaw) 8:269-279.
25. Krzanowska, H. 1962. Acta Biol. Cracov. Ser. Zool. 5:279-290.
26. Krzanowska, H. 1964. Folia Biol. (Warsaw) 12:231-244.
27. Krzanowska, H. 1976. Genet. Res. 28:189-198.
28. MacArthur, J. W. 1944. Am. Nat. 78:142-157.
29. McCarthy, J. C. 1965. Genetics 51:217-222.
30. Nash, D. J., et al. 1964. Am. Zool. 4:404-405.
31. Nobunaga, T. 1973. Lab. Anim. Sci. 23:803-811.
32. Parkening, T. A., and M. C. Chang. 1973. Biol. Reprod. 15:647-653.
33. Poiley, S. M. 1972. Lab. Anim. Sci. 22:759-779.
34. Russell, E. S., and F. A. Lawson. 1959. J. Hered. 50: 19-25.
35. Russell, E. S., et al. 1970. Blood 35:838-850.
36. Schlager, G. 1968. J. Hered. 59:171-174.
37. Schlager, G., and T. H. Roderick. 1968. Ibid. 59:363-365.
38. Sneyd, J. G. T. 1964. J. Endocrinol. 28:163-172.
39. Staats, J. 1976. Cancer Res. 36:4333-4377.

continued

40. Stutman, O. 1974. Fed. Proc. Fed. Am. Soc. Exp. Biol. 33:2028-2032.
41. Takahashi, K. W. 1971. Jpn. J. Anim. Reprod. 17:47-54.
42. Takahashi, K. W., and T. Nobunaga. 1973. Ibid. 19: 94-98.
43. Wakasugi, N. 1973. J. Reprod. Fertil. 33:283-291.
44. Whitten, W. K. 1973. Ibid., Suppl. 19:405-410.

10. NORMAL EMBRYO CHARACTERISTICS: MOUSE

Data in brackets refer to the column heading in brackets.

Stage	Age	Crown-to-Rump Length[1], mm[1] [Somites]	Form	Digestive, Endocrine, Respiratory & Lymphatic Systems	Vascular System	Nervous System & Sensory Organs	Urogenital System
1	1-20 h	....	One-celled egg	....	....	....	....
2	20-24 h	80-100[2]	Two-celled egg	....	....	....	....
3	2 d	....	4- to 16-cell morula	....	....	....	....
4	3 d	....	16- to 43-cell morula-blastocyst	....	....	....	....
	3½ d	....	Determination of inner-cell mass & trophoblast	....	....	....	....
5	4 d	....	Free blastocyst without zona pellucida	....	....	....	....
6	4½ d	....	Implantation	Appearance of embryonic endoderm below formative cells (embryonic ectoderm)	....	....	....
7	5 d	....	Compact egg cylinder	....	Ectoplacental cone develops	....	....
8	6 d	....	Proamniotic cavity	Reichert's membrane forming	Cone fills with maternal blood	....	....
9	6½ d	....	Embryonic axis determined	....	....	....	....
10	7 d	....	Amnion formation; primitive streak	....	....	....	....
11	7½ d	....	Presomite germ disc; allantois appearing	Blood islets in yolk sac; foregut pocket present	....	Neural plate	....
12	8 d	0.8 [1-7]	Lordotic flexure	Posterior intestinal portal appears as slight depression	1st aortic arch; allantois contacts cone in older specimens	Neural groove; otic placode	Germ cells near hindgut pocket
13	8½ d	[8-12]	Rotation (→ kyphotic flexure)	Thyroid rudiment; 2nd pharyngeal pouch; hepatic diverticulum	Paired heart primordia fusing anteriorly	Neural folds closing, beginning at cervico-cranial boundary	Pronephros

[1] Unless otherwise indicated. [2] Value is diameter given in μm.

continued

10. NORMAL EMBRYO CHARACTERISTICS: MOUSE

Stage	Age	Crown-to-Rump Length, mm [Somites]	Form	Digestive, Endocrine, Respiratory & Lymphatic Systems	Vascular System	Nervous System & Sensory Organs	Urogenital System
14	9 d	1.2-2.5 [13-20]	....	Oral plate ruptures; vitelline duct open; wide hypophyseal pouch	Paired umbilical, vitelline & cardinal veins; heart begins to beat; 3 paired aortic arches	Anterior neuropore; optic vesicle; otic pit; olfactory placode	Pronephric duct still solid
15	9½ d	1.8-3.3 [21-29]	Forelimb bud	Lung primordia; pancreas evaginations; vitelline duct closed	Heart not yet divided into right & left parts; dorsal aortae fused	Posterior neuropore; otic vesicle	....
16	10 d	3.3-3.9 [30-34]	Hindlimb bud & tail bud	Primary bronchi	....	Lens placode	Wolffian ducts contact cloaca in older specimens
17	10½ d	3.5-4.9 [35-39]	Tail	Umbilical loop; cloacal membrane	Aorta abdominalis unpaired; 6th aortic arch	Deep lens pit	Mesonephric tubules
18	11 d	5-6 [40-44]	....	Narrow hypophyseal pouch	Bulbar ridges; spleen primordium	Lens vesicle closing; rims of olfactory placode fusing	Distinct genital ridge
19	11½ d	6-7 [>45]	Forefoot plate	Bucco-nasal membrane	Partitioned atrium; unpaired ventricle	Lens vesicle detached	Ureteric buds
20	12 d	7-9	Hindfoot plate	Hypophyseal pouch closed; tongue primordium; thymus & parathyroid primordium	Partition of arterial trunk begins	Pineal body evaginates	Sexual differentiation of gonads in older specimens
21	13 d	9-11	Forefoot plate indented	Palatine processes vertical; dental laminae	Aortic & pulmonary trunks separated	Lens solid	Cloaca subdivided
22	14 d	11-12	Fingers separate distally	Palatine processes elevating	Interventricular septum closed	Ganglionic cells of retina	Ureter opens separately from Wolffian duct into urogenital sinus
23	15 d	12-14	Toes separate	Small intestine with villi; palatine processes fused	Coronary vessels	....	....
24	16 d	14-17	Skin becomes wrinkled	Reposition of umbilical hernia (Digestive & Vascular) Adrenal glands contain scattered medullary cells		Eyelids fusing	Large central glomeruli in kidney
25	17 d	17-20	Fingers & toes joined together	Alveolar ducts of lung	....	Ciliary body delineated	....

continued

10. NORMAL EMBRYO CHARACTERISTICS: MOUSE

Stage	Age	Crown-to-Rump Length, mm [Somites]	Form	Digestive, Endocrine, Respiratory & Lymphatic Systems	Vascular System	Nervous System & Sensory Organs	Urogenital System
26	18 d	19.5-22.5	Long whiskers	Thyroid has many colloid-filled follicles; pancreatic islets of Langerhans	....	Iris & ciliary body	Solid cords of prostate cells
27	19 d	23-27	Birth	Cortex of thymus subdivided into 2 zones	Ductus arteriosus shrinks; foramen ovale closes functionally	....	Testis cords still solid

Contributor: K. Theiler

References

1. Jackson Laboratory. 1975. Biology of the Laboratory Mouse. Ed. 2. Dover, New York.
2. Theiler, K. 1972. The House Mouse. Springer-Verlag, New York. pp. 26-28.

11. COMPARATIVE AGES: MOUSE AND OTHER MAMMALS

Ages: **Rat, Hamster,** and **Rabbit**—Observed ages from time of mating of rat and rabbit were reduced by Witschi by the estimated number of hours necessary for the establishment of contact between egg and sperm (for rat, 9 hours; for hamster, 8 hours; for rabbit, 14 hours). Therefore the estimated embryonic age differs considerably from the values given by Gottschewski [1]. The two-cell stage of the rabbit, according to Gottschewski, is 24 hours (post-coitum). Figures in heavy brackets are reference numbers.

Developmental Characteristics	Stages						Ages, d[1]			
	Man				Mouse					
	Carnegie, after O'Rahilly [3]	After Streeter [4]								
		Horizon	Age, d[1]	Length mm	After Witschi [6]	After Theiler [5]	Mouse	Rat	Hamster	Rabbit
One-celled egg	1	I	....		1	1	1-20 h	...	...	...
Beginning of cleavage				0.17	2[2]			24 h	16 h	8 h
	2	II	....		3[3]	2	20-24 h	50 h	40 h	11 h
Egg cleavage				0.18[4,5]	4-5	3	2			
Advanced cleavage						4	3			
Early blastocyst			....	0.13[4,6]	6			4	...	...
	3	III				5	4			
Free blastocyst					7			5	...	3½
Implantation	4-5	IV	(6-7)	....	8	6	4½	6	4½	4
Formation of egg cylinder	5	V	....	0.1-0.2[7]	9-10	7	5	7	...	...
Differentiation of egg cylinder[8]						8	6			
					11			7¾		
Advanced endometrial reaction	6	VI-VII	(8-14)	0.2[7]		9	6½			
Primitive streak	7	VIII	(14-17)	0.4[7]	12	10	7	8½	6½-7	6½-7
Neural plate; presomite stage	8	IX	(18-20)	1.0-1.5[7]	13	11	7½	9	...	...

[1] Unless otherwise specified. [2] 2 cells. [3] 4 cells. [4] Fresh. [5] With zona pellucida. [6] Without zona pellucida. [7] Fixed. [8] In rodents.

continued

11. COMPARATIVE AGES: MOUSE AND OTHER MAMMALS

Developmental Characteristics	Stages						Ages, d [1]			
	Man				Mouse		Mouse	Rat	Hamster	Rabbit
	Carnegie after O'Rahilly [3]	After Streeter [4]			After Witschi [6]	After Theiler [5]				
		Horizon	Age, d [1]	Length mm						
First somites	9-10	X	(21-23)	1.5-3.5 [7]	14	12	8	9½	...	...
Early somites (up to 12)					15	13	8½	10		
Anterior neuropore	11	XI	(23-25)	2.5-4.5 [7]	16	14	9	10½	8	9
Forelimb bud	12	XII	(25-27)	3-5 [7]	17-18	15	9½	11½	8½	9½
Hindlimb bud	13	XIII	(27-29)	4-5 [7]	19-20	16	10	12	...	...
Deep lens pit	14	XIV	(29-32) [9]	5-7 [7]	21	17	10½			
Posterior tail somites [8]	14-15	XIV-XV	33	7-9 [7]	22-23	18	11	13	9	10
Lens vesicle	16	XVI	(33-37)	8-11 [7]	24	19	11½			
Forefoot ⟨Hand-⟩ plate & hind-foot ⟨foot-⟩ plate demarcated	17	XVII	(37-41)	11-14 [7]	25	20	12	13½	13	13½
Sex differentiation of gonads	18-19	XVII-XIX	(41-48)	13-20 [7]	26	21	13	...	...	...
Fingers separate distally	20-23	XX-XXIII	(48-56)	20-31 [7]	......	22	14			
Toes separate	.....	....	2 mo	....	......	23	15			
Reposition of umbilical hernia						24	16			
Eyelids closed					35	25-26	17	17	...	19
Birth					Birth	27	19	22	16	32

[1] Unless otherwise specified. [7] Fixed. [8] In rodents. [9] Human ages adjusted according to Olivier & Pineau [2].

Contributor: K. Theiler

References

1. Gottschewski, G. H. M., and W. Zimmermann. 1973. Die Embryonalentwicklung des Hauskaninchens, Normogenese und Teratogenese. Schaper, Hannover.
2. Olivier, G., and H. Pineau. 1962. Bull. Assoc. Anat. 47:573-576.
3. O'Rahilly, R. 1973. Carnegie Inst. Wash. Publ. 631.
4. Streeter, G. L. 1942. Ibid. 541:213-245.
5. Theiler, K. 1972. The House Mouse. Springer-Verlag, Berlin and New York.
6. Witschi, E. 1956. Development of Vertebrates. Saunders, London and New York.

12. EXISTING *T/t* MUTANT STOCKS: MOUSE

Holder: For full name, *see* list of HOLDERS at front of book; Common = holdings by more than four laboratories.

Part I. Dominant Mutations

All mutants named are semidominant homozygous lethals.

Mutant Symbol	Mutant Name	Gestational Day of Death	Heterozygous Phenotype (*T*/+)	Derivation	Comments	Holder	Reference
T	Brachyury	10¾	Short tail	Spontaneous mutation	Original mutation	Common	1
T^c	Curtailed	10¾	Short tail	Radiation-induced	Embryo effects more severe than for *T*	HAR	5
T^{hp}	Hairpin-tail	7	Short tail	Spontaneous mutation	Maternal effect; elicits pseudodominance of quaking (*qk*)	HAR, LDS, SKI	3

continued

12. EXISTING T/t MUTANT STOCKS: MOUSE

Part I. Dominant Mutations

Mutant Symbol	Mutant Name	Gestational Day of Death	Heterozygous Phenotype ($T/+$)	Derivation	Comments	Holder	Reference
T^{Or}	T-Oak Ridge	7	Short tail	Radiation-induced	3 of this class exist: T^{Or1}, T^{Or2}, T^{Or6}	SKI	2
T^{Orl}	T-Orleans	7	Short tail	Spontaneous mutation	Elicits pseudodominance of quaking (*qk*)	ORL, SKI	4

Contributor: Karen Artzt

References

1. Bennett, D. 1975. Cell 6:441-454.
2. Bennett, D., et al. 1975. Genet. Res. 26:95-108.
3. Johnson, D. R. 1975. Ibid. 24:207-213.
4. Moutier, R. 1973. Mouse News Lett. 49:42.
5. Searle, A. G. 1966. Genet. Res. 7:86-95.

Part II. Recessive Mutations

The nomenclature of recessive *t*-alleles is confusing. Dunn's laboratory and others originally used serial numerical superscripts (e.g., t^0, t^1, t^2, etc.) to identify different alleles of independent origin as they were detected in laboratory stocks or arose from pre-existing mutants. When the first *t*-alleles were taken from wild populations, they were designated t^{w1}, t^{w2}, etc., to indicate their wild origin. As these wild-derived, new alleles in turn generated new forms of *t*-mutants in the laboratory, the new alleles were also designated t^w. Later, t^h and t^{AE} were used, respectively, to indicate alleles arising and studied by Lyon at Harwell and by Gluecksohn-Waelsch at Albert Einstein College of Medicine.

Lethality [Sterility]	Complementation Group	Member Haplotype [Viability, %]	Gestational Day of Death	Heterozygous Phenotype (T/t)	Recombination	Transmission Ratio	Holder	Reference
Homozygous lethal	t^0	t^0, (t^1)[1], t^6	5½	Tailless	Suppressed	80%[2]	Common	1
		t^{h7}	5½	Normal tail	Suppressed	Normal	HAR	5
		t^{h17}	5½	Short tail	Fn[3]	Fn[3]	HAR	6
		t^{h18}	5½	Short tail	Enhanced	Normal	HAR	5
		t^{h20}	5½	Tailless	Fn[3]	Fn[3]	HAR	6
	t^9	t^4	7½-9	Tailless	Normal	Normal[4]	ALB	1
		t^{w18}	7½-9	Tailless	Normal	Normal[4]	Common	1
	t^{12}	t^{12}, t^{w32}	3	Tailless	Suppressed	75%[2]	Common	1
	t^{w1}	t^{w1}, t^{w12}	9-birth	Tailless	Suppressed	95%	Common	1
		t^{w71}	9-birth	Tailless	Suppressed	95%	SKI	1
	t^{w5}	t^{w5}	5	Tailless	Suppressed	90%	Common	1
		t^{w75}	5	Tailless	Suppressed	90%	SKI	1
	t^{w73}	t^{w73}	5-6	Tailless	Suppressed	90%[2]	PAS, SKI	1
Homozygous semilethal [♂ sterile]	..	t^{w2} [51]	....	Tailless	Suppressed	95%	PAS, SKI	1
		t^{w8} [12]	....	Tailless	Suppressed	76%	SKI	1

[1] t^0 & t^1 do not represent independent mutations but are actually the same mutation; different designations are due to a laboratory mix-up [ref. 7]. [2] Average ratio; some males of this genotype, however, may have ratios as low as 30%. [3] *See* reference 7. [4] Average ratio. Some males have very heterogeneous ratios, with some ratios being as low as 25% and others as high as 90%; variation indicates some abnormality in transmission of these mutants through sperm, since $T/+$ males, for example, do not exhibit this diversity.

continued

12. EXISTING T/t MUTANT STOCKS: MOUSE

Part II. Recessive Mutations

Lethality [Sterility]	Complementation Group	Member Haplotype [Viability, %]	Gestational Day of Death	Heterozygous Phenotype (T/t)	Recombination	Transmission Ratio	Holder	Reference
Homozygous viable 5/	..	t^{h2} 6/	....	Tailless	Normal	Low	HAR	5
		t^{3} 7/	....	Tailless	Normal	Normal	ALB	4
		t^{AE5} 8/	....	Tailless	....	....	ALB	8
		t^{vSKI} 9/	....	Tailless	Normal	Low to normal	SKI	2,3

5/ Homozygous viable alleles (t^v) are by far the most frequent type obtained from known t-mutants by recombination. It is difficult to discriminate among them, and not all have been well studied. They are divided here into several groups, primarily on the basis of the transmission ratio. 6/ Derived from t^6. 7/ Derived from t^1. 8/ Homozygous short tail derived from t^{11}. 9/ SKI maintains the following viable mutations: t^{38} & t^{46} from t^0; t^{52} from t^{12}; t^{w82} from t^{w32}; t^{w84} from $t^{w12}tf$; t^{w86} from t^{w12}; t^{w88} from t^{w8}; t^{w90} & t^{w91} from t^{w5}; t^{w92} from t^{w18}.

Contributor: Karen Artzt

References

1. Bennett, D. 1975. Cell 6:441-454.
2. Bennett, D. Unpublished. Cornell Univ. Medical College, New York, 1977.
3. Bennett, D., et al. 1976. Genetics 83:361-372.
4. Dunn, L. C., and S. Gluecksohn-Waelsch. 1953. Ibid. 38:261-271.
5. Lyon, M. F., and R. Meredith. 1964. Heredity 19: 301-327.
6. Lyon, M. F., and K. Bechtol. Unpublished. Medical Research Council, Harwell, England, 1977.
7. Silagi, S. 1962. Dev. Biol. 5:35-67.
8. Vojtiskova, M., et al. 1976. J. Embryol. Exp. Morphol. 36:443-451.

13. MUTANT GENE EFFECTS: MOUSE

Part I. Congenital Malformations

Congenital malformations are here defined as gross structural abnormalities of systems, organs, or parts, which are present, or presumably present, at birth or become manifested soon after. In all instances, gastrointestinal defects are combined with other defects. For genes producing gastrointestinal defects, *see* entries 66, 71, 75, 79, 97, 98, 100, 105, 111, and 112. **Holder:** For full name, *see* list of HOLDERS at front of book.

Entry No.	Gene Symbol	Gene Name	Chromosome	Holder	Reference
Skeletal Malformations Only					
Axial Except Skull					
1	*dm*	Diminutive	2	JAX	178
2	*Fn*	Funny tail	....	....	192
3	*gt*	Gyre-tail	....	....	18
4	*Hi*	Hare tail	....	....	51
5	*Ht*	High-tail	15	OAK	75
6	*Kb*	Knobbly	17	HAR	139
7	*kt*	Kimbo-tail	....	....	191
8	*Kw*	Kinky-waltzer	....	OAK	76
9	*Mv*	Malformed vertebra	....	....	186
10	*pr*	Porcine tail	....	....	145

continued

13. MUTANT GENE EFFECTS: MOUSE

Part I. Congenital Malformations

Entry No.	Gene Symbol	Gene Name	Chromo-some	Holder	Reference
11	*pu*	Pudgy	7	JAX, OAK	94
12	*rg*	Rotating	....	....	35
13	*rh*	Rachiterata	2	JAX	195
14	*sb*	Stub	....	....	46
15	*sno*	Snubnose	4	....	104
16	*spi*	Spiral tail	....	....	183
17	*stb*	Stubby	2	JAX, LDS	134
18	*T*	Brachyury	17	Common	4,125
19	T^{h}	T-Harwell	17	....	4,125
20	T^{Or}	T-Oak Ridge	17	....	4
21	T^{Orl}	T-Orleans	17	....	4
22	*tk*	Tail-kinks	9	CIN, CON, HAR, JAX, LUC, OAK, PRA	89
23	*Tz*	Tail-zigzagged	....	....	65
24	Un^{s}	Undulated-short-tail	2	MOS	50
25	*us*	Urogenital syndrome	13	JAX	133
26	*vt*	Vestigial tail	11	CAM, HAR, SYD	92,97,102
Appendicular					
27	*bp*	Brachypodism	2	Common	132
28	*cl*	Clubfoot	....	....	163
29	*Dsy*	Dominant syndactylism	....	....	173
30	*fl*	Flipper-arm	....	OAK	76
31	*fld*	Forelimb deformity	....	....	98
32	*Hm*	Hammer-toe	5	KAN	76
33	*hop*	Hop-sterile	....	LDS	122
34	*Hx*	Hemimelic extra toes	5	JAX	41
35	*Hyp*	Hypophosphatemia	X	JAX	42
36	*jt*	Joined toes	....	CIN	23
37	*oc*	Osteosclerotic	19	JAX	76
38	*Po*	Postaxial polydactyly	....	....	150
39	*Ps*	Polysyndactyly	4	HAR, JAX	118
40	*py*	Polydactyly	1	BRN, CAM, YUR	107,154
41	*sy*	Shaker-with-syndactylism	18	HAL, JAX	91,95,101
42	sy^{fp}	Fused phalanges	18	JAX	114
43	*To*	Tortoiseshell	X	ARG, JAX, KAN, NMI	38
44	*tu*	Toe-ulnar	....	....	22
45	*Ul*	Ulnaless	2	HAR, YUR	149
Systemic					
46	*am*	Amputated	8	LMX	76
47	*Bst*	Belly spot and tail	....	JAX	172
48	*can*	Cartilage anomaly	....	LDS	123
49	*cn*	Achondroplasia	4	JAX, LDS	134
50	*Gy*	Gyro	X	HAR, JAX, LUC	76
51	*lu*	Luxoid	9	ANN, JAX	75
52	Mo^{dp}	Dappled	X	ARG, JAX	155
53	*stm*	Stumpy	....	CAM, LDS	199
54	*sm*	Syndactylism	....	JAX	91
55	T^{hp}	Hairpin-tail	17	JAX, LDS	76,120,121
56	*tl*	Non-erupted teeth	....	OAK	76
57	*un*	Undulated	2	CAM, HAR, JAX, OAK	84,202
Skeletal Plus Associated Malformations					
Axial					
58	*Bn*	Bent-tail	X	ARG, HAR, JAX, NMI	64,90

continued

13. MUTANT GENE EFFECTS: MOUSE

Part I. Congenital Malformations

Entry No.	Gene Symbol	Gene Name	Chromo-some	Holder	Reference
59	......	Coiled	....	....	125
60	*Cd*	Crooked	....	LUC	81,148
61	*ct*	Curly tail	....	....	88
62	*cy*	Crinkly-tail	4	CAM	198
63	*f*	Flexed-tail	13	ARG, EDI, JAX, KAN	116,126
64	*Fu^ki^*	Kinky	17	ALB, ANN, COR, MIS, PAS	20,69
65	*gr*	Grizzled	10	EDI, HAR, JAX, OAK	6
66	*Hk*	Hook	8	JAX	106
67	*jg*	Jagged-tail	5	JAX	76
68	*Lp*	Loop-tail	1	JAX, NIJ, SHA	125,180
69	......	Lumbarless	....	....	71
70	......	Pigtail	....	....	29
71	*Pt*	Pintail	4	JAX, KAN, NCI	105
72	*Q*	Quinky	8	....	76
73	*Rf*	Rib fusion	....	JAX	143,185
74	*sc*	Screw-tail	....	....	141
75	*Sd*	Danforth's short tail	2	Common	48,68,86,93
76	*Sh-3*	Shaker-3	....	....	170
77	*Sp*	Splotch	1	HAR, JAX, KAN, OAK	1,76,125
78	*st*	Shaker-short	....	....	7,44,125
79	T^c	Curtailed	17	HAR	4,165
80	*Ta*	Tabby	X	Common	52
81	*tc*	Truncate	6	JAX, ZUR	182
82	*ur*	Urogenital	....	....	47,56
Appendicular					
83	*bl*	Blebbed	5	HAR	156
84	*eb*	Eye-blebs	10	JAX	76
85	*fh*	Fetal hematoma	....	....	24,25
86	*Hd*	Hypodactyly	6	JAX	112
87	*heb*	Head-bleb	....	JAX	193
88	*hpy*	Hydrocephalic polydactyl	6	CIN	103
89	*ld*	Limb-deformity	2	FRA, JAX, OAK	30,76
90	*Mp*	Micropinna-microphthalmia	....	JAX, OAK	158
91	*Os*	Oligosyndactylism	8	Common	91,124
92	*Sig*	Sightless	6	HAR, JAX, LUC	165
Systemic					
93	*abn*	Abnormal	....	....	45
94	*ch*	Congenital hydrocephalus	13	JAX	79,82,87,96
95	*cho*	Chondrodysplasia	....	COR	167
96	*de*	Droopy-ear	12	EDI, HAR	31
97	*Dh*	Dominant hemimelia	1	HAR, JAX, PAS	76,77
98	*Ds*	Disorganization	14	CIN, JAX, PAS	109,110
99	*fi*	Fidget	2	CAM, HAR, JAX	83,188
100	*Fu*	Fused	17	Common	49,161,184
101	*lst*	Strong's luxoid	2	DET	61,179
102	*lx*	Luxate	5	EDI, JAX, KAN	13-15
103	*my*	Blebs	3	JAX	16,17,125,135,136
104	*ol*	Oligodactyly	7	HAL	63,100,101
105	*pad*	Paddle	....	....	151
106	*pc*	Phocomelic	....	ALB	56,72
107	*pf*	Pupoid-fetus	4	LMX	76
108	*px*	Postaxial hemimelia	6	HAR	76

continued

13. MUTANT GENE EFFECTS: MOUSE

Part I. Congenital Malformations

Entry No.	Gene Symbol	Gene Name	Chromo-some	Holder	Reference
109	*px^r*	Postaxial hemimelia-Russell	6	OAK	164
110	*se*	Short-ear	9	Common	78
111	*sho*	Shorthead	....	....	57,58
112	*srn*	Siren	....	NHA	108
113	*t*[1]	....	17	v. MNL 56	4,125
114	*Ts*	Tail short	....	JAX, LUC	34,147
115	*Xt*	Extra-toes	13	HAR, JAX, LUC, PAS	117
116	*Xt^bph*	Brachyphalangy	13	....	119
CNS Malformations Only					
117	*ac*	Absent corpus callosum	....	....	125,130
118	*ax*	Ataxia	18	JAX, MAS, PAS	28,170
119	*cb*	Cerebral degeneration	....	LUC	37
120	*crn*	Cranioschisis	....	CIN, NIDR	10
121	*hy-1*	Hydrocephalus-1	2	....	27,125
122	*hy-2*	Hydrocephalus-2	....	....	125,203
123	*hy-3*	Hydrocephalus-3	8	JAX	83,125
124	*hy-like*	Hydrocephalus-like	....	....	125
125	*la*	Leaner	....	....	170
126	....	Leukencephalosis	....	....	55,125
127	*Lc*	Lurcher	6	HAR	181
128	*oh*	Obstructive hydrocephalus	....	....	9
129	....	Pseudencephaly	....	....	8,125
130	*rl*	Reeler	5	HVD, ILL, JAX, PAS	99
131	*sg*	Staggerer	9	JAX, HVD, MAS, OAK, PAS	169
132	*sh-1*	Shaker-1	7	CAM, HAR, JAX, OAK	204
133	*Sp^d*	Delayed-splotch	1	JAX	39
134	*sw*	Swaying	15	HVD, JAX	168
135	*wv*	Weaver	....	HVD, ILL, JAX, PAS	170
136	*xn*	Exencephaly	....	CAM	131
CNS Plus Other Non-skeletal Defects[2]					
137	*bh*	Brain hernia	7	....	2
138	....	Double toe	....	....	73
139	*dr*	Dreher	1	CIN, HAL, LUC	54,125
140	*Ph*	Patch	5	HAR, JAX, KAN, LUC	96
141	*Ph^e*	Patch-extended	5	LUC	190
142	*vi*	Visceral inversion	....	....	187
143	*vl*	Vacuolated lens	1	JAX	40
Eye Defects Alone or Combined with Other Defects[3]					
144	*ak*	Aphakia	19	JAX	194
145	*Bld*	Blind	15	ROC	200
146	*bs*	Blind-sterile	....	OAK	197
147	*Cm*	Coloboma	2	JAX, PAS	165
148	*Dey*	Dickie's small eye	....	JAX	196
149	*Eo*	Eye opacity	....	JAX	76

[1] 111 haplotypes as of 1975; all but 2 produce short-tail or taillessness in *T/t* heterozygotes (*see* Part III). [2] For other genes causing CNS defects, *see* entries 58-65, 68-70, 72, 73, 75-79, 81, 88, 92, 94, 98-101, 103, 113-116, & 189.

[3] For other genes causing eye defects, *see* entries 60, 80, 83-85, 90, 92, 98-101, 103, 111, 115, 137, 138, 143, 166, 184, & 189.

continued

13. MUTANT GENE EFFECTS: MOUSE

Part I. Congenital Malformations

Entry No.	Gene Symbol	Gene Name	Chromo-some	Holder	Reference
150	*ey-1*	Eyeless-1	....	BRN, PAS	26,85
151	*ey-2*	Eyeless-2	....	BRN, PAS	26,85
152	*gp*	Gaping lids	....	BRC	129
153	*Ie*	Eye-ear reduction	X	OAK	115
154	*j*	Jaw-lethal	....	....	125
155	*lg*	Lid gap	....	BRC, OAK	76
156	lg^{ga}	Ophthalmatrophy	....	....	66
157	lg^{stn}	Slit-lid	....	....	176
158	*lo*	Lids open	....	....	76
159	*mi*	Microphthalmia	6	HAL, JAX, LUC, OAK	76,101
160	mi^{ew}	Eyeless white	6	....	201
161	Mi^{or}	Microphthalmia-Oak Ridge	6	OAK	177
162	mi^{rw}	Red-eyed white	6	....	171
163	Mi^{wh}	White	6	EDI, HAR, JAX, KAN	76
164	*o*	Eyelids-open-at-birth	....	....	85
165	*oe*	Open eyelids	11	BRC, CIN	142
166	*oel*	Open-eyelids with cleft palate	....	....	76
167	*or*	Ocular retardation	....	JAX, LUC	189
168	*po*	Palpebra operta [*sic*]	....	....	70
169	*Sey*	Small eye	....	EDI	162
170	*sq*	Squint	....	....	12
171	*wa-1*	Waved-1	6	BRN, HAR, JAX, PAS	5
172	*wa-2*	Waved-2	11	Common	12,76
Ear Defects Alone or Combined with Other Defects[4/]					
173	*Dc*	Dancer	19	HAR, JAX	36
174	*Dre*	Dominant reduced ear	4	OAK	127
175	*Em*	Ear malformed	....	....	66
176	*ft*	Flaky tail	12	JAX, OAK	76
177	*jv*	Jackson waltzer	....	JAX	43
178	*kr*	Kreisler	2	HAL, JAX, OAK	170
179	*mu*	Muted	13	HAR	76
180	*pa*	Pallid	2	Common	137
181	*pg*	Pygmy	10	EDI, NCT	87
182	*Pv*	Pivoter	....	LMX	146
183	*Rp*	Reduced pinna	....	....	76
184	*sh-2*	Shaker-2	11	CAM, JAX	32
185	*Tw*	Twirler	18	CAM, EDI, HAR, JAX	138
186	*ub*	Unbalanced	....	....	170
187	v^{df}	Deaf	10	LUC	33
188	*wi*	Whirler	4	JAX, OAK	76
189	*Wt*	Waltzer-type	....	OAK	174,175
Urogenital Defects Alone or Combined with Other Defects[5/]					
190	*at*	Atrichosis	10	JAX	111
191	*Br*	Brachyrrhine	....	HAR	166
192	Sl^{cg}	Cloud-gray	10	....	128
193	*Sxr*	Sex reversal	....	GLA, HAR	21
194	*Tfm*	Testicular feminization	X	GLA, HAR, HVD, PAT	140

[4/] For other genes causing ear defects, *see* entries 74, 78, 90, 96, 98, 99, 103, 113-115, 139, & 148. [5/] For other genes causing urogenital defects, *see* entries 67, 68, 71, 75, 82-84, 86, 89, 94, 97, 98, 100, 102-104, 108-112, & 115.

continued

Part I. Congenital Malformations

Entry No.	Gene Symbol	Gene Name	Chromosome	Holder	Reference
Orofacial Defects[6]					
195	*di*	Duplicate incisors	....	....	76
196	*It*	Irregular teeth	X	OAK	159
197	*mdg*	Muscular dysgenesis	....	ALB	71,152
198	*op*	Osteopetrosis	12	JAX	144
199	p^{cp}	p-cleft palate	7	HAR	157
Skin & Hair Defects[7]					
200	*ab*	Asebia	19	GUE, JAX, PSC, RMD	67
201	*ba*	Bare	....	....	160
202	*cr*	Crinkled	13	EDI, JAX, PSC	76
203	*N*	Naked	15	Common	85
204	*nu*	Nude	11	Common	59,153
205	Ra^{op}	Opossum	2	JAX, PSC	74
206	*Sha*	Shaven	15	DUB	60
Blebs & Edema[8]					
207	*Ra*	Ragged	2	CAM, EDI, HAR, JAX, OAK, PAS	19
Congenital Dwarfism, Proportionate & Disproportionate[9]					
208	*bal*	Balding	....	JAX	62
209	*sla*	Sex-linked anemia	X	EDI, JAX, NYB	53
Miscellaneous Defects[10]					
210	*cri*	Cribriform degeneration	4	HVD, ILL, JAX	80
211	*hf*	Hepatic fusion	7	JAX	11
212	*iv*	Situs inversus viscerum	....	JAX	113
213	*mn*	Miniature	15	CAM	3

[6] For other genes causing orofacial defects, *see* the following: for cleft lip, cleft palate, or both—entries 82, 93, 95, 98, 105, 106, 111, 113, 116, 140, 154, 166, 173, 185, 197, & 199; congenital open eyelid—65, 85, 87, 92, 98, 101, 103, 111, 113, 145, 150, 153, 155, 156, 158, 164-166, 168, & 170-172; abnormal or absent teeth—37, 60, 98, 106, 161, 195, 196, & 198; other orofacial defects—80 & 191. [7] For other genes causing skin & hair defects, *see* entries 80, 98, 99, 107, 108, & 115. [8] For other genes causing blebs or edema, *see* entries 46, 55, 83, 85, 87, 116, 140, 141, 197, & 205. [9] For other genes causing congenital dwarfism, *see* entries 1, 4, 5, 11-14, 25, 35, 41, 48, 53, 60, 65, 68, 82, 90, 95, 106, 110, 111, 113, 114, 142, 149, 163, 174, 181, 191, 205, 207, & 213. [10] Other genes causing miscellaneous defects: *stb* (defect: stubby head), entry 17; *gr* (short snout), 65; *Lp* (short umbilical cord, thoracic & umbilical eventration), 68; T^c (hindlimb paralysis), 79; *Os* (abnormal limb muscles), 91; *Dh* (spleen absent), 97; *Ds* (gastro- and thoracoschisis, abnormalities of the diaphragm, exstrophy of bladder, etc.), 98; *fi* (lachrymal glands absent), 99; *lst* (posteriorly shifted umbilicus, abnormal posteroventral body wall), 101; *my* (situs inversus, otocephaly), 103; *ol* (small abnormal spleen), 104; *se* (diaphragmatic hernia, ectopic renal artery), 110; *t* (distended pericardium), 113; *Xt* (duplicated ectopic adrenal), 115; *bh* (foreshortened head), 137; *Ph* (wide short skull), 140; Ph^e (wide short skull), 141; *ak* (large lachrymal glands), 144; *j* (short head), 154; *sq* (reduced eyeball muscles), 170; *Br* (short snout), 191; *Tfm* (female-type teats in male), 194; *mdg* (absent or abnormal striated muscle), 197; & *nu* (thymus absent), 204.

Contributor: Harold Kalter

References

1. Auerbach, R. 1954. J. Exp. Zool. 127:305-329.
2. Bennett, D. 1959. J. Hered. 50:265-268.
3. Bennett, D. 1961. Ibid. 52:95-98.
4. Bennett, D. 1975. Cell 6:441-454.
5. Bennett, J. H., and G. A. Gresham. 1956. Nature (London) 178:272-273.

continued

6. Bloom, J. L., and D. S. Falconer. 1966. Genet. Res. 7:159-167.
7. Bonnevie, K. 1935. Erbarzt 2:145-150.
8. Bonnevie, K. 1936. Skr. Nor. Vidensk. Akad. Oslo I, 9:1-39.
9. Borit, A., and R. L. Sidman. 1972. Acta Neuropathol. 21:316-331.
10. Brown, K. S. Unpublished. N.I.D.R., National Institutes of Health, Bethesda, MD, 1977.
11. Bunker, L. E. 1959. J. Hered. 50:40-44.
12. Butler, L., and D. A. Robertson. 1953. Ibid. 44:13-16.
13. Carter, T. C. 1951. J. Genet. 50:277-299.
14. Carter, T. C. 1953. Ibid. 51:442-457.
15. Carter, T. C. 1954. Ibid. 52:1-35.
16. Carter, T. C. 1956. Ibid. 54:311-326.
17. Carter, T. C. 1959. Ibid. 56:401-436.
18. Carter, T. C., and R. S. Phillips. 1952. In A. Haddow, ed. Biological Hazards of Atomic Radiation. Clarendon, Oxford. pp. 73-81.
19. Carter, T. C., and R. S. Phillips. 1954. J. Hered. 45:151-154.
20. Caspari, E., and P. R. David. 1940. Ibid. 31:427-431.
21. Cattanach, B. M., et al. 1971. Cytogenetics 10:318-337.
22. Center, E. M. 1955. J. Hered. 46:144-148.
23. Center, E. M. 1966. Genet. Res. 8:33-40.
24. Center, E. M. 1971. Genetics 68:s9.
25. Center, E. M. 1977. Genet. Res. 29:147-157.
26. Chase, H. B. 1942. Genetics 27:339-348.
27. Clark, F. H. 1932. Proc. Natl. Acad. Sci. USA 18:654-656.
28. Coggeshall, R. E., et al. 1961. Fed. Proc. Fed. Am. Soc. Exp. Biol. 20:330.
29. Crew, F. A. E., and C. Auerbach. 1941. J. Genet. 41:267-274.
30. Cupp, M. B. 1958. Mouse News Lett. 19:37.
31. Curry, G. A. 1959. J. Embryol. Exp. Morphol. 7:39-65.
32. Deol, M. S. 1954. J. Genet. 52:562-588.
33. Deol, M. S. 1956. J. Embryol. Exp. Morphol. 4:190-195.
34. Deol, M. S. 1961. Proc. R. Soc. London B 155:78-95.
35. Deol, M. S., and M. M. Dickie. 1967. J. Hered. 58:69-74.
36. Deol, M. S., and P. W. Lane. 1966. J. Embryol. Exp. Morphol. 16:543-558.
37. Deol, M. S., and G. M. Truslove. 1963. Proc. 11th Int. Congr. Genet. 1:183-184.
38. Dickie, M. M. 1954. J. Hered. 45:158, 190.
39. Dickie, M. M. 1964. Ibid. 55:97-101.
40. Dickie, M. M. 1967. Mouse News Lett. 36:39.
41. Dickie, M. M. 1968. Ibid. 38:24.
42. Dickie, M. M. 1969. Ibid. 41:30.
43. Dickie, M. M., and M. S. Deol. 1966. Ibid. 35:31.
44. Dunn, L. C. 1934. Proc. Natl. Acad. Sci. USA 20:230-232.
45. Dunn, L. C., and D. Bennett. 1970. Mouse News Lett. 43:57.
46. Dunn, L. C., and S. Gluecksohn-Schoenheimer. 1942. J. Hered. 33:235-239.
47. Dunn, L. C., and S. Gluecksohn-Schoenheimer. 1947. J. Exp. Zool. 104:25-51.
48. Dunn, L. C., et al. 1940. J. Hered. 31:343-348.
49. Dunn, L. C., et al. 1954. J. Genet. 53:383-391.
50. Egorov, I. K. 1967. Mouse News Lett. 36:57.
51. Egorov, I. K. 1969. Ibid. 41:47.
52. Falconer, D. S. 1953. Z. Indukt. Abstamm. Vererbungsl. 8:210-219.
53. Falconer, D. S., and J. H. Isaacson. 1962. Genet. Res. 3:248-250.
54. Falconer, D. S., and U. Sierts-Roth. 1951. Z. Indukt. Abstamm. Vererbungsl. 84:71-73.
55. Fischer, H. 1959. Z. Menschl. Vererbungsl. Konstitutionsl. 35:46-70.
56. Fitch, N. 1957. J. Exp. Zool. 136:329-361.
57. Fitch, N. 1961. J. Morphol. 109:141-149.
58. Fitch, N. 1961. Ibid. 109:151-157.
59. Flanagan, S. P. 1966. Genet. Res. 8:295-310.
60. Flanagan, S. P., and J. H. Isaacson. 1967. Ibid. 9:99-110.
61. Forsthoefel, P. F. 1962. J. Morphol. 110:391-420.
62. Fox, S., and E. M. Eicher. 1977. Mouse News Lett. 56:40.
63. Freye, H. 1954. Wiss. Z. Martin-Luther-Univ. 3:801-824.
64. Garber, E. D. 1952. Proc. Natl. Acad. Sci. USA 38:876-879.
65. Gates, A. H. 1967. Mouse News Lett. 36:52.
66. Gates, A. H. 1968. Ibid. 39:37.
67. Gates, A. H., and M. Karasek. 1965. Science 148:1471-1472.
68. Gluecksohn-Schoenheimer, S. 1943. Genetics 28:341-348.
69. Gluecksohn-Schoenheimer, S. 1949. J. Exp. Zool. 110:47-76.
70. Gluecksohn-Waelsch, S. 1963. Mouse News Lett. 28:13.
71. Gluecksohn-Waelsch, S. 1963. Science 142:1269-1276.
72. Gluecksohn-Waelsch, S., et al. 1956. J. Morphol. 99:465-479.
73. Green, E. L. 1964-65. Annu. Rep. Jackson Lab. 36:67-68.
74. Green, E. L., and S. J. Mann. 1961. J. Hered. 52:223-227.
75. Green, M. C. 1955. Ibid. 46:91-99.
76. Green, M. C. 1966. In E. L. Green, ed. Biology of the Laboratory Mouse. Ed. 2. McGraw-Hill, New York. pp. 87-150.
77. Green, M. C. 1967. Dev. Biol. 15:62-89.
78. Green, M. C. 1968. J. Exp. Zool. 167:129-150.
79. Green, M. C. 1970. Dev. Biol. 23:585-608.
80. Green, M. C. 1971. Mouse News Lett. 45:29.
81. Grewal, M. S. 1962. J. Embryol. Exp. Morphol. 10:202-211.
82. Grüneberg, H. 1943. J. Genet. 45:1-21.
83. Grüneberg, H. 1943. Ibid. 45:22-28.

continued

84. Grüneberg, H. 1950. Ibid. 50:142-173.
85. Grüneberg, H. 1952. The Genetics of the Mouse. Ed. 2. Nijhoff, The Hague.
86. Grüneberg, H. 1953. J. Genet. 51:317-326.
87. Grüneberg, H. 1953. Ibid. 51:327-358.
88. Grüneberg, H. 1954. Ibid. 52:52-62.
89. Grüneberg, H. 1955. Ibid. 53:536-550.
90. Grüneberg, H. 1955. Ibid. 53:551-562.
91. Grüneberg, H. 1956. Ibid. 54:113-145.
92. Grüneberg, H. 1957. Ibid. 55:181-194.
93. Grüneberg, H. 1958. J. Embryol. Exp. Morphol. 6: 124-148.
94. Grüneberg, H. 1961. Genet. Res. 2:384-393.
95. Grüneberg, H. 1962. Ibid. 3:157-166.
96. Grüneberg, H., and G. M. Truslove. 1960. Ibid. 1:69-90.
97. Grüneberg, H., and G. A. de S. Wickramaratne. 1974. J. Embryol. Exp. Morphol. 31:207-222.
98. Guénet, J. L., and M. Mercier-Balaz. 1975. Mouse News Lett. 53:57.
99. Hamburgh, M. 1963. Dev. Biol. 8:165-185.
100. Hertwig, H. 1939. Erbarzt 6:41-43.
101. Hertwig, P. 1942. Z. Indukt. Abstamm. Vererbungsl. 80:220-246.
102. Heston, W. E. 1951. J. Hered. 42:71-74.
103. Hollander, W. F. 1966. Am. Zool. 6:588-589.
104. Hollander, W. F. 1976. Am. J. Anat. 146:173-180.
105. Hollander, W. F., and L. C. Strong. 1951. J. Hered. 42:179-182.
106. Holman, S. P. 1951. Ibid. 42:305-306.
107. Holt, S. B. 1945. Ann. Eugen. 12:220-249.
108. Hoornbeek, F. K. 1970. Teratology 3:7-10.
109. Hummel, K. P. 1958. J. Exp. Zool. 137:389-423.
110. Hummel, K. P. 1959. Pediatrics 23:212-221.
111. Hummel, K. P. 1966. Mouse News Lett. 34:31.
112. Hummel, K. P. 1970. J. Hered. 61:219-220.
113. Hummel, K. P., and D. B. Chapman. 1959. Ibid. 50: 9-13.
114. Hummel, K. P., and D. B. Chapman. 1971. Mouse News Lett. 45:28.
115. Hunsicker, P. 1974. Ibid. 50:51.
116. Hunt, H. R., et al. 1933. Genetics 18:335-366.
117. Johnson, D. R. 1967. J. Embryol. Exp. Morphol. 17: 543-581.
118. Johnson, D. R. 1969. Ibid. 21:285-294.
119. Johnson, D. R. 1969. Genet. Res. 13:275-280.
120. Johnson, D. R. 1974. Genetics 76:795-805.
121. Johnson, D. R. 1975. Genet. Res. 24:207-213.
122. Johnson, D. R., and D. M. Hunt. 1971. J. Embryol. Exp. Morphol. 25:223-236.
123. Johnson, D. R., and J. M. Wise. 1971. Ibid. 25:21-32.
124. Kadam, K. M. 1962. Genet. Res. 3:139-156.
125. Kalter, H. 1968. Teratology of the Central Nervous System. Univ. Chicago Press, Chicago.
126. Kamenoff, R. J. 1935. J. Morphol. 58:117-155.
127. Kelly, E. M. 1968. Mouse News Lett. 38:31.
128. Kelly, E. M. 1974. Ibid. 50:52.
129. Kelton, D. E., and V. Smith. 1964. Genetics 50:261-262.
130. King, L. S., and C. E. Keeler. 1932. Proc. Natl. Acad. Sci. USA 525-528.
131. Knights, P. J. 1972. Mouse News Lett. 47:24.
132. Landauer, W. 1952. J. Hered. 43:293-298.
133. Lane, P. W. 1973. Mouse News Lett. 49:32.
134. Lane, P. W., and M. M. Dickie. 1968. J. Hered. 59: 300-308.
135. Little, C. C., and H. J. Bagg. 1923. Am. J. Roentgenol. 10:975-989.
136. Little, C. C., and H. J. Bagg. 1924. J. Exp. Zool. 41: 45-91.
137. Lyon, M. F. 1953. J. Genet. 51:638-650.
138. Lyon, M. F. 1958. J. Embryol. Exp. Morphol. 6: 105-116.
139. Lyon, M. F. 1977. Mouse News Lett. 56:37.
140. Lyon, M. F., and S. G. Hawkes. 1970. Nature (London) 227:1217-1219.
141. MacDowell, E. C., et al. 1942. J. Hered. 33:439-449.
142. Mackensen, J. A. 1960. Ibid. 51:188-190.
143. Mackensen, J. A., and L. C. Stevens. 1960. Ibid. 51: 264-268.
144. Marks, S. C., Jr., and P. W. Lane. 1976. Ibid. 67:11-18.
145. McNutt, W. 1969. Anat. Rec. 163:340.
146. Meredith, R. 1969. Mouse News Lett. 40:36.
147. Morgan, W. C. 1950. J. Hered. 41:208-215.
148. Morgan, W. C. 1954. J. Genet. 52:354-373.
149. Morris, T. 1967. Mouse News Lett. 36:34.
150. Nakamura, A., et al. 1963. Annu. Rep. Natl. Inst. Genet. Jpn., 1962, 13:31.
151. Nash, D. J. 1969. Mouse News Lett. 40:20.
152. Pai, A. C. 1965. Dev. Biol. 11:82-92.
153. Pantelouris, E. M., and J. Hair. 1970. J. Embryol. Exp. Morphol. 24:615-625.
154. Parsons, P. A. 1958. Heredity 12:77-95.
155. Phillips, R. J. 1961. Genet. Res. 2:290-295.
156. Phillips, R. J. 1970. Mouse News Lett. 42:26.
157. Phillips, R. J. 1975. Ibid. 53:29.
158. Phipps, E. L. 1964. Ibid. 31:41.
159. Phipps, E. L. 1969. Ibid. 40:41.
160. Randelia, H. P., and L. D. Sanghvi. 1961. Genet. Res. 2:283-289.
161. Reed, S. C. 1937. Genetics 22:1-13.
162. Roberts, R. C. 1966. Mouse News Lett. 35:24-25.
163. Robins, M. W. 1959. J. Hered. 50:188-192.
164. Russell, L. B. 1972. Mouse News Lett. 47:61.
165. Searle, A. G. 1965. Ibid. 33:29.
166. Searle, A. G. 1966. Ibid. 35:27.
167. Seegmiller, R., et al. 1971. J. Cell Biol. 48:580-593.
168. Sidman, R. L. 1968. In F. D. Carlson, ed. Physiological and Biochemical Aspects of Nervous Integration. Prentice-Hall, Englewood Cliffs, NJ. pp. 163-193.
169. Sidman, R. L., et al. 1962. Science 137:610-612.
170. Sidman, R. L., et al. 1965. Catalog of the Neurological Mutants of the Mouse. Harvard Univ. Press, Cambridge.
171. Southard, J. L. 1974. Mouse News Lett. 51:23.
172. Southard, J. L., and E. M. Eicher. 1977. Ibid. 56:40.
173. Steele, M. S. H. 1960. Ibid. 23:58.

continued

13. MUTANT GENE EFFECTS: MOUSE

Part I. Congenital Malformations

174. Stein, K. F., and S. H. Filosa. 1969. Dev. Biol. 19: 358-367.
175. Stein, K. F., and S. A. Huber. 1960. J. Morphol. 106: 197-203.
176. Stein, K. F., and C. N. Kettyle. 1973. Teratology 8: 51-54.
177. Stelzner, K. F. 1966. Mouse News Lett. 34:41.
178. Stevens, L. C., and J. A. Mackensen. 1958. J. Hered. 49.153-160.
179. Strong, L. C., and L. B. Hardy. 1956. Ibid. 47:277-284.
180. Strong, L. C., and W. F. Hollander. 1949. Ibid. 40: 329-334.
181. Swisher, D. A., and D. B. Wilson. 1975. Anat. Rec. 181:489.
182. Theiler, K. 1959. Am. J. Anat. 104:319-343.
183. Theiler, K. 1968. Mouse News Lett. 38:41.
184. Theiler, K., and S. Gluecksohn-Waelsch. 1956. Anat. Rec. 125:83-104.
185. Theiler, K., and L. C. Stevens. 1960. Am. J. Anat. 106:171-183.
186. Theiler, K., et al. 1975. Anat. Embryol. 147:161-166.
187. Tihen, J. A., et al. 1948. J. Hered. 39:29-31.
188. Truslove, G. M. 1956. J. Genet. 54:64-86.
189. Truslove, G. M. 1962. J. Embryol. Exp. Morphol. 10:652-660.
190. Truslove, G. M. 1977. Genet. Res. 29:183-186.
191. Van Pelt, A. 1965. Mouse News Lett. 32:33.
192. Van Valen, P. 1964. Ibid. 31:39.
193. Varnum, D. S., and S. Fox. 1976. Ibid. 55:16.
194. Varnum, D. S., and L. C. Stevens. 1968. J. Hered. 59:147-150.
195. Varnum, D. S., and L. C. Stevens. 1970. Mouse News Lett. 43:34.
196. Varnum, D. S., and L. C. Stevens. 1974. Ibid. 50:43.
197. Varnum, D. S., and L. C. Stevens. 1977. Ibid. 56:40.
198. Wallace, M. E. 1971. Ibid. 41:18.
199. Wallace, M. E. 1973. Ibid. 49:23.
200. Watson, M. L. 1968. J. Hered. 59:60-64.
201. Wolfe, H. G., and G. Miner. 1969. Mouse News Lett. 40:32.
202. Wright, M. E. 1947. Heredity 1:137-141.
203. Zimmerman, K. 1933. Z. Indukt. Abstamm. Vererbungsl. 64:176-180.
204. Zimmerman, K. 1935. Erbarzt 2:119-120.

Part II. Lethal and Semilethal Genes

Lethal genes are defined as causing death in all homozygotes for recessive alleles, heterozygotes for semidominant alleles, and hemizygous males, prenatally or before roughly the first day of age. Semilethal genes are defined as causing death in some homozygotes for recessive alleles, some heterozygotes for semidominant alleles, and some hemizygous males, prenatally or before roughly the first day of age. For other prenatal and neonatal lethal and semilethal genes, *see* Part I, entries, 2, 5, 6, 9, 19-21, 23, 24, 39, 43, 46, 52, 55, 58-60, 64, 65, 68, 72-77, 79, 82, 83, 86, 87, 91-95, 97, 98, 100, 102, 103, 105-107, 111, 112, 114-116, 120, 124, 133, 136, 140-142, 145, 149, 154, 166, 169, 173, 183, 185, 189, 196, 197, 205, & 207.

Entry No.	Gene Symbol	Gene Name	Time of Death	Chromosome	Holder	Reference
1	A^y	Yellow	Preimplantation	2	Common	3
2	a^x	Lethal non-agouti	Prenatal	2	NCT, OAK	3
3	a^l	Non-agouti lethal	Probably prenatal	....	HAR	8
4	*Er*	Repeated epilation	Probably midgestation	....	JAX, OAK, PAS	3,6
5	*Fk*	Fleck	Possibly prenatal	....	STK	10
6	*ja*	Jaundiced	Neonatal	8	JAX	3
7	*Lc*	Lurcher	Neonatal	6	HAR	3
8	*Mo*	Mottled	Hemizygote: midgestation; heterozygote: pre- and neonatal semilethal	X	....	3
9	*om*	Ovum mutant	Just before and during implantation	....	....	12
10	*Rw*	Rump-white	Probably prenatal	5	HAR, PAS	3
11	*Sk*	Scaly	Possibly prenatal	2	JAX, OAK	3
12	*Sl*	Steel	15th-16th day of gestation	10	ARG, EDI, JAX, OAK	2,9
13	*sph*	Spherocytic anemia	Perinatal	....	BRC, HAL, JAX	3
14	*Sta*	Autosomal striping	Probably prenatal	....	OAK	1
15	*Str*	Striated	Approx. 11th day of gestation	X	HAR	3
16	*Tsk*	Tight skin	8th-9th day of gestation	2	JAX	5

continued

13. MUTANT GENE EFFECTS: MOUSE

Part II. Lethal and Semilethal Genes

Entry No.	Gene Symbol	Gene Name	Time of Death	Chromosome	Holder	Reference
17	*Ve*	Velvet coat	Probably prenatal	15	JAX, OAK, PAS	11
18	*Wc*	Waved coat	Prenatal; ? age	14	JAX	4
19	*Ym*	Yellow mottling	? age	X	OAK	7

Contributor: Harold Kalter

References

1. Bangham, J. W. 1968. Mouse News Lett. 28:31.
2. Bennett, D. 1956. J. Morphol. 98:199-233.
3. Green, M. C. 1966. In E. L. Green, ed. Biology of the Laboratory Mouse. Ed. 2. McGraw-Hill, New York. pp. 87-150.
4. Green, M. C. 1972. Mouse News Lett. 47:36.
5. Green, M. C., et al. 1976. J. Pathol. 82:493-512.
6. Guénet, J. L. 1976. Mouse News Lett. 54:51.
7. Hunsicker, P. 1968. Ibid. 38:31.
8. Phillips, R. J. 1976. Ibid. 55:14.
9. Sarvella, P. A., and L. B. Russell. 1956. J. Hered. 47: 123-128.
10. Sheridan, W. 1968. Mutat. Res. 5:323-328.
11. Stieler, C., and W. F. Hollander. 1972. J. Hered. 63: 212-213.
12. Wakasugi, N. 1974. J. Reprod. Fertil. 41:85-96.

Part III. Lethal *t* Haplotypes

Lethality Group—Lethality schedule for homozygotes: t^0 die shortly after implantation; t^9 die after formation of abnormal primitive streak; t^{12} die upon reaching morula stage; t^{w1} die after beginning to form neural tube; t^{w5} die after forming egg cylinder; t^{w73} die during process of implanting in uterus.

Entry No.	Gene Symbol	Gene Name	Lethality Group
1	$t^{0 (or\ 1)}$	t-0	t^0
2	t^4	t-4	t^9
3	t^6	t-6	t^0
4	t^9	t-9	t^9
5	t^{12}	t-12	t^{12}
6	t^{30}	t-30	t^0
7	t^{h7}	t-Harwell-7	t^0
8	t^{h13}	t-Harwell-13	t^0
9	t^{h16}	t-Harwell-16	t^0
10	t^{h18}	t-Harwell-18	t^0
11	t^{w1}	t-wild-1	t^{w1}
12	t^{w2}	t-wild-2	semilethal
13	t^{w3}	t-wild-3	t^{w1}
14	t^{w5}	t-wild-5	t^{w5}
15	t^{w6}	t-wild-6	t^{w5}
16	t^{w8}	t-wild-8	semilethal
17	t^{w10}	t-wild-10	t^{w5}
18	t^{w11}	t-wild-11	t^{w5}
19	t^{w12}	t-wild-12	t^{w1}
20	t^{w13}	t-wild-13	t^{w5}
21	t^{w14}	t-wild-14	t^{w5}
22	t^{w15}	t-wild-15	t^{w5}
23	t^{w16}	t-wild-16	t^{w5}
24	t^{w17}	t-wild-17	t^{w5}
25	t^{w18}	t-wild-18	t^{w9}
26	t^{w20}	t-wild-20	t^{w1}
27	t^{w21}	t-wild-21	t^{w1}
28	t^{w30}	t-wild-30	t^9
29	t^{w32}	t-wild-32	t^{12}
30	t^{w36}	t-wild-36	semilethal
31	t^{w37}	t-wild-37	t^{w5}
32	t^{w38}	t-wild-38	t^{w5}
33	t^{w39}	t-wild-39	t^{w5}
34	t^{w41}	t-wild-41	t^{w5}
35	t^{w46}	t-wild-46	t^{w5}
36	t^{w47}	t-wild-47	t^{w5}
37	t^{w49}	t-wild-49	semilethal
38	t^{w52}	t-wild-52	t^9
39	t^{w71}	t-wild-71	t^{w1}
40	t^{w72}	t-wild-72	t^{w1}
41	t^{w73}	t-wild-73	t^{w73}
42	t^{w74}	t-wild-74	t^{w5}
43	t^{w75}	t-wild-75	t^{w5}
44	t^{w80}	t-wild-80	t^{w5}
45	t^{w81}	t-wild-81	t^{w5}

Contributor: Harold Kalter

Reference: Bennett, D. 1975. Cell 6:441-454.

14. MUTANT GENES AFFECTING DEVELOPMENT OF ANEMIA AND ERYTHROCYTE PRODUCTION: MOUSE

Holder: For full names, *see* list of HOLDERS at front of book; Common = holdings by more than seven laboratories.

Locus & Alleles	Gene Name	Chromosome [Linkage Group]	Characteristics & Gene Effects	Strain Distribution	Holder	Reference
an	Anemia (Hertwigs)	4 [VIII]	Normochromic macrocytic anemia, with deficiencies in both erythroid & myeloid precursors	C57BL/6J-*an* B^{lt}; WB/Re-*an* B^{lt}	HAL, JAX	4,8
dm	Diminutive	2 [V]	Macrocytic anemia; dwarfism	C57BL/6J-*dm*; WB/Re-*dm*	JAX	1,4, 8
f	Flexed-tail	13 [XIV]	Transitory siderocytic anemia caused by disturbance of hemopoietic function of fetal liver; color or spotting effects; tail & other appendages affected	*a/a f/f je/+ ru/ru*; FL/1Re *a/a f/f*; FL/1Re *a/a f/+ W/+*	ARG, EDI, JAX, KAN	1,4, 8
ha	Hemolytic anemia	....	Severe hemolytic disease; homozygous lethality or sublethality; reproductive organs affected—sterility	C57BL/6J-*ha*; WB/Re-*ha*	JAX	1,4, 8
hbd	Hemoglobin deficit	....	Hematological	....	HAL	9
ja	Jaundiced	....	Severe hemolytic disease; homozygous lethality or sublethality; reproductive organs affected—sterility	C57BL/6J-*ja*; WB/Re-*ja*	JAX	1,4, 8
lst	Strong's luxoid	2 [V]	Temporary postnatal anemia due to bleeding at umbilicus; skeleton, & tail & other appendages affected	....	DET	4
mk	Microcytic anemia	15 [VI]	Microcytic hypochromic anemia	MK/Re *mk*/+; C57BL/6J-*mk*; SEC/1Re-*mk*; WB/Re-*mk*	JAX	1,4, 8
nb	Normoblastic anemia	....	Severe hemolytic disease; homozygous lethality or sublethality; reproductive organs affected—sterility	C57BL/6J-*nb*; WB/Re-*nb*	JAX	1,5, 8
Sl[1]	Steel	10 [IV, X]	Macrocytic anemia, with normal stem cells but an inhibiting environment preventing their normal differentiation; color or spotting effects; homozygous lethality or sublethality; reproductive organs affected—sterility	WC/Re-*Sl*; C57BL/6By-*CaSl*; C57BL/6J-Ca^J *Sl Hm*; LT/ChReSv-*a/a* B^{lt}/B^{lt} *Sl*/+	ARG, EDI, JAX, OAK	1,2, 4,8
Sl^{con}	Contrasted	10 [IV, X]	Color or spotting effects; hematological effects; other biochemical effects	....	HAR	3,6
Sl^{d}	Steel-Dickie	10 [IV, X]	Color or spotting effects; hematological effects; other biochemical effects	C57BL/6J-Sl^d; 129/Sv-Sl^d/+ $c^+ p^+/c^+ p^+$; WB/Re-Sl^d	KAN, PAS	1,4, 8
Sl^{gb}	Grizzle-belly	10 [IV, X]	Color or spotting effects; hematological effects; homozygous lethality or sublethality	....	KAN	4
Sl^{so}	Sooty	10 [IV, X]	Color or spotting effects; hematological effects; homozygous lethality or sublethality	....	....	4
sla	Sex-linked anemia	X [XX]	Defective iron utilization	C57BL/6J-*sla*; C57BL/6J-*sla* Mo^{to}; WB/Re-*sla*	EDI, JAX, NYB	1,4, 8
sph	Spherocytic anemia	....	Severe hemolytic disease; homozygous lethality or sublethality; reproductive organs affected—sterility	C57BL/6J-*sph*; WB/Re-*sph*	HAL, JAX	1,4, 7,8

[1] Another allele, not listed below, is Sl^J.

continued

14. MUTANT GENES AFFECTING DEVELOPMENT OF ANEMIA AND ERYTHROCYTE PRODUCTION: MOUSE

Locus & Alleles	Gene Name	Chromosome [Linkage Group]	Characteristics & Gene Effects	Strain Distribution	Holder	Reference
Ts	Tail-short	....	Prenatal anemia which disappears at birth and can be traced to deficiency of blood islands in yolk sac of 8-d embryos; skeleton, & tail & other appendages affected	*b/b Ts/+*	JAX, LUC	4
W[2]	Dominant spotting	5 [XVII]	Macrocytic anemia characterized by defect of erythropoietic stem cells; color or spotting effects; homozygous lethality or sublethality	C57BL/6J-*W*; WB/Re *a/a W/+*; WC/Re *a/a W/+*; WH/Re *a/a W/+*; WK/Re *a/a b/b W/+*; FL/1Re-*W*	CAM, JAX, KAN, NYH, PAT	1,2, 4,8
W^a	Ames dominant spotting	5 [XVII]	Color or spotting effects; hematological effects; homozygous lethality or sublethality	....	ARG, KAN	4
W^b	Ballantyne's spotting	5 [XVII]	Color or spotting effects; hematological effects; homozygous lethality or sublethality	....	MIS	4
W^{2J}	Dominant spotting-2J	5 [XVII]	Color or spotting effects; hematological effects; homozygous lethality or sublethality	C57BL/6J-W^{2J}	JAX	4,5
W^{pw}	Panda-white	5 [XVII]	Color or spotting effects; hematological effects; homozygous lethality or sublethality	....	....	4
W^s	Strong's dominant spotting	5 [XVII]	Color or spotting effects; hematological effects; homozygous lethality or sublethality	....	....	10
W^v	Viable dominant spotting	5 [XVII]	Color or spotting effects; hematological effects; homozygous lethality or sublethality	C57BL/6J-W^v; C57BL/6J-*Ph* +/+ W^v; C57BL/6J-*lx* W^v; WB/Re-*lx* W^v; MWT/Le a^t/a^t $Mi^{wh}/+$ $W^v/+$ *T*/+; C57BL/6By-M^{wh} W^v *T*; 129/Sv-*ter* $W^v/+$	Common	1,4, 8

[2] Other alleles not listed are W^x & W^j.

Contributor: Seldon E. Bernstein

References

1. Bernstein, S. 1969. NAS-NRC Publ. 1724:9-33.
2. Bernstein, S. E., et al. 1968. Ann. N.Y. Acad. Sci. 149:475-485.
3. Cattanach, B. M., et al. 1974. Mouse News Lett. 50: 39-42.
4. Green, E. L., ed. 1966. Biology of the Laboratory Mouse. Ed. 2. McGraw-Hill, New York.
5. Lane, P. W., et al. 1969. Mouse News Lett. 40:28-30.
6. Lyon, M. F., et al. 1968. Ibid. 38:22.
7. Miller, J. R., et al. 1976. Ibid. 55:9.
8. Russell, E. S. 1970. In A. S. Gordon, ed. Regulation of Hematopoiesis. Appleton-Century-Crofts, New York. v. 1, pp. 649-676.
9. Schmidt, R., et al. 1969. Mouse News Lett. 40:24.
10. Strong, L. C., and W. F. Hollander. 1953. J. Hered. 44:41-44.

15. MUTANT GENES AFFECTING MUSCLE DEVELOPMENT: MOUSE

Gene Symbol	Gene Name	Chromosome	Characteristics	Reference
dt	Dystonia musculorum	1	Severe sensory ataxia; atrophy of primary sensory neurons beginning in axon and spreading to cell body	5,16

continued

15. MUTANT GENES AFFECTING MUSCLE DEVELOPMENT: MOUSE

Gene Symbol	Gene Name	Chromosome	Characteristics	Reference
dy	Dystrophia muscularis	10	Progressive weakness and paralysis beginning at ∿2 wk; defective myelination of dorsal & ventral spinal roots & proximal parts of peripheral nerves; muscle defect probably secondary to neural defect	2,9,11, 14,15
dy^{2J}	Dystrophia muscularis-2J	10	Similar clinically & histologically to *dy* when in same genetic background; similar to *dy* in abnormalities of nerve roots	8,17
lz	Lizard	15	Partial paralysis of hind limbs; marked tremor of whole musculature; preliminary studies suggest muscular rather than nervous origin	10
mdg	Muscular dysgenesis	....	Complete failure of myoblasts to differentiate into striated myotubes in skeletal muscle in vivo; myoblasts develop further in culture but do not become contractile; electrical properties of cultured cells normal despite absence of contractility	12,13, 18
med	Motor end-plate disease	15	Progressive fatal muscular weakness due to abnormal motor end-plates which result in functionally denervated muscle fibers; proximal muscles most severely affected	4
myd	Myodystrophy	8	Diffuse progressive myopathy with focal lesions in all skeletal muscles; life-span considerably reduced	7
Swl	Sprawling	....	Ataxia chiefly affecting hindlimbs, detectable at 12 d postnatal; deficiency of central & peripheral axons of sensory ganglion cells, & of muscle spindles & sensory receptors of skin & joints	3
wr	Wobbler	....	Progressive motor denervation affecting forequarters most severely; degeneration of motor neurons of brainstem & cervical spinal cord, with broad spectrum of cytopathological changes	1,6

Contributor: Margaret C. Green

References

1. Andrews, J. M. 1975. J. Neuropathol. Exp. Neurol. 34:12-27.
2. Brimijoin, S., and C. Jablecki. 1976. Exp. Neurol. 53: 454-464.
3. Duchen, L. W. 1975. Neuropathol. Appl. Neurobiol. 1:89-101.
4. Duchen, L. W., and E. Stefani. 1971. J. Physiol. (London) 212:535-548.
5. Duchen, L. W., and S. J. Strich. 1964. Brain 87:367-378.
6. Duchen, L. W., and S. J. Strich. 1968. J. Neurol. Neurosurg. Psychiatr. 31:535-542.
7. Lane, P. W., et al. 1976. J. Hered. 67:135-138.
8. MacPike, A. D., and H. Meier. 1976. Proc. Soc. Exp. Biol. Med. 151:670-672.
9. Madrid, R. E., et al. 1975. Nature (London) 257:319-321.
10. Morris, T. 1966. Mouse News Lett. 34:27.
11. Neerunjun, J. S., et al. 1976. Exp. Neurol. 52:556-564.
12. Pai, A. C. 1965. Dev. Biol. 11:93-109.
13. Powell, J. A., and D. M. Fambrough. 1973. J. Cell. Physiol. 82:21-38.
14. Staats, J. 1965. Z. Versuchstierk. 6:56-68.
15. Stirling, C. A. 1975. J. Anat. 119:169-180.
16. Thornburg, L. P., and J. S. Hanker. 1975. Acta Neuropathol. 32:91-101.
17. Weinberg, H. J., et al. 1975. Brain Res. 88:532-537.
18. Yao, M.-L., and F. B. Essien. 1975. Dev. Biol. 45:166-175.

16. MUTANT GENES AFFECTING DEVELOPMENT OF THE IMMUNE SYSTEM: MOUSE

Data are for single gene mutations having effects on the immune system. The association between observed immunological dysfunction and other phenotypic characteristics is not well understood. To facilitate comprehension of gene action, non-immunological characteristics are also listed. Holders of these mutant stocks are listed in reference 64, or in the Lists of Mutations and Mutant Stocks of the Mouse, which may be obtained by contacting the Animal

continued

Resources Department, Jackson Laboratory, Bar Harbor, Maine 04609. **Abbreviations**: Con-A = concanavalin-A; GMuLV = Gross murine leukemia virus; GVH = graft versus host; IgG (or IgG_{2a}) = immunoglobulin G (or G_{2a}); IgM = immunoglobulin M; Lps = bacterial lipopolysaccharide; PFC = plaque-forming cell; PHA = phytohemagglutinin; SRBC = sheep erythrocytes; SSS-III = type III pneumococcal polysaccharide.

Gene Symbol	Gene Name	Chromosome	Phenotypic Characteristics	Reference
A^{vy}	Viable yellow	2	Decreased reactivity of spleen cells in GVH assay	23
			Increased susceptibility to hepatomas, mammary tumors, cholangiomas	32,71
			Obesity	32
A^{y}	Yellow	2	Decreased reactivity of spleen cells in GVH assay	23
			Increased susceptibility to reticular neoplasms, hepatomas, mammary tumors, lung tumors	10,30,31
			Obesity	31
bg	Beige	13	Defective granulocyte chemotaxis; impaired intracellular killing of *Staphylococcus aureus*	22
			Increased susceptibility to spontaneous pneumonitis	37
			Increased susceptibility to experimental infection with *Candida albicans* & *Streptococcus pneumoniae*[1]; elevated levels of serum IgM	20
			Presence of giant lysosomal granules in all granule-containing cells	22
			Homologue for Chediak-Higashi syndrome in man	48
df	Ames dwarf	11	Involution of thymus after weaning; lymphopenia	16
			Decreased reactivity of spleen cells in GVH assay	16
			Depressed blastogenic response of spleen cells to PHA & Con-A	18
			Reduced humoral response to SRBC	16,18
			Pituitary hormone deficiency	4
Dh	Dominant hemimelia	1	Congenital absence of spleen	25,63
			Hypertrophic lymph nodes	41
			Leukocytosis	38
			Reduced humoral response to SRBC	5,39,40
			Decreased rate of intravascular carbon clearance	39
			Hind limb defects & widespread defects of urogenital & digestive systems	25,63
dw	Dwarf (Snell's)	16	Involution of thymus after weaning; lymphopenia	2,21
			Decreased levels of thymic hormone in serum	52
			Decreased reactivity of spleen cells in GVH assay	17
			Reduced humoral response to SRBC	2,17,21
			Depressed humoral response to *Brucella* antigen	17
			Reduced contact sensitivity response to picryl chloride	3
			Decreased susceptibility to autoimmunity	52
			Pituitary hormone deficiency	21
hr	Hairless	14	Thymic cortical atrophy at 6 mo	27
			Decreased reactivity of spleen cells in GVH assay	34,67
			Impaired cellular immune response to SRBC & GMuLV	68
			Increased titer of GMuLV	28
			Increased susceptibility to leukemia	43
			Decreased humoral response to tetanus toxoid	27
			Reduced contact sensitivity response to picryl chloride	67
lh	Lethargic	2	Decreased size of thymus & spleen	14
			Involution of thymus by 25 d	12,15
			Decreased size & number of Peyer's patches	15
			Lymphopenia	14
			Reduced reactivity of spleen cells in GVH assay; delayed allograft rejection	13
			Sluggish behavior	11,14
			High mortality before 45 d	14

[1] Synonym: *Diplococcus pneumoniae*.

continued

Gene Symbol	Gene Name	Chromosome	Phenotypic Characteristics	Reference
lpr	Lymphoproliferation	?	Massive lymph node enlargement by 40 wk	46
			Hyperimmunoglobulinemia; thymocytotoxic autoantibodies; immune complex nephritis	47
Lps	Lps response[2]	4	Impaired humoral & mitogenic response to Lps	24,72
			Increased resistance to endotoxemia	44
			Defective tumoricidal capacity of macrophages	58
me	Motheaten	6	Splenomegaly; small thymus; reduced size & number of Peyer's patches; depletion of lymphocytes from B-dependent areas of lymphoid tissues	26
			Decreased number of splenic B-cells	69
			Non-reactivity of spleen cells in GVH assay	66
			Severely depressed humoral response to SRBC	66
			Reduced DNA synthetic & antibody-forming PFC responses to B-cell mitogens & thymus-independent antigens	70
			Polyclonal hyperimmunoglobulinemia; antithymocyte autoantibody; anti-DNA autoantibody; immune complex nephritis	69
			Decreased life-span	26
nu	Nude[3]	11	Congenital absence of thymus	49,51
			Severe reduction in numbers of T-cells	56
			Lymphopenia	73
			Depressed humoral response to SRBC	9,50,54,73
			Decreased humoral response to bacteriophage T4	36
			Impaired IgG antibody response to *Brucella abortus*	8
			Inability to reject allogeneic or xenogeneic skin grafts	42,50,57
			Reduced contact sensitivity response to oxazolone	53,54
			Hypoimmunoglobulinemia	8,55
			Antinuclear autoantibody; glomerulonephritis	45
			Decreased life-span	73
nu^{str}	Streaker[4]	11	Congenital absence of thymus	19,69
			Severe reduction in numbers & function of T-cells	69
ob	Obese	6	Decreased humoral response to SRBC; delayed allograft rejection; decreased level of serum IgG_{2a}	65
			Obesity; hyperinsulinemia	7,29
Xid	X-linked immunodeficiency	X	Decreased numbers of B-cells	61
			Intrinsic defect in B-cell development	62
			Subnormal incidence of colony-forming B-cells in lymphoid tissues	35
			Absence of subclass of mature B-cells	33
			Reduced IgM antibody response to SSS-III	1
			Reduced antibody response to synthetic double-stranded RNA	60
			Impaired humoral response to thymus-independent antigens	6

[2] The codominant mutation to lipopolysaccharide unresponsiveness occurred in the Animal Resources Colony of C3H/HeJ mice at the Jackson Laboratory. [3] For a more comprehensive reference list on nude mice, consult reference 59. [4] The streaker mutation occurred in the AKR/J strain in 1974. At present, the difference in phenotypic expression between streaker and nude mutations do not appear any greater than reported variation among different experiments with *nu*.

Contributor: Leonard D. Shultz

References

1. Amsbaugh, D. F., et al. 1972. J. Exp. Med. 136:931-949.
2. Baroni, C. 1967. Experientia 23:282-283.
3. Baroni, C. D., et al. Nature (London) New Biol. 237: 219-220.
4. Bartke, A. 1964. Anat. Rec. 149:225-236.
5. Battisto, J. R., et al. 1969. Nature (London) 222: 1196-1198.
6. Berning, A., et al. 1978. Fed. Proc. Fed. Am. Soc. Exp. Biol. (in press).

continued

7. Coleman, D. L. 1978. Diabetologia 14:141-148.
8. Crewther, P., and N. L. Warner. 1972. Aust. J. Exp. Biol. Med. Sci. 50:625-635.
9. Croy, B. A., and D. Osaba. 1973. Cell. Immunol. 9: 306-318.
10. Deringer, M. K. 1970. J. Natl. Cancer Inst. 45:1205-1210.
11. Dickie, M. M. 1964. Mouse News Lett. 30:30.
12. Dung, H. C. 1976. Am. J. Anat. 147:255-264.
13. Dung, H. C. 1977. Transplantation 23:39-43.
14. Dung, H. C., and R. H. Swigart. 1971. Texas Rep. Biol. Med. 29:273-278.
15. Dung, H. C., and R. H. Swigart. 1972. Ibid. 30:23-39.
16. Duquesnoy, R. J. 1972. J. Immunol. 108:1578-1590.
17. Duquesnoy, R. J., et al. 1970. Proc. Soc. Exp. Biol. Med. 133:201-206.
18. Duquesnoy, R. J., et al. 1975. Am. Zool. 15:167-174.
19. Eicher, E. M. 1976. Mouse News Lett. 54:40.
20. Elin, R. J., and S. M. Wolff. 1973. Fed. Proc. Fed. Am. Soc. Exp. Biol. 32:1030.
21. Fabris, N., et al. 1971. Clin. Exp. Immunol. 9:209-225.
22. Gallin, J. I., et al. 1974. Blood 43:201-206.
23. Gasser, D. L., and T. Fischgrund. 1973. J. Immunol. 110:305-308.
24. Glode, L., and D. L. Rosenstreich. 1976. Ibid. 117: 2061-2066.
25. Green, M. C. 1967. Dev. Biol. 15:62-89.
26. Green, M. C., and L. D. Shultz. 1975. J. Hered. 66: 250-258.
27. Heiniger, H. J., et al. 1974. Cancer Res. 34:201-211.
28. Heiniger, H. J., et al. 1976. J. Natl. Cancer Inst. 56: 1073-1074.
29. Herberg, L., and D. L. Coleman. 1977. Metabolism 26:59-99.
30. Heston, W. E., and M. K. Deringer. 1947. J. Natl. Cancer Inst. 7:463-465.
31. Heston, W. E., and G. Vlahakis. 1961. Ibid. 26:969-983.
32. Heston, W. E., and G. Vlahakis. 1968. 40:1161-1168.
33. Huber, B., et al. 1977. J. Exp. Med. 145:10-20.
34. I'anson, V. A., and D. L. Gasser. 1973. J. Immunol. 111:1604-1606.
35. Kincade, P. W. 1977. J. Exp. Med. 145:249-263.
36. Kindred, B., et al. 1971. Eur. J. Immunol. 1:59-61.
37. Lane, P. W., and E. D. Murphy. 1972. Genetics 72: 451-460.
38. Lozzio, B. B. 1972. Am. J. Physiol. 222:290-295.
39. Lozzio, B. B., and M. V. Burns. 1972. J. Reticuloendothel. Soc. 11:429-430.
40. Lozzio, B. B., and L. B. Wargon. 1974. Immunology 27:167-178.
41. Machado, E. A., and B. B. Lozzio. 1976. Am. J. Pathol. 85:515-518.
42. Manning, D. D., et al. 1973. J. Exp. Med. 138:488-494.
43. Meier, H., et al. 1969. Proc. Natl. Acad. Sci. USA 63: 759-766.
44. Moeller, G. R., et al. 1978. J. Immunol. 120:116-123.
45. Morse, H. C., et al. 1974. Ibid. 113:688-697.
46. Murphy, E. D., and J. B. Roths. 1977. Proc. Am. Assoc. Cancer Res. 18:157.
47. Murphy, E. D., and J. B. Roths. 1978. Proc. 16th Int. Congr. Hematol., pp. 58-61.
48. Oliver, C., and E. Essner. 1973. J. Histochem. Cytochem. 21:218-228.
49. Pantelouris, E. M. 1968. Nature (London) 217:370-371.
50. Pantelouris, E. M. 1971. Immunology 20:247-252.
51. Pantelouris, E. M., and J. Hair. 1970. J. Embryol. Exp. Morphol. 24:615-623.
52. Pelletier, M., et al. 1976. Immunology 30:783-788.
53. Pritchard, H., and H. S. Micklem. 1972. Clin. Exp. Immunol. 10:151-161.
54. Pritchard, H., and H. S. Micklem. 1974. Proc. 1st Int. Workshop Nude Mice, pp. 127-139.
55. Pritchard, H., et al. 1973. Clin. Exp. Immunol. 13: 125-128.
56. Raff, M. C., and H. H. Wortis. 1970. Immunology 18: 931-942.
57. Reed, N. D., and D. D. Manning. 1973. Proc. Soc. Exp. Biol. Med. 143:350-353.
58. Ruco, L. P., and M. S. Meltzer. 1978. J. Immunol. 120:329-334.
59. Rygaard, J., and C. O. Povlsen, ed. 1977. Bibliography of the Nude Mouse, 1966-1976. Fischer, Stuttgart.
60. Scher, I., et al. 1973. J. Immunol. 110:1396-1401.
61. Scher, I., et al. 1975. J. Exp. Med. 141:788-803.
62. Scher, I., et al. 1975. Ibid. 142:637-650.
63. Searle, A. G. 1964. Genet. Res. 5:171-197.
64. Searle, A. G., ed. 1978. Mouse News Lett. 58.
65. Shultz, L. D. Unpublished. Jackson Laboratory, Bar Harbor, ME, 1978.
66. Shultz, L. D., and M. C. Green. 1976. J. Immunol. 116:936-943.
67. Shultz, L. D., et al. 1974. Annu. Rep. Jackson Lab. 45:19.
68. Shultz, L. D., et al. 1975. Ibid. 47:25.
69. Shultz, L. D., et al. 1978. Comparative and Developmental Aspects of Immunity and Disease. Pergamon, New York. In press.
70. Sidman, C. L. 1977. Fed. Proc. Fed. Am. Soc. Exp. Biol. 36:1302.
71. Vlahakis, G., and W. E. Heston. 1971. J. Natl. Cancer Inst. 46:677-683.
72. Watson, J., et al. 1978. J. Immunol. 120:422-424.
73. Wortis, H. H. 1971. Clin. Exp. Immunol. 8:305-317.

17. MUTANT GENES AFFECTING COLOR: MOUSE

Data are for all the known mutant genes of the mouse which influence coat color. Many of these mutant genes also influence the color of the eye; some control the biosynthesis of melanin, some affect various attributes of the

continued

pigment granule, and still others affect melanoblast survival. **Origin**: 1 = spontaneous; 2 = radiation-induced. **Holder**: For full name, *see* list of HOLDERS at front of book; Common = holdings by more than seven laboratories.

Gene Symbol	Gene Name	Origin	Chromosome No.	Holder	Reference
A^{vy}	Viable yellow	1-in C3H/HeJ	2	CAL, JAX, LIL, NCI, NCT	8
A^{y}	Yellow	Old mutant	2	Common	8,9
A^{w}	White-bellied agouti	Old mutant	2	Common	8,9
A^{i}	Intermediate agouti	1-in C57BL/6J	2	JAX	8
A^{iy}	Intermediate yellow	1-in C3H/HeJ	2	JAX	22
A^{s}	Agouti-suppressor	2	2	HAR, NCT, YUR	36
A^{sy}	Sienna yellow	1-in C57BL/6J	2	JAX	25
a^{u}	Agouti umbrous	2	2	HAR	37
a^{t}	Black and tan	Old mutant	2	Common	8,9
a^{td}	Tanoid	1-in C57BL/6J	2	JAX, NCT	8
a^{da}	Non-agouti dark belly	2	2	HAR	37
a^{x}	Lethal non-agouti	2	2	NCT, OAK	8
a	Non-agouti	Old mutant	2	Common	8,9
a^{e}	Extreme non-agouti	2	2	Common	8
a^{m}	Mottled agouti	2	2	NCT, OAK	40
a^{l}	Non-agouti lethal	2	2	HAR	37
ash	Ashen	1-in C3H/DiSn	9	JAX	20
b	Brown	Old mutant	4	Common	8,9
b^{c}	Cordovan	2(?)-recognized in (C57BL/10 x DBA/1)F_1	4	HAR, JAX	8
B^{lt}	Light	1-in C58	4	BRN, HAR, JAX, KAN, PAS	8,9
B^{w}	White-based brown	2	4	OAK	43
bf	Buff	1-in C57BL/6J	5	HAR, JAX	8
bg	Beige	1(in YZ57/ch) and 2	13	CON, HAR, HVD, JAX, MIC, NCT, OAK	8
bg^{J}	Beige-Jackson	2-in C57BL/6J	13	JAX	21
Bst	Belly spot and tail	1-in C57BL/KsJ	....	JAX	17
bt	Belted	1-in DBA	15	Common	8,9
bt^{J}	Belted-Jackson	1-in CBA/J	15	JAX	34
bt-2	Belted-2	Induced by *N*-methyl-*N'*-nitro-*N*-nitrosoguanidine	....	OAK	47
c	Albino	Old mutant	7	Common	8,9
c^{ch}	Chinchilla	Old mutant	7	Common	8,9
c^{e}	Extreme dilution	Mutation in wild	7	Common	8,9
c^{h}	Himalayan	1-recognized in AKD2F_1/J	7	CAM, EDI, HVD, JAX, MCM, MIC, PAT	8
c^{P}	Platinum	1-recognized in AKD2F_1/J	7	HVD, JAX	22
c^{m}	Chinchilla-mottled	2	7	HAR	33
c^{i}	Intense chinchilla	1-in chinchilla stock	7	....	9
d	Dilute	Old mutant	9	Common	8,9
d^{l}	Dilute-lethal	1-in C57BL/Gr	9	HAR, JAX, MAS, OAK, PAS	8
d^{s}	Slight dilution	1-recognized in B6D2F_1/J	9	JAX	8
d^{15}	Dilute-15	2	9	MAS, OAK	8
da	Dark	1-in CBA/Fa	7	....	8
Dfp	Dark foot pads	2	....	OAK	42,44
dp	Dilution-Peru	1	15	CAM	55,56

continued

Gene Symbol	Gene Name	Origin	Chromosome No.	Holder	Reference
E^{so}	Sombre	1-in C3H	8	Common	8,48
E^{tob}	Tobacco darkening	1	8	BON	54
e	Recessive yellow	1-in C57BL/6Ha	8	ARG, HAR, JAX, MIS	12
ep	Pale ear	1-in C3HeB/FeJ	19	HAR, JAX, OAK, PAS	19
f	Flexed-tail	1	13	ARG, EDI, JAX, KAN	8,9
Fk	Fleck	1-in CBA	....	STK	51
Fkl	Freckled	1	14	HAR	32
Ga	Graying with age	Mutation in wild	....	....	15
gl	Gray-lethal	1	10	JAX, LUC, OAK	8,9
gm	Gunmetal	1-in C57BL/6J	14	JAX	8
gr	Grizzled	1	10	EDI, HAR, JAX, OAK	1
le	Light ear	1-in C3H/HeJ	5	HVD, JAX, KAN	19
ln	Leaden	1-in C57BR	1	Common	8,9
ls	Lethal spotting	1-in a C57BL-a^t subline	2	HAR, JAX	8
m	Misty	1-in DBA/J	4	HAR, HVD, JAX, KAN, MIC, OAK	8,9
md	Mahoganoid	1-in C3H/HeJ	16	JAX	8
mg	Mahogany	1	2	JAX, LAC	8
mh	Mocha	1-in C57BL/6J-*pi*	10	JAX	18
mi	Microphthalmia	2?	6	HALLE, JAX, LUC, OAK	8,9
mi^{bw}	Black-eyed white	1-in C3H	6	KAN	8
mi^{rw}	Red-eyed white	1-in CBA/J	6	....	29
mi^{sp}	Microphthalmia-spotted	1-in a C57BL/6J-Mi^{wh} stock	6	JAX, KAN	8
mi^{ws}	White-spot	2(?)-in C57BL/6	6	....	8,14
Mi^{b}	Microphthalmia-brownish	1-recognized in (101 x C3H)F_1	6	OAK	41
Mi^{or}	Microphthalmia-Oak Ridge	2	6	OAK	41
Mi^{wh}	White	2?-recognized in (C57BL x DBA)F_1	6	EDI, HAR, JAX, KAN, LUC, OAK	8,9
Mo	Mottled	1	X	....	8
Mo^{blo}	Blotchy	1	X	Common	3,8
Mo^{br}	Brindled	1-in C57BL	X	Common	8
Mo^{dp}	Dappled	2	X	ARG, HAR	8
Mo^{vbr}	Viable-brindled	1	X	ARG, GLA, HAR	2
mu	Muted	1	13	HAR	30
nc	Non-agouti curly	Mutagenesis experiment using caffeine	....	HAR	50
Och	Ochre	2	4	HAR	4
p	Pink-eyed dilution	Old mutant	7	Common	8,9
p^{bs}	*p*-black-eyed sterile	2	7	HAR, KAN	33
p^{cp}	*p*-cleft palate (formerly p^{11})	2	7	HAR	6
p^{d}	Dark pink-eye	2	7	HAR, PAS	8,9
p^{dn}	*p*-darkening	Induced by ethyl methanesulfonate	7	HAR	38
p^{m1}	Pink-eyed mottled-1	2	7	OAK	40
p^{m2}	Pink-eyed mottled-2	2	7	OAK	40
p^{r}	Japanese ruby	Japanese waltzing mice	7	....	8,9
p^{s}	*p*-sterile	2	7	....	8
p^{un}	Pink-eyed unstable	1-in C57BL/6	7	HAR, JAX, KAN	35
p^{x}	*p*-extra dark	1-in C3H	7	HAR	38
pa	Pallid	Mutation in wild	2	Common	8,9
pe	Pearl	1-in C3H/He	13	HAR, JAX, MIC, OAK, PAS	8
Ph	Patch	1-in C57BL	5	HAR, JAX, KAN, LUC	10

continued

17. MUTANT GENES AFFECTING COLOR: MOUSE

Gene Symbol	Gene Name	Origin	Chromosome No.	Holder	Reference
Ph^e	Patch-extended	1-from a *Ph/+* female	5	LUC	53
Rn	Roan	1	14	JAX	22
rs	Recessive spotting	1-in C3H/HeJ	5	JAX	23
ru	Ruby-eye	1-in a silver-piebald stock	19	Common	8,9
ru-2	Ruby-eye-2	1-in C57BL/6	7	HAR, JAX, OAK	7
$ru\text{-}2^{hz}$	Haze	1-in DBA/2J	7	JAX	8,28
$ru\text{-}2^{mr}$	Maroon	1	7	CAM, HAR, JAX, MIC	8
$ru\text{-}2^{r}$	Ruby-eye-2^r	1-in C57BL/10	7	OAK	47
Rw	Rump-white	2	5	HAR, PAS	49
s	Piebald	Old mutant	14	Common	8,9
s^l	Piebald-lethal	1-recognized in (C3H/HeJ x C57BL/6J)F_2	14	CON, OAK	8
sea	Sepia	1-in C57BL/6J	1	JAX	52
si	Silver	Old mutant	10	BRN, JAX, KAN, OAK	8,9
Sl	Steel	1-in C3H	10	ARG, EDI, JAX, OAK	8
Sl^{cg}	Cloud-gray	2	10	....	45,46
Sl^{con}	Contrasted	2	10	HAR	5,31
Sl^d	Steel-Dickie	1-in DBA/2J	10	HAR, JAX, KAN, PAS	8
Sl^{gb}	Grizzle-belly	1-in a belted white female	10	KAN	8
Sl^m	Steel-Miller	1?-in C57BL/6	10	....	13
Sl^{so}	Sooty	1?-in C57BL/6	10	....	8
sl^{du}	Dusty	1-in C3H/HeJ	10	JAX	27
slt	Slaty	1-in noninbred stock	14	JAX	26
Sp	Splotch	1-in C57BL	1	HAR, JAX, KAN, OAK	8,9
Sp^d	Delayed-splotch	1-in C57BL/6J	1	JAX	8
te	Light-head	1	5	....	9
To	Tortoiseshell	1-in an obese stock	X	ARG, JAX, KAN, NMI	8
tp	Taupe	1-in C57BL/10	7	JAX, MIC, OAK	8
U	Umbrous	1	....	JAX	8,9
uw	Underwhite	1-in C57BL/6J	15	HAR, JAX	8,24
Va	Varitint-waddler	1-recognized in descendants of C57BL x C57BR cross	12	Common	8,9
Va^J	Varitint-waddler-Jackson	1	12	JAX	16
W	Dominant spotting	Old mutant	5	CAM, JAX, KAN, NYH, PAT	8,9
W^x	Dominant spotting-x	1-in C3H/J	5	JAX	39
W^a	Ames dominant spotting	2	5	ARG, KAN	8
W^b	Ballantyne's spotting	1-in C57BL/St	5	MIS	8
W^f	W-fertile	1-in C3H/He	5	PAS	11
W^j	Jay's dominant spotting	1-in C3H	5	JAX	39
W^{pw}	Panda-white	1	5	....	46
W^v	Viable dominant spotting	1-in an *a/a si/si* stock	5	Common	8,9
Ym	Yellow mottling	1	X	OAK	42

Contributors: Willys K. Silvers, Timothy W. Poole, and Gwendolyn Murphy

References

1. Bloom, J. L., and D. S. Falconer. 1966. Genet. Res. 7:159-167.
2. Cattanach, B. M., et al. 1969. Ibid. 14:223-235.
3. Cattanach, B. M., et al. 1972. Mouse News Lett. 47: 31-35.
4. Cattanach, B. M., et al. 1973. Ibid. 48:30-32.
5. Cattanach, B. M., et al. 1974. Ibid. 50:39-42.
6. Cattanach, B. M., et al. 1975. Ibid. 53:27-31.
7. Eicher, E. 1970. Genetics 64:495-510.
8. Green, E. L., ed. 1966. Biology of the Laboratory Mouse. Ed. 2. McGraw-Hill, New York.
9. Grüneberg, H. 1952. The Genetics of the Mouse. Ed. 2. M. Nijhoff, The Hague.
10. Grüneberg, H., and G. M. Truslove. 1960. Genet. Res. 1:69-90.
11. Guénot, J. L., et al. 1975. Mouse News Lett. 53:53-58.

continued

12. Hauschka, T. S., et al. 1968. J. Hered. 59:339-341.
13. Hollander, W. F. 1964. Mouse News Lett. 30:29.
14. Hollander, W. F. 1968. Genetics 60:189(Abstr.).
15. Kirby, G. C. 1974. J. Hered. 65:126-128.
16. Lane, P. W. 1972. Ibid. 63:135-140.
17. Lane, P. W. 1977. Mouse News Lett. 56:40.
18. Lane, P. W., and M. S. Deol. 1974. J. Hered. 65:362-364.
19. Lane, P. W., and E. L. Green. 1967. Ibid. 58:17-20.
20. Lane, P. W., and J. E. Womack. 1977. Mouse News Lett. 57:18-23.
21. Lane, P. W., et al. 1962. Ibid. 26:35-38.
22. Lane, P. W., et al. 1966. Ibid. 34:29-32.
23. Lane, P. W., et al. 1966. Ibid. 35:31-33.
24. Lane, P. W., et al. 1968. Ibid. 39:27-28.
25. Lane, P. W., et al. 1969. Ibid. 41:29-32.
26. Lane, P. W., et al. 1972. Ibid. 47:36-51.
27. Lane, P. W., et al. 1973. Ibid. 49:31-33.
28. Lane, P. W., et al. 1974. Ibid. 50:42-44.
29. Lane, P. W., et al. 1974. Ibid. 51:23-25.
30. Lyon, M. F., and R. Meredith. 1969. Genet. Res. 14: 163-166.
31. Lyon, M. F., et al. 1968. Mouse News Lett. 38:22.
32. Lyon, M. F., et al. 1969. Ibid. 40:25-26.
33. Lyon, M. F., et al. 1970. Ibid. 42:25-28.
34. Mayer, T. C., and E. Maltby. 1965. Dev. Biol. 9:269-286.
35. Melvold, R. W. 1971. Mutat. Res. 12:171-174.
36. Phillips, R. J. S. 1966. Genetics 54:485-495.
37. Phillips, R. J. S. 1976. Mouse News Lett. 55:14.
38. Phillips, R. J. S. 1977. Ibid. 56:37-39.
39. Russell, E. S., et al. 1957. J. Hered. 48:119-123.
40. Russell, L. B. 1964. Role of Chromosomes in Development. Academic, New York.
41. Russell, L. B., et al. 1966. Mouse News Lett. 34:40-42.
42. Russell, L. B., et al. 1968. Ibid. 38:31-32.
43. Russell, L. B., et al. 1969. Ibid. 40:41-42.
44. Russell, L. B., et al. 1970. Ibid. 43:59-60.
45. Russell, L. B., et al. 1972. Ibid. 47:56-61.
46. Russell, L. B., et al. 1974. Ibid. 50:51-53.
47. Russell, L. B., et al. 1975. Ibid. 52:46-47.
48. Searle, A. G. 1968. J. Hered. 59:341-342.
49. Searle, A. G., and G. M. Truslove. 1970. Genet. Res. 15:227-235.
50. Searle, A. G., et al. 1971. Mouse News Lett. 45:23-25.
51. Sheridan, W. 1968. Mutat. Res. 5:323-328.
52. Sweet, H. O., and P. W. Lane. 1977. Mouse News Lett. 57:18-23.
53. Truslove, G. M. 1977. Genet. Res. 29:183-186.
54. Von Lehmann, E. 1974. Mouse News Lett. 50:27.
55. Wallace, M. E. 1974. Ibid. 50:28-29.
56. Wallace, M. E., et al. 1971. Ibid. 44:16-18.

18. MUTANT GENES AFFECTING SKIN AND HAIR: MOUSE

Holder: For full names, *see* list of HOLDERS at front of book; Common = holdings by more than seven laboratories.

Gene Symbol	Gene Name	Chromosome	Holder	Reference
ab	Asebia	19	GUE, JAX, PSC, RMD	13
Al	Alopecia	11	CAM, PSC	13
alp[1/]	Alopécie	....	....	28
ao	Apampischo	....	....	52
ap	Alopecia periodica	....	....	13
at	Atrichosis	10	JAX	23,25
ba	Bare	....	....	13
Bpa	Bare patches	X	HAR	42,43
Ca	Caracul	15	Common	13
Ca^d	Directional caracul	15	HAR, OAK, PAS	45
cr	Crinkled	13	EDI, JAX, PSC	13,38
cv	Calvino	....	....	13
cw	Curly whiskers	9	ANN, HAR, JAX, OAK, PRA	10
Den	Denuded	....	JAX	47
dep	Depilated	4	JAX	37
dl	Downless	10	CAM, HAR, JAX	13,48
Eh	Hairy ears	15	JAX, OAK, PAS	1
Er	Repeated epilation	....	JAX, OAK, PAS	13
fd	Fur deficient	9	JAX	30
fr	Frizzy	7	CAM, HAR, JAX, OAK	13

1/ Provisional gene symbol.

continued

18. MUTANT GENES AFFECTING SKIN AND HAIR: MOUSE

Gene Symbol	Gene Name	Chromosome	Holder	Reference
Frl[1/]	Fur-loss	14	OAK	44
fs	Furless	13	JAX, PSC	13
ft	Flaky tail	12	JAX, OAK	29
fz	Fuzzy	1	Common	13,36
fz^{fy}	Frowzy	1	OAK	13,22
fzt	Fuzzy tail	....	JAX	53
go	Angora	5	HAR, JAX, PSC, TXA, YUR	13
Gs	Greasy	X	ARG, HAR, OAK	13
Hct	Single constriction	....	....	35
hid	Hair interior defect	....	JAX	51
hl	Hair-loss	15	KAN	13
hr	Hairless	14	Common	13
hr^{ba}	Bald	14	....	13
hr^{rh}	Rhino	14	JAX, PSC	13
ic	Ichthyosis	1	JAX	19
lt	Lustrous	11	JAX	13
ma	Matted	12	HAR, JAX, OAK	13
mc	Marcel	5	....	13,18
me	Motheaten	6	JAX	16
Mo	Mottled	X	Common	13,26
Mo^{blo}	Blotchy	X	Common	13,34
Mo^{br}	Brindled	X	Common	13,20
Mo^{dp}	Dappled	X	ARG, HAR	13
Mo^{vbr}	Viable-brindled	X	ARG, GLA, HAR	3
N	Naked	15	Common	13
nc	Non-agouti curly	....	HAR	41
nu	Nude	11	Common	11,40
nu^{str}[1/]	Streaker	11	JAX	9
olt	Oligotriche	....	ORL	39
pf	Pupoid-fetus	4	LMX	13
pk	Plucked	....	JAX, PSC	13
Ra	Ragged	2	CAM, EDI, HAR, JAX, OAK, PAS	13
Ra^{op}	Opossum	2	JAX, PSC	13
rc	Rough coat	....	JAX	5
Re	Rex	11	Common	13
Re^{wc}	Wavy coat	11	HAR	33
ro	Rough	2	CAM	13
rst	Rosette	....	CAM, HAR	54
sa	Satin	13	BRN, HAR, JAX, NCT, OAK	13
sal	Satin-like	....	OAK	31
sch	Scant hair	9	JAX	24
sf	Scurfy	X	OAK	13
Sha	Shaven	15	DUB	12,13
Sk	Scaly	2	JAX, OAK	13
Slk	Sleek	....	HAR	2
soc	Soft coat	12	JAX	6
spc	Sparse coat	14	JAX	15
spf	Sparse-fur	X	ARG, HAR, JAX, OAK	13
spf^{ash}	Abnormal skin & hair	X	....	8
Str	Striated	X	HAR	13,20
Ta	Tabby	X	Common	4,13,21,49
Ta^{J}	Tabby-Jackson	X	JAX	13

[1/] Provisional gene symbols.

continued

18. MUTANT GENES AFFECTING SKIN AND HAIR: MOUSE

Gene Symbol	Gene Name	Chromosome	Holder	Reference
tf	Tufted	17	Common	13
thd	Tail hair depletion	....	JAX	32
thf	Thin fur	17	ANN	27
Tsk	Tight skin	2	JAX	17
Ve	Velvet coat	15	JAX, OAK, PAS	50
wa-1	Waved-1	6	BRN, HAR, JAX, PAS	13
wa-2	Waved-2	11	Common	13
Wc	Waved coat	14	JAX	7,14
we	Wellhaarig	2	CAM, HAR, JAX, OAK	13
Wfl[1]	Wavy fur	....	....	46

[1] Provisional gene symbols.

Contributor: Thomas C. Mayer

References

1. Bangham, J. W. 1965. Mouse News Lett. 33:68.
2. Cattanach, B. M. 1975. Ibid. 53:29.
3. Cattanach, B. M., et al. 1969. Genet. Res. 14:223-235.
4. Claxton, J. H. 1967. Ibid. 10:161-171.
5. Dickie, M. M. 1966. Mouse News Lett. 34:30.
6. Dickie, M. M. 1968. Ibid. 38:24.
7. Diwan, S., and L. C. Stevens. 1974. Ibid. 51:24-25
8. Doolittle, D. P., et al. 1974. J. Hered. 65:194-195.
9. Eicher, E. M. 1976. Mouse News Lett. 54:40.
10. Falconer, D. S., and J. H. Isaacson. 1966. Genet. Res. 8:111-113.
11. Flanagan, S. P. 1966. Ibid. 8:295-309.
12. Flanagan, S. P., and J. H. Isaacson. 1967. Ibid. 9:99-110.
13. Green, E. L., ed. 1966. Biology of the Laboratory Mouse. Ed. 2. McGraw-Hill, New York.
14. Green, M. C. 1972. Mouse News Lett. 47:36.
15. Green, M. C. 1974. Ibid. 51:23.
16. Green, M. C., and L. D. Shultz. 1975. J. Hered. 66: 250-258.
17. Green, M. C., and H. O. Sweet. 1973. Mouse News Lett. 48:34.
18. Green, M. C., and E. F. Woodworth. 1967. Ibid. 37: 34.
19. Green, M. C., et al. 1974. J. Embryol. Exp. Morphol. 32:715-721.
20. Grüneberg, H. 1969. Ibid. 22:145-179.
21. Grüneberg, H. 1971. Ibid. 25:1-19.
22. Hollander, W. F. 1966. Mouse News Lett. 35:30.
23. Hummel, K. P., and D. B. Chapman. 1966. Ibid. 34: 31.
24. Hummel, K. P., and D. B. Chapman. 1971. Ibid. 45: 28.
25. Hummel, K. P., and D. B. Chapman. 1971. Ibid. 45: 29.
26. Hunt, D. M. 1974. Nature (London) 249:852-854.
27. Key, M., and W. F. Hollander. 1972. J. Hered. 63: 97-98.
28. Kobozieff, N., et al. 1970. Ann. Genet. Sel. Anim. 2:111-117.
29. Lane, P. W. 1972. J. Hered. 63:135-140.
30. Lane, P. W., and H. O. Sweet. 1973. Mouse News Lett. 48:34-35.
31. Larsen, M. 1972. Ibid. 47:61.
32. Les, E. P., and J. B. Roths. 1975. Ibid. 53:34.
33. Lyon, M. F. 1968. Ibid. 39:25.
34. Lyon, M. F. 1972. Ibid. 47:34.
35. Mann, S. J., and W. E. Straile. 1972. Genetics 70:631-637.
36. Mayer, T. C., et al. 1974. J. Embryol. Exp. Morphol. 32:707-713.
37. Mayer, T. C., et al. 1976. Genetics 84:59-65.
38. Mayer, T. C., et al. 1977. Dev. Biol. 55:397-401.
39. Moutier, R. 1974. Mouse News Lett. 50:53.
40. Pantelouris, E. M. 1968. Nature (London) 217:370-371.
41. Phillips, R. J. S. 1971. Mouse News Lett. 45:25.
42. Phillips, R. J. S., and M. H. Kaufman. 1974. Genet. Res. 24:27-41.
43. Phillips, R. J. S., et al. 1973. Ibid. 22:91-99.
44. Raymer, D. 1971. Mouse News Lett. 45:39.
45. Russell, L. B. 1969. Ibid. 40:41.
46. Russell, L. B. 1970. Ibid. 43:59.
47. Snell, G. D., and H. P. Bunker. 1968. Ibid. 39:28.
48. Sofaer, J. A. 1973. Dev. Biol. 34:289-296.
49. Sofaer, J. A. 1974. Genet. Res. 23:219-225.
50. Stieler, C., and W. F. Hollander. 1972. J. Hered. 63: 212-214.
51. Trigg, M. J. 1972. J. Zool. 168:165-198.
52. Van Pelt, A. F., et al. 1969. J. Hered. 60:78.
53. Varnum, D. S. 1976. Mouse News Lett. 55:17.
54. Wallace, M. E. 1971. Ibid. 44:18.

19. BIOCHEMICAL VARIATION: MOUSE

Part I. Proteins Controlled and Tissue Distribution

Gene Name: NDV = Newcastle disease virus. **Protein or Enzyme Activity:** NAD = nicotinamide-adenine dinucleotide; NAD^+ = nicotinamide-adenine dinucleotide, oxidized form; NAD(P)H = nicotinamide-adenine dinucleotide, reduced form, or nicotinamide-adenine dinucleotide phosphate, reduced form. Data in brackets refer to the column heading in brackets.

Gene Symbol ⟨Synonym⟩	Gene Name	Protein or Enzyme Activity ⟨Synonym⟩	Chromosome	Tissue	Reference
Adh-1	Alcohol dehydrogenase	Alcohol dehydrogenase	....	Liver, kidney	149
Ags	α-Galactosidase	α-Galactosidase	X	Brain, liver, kidney	96
Tag	Temporal α-galactosidase	α-Galactosidase	Autosomal	Liver	95
Ah ⟨*Ahh*⟩	Aromatic hydrocarbon responsiveness	Receptor protein	....	Liver	7,130,138,163, 164
Alb-1	Albumin	Albumin	5	Serum	110,127
Amy-1	Salivary α-amylase	α-Amylase	12	Saliva	83
Amy-2	Pancreatic α-amylase	α-Amylase	12	Pancreas	83
Aox	Aldehyde oxidase	Aldehyde oxidase	1	Liver	63,172
Apk ⟨*Acp-2*⟩	Acid phosphatase—kidney	Acid phosphatase	10	Kidney	180,181
Apl ⟨*Acp-1*⟩	Acid phosphatase—liver	Acid phosphatase	17	Liver	91,92,182
As-1	Arylsulfatase B	Arylsulfatase B	....	Kidney	25,26
Asr-1	Arylsulfatase B regulation	Arylsulfatase B	....	Liver	25,26
Bge	β-Galactosidase—electrophoresis	β-Galactosidase	9	Liver, kidney	4
Bgs	β-Galactosidase activity	β-Galactosidase	9	Liver, kidney, spleen, brain	4,120
Bgt	β-Galactosidase, temporal	β-Galactosidase	9	Liver	4,120
c[1]	Albino	Monophenol monooxygenase ⟨Tyrosinase⟩[2]	7	Skin	20
c[3]	Albino	Multiple proteins	7	Liver	46,47,58,62, 64,81,165, 166
Car-1	Carbonic anhydrase-1	Carbonate dehydratase ⟨Carbonic anhydrase⟩	3	Erythrocytes	43,157
Car-2	Carbonic anhydrase-2	Carbonate dehydratase ⟨Carbonic anhydrase⟩	3	Erythrocytes	43,157
Ce-1	Catalase degradation	Catalase	....	Liver	56,57,71,136, 137
Ce-2	Kidney-catalase	Catalase	17	Kidney	74
Coh	Coumarin hydroxylase	*p*-Coumarate 3-monooxygenase ⟨Coumarin hydroxylase⟩	7	Liver	161,186
Cs-1	Catalase	Catalase	2	Liver, kidney, erythrocytes	33,48,57,66,73
d	Dilute	Phenylalanine 4-monooxygenase ⟨Phenylalanine hydroxylase⟩[4]	9	Skin	19,187
Dip-1 ⟨*Pep-c*⟩	Dipeptidase	Dipeptidase	1	Kidney	72,94,140,157
Eg	Endoplasmic β-glucuronidase	Egasyn[5]	8	Liver, kidney, lung	55,97,159,167
Erp-1	Erythrocytic protein-1	Erythrocytic protein	....	Erythrocytes	85

[1] Order of monophenol monooxygenase ⟨tyrosinase⟩ activity: $C > c^{ch} > c^{h} > c^{e} > c$. [2] Activity is also affected by the brown, agouti, and pink-eyed dilution loci. [3] A series of radiation-induced deletions of varying extent. [4] The *d* locus may not govern phenylalanine 4-monooxygenase ⟨phenylalanine hydroxylase⟩. [5] Egasyn levels are additive in heterozygotes; the presence of endoplasmic β-glucuronidase is dominant.

continued

19. BIOCHEMICAL VARIATION: MOUSE

Part I. Proteins Controlled and Tissue Distribution

Gene Symbol ⟨Synonym⟩	Gene Name	Protein or Enzyme Activity ⟨Synonym⟩	Chromosome	Tissue	Reference
Es-1	Esterase-1	Esterase	8	Serum, kidney	125,126,134, 157
Es-2	Esterase-2	Esterase	8	Kidney, serum	125,132,133, 140
Es-3	Esterase-3	Esterase	11	Kidney, erythrocytes	139,141
Es-5	Esterase-5	Esterase	8	Serum	3,126,157
Es-6	Esterase-6	Esterase	8	Kidney	128,157,178
Es-7	Esterase-7	Esterase	8	Erythrocytes	11
Es-8	Esterase-8	Esterase	7	Erythrocytes	14
Es-9	Esterase-9	Esterase	8	Kidney, liver	148
Es-10	Esterase-10	Esterase[6]	14	Erythrocytes, kidney	124,157,177, 183
Es-11	Esterase-11	Esterase	8	Liver, kidney	123
Es-12	Esterase-12	Esterase	....	Serum	162
For-1	Formamidase-1	Formamidase	Autosomal	Liver	23
Gdc-1	NAD-α-glycerol phosphate dehydrogenase	Glycerol-3-phosphate dehydrogenase (NAD^+) ⟨NAD-α-glycerol phosphate dehydrogenase⟩	15	Heart, kidney, brain	41,42,86-88
Gdr-1; Gdr-2	Glucose-6-phosphate dehydrogenase regulator	Glucose-6-phosphate dehydrogenase	Autosomal	Erythrocytes	78
Glk	Galactokinase	Galactokinase	11	Erythrocytes	102
Got-1	Glutamate oxaloacetate transaminase (soluble)-1	Soluble aspartate aminotransferase ⟨Soluble glutamate-oxaloacetate transaminase⟩	19	Kidney, liver	10,11,44
Got-2	Glutamate oxaloacetate transaminase (mitochondrial)-2	Mitochondrial aspartate aminotransferase ⟨Mitochondrial glutamate-oxaloacetate transaminase⟩[7]	8	Kidney, liver	28
Gpd-1	Glucose-6-phosphate dehydrogenase, autosomal	Glucose-6-phosphate dehydrogenase ⟨Hexose-6-phosphate dehydrogenase⟩	4	Kidney	11,79,140,142, 157
Gpi-1	Glucose phosphate isomerase	Glucosephosphate isomerase	7	Erythrocytes	8,27,77,157
Gpt-1	Glutamate pyruvate transaminase	Alanine aminotransferase ⟨Glutamate-pyruvate transaminase⟩	15	Liver, small intestine, stomach, heart	15,42
Gr-1	Glutathione reductase	Glutathione reductase (NAD(P)H)	8	Kidney	54,108
Gur[8]	β-Glucuronidase regulation	β-Glucuronidase	5	Kidney	119,160
Gus ⟨*G*⟩	β-Glucuronidase	β-Glucuronidase	5	Liver, kidney, brain, spleen	90,93,103,104, 116-119,156
Gut	β-Glucuronidase, temporal	β-Glucuronidase	5	Liver	117,119
Hao-1	α-Hydroxyacid oxidase	L-2-Hydroxyacid oxidase	2	Liver	40
Hba	Hemoglobin α-chain	Hemoglobin, α-chain	11	Erythrocytes	131,144,145, 157
Hbb	Hemoglobin β-chain	Hemoglobin, β-chain	7	Erythrocytes	77,80,105,135, 144,157

[6] Specific for the substrate 4-methylumbelliferyl acetate. [7] Original publication mistitled enzyme name. [8] May be the same as the *O* locus.

continued

19. BIOCHEMICAL VARIATION: MOUSE

Part I. Proteins Controlled and Tissue Distribution

Gene Symbol ⟨Synonym⟩	Gene Name	Protein or Enzyme Activity ⟨Synonym⟩	Chromosome	Tissue	Reference
Hby	Hemoglobin embryonic y-chain	Hemoglobin, embryonic y-chain	7	Erythrocytes	60,158
Hex-1	β-D-*N*-Acetylhexosaminidase	β-*N*-Acetylhexosaminidase	....	Liver	45
His	Histidase	Histidine ammonia-lyase ⟨Histidase⟩	....	Liver	5,82
Id-1	Isocitrate dehydrogenase	Soluble isocitrate dehydrogenase	1	Kidney	69,79,157
If-1	Interferon-1 (NDV-induced)	Interferon	Autosomal	Serum	29-31,157
If-2	Interferon-2	Interferon	....	Serum	31
Ipo-1	Indophenol oxidase-1 [9]	Cytochrome *c* oxidase ⟨Indophenol oxidase⟩	....	Liver	149
Lap-1	Leucine arylaminopeptidase	Leucine aminopeptidase	9	Intestine	184
Ldr-1	Lactate dehydrogenase regulator	Lactate dehydrogenase	6	Erythrocytes	61,79,140,150, 157
Lv	δ-Aminolevulinate dehydratase	Porphobilinogen synthase ⟨δ-Aminolevulinate dehydratase⟩	4	Liver	21,22,39,146
Map-1	α-Mannosidase processing-1	α-Mannosidase processing	5	Liver	35,36
Map-2	α-Mannosidase processing-2	α-Mannosidase processing	17	Liver	36
Mod-1 ⟨*Mdh-1*⟩	Malic enzyme, supernatant	Soluble malate dehydrogenase (decarboxylating) ⟨Soluble malic enzyme⟩	9	Kidney, liver	70,152,157
Mod-2	Malic enzyme, mitochondrial	Mitochondrial malate dehydrogenase (decarboxylating) ⟨Mitochondrial malic enzyme⟩	7	Kidney	152,157
Mor-1	Malate dehydrogenase, mitochondrial	Mitochondrial malate dehydrogenase	5	Kidney	152,185
Mpi-1	Mannose phosphate isomerase	Mannosephosphate isomerase	9	Kidney	109
Mup-1	Major urinary protein-1	Major urinary protein	4	Urine	50-53,75,157, 173
Np-1	Nucleoside phosphorylase-1	Purine nucleoside phosphorylase	14	Liver, kidney, erythrocytes	11,179
O [10]	Operator	β-Glucuronidase	5	Kidney	37,38,114
Pd	Pyrimidine degradation	Pyrimidine	....	Liver	24,147
Pgd	6-Phosphogluconate dehydrogenase	Phosphogluconate dehydrogenase	4	Kidney	9
Pgk-1	Phosphoglycerate kinase-1	Phosphoglycerate kinase	X	Kidney, liver, erythrocytes	12,89,111
Pgk-2	Phosphoglycerate kinase-2	Phosphoglycerate kinase	17	Testes	16,168
Pgm-1	Phosphoglucomutase-1	Phosphoglucomutase	5	Erythrocytes	101,151,157
Pgm-2	Phosphoglucomutase-2	Phosphoglucomutase	4	Kidney, erythrocytes, heart	13
Phk	Phosphorylase kinase	Phosphorylase kinase	X	Muscle	67,76,98
Pre-1 ⟨*Pre*⟩	Prealbumin protein-1	Prealbumin protein	....	Serum	65,155,174, 175
Pre-2 ⟨*Pre-4*⟩	Prealbumin protein-2	Prealbumin protein	....	Serum	65,174,175

[9] This may be similar to human superoxide dismutase. [10] May be the same locus and alleles as *Gur*.

continued

Part I. Proteins Controlled and Tissue Distribution

Gene Symbol ⟨Synonym⟩	Gene Name	Protein or Enzyme Activity ⟨Synonym⟩	Chromosome	Tissue	Reference
Pro-1	Proline oxidase, mitochondrial	Mitochondrial pyrroline-5-carboxylate reductase ⟨Mitochondrial proline dehydrogenase⟩	....	Liver	2
Prt-1	Proteinase-1	Proteinase (trypsin)	....	Pancreas	169,170
Prt-2	Proteinase-2	Proteinase (chymotrypsin)	8	Pancreas	169-171
Prt-3	Proteinase-3	Proteinase (trypsin activity)	....	Pancreas	170
Sas-1	Antigenic serum substance	Antigenic serum substance	....	Serum	188
Slp	Sex-limited serum protein	Sex-limited serum protein [11]	17	Serum	68,121
spf	Sparse-fur	Ornithine carbamoyltransferase ⟨Ornithine transcarbamylase⟩	X	Liver	32
Ss	Serum antigen	Mouse serum antigen (globulin)	17	Serum	68,122,157
Svp-1	Seminal vesicle protein-1	Seminal vesicle protein	2	Seminal vesicles	129
Svp-2	Seminal vesicle protein-2	Seminal vesicle protein	....	Seminal vesicles	100
Tfm	Testicular feminization	Androgen receptor protein	X	Testes, kidney	1,6,59,99,112, 113,115
Trf	Transferrin	Transferrin	9	Serum	18,153,154, 157
No gene symbol [12]	No gene name	Alkaline phosphatase (polygenic)	....	Intestine	106,107
		Esterase [13]	....	Serum	84
		Noradrenalin *N*-methyltransferase ⟨Phenylethanolamine *N*-methyl transferase⟩	....	Adrenal gland	17
		Renin activity (regulatory) [14]	....	Submaxillary gland	176
		Tryptophan 5-monooxygenase ⟨Tryptophan hydroxylase⟩ [13]	....	Brain	34
		Xanthine oxidase ⟨Xanthine dehydrogenase⟩ [13]	....	Liver	49

[11] Protein is only expressed in males. [12] No loci have yet been discovered. [13] Only inter-strain differences observed. [14] Androgen-inducible; both basal and induced levels affected.

Contributors: Verne M. Chapman, Kenneth Paigen, Linda Siracusa, and James E. Womack.

References: *See* **Part II.**

Part II. Phenotypes and Strain Distribution

Phenotypic Expression: TCDD = 2,3,7,8-tetrachlorodibenzo-*p*-dioxin; K_m = Michaelis constant. **Strains:** V.I.S. = various inbred strains. Data in brackets refer to the column heading in brackets.

Gene Symbol ⟨Synonym⟩	Mode of Inheritance	Allele	Phenotypic Expression [Relative Enzyme Activity]	Strain	Reference
Adh-1	Codominant	*a*	Electrophoretic variation	C57BL, C3H, V.I.S.	149
		b	Electrophoretic variation	*Mus musculus musculus, M. musculus domesticus* (Denmark)	
Ags	Codominant	*h*	Thermostable	C3H/HeJ, V.I.S.	96
		m	Thermolabile	*M. musculus molossinus*	

continued

Part II. Phenotypes and Strain Distribution

Gene Symbol ⟨Synonym⟩	Mode of Inheritance	Allele	Phenotypic Expression [Relative Enzyme Activity]	Strain	Reference
Tag	Additive	......	[High: 2]	C57BL/6Ha, C57BL/6J, C57BL/6J-*bg*, C57BL/10J, C57BR/cdJ, C58/J, CE/J, MK/Re, 129/J	95
		......	[Low: 1]	A/HeJ, A/J, AU/SsJ, CBA/CaJ, CBA/J, C3H/HeJ, DBA/1J, DBA/2J, FL/1Re, LG/J, MA/J	
Ah ⟨*Ahh*⟩	*b* dominant	*b*	Responsive to TCDD, higher affinity for TCDD [K_m low]	A/HeJ, A/J, AU/SsJ, A2G, BALB/cJ, CBA/J, C3H/HeJ, C3H$_e$B/FeJ, C57BL/6J, C57BL/10ScSnA, C57L/J, C58/J, GRS/A, LIS/A, LTS/A, MAS/A, O20/A, PL/J, RIII/SeA, SM/J, STS/A, WLL/BrA	7,130,138, 163,164
		d	Non-responsive to TCDD, lesser affinity for TCDD [K_m high]	AKR/J, DBA/1J, DBA/2J, DD/HeA, I/LnJ, LP/J, RF/J, SJL/J, ST/bJ, SWR/J, 129/J	
Alb-1	Codominant	*a*	Electrophoretic variation	V.I.S.	110,127
		c	Electrophoretic variation	*M. musculus domesticus* (Ontario)	
Amy-1	Codominant	*a*	Electrophoretic variation	A/HeJ, AKR, AKR/J, AU/SsJ, BALB/cJ, BDP/J, BUB/BnJ, CBA/J, C3H, C57BL/6J, C57BR/J, C57L/J, C58/J, DBA, DBA/1J, DBA/2J, IS/Cam, LG/J, LP/J, MA/J, P/J, PL/J, RF/J, RIII/2J, SEA/Gn, SEC/1ReJ, SF/Cam, SJL/J, SK/Cam, SM/J, ST/aEh, SWR/J, T6, 129/J, wild Danish mice	83
		b	Electrophoretic variation	CE/J, DE/J, MOR/Cv, YBR/Cv, *M. musculus castaneus*, single male from fancier	
Amy-2	Codominant	*a*	Electrophoretic variation	A/HeJ, AKR/J, AU/SsJ, BALB/cJ, BDP/J, BUB/BnJ, CBA/J, C3H/HeJ, C57BL/6J, C57BR/J, C57L/J, C58/J, DBA/1J, DBA/2J, LG/J, LP/J, MA/J, P/J, PL/J, RF/J, RIII/2J, SEA/Gn, SEC/1ReJ, SF/Cam, SJL/J, SM/J, SWR/J, 129/J	83
		b	Electrophoretic variation	CE/J, DE/Cv, IS/Cam, MOR/Cv, SK/Cam, YBR/Cv	
Aox	Additive	*a*	[High: 10; (K_m = 1X)]	C57BL/6J	63,172
		b	[Low: 1; (K_m = 2X)]	DBA/2J	
Apk (Acp-2)	Codominant	*a*	Electrophoretic variation	V.I.S.	180,181
		m	Electrophoretic variation	*M. musculus molossinus*	
Apl ⟨*Acp-1*⟩	Additive	......	Present	SWR/J	91,92,182
		......	Absent	SM/J	
As-1	Additive	*a*	Thermostable [High: 2]	BALB/cJ, C3H/HeJ, DBA/2J, SWR/J	25,26
		b	Thermolabile [Low: 1]	A/HeJ, C57BL/6J	
Asr-1	Additive	*a*	[High: 2]	C57BL/6J	25,26
		b	[Low: 1]	A/HeJ, C3H/HeJ	
Bge	Codominant	*b*	Electrophoretic variation	AE/Wf, AG/JNt, AKR/J, AU/SsJ, BA/Nmg, BDP/J, BRSUNT/N, BRSUNT/St, BRVR/Pl, BSVS/Pl, BUA/Hn, BUB/BnJ, CBA/J, CFCW/Rl, CPB-Mo/Cpb, CPB-R/Cpb, C3H/St, C57BL/KsJ, C57BL/6ByJ, C57BL/6J, C57BL/6J-bg^J/bg^J, C57BL/6J-*le rd*, C57BL/10J, C57BR/cdJ, C57$_e$/Ha, C57L/J, C58/J, DBA/LiHa, DBA/1J, DBA/2J, F/St, GR/A, GR/N, I/StN, JBT/Jd, K/Gh, KE/Kw, KR/Nga, LG/J, LIS/A, LP/J, LTS/A, MAS/A, N/St, NH/N, NLC/Rd, NZB/BlNJ, NZC/Bl, NZO/L, NZW/N, NZY/Bl, O20/A, P/J, PH/Re, PHL/We, PL/J, RF/J, RFM/Un, RIII/2J, SEC/1ReJ, SM/J, ST/bJ, STR/N, STR/1N, STS/A, SWR/J, TSI/A, WB/ReJ, WLL/BrA, XVII/Rd, YBR/HaCv, YS/Wf, 101/Rl, 129/J	4

continued

Part II. Phenotypes and Strain Distribution

Gene Symbol ⟨Synonym⟩	Mode of Inheritance	Allele	Phenotypic Expression [Relative Enzyme Activity]	Strain	Reference
		a	Electrophoretic variation	A/HeJ, A/J, AL/N, BALB/cByJ, BALB/cJ, CE/J, C3H/HeJ, DD/He, DE/J, DHS/Mk, DW/J-*dw*/+, GR/Rl, I/St, KSB/Nga, MA/MyJ, MK/Re, NGP/N, S/Gh	
Bgs	Additive	*h*	[High: 2]	A/HeJ, A/J, BALB/cJ, CE/J, C3H/HeJ, C57BL/6J, C57BL/10J, C57BR/cdJ, C57$_e$/Ha, C57L/J, C58/J, DE/J, DW/J, FL/1Re, HRS/J, MA/J, MK/Re, PH/Re, WB/Re, YBR/WiHa	4,120
		d	[Low: 1]	AKR/J, AU/SsJ, BDP/J, BUB/BnJ, CBA/CaJ, CBA/J, C57BL/KsJ, DBA/1J, DBA/2J, I/FnLn, LG/J, LP/J, NZB/BlNJ, P/J, PL/J, RF/J, RIII/2J, SEC/1ReJ, SJL/J, SM/J, ST/bJ, SWR/J, WC/Re, WH/Re, WK/Re, 129/J	
Bgt	Additive	*b*	[High: 2]	BA/Nmg, C57BL/6ByJ, C57BL/6J, C57BL/6J-*bg*^{J}/*bg*^{J}, C57BL/6J-*le rd*, C57BL/10J, C57BR/cdJ, C57$_e$/Ha, C57L/J, C58/J, DD/He, GR/Rl, MK/Re, NGP/N, S/Gh, WB/ReJ, YS/Wf	4,120
		d	[Low: 1]	A/HeJ, A/J, AE/Wf, AG/JNt, AKR/J, AL/N, AU/SsJ, BALB/cByJ, BALB/cJ, BDP/J, BRSUNT/N, BRSUNT/St, BRVR/Pl, BSVS/Pl, BUA/Hn, BUB/BnJ, CBA/J, CE/J, CFCW/Rl, CPB-Mo/Cpb, CPB-R/Cpb, C3H/HeJ, C3H/St, C57BL/KsJ, DBA/LiHa, DBA/1J, DBA/2J, DE/J, DHS/Mk, DW/J-*dw*/+, F/St, GR/A, GR/N, I/St, I/StN, JBT/Jd, K/Gh, KE/Kw, KR/Nga, KSB/Nga, LG/J, LIS/A, LP/J, LTS/A, MA/J, MAS/A, N/St, NH/N, NLC/Rd, NZB/BlNJ, NZC/Bl, NZO/L, NZW/N, NZY/Bl, O20/A, P/J, PH/Re, PHL/We, PL/J, RF/J, RFM/Un, RIII/2J, SEC/1ReJ, SM/J, ST/bJ, STR/N, STR/1N, STS/A, SWR/J, TSI/A, WLL/BrA, XVII/Rd, YBR/HaCv, 101/Rl, 129/J	
c	Additive	*C*	Colored—[Normal]	....	20
		c^{ch}	Chinchilla [Low]	....	
		c^h	Himalayan; [Thermolabile]	....	
		c^e	Extreme dilution [Low]	....	
		c	Albino [Low]	....	
	Recessive	c^{14Cos}, c^{65K}, c^{112K}	Albino (Oak Ridge) 1/	....	46,47,58, 62,64,81, 165,166
		c^{3H}, c^{6H}, c^{25H}	Albino (Harwell) 1/	....	
Car-1	Codominant	*a*	Electrophoretic variation	A/HeJ, A/J, ABP/Le, AU/SsJ, BALB/cBy, BALB/cJ, BDP/J, BUB/BnJ, B10.D2/nSn, CBA/CaJ, CBA/H-T6, CBA/J, CE/J, C3H/HeJ, C3H$_e$B/FeJ, C57BL/KsJ, C57BL/6By, C57BL/6J, C57BL/10J, C57BL/10Sn, C57BR/cdJ, C57L/J, C58/J, DBA/1J, DBA/2J, DBA/2DeJ, DW/J, FL/1Re, FL/2Re, FL/4Re, FS/Ei, HRS/J, I/LnReJ, LG/J, LP/J, LT/ChReSv, MA/J, MA/MyJ, MK/Re, MWT/Ww, NZB/BlNJ, P/J, PERU, PL/J, PRO/Re, RF/J, RIII/2J, ROP/Gn, SEA/Gn, SEC/1ReJ, SF/Cam, SJL/J, ST/bJ, SWR/J, 129/J, 129/ReJ	43,157
		b	Electrophoretic variation	IS/Cam, SK/Cam, SM/J, WB/J, WC/Re, WK/Re, *M. musculus castaneus, M. musculus molossinus*	

1/ These are radiation-induced mutations, resulting from chromosomal deletions.

continued

Part II. Phenotypes and Strain Distribution

Gene Symbol ⟨Synonym⟩	Mode of Inheritance	Allele	Phenotypic Expression [Relative Enzyme Activity]	Strain	Reference
Car-2	Codominant	*a*	Electrophoretic variation	ABP/Le, AKR/J, BDP/J, B10.D2/nSn, CBA/CaJ, CBA/H-T6, C57BL/KsJ, C57BL/6By, C57BL/6J, C57BL/10J, C57BL/10Sn, C58/J, DBA/1J, FS/El, I/LnReJ, IS/Cam, LG/J, LP/J, MA/J, MA/MyJ, MK/Re, MWT/Ww, NZB/BlNReJ, P/J, PERU, PRO/Re, RF/J, ROP/Gn, SEA/Gn, SF/Cam, WB/J, WC/Re, WK/Re, 129/J, 129/ReJ, *M. musculus castaneus, M. musculus molossinus*	43,157
		b	Electrophoretic variation	A/HeJ, A/J, AU/SsJ, BALB/cBy, BALB/cJ, BUB/BnJ, CBA/J, CE/J, C3H/HeJ, $C3H_eB$/FeJ, C57BR/cdJ, C57L/J, DBA/2DeJ, DBA/2J, DW/J, FL/1Re, FL/2Re, FL/4Re, HRS/J, LT/ChReSv, PH/Re, PL/J, RIII/2J, SEC/1ReJ, SJL/J, SM/J, ST/bJ, SWR/J	
Ce-1	*c* dominant	*b*	[High: 2 (slow degradation)]	BALB/cDe, C3H/He, $C3H_f$/He, C57BL/An, C57BL/Ha, C57BL/He, YBR/He	56,57,71, 136,137
		c	[Low: 1 (fast degradation)]	C57BL/6, C57BL/10, C57BR/cdJ, C57L/J, C58/J, DBA/2	
Ce-2	*a* dominant	*a*	Electrophoretic variation (processing)	C57BL/6J, (C3H x C57BL/6)F_1	74
		b	Electrophoretic variation (processing)	C3H	
Coh	Additive	......	[High: 3]	DBA/2J	161,186
		......	[Low: 1]	AKR/J, C3H/HeJ, C57BL/6J	
Cs-1	Additive	*a*	[High: 2]	BALB/c, DBA/2	33,48,57, 66,73
		b	[Low: 0.02-0.20]	....	
		c	[Low: 0.02-0.20]	....	
		d	[Low: 0.02-0.20]	....	
		e	[Low: 0.02-0.20]	....	
		f	[Low: 0.02-0.20]	....	
		g	[Low: 1]	C57BL/An, C57BL/Ha, C57BL/6J	
d	Additive	*D*	[High: 1]	C57BR/cdJ	19,187
		d	[Low: 0.5]	DBA/1J, DBA/2J	
		d^l	[Lowest: 0.14]	Research colony of D. Kelton	
Dip-1 ⟨*Pep-c*⟩	Codominant	*a*	Electrophoretic variation	BALB/cJ, C57BL/6J, C57BL/10J, C57BR/cdJ, C58/J, HRS/J, LP/J, SEA/Gn, SEC/1ReJ	72,94,140, 157
		b	Electrophoretic variation	A/HeJ, AKR/Cum, AKR/FuA, AKR/FuRdA, AKR/J, AKR/Lw, AU/SsJ, A2G, BDP/J, BUB/BnJ, CBA/J, CE/J, C3H/HeA, $C3H_eB$/FeJ, DBA/2J, DE/J, DW/J, GRS/A, I/LnJ, LG/J, LIS/A, LTS/A, MA/J, MAS/A, O20, P/J, PH/Re, PL/J, RF/J, RIII/2J, SJL/J, SM/J, ST/bJ, STS/A, SWR/J, 129/J	
		c	Electrophoretic variation	DD/HeA, WLL/BrA	
Eg	*a* dominant	*a*	Present	C57BL/6J, V.I.S.	55,97,159, 167
		o	Absent	KE/Kw, LIS/A, LTS/A, STS/A, YBR/Cv	
Erp-1	Codominant	*a*	Electrophoretic variation	BALB/c, CBA, C3H, C57BL, DBA, SWR, most wild mice	85
		b	Electrophoretic variation	Wild mice	
Es-1	Codominant	*a*	Electrophoretic variation	BP/Ww, C57BL/6J, C57BL/10, C57BR/cdJ, C57L/J, MWT/Ww, ROP/Gn, WN	125,126, 134,157

continued

Part II. Phenotypes and Strain Distribution

Gene Symbol (Synonym)	Mode of Inheritance	Allele	Phenotypic Expression [Relative Enzyme Activity]	Strain	Reference
		b	Electrophoretic variation	A/He, AKR/Cum, AKR/Fu, AKR/J, AKR/Lw, AU/SsJ, AY/Nga, A2G, BALB/cJ, BDP/J, BUB/BnJ, CBA/J, CE/J, CS, C3H/He, C3H$_e$B/FeJ, C58/J, DBA/2, DD/HeA, DDK, DE/J, DW/J, GRS/A, HRS/J, I/LnJ, IS/Cam, ITES, IITES, IXBL, KK, KR, KSB, LG/J, LIS/A, LP/J, LTS/A, MA/J, MAS/A, NBR, NC, NZB/Hz, O20, P/J, PH/Re, PL/J, RF/J, RIII/SeA, RIII/2J, SEA/Gn, SEC/1ReJ, SF/Cam, SIIT, SJL/J, SK/Cam, SM/J, ST/a, ST/bJ, STS/A, SWR/J, WLL/BrA, XLII/Orl, 129/J	
Es-2	Codominant	*a*	Electrophoretic variation	IS/Cam, ITES, KK, KR, KSB, RFM/Un, SIIT, SK/Cam	125,132, 133,140
		b	Electrophoretic variation	A/HeJ, AKR/J, AU/SsJ, AY, BALB/cJ, BDP/J, BP/Ww, BUB/BnJ, CBA/J, CE/J, CS, C3H/HeJ, C3H$_e$B/FeJ, C57BL/6J, C57BR/cdJ, C58/J, DBA/2J, DDK, DE/J, DW/J, GR, HRS/J, I/LnJ, IITES, IXBL, LG/J, LP/J, MA/J, MWT/Ww, NBR, NC, NMRI, NZB/Hz, P/J, PH/Re, RF/J, ROP/Gn, RIII/2J, SEA/Gn, SEC/1ReJ, SF/Cam, SJL/J, SM/J, ST/a, ST/bJ, SWR/J, WN, 129/J	
		c	Electrophoretic variation	PL/J	
Es-3	Codominant	*a*	Electrophoretic variation	BALB/cJ, BDP/J, BP/Ww, C57BL/6J, C57BL/10, C57BR/cdJ, C57L/J, DW/J, HRS/J, MWT/Ww, NZB/Hz, P/J, ROP/Gn, SEA/Gn, SEC/1ReJ	139,141
		b	Electrophoretic variation	AKR/FuA, GRS/A, RF/J, ST/a, ST/bJ	
		c	Electrophoretic variation	A/HeJ, AKR/Cum, AKR/FuRdA, AKR/J, AKR/Lw, AU/SsJ, A2G, BUB/BnJ, CBA/J, CE/J, C3H/He, C3H$_e$B/FeJ, C58/J, DBA/1J, DBA/2J, DD/HeA, DE/J, I/LnJ, IS/Cam, LIS/A, LG/J, LP/J, LTS/A, MA/J, MAS/A, O20, PH/Re, PL/J, RIII/2J, SF/Cam, SJL/J, SK/Cam, SM/J, STS/A, SWR/J, WLL/BrA, 129/J	
Es-5	Codominant	*a*	Absent	DD/HeA, wild mice	3,126,157
		b	Present	A/BrA, A/SnA, AKR/FuA, AKR/FuRdA, A2G, BALB/cByA, BALB/cCrgl, BALB/cHeA, CBA/J, C3H/DiA, C3H/He, C57BL/6ByA, C57BL/10, DBA/HeA, DBA/LiA, GRS/A, IF/BrA, LIS/A, LTS/A, MAS/A, O20/A, RIII/SeA, STS/A, WLL/BrA	
Es-6	*b* dominant	*a*	Electrophoretic variation (multiple-banded pattern)	A/HeJ, AKR/J, AU/SsJ, BALB/cJ, BDP/J, BUB/BnJ, CBA/J, CE/J, C3H$_e$B/FeJ, C57BL/6J, C57BL/10, C57BR/cdJ, C58/J, DBA/2J, DE/J, DW/J, HRS/J, LG/J, LP/J, MA/MyJ, NZB/Hz, P/J, PH/Re, PL/J, RF/J, ROP/Gn, RIII/2J, SEA/Gn, SEC/1ReJ, SJL/J, SM/J, ST/bJ, SWR/J, 129/J	128,157, 178
		b	Electrophoretic variation (double-banded pattern)	SK/Cam, *M. musculus molossinus*	
Es-7	Codominant	*b*	Electrophoretic variation	V.I.S.	11
		a,c	Electrophoretic variation	*M. musculus castaneus*	
Es-8	Codominant	*a*	Electrophoretic variation	A/HeJ, AKR/J, AU/SsJ, BDP/J, BUB/BnJ, CBA/J, CE/J, C3H/HeJ, C57BL/6J, C57BR/cdJ, C57L/J, C58/J, DBA/2J, DE/J, DW/J, HRS/J, LG/J, LP/J, MA/J, P/J, PH/ReJ, PL/J, RF/J, ROP/J, SEA/GnJ, SEC/1ReJ, SJL/J, SWR/J, 129/J	14
		b	Electrophoretic variation	*M. musculus castaneus*	

continued

Part II. Phenotypes and Strain Distribution

Gene Symbol ⟨Synonym⟩	Mode of Inheritance	Allele	Phenotypic Expression [Relative Enzyme Activity]	Strain	Reference
Es-9	Codominant	*a*	Electrophoretic variation	C57BL/10Sn, NMRI/Lac	148
		b	Electrophoretic variation	*M. musculus molossinus*	
Es-10	Codominant	*a*	Electrophoretic variation	A/HeJ, A2G/Lac, BALB/cBy, BALB/cGr, BALB/cJ, C57BL/Gr, C57BL/KsJ, C57BL/6By, C57BL/6J, C57BL/10J, C57BL/10Sn, C57BR/cdJ, C57BR/cdLac, C57L/Lac, DW/J, FL/1Re, FL/2Re, FL/4Re, HRS/J-+/+, IS/Cam, MA/MyJ, MK/Re, P/J, PRO/Re, ROP/Gn, SEC/1ReJ, ST/bJ, SWR/J, WB/J, WC/Re, *M. musculus castaneus*	124,157, 177,183
		b	Electrophoretic variation	AKR/J, CBA/CaJ, CBA/J, CE/J, C3H/HeJ, C3H$_e$B/FeJ, DBA/1J, DBA/2J, HRS/J-*hr/hr*, I/LnReJ, LG/J, MWT/Le, NZB/BlNRe, PERU, PH/Re, SF/Cam, SJL/J, SK/Cam, SM/J, WK/Re, 129/J, 129/ReJ	
		c (b)[2]	Electrophoretic variation	BUB/BnJ, *M. musculus molossinus, M. musculus musculus* (Denmark)	
Es-11	Codominant	*a*	Electrophoretic variation	A2G/Lac, BALB/cGr, CBA/Cam, CE/Lac, C3H/HeLac, C57BL/Gr, C57BR/cdLac, C57L/Lac, DBA/1, F/StLac, ICFW/Lac, NMRI/Lac, Schneider, SM/J, TO, 129/RrLac	123
		b	Electrophoretic variation	IS/Cam (also wild stock, Peru-Emaus, and feral mice)	
Es-12	....	*a*	Electrophoretic variation	C3H/HeJ, C57BL/6J, DBA/2J, SWR/J	162
		b	Electrophoretic variation	AKR/J, C57BR/cdJ, C57L/J	
For-1	Additive	*a*	Thermostable [High: 3]	C3H$_f$/Rl	23
		b	Thermolabile [Low: 1]	C57BL/10ScSn	
Gdc-1	Codominant	*b*[3]	Electrophoretic variation; thermostable	A/J, AKR/J, BUB/BnJ, C3H/HeJ, C3H$_e$B/FeJ, C57BL/KsJ, C57BL/6J, C57BR/cdJ, DBA/2J, LP/J, RF/J, SJL/J, SWR/J	41,42,86-88
		c[3]	Electrophoretic variation; thermolabile	BALB/cJ, CBA/J	
		d	Electrophoretic variation; thermostable	*M. musculus castaneus*	
Gdr-1; Gdr-2	Additive	......	[High: 3]	A/HeJ, A/J	78
		......	[Intermediate: 2]	AKR/J, BALB/cJ, CBA/J, C3H/HeJ, C57BL/6J, C58/J, DBA/2J, RF/J, SEC/1ReJ, SJL/J, SWR/J, 129/J	
		......	[Low: 1][4]	C57BR/cdJ, C57L/J	
Glk	Additive	*a*	[Low: 1]	BALB/cJ, BDP/J, CE/J, C57BL/6J, C57BL/10J, C57BR/cdJ, C57L/J, C58/J, DBA/1J, DBA/2J, I/LnJ, LG/J, MA/MyJ, NZB/BlNJ, P/J, PL/J, RIII/2J, SJL/J, SM/J, SWR/J, 129/J	102
		b	[High: 2]	A/HeJ, A/J, AKR/J, CBA/J, C3H/HeJ, LP/J, ST/bJ	
Got-1	Codominant	*a*	Electrophoretic variation	C57BL/6J, SWR/J, V.I.S.	10,11,44
		b	Electrophoretic variation	*M. musculus castaneus* (Thailand)	
Got-2	Codominant	*a*	Electrophoretic variation	IS/Cam, PERU, SK/Cam, SWR/J	28
		b	Electrophoretic variation	A/HeJ, AKR/J, AU/SsJ, BALB/cJ, BUB/BnJ, CBA/J, CE/J, C3H$_e$B/FeJ, C57BL/6J, C57BR/cdJ, C58/J, DBA/2J, DE/J, DW/J, LP/J, MA/J, P/J, PL/J, RF/J, RIII/2J, SEA/Gn, SJL/J, SM/J, ST/bJ, 129/J	

[2] Allele "*c*" is referred to as "*b*" in references 177 & 183. [3] Alleles *b* and *c* are electrophoretically the same but differ from *d*. [4] Low activity is probably the result of reduced enzyme stability during red cell aging.

continued

Part II. Phenotypes and Strain Distribution

Gene Symbol ⟨Synonym⟩	Mode of Inheritance	Allele	Phenotypic Expression [Relative Enzyme Activity]	Strain	Reference
Gpd-1	Codominant	*a*	Electrophoretic variation	BDP/J, BP/Ww, CE/J, C57BL/6J, C57BR/cdJ, C57L, C58/J, DW/J, HRS/J, IXBL, KK, MWT/Ww, O20, P/J, PL/J, RF/J, ROP/Gn, SEA/Gn, SF/Cam, SK/Cam, WLL/BrA, WN, 129/J	11,79,140, 142,157
		b	Electrophoretic variation	A/HeJ, AKR/Cum, AKR/FuA, AKR/FuRdA, AKR/J, AKR/Lw, AU/SsJ, BALB/cJ, BUB/BnJ, CBA/J, C3H/HeOrl, C3H$_e$B/FeJ, DBA/2J, DD/HeA, DE/J, GRS/A, I/LnJ, IS/Cam, KR, KSB, LG/J, LIS/A, LP/J, LTS/A, MA/J, MAS/A, NZB/Hz, PH/Re, RIII/2J, SEC/1ReJ, SJL/J, SM/J, ST/a, ST/bJ, STS/A, SWR/J, XLII/Orl	
		c	Electrophoretic variation	A2G/A, *M. musculus castaneus*	
Gpi-1	Codominant	*a*	Electrophoretic variation	A/HeJ, AKR/Cum, AKR/FuA, AKR/FuRdA, AKR/J, AKR/Lw, A2G, BALB/cJ, BDP/J, CE/J, C57BR/cdJ, C58/J, DBA/2J, DD/HeA, DE/J, DW/J, HRS/J, IS/Cam, LG/J, LP/J, MA/J, MWT/Ww, NZB/Hz, P/J, PH/Re, PL/J, RF/J, RIII/2J, ROP/Gn, SEA/Gn, SEC/1ReJ, SJL/J, SK/Cam, SM/J, ST/bJ, WLL/BrA, 101/H, 129/J	8,27,77, 157
		b	Electrophoretic variation	AU/SsJ, BP/Ww, BUB/BnJ, CBA/J, C3H/HeA, C3H$_e$B/FeJ, C57BL/6J, GRS/A, JU, LIS/A, LTS/A, MAS/A, SF/Cam, STS/A, SWR/J	
Gpt-1	Codominant	*a*	Electrophoretic variation	A/HeJ, A/J, ABP/Le, AKR/J, AU/SsJ, BALB/cJ, BDP/J, BUB/BnJ, B10.D2/nSn, CBA/CaJ, CBA/H-T6, CBA/J, CE/J, C3H/HeJ, C3H$_e$B/FeJ, C57BL/KsJ, C57BL/6J, C57BL/10J, C57BL/10Sn, C57BR/cdJ, C57L/J, C58/J, DBA/1J, DBA/2DeJ, DBA/2J, DE/J, DW/J, FL/1Re, FL/2Re, FL/4ReII, FS/Ei, HRS/J, I/LnReJ, IS/CamEi, JE, LG/J, LP/J, LT/ChReSv, MK/Re, MWT/Ww, P/J, PERU-Atteck, PH/Re, PL/J, PRO/J, RF/J, ROP/Gn, RSV/Le, SEA/GnJ, SEC/1ReJ, SJL/J, SK/CamEi, SM/J, ST/bJ, STX/Le, SWR/J, TF/Gn, V/Le, WB/Re, WC/Re, WK/Re, 129/J, 129/ReJ, 129/Sv-*ter*	15,42
		b	Electrophoretic variation	MA/J, MA/MyJ, NZB/BlNReJ, RIII/2J, SF/CamEi, *M. musculus castaneus, M. musculus molossinus*	
		c	Electrophoretic variation	*M. musculus castaneus, M. musculus molossinus*	
Gr-1	Codominant	*a*	Electrophoretic variation	A/HeJ, AKR/J, AU/SsJ, BALB/cJ, BDP/J, CBA/J, CE/J, C3H/HeJ, C57BL/6J, C57BL/10J, C57BR/cdJ, DBA/2J, DW/J, LG/J, MA/J, MWT/Ww, NZB/J, PH/Re, PL/J, RF/J, RIII/2J, SEA/GnJ, SEC/1ReJ, SM/J, ST/bJ, 129/J	54,108
		b	Electrophoretic variation	ABP/J, BUB/BnJ, SJL/J, SWR/J	
Gur	Additive (*cis*-acting)	*a*	High inducibility	A/HeJ, A/J, AU/SsJ, BALB/cJ, DBA/Ha, DBA/LiHa, EBT/Ha, HRS/J, LG/J, SEA/GnJ, SEC/1ReJ, SM/J	119,160
		b	Low inducibility	BBT/Ha, BDP/J, BUB/BnJ, CBA/CaJ, CE/J, C57BL/6J, C57BL/10J, C57BR/cdJ, C57L/J, C58/J, DBA/1J, DBA/2J, JBT/Jd, KSB/Nga, LIS/A, LP/J, LTS/A, MA/MyJ, MAS/A, NZB/BlNJ, O20/A, P/J, PHL/We, PL/J, RF/J, STS/A, SWR/J, 129/J	
Gus ⟨*G*⟩	Codominant	*a*	Electrophoretic variation	A/HeJ, A/J, A/St, AE/Wf, BALB/cByJ, BALB/cJ, DBA/LiHa, DE/J, LG/J, S/Gh, SEA/GnJ, SEC/1ReJ, SM/J	90,93,103, 104,116-119,156

continued

Part II. Phenotypes and Strain Distribution

Gene Symbol (Synonym)	Mode of Inheritance	Allele	Phenotypic Expression [Relative Enzyme Activity]	Strain	Reference
		b	Electrophoretic; thermostable (standard)	AU/SsJ, BDP/J, BRVR/Pl, BSVS/Pl, BUA/Hn, BUB/BnJ, CE/J, CPB-R/Cpb, C3H/St, C57BL/6J, C57BL/6J-*bg*[J]/*bg*[J], C57BL/10J, C57$_e$/IIa, C57L/J, C58/J, DBA/1J, DBA/2J, DD/He, DHS/Mk, JBT/Jd, KR/Nga, KSB/Nga, LIS/A, LP/J, LTS/A, MAS/A, NZB/BlNJ, NZC/Bl, NZY/Bl, O20/A, P/J, PHL/We, PL/J, RF/J, RFM/Un, RIII/2J, STS/A, SWR/J, YS/Wf, 129/J	
		h	Thermolabile	AKR/J, CBA/J, C3H/HeJ, C57BL/6J-*le rd*, GR/Rl, K/Gh	
Gut	....	*a*	[High: 1]	A/J, A/St, AU/SsJ, BALB/cJ, BUB/BnJ, CE/J, C57BL/6J, C57BL/10J, C57BR/cdJ, C57L/J, C58/J, DBA/2J, LG/J, LP/J, MA/J, P/J, PL/J, RF/J, SEA/GnJ, SJL/J, SM/J, ST/bJ, SWR/J, WB/ReCz, 129/J	117,119
		h	[Low: 0.05-0.15]	AKR/J, CBA/J, C3H/HeJ	
Hao-1	Codominant	*a*	Electrophoretic variation	BALB/c, CBA/H, C57BL/KaLw	40
		b	Electrophoretic variation	NZC	
Hba	Codominant	*a*	Solubility[5]	A/HeJ, AKR/J, BP/Ww, CE/J, CXBE, CXBG, CXBI, CXBJ, C57BL/6J, C57BL/10J, C57BR/cdJ, C57L/J, DBA/2J, MA/J, MWT, RF/J, WK/Re, 129/J	131,144, 145,157
		b	Solubility[5]	BALB/cJ, CXBD, CXBH, CXBK, C58/J, DW/J, HRS/J, LT/Re, SEC/1ReJ	
		c	Solubility[5], chromatography	BDP/J, C3H/HeJ, C3H$_e$B/FeJ, DE/J, FL/2Re, LS/Le, MK/Re, P/J, SEA/Gn, SJL/J, ST/bJ, SWR/J, TW, WB/Re, WC/Re, WH/Re	
		d	Solubility[5]	PH/Re, ROP/Gn, SM/J	
		Not *c*	Chromatography	AU/SsJ, BUB/BnJ, CBA/J, DBA/1, GRS, I/LnJ, LG/J, LP/J, NZB/BlN, PL/J	
Hbb	Codominant	*d*	Electrophoretic variation	A/HeJ, A/J, AB, AKR/Cum, AKR/FuA, AKR/FuRdA, AKR/J, AKR/Lw, AY, A2G, BALB/cJ, BDP/J, BUB/BnJ, CBA/CaJ, CBA/J, CXBG, CXBI, C3H/He, C3H$_e$B/FeJ, DBA/1, DBA/2J, DD/HeA, FL/1Re, FL/2Re, I/LnJ, IF/BrA, IS/Cam, JU/Fa, LG/J, LIS/A, LTS/A, MK/Re, NZB/BlNJ, O20, P/J, PL/J, PRO/Re, RF/J, RIII/SeA, SB/Le, SEA/Gn, SF/Cam, SK/Cam, ST/bJ, STS/A, WB/Re, WC/Re, WH/Re, WLHR/J, WLL/BrA, XLII/Orl, 129/J	77,80,105, 135,144, 157
		p	Electrophoretic variation	AU/SsJ	
		s	Electrophoretic variation	BP/Ww, CE/J, CS, CXBD, CXBE, CXBH, CXBJ, CXBK, C57BL/Ks, C57BL/6J, C57BL/10, C57BR/cd, C57L, C58, DDI, DDK, DE/J, DU/Ei, DW/J, FSB/Gn, HRS/A, HRS/J, IDT, ITES, IITES, IXBL, KK, KR, KSA, KSB, LP/J, LS/Le, LT/Re, MA/J, MAS/A, MWT/Ww, NBR, NC, PH/Re, RIII/2J, ROP/Gn, SEC/1ReJ, SIIT, SJL/J, SM/J, SWR/J, TF/Gn, TW, UW, WK/Re, WN	
Hby	Codominant	y^1	Electrophoretic variation	C57BL/KsJ, C57L/J, C58/J, DW/J, FSB/Gn, LP/J, LS/Le, MA/MyJ, PH/Re, RIII/2J, SEC/1Re, SJL/J, SWR/J	60,158
		y^2	Electrophoretic variation	A/J, AKR/J, AU/SsJ, BALB/cJ, BDP/J, BUB/BnJ, CBA/CaJ, CBA/J, DBA/1J, FL/1Re, I/LnRe, IS/CamEi, LG/J, MK/Re, NZB/BlNJ	

[5] Solubility test applies only to those strains that are *Hbb*[s]/*Hbb*[s].

continued

Part II. Phenotypes and Strain Distribution

Gene Symbol ⟨Synonym⟩	Mode of Inheritance	Allele	Phenotypic Expression [Relative Enzyme Activity]	Strain	Reference
Hex-1	Codominant	*a*	Electrophoretic variation	A/HeHa, AE, AKR/Sm, AL/N, BA, BALB/c, BDP, BRS/SE, BRSUNT/N, BRVR, BSVS, BUA, CBA/J, CE/J, C3H/StHa, C57BL/Ks, C57BL/6J, C57L/J, C58/J, DBA/1J, DBA/2J, DD, DE/Cv, DW, F/St, FL/1Re, GR/N, GR/Rl, HA/Icr (random-bred), I/AnN, I/St, I/StN, ICR/Ha, JBT/Jd, K, KE, KP, L/St, LG/J, LIS/A, LP/J, LTS, MAS/J, MK/Re, MT, N/St, NGP/N, NH/LwN, NLC, NZC, NZO, NZW/BlN, NZY, O20, P/J, PH/Re, PHL, PL/J, POLY/St, RUS, S/Gn, SEC/1ReJ, SJL/J, STAR/N, STR/N, STS/A, SWR/J, TSI/A, WLL/BrA, X/Gf, XVII, YBR/Cv, YS, 129/Rr	45
		b	Electrophoretic variation	A/J, BUB/BnJ, DM, some *M. musculus castaneus*	
		c	Electrophoretic variation	CS, KSB, RIII, *M. musculus molossinus*	
His	Additive	......	[High: 3 (normal)]	C57BL/6J	5,82
		......	[Low: 1]	Cambridge stock derived from wild mice trapped in Peru in 1962	
Id-1	Codominant	*a*	Electrophoretic variation	A/HeJ, A2G, BALB/cJ, BP/Ww, CE/J, C3H/He, $C3H_eB$/FeJ, C57BL/6J, C57BL/10J, C58/J, DD/HeA, DDK, DE/J, DW/J, GRS/A, HRS/J, I/LnJ, KK, LG/J, LIS/A, LP/J, LTS/A, MA/J, MAS/A, MWT/Ww, NZB/Hz, O20, RF/J, RIII/2J, SEA/Gn, SEC/1ReJ, SF/Cam, ST/bJ, STS/A, SWR/J, WLL/BrA, WN, 129/J	69,79,157
		b	Electrophoretic variation	AKR/Cum, AKR/FuA, AKR/FuRdA, AKR/J, AKR/Lw, AU/SsJ, AY/Nga, BDP/J, BUB/BnJ, CBA/J, CS, C57BR/cdJ, DBA/1J, DBA/2J, IS/Cam, ITES, IITES, KR, KSB, NBR, NC, P/J, PH/Re, PL/J, RIII/SeA, ROP/Gn, SIIT, SJL/J, SM/J, XLII/Orl	
If-1	Not clear	*l*	[Low:1]	A/HeJ, A/J, AKR/FuA, BALB/cBy, BALB/cGif, B6.C-*H-28*, CBA/J, CE/J, CXBD, CXBE, CXBI, CXBJ, C3H/HeJ, C3H/HeLac, DBA/1J, DD/HeA, O20, ST/bJ	29-31,157
		h	[High: 7]	AKR/FuRdA, AKR/J, A2G, B10.D2/nSn, B10.D2/oSn, CBA/BrA, CXBG, CXBH, CXBK, C57BL/KsJ, C57BL/Lac, C57BL/6J, C57BL/10J, C57BL/10Sn, C57BR/cdJ, C58/J, DBA/2J, GRS/A, LTS/A, MAS/A, RF/J, STS/A, WLL	
If-2	*l* dominant	*l*	[Low]	BALB/c, CXBD, CXBH, CXBK	31
		h	[High]	B6.C-*H-28*, CXBE, CXBG, CXBI, CXBJ, C57BL/6	
Ipo-1	....	*a*	Electrophoretic variation	*M. musculus domesticus*	149
		b	Electrophoretic variation	*M. musculus musculus*	
Lap-1	....	*a*	Present	C57BL/KsJ, C57BL/6J, C57BL/10J, C57BR/cdJ, C57L/J, C58/J, LT/ChReSv, PRO/Re, RIII/2J	184
		b	Absent	A/HeJ, A/J, AKR/J, AU/SsJ, BALB/cJ, BDP/J, BUB/BnJ, CBA/CaJ, CBA/J, CE/J, C3H/HeJ, $C3H_eB$/FeJ, DBA/1J, DBA/2J, HRS/J, LG/J, LP/J, MA/MyJ, NZB/BlNRe, P/J, PERU, PL/J, RF/J, SEA/GnJ, SEC/1Re, SF/Cam, SJL/J, SM/J, ST/bJ, SWR/J, 129/J, *M. musculus castaneus, M. musculus molossinus*	

continued

Part II. Phenotypes and Strain Distribution

Gene Symbol ⟨Synonym⟩	Mode of Inheritance	Allele	Phenotypic Expression [Relative Enzyme Activity]	Strain	Reference
Ldr-1	*b* dominant	*a*	Electrophoretic variation; B subunit absent	A/HeJ, AKR/FuA, AKR/FuRdA, AKR/J, AU/SsJ, A2G, BALB/cJ, BDP/J, BP/Ww, BUB/BnJ, CBA/J, CE/J, $C3H_eB$/FeJ, C57BL/6J, C57BR/cdJ, C57L/J, C58/J, DBA/1J, DBA/2J, DD/HeA, GRS/A, HRS/J, I/LnJ, LG/J, LIS/A, MA/J, MAS/A, MWT/Ww, NZB/Hz, O20, P/J, PL/J, RF/J, RIII/2J, ROP/Gn, SEA/Gn, SEC/1ReJ, SF/Cam, SJL/J, SM/J, ST/bJ, STS/A, WLL/BrA, 129/J	61,79,140, 150,157
		b	Electrophoretic variation; B subunit present	DE/J, DW/J, IS/Cam, LP/J, LTS/A, PH/Re, SK/Cam, SWR/J	
Lv	Additive	*a*	[High: 3 (faster synthesis)]	A/HeJ, AKR/J, BALB/cJ, CBA/J, CE/J, C3H/He, $C3H_eB$/FeJ, C57BR/cdJ, C57L, DBA/1J, DBA/2J, HRS/J, MA/J, SJL/J, ST/bJ, SWR/J	21,22,39, 146
		b	[Low: 1 (slow synthesis)]	C57BL/6J, RF/J, SM/J	
		c	[Intermediate: 2]	BUB/BnJ, C57BL/10, C58/J, LP/J, PL/J, SEC/1ReJ, 129/J	
Map-1	*b* dominant	*b*	Electrophoretic variation	A/HeHa, A/J, AE/Wf, AG/JNt, AKR/Sn, AL/N, AU/SsJ, BA/Nmg, BALB/cJ, BDP/J BRS/St, BRSUNT/N, BRVR/Pl, BRVS/Pl, BUA/Hn, BUB/BnJ, CBA/CaJ, CE/J, CFCW/Rl, CPB-Mo/Cpb, CPB-R/Cpb, C3H/St, C3H/StHa, C57BL/6J, C57BL/6J-*bg*^{J}/*bg*^{J}, C57BL/6J-*ot*/+, C57BL/10J, C57BL/10Sn, C57L/J, C58/J, DBA/1J, DBA/2Ha, DBA/2J, DD/He, DE/J, DHS/Mk, DW/J-*dw*/+, F/St, GR/A, GR/N, HA/Icr (random-bred), I/AnN, I/LnJ, I/St, I/StN, ICR/Ha, KE/Kw, KP/Kw, KR/Nga, KSB/Nga, L/St, LG/J, LIS/A, LP/J, LTS/A, MAS/A, MT/Mk, N/St, NC/Nga, NGP/N, NH/LwN, NLC/Rd, NZC/Bl, NZW/BlN, PH/Re, PHL/We, PL/J, POLY/St, RF/J, RFM/Un, RIII/2J, RUS/Rl, S/Gh, SEC/1ReJ, SJL/J, STAR/N, STR/N, STS/A, SWM/Ms, SWR/J, TSI/A, WB/Re, WB/ReCz, WC/Re, WH/Re, WLL/BrA, X/Gf, XVII/Rd, YBR/Cv, YS/Wf, 101/Rl, 129/RrJ	35,36
		h	Electrophoretic variation	AKR/J, CBA/J, C3H/HeJ, C57BL/6-*le rd Gus*^{h}	
Map-2	*b* dominant	*b*	Electrophoretic variation	C57BL/6J	36
		sm	Electrophoretic variation	SM/J	
Mod-1 ⟨*Mdh-1*⟩	Codominant	*a*	Electrophoretic variation	A/HeJ, AKR/Cum, AKR/FuA, AY, A2G, BALB/cJ, BDP/J, BP/Ww, BUB/BnJ, CE/J, CS, C3H/HeOrl, $C3H_eB$/FeJ, DBA/2J, DD/HeA, DDK, DE/J, DW/J, GRS/A, IS/Cam, ITES, IXBL, KK, KR, KSB, LG/J, LP/J, LTS/A, MA/J, MAS/A, MWT/Ww, NBR, NC, O20, PL/J, RF/J, ROP/Gn, SEC/1ReJ, SF/Cam, SIIT, SM/J, SWR/J, WLL/BrA, 129/J	70,152, 157
		b	Electrophoretic variation	AKR/FuRdA, AKR/J, AKR/Lw, AU/SsJ, CBA/J, C57BL/6J, C57BR/cdJ, C58/J, HRS/J, I/LnJ, LIS/A, NZB/Hz, P/J, PH/Re, RIII/2J, SEA/Gn, SK/Cam, ST/bJ, STS/A, XLII/Orl	
Mod-2	Codominant	*a*	Electrophoretic variation	A/BrA, AKR/FuA, A2G, CBA/BrA, C57BL/LiA, DBA/HeA, GRS/A, LIS/A, LTS/A, SM/J	152,157

continued

Part II. Phenotypes and Strain Distribution

Gene Symbol ⟨Synonym⟩	Mode of Inheritance	Allele	Phenotypic Expression [Relative Enzyme Activity]	Strain	Reference
		b	Electrophoretic variation	A/HeJ, AU/SsJ, BALB/cJ, CBA/CaJ, CBA/J, C3H/HeJ, C57BL/6J, C57BR/cdJ, DBA/2J, DE/J, JU, LG/J, MA/J, NZB/Hz, PL/J, RIII/2J, SJL/J, SWR/J, 101, 129/J	
Mor-1	Codominant	*a*	Electrophoretic variation	C57BL/6J, SJL/J, SM/J, SWR/J	152,185
		b	Electrophoretic variation	MOR/Cv (inbred strain derived from feral mice from Ohio)	
Mpi-1	Codominant	*a*	Electrophoretic variation	MA/J, YBR/Cv, *M. musculus castaneus*	109
		b	Electrophoretic variation	A/HeJ, AKR/J, AU/SsJ, BALB/cJ, BDP/J, BUB/BnJ, CBA/J, CE/J, C57BL/6J, C57BR/cdJ, C57L/J, DBA/2J, DE/J, DW/J, LG/J, LP/J, MWT/Ww, P/J, PH/Re, PL/J, RF/J, ROP/GnJ, SEA/GnJ, SEC/1ReJ, SJL/J, SM/J, ST/bJ, SWR/J	
Mup-1	Codominant	*a*	Electrophoretic variation	A/He, AKR/Cum, AKR/FuRdA, AKR/Lw, AL/N, A2G, BALB/cBy, BALB/cHf, BL/De, BRSUNT/N, CBA_f/Lw, CE/J, CFCW/Rl, CXBE, CXBG, CXBH, CXBJ, C3H/HeA, $C3H_f$/De, C57BR/cdJ, DBA/1J, $DBA/2_e$De, DD/He, F, FL, I/An, LIS/A, LP/J, MA/J, NBL/N, NH/Lw, RF/Lw, RIII/SeA, ST/bJ, STR/N, STR/1N, SWR/De, WB, WC, WLL/BrA, 129/J	50-53,75, 157,173
		b	Electrophoretic variation	AKR/FuA, CXBD, CXBI, CXBK, C57BL/An, C57BL/He, C57BL/6Hf, C57BL/6J, C57BL/10N, C57L/He, C58/Lw, GRS/A, HR/De, LTS/A, MAS/A, O20, P/J, POLY, RIII/An, STS/A, YBR/He	
Np-1	Codominant	*a*	Electrophoretic variation	V.I.S.	11,179
		b	Electrophoretic variation	*M. musculus castaneus, M. musculus molossinus*	
O	....	*t*	[High]	Unknown	37,38,114
		s	[Low]	Unknown	
Pd	Additive	*a*	[Slow degradation]	BALB/cJ, BDP/J, $C3H_e$B/FeJ, C57BL/6J, C57BL/10, C57BR/cdJ, C58/J, LP/J, PL/J, SWR/J, 129/J	24,147
		b	[Fast degradation]	SJL/J	
Pgd	Codominant	*a*	Electrophoretic variation	*M. musculus musculus* (Denmark)	9
		b	Electrophoretic variation	A/HeJ, AKR/J, AU/SsJ, BDP/J, BUB/J, CBA/J, CE/J, C3H/HeJ, C57BL/6J, C57BR/cdJ, C57L/J, C58/J, DBA/2J, DE/J, HRS/J, LG/J, LP/J, MA/J, P/J, PH/ReJ, PL/J, ROP/J, SEA/GnJ, SEC/1ReJ, SJL/J, SWR/J, 129/J	
Pgk-1	Codominant	*a*	Electrophoretic variation	*M. musculus musculus* (Denmark)	12,89,111
		b	Electrophoretic variation	A/HeJ, A/J, AKR/J, AU/SsJ, BALB/cJ, BDP/J, CBA/CaGnJ, CBA/CaJ, CBA/H-T6, CE/J, C3H/HeJ, $C3H_e$B/FeJ, C57BL/KsJ, C57BL/6J, C57BL/10J, C57BR/cdJ, C57L/J, C58/J, DBA/1J, DBA/2DeJ, DBA/2J, DE/J, I/LnRe, LG/J, MA/J, MWT/WeGn, NZB/BlNJ, P/J, PL/J, RF/J, RIII/2J, ROP/Gn, SEA/GnJ, SEC/1ReJ, SJL/J, SM/J, ST/bJ, SWR/J, WB/Re, 129/J, 129/Sv-*ter, M. musculus molossinus*	
Pgk-2	....	......	Electrophoretic variation; thermolabile	BALB/c, C57BL	16,168
		......	Electrophoretic variation; thermostable	C3H	

continued

Part II. Phenotypes and Strain Distribution

Gene Symbol (Synonym)	Mode of Inheritance	Allele	Phenotypic Expression [Relative Enzyme Activity]	Strain	Reference
Pgm-1	Codominant	*a*	Electrophoretic variation	A/HeJ, AKR/Cum, AKR/FuA, AKR/FuRdA, AKR/J, AKR/Lw, BALB/cJ, BP/Ww, BUB/BnJ, CBA/J, CE/J, C57BL/6J, C57BR/cdJ, C57L/J, C58/J, DE/J, DW/J, HRS/J, LG/J, LP/J, MWT/Ww, RF/J, RIII/2J, ROP/Gn, SEA/Gn, SEC/1Re, SK/Cam, ST/a, ST/bJ, SWR/J, 129/J	101,151, 157
		b	Electrophoretic variation	AU/SsJ, A2G, BDP/J, CBA/BrA, C3H/He, $C3H_eB$/FeJ, DBA/1J, DBA/2J, DD/HeA, GRS/A, I/LnJ, IS/Cam, LIS/A, LTS/A, MA/J, MAS/A, P/J, PH/Re, PL/J, SJL/J, STS/A, SWR/J	
		c	Electrophoretic variation	C3H/HeNWe	
		d	Electrophoretic variation	129/ReWl	
Pgm-2	Codominant	*a*	Electrophoretic variation	A/HeJ, AKR/J, AU/SsJ, BALB/cJ, BDP/J, BP/Ww, BUB/BnJ, CBA/J, CE/J, $C3H_eB$/FeJ, C57BL/6J, C57BR/cdJ, C58/J, DBA/2J, DE/J, DW/J, GRS, HRS/J, I/LnJ, IS/Cam, LG/J, LP/J, MA/J, MWT/Ww, NZB/Hz, P/J, PH/Re, PL/J, RF/J, RIII/2J, ROP/Gn, SEA/Gn, SEC/1ReJ, SJL/J, SK/Cam, ST/a, ST/bJ, SWR/J, 129/J	13
		b	Electrophoretic variation	SM/J	
Phk	Intermediate[6]	*a*	Present	CBA/J, C57BL/FnLn, C57BL/6J	67,76,98
		b	Absent	I/FnLn, I/LnJ	
Pre-1 ⟨*Pre*⟩	Codominant	*a*	Electrophoretic variation	AKR/J, BALB/cJ, BDP/J, BUB/BnJ, CBA/CaJ, CE/J, DBA/1J, DBA/2J, I/LnJ, LT/Re, NZB/BlNJ, P/J, RF/J, RIII/2J, SEA/GnJ, SEC/1ReJ, SJL/J, ST/bJ, WB/Re, 129/J	65,155, 174,175
		b	Electrophoretic variation	SWR/J	
		c	Electrophoretic variation	C3H/HeJ	
Pre-2 ⟨*Pre-4*⟩	*b* dominant	*a*	Absent	AKR/J, BALB/cJ, CBA/J, C57BR/cdJ, DBA/2J, LP/J, RF/J, SEC/1ReJ, SWR/J	65,174, 175
		b	Present	A/HeJ, A/J, CE/J, C3H/HeJ, C57BL/6J	
Pro-1	Additive	*a*	Present	C57BL/6J, DBA/2J, ST/bJ, 129/ReJ	2
		b	Absent	PRO/Re	
Prt-1	....	*a*	Present	A/He, BALB/cCrgl, BALB/cJ, CBA/J, C3H/He, C57BL/6J, DBA/2, KK	169,170
		b	Absent	AU/SsJ, BS, CS, C57L, DDK, KR, KSA, KSB, NBC, NC, PON	
Prt-2	Codominant	*a*	Electrophoretic variation	A/He, AU/SsJ, BALB/cCrgl, BALB/cJ, BS, CBA/J, CS, C3H/He, C57BL/6J, C57L, DBA/2, DDK, KK, KR, KSA, KSB, NBC, NC, PON	169-171
		b	Electrophoretic variation	Mol-A, Mol-O (inbred strains derived from feral mice in Japan)	
Prt-3	Additive	*a*	[Low]	A/He, AU/SsJ, BALB/cCrgl, BALB/cJ, BS, CBA/J, CS, C57BL/6J, DBA/2, KR, NC, PON	170
		b	[High]	Mol-A	

[6] Additivity due to X-chromosome expression in a syncytial tissue.

continued

Part II. Phenotypes and Strain Distribution

Gene Symbol ⟨Synonym⟩	Mode of Inheritance	Allele	Phenotypic Expression [Relative Enzyme Activity]	Strain	Reference
Sas-1	*a* dominant	*a*	Present	AKR/J, BALB/cJ, BDP/J, C57BL/6J, C57BL/10, C57BR/cdJ, C57L, C58/J, LP/J, SWR/J, 129/J	188
		o	Absent	CBA/J, CE/J, C3H/He, DBA/1, DBA/2, DE/J, PL/J, SJL/J, SM/J, ST/bJ	
Slp	*a* dominant	*a*	Present	BALB/cJ, BDP/J, B10.A(2R), B10.D2, B10.D2(M504), B10.D2(R103), B10.LP, B10.WB, C3H.OL, DBA/2J, P/J, YBR	68,121
		o	Absent	AKR/J, B10.A(4R), B10.AKM, B10.BR, B10.HTT, CBA/J, C3H/He, C57BL/6J, C57BL/10J, C57BR/cdJ, C57L, C58/J, DBA/1J, MA/J, PL/J, RF/J	
spf	Intermediate	*spf*$^+$	[Normal: pH 7.6 = 1.0; pH 10 = 1.0]	C57BL	32
		spf	[Low activity: pH 7.6 = 0.2; pH 10 = 2.0]	Oak Ridge SPFCP strain; maintained on C57BL genetic background for mating purposes	
Ss	Additive	*h*	[High: 1]	A/HeJ, A/J, AKR/FuA, A2G, BALB/cJ, B10.A(2R), B10.A(4R), B10.A(5R), B10.D2, B10.D2(M504), B10.D2(R103), B10.PL, B10.WB, CBA/BrA, C3H.OH, C57BL/6J, C57BL/10J, C57L, DBA/1J, DBA/2J, E/Gw, GRS/A, IF/BrA, LIS/A, LTS/A, MAS/A, O20, RIII/SeA, S/Gw, STOLI/Lw, STS/A, WB/Re, WC/Re, WLL/BrA, YBR/HeWiHa	68,122, 157
		l	[Low: 0.05]	AKR/FuRdA, AKR/J, B10.BR, B10.HTT, CBA/J, C3H/HeJ, C3H.OL, C57BR/cdJ, RF/J, ST/bJ	
Svp-1	Codominant	*a*	Electrophoretic variation	AKR/J, C3H/He, DBA/2Orl, XLII/Orl, 129/J	129
		b	Electrophoretic variation	A/Orl, AL, BALB/cJ, CBA/J, C57BL/6J, C57BL/10, C57BR/cdOrl	
Svp-2	Codominant	*a*	Electrophoretic variation	C57BL/6Orl	100
		b	Electrophoretic variation	AKR/Orl, CBA/Orl, C57BR/cdOrl, XLII/Orl	
		c	Electrophoretic variation	A/Orl, BALB/cOrl, CBA/H-T6, C3H/HeOrl, DW/Orl	
Tfm	....	*Tfm*$^+$	[Normal]	V.I.S.	1,6,59,99, 112,113, 115
		Tfm	[Low-absent]	Non-inbred	
Trf	Codominant	*a*	Electrophoretic variation	AU/SsJ, CBA/BrA, CBA/Fa, CBA/J, CBA/Lac, XLII/Orl, YBR/Cv	18,153, 154,157
		b	Electrophoretic variation	A/Fa, A/HeJ, A/J, A/Lac, A/Orl, AKR/Cum, AKR/J, AKR/Orl, A2G, BALB/cJ, BALB/cOrl, CE/Lac, CS, C3H/HeOrl, $C3H_eB$/FeJOrl, C57BL/6J, C57BL/6Orl, C57BL/10Gn-*lu*, C57BR/cdJ, C57BR/cdOrl, C57L/Lac, DBA/1J, DBA/2J, DBA/2Orl, DD/HeA, DDK, GRS/A, IF/BrA, ITES, JU/Fa, KK, KL/Fa, KR, KSA, KSB, LG/J, LIS/A, LTS/A, MAS/A, NC, O20, RIII/Fa, S, SIIT, ST/a, ST/b, STS/A, SWR, TB/Orl, WB/Re, WC/Re, WLL/Br	
Alkaline phosphatase [7]	....	......	[High: 3]	SWR/J	106,107
		......	[Low: 1]	LAS	

[7] No gene symbol yet assigned.

continued

19. BIOCHEMICAL VARIATION: MOUSE

Part II. Phenotypes and Strain Distribution

Gene Symbol ⟨Synonym⟩	Mode of Inheritance	Allele	Phenotypic Expression [Relative Enzyme Activity]	Strain	Reference
Esterase [7]	....	......	[High: 2]	C3H/St	84
		......	[Low: 1]	C57BL/J	
Noradrenalin *N*-methyltransferase ⟨Phenylethanolamine *N*-methyltransferase⟩ [7]	Additive	......	[High: 2 (slow degradation)]	BALB/cJ	17
		......	[Low: 1 (fast degradation)]	BALB/cN	
Renin activity [7]	Additive	......	[High: 100]	SWR/J	176
		......	[Low: 1]	C57BL/10J	
Tryptophan 5-monooxygenase ⟨Tryptophan hydroxylase⟩ [7]	Additive	......	[High: 1.3]	C57BL/6Bg, C57BL/10Bg	34
		......	[Low: 1]	DBA/1Bg	
Xanthine oxidase ⟨Xanthine dehydrogenase⟩ [7]	....	......	[High: 2]	JK	49
		......	[Low: 1]	C3H/St	

[7] No gene symbol yet assigned.

Contributors: Verne M. Chapman, Kenneth Paigen, Linda Siracusa, and James E. Womack

References

1. Attardi, B., and S. Ohno. 1974. Cell 2:205-212.
2. Blake, R. L., et al. 1976. Biochem. Genet. 14:739-757.
3. Bouw, J., and L. van Zutphen. 1976. Mouse News Lett. 54:30-31.
4. Breen, G., et al. 1977. Genetics 85:73-84.
5. Bulfield, G., and H. Kacser. 1974. Arch. Dis. Child. 49:545-552.
6. Bullock, L. P., and C. W. Bardin. 1974. Endocrinology 94:746-756.
7. Burki, K., et al. 1975. Biochem. Genet. 13:417-433.
8. Carter, N. D., and C. W. Parr. 1967. Nature (London) 216:511.
9. Chapman, V. M. 1975. Biochem. Genet. 13:849-856.
10. Chapman, V. M., and F. H. Ruddle. 1972. Genetics 70:299-305.
11. Chapman, V. M., and F. H. Ruddle. 1973. Mouse News Lett. 48:44-45.
12. Chapman, V. M., and T. B. Shows. 1976. Nature (London) 259:665-667.
13. Chapman, V. M., et al. 1971. Biochem. Genet. 5:101-110.
14. Chapman, V. M., et al. 1974. Ibid. 11:347-358.
15. Chen, S.-H., and R. P. Donahue. 1973. Ibid. 10:23-28.
16. Cheng, M., and E. Eicher. 1976. Mouse News Lett. 54:41.
17. Ciaranello, R. D., and J. Axelrod. 1973. J. Biol. Chem. 248:5616-5623.
18. Cohen, B. L. 1960. Genet. Res. 1:431.
19. Coleman, D. L. 1960. Arch. Biochem. Biophys. 91:300-306.
20. Coleman, D. L. 1962. Ibid. 96:562-568.
21. Coleman, D. L. 1966. J. Biol. Chem. 241:5511-5517.
22. Coleman, D. L. 1971. Science 173:1245-1246.
23. Cummings, R. B. 1977. Mouse News Lett. 56:53-54.
24. Dagg, C. P., et al. 1964. Genetics 49:979-989.
25. Daniel, W. L. 1976. Ibid. 82:477-491.
26. Daniel, W. L. 1976. Biochem. Genet. 14:1003-1018.
27. DeLorenzo, R. J., and F. H. Ruddle. 1969. Ibid. 3:151-162.
28. DeLorenzo, R. J., and F. H. Ruddle. 1970. Ibid. 4:259-273.
29. DeMaeyer, E., and J. DeMaeyer-Guignard. 1969. J. Virol. 3:506-512.
30. DeMaeyer, E., et al. 1973. Mouse News Lett. 48:39.
31. DeMaeyer, E., et al. 1974. J. Genet. Virol. 23:209-211.
32. DeMars, R., et al. 1976. Proc. Natl. Acad. Sci. USA 73:1693-1697.
33. Dickerman, R. C., et al. 1968. J. Hered. 59:177-178.
34. Diez, J. A., et al. 1976. Brain Res. 109:413-417.
35. Dizik, M., and R. W. Elliott. 1977. Biochem. Genet. 15:31-46.
36. Dizik, M., and R. W. Elliott. 1977. Mouse News Lett. 56:58.
37. Dofuku, R., et al. 1971. Nature (London) New Biol. 232:5-7.
38. Dofuku, R., et al. 1971. Ibid. 234:259-261.
39. Doyle, D., and R. T. Schimke. 1969. J. Biol. Chem. 244:5449-5459.
40. Duley, J., and R. S. Holmes. 1974. Genetics 76:93-97.
41. Eicher, E. M. 1976. Mouse News Lett. 54:41.
42. Eicher, E. M., and J. E. Womack. 1977. Biochem. Genet. 15:1-8.

continued

43. Eicher, E. M., et al. 1976. Ibid. 14:651-660.
44. Eicher, E. M., et al. 1977. Mouse News Lett. 56:42.
45. Elliott, R. W. Unpublished. Roswell Park Memorial Institute, Buffalo, NY, 1977.
46. Erickson, R. P., et al. 1968. Proc. Natl. Acad. Sci. USA 59:437-444.
47. Erickson, R. P., et al. 1974. Nature (London) 248: 416-418.
48. Feinstein, R. N., et al. 1966. Genetics 53:923-933.
49. Figge, F. H. J., and L. C. Strong. 1941. Cancer Res. 1:779-784.
50. Finlayson, J. S., et al. 1963. J. Natl. Cancer Inst. 31: 91-97.
51. Finlayson, J. S., et al. 1964. Science 149:981-982.
52. Finlayson, J. S., et al. 1968. Biochem. Genet. 2:127-140.
53. Finlayson, J. S., et al. 1974. Ibid. 11:325-335.
54. Frels, W. I., and V. M. Chapman. 1977. Mouse News Lett. 56:58.
55. Ganschow, R., and K. Paigen. 1967. Proc. Natl. Acad. Sci. USA 58:938-945.
56. Ganschow, R. E., and R. T. Schimke. 1969. J. Biol. Chem. 244:4649-4658.
57. Ganschow, R. E., and R. T. Schimke. 1970. Biochem. Genet. 4:157-167.
58. Garland, R. C., et al. 1976. Proc. Natl. Acad. Sci. USA 73:3376-3380.
59. Gehring, U., and G. M. Tomkins. 1974. Cell 3:59-64.
60. Gilman, J. G., and O. Smithies. 1968. Science 160: 885-886.
61. Glass, R. D., and D. Doyle. 1972. Ibid. 176:180-181.
62. Gluecksohn-Waelsch, S. 1975. Dev. Biol. 45:369-371.
63. Gluecksohn-Waelsch, S., et al. 1967. J. Biol. Chem. 242:1271-1273.
64. Gluecksohn-Waelsch, S., et al. 1974. Proc. Natl. Acad. Sci. USA 71:825-839.
65. Green, M. 1976. Mouse News Lett. 54:3.
66. Grieshaber, C. K., and H. A. Hoffman. 1975. Biochem. Genet. 13:447-456.
67. Gross, S. R., et al. 1975. Ibid. 13:567-584.
68. Hansen, T. H., et al. 1974. Ibid. 12:281-293.
69. Henderson, N. S. 1965. J. Exp. Zool. 158:263-274.
70. Henderson, N. S. 1966. Arch. Biochem. Biophys. 117: 28-33.
71. Heston, W. E., Jr., et al. 1965. Genet. Res. 6:387-397.
72. Hilgers, J. Unpublished. The Netherlands Cancer Institute, Amsterdam, 1977.
73. Hoffman, H. A., and C. K. Grieshaber. 1974. J. Hered. 65:277-279.
74. Hoffman, H. A., and C. K. Grieshaber. 1976. Biochem. Genet. 14:59-66.
75. Hudson, D. M., et al. 1967. Genet. Res. 10:195-198.
76. Huijing, F., et al. 1973. Biochem. Genet. 9:193-196.
77. Hutton, J. J. 1969. Ibid. 3:507-515.
78. Hutton, J. J. 1971. Ibid. 5:315-331.
79. Hutton, J. J., and T. H. Roderick. 1970. Ibid. 4:339-350.
80. Hutton, J. J., et al. 1962. Proc. Natl. Acad. Sci. USA 48:1505-1513.
81. Jagiello, G., et al. 1976. Chromosoma 58:377-386.
82. Kacser, H., et al. 1973. Nature (London) 244:77-79.
83. Kaplan, R. D., et al. 1973. J. Hered. 64:155-157.
84. Khanolkar, V. R., and R. G. Chitre. 1942. Cancer Res. 2:567-570.
85. Kirby, G. C. 1974. Anim. Blood Groups Biochem. Genet. 5:153.
86. Kozak, L. P. 1972. Proc. Natl. Acad. Sci. USA 69: 3170-3174.
87. Kozak, L. P. 1974. Biochem. Genet. 12:69-80.
88. Kozak, L. P., and K. J. Erdelsky. 1975. J. Cell Physiol. 85:437-447.
89. Kozak, L. P., et al. 1974. Biochem. Genet. 11:41-47.
90. Lalley, P. A., and T. B. Shows. 1974. Science 185: 442-444.
91. Lalley, P. A., and T. B. Shows. 1974. Genetics 77: s38.
92. Lalley, P. A., and T. B. Shows. 1977. Ibid. 87:305-317.
93. Law, L. W., et al. 1952. J. Natl. Cancer Inst. 12:909-916.
94. Lewis, W. H. P., and G. M. Truslove. 1969. Biochem. Genet. 3:493-498.
95. Lusis, A. J., and K. Paigen. 1975. Cell 6:371-378.
96. Lusis, A. J., and J. D. West. 1976. Biochem. Genet. 14:849-855.
97. Lusis, A. J., et al. 1977. Ibid. 15:115-122.
98. Lyon, J. B., Jr. 1970. Ibid. 4:169-185.
99. Lyon, M. F., and S. G. Hawkes. 1970. Nature (London) 227:1217-1219.
100. Mentier, R., et al. 1971. Exp. Anim. (Paris) 4:7.
101. Miner, G. D., and H. G. Wolfe. 1972. Biochem. Genet. 7:247-252.
102. Mishkin, J. D., et al. 1976. Ibid. 14:635-640.
103. Morrow, A. G., et al. 1949. J. Natl. Cancer Inst. 10: 657-661.
104. Morrow, A. G., et al. 1949. Ibid. 10:1199-1203.
105. Morton, J. R. 1966. Genet. Res. 7:76-85.
106. Nayudu, P. R. V., and F. Moog. 1966. Science 152: 656-657.
107. Nayudu, P. R. V., and F. Moog. 1967. Biochem. Genet. 1:155-170.
108. Nichols, E. A., and F. H. Ruddle. 1975. Ibid. 13: 323-330.
109. Nichols, E. A., et al. 1973. Ibid. 8:47-53.
110. Nichols, E. A., et al. 1975. Ibid. 13:551-556.
111. Nielsen, J. T., and V. M. Chapman. Unpublished. Roswell Park Memorial Institute, Buffalo, NY, 1977.
112. Ohno, S., and M. F. Lyon. 1970. Clin. Genet. 1:121-127.
113. Ohno, S., et al. 1971. Ibid. 2:128-140.
114. Ohno, S., et al. 1971. Hereditas 69:107-124.
115. Ohno, S., et al. 1973. Nature (London) New Biol. 245:92-93.
116. Paigen, K. 1961. Exp. Cell Res. 25:286-301.
117. Paigen, K. 1961. Proc. Natl. Acad. Sci. USA 47: 1641-1649.
118. Paigen, K., and W. K. Noell. 1961. Nature (London) 190:148-150.
119. Paigen, K., et al. 1975. J. Cell Physiol. 85:379-392.
120. Paigen, K., et al. 1976. Cell 9:533-539.

continued

Part II. Phenotypes and Strain Distribution

121. Passmore, H. C., and D. C. Shreffler. 1968. Genetics 60:210-211(Abstr.).
122. Passmore, H. C., and D. C. Shreffler. 1970. Biochem. Genet. 4:351-366.
123. Peters, J. 1977. Mouse News Lett. 56:45.
124. Peters, J., and H. R. Nash. 1976. Biochem. Genet. 14:119-125.
125. Petras, M. L. 1963. Proc. Natl. Acad. Sci. USA 50: 112-116.
126. Petras, M. L., and F. G. Biddle. 1967. Can. J. Genet. Cytol. 9:704-710.
127. Petras, M. L., and I. A. McLaren. 1976. Biochem. Genet. 14:67-73.
128. Petras, M. L., and P. Sinclair. 1969. Can. J. Genet. Cytol. 11:97-102.
129. Platz, R. D., and H. G. Wolfe. 1969. J. Hered. 60: 187-192.
130. Poland, A., et al. 1976. J. Biol. Chem. 251:4936-4946.
131. Popp, R. A. 1962. J. Hered. 53:73-80.
132. Popp, R. A. 1965. Ibid. 56:107-108.
133. Popp, R. A. 1967. Ibid. 58:186-188.
134. Popp, R. A., and D. M. Popp. 1962. Ibid. 53:111-114.
135. Popp, R. A., and W. St. Amand. 1960. Ibid. 51:141-144.
136. Rechcigl, M., and W. E. Heston, Jr. 1967. Biochem. Biophys. Res. Commun. 27:119-124.
137. Rechcigl, M., Jr., and W. E. Heston, Jr. 1963. J. Natl. Cancer Inst. 30:855-864.
138. Robinson, J. R., et al. 1974. J. Biol. Chem. 249: 5851-5859.
139. Roderick, T. H., et al. 1970. J. Hered. 61:278-279.
140. Roderick, T. H., et al. 1971. Biochem. Genet. 5: 457-466.
141. Ruddle, F. H., and T. H. Roderick. 1965. Genetics 51:445-454.
142. Ruddle, F. H., et al. 1968. Ibid. 58:599-606.
143. Ruddle, F. H., et al. 1969. Ibid. 62:393-399.
144. Russell, E. S., and E. C. McFarland. 1974. Ann. N. Y. Acad. Sci. 241:25-38.
145. Russell, E. S., et al. 1972. Biochem. Genet. 7:313-330.
146. Russell, R. L., and D. L. Coleman. 1963. Genetics 48:1033-1039.
147. Sanno, Y., et al. 1970. J. Biol. Chem. 245:5668-5676.
148. Schollen, J., et al. 1975. Biochem. Genet. 13:369-377.
149. Selander, R. K., et al. 1969. Evolution 23:379-390.
150. Shows, T. B., and F. H. Ruddle. 1968. Proc. Natl. Acad. Sci. USA 61:574-581.
151. Shows, T. B., et al. 1969. Biochem. Genet. 3:25-35.
152. Shows, T. B., et al. 1970. Ibid. 4:707-718.
153. Shreffler, D. C. 1960. Proc. Natl. Acad. Sci. USA 46:1378-1384.
154. Shreffler, D. C. 1963. J. Hered. 54:127-129.
155. Shreffler, D. C. 1964. Genetics 49:629-634.
156. Sidman, R. L., and M. C. Green. 1965. J. Hered. 56: 23-29.
157. Staats, J. 1976. Cancer Res. 36:4333-4377.
158. Stern, R. H., et al. 1976. Biochem. Genet. 14:373-382.
159. Swank, R. T., and K. Paigen. 1973. J. Mol. Biol. 77: 371-389.
160. Swank, R. T., et al. 1973. Ibid. 81:225-243.
161. Taylor, B. A. 1975. Mouse News Lett. 52:37.
162. Taylor, B. A. 1977. Ibid. 56:41.
163. Thomas, P. E., and J. J. Hutton. 1973. Biochem. Genet. 8:249-257.
164. Thomas, P. E., et al. 1972. Ibid. 6:157-168.
165. Thorndike, J., et al. 1973. Ibid. 9:25-39.
166. Tomino, S., and K. Paigen. 1975. J. Biol. Chem. 250: 1146-1148.
167. Trigg, M. J., and S. Gluecksohn-Waelsch. 1973. J. Cell Biol. 58:549-563.
168. VandeBerg, J. L., et al. 1973. Nature (London) New Biol. 243:48-50.
169. Watanabe, T., and T. Tomita. 1974. Biochem. Genet. 12:419-428.
170. Watanabe, T., et al. 1976. Ibid. 14:697-707.
171. Watanabe, T., et al. 1976. Ibid. 14:999-1002.
172. Watson, J. G., et al. 1972. Ibid. 6:195-204.
173. Wilcox, F. H. 1975. Ibid. 13:243-246.
174. Wilcox, F. H. 1975. J. Hered. 66:19-22.
175. Wilcox, F. H., and D. C. Shreffler. 1976. Ibid. 67: 113.
176. Wilson, C. M., et al. 1977. Proc. Natl. Acad. Sci. USA 74:1185-1189.
177. Womack, J. E. 1975. Mouse News Lett. 53:35.
178. Womack, J. E. 1975. Biochem. Genet. 13:311-322.
179. Womack, J. E. 1977. Ibid. 15:347.
180. Womack, J. E. Unpublished. Jackson Laboratory, Bar Harbor, ME, 1977.
181. Womack, J. E., and S. Belsky. 1975. Mouse News Lett. 52:37.
182. Womack, J. E., and E. M. Eicher. 1975. Ibid. 54:41.
183. Womack, J. E., and M. Sharp. 1976. Genetics 82: 665-675.
184. Womack, J. E., et al. 1975. Biochem. Genet. 13: 511-518.
185. Womack, J. E., et al. 1975. Ibid. 13:519-526.
186. Wood, A. W., and A. H. Conney. 1974. Science 185: 612-614.
187. Woolf, L. I., et al. 1970. Biochem. Genet. 119:895-903.
188. Wortis, H. H. 1965. Genetics 52:267-273.

Part I.

The enzyme or other activity represented by each symbol can be found in Part II. At the bottom of the map are symbols volving variant mice.

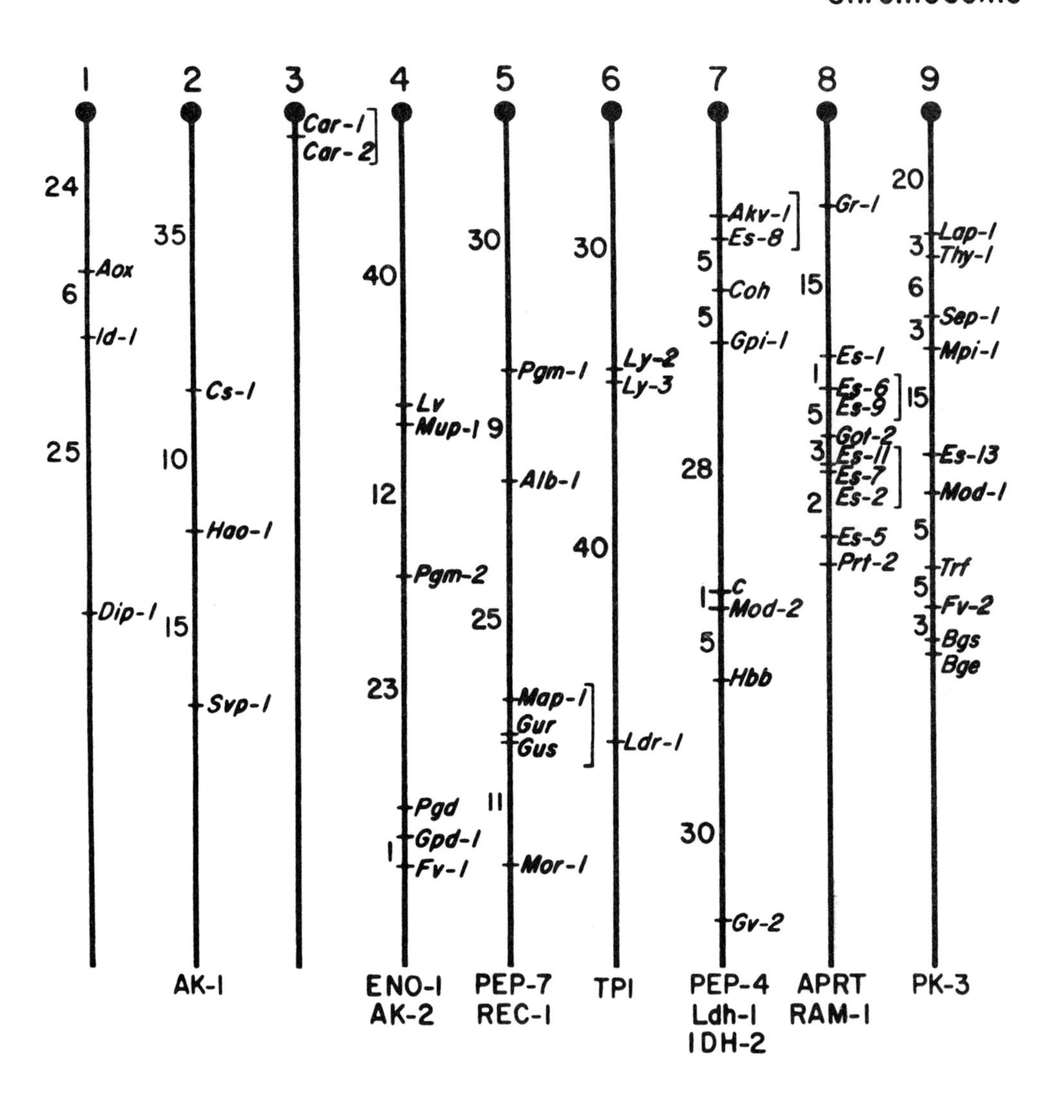

Contributor: James E. Womack

Map

for proteins coded by genes that have been assigned to a chromosome by somatic cell genetics or other methods not in-

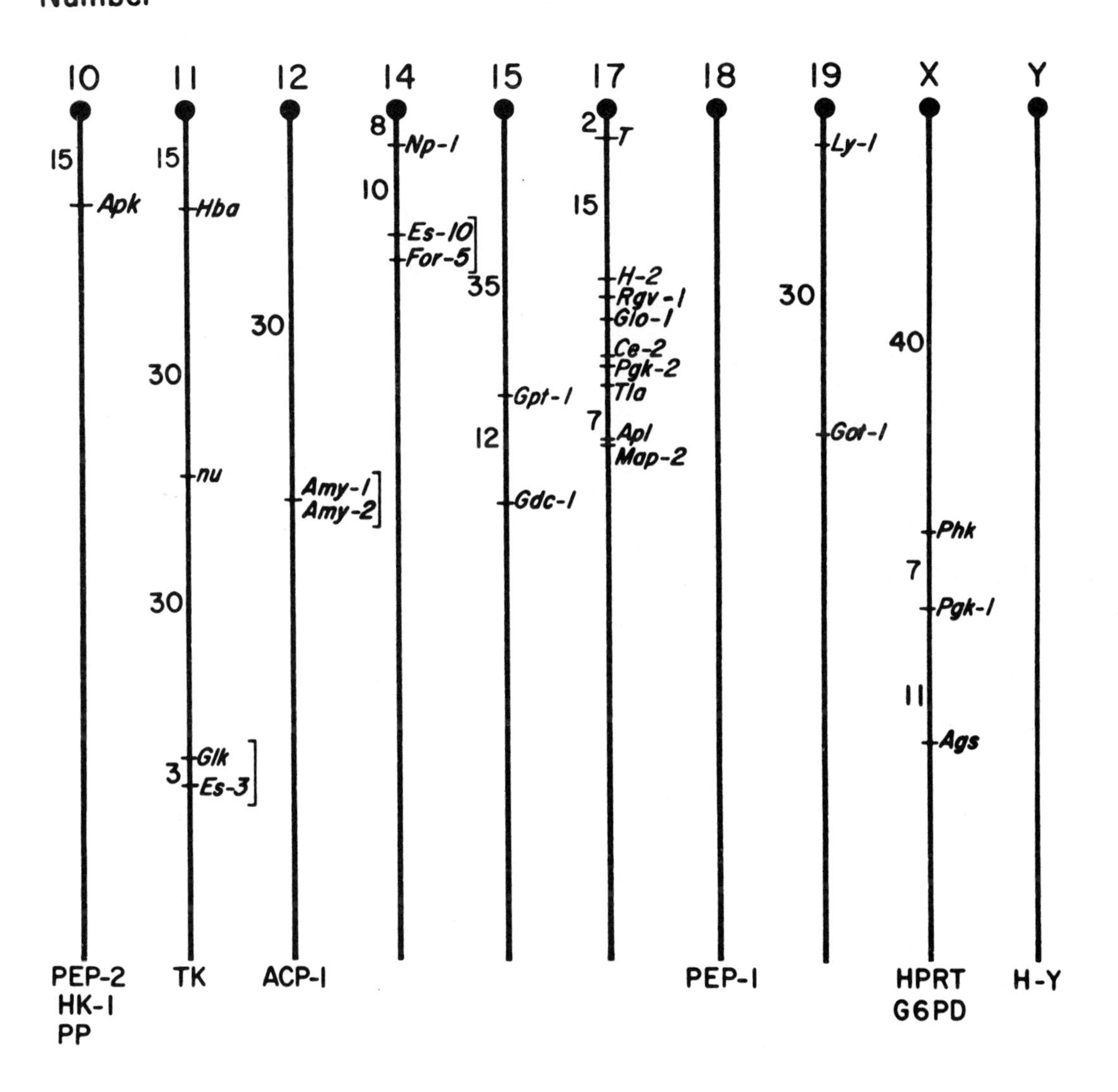

continued

20. BIOCHEMICAL LOCI: MOUSE

Part II. References

Data in brackets refer to the column heading in brackets.

Gene Symbol [Protein Symbol]	Protein or Activity ⟨Synonym⟩	Reference
Chromosome 1		
Aox	Aldehyde oxidase	89
Id-1	Soluble isocitrate dehydrogenase	34,40
Dip-1	Dipeptidase	49
Chromosome 2		
Cs-1	Catalase	26,31
Hao-1	L-2-Hydroxyacid oxidase ⟨α-Hydroxy acid oxidase⟩	18
Svp-1	Seminal vesicle protein	67
[AK-1]	Adenylate kinase	29
Chromosome 3		
Car-1	Carbonate dehydratase ⟨Carbonic anhydrase⟩	21
Car-2	Carbonate dehydratase ⟨Carbonic anhydrase⟩	21
Chromosome 4		
Lv	Porphobilinogen synthase ⟨δ-Aminolevulinate dehydrogenase⟩	39
Mup-1	Major urinary protein	27
Pgm-2	Phosphoglucomutase	78
Pgd	6-Phosphogluconate dehydrogenase	7
Gpd-1	Glucose-6-phosphate dehydrogenase ⟨Hexose 6-phosphate dehydrogenase⟩	40,73
Fv-1	Murine type C virus	71,85
[ENO-1]	Enolase	48
[AK-2]	Adenylate kinase	48
Chromosome 5		
Pgm-1	Phosphoglucomutase	78
Alb-1	Albumin	60
Map-1	α-Mannosidase processing	16,17
Gur	Glucuronidase regulation	83
Gus	β-Glucuronidase	30
Mor-1	Mitochondrial malate dehydrogenase	79,95
[PEP-7]	Peptidase	48
[REC-1]	Ecotropic virus replication	32
Chromosome 6		
Ly-2	Lymphocyte antigen	4
Ly-3	Lymphocyte antigen	5
Ldr-1	Lactate dehydrogenase regulator	40,77
[TPI]	Triosephosphate isomerase	56
Chromosome 7		
Akv-1	AKR type C virus	72
Es-8	Esterase	10
Coh	*p*-Coumarate 3-monooxygenase ⟨Coumarin hydroxylase⟩	84,98
Gpi-1	Glucosephosphate isomerase	14,40
c[1/]	Monophenol monooxygenase ⟨Tyrosinase⟩	12,33
Mod-2	Mitochondrial malate dehydrogenase (decarboxylating) ⟨Mitochondrial malic enzyme⟩	79
Hbb	Hemoglobin β-chain	38
Gv-2	G_{IX} cell surface antigen	82
[PEP-4]	Peptidase	48
Ldh-1	Lactate dehydrogenase-A	48,81
[IDH-2]	Mitochondrial isocitrate dehydrogenase	48
Chromosome 8		
Gr-1	Glutathione reductase	58
Es-1	Esterase	69
Es-6	Esterase	66,90
Es-9	Esterase	75
Got-2	Mitochondrial aspartate aminotransferase ⟨Mitochondrial glutamate oxaloacetate transaminase⟩	15
Es-11	Esterase	64
Es-7	Esterase	8
Es-2	Esterase	74
Es-5	Esterase	65
Prt-2	Proteinase (chymotrypsin)	87,88
[APRT]	Adenine phosphoribosyltransferase	45
[RAM-1]	Amphotropic virus replication	32
Chromosome 9		
Lap-1	Aminopeptidase (cytosol) ⟨Leucine aminopeptidase⟩	94
Thy-1	Thymus cell antigen (theta)	2,41
Sep-1	Serum protein	23
Mpi-1	Mannosephosphate isomerase	59
Es-13	Esterase	97
Mod-1	Soluble malate dehydrogenase (decarboxylating) ⟨Soluble malic enzyme⟩	79
Trf	Transferrin	80
Fv-2	Murine type C virus	51
Bgs	β-Galactosidase	6,62,76

1/ Albino.

continued

Part II. References

Gene Symbol [Protein Symbol]	Protein or Activity ⟨Synonym⟩	Reference
Bge	β-Galactosidase	6
[PK-3]	Mitochondrial pyruvate kinase	48
	Chromosome 10	
Apk	Acid phosphatase	91
[PEP-2]	Peptidase	29
[HK-1]	Hexokinase	48
[PP]	Pyrophosphate	48
	Chromosome 11	
Hba	Hemoglobin, α-chain	38,68
nu	....	28
Glk	Galactokinase	44,54, 57
Es-3	Esterase	70
[TK]	Thymidine kinase	44,54
	Chromosome 12	
Amy-1	α-Amylase	19,42
Amy-2	α-Amylase	19,42
[ACP-1]	Acid phosphatase	29
	Chromosome 14	
Np-1	Nucleoside phosphorylase	8,96
Es-10	Esterase	63,93, 96
For-5	Formamidase	13
	Chromosome 15	
Gpt-1	Alanine aminotransferase ⟨Glutamate pyruvate transaminase⟩	20
Gdc-1	Glycerol-3-phosphate dehydrogenase (NAD^+) ⟨NAD-α-glycerol phosphate dehydrogenase⟩	8,19,46
	Chromosome 17	
T	Embryonic cell surface antigen	1
H-2	Cell surface antigen	43
Rgv-1	Gross virus resistance	50
Glo-1	Lactoyl-glutathione lyase ⟨Glyoxalase⟩	55,56
Ce-2	Catalase	35,36
Pgk-2	Phosphoglycerate kinase	9,11,86
Tla	Thymic leukemia antigen	3
Apl	Acid phosphatase	47,92
Map-2	α-Mannosidase processing	17
	Chromosome 18	
[PEP-1]	Peptidase	29
	Chromosome 19	
Ly-1	Lymphocyte antigen	4
Got-1	Soluble aspartate aminotransferase ⟨Soluble glutamate oxaloacetate transaminase⟩	8,15,22
	X Chromosome	
Phk	Phosphorylase kinase	37,53
Pgk-1	Phosphoglycerate kinase	9,61
Ags	α-Galactosidase	9,45,52
[HPRT]	Hypoxanthine phosphoribosyltransferase	25,45
[G6PD]	Glucose-6-phosphate dehydrogenase	9
	Y Chromosome	
[H-Y]	Cell surface antigen	24

Contributor: James E. Womack

References

1. Bennett, D. 1975. Cell 6:441-454.
2. Blankenhorn, E. P., and T. C. Douglas. 1972. J. Hered. 63:259-263.
3. Boyse, E. A., and L. J. Old. 1969. Annu. Rev. Genet. 3:269-290.
4. Boyse, E. A., et al. 1968. Proc. R. Soc. London B170: 175-193.
5. Boyse, E. A., et al. 1971. Transplantation 11:351-353.
6. Breen, G., et al. 1977. Genetics 85:73-84.
7. Chapman, V. M. 1975. Biochem. Genet. 13:849-856.
8. Chapman, V. M., and F. H. Ruddle. 1973. Mouse News Lett. 48:45.
9. Chapman, V. M., and T. B. Shows. 1976. Nature (London) 259:665-667.
10. Chapman, V. M., et al. 1974. Biochem. Genet. 11: 347-358.
11. Cherry, M., and W. M. Eicher. 1976. Mouse News Lett. 54:41.
12. Coleman, D. L. 1962. Arch. Biochem. Biophys. 96: 562-568.
13. Cumming, R. B., et al. 1978. Biochem. Genet. 16:(in press).
14. DeLorenzo, R. J., and F. H. Ruddle. 1969. Ibid. 3: 151-162.
15. DeLorenzo, R. J., and F. H. Ruddle. 1970. Ibid. 4: 259-273.
16. Dizik, M., and R. W. Elliott. 1977. Ibid. 15:31-46.
17. Dizik, M., and R. W. Elliott. 1977. Mouse News Lett. 56:58.
18. Duley, J., and R. S. Holmes. 1974. Genetics 76:93-97.
19. Eicher, E. M. 1976. Mouse News Lett. 54:41.

continued

Part II. References

20. Eicher, E. M., and J. E. Womack. 1977. Biochem. Genet. 15:1-8.
21. Eicher, E. M., et al. 1976. Ibid. 14:651-660.
22. Eicher, E. M., et al. 1977. Mouse News Lett. 56:42.
23. Eicher, E. M., et al. 1978. (In preparation).
24. Eichwald, E. J., and C. R. Silmser. 1955. Transplant. Bull. 2:148-149.
25. Epstein, C. J. 1972. Science 175:1467-1468.
26. Feinstein, R. N. 1970. Biochem. Genet. 4:135-155.
27. Finlayson, J. S., et al. 1969. Genet. Res. 14:329-331.
28. Flanagan, S. P. 1966. Ibid. 8:295-309.
29. Francke, U., et al. 1977. Cytogenet. Cell Genet. 19: 57-84.
30. Ganschow, R. E., and B. G. Bunker. 1970. Biochem. Genet. 4:127-133.
31. Ganschow, R. E., and R. T. Schimke. 1970. Ibid. 4: 157-167.
32. Gazdar, A. F., et al. 1977. Cell 11:949-956.
33. Haldane, J. B. S., et al. 1915. J. Genet. 5:133-135.
34. Henderson, N. S. 1965. J. Exp. Zool. 158:263-273.
35. Hoffman, H. A., and C. K. Grieshaber. 1976. Biochem. Genet. 14:59-66.
36. Hoffman, H. A., and C. K. Grieshaber. 1977. Mouse News Lett. 56:51.
37. Huijing, F., et al. 1973. Biochem. Genet. 9:193-196.
38. Hutton, J. J. 1969. Ibid. 3:507-515.
39. Hutton, J. J., and D. L. Coleman. 1969. Ibid. 3:517-523.
40. Hutton, J. J., and T. H. Roderick. 1970. Ibid. 4:339-350.
41. Itakura, K., et al. 1972. Transplantation 13:239-243.
42. Kaplan, R. D., et al. 1973. J. Hered. 64:155-157.
43. Klein, J., and D. C. Shreffler. 1971. Transplant. Rev. 6:3-29.
44. Kozak, C., and F. H. Ruddle. 1977. Somat. Cell Genet. 3:121-133.
45. Kozak, C., et al. 1975. Ibid. 1:371-382.
46. Kozak, L. P., and K. J. Erdelsky. 1975. J. Cell. Physiol. 85:437-447.
47. Lalley, P. A., and T. B. Shows. 1977. Genetics 87: 305-317.
48. Lalley, P. A., et al. Unpublished. 4th Int. Workshop Hum. Gene Mapp. Winnipeg, 1977.
49. Lewis, W. H. P., and G. M. Truslove. 1969. Biochem. Genet. 3:493-498.
50. Lilly, F. 1966. Genetics 53:529-539.
51. Lilly, F. 1970. J. Natl. Cancer Inst. 45:163-169.
52. Lusis, A. J., and J. D. West. 1976. Biochem. Genet. 14:849-855.
53. Lyon, J. B., Jr. 1970. Ibid. 4:169-185.
54. McBreen, P., et al. 1977. Cytogenet. Cell Genet. 19: 7-13.
55. Meo, T., et al. 1977. Science 198:311-313.
56. Minna, J., et al. 1978. Somat. Cell Genet. 4:241-252.
57. Mishkin, J. D., et al. 1976. Biochem. Genet. 14:635-640.
58. Nichols, E. A., and F. H. Ruddle. 1975. Ibid. 13:323-329.
59. Nichols, E. A., et al. 1973. Ibid. 8:43-53.
60. Nichols, E. A., et al. 1975. Ibid. 13:551-555.
61. Nielsen, J. T., and V. M. Chapman. 1977. Genetics 87: 319-325.
62. Paigen, K., et al. 1976. Cell 9:533-539.
63. Peters, J., and H. R. Nash. 1976. Biochem. Genet. 14: 119-125.
64. Peters, J., and H. R. Nash. 1977. Ibid. 15:217-226.
65. Petras, M. L., and F. G. Biddle. 1967. Can. J. Genet. Cytol. 9:704-710.
66. Petras, M. L., and P. Sinclair. 1969. Ibid. 11:97-102.
67. Platz, R. D., and H. G. Wolfe. 1969. J. Hered. 60:187-192.
68. Popp, R. A. 1969. Ibid. 60:126-133.
69. Popp, R. A., and D. M. Popp. 1962. Ibid. 53:111-114.
70. Roderick, T. H., et al. 1970. Ibid. 61:278-279.
71. Rowe, W. P., and H. Sato. 1973. Science 180:640-641.
72. Rowe, W. P., et al. 1972. Ibid. 178:860-862.
73. Ruddle, R. H., et al. 1968. Genetics 58:599-606.
74. Ruddle, R. H., et al. 1968. Ibid. 62:393-399.
75. Schollen, J., et al. 1975. Biochem. Genet. 13:369-377.
76. Seyedyazdani, I., and L. G. Ludin. 1973. J. Hered. 64:295-296.
77. Shows, T. B., and F. H. Ruddle. 1968. Science 160: 1356-1357.
78. Shows, T. B., et al. 1969. Biochem. Genet. 3:25-35.
79. Shows, T. B., et al. 1970. Ibid. 4:707-718.
80. Shreffler, D. C. 1963. J. Hered. 54:127-129.
81. Soares, E. R. 1977. Mouse News Lett. 57:33.
82. Stockert, E., et al. 1972. Science 178:862-863.
83. Swank, R. T., et al. 1973. J. Mol. Biol. 81:225-243.
84. Taylor, B. A. 1975. Mouse News Lett. 52:37.
85. Taylor, B. A., et al. 1977. J. Virol. 23:106-109.
86. Vanderberg, J. L., et al. 1973. Nature (London) New Biol. 243:48-50.
87. Watanabe, T., and T. Tomita. 1974. Biochem. Genet. 12:419-428.
88. Watanabe, T., et al. 1976. Ibid. 14:999-1002.
89. Watson, J. G., et al. 1972. Ibid. 6:195-204.
90. Womack, J. E. 1975. Ibid. 13:311-322.
91. Womack, J. E., and S. Auerbach. 1978. Ibid. 16:239-245.
92. Womack, J. E., and E. M. Eicher. 1977. Mol. Gen. Genet. 155:315-317.
93. Womack, J. E., and M. Sharp. 1976. Genetics 82:665-675.
94. Womack, J. E., et al. 1975. Biochem. Genet. 13:511-518.
95. Womack, J. E., et al. 1975. Ibid. 13:519-525.
96. Womack, J. E., et al. 1977. Ibid. 15:347-355.
97. Womack, J. E., et al. 1978. Ibid. 16:(in press).
98. Wood, A. W., and A. H. Conney. 1974. Science 185: 612-614.

Phenotypic characteristics listed below are not meant to be inclusive, but rather to highlight characteristics most likely to be associated with endocrine functions. Differences in endocrine parameters reported between strains were included when appropriate data on F_1 and F_2 or backcross individuals established the genetic bases for such differences. Mutant genes associated with dwarfism were included when endocrinologic data were available. For institutions maintaining these mutant stocks, consult references 23 and 35. **Phenotypic Characteristics:** GH = growth hormone; PRL = prolactin. Data in brackets refer to the column heading in brackets.

Gene Symbol	Gene Name [Allele]	Chromosome	Phenotypic Characteristics	Reference
A^y	Yellow [Many]	2	Obesity, islets of Langerhans cell hypertrophy, elevated plasma insulin, adrenal enlargement	18
ald-1	Adrenocortical lipid depletion-1	1	Depleted esterified cholesterol stores from adrenal cortex (strain AKR)	1,7
ald-2	Adrenocortical lipid depletion-2	....	Depleted esterified cholesterol stores from adrenal cortex; decreased corticosterone produced in vitro (strain DBA/2)	8
Bfo	Bell-flash ovulation	4	Enhanced ovulation after exogenous gonadotropin	17
Cpl-1	Plasma corticosterone level-1	....	High ($Cpl\text{-}1^h$) or low ($Cpl\text{-}1^l$) levels of endogenous corticosterone in normal, untreated mice	16
Cpl-2	Plasma corticosterone level-2	....	Epistatic, and not allelic, to *Cpl-1*	16
d^l	Dilute-lethal	9	Neurological effects at early age; elevated adrenal epinephrine & norepinephrine; decreased urinary epinephrine, but normal norepinephrine	9
db	Diabetes [Adipose, db^{ad}]	4	Obesity, hyperglycemia, temporary hyperinsulinemia (dependent on genetic background)	18
df	Ames dwarf	11	Pituitary hormone deficient (GH, PRL absent)	3,34
du	Ducky [Torpid, du^{td}]	9	Low fertility in ♀; pituitary morphologically described as deficient in PRL-producing cells	10,40
dw	Dwarf[1/] [Dwarf-J; dw^J]	....	Pituitary hormone deficient (GH, PRL absent; thyrotropin ⟨TSH⟩, luteinizing hormone ⟨LH⟩, follicle-stimulating hormone ⟨FSH⟩ reduced)	24
Ex	Earlier X-zone degeneration	7	Advances time of adrenal X-zone degeneration in ♀ (endocrine function of X-zone unknown)	38
Ezg	Extent of zona glomerulosa	....	Reduced zona glomerulosa in adrenal cortex (PERU strain); possible elevated plasma ACTH; plasma aldosterone not affected	36,39
fat	Fat	....	Obese by 4-5 wk; hyperglycemia infrequent; hyperinsulinemia; moderate increase in number & size of islets of Langerhans	19
gl	Gray-lethal	10	Decreased serum Ca & P; elevation of serum alkaline phosphatase & of Ca in bone ash; increased calcitonin-like activity in plasma	29,30
Hnl	Hypothalamic norepinephrine level	....	High level of endogenous hypothalamic norepinephrine (strain BALB/cBy)	14
Hom-1	Androgen hormone metabolism-1	17	Smaller relative weights of androgen target glands (seminal vesicles, thymus, axillary lymph nodes)	20
lh	Lethargic	2	Neurological symptoms, thymic involution, elevated serum corticosterone	11
lit	Little	6	Reduced growth rate; pituitary deficient in GH & PRL; plasma deficient in GH	4,13
nu	Nude [Nude-streaker, nu^{str}]	11	Serum thyroxine reduced, enlarged adrenal X-zone	12,31,37
ob	Obese	6	Obesity, hyperglycemia, hyperinsulinemia, adrenal hypertrophy	18

1/ Synonym: Snell's dwarf.

continued

21. MUTANT GENES WITH ENDOCRINE EFFECTS: MOUSE

Gene Symbol	Gene Name [Allele]	Chromosome	Phenotypic Characteristics	Reference
p	Pink-eyed dilution [Many]	7	Variable sterility in both sexes associated with different alleles; anterior pituitary reduced to half normal size but morphologically normal; thyroid incorporation of iodine markedly reduced	21,26,41
pg	Pygmy	10	Reduced body size; peripheral resistance to exogenous GH	22,32
Sl	Steel [Many]	10	Germ cells deficient (both sexes); ovarian follicular structures deficient; elevated pituitary & plasma gonadotropins	5,33
Spl	Plasma serotonin level	9	High plasma serotonin levels	15
Tfm	Testicular feminization	X	Chromosomally XY ♂, but phenotypically ♀; deficient in androgen receptors in target tissue cytosol, normal estrogen receptors (brain)	2,6,25
W	Dominant spotting [Many]	5	Germ cells deficient (both sexes); ovarian follicular structures deficient (degree of deficiencies dependent on allele); W^v/W^x, elevated plasma gonadotropins	27,28

Contributor: Wesley G. Beamer

References

1. Arnesen, K. 1955. Acta Endocrinol. (Copenhagen) 18:396-401.
2. Attardi, B., et al. 1976. Endocrinology 98:864-874.
3. Bartke, A. 1965. Gen. Comp. Endocrinol. 5:418-426.
4. Beamer, W. G., and E. M. Eicher. 1976. J. Endocrinol. 71:37-45.
5. Beamer, W. G. Unpublished. Jackson Laboratory, Bar Harbor, ME, 1977.
6. Bullock, L. P., and C. W. Bardin. 1974. Endocrinology 94:746-756.
7. Doering, C. H., et al. 1972. Ibid. 90:93-101.
8. Doering, C. H., et al. 1973. Biochem. Genet. 8:101-111.
9. Doolittle, C. H., and H. Rauch. 1965. Biochem. Biophys. Res. Commun. 18:43-47.
10. Dung, H. C. 1975. J. Reprod. Fertil. 45:91-99.
11. Dung, H. C. 1976. Am. J. Anat. 147:255-263.
12. Eicher, E. M. 1976. Mouse News Lett. 54:40.
13. Eicher, E. M., and W. G. Beamer. 1976. J. Hered. 67: 87-91.
14. Eleftheriou, B. E. 1974. Brain Res. 70:538-540.
15. Eleftheriou, B. E., and D. W. Bailey. 1972. J. Endocrinol. 55:225-226.
16. Eleftheriou, B. E., and D. W. Bailey. 1972. Ibid. 55: 415-420.
17. Eleftheriou, B. E., and M. B. Kristal. 1974. J. Reprod. Fertil. 38:41-47.
18. Herberg, L., and D. L. Coleman. 1977. Metabolism 26:59-99.
19. Hummel, K., and D. L. Coleman. 1974. Mouse News Lett. 50:43.
20. Ivanyi, P., et al. 1972. Folia Biol. (Prague) 18:81-97.
21. Johnson, D. R., and D. M. Hunt. 1975. J. Reprod. Fertil. 42:51-58.
22. King, J. W. B. 1955. J. Genet. 53:487-497.
23. Lane, P. W. 1976. Lists of Mutations and Mutant Stocks of the Mouse. Jackson Laboratory, Bar Harbor.
24. Lewis, U. J. 1967. Mem. Soc. Endocrinol. 15:179-192.
25. Lyon, M. F., and S. G. Hawkes. 1970. Nature (London) 227:1217-1219.
26. Melvold, R. W. 1974. Genet. Res. 23:319-325.
27. Mintz, B., and E. S. Russell. 1957. J. Exp. Zool. 134: 207-237.
28. Murphy, E. D., and W. G. Beamer. 1973. Cancer Res. 33:721-727.
29. Murphy, H. M. 1968. Genet. Res. 11:7-14.
30. Murphy, H. M. 1972. J. Endocrinol. 53:139-150.
31. Pierpaoli, W., and E. Sorkin. 1972. Nature (London) New Biol. 238:282-285.
32. Rimoin, D. L., and L. Richmond. 1972. J. Clin. Endocrinol. Metab. 35:467-468.
33. Russell, E. S., and S. E. Bernstein. 1966. In E. L. Green, ed. Biology of the Laboratory Mouse. Ed. 2. McGraw-Hill, New York. pp. 351-372.
34. Schaible, R., and J. W. Gowen. 1961. Genetics 46: 896.
35. Searle, A. G., ed. 1977. Mouse News Lett. 56:4-23, 24-28.
36. Shire, J. G. M. 1969. Endocrinology 85:415-422.
37. Shire, J. G. M. 1973. Mouse News Lett. 48:28.
38. Shire, J. G. M., and S. G. Spickett. 1968. Gen. Comp. Endocrinol. 11:355-365.
39. Shire, J. G. M., and J. Stewart. 1972. J. Endocrinol. 55:185-193.
40. Snell, G. D. 1955. J. Hered. 46:27-29.
41. Wolfe, H. G. 1971. Biol. Reprod. 4:161-173.

22. VARIANTS OF COMPLEMENT AND OTHER PROTEINS: MOUSE

Part I. Complement

Data in brackets refer to the column heading in brackets.

Protein	Nature of Variation	Locus Symbol [Locus Name]	Chromo-some	Allele	Strain Distribution	Refer-ence
C1, C2, C4	Quantitative difference	....	17	"High"	B10.A, B10.D2, B10.M, B10.S, C3H.OH, C3H.SW, C57BL/10	7
				"Low"	B10.AKM, B10.BR, B10.HTT, C3H, C3H.OL	
C3	Quantitative difference	....	17	"High"	A.AL, A.CA, AKR, B10.BR, B10.M, B10.S, C3H, C3H.OL	3,4
				"Low"	A/Sn, B10.D2, C3H.B10, C3H.OH, C57BL/10, DBA/2	
C4[1]	Quantitative difference	*Ss* [Serum substance]	17	*h*[2]	A/He, A/J, A/Sn, BALB/c, BDP, C57BL/6, C57BL/10, C57L, DA, DBA/2, F/St, I/Ao, JK/St, LP, P, SEC, SJL, SWR, WB/Re, YBR/He, 129	9,11
				l	AKR, CBA, CE, CHI, C3H, C57BR, C58, FL/2, L/St, MA, RF, ST/b	
	Allotype, presence-absence	*Slp* [Sex-limited protein]	17	*a*[2]	A/He, A/J, A/Sn, BALB/c, BDP, DBA/2, F/St, I/Ao, JK/St, P, SJL, YBR/He	9,11
				o	AKR, CBA, C3H, C57BL/6, C57BL/10, C57BR, C57L, C58, DA, FL/2, MA, RF, SWR, WB/Re, 129	9,11
				w	B10.WR7, C3H.WSlp	8
C5	Presence-absence	*Hc* [Hemolytic complement]	2	*1*	BALB/c, BDP, BRVR, BSVS, BUB, CBA, CHI/St, C3H/He, C57BL/6, C57BL/10, C57BR/cd, C57L, C58, DBA/1, F/St, HR/De, LP/J, MA/J, NZB, P/J, PHL, PL/J, RIII/An, RIII/2J, SEC, SJL, SM/J, YBR/He, 129/J	1,2
				0	A/He, A/J, AKR, AU, BRSUNT, CE/J, DBA/2, DE/J, PHH, RF/J, ST/b, SWR	
Complement receptor	Quantitative difference	*Crl-1* [Complement receptor lymphocytes]	17	"High"	A/He, A/Sn, AKR, B10.A, B10.A(5R), SJL	5,6
				"Low"	A.BY, A.CA, A.SW, AL, BALB/c, B10.A(2R), B10.A(4R), B10.D2, C3H/He, C57BL/6, C57BL/10, C57BR, DBA/2	

[1] For this protein, only *standard* strains are given; for distribution in *congenic* strains, consult reference 9. [2] Represents broad *class* of alleles; for details, consult reference 10.

Contributor: Donald C. Shreffler

References

1. Cinader, B., et al. 1964. J. Exp. Med. 120:897-924.
2. Erickson, R. P., et al. 1964. J. Immunol. 92:611-615.
3. Ferreira, A., and V. Nussenzweig. 1975. J. Exp. Med. 141:513-517.
4. Ferreira, A., and V. Nussenzweig. 1976. Nature (London) 260:613-615.
5. Gelfand, M. C., et al. 1974. J. Exp. Med. 139:1125-1141.
6. Gelfand, M. C., et al. 1974. Ibid. 139:1142-1153.
7. Goldman, M. B., and J. N. Goldman. 1976. J. Immunol. 117:1584-1588.
8. Hansen, T. H., and D. C. Shreffler. 1976. Ibid. 117:1507-1513.
9. Klein, J. 1975. Biology of the Mouse Histocompatibility-2 Complex. Springer-Verlag, New York. pp. 496-498.
10. Shreffler, D. C. 1976. Transplant. Rev. 32:140-167.
11. Shreffler, D. C., and H. C. Passmore. 1971. In A. Lengerova and M. Vojtiskova, ed. Immunogenetics of the H-2 System. Karger, Basel. pp. 58-68.

continued

22. VARIANTS OF COMPLEMENT AND OTHER PROTEINS: MOUSE

Part II. Other Proteins

Protein	Nature of Variation	Locus	Chromo-some	Allele	Strain Distribution	Refer-ence
Albumin	Conforma-tional	*Acf-1*	..	*a*	A/He, A/J, AU, CE, C3H, $C3H_eB$/Fe, C57BL/6, C58, DE, LP, NZB	8,9
				b	AKR, BALB/c, BUB, CBA, DBA/2, LG, MA, P, PL, RF, SEC, SJL, SM, ST/b, SWR, 129	
	Electro-phoretic	*Alb-1*	5	*a*	A/He, AKR, AU, BALB/c, BDP, BUB, CBA, CE, C3H, C57BL/6, C57BL/10, C57BR/cd, C57L, DBA/2, DW, LG, LP, MA, NZB, P, PL, ROP, RIII, SEA, SEC, SJL, SM, ST/b, SWR, 129	4
				c	ALB-1C[1]	
Esterase	Electro-phoretic	*Es-1*	8	*a*	C57BL, C57BR, C57L	6
				b	A/He, AKR, AU, BALB/c, BDP, BUB, CBA, CE, C3H/He, $C3H_eB$/Fe, C58, DBA/1, DBA/2, DE, DW, HRS, IS/Cam, LG, LP, MA, NZB, P, PH/Re, PL, RF, RIII/2J, SEA, SEC, SF/Cam, SJL, SK/Cam, SM, ST/b, 129	
		Es-2	8	*a*	IS/Cam[1], SK/Cam[1]	6
				b	A/He, AKR, AU, BALB/c, BDP, BUB, CBA, CE, C3H/He, $C3H_eB$/Fe, C57BL/6, C57BR/cd, C57L, C58, DBA/1, DBA/2, DE, DW, HRS, LG, LP, MA, NZB, P, PH/Re, RF, RIII/2J, SEA, SEC, SF/Cam, SJL, SM, ST/b, SWR, 129	
				c	PL	
Prealbu-min	Electro-phoretic	*Pre-1*	..	*a*	AKR, BALB/c, B10-*Pre^a*, CBA, DBA/1, DBA/2, RF, SEC, WB	7,9,10
				b	CE, C57BR/cd, LP, SWR	
				c	A/He, A/J, C3H/He, C57BL/6	
	Presence-absence	*Pre-2*	..	*a*	AKR, BALB/c, CBA, C57BR/cd, DBA/2, LP, RF, SEC, SWR	9,10
				b	A/He, A/J, CE, C3H, C57BL/6	
Seminal vesicle protein	Electro-phoretic	*Svp-1*	2	*a*	AKR, C3H, PHH, PHL, 129	5
				b	A/J, AL, BALB/c, CBA, C57BL/6, C57BL/10, C57BR/cd	
		Svp-2	..	*b*	C57BR/cd	2
				c	A/J	
Serum antigen	Antigenic	*Mud*	..	*1*	AKR, BALB/c, BDP, C57BL/6, C57BL/10, C57BR/cd, C57L, C58, HRS, LP, MA, P, PHH, SEA, SEC, ST/b, SWR, 129	3
				2	A/He, A/J, CBA, CE/J, DBA/1, DBA/2, DE, PL, RF, RIII, SJL, SM	
		Sas-1	..	*a*	AKR, AL, BALB/c, BDP, C57BL/6, C57BL/10, C57BR/cd, C57L, C58, LP, RIII, ST, SWR, 101, 129	11
				o	A/J, CBA, CE, C3H/He, DA, DBA/1, DBA/2, DD, DE, NZB, PL, RF, SJL, SM, ST/b	
Trans-ferrin	Electro-phoretic	*Trf*	9	*a*	B10-*Trf^a*, CBA	1
				b	A/He, A/J, AKR, A2G, BALB/c, C3H, C57BL/6, C57BL/10, C57BR/cd, C57L, CE, DBA/1, DBA/2, KL, RF, RIII, WB, WC	

[1] Wild-derived stocks.

Contributor: Donald C. Shreffler

References

1. Cohen, B. L., and D. C. Shreffler. 1961. Genet. Res. 2:306-308.
2. Moutier, R., et al. 1971. Exp. Anim. 4:7-18.
3. Naylor, D. H., and B. Cinader. 1970. Int. Arch. Allergy Appl. Immunol. 39:511-539.
4. Nichols, E. A., et al. 1975. Biochem. Genet. 13:551-555.
5. Platz, R. D., and H. G. Wolfe. 1969. J. Hered. 60:187-192.
6. Roderick, T. H., et al. 1971. Biochem. Genet. 5:457-466.
7. Shreffler, D. C. 1964. Genetics 49:629-634.
8. Wilcox, F. H. 1973. Biochem. Genet. 10:69-78.
9. Wilcox, F. H. 1975. J. Hered. 66:19-22.
10. Wilcox, F. H., and D. C. Shreffler. 1976. Ibid. 67:113.
11. Wortis, H. H. 1965. Genetics 52:267-273.

23. HEMOGLOBIN GENES: MOUSE

Adult hemoglobins are tetramers: $\alpha_2\beta_2$. A homogeneous form of hemoglobin is produced in mice carrying the single β-chain allele, Hbb^s, whereas two electrophoretically distinguishable forms of adult hemoglobin appear in mice having Hbb^d or Hbb^p. Each of the latter alleles is composed of two closely-linked β-chain genes; each controls the synthesis of a different β-chain polypeptide. $Hbb^{s(dp)}$ symbolizes a radiation-induced duplication of a segment of chromosome 7 including Hbb^s. Products of known alleles at the α-chain locus are electrophoretically indistinguishable. A single α-chain polypeptide is produced by mice carrying Hba^a, but two kinds of α-chain polypeptides are found in adult mice carrying Hba^b, Hba^c, or Hba^d because each of these alleles is composed of two closely-linked α-chain genes. $Hba^{(df)}$ symbolizes a radiation-induced α-chain deficiency. Embryonic hemoglobins are tetramers: E_I is x_2y_2, E_{II} is α_2y_2, E_{III} is α_2z_2. **Holder**: For full name, *see* list of HOLDERS at front of book; Common = holdings by many laboratories. Data in brackets refer to the column heading in brackets.

Gene Symbol	Gene Name [Time of Expression]	Chromosome	Allele	Prototype Strain	Holder	Reference
Hba	Hemoglobin α-chain [Adult]	11	Hba^a	C57BL	Common	7,11
			Hba^b	BALB/c, SEC	Common	7,11
			Hba^c	C3H/AnCum	CUM	4,7
			Hba^d	CBA, C3H.B	Common	4,7
			$Hba^{(df)}$	27HB/Rl, 352HB/Rl	RL	12
Hbb	Hemoglobin β-chain [Adult]	7	Hbb^d	BALB/c, C3H	Common	9,10
			Hbb^p	AU/SsJ	JAX	6,11
			Hbb^s	C57BL, SEC	Common	8,10
			$Hbb^{s(dp)}$	86HB/Rl	RL	12
Hbx	Hemoglobin x-chain [Embryonic]	....	....	C3H, C57BL	Common	1,5
Hby	Hemoglobin y-chain [Embryonic]	7	Hby^a	C57BL	Common	1-3,13
			Hby^b	C3H, DBA/2	Common	1-3,13
Hbz	Hemoglobin z-chain [Embryonic]	....	....	C3H, C57BL	Common	1,14

Contributor: Raymond A. Popp

References

1. Fantoni, A., et al. 1967. Science 157:1327-1329.
2. Gilman, J. G. 1976. Biochem. J. 155:231-241.
3. Gilman, J. G., and O. Smithies. 1968. Science 160: 885-886.
4. Hilse, K., and R. A. Popp. 1968. Proc. Natl. Acad. Sci. USA 61:930-936.
5. Melderis, H., et al. 1974. Nature (London) 250:774-776.
6. Morton, J. R. 1966. Genet. Res. 7:76-85.
7. Popp, R. A. 1969. J. Hered. 60:126-133.
8. Popp, R. A. 1973. Biochim. Biophys. Acta 303:52-60.
9. Popp, R. A., and E. G. Bailiff. 1973. Ibid. 303:61-67.
10. Popp, R. A., and W. St. Amand. 1960. J. Hered. 51: 141-144.
11. Russell, E. S., and E. C. McFarland. 1974. Ann. N.Y. Acad. Sci. 241:25-38.
12. Russell, L. B., et al. 1976. Proc. Natl. Acad. Sci. USA 73:2843-2846.
13. Stern, R. H., et al. 1976. Biochem. Genet. 14:373-381.
14. Vulpis, G., and A. Bank. 1970. Biochim. Biophys. Acta 207:390-394.

24. URINARY PROTEINS: MOUSE

The urine of normal inbred [ref. 5, 22], random-bred [ref. 3], and wild [ref. 9] mice contains a considerable quantity of protein [ref. 13, 14]. Male mice excrete more than do females [ref. 5, 22, 24]. The concentration in the urine of sexually mature males is on the order of 5 mg/ml, but the value depends, inter alia, on strain, age, and dietary protein level; the concentration in the urine of females averages one-sixth to one-half of this figure. Mouse urine also contains much low-molecular-weight peptide. Chemical analyses of urine without prior fractionation (e.g., dialysis, ultrafiltration, precipitation, or gel chromatography) can therefore yield spuriously high values for protein.

The urinary proteins of inbred laboratory mice (*Mus musculus*) include uromucoid [ref. 7], analogous to the Tamm-Horsfall mucoprotein of human urine; variable, though small, quantities of α- and β-globulins [ref. 2, 7]; and a family of prealbumins (molecular weight, 17,500) known collectively as major urinary protein (MUP). MUP consists of components 1, 2, and 3, numbered in order of increasing

continued

electrophoretic mobility toward the anode [ref. 5]; the distribution is subject to both hormonal and genetic control. Mice of a given inbred strain excrete either 1 and 3 or 2 and 3; males and females of a single strain show the same components, but the ratio may vary with the strain [ref. 5]. Assignment of the phenotype is also affected by the resolving power of the electrophoretic system used. At high resolution, for example, a small amount of protein with the mobility of component 1 is seen in the urine of males from 2,3 strains [ref. 5, 12]; at lower resolution, 2 appears as an elongated component designated 2′ [ref. 5].

Earlier studies showed that the 1,2 variant was controlled by a locus [ref. 12] on chromosome 4, situated near brown (*b*) and pintail (*Pt*) approximately 5 centimorgans from *b* [ref. 8, 12]. Recent work [ref. 20] revealed that after induction by testosterone injection into female mice [ref. 6], all three components are present in the urine of all strains, and that the relative proportions are controlled by this locus. The locus was initially designated *Mup-a*, and the codominant alleles controlling 1 and 2, respectively, were designated *Mup-a¹* and *Mup-a²* [ref. 12]. It was subsequently suggested that the designations be changed to *Mup-1ᵃ* and *Mup-1ᵇ* [ref. 10]. The codominant nature of the control has been confirmed, but the recent data cited above suggest that *Mup-a* is a regulatory locus [ref. 20]. Since MUP is synthesized in the liver [ref. 6] and circulates in the plasma [ref. 6, 18] before being excreted, phenotyping may be possible by electrophoresis of serum [ref. 16, 17] as well as of urine.

In addition to the strains listed in the table, strains BUB/Bn [ref. 21] and CO_{13} [ref. 16] have been examined, but their MUP patterns do not coincide with those of either *Mup-a¹* or *Mup-a²*. A similar situation has been reported recently for other strains by A. Groen, 1977, Lab. Anim. 11:209-214.

Strain	Reference
Inbred Strains & Substrains Carrying *Mup-a¹* ⟨*Mup-1ᵃ*⟩	
A/HeJ	21
A/HeN	4,10
A/J	21
AKR/Cum	1
AKR/FuRdA	1
AKR/J	1
AKR/LwN	4,10
AL/N	4,10
AU/SsJ	21
A2G/A	11
BALB/cAnN	5
BALB/cBy	19
BALB/cHf	10
BL/De	10,12
BRSUNT/N	15
BXH-2	23
BXH-3	23
BXH-4	23
BXH-6	23
BXH-7	23
BXH-8	23
BXH-9	23
BXH-12	23
BXH-14	23
CBA_f/Lw	10
CE/J	10,21
CFCW/Rl	10
CFW/N	4
CXBE	15
CXBG	15
CXBH	15
CXBJ	15
C3H/HeJ	21
C3H/Lw_f	5
$C3H_eB$/FeJ	21
$C3H_f$/De	10
C57BR/cdJ	5,10
DBA/1J	10,21
DBA/2_eDe	10
DBA/2Lw_f	5
DBA/2N	12
DD/He	10
E-B+	5
F	5
FL	15
HRS/J	21
I/An	5,10
I/Ln	21
LG/J	21
LP/J	10,21
LTS/A	11
MA/J	10
MA/My	21
N	5
NBL/N	12
NH/Lw	5,10
NZB/Hz	21
PBR	5
RF/J	21
RF/Lw	10
RIII/SeA	11
SEA/Gn	21
SEC/1ReJ	21
SJL/J	4,21
SM	12
ST/bJ	10,21
STR/N	4,10
STR/1N	10
SWR/De	10
WB	15,21
WC	15,21
WH/Re	21
WK/Re	21
WLL/BrA	11
129/J	5,10
Inbred Strains & Substrains Carrying *Mup-a²* ⟨*Mup-1ᵇ*⟩	
AKR/FuA	1
BDP/J	21
BXH-5	23
BXH-10	23
BXH-11	23
BXH-19	23
B10.D2/n	5
B10.D2/o	5
CXBD	15
CXBI	15
CXBK	15
C57BL/An	10
C57BL/He	10
C57BL/Ka	5
C57BL/KsJ	21
C57BL/Rij	16
C57BL/6Hf	10
C57BL/6J-*Pt*/+	8
C57BL/6N	12
C57BL/10J	21
C57BL/10N	10
C57BL/10Sc	5
C57L/He	5,10
C57L/J	4
C58/J	21
C58/Lw	5,10
GRS/A	11

continued

24. URINARY PROTEINS: MOUSE

Strain	Reference	Strain	Reference	Strain	Reference
HR/De	10	O20	11	RIII/2J	21
LIS/A	11	P/J	10,21	STS/A	11
LT/Re	21	POLY-2	5	YBR-A^y/a	4
MAS/A	11	PRO/Re	21	YBR-*a/a*	4
MWT/Ww	21	RIII/An	10	YBR/He	10

Contributor: J. S. Finlayson

References

1. Acton, R. T., et al. 1973. Nature (London) New Biol. 245:8-10.
2. Fahey, J. L., and M. Potter. 1959. Nature (London) 184:654-655.
3. Finlayson, J. S., and C. A. Baumann. 1958. Am. J. Physiol. 192:69-72.
4. Finlayson, J. S., and M. Potter. Unpublished. Food and Drug Administration, Bureau of Biologics, Bethesda, MD, 1978.
5. Finlayson, J. S., et al. 1963. J. Natl. Cancer Inst. 31: 91-107.
6. Finlayson, J. S., et al. 1965. Science 149:981-982.
7. Finlayson, J. S., et al. 1968. Biochem. Genet. 2:127-140.
8. Finlayson, J. S., et al. 1969. Genet. Res. 14:329-331.
9. Finlayson, J. S., et al. 1974. Biochem. Genet. 11:325-335.
10. Hoffman, H. A. 1970. Proc. Soc. Exp. Biol. Med. 135: 81-83.
11. Hoffman, H. A. Unpublished. National Institutes of Health, Bethesda, MD, 1978.
12. Hudson, D. M., et al. 1967. Genet. Res. 10:195-198.
13. Parfentjev, I. A. 1932. Proc. Soc. Exp. Biol. Med. 29: 1285-1286.
14. Parfentjev, I. A., and W. A. Perlzweig. 1933. J. Biol. Chem. 100:551-555.
15. Potter, M., et al. 1973. Genet. Res. 22:325-328.
16. Reuter, A. M., et al. 1968. Comp. Biochem. Physiol. 25:921-928.
17. Reuter, A. M., et al. 1968. Eur. J. Biochem. 5:233-238.
18. Rümke, P., and P. J. Thung. 1964. Acta Endocrinol. (Copenhagen) 47:156-164.
19. Szoka, P. Unpublished. Roswell Park Memorial Institute, Buffalo, NY, 1978.
20. Szoka, P., and K. Paigen. 1977. Fed. Proc. Fed. Am. Soc. Exp. Biol. 36:648.
21. Taylor, B. A. Unpublished. Jackson Laboratory, Bar Harbor, ME, 1978.
22. Thung, P. J. 1962. Acta Physiol. Pharmacol. Neerl. 10:248-261.
23. Watson, J., et al. 1977. J. Immunol. 118:2088-2093.
24. Wicks, L. F. 1941. Proc. Soc. Exp. Biol. Med. 48: 395-400.

25. IMMUNOGLOBULIN ALLOTYPES: MOUSE

Figures in heavy brackets are reference numbers.

Part I. Immunoglobulin Classes and Corresponding Loci

P & L refers to nomenclature of Potter and Lieberman [10, 11]. **H, et al.** refers to nomenclature of Herzenberg, et al. [5, 6].

Chain Classes	Chain Designation	Gene Locus Designation		
		P & L	H, et al.	Proposed
Heavy Chains				
IgG_{2a} [2,13]	γ_{2a}	*IgG* [10]	*Ig-1* [6]	*IgG2a*
IgG_{2b} [2]	γ_{2b}	*IgH* [10]	*Ig-3* [6]	*IgG2b*
IgG_1 [2]	γ_1	*IgF* [10]	*Ig-4* [6]	*IgG1*
IgG_3 [4]	γ_3	....	....	......
IgA [2]	α	*IgA* [10]	*Ig-2* [6]	*IgA*
IgD [3]	δ	....	*Ig-5* [3]	*IgD*
IgM [2]	μ	*IgM* [8]	*Ig-6* [7]	*IgM*
IgE [12]	ϵ	....	....	......
Light Chains				
Kappa [9]	κ	....	....	......
Lambda 1 [14]	$\lambda 1$	....	....	......
Lambda 2 [1]	$\lambda 2$	....	....	......

Contributor: Rose Lieberman

continued

25. IMMUNOGLOBULIN ALLOTYPES: MOUSE

Part I. Immunoglobulin Classes and Corresponding Loci

References

1. Dugan, E. S., et al. 1973. Biochemistry 12:5400-5416.
2. Fahey, J. L., et al. 1964. J. Exp. Med. 120:223-243.
3. Goding, J. W., et al. 1976. Proc. Natl. Acad. Sci. USA 73:1305-1309.
4. Grey, H. M., et al. 1971. J. Exp. Med. 133:289-304.
5. Herzenberg, L. A., et al. 1967. Cold Spring Harbor Symp. Quant. Biol. 32:181-186.
6. Herzenberg, L. A., et al. 1968. Annu. Rev. Genet. 2: 209-244.
7. Herzenberg, L. A., et al. 1976. Cold Spring Harbor Symp. Quant. Biol. 41:33-45.
8. Mage, R., et al. 1973. The Antigens. Academic, New York. v. 1, pp. 299-376.
9. Potter, M. 1977. Adv. Immunol. 25:141-211.
10. Potter, M., and R. Lieberman. 1967. Ibid. 7:91-145.
11. Potter, M., and R. Lieberman. 1967. Cold Spring Harbor Symp. Quant. Biol. 32:187-202.
12. Prouvast-Danon, A., et al. 1975. Immunology 29:151-162.
13. Robinson, E. A., and E. Appella. 1977. Proc. Natl. Acad. Sci. USA 74:2465-2469.
14. Weigert, M., et al. 1970. Nature (London) 28:1045-1047.

Part II. Alleles of IgC_H Genes

Two nomenclature systems are currently in use to designate the immunoglobulin allotypes of mice and the allelic forms of the genes controlling the C_H regions. Potter and Lieberman [6] designate allotypic determinants by an upper case letter indicating the immunoglobulin class, and an arabic superscript indicating the specific determinant (e.g., G^1). Unassigned determinants appear as numerals without any class designation (e.g., 18) until the class designation can be determined. An italicized upper case letter indicates the genetic locus, and is the same letter that indicates the immunoglobulin class. The allele of that locus is designated by the upper case letter, together with arabic superscripts showing all of the determinants on the C_H region controlled by that allele (e.g., $G^{1,6,7,8,26}$).

Herzenberg, et al. [2] designate allotypic determinants by a numeral code (e.g., 1.1). The number before the decimal point indicates the class of the H chain, and the number after the decimal point indicates the specificity or determinant. The allele is shown by an arabic number representing the locus, and a lower case superscript indicates the allele at the locus (e.g., $Ig\text{-}1^a$).

One important difference between the two systems is that Potter and Lieberman [6] use consecutive numbers to identify determinants regardless of the class on which that determinant appears. Thus, a determinant which appears on two different classes is given the same number (e.g., G^4 and H^4 found on IgG and IgH proteins). Herzenberg's [2] system starts with determinant 1 for each class and uses consecutive numbers for additional determinants of that class (e.g., 1.1 and 2.1 specificities are unrelated).

Although many of the same determinants are identified in both systems, there are also determinants which are different for each system. Spring and Nisonoff [8] have identified an IgG_{2a} antigen-binding fragment ⟨Fab⟩ (Fd) with allotypic determinants linked to the IgC_H region, and we have taken the liberty of designating them G^{26} and G^{27}. G^{26} is identified by the percent of ^{125}I Fab fragments bound using guinea-pig anti-AKR Fab. G^{27} is identified by the percent of ^{125}I Fab fragments bound using anti-C57BL Fab. IgD and IgM specificities have been identified on lymphocyte membranes by Goding, et al. [1], and Herzenberg, et al. [4], and two allelic genes for each class have been identified by Herzenberg, et al. [4]: $5^{a,c,d,e,f,g,h}$ and 5^b for IgD; $6^{a,c,d,e,f,g,h}$ and 6^b for IgM.

System of P & L refers to nomenclature of Potter and Lieberman [6, 7]. **System of H, et al.** refers to nomenclature of Herzenberg, et al. [2, 3].

System of P & L: Allele	System of H, et al.		Prototype Strain
	Allele	Determinant	
$G^{1,6,7,8,26}$	1^a	1, 2, 6, 7, 8, 10, 12	BALB/c
$G^{2,27}$	1^b	4, 7	C57BL/6

continued

25. IMMUNOGLOBULIN ALLOTYPES: MOUSE

Part II. Alleles of IgC_H Genes

System of P & L: Allele	System of H, et al.		Prototype Strain
	Allele	Determinant	
$G^{3,8}$	1^c	2, 3, 7	DBA/2
$G^{4,6,7,8,26}$	1^d	1, 2, 5, 7, 12	AKR
$G^{4,6,7,8,26}$	1^e	1, 2, 5, 6, 7, 8, 12	A/J
$G^{5,7,8,26}$	1^f	1, 2, 8, 11	CE/J
$G^{3,8,26}$	1^g	2, 3	RIII/2J
$G^{1,6,7,8}$	1^h	1, 2, 6, 7, 10, 12	SEA/J
$G^{2+1,6,7,8}$ [5]	....	....	Kyushu (wild)[1]
$G^{3+5,7,8}$ [5]	....	....	Kitty Hawk (wild)[1]
$G^{3+5,8}$ [5]	....	....	Kitty Hawk (wild)[1]
$H^{9,11,22}$	$3^{a,c,h}$	1, 2, 4, 7, 8	BALB/c, DBA/2, SEA
$H^{9,16,22}$	3^b	4, 7, 8, 9	C57BL/6
$H^{4,23}$	3^d	1, 3, 7, 8	AKR/J
$H^{4,23}$	3^e	1, 3, 7	A/J
$H^{9,11}$	3^f	1, 2, 3, 4	CE
$H^{9,11}$	3^g	1, 2, 4 (?)	RIII
$F^{8,19}$	$4^{a,c,d,e,f,g,h}$	1	All except C57BL/6
F^-	4^b	2	C57BL/6
$A^{12,13,14}$	$2^{a,h}$	2, 3, 4	BALB/c, SEA
A^{15}	2^b	....	C57BL/6
Not yet assigned	$2^{c,g}$	1	DBA/2, RIII
$A^{13,17}$	$2^{d,c}$	3	A/J, AKR
A^{14}	2^f	4	CE
D^+	$5^{a,c,d,e,f,g,h}$	1 [4]	All except C57BL/6
D^-	5^b	2 [4]	C57BL/6
M^+	$6^{a,c,d,e,f,g,h}$	1 [4]	All except C57BL/6
M^-	6^b	2 [4]	C57BL/6
10[2]	....	....	DD
18[2]	....	....	AKR, AL
20[2]	....	....	DBA/2
21[2]	....	....	RIII
24[2]	....	....	BL
25[2]	....	....	CE, RIII

[1] Wild mice. [2] Class not yet determined.

Contributor: Rose Lieberman

References

1. Goding, J. W., et al. 1976. Proc. Natl. Acad. Sci. USA 73:1305-1309.
2. Herzenberg, L. A., et al. 1967. Cold Spring Harbor Symp. Quant. Biol. 32:181-186.
3. Herzenberg, L. A., et al. 1968. Annu. Rev. Genet. 2: 209-244.
4. Herzenberg, L. A., et al. 1976. Cold Spring Harbor Symp. Quant. Biol. 41:33-45.
5. Lieberman, R., and M. Potter. 1969. J. Exp. Med. 130: 519-541.
6. Potter, M., and R. Lieberman. 1967. Adv. Immunol. 7:91-145.
7. Potter, M., and R. Lieberman. 1967. Cold Spring Harbor Symp. Quant. Biol. 32:187-202.
8. Spring, S. B., and A. Nisonoff. 1974. J. Immunol. 113: 470-478.

continued

Part III. Determinants Assigned to Specific IgC_H Loci

Data show correspondence between determinants identified by different investigators. **P & L** refers to nomenclature of Potter and Lieberman [3, 4]. **H, et al.** refers to nomenclature of Herzenberg, et al. [1, 2].

Locus	Determinant According to P & L	Determinant According to H, et al.
G	G^{1}	1.10
	G^{2}	1.4
	G^{3}	1.3
	G^{4}	.....
	G^{5}	1.11
	G^{6}	1.12
	G^{7}	1.1
	G^{8}	.....
	G^{1+2}	.....
	G^{3+5}	.....
	....	1.2
	....	1.5
	....	1.6
	....	1.7
	....	1.8
	G^{26}	Fab[1/]
	G^{27}	Fab[1/]
H	H^{4}	.....
	H^{9}	3.4
	H^{11}	3.2
	H^{16}	3.9
	H^{22}	.....
	H^{23}	.....
	....	3.1
	....	3.3
	....	3.7
	....	3.8
F	F^{8}	.....
	F^{19}	4.1
	F^{-}	4.2
A	A^{12}	2.2
	A^{13}	2.3
	A^{14}	2.4
	A^{15}	.....
	A^{17}	.....
	....	2.1
D	....	5.1
	....	5.2
M	....	6.1
	....	6.2
Unassigned	10	.....
	18	.....
	20	.....
	21	.....
	24	.....
	25	.....

[1/] Identified by Spring and Nisonoff [5].

Contributor: Rose Lieberman

References

1. Herzenberg, L. A., et al. 1967. Cold Spring Harbor Symp. Quant. Biol. 32:181-186.
2. Herzenberg, L. A., et al. 1968. Annu. Rev. Genet. 2: 209-244.
3. Potter, M., and R. Lieberman. 1967. Adv. Immunol. 7:91-145.
4. Potter, M., and R. Lieberman. 1967. Cold Spring Harbor Symp. Quant. Biol. 32:187-202.
5. Spring, S. B., and A. Nisonoff. 1974. J. Immunol. 113: 470-478.

Part IV. Heavy Chain Linkage Groups in Inbred and Wild Mice

Data are from reference 3. For explanations of nomenclature, *see* headnote to Part II. *D* and *M* are lymphocyte membrane immunoglobulin markers. IgD was described by Goding, et al. [1]. Herzenberg, et al., [2] described IgD and IgM membrane markers linked to the allotype locus.

Antigenic Determinants for Locus							Strain
G	*H*	*F*	*A*	*D*	*M*	Unassigned	
1, 6, 7, 8, 26[1/]	9, 11, 22	8, 19	12, 13, 14	+	+	......	BALB/c, BDP[2/], BRSUNT[2/], BSL[2/], C57L, C58, P/J, ST[2/], STR/N, 129
1, 6, 7, 8	9, 11, 22	8, 19	12, 13, 14	+	+	......	CBA, C3H, C57BR, MA, PL, SEA, SEC
1, 6, 7, 8	9, 11, 22	8, 19	12, 13, 14	+	+	10	DD[2/]
2, 27[1/]	9, 16, 22	–	15	–	–	......	C57BL, C57BL/6, C57BL/10, HR[2/], LP, NBL[2/], SJL, SM, STR/1[2/]

[1/] G^{26} and G^{27} are IgG_{2a} markers described by Spring and Nisonoff [4], found on antigen-binding fragment ⟨Fab⟩ (Fd), and linked to the allotype. [2/] Not tested by Spring and Nisonoff [4] for G^{26} and G^{27}.

continued

25. IMMUNOGLOBULIN ALLOTYPES: MOUSE

Part IV. Heavy Chain Linkage Groups in Inbred and Wild Mice

Antigenic Determinants for Locus							Strain
G	*H*	*F*	*A*	*D*	*M*	Unassigned	
3, 8, 26[1/]	9, 11	8, 19	–	+	+	21, 25	BSVS[2/], DA[2/], FZ[2/], RIII
3, 8	9, 11	8, 19	–	+	+	21, 25	SWR
3, 8	9, 11	8, 19	–	+	+	20	DBA/1, DBA/2[2/], I[2/], RF, STOLI[2/], YBR
4, 6, 7, 8, 26[1/]	4, 23	8, 19	13, 17	+	+	10	A, NZB, NZW[2/]
4, 6, 7, 8, 26[1/]	4, 23	8, 19	13, 17	+	+	10, 18	AKR, AL[2/]
4, 6, 7, 8	4, 23	8, 19	13, 17	+	+	10, 24	BL[2/]
5, 7, 8, 26[1/]	9, 11	8, 19	14	+	+	25	CE, DE[2/], NH[2/]
2+1, 6, 7, 8	9, 16	–	15	...	...	......	KY (Kyushu wild mouse *Mus musculus molossinus*)[2/]
3+5, 7, 8	9, 11	8, 19	–	...	...	......	KH-1 (Kitty Hawk wild mouse)[2/]
3+5, 8	9, 11	8, 19	–	...	...	......	KH-2 (Kitty Hawk wild mouse)[2/]

1/ G^{26} and G^{27} are IgG_{2a} markers described by Spring and Nisonoff [4], found on antigen-binding fragment ⟨Fab⟩ (Fd), and linked to the allotype. 2/ Not tested by Spring and Nisonoff [4] for G^{26} and G^{27}.

Contributor: Rose Lieberman

References

1. Goding, J. W., et al. 1976. Proc. Natl. Acad. Sci. USA 73:1305-1309.
2. Herzenberg, L. A., et al. 1976. Cold Spring Harbor Symp. Quant. Biol. 41:33-45.
3. Mage, R., et al. 1973. The Antigens. Academic, New York. v. 1, pp. 299-376.
4. Spring, S. B., and A. Nisonoff. 1974. J. Immunol. 113: 470-478.

Part V. Origin and Characteristics of Allotype Congenic Strains

Donor Strain is the strain source of introduced IgC_H. N = number of introgressive backcrosses. F = number of generations of subsequent brother by sister mating.

Congenic Strain	Inbred Background	Donor Strain	N	F	IgC_H	Presumed IgV_H	Reference
BAB-14	BALB/c	C57BL/Ka	14	10	C57BL/Ka	BALB/c	4
BALB.Ig^b	BALB/c	C57BL/6	10	....	C57BL	C57BL	5
BALB.Ig^c	BALB/c	DBA/2	10	....	DBA/2	DBA/2	5
BALB.Ig^d	BALB/c	AKR	10	....	AKR	AKR	5
BALB.Ig^f	BALB/c	CE	10	....	CE	CE	5
BALB.Ig^g	BALB/c	RIII	10	....	RIII	RIII	5
B10.D2b.Ig^e	B10.D2	NZB	20	....	NZB	NZB	5
BC-8	C57BL/Ka	BALB/c	8	10	BALB/c	BALB/c	2
CAL-9	BALB/c	AL/N	9	15	AL	AL	3
CAL-20	BALB/c	AL/N	20	11	AL	AL	3
CB-20	BALB/c	C57BL/Ka	20	13	C57BL/Ka	C57BL/Ka	2
C57BL/6.Ig^e	C57BL/6	NZB	20	....	NZB	NZB	5
NZB.Ig^b	NZB	C57BL/6	8	....	C57BL/6	C57BL/6	5
SJA	SJL	BALB/c	8	8	BALB/c	BALB/c	1

Contributor: Rose Lieberman

continued

25. IMMUNOGLOBULIN ALLOTYPES: MOUSE

Part V. Origin and Characteristics of Allotype Congenic Strains

References

1. Herzenberg, L. A., et al. 1976. Cold Spring Harbor Symp. Quant. Biol. 41:33-45.
2. Lieberman, R., et al. 1976. J. Immunol. 117:2105-2111.
3. Mage, R., et al. 1973. The Antigens. Academic, New York. v. 1, pp. 299-376.
4. Riblet, R., et al. 1975. Eur. J. Immunol. 5:775-777.
5. Warner, N. L. Unpublished. Univ. New Mexico School Medicine, Albuquerque, 1978.

26. STRAIN DISTRIBUTION OF *If-1* ALLELES: MOUSE

If-1 is a locus with a quantitative effect on circulating interferon induced by Newcastle disease virus. Two alleles are known: $If\text{-}1^h$ (high producer), $If\text{-}1^l$ (low producer). The difference between low and high is tenfold. The absolute levels of interferon are a function of the amount of virus used to induce the interferon. *If-1* is linked to *H-28*.

Allele	Strain
$If\text{-}1^h$	AKR/FuRdA
	AKR/J
	A2G/GA
	BIMA/A
	BIR/A
	B10.D2/nSn
	B10.D2/oSn
	CBA/BrA
	CXBG/By
	CXBH/By
	CXBK/By
	C57BL/Lac
	C57BL/LiA
	C57BL/MHeA
	C57BL/6By
	C57BL/6ByA
	C57BL/6J
	C57BL/10J
	C57BL/10ScSnA
	C57BL/10Sn
	DBA/HeA
	DBA/2J
	GRS/A
	LTS$_f$/A
	MAS/A
	STS/A
	TS1/A
	WLL/BrA
	WLL/Br$_e$BA
$If\text{-}1^l$	A/BrA
	A/HeJ
	A/J
	BALB/cBy
	BALB/cCrglA
	BALB/cHeA
	BALB/cJ
	CE/J
	CXBD/By
	CXBE/By
	CXBI/By
	CXBJ/By
	C3H/He$_f$A
	C3H/HeJ
	C3H/J
	C3H/Lac
	DBA/He$_f$A
	DBA/LiA
	DBA/Li$_f$A
	DBA/1J
	HW81/By
	HW94/By
	HW97/By
	IF/BrA
	O20/A
	P/A
	RIII/Se$_f$A

Contributor: Edward De Maeyer

General References

1. DeMaeyer, E., and J. DeMaeyer-Guignard. 1969. J. Virol. 3:506-512.
2. DeMaeyer, E., et al. 1975. Immunogenetics 1:438-443.
3. DeMaeyer, E., et al. 1975. Ibid. 2:151-160.
4. Hilger, J. Unpublished. The Netherlands Cancer Institute, Amsterdam, 1977.

27. IMMUNE RESPONSE GENES, EXCEPT *Ir-1*: MOUSE

Gene responses expressed in F1 hybrids as dominant, co-dominant, or intermediate are designated as "dominant." Inheritance of high response to all antigens is "dominant," except as otherwise indicated.

Antigen	Antigen Symbol	Gene Symbol	Linkage	Comments	Reference
Dextran, α-1,3-linked	Dextran	V_H-*DEX*	Linked with Ig allotype loci	Believed to be property of an immunoglobulin variable region ⟨V_H⟩ gene	2,15

continued

27. IMMUNE RESPONSE GENES, EXCEPT *Ir-1*: MOUSE

Antigen	Antigen Symbol	Gene Symbol	Linkage	Comments	Reference
DNA, denatured	DNA	....	X chromosome	....	12
Erythrocyte alloantigen 1	Ea-1	*Ir-2*[1/]	Chromosome 2	Also affected by *H-2*-linked *Ir*-gene(s)	5-7
H-2.2 alloantigen	H-2.2	*Ir-4*	....	Also affected by *H-2*-linked *Ir*-gene(s)	9,10
H-2.28 alloantigen	H-2.28	....	....	....	8
H-2.5 alloantigen	H-2.5	....	....	....	14
Hen albumen lysozyme loop region-poly(DL-alanine)	Loop-A—Loop	....	....	....	11
Lipopolysaccharides, bacterial	LPS	....	....	Low response found in C3H/HeJ substrain. Other lines of C3H are responders.	19
Poly(L-Tyr,Glu)-poly(L-Pro)-poly(L-Lys)	(T,G)-Pro-L	*Ir-3*	....	....	13
Streptococcal Group A polysaccharide	A-CHO	*Ir-A-CHO*	....	....	4
	GAC	*Ir-GAC*	....	Also *H-2* and Ig-allotype-linked effects. Strain distribution different from *Ir-A-CHO*.	3
Theta alloantigen 1	Thy 1.1	*Ir-5*	Chromosome 17	Also affected by *H-2*-linked *Ir*-gene(s)	20
Various thymus-independent antigens	....	....	X chromosome	Defective response found in substrain CBA/N	1,16-18

[1/] Inheritance of high response to antigen is recessive.

Contributor: Howard C. Passmore

References

1. Amsbaugh, D. F., et al. 1972. J. Exp. Med. 136:931-949.
2. Blomberg, B., et al. 1972. Science 177:178-180.
3. Briles, D. E., et al. 1977. Immunogenetics 4:381-392.
4. Cramer, M., and D. G. Braun. 1975. Eur. J. Immunol. 5:823-830.
5. Gasser, D. L. 1969. J. Immunol. 103:66-70.
6. Gasser, D. L., and D. C. Shreffler. 1972. Nature (London) New Biol. 235:155-156.
7. Gasser, D. L., and D. C. Shreffler. 1974. Immunogenetics 1:133-140.
8. Hansen, T. H., et al. 1977. J. Immunol. 118:1403-1408.
9. Lilly, F., et al. 1971. Immunogenet. H-2 Syst. 1970 Proc. Symp., pp. 197-199.
10. Lilly, F., et al. 1973. Transplant. Proc. 5:193-196.
11. Maron, E., et al. 1973. J. Immunol. 111:101-105.
12. Mozes, E., and S. Fuchs. 1974. Nature (London) 249: 167-169.
13. Mozes, E., et al. 1969. J. Exp. Med. 130:1263-1278.
14. Peters, K., et al. 1974. Immunogenetics 1:141-157.
15. Riblet, R., et al. 1975. Eur. J. Immunol. 5:775-777.
16. Scher, I., et al. 1973. J. Immunol. 110:1396-1401.
17. Scher, I., et al. 1975. J. Exp. Med. 141:788-803.
18. Scher, I., et al. 1975. Ibid. 142:637-650.
19. Watson, J., and R. Riblet. 1974. Ibid. 140:1147-1161.
20. Zaleski, M. B., and J. Klein. 1974. J. Immunol. 113: 1170-1177.

28. CELLULAR ALLOANTIGENS: MOUSE

Data in brackets refer to the column heading in brackets.

Locus ⟨Synonym⟩ [Chromosome]	Alleles ⟨Synonym⟩	Specificity ⟨Synonym⟩	Strain Distribution	Tissue Distribution	Reference
Ea-1 [8]	*a*	1 ⟨A⟩	Wild mice	Present: erythrocytes	8,13,14,16,17
	b	2 ⟨B⟩	Wild mice		
	c ⟨*o*⟩	....	CBA/J, C3H/He, C57BL, C57BR/cdJ, DBA/2J, SEC/1ReJ, YBR		

continued

Locus ⟨Synonym⟩ [Chromosome]	Alleles ⟨Synonym⟩	Specificity ⟨Synonym⟩	Strain Distribution	Tissue Distribution	Reference
Ea-2 ⟨*H-14*⟩	*1* ⟨*a*⟩	1 ⟨R⟩	F/St, RF/J, RFM	Present: brain, erythrocytes, kidney, liver, lung, lymph nodes, spleen, testis, thymus Absent: epidermal cells	8,10,14-17
	2 ⟨*b*⟩	2 ⟨Z⟩	A/Sn, AKR, BALB/c, BDP, BUB/BnJ, CBA/J, CE/J, C3H/He, C3H/St, C57BL/Ks, C57BL/6, C57BL/10, C57BR/cd, C57L, C58, DA/HuSn, DBA/1, DBA/2, FL/2Re, HRS, HTG, HTI, I, JK, LP, MA/J, NZB, PL, SEC/1Re, SJL, SM/J, ST/bJ, SWR, WB/Re, YBR, 129		
Ea-3	*a*	1	C57L	Present: erythrocytes	6,8,14, 16,17
	b	....	A/Mv, BALB/cDe, C3H/HeDiSn, C57BL/He, C57BL/10ScSn		
Ea-4	*a*	1	A, AKR, BALB/c, BDP, BUB/Bn, CBA/J, CE/J, CXBD, CXBH, C3H/He, C57BR/cd, C57L, C58, DA/HuSn, DBA/1, DBA/2, F/St, FL/2Re, HRS, HTG, I, LP, MA/J, NZB, PL, RF/J, RIII, SEC/1Re, SJL, SM/J, ST/bJ, SWR, WB/Re, 129	Present: erythrocytes, kidney, lung, lymph nodes, ? thymus Absent: brain, heart, liver, muscle, spleen, testis, uterus	8,9,12, 14,16, 17
	b	2	CXBE, CXBG, CXBI, CXBJ, CXBK, C57BL, C57BL/Ks		
Ea-5 ⟨*H-5*⟩	*a*	1	A, C3H/St, F/St, I, YBR, 129	Present: brain, erythrocytes, kidney, lung, muscle, spleen, testis, ? liver	2,8,14, 16,17
	b	....	C3H/He, C57BL, C58, DBA/1, DBA/2, JK, RF/J, RIII		
Ea-6 ⟨*H-6*⟩ [2]	*a*	1	A, AKR, CBA/J, CXBE, CXBH, CXBI, CXBJ, CXBK, C3H/He, C57BL/6, C58, F/St, I, YBR, 129	Present: brain, erythrocytes, feces, gut, kidney, lung, muscle, spleen, testis, uterus, ? heart	2,8,14, 16,17
	b	2	BALB/c, CXBD, CXBG, C3H/St, DBA/1, DBA/2, JK, RF/J		
Ea-7	*a*	1	BUB/Bn, CBA/J, C3H/He, C58, LP, 129	Present: erythrocytes	1,8,14, 16-18
	b	2	A, AKR, BALB/c, BDP, CE/J, C57BL/6, C57BR/cd, C57L, DA/HuSn, DBA/1, DBA/2, FL/2Re, HRS, HTG, I, MA/J, PL, RF/J, RIII, SEC/1Re, SJL, SM/J, ST/bJ, SWR, WB/Re		
H-Y [Y]	No serological polymorphism detected; ♂, but not ♀, from all strains tested express antigen		A, AKR, BALB/c, $C3H_f$/Bi, C57BL/6	Present: bone marrow, epidermal cells, lymph nodes, sperm, spleen, thymocytes	7,11,16
Mph-1 [7]	*a*	1	F/St, I/StDa	Present: peritoneal exudate cells (macrophages) Absent: erythrocytes, lymph nodes	3,4,16, 17
	b	2	A/J, AKR, BALB/c, CBA/Ca, C3H/He, C57BL/6-*Tla*, C57BL/10ScSn, DBA/1, DBA/2, NZB, NZW, SWR, 129		
Sk	*1*	1	A, AKR, CBA/H, C3H/An	Present: brain, epidermal cells Absent: lymph nodes, spleen, thymocytes	11,16
	2	2	C57BL/6, C57BL/10		
T [17]	*T*	Not designated	BALB/cHu-T^J/+	Present: sperm, testicular cells Absent: brain, epidermal cells, lymphocytes, thymocytes	5,16
	+	....	BALB/cHu		

Contributor: Donal B. Murphy

continued

References

1. Amos, D. B. 1959. Can. Cancer Conf. 3:241-258.
2. Amos, D. B., et al. 1963. Transplantation 1:270-283.
3. Archer, J. R. 1975. Genet. Res. 26:213-219.
4. Archer, J. R., and D. A. L. Davies. 1974. J. Immunogenet. 1:113-123.
5. Bennett, D., et al. 1972. Proc. Natl. Acad. Sci. USA 69:2076-2080.
6. Egorov, I. K. 1965. Genetika 6:80.
7. Goldberg, E. H., et al. 1971. Nature (London) 232: 478-480.
8. Klein, J. 1975. Biology of the Mouse Histocompatibility-2 Complex. Springer-Verlag, New York. pp. 151-177.
9. Klein, J., and J. Martinkova. 1968. Folia Biol. (Prague) 14:237-238.
10. Popp, D. M. 1967. Transplantation 5:290-299.
11. Scheid, M., et al. 1972. J. Exp. Med. 135:938-955.
12. Shreffler, D. C. 1966. Genetics 54:362.
13. Singer, M. F., et al. 1964. Ibid. 50:285-286.
14. Snell, G. D., and M. Cherry. 1972. In P. Emmelot and P. Bentvelzen, ed. RNA Viruses and Host Genome in Oncogenesis. North-Holland, Amsterdam. pp. 221-228.
15. Snell, G. D., et al. 1967. Transplantation 5:481-491.
16. Snell, G. D., et al. 1976. Histocompatibility. Academic, New York. pp. 67-90.
17. Staats, J. 1976. Cancer Res. 36:4333-4377.
18. Stimpfling, J. H., and G. D. Snell. 1968. Transplantation 6:468-475.

29. CELLULAR ALLOANTIGENS IN CONGENIC LINES: MOUSE

Producer: For full name, *see* list of HOLDERS at front of book.

Differential Locus & Allele ⟨Synonym⟩	Congenic Line	Inbred Partner	Donor Strain	Producer
Erythrocyte Alloantigen Loci				
Ea-2^a ⟨*R, rho, H-14*⟩	B6.RIII(76NS)	C57BL/6	RIII/WyJ	SN
	B10.F-*Ea-2*^{a}	C57BL/10	F/St	PP
	B10.RIII(R)	C57BL/10Sn	RIII/WyJ	SG
	B10.RIII(72NS)	C57BL/10Sn	RIII/WyJ	SN
	C3H-*Ea-2*^{a}	C3H/JSf	RIII/WyJ	SF
	C57BL/6-*Ea-2*^{a}	C57BL/6JBoy	B10.F-*Ea-2*^{a}	BOY
	C57BL/10-*Ea-2*^{a} *Ea-7*^{a}	C57BL/10Sn	RIII/WyJ	SN
Ea-3^a ⟨λ⟩	B10.L	C57BL/10SnEg	C57L/J	EG
Ea-4^b ⟨*BL, D*⟩	C3H-*Ea-4*^{b}	C3H/JSf	C57BL/10J	SF
Ea-6^a ⟨*H-6*^{a}⟩	C3H/Bi-*H-6*^{a}	C3H$_f$/Bi	C3H$_f$/An	LI
Ea-6^b ⟨*H-6*^{b}⟩	C3H/Bi-*H-6*^{b}	C3H$_f$/Bi	C57BL/6J	LI
Ea-7^a ⟨*T*⟩	B10.129(5M)	C57BL/10Sn	129/J	SN
	C57BL/10-*Ea-2*^{a} *Ea-7*^{a}	C57BL/10Sn	RIII/WyJ, 129/J	SN
Lymphocyte Alloantigen Loci				
Ly-1^a ⟨*Ly-A, mu*⟩	C57BL/6-*Ly-1*^{a}	C57BL/6JBoy	C3H	BOY
	C57BL/6-*Ly-1*^{a}	C57BL/6J	Mixed	SN
Ly-2^a ⟨*Ly-B*⟩	B6.PL(75NS)	C57BL/6J	PL/J	SN
	C57BL/6-*Ly-2*^{a}	C57BL/6JBoy	C3H	BOY
Ly-2^a, *Ly-3*^{a}	B6.PL(76NS)	C57BL/6J	PL/J	SN
	C57BL/6-*Ly-2*^{a} *Ly-3*^{a}	C57BL/6JBoy	RF	BOY
Thymocyte Alloantigen Locus				
Thy-1^a ⟨θAKR⟩	A-*Thy-1*^{a}	A/JBoy	A.SW & AL/N	BOY
	B6.PL(74NS)	C57BL/6	PL/J	SN
Thymus-Leukemia Antigen Locus				
Tla^a	C57BL/6-*Tla*^{a}	C57BL/6JBoy	A/JBoy	BOY
Tla^b	A-*Tla*^{b}	A/JBoy	C57BL/6JBoy	BOY

Contributor: Donald C. Shreffler

Reference: Klein, J. 1973. Transplantation 15:137-153.

In addition to the loci described below, the specificity Lyb-3 also has been recently defined [ref. 12]. It is a B cell marker, present in all strains tested except the mutant CBA/N. **Tissue Distribution**: MuLV = murine leukemia virus. **Functional Distribution**: DTH = delayed type hypersensitivity; T_{DTH} = DTH T cells; T_H = helper T cells; T_K = killer T cells; T_S = suppressor T cells; IgM = immunoglobulin M. *Symbols:* T = thymus or thymus-derived cells; B = bone marrow-derived cells. Data in brackets refer to the column heading in brackets. Synonyms in broken brackets give alternate new nomenclature.

Locus ⟨Synonym⟩ [Chromosome No. (Linkage Group)]	Allele	Specificity	Strain Distribution	Tissue Distribution	Functional Distribution	Reference
Ala-1	*Ala-1*[a]	Ala-1.1	A, BALB/c, CBA/J, CE, C3H/He	T cells: extra-thymic; activated T cells	Activated T cells; T_H; T_K; IgM antibody-forming cells	8
	Ala-1[b]	Ala-1.2	AKR, C57BL/6, C57BR/cd, C57L, C58, DBA/2, SJL, 129			
Gix	......	Gix+	A, AKR, CE, C3H/He, C58, DBA/2, SJL, 129	Thymus; other tissues (age-, strain-, and MuLV-dependent)	....	3, 13, 28
		Gix−	BALB/c, CBA/J, C57BL/6, C57BR/cd, DBA/1, RF			
Lyb-2 [4 (VIII)]	*Lyb-2*[a]	Lyb-2.1	BUB/Bn, CBA/J, $C3H_f$/Bi, C57BR, C57L, C58, DBA/1, DBA/2, HSFS/N (Swiss), I, SWR	B cells	B cells	24
	Lyb-2[b]	Lyb-2.2	A, BALB/c, CBA/H-T6, C3H/An, C57BL/6, HRS, MA/MyJ, RIII/2J, SEC/1ReJ, 129			
	Lyb-2[c]	Lyb-2.3	AKR, BDP, CE, GR/A, RF, SEA/GnJ, SJL			
Ly-1 ⟨*Lyt-1*⟩ [19 (XII)]	*Ly-1*[a]	Ly-1.1	CBA/H-T6, CBA/J, C3H/An, C3H/He, $C3H_f$/Bi, DBA/1, DBA/2, FL/2Re, JK, MA/MyJ	T cells: intra-thymic, extra-thymic	T_H; T_K; T_{DTH}; T_S; precursor cells	2,5-7, 11, 22, 31
	Ly-1[b]	Ly-1.2	A, AKR, BALB/c, BDP, BUB/Bn, CE, C57BL/6, C57BR, C57BR/cd, C57L, C58, DA, GR/A, HRS, HSFS/N (Swiss), I, LP, NZB, PL, RF, RIII/2J, SEA/GnJ, SEC/1ReJ, SJL, SM/J, ST/bJ, STS/A, SWR, WB, 129			
Ly-2 ⟨*Lyt-2*⟩ [6 (XI)]	*Ly-2*[a]	Ly-2.1	AKR, BDP, CBA/H-T6, CBA/J, CE, C3H/An, C3H/He, $C3H_f$/Bi, C58, DBA/1, DBA/2, GR/A, HSFS/N (Swiss), I, JK, MA/MyJ, PL, RF, SM/J	T cells: intra-thymic, extra-thymic	T_K; T_S for antibody formation	2,5, 7
	Ly-2[b]	Ly-2.2	A, BALB/c, BUB/Bn, C57BL/6, C57BR, C57BR/cd, C57L, DA, FL/2Re, HRS, LP, NZB, RIII/2J, SEA/GnJ, SEC/1ReJ, SJL, ST/bJ, STS/A, SWR, WB, 129			
Ly-3 ⟨*Lyt-3*⟩ [6 (XI)[1]]	*Ly-3*[a]	Ly-3.1	AKR, C58, HSFS/N (Swiss), PL, RF	T cells: intra-thymic, extra-thymic	T_K; T_S for antibody formation	2,5, 7
	Ly-3[b]	Ly-3.2	A, BALB/c, BDP, CBA/H-T6, CBA/J, CE, C3H/An, C3H/He, $C3H_f$/Bi, C57BL/6, C57BR, C57BR/cd, C57L, DA, DBA/1, DBA/2, FL/2Re, GR/A, HRS, I, LP, MA/MyJ, NZB, RIII/2J, SEA/GnJ, SEC/1ReJ, SJL, SM/J, ST/bJ, STS/A, SWR, WB, 129			
Ly-4 ⟨*Lyb-1*⟩ [2 (V)[2]]	*Ly-4*[a]	Ly-4.1	A, AKR, BALB/c, BDP, CBA/J, CE, C3H/An, C3H/He, $C3H_f$/Bi, C58, DA, DBA/1, DBA/2, LP, NZB, PL, RF, SJL, SWR, 129	B cells	B cells; antibody-forming cells	15-17, 26
	Ly-4[b]	Ly-4.2	C57BL/6, C57BR, C57BR/cd, C57L, MA/MyJ, WB			

[1] *Ly-2* and *Ly-3* are closely linked. [2] Linked to *H-3*.

continued

Locus ⟨Synonym⟩ [Chromosome No. (Linkage Group)]	Allele	Specificity	Strain Distribution	Tissue Distribution	Functional Distribution	Reference
Ly-5 ⟨*Lyt-4*⟩	*Ly-5*^{a}	Ly-5.1	A, AKR, BALB/c, BDP, BUB/Bn, CBA/H-T6, CBA/J, CE, C3H/An, C3H/He, C3H$_f$/Bi, C57BL/6, C57BR, C57BR/cd, C57L, C58, DBA/1, DBA/2, GR/A, HRS, HSFS/N (Swiss), I, LP, MA/MyJ, NZB, PL, RF, SEA/GnJ, SEC/1ReJ, ST/bJ, SWR, WB, 129	T cells: intrathymic, extrathymic. Other tissues under investigation.	Probably all T cells; some B cells, & possibly other cells	14, 32, 33
	Ly-5^{b}	Ly-5.2	DA, RIII/2J, SJL, STS/A			
Ly-6 ⟨*Lyt-5*⟩	*Ly-6*^{a}	Ly-6.1	A, BALB/c, CBA/J, CE, C3H/An, C3H/He, LP, NZB	T cells: extrathymic. Kidney.	T$_K$	18, 19
	Ly-6^{b}	Ly-6.2	AKR, BDP, C57BL/6, C57BR, C57BR/cd, C57L, C58, DA, DBA/1, DBA/2, MA/MyJ, PL, RF, SJL, SWR, WB, 129			
Ly-7	*Ly-7*^{a}	....	C57BL/6, C58	B cells; ? T cells	B cells; antibody-forming cells	19
	Ly-7^{b}	Ly-7.2	A, AKR, BALB/c, BDP, CBA/J, CE, C3H/An, C3H/He, C57BR, C57BR/cd, C57L, DA, DBA/1, DBA/2, LP, MA/MyJ, NZB, PL, RF, SJL, SWR, WB, 129[3]			
Ly-8	*Ly-8*^{a}	Ly-8.1	A, BALB/c, BDP[3], CBA/J[3], C3H/An, C3H/He, DBA/1[3], PL, RF	T cells; B cells	....	10
	Ly-8^{b}	Ly-8.2	AKR, C57BL/6, C57BR, C57BR/cd, C57L, DBA/2, SJL, SM/J, SWR			
Mls [1 (XIII)]	......	Mls^{++}	AKR, CBA/J, DBA/2	B cells; macrophages	Antibody-forming cells	30
		Mls^{+}	BALB/c, C57BL/6, C57BR, DBA/1, 129			
		Mls^{-}	A, CE, C3H/An, C3H/He, RF, SJL			
Mph-1 [7 (I)]	......	Mph-1^{+}	A, AKR, BALB/c, C3H/He, C57BL/6, DBA/1, DBA/2, RF, 129	Peritoneal exudate cells (? macrophages)	....	1
		Mph-1^{-}	....			
Pca-1	......	Pca-1^{+}	A, AKR, BALB/c, CE, C3H/An, C3H/He, GR/A, HRS, HSFS/N (Swiss), MA/MyJ, NZB, PL, RF, SJL, STS/A, SWR	B cells; plasma cells; myelomas. Brain; liver; kidney.	Antibody-forming cells, not T cells	29
		Pca-1^{-}	C57BL/6, C57BR/cd, C57L, C58, DBA/2, 129			
Qa-1 [17 (IX)]	......	Qa-1^{+}	A, C57BR/cd, C58, PL, SJL	T cells	....	27
		Qa-1^{-}	AKR, BALB/c, C3H/He[3], C57BL/6, DBA/2, RF, 129			
Qa-2 [17 (IX)]	......	Qa-2^{+}	A, BALB/c, C57BL/6, C57L, DBA/1, DBA/2, 129	....	....	9
		Qa-2^{-}	AKR, CBA/J, C3H/He, C57BR/cd, C58, RF			
Thy-1 [9 (II)]	*Thy-1*^{a}	Thy-1.1	AKR, BDP, MA/MyJ, PL, RF, RIII/2J, SEA/GnJ	T cells: intrathymic, extrathymic. Brain; skin.	All T cells	21, 23
	Thy-1^{b}	Thy-1.2	A, BALB/c, CBA/H-T6, CBA/J, CE, C3H/An, C3H/He, C3H$_f$/Bi, C57BL/6, C57BR, C57BR/cd, C57L, C58, DA, DBA/1, DBA/2, FL/2Re, GR/A, HRS, HSFS/N (Swiss), I, LP, NZB, SEC/1ReJ, SJL, SM/J, ST/bJ, STS/A, SWR, WB, 129			

[3] Assignment possible, but not definite.

continued

Locus ⟨Synonym⟩ [Chromosome No. (Linkage Group)]	Allele	Specificity	Strain Distribution	Tissue Distribution	Functional Distribution	Reference
Tla [17 (IX)][4/]	*Tla*[a]	Tla-1,2,3	A, BDP, BUB/Bn, C57BR, C57BR/cd, C58, HSFS/N (Swiss), NZB, PL, SJL, SWR	T cells: intra-thymic. Skin.	Immature T cells	4,20
	Tla[b]	Tla-0	AKR, CBA/H-T6, CBA/J, CE, C3H/An, C3H/He, $C3H_f$/Bi, C57BL/6, DA, DBA/1, FL/2Re, HRS, I, JK, MA/MyJ, RF, RIII/2J, SM/J, ST/bJ, WB			
	Tla[c]	Tla-2	BALB/c, C57L, DBA/2, GR/A, LP, SEA/GnJ, SEC/1ReJ, STS/A, 129			
X-1	......	$X\text{-}1^+$	AKR, C58, NZB, 129	?	....	25
		$X\text{-}1^-$	A, AKR/Cum, BALB/c, C3H/He, C57BL/6, DBA/2, SJL			

4/ Near *H-2D*.

Contributors: Ian F. C. McKenzie and Terry Potter; Fung-Win Shen and Edward A. Boyse

References

1. Archer, J. R., and D. A. L. Davies. 1974. J. Immunogenet. 1:113-123.
2. Beverley, P. C. L., et al. 1976. Nature (London) 262: 495-497.
3. Boyse, E. A. 1977. Immunol. Rev. 33:125-145.
4. Boyse, E. A., and L. J. Old. 1969. Annu. Res. Genet. 3:269-290.
5. Boyse, E. A., et al. 1968. Proc. R. Soc. London B170: 175-193.
6. Cantor, H., and E. A. Boyse. 1975. J. Exp. Med. 141: 1376-1389.
7. Cantor, H., and E. A. Boyse. 1975. Ibid. 141:1390-1399.
8. Feeney, A. J., and V. Hammerling. 1976. Immunogenetics 3:369-379.
9. Flaherty, L. 1976. Ibid. 3:533-539.
10. Frelinger, J. A., and D. B. Murphy. 1976. Ibid. 3:481-487.
11. Huber, B., et al. 1976. J. Exp. Med. 143:1334-1339.
12. Huber, B., et al. 1977. Ibid. 145:10-20.
13. Ikeda, H., et al. 1973. Ibid. 137:1103-1108.
14. Komuro, K., et al. 1975. Immunogenetics 1:452-456.
15. McKenzie, I. F. C. 1975. J. Immunol. 114:856-862.
16. McKenzie, I. F. C., and G. D. Snell. 1975. Ibid. 114: 848-855.
17. McKenzie, I. F. C., et al. 1975. Isr. J. Med. Sci. 11: 1278-1284.
18. McKenzie, I. F. C., et al. 1977. Immunogenetics 5: 25-32.
19. McKenzie, I. F. C., et al. 1977. Transplant. Proc. 9: 667-669.
20. Owen, J. J. T., and M. C. Raff. 1970. J. Exp. Med. 132:1216-1232.
21. Raff, M. C. 1970. Immunology 19:637-650.
22. Ramshaw, I. A. Unpublished. Australian National University, Dep. Microbiology, Canberra, 1977.
23. Reif, A. E., and J. M. V. Allen. 1964. J. Exp. Med. 120:413-433.
24. Sato, H., and E. A. Boyse. 1976. Immunogenetics 3: 565-572.
25. Sato, H., et al. 1973. J. Exp. Med. 138:593-606.
26. Snell, G. D., et al. 1973. Proc. Natl. Acad. Sci. USA 70:1108-1111.
27. Stanton, T. H., and E. A. Boyse. 1976. Immunogenetics 3:525-531.
28. Stockert, E., et al. 1971. J. Exp. Med. 133:1334-1355.
29. Takahashi, T., et al. 1970. Ibid. 131:1325-1341.
30. Tonkonogy, S. L., and H. J. Winn. 1976. J. Immunol. 116:835-841.
31. Vadas, M. A., et al. 1976. J. Exp. Med. 144:10-19.
32. Woody, J., et al. 1977. J. Immunol. 119:1739-1743.
33. Scheid, M. P. Unpublished. Memorial Sloan-Kettering Cancer Center, New York, 1978.

31. HISTOCOMPATIBILITY SYSTEMS, EXCEPT *H-2*: MOUSE

Part I. Standard Loci

Strain includes both background and congenic strains possessing the given allele. Most strains are maintained by The Jackson Laboratory, unless otherwise indicated. For information on holders, *see* reference 16.

continued

31. HISTOCOMPATIBILITY SYSTEMS, EXCEPT *H-2*: MOUSE

Part I. Standard Loci

Locus	Chromosome	Allele	Strain	Reference
H-1	7	*a*	B10.C3H-*H-1*^{a} [1], B10.D2-*H-1*^{a}, C3H/Sn	2,5,6,7,13
		b	B6.C-*H-1*^{b}, B10.A-*H-1*^{b}, B10.AK-*H-1*^{b}, B10.C-*H-1*^{b}, B10.K-*H-1*^{b} [1], B10.129-*H-1*^{b}, C3H.K	
		c	C57BL/6J, C57BL/10Sn	
		?	B10.CE(62NX) [1], B10.P(61NX) [1], B10.WB(66NX) [1]	
H-3	2	*a*	B10-*H-3*^{a} *we*, C57BL/6By, C57BL/10Sn	8,11,14,17
		b	B10-*H-2*^{a} *H-3*^{b} [2], B10-*H-2*^{d} *H-3*^{b} [2], B10-*H-3*^{b} *we*, B10.LP, B10.LP-*a*, B10.VW	
		c	B10.C-*H-3*^{c} [1]	
		d	B10.KR-*H-3*^{d} [1]	
		e	B10-*pa H-3*^{e} *a*^{t} [1]	
H-4	7	*a*	C57BL/6By, C67BL/10Sn	7,13,18
		b	B10-*H-2*^{a} *H-4*^{b} [2], B10-*H-2*^{d} *H-4*^{b} [2], B10.129-*H-4*^{b}	
H-7	9	*a*	C57BL/6By, C57BL/10Sn	2,7,12,19
		b	B6.C-*H-7*^{b}, B10.C-*H-7*^{b}, B10-*H-2*^{a} *H-7*^{b} [2], B10-*H-2*^{d} *H-7*^{b} [2]	
H-8	....	*a*	C57BL/6By, C57BL/10Sn	7,12
		b	B10.D2-*H-8*^{b}	
		c	B6.C-*H-8*^{c}	
H-9	....	*a*	C57BL/6By, C57BL/10Sn	7,12
		b	B10.C-*H-9*^{b}	
H-10	....	*a*	C57BL/10Sn	7,12
		b	B10.129-*H-10*^{b}	
H-11	....	*a*	C57BL/10Sn	7,12
		b	B10.129-*H-11*^{b}	
		?	B10.D2-*H-11?*	
H-12	....	*a*	C57BL/10Sn	7,12,15
		b	B10.129-*H-12*^{b}	
H-13	2	*a*	C57BL/10Sn	5,7,14
		b	B10.CE-*H-13*^{b}, B10.LP, B10.LP-*H-13*^{b} *A*^{w}	
		c	B10.C3H-*H-13*^{c}, B10.KR-*H-13*^{c} *A*^{w} [1]	
H-15	4	*b*	C57BL/6By	2
		c	B6.C-*H-15*^{c}	
H-16	4	*b*	C57BL/6By	2
		c	B6.C-*H-16*^{c}	
H-17	....	*b*	C57BL/6By	2
		c	B6.C-*H-17*^{c}	
H-18	4	*b*	C57BL/6By	2
		c	B6.C-*H-18*^{c}	
H-19	8	*b*	C57BL/6By	2
		c	B6.C-*H-19*^{c}	
H-20	....	*b*	C57BL/6By	2
		c	B6.C-*H-20*^{c}	
H-21	....	*b*	C57BL/6By	2
		c	B6.C-*H-21*^{c}	
H-22	7	*b*	C57BL/6By	2
		c	B6.C-*H-22*^{c}	
H-23	....	*b*	C57BL/6By	2
		c	B6.C-*H-23*^{c}	
H-24	....	*b*	C57BL/6By	2
		c	B6.C-*H-24*^{c}	

[1] Holder: Ralph J. Graff, The Jewish Hospital of St. Louis, St. Louis, Missouri 63110. [2] Holder: Peter Wettstein, Dept. of Microbiology, USC Medical School, Los Angeles, California 90033.

continued

31. HISTOCOMPATIBILITY SYSTEMS, EXCEPT *H-2*: MOUSE

Part I. Standard Loci

Locus	Chromosome	Allele	Strain	Reference
H-25	....	*b*	C57BL/6By	2
		c	B6.C-$H\text{-}25^{c}$	
H-26	....	*b*	C57BL/6By	2
		c	B6.C-$H\text{-}26^{c}$	
H-27	....	*b*	C57BL/6By	2
		c	B6.C-$H\text{-}27^{c}$	
H-28	....	*b*	C57BL/6By	2
		c	B6.C-$H\text{-}28^{c}$	
H-29	....	*b*	C57BL/6By	2
		c	B6.C-$H\text{-}29^{c}$	
H-30	....	*b*	C57BL/6By	2
		c	B6.C-$H\text{-}30^{c}$	
H-31	17	*a*	A/JBoy[3/], C57BL/6JBoy-Tla^{a}[3/]	4
		b	A-Tla^{b}, C57BL/6JBoy[3/]	
		c	B6.AK-$H\text{-}2^{k}$ Tla^{b}[3/]	
H-32	17	*a*	A/JBoy[3/]	4
		b	A/JBoy-Tla^{b}[3/], C57BL/6JBoy[3/], C57BL/6JBoy-Tla^{a}[3/]	
H-33	17	*a*	BALB.TTF-$H\text{-}33^{a}/H\text{-}33^{b}$	3
		b	BALB/cBoy	
H-34	....	*b*	C57BL/6By	2
		c	B6.C-$H\text{-}34^{c}$	
H-35	....	*b*	C57BL/6By	2
		c	B6.C-$H\text{-}35^{c}$	
H-36	....	*b*	C57BL/6By	2
		c	B6.C-$H\text{-}36^{c}$	
H-37	....	*b*	C57BL/6By	2
		c	B6.C-$H\text{-}37^{c}$	
H-38	....	*b*	C57BL/6By	2
		c	B6.C-$H\text{-}38^{c}$	
H-X	X	*b*	C57BL/6By	1,9
		c	BALB/cBy	
		l	LG/Ckc	
H-Y	Y	*a*	A/J	9,10
		b	C57BL/6	

[3/] Holder: Edward Boyse, Memorial Sloan-Kettering Cancer Center, 1275 York Avenue, New York, N.Y. 10021.

Contributor: Ralph J. Graff

References

1. Bailey, D. W. 1963. Transplantation 1:70-74.
2. Bailey, D. W. 1975. Immunogenetics 2:249-256.
3. Flaherty, L. 1975. Ibid. 2:325-329.
4. Flaherty, L., and S. S. Wachtel. 1975. Ibid. 2:81-85.
5. Graff, R. J. Unpublished. The Jewish Hospital of St. Louis, St. Louis, MO, 1977.
6. Graff, R. J., and G. D. Snell. 1968. Transplantation 6:598-617.
7. Graff, R. J., et al. 1966. Ibid. 4:425-437.
8. Graff, R. J., et al. 1973. Transplant. Proc. 5:299-302.
9. Hildemann, W. H. 1970. Transplant. Rev. 3:5-21.
10. Hildemann, W. H., and E. L. Cooper. 1967. Transplantation 5:707-720.
11. Snell, G. D., and H. P. Bunker. 1964. Ibid. 2:743-751.
12. Snell, G. D., and H. P. Bunker. 1965. Ibid. 3:235-252.
13. Snell, G. D., and L. C. Stevens. 1961. Immunology 4: 366-379.
14. Snell, G. D., et al. 1967. Transplantation 5:492-503.
15. Snell, G. D., et al. 1971. Ibid. 11:525-530.
16. Staats, J. 1976. Cancer Res. 36:4333-4377.
17. Wettstein, P. J. Unpublished. Univ. Southern California Medical School, Dep. Microbiology, Los Angeles, 1977.
18. Wettstein, P. J., and G. Haughton. 1974. Transplantation 17:513-517.
19. Wettstein, P. J., and G. Haughton. 1977. Immunogenetics 4:65-78.

continued

31. HISTOCOMPATIBILITY SYSTEMS, EXCEPT *H-2*: MOUSE

Part II. Temporary Designations for New Loci

Eight strains differing from C57BL/6J at various mutant marker genes affecting morphological and behavioral traits have been shown to be histoincompatible with C57BL/6J at histocompatibility loci linked to the marker genes. The *H* genes have been temporarily symbolized by the nearby marker genes. *H(js)*, *H(lt)*, and *H(tn)* have not been proven to be distinct from one another. Likewise, *H(go)* and *H(pi)* have not been proven to be distinct from one another.

Locus	Chromosome	Strain Possessing Allele	
		Background	Congenic
H(Eh)	15	C57BL/6J	C57BL/6-*Eh*
H(ep)	19	C57BL/6J	C57BL/6-*ep*
H(go)	5	C57BL/6J	C57BL/6-*go*
H(js)	11	C57BL/6J	C57BL/6-*js*

Locus	Chromosome	Strain Possessing Allele	
		Background	Congenic
H(ln)	1	C57BL/6J	C57BL/6-*ln*
H(lt)	11	C57BL/6J	C57BL/6-*lt*
H(pi)	5	C57BL/6J	C57BL/6-*pi*
H(tn)	11	C57BL/6J	C57BL/6-*tn*

Contributor: Ralph J. Graff

Reference: Bailey, D. W., and H. P. Bunker. 1972. Mouse News Lett. 47:18.

Part III. Histocompatibility Typing Results

Typing was accomplished by histogenetic techniques, generally by F_1 testing.

Strains	*H-1*	*H-3*	*H-4*	*H-7*	*H-8*	*H-9*	*H-12*	*H-13*
A/He	Not *a* or *c*	Not *a* or *b*	Not *a* or *b*	*b*	Not *a* or *b*	*a*	Not *a* or *b*	*a*
A/Sn	*b*[1/]	Not *a* or *b*	Not *a* or *b*	*b*	Not *a* or *b*	*a*	*a*	*a*
AKR/Sn	*b*[1/]	Not *a* or *b*	*a*	Not *a* or *b*	Not *a* or *b*	Fn[2/]	Not *a* or *b*	*a*
BALB/cJ	*b*[1/]	*c*[1/]	Not *a* or *b*	*b*[1/]	*c*[1/]	*b*	*a*	Not *a* or *b*
BDP/J	*a*	Not *a* or *b*	*b*	*a*	Not *a* or *b*	Not *a* or *b*	*a*	Not *a* or *b*
CBA/Ca	*a*	Not *a* or *b*	*a*	Not *a* or *b*	Not *a* or *b*	*b*	Not *a* or *b*	*a*
CBA/J	*a*	Not *a* or *b*	*a*	Not *a* or *b*	Not *a* or *b*	*b*	Not *a* or *b*	*a*
CE/J	Not *a*, *b*, or *c*[1/]	*b*	Not *a* or *b*	Not *a* or *b*	Not *a* or *b*	*a*	*a*	*b*[1/]
C3H/Bi	*a*	*b*	*a*	Not *a* or *b*	Not *a* or *b*	*a*	*a*	*a*
C3H/Sn	*a*[1/]	*b*	*a*	*b*	Not *a* or *b*	*a*	*a*	*c*[1/]
$C3H_f$/An	*a*	*b*	*a*	*b*	Not *a* or *b*	*b*	Not *a* or *b*	*a*
C57BL/Ks	Not *a* or *c*	*a*	*a*	*a*	Not *a* or *b*	Not *a* or *b*	Not *a* or *b*	*a*
C57BL/6J	*c*	*a*	*a*	*a*	*a*	Not *a* or *b*	*a*	*a*
C57BL/10Sn	*c*[1/]	*a*[1/]	*a*[1/]	*a*[1/]	*a*[1/]	*a*[1/]	*a*[1/]	*a*[1/]
C57BR/cdJ	*c*	*a*	*a*	*a*	*a*	*a*	*a*	*a*
C57L/J	*c*	*a*	*a*	*a*	*a*	*a*	*a*	*a*
C58/J	*c*	*b*	*a*	*a*	*a*	Fn[2/]	*a*	*b*
DA/Hu	Not *a* or *c*	Not *a* or *b*	Not *a* or *b*	*b*	Not *a* or *b*	*b*	Not *a* or *b*	*a*
DBA/1Sn	*a*	*b*	Not *a* or *b*	*a*	*b*	Fn[2/]	*a*	*b*
DBA/2J	*a*[1/]	Not *a* or *b*	Not *a* or *b*	*a*	*b*[1/]	Fn[2/]	*a*	*b*
FL/2Re	Not *a* or *c*	Not *a* or *b*	Not *a* or *b*	*b*	Not *a* or *b*	*a*	*a*	*a*
LP/J	Not *a* or *c*	*b*[1/]	Not *a* or *b*	Not *a* or *b*	*b*	*b*	Not *a* or *b*	*b*[1/]

1/ Indicates typing accomplished by producing congenic strains. 2/ Indeterminate results. Strains type positive for both *a* and *b* alleles.

continued

31. HISTOCOMPATIBILITY SYSTEMS, EXCEPT *H-2*: MOUSE

Part III. Histocompatibility Typing Results

Strains	*H-1*	*H-3*	*H-4*	*H-7*	*H-8*	*H-9*	*H-12*	*H-13*
PL/J	Not *a* or *c*	Not *a* or *b*	Not *a* or *b*	Not *a* or *b*	Not *a* or *b*	*a*	*a*	Not *a* or *b*
RF/J	Not *a* or *c*	Not *a* or *b*	Not *a* or *b*	Not *a* or *b*	Not *a* or *b*	Not *a* or *b*	Not *a* or *b*	*a*
SJL/J	Not *a* or *c*	Not *a* or *b*	*a*	Not *a* or *b*	Not *a* or *b*	Not *a* or *b*	Not *a* or *b*	Not *a* or *b*
SWR/J	Not *a* or *c*	*b*	Not *a* or *b*	Not *a* or *b*	Not *a* or *b*	Fn[2/]	*b*	*a*
WB/Re	Not *a, b,* or *c*[1/]	*a*	Not *a* or *b*	Not *a* or *b*	Not *a* or *b*	*a*	Not *a* or *b*	Not *a* or *b*
WC/Re	*a*(?)	*a*	Not *a* or *b*	Not *a* or *b*	Not *a* or *b*	Not *a* or *b*	Not *a* or *b*	*a*
WK/Re	Not *a* or *c*	Not *a* or *b*	*a*	*a*	Not *a* or *b*	Not *a* or *b*	Not *a* or *b*	*a*
129/Sn	*b*[1/]	*b*[1/]	*b*[1/]	*a*	*b*	Fn[2/]	*b*	*b*[1/]

[1/] Indicates typing accomplished by producing congenic strains. [2/] Indeterminate results. Strains type positive for both *a* and *b* alleles.

Contributor: Ralph J. Graff

Reference: Graff, R. J. 1970. Transplant. Proc. 2:15-23.

32. HISTOCOMPATIBILITY LINKAGE MAP: MOUSE

Chromosome numbers are represented by Arabic numerals at one end of the chromosome, while linkage group numbers are represented by Roman numerals at the opposite end. The smaller Arabic numerals to the left of the chromosome/linkage group symbol represent map distances, in centimorgans, between loci. Brackets indicate that the order within the bracketed group is unknown. Knobs indicate the location of those centromeres that are known.

MARKER GENES

Gene Symbol	Gene Name	Description
a	Non-agouti	Coat color gene
b	Brown	Coat color gene
c	Albino	Coat color gene
Eh	Hairy ears	....
ep	Pale ears	Ears and tail are pale
Es-1	Serum esterase-1	....
Fu	Fused	Shortened and kinked tail
go	Angora	Extra length of guard hairs
Gpd-1	Glucose-6-phosphate dehydrogenase	....
Gpi-1	Glucose phosphate isomerase	....
Hbb	Hemoglobin β-chain	....
js	Jackson shaker	....
ln	Leaden	Coat color gene
p	Pink-eyed dilution	Coat color gene
pa	Pallid	Coat color gene
pi	Pirouette	Degenerative changes in inner ear
tf	Tufted	Repeated waves of hair loss and regrowth
Tla	Thymus-leukemia antigen	....
un	Undulated	Dorsal kyphosis and kinked tail
we	Wellhaarig	Wavy coat

continued

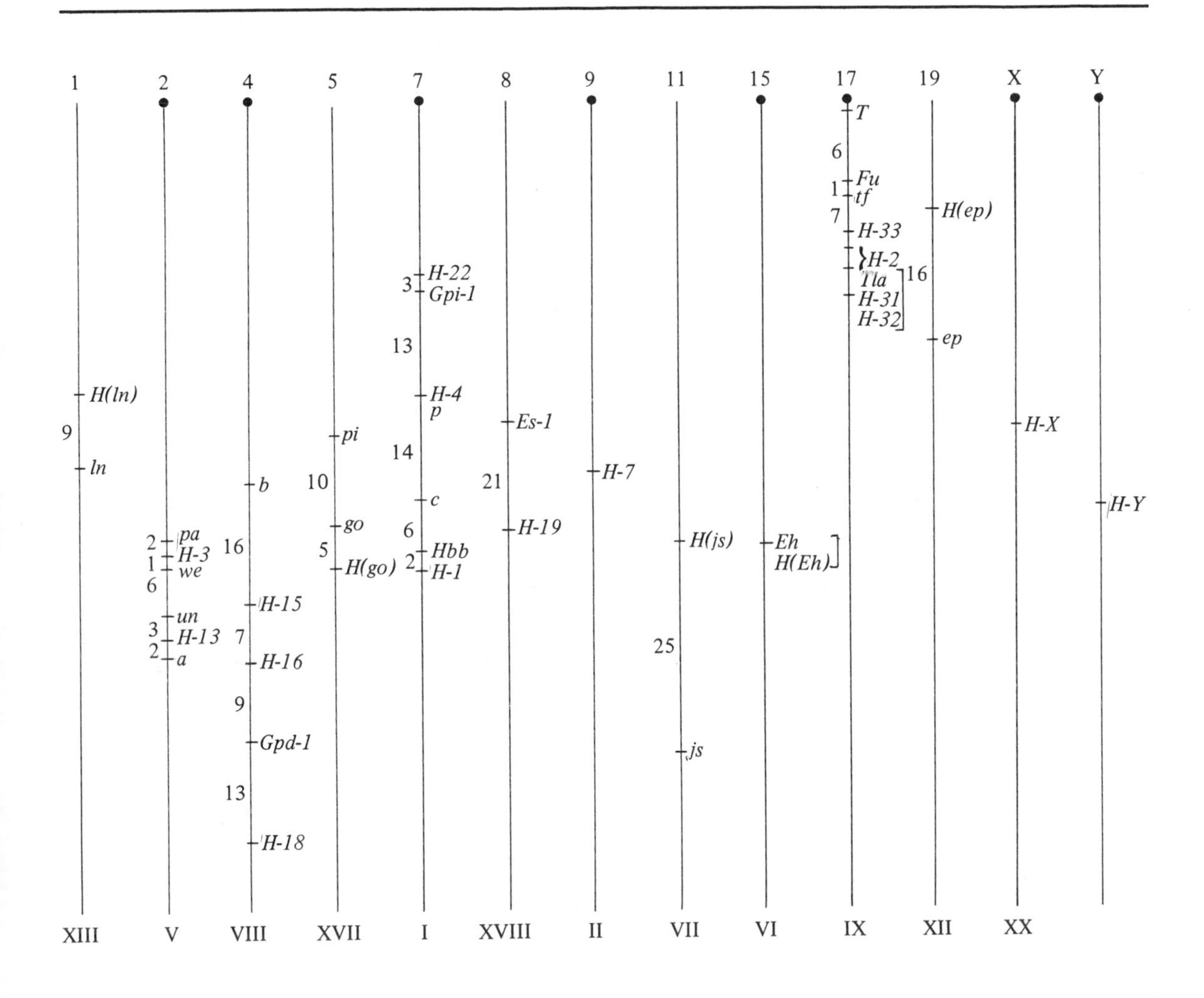

Contributor: Ralph J. Graff

Reference: Womack, J. E. 1976. Mouse News Lett. 55:6.

33. STANDARD INDEPENDENT *H-2* HAPLOTYPES AND MARKERS OF THE *H-2* GENE COMPLEX: MOUSE

Other Major Strains: For further listings, consult reference 2. **Primary Serological Markers:** Markers for *K* and *D* are private *H-2* specificities (e.g., H-2K.33, H-2D.2, etc.); for *G*, the H-2.7 antigen; for *I*, the private Ia specificities (e.g., Ia.20, Ia.11, etc.); for *S*, Ss^h vs. Ss^l, and *Slp+* vs. *Slp−*; for *TL*, the specificities Tla.1, 2, 3; + = presence; − = absence; ? = marker unknown.

Haplotype	Type Strain	B10 Congenic Strain	Other Major Strains	Primary Serological Markers							
				K	*I-A*	*I-C*	*S* *Ss*	*S* *Slp*	*G*	*D*	*TL*
b	C57BL/10ScSn	....	C57BL/6, C57L, 129, LP, C3H.SW, A.BY	33	20	..	h	–	–	2	–

continued

33. STANDARD INDEPENDENT *H-2* HAPLOTYPES AND MARKERS OF THE *H-2* GENE COMPLEX: MOUSE

Haplotype	Type Strain	B10 Congenic Strain	Other Major Strains	Primary Serological Markers							
				K	*I-A*	*I-C*	*S* *Ss*	*S* *Slp*	*G*	*D*	*TL*
d	DBA/2J	B10.D2/nSn	BALB/c, C57BL/Ks, SEA, SEC, YBR, NZB[1]	31	11	23	h	+	–	4	2
f	A.CA/Sn	B10.M/Sn	RFM	26	14	..	h	–	7	9	1, 2, 3
j	JK/St	B10.WB/Sn	C3H.JK, I/St, N/St	15	?	?	h	–	7	2	–
k	C3H/DiSn	B10.BR/Sg[1]	CBA, AKR, CE, C57BR[1], C58[1], FL, CHI, RF, MA, ST, 101	23	2	22	1	–	7[2]	32	–
p	P/J	B10.P/Sg	C3H.NB, BDP	16	?	21	h	+	7	22	?
q	DBA/1J	B10.Q/Sg	C3H.Q, B10.G, SWR[1], AU, BUB	17	10	?	h	–	–	30	–
r	RIII/2J	B10.RIII/Sn	LP.RIII	18	19	7	h	–	–	?	–
s	A.SW/Sn	B10.S/Sg	SJL[1]	19	4	–	h	+	7	12	–
u	PL/J	B10.PL/Sn	....	20	?	?	h	+	–	4	1, 2, 3
v	SM/J	B10.SM/Sn	....	21	?	?	h	–	–	?	–

[1] This strain differs from the Type Strain in Tla Serological Marker. Marker is Tla.1, 2, 3. [2] Weakly positive or negative by hemagglutination, positive by absorption.

Contributor: Donald C. Shreffler

General References

1. David, C. S. 1976. Transplant. Rev. 30:299-322.
2. Klein, J. 1975. Biology of the Mouse Histocompatibility-2 Complex. Springer-Verlag, New York.
3. Shreffler, D. C., and C. S. David. 1975. Adv. Immunol. 20:125-195.
4. Snell, G. D., et al. 1976. Histocompatibility. Academic, New York.

34. *H-2* CONGENIC LINES CARRYING STANDARD HAPLOTYPES: MOUSE

Producer: For full name, *see* list of HOLDERS at front of book.

Differential Locus	Congenic Line ⟨Synonym⟩	Inbred Partner	Donor Strain	Producer
$H\text{-}2^a$	B10.A	C57BL/10Sn	A/WySnSg	SG
	C3H.A/Ha	$C3H_f$/HeHa	A/HeHa	HA
$H\text{-}2^b$	A.BY/Kl	A/WySn	Brachyury	SN-KL
	A.BY/Sn	A/WySn	Brachyury	SN
	AKR.B6/1 ⟨AKR-$H\text{-}2^b$⟩	AKR/JBoy	C57BL/6JBoy (♂)	BOY
	AKR.B6/2 ⟨AKR-$H\text{-}2^b$⟩	AKR/JBoy	C57BL/6JBoy (♀)	BOY
	BALB.B10 ⟨BALB/c-$H\text{-}2^b$⟩	BALB/cJ	C57BL/10Sn	LI
	C3H/Bi-$H\text{-}2^b$	$C3H_f$/Bi	C57BL/6J	LI
	C3H.B10	C3H/JSf	C57BL/10J	SF
	C3H.SW	C3H/HeDiSn	Swiss	SN
	D1.LP	DBA/1J	LP/J	SN
$H\text{-}2^d$	B6.C/a-$H\text{-}2^d$	C57BL/6By	BALB/cBy	BY
	B6.C/b-$H\text{-}2^d$	C57BL/6By	BALB/cBy	BY
	B6.C/c-$H\text{-}2^d$	C57BL/6By	BALB/cBy	BY
	B10.D2/n[1]	C57BL/10Sn	B10.D2/o	SN
	B10.D2/o[2]	C57BL/10Sn	DBA/2	SN
	D1.C	DBA/1J	BALB/c	SN
$H\text{-}2^f$	A.CA/Kl	A/WySn	Caracul	SN-KL
	A.CA/Sn	A/WySn	Caracul non-inbred	SN
	B10.M	C57BL/10Sn	Non-inbred	SN

[1] Carries Hc^1. [2] Carries Hc^0.

continued

34. *H-2* CONGENIC LINES CARRYING STANDARD HAPLOTYPES: MOUSE

Differential Locus	Congenic Line ⟨Synonym⟩	Inbred Partner	Donor Strain	Producer
$H\text{-}2^j$	BALB.I	BALB/cJ	I/St	LI
	C3H.JK	C3H/HeDiSn	JK	SN
$H\text{-}2^{ja}$	B10.WB(69NS)	C57BL/10Sn	WB/Re	SN
$H\text{-}2^k$	BALB.K ⟨BALB.C3H; BALB/c-$H\text{-}2^k$⟩	BALB/cJ	C3H	LI
	B6.C3H ⟨C57BL/6-$H\text{-}2^k$⟩	C57BL/6J	$C3H_f$/Bi	LI
	B10.AKR ⟨C57BL/10Sn-$H\text{-}2^k$⟩	C57BL/10Sn	AKR	LI
	B10.BR	C57BL/10Sn	C57BR/cd	SG
	B10.CBA	C57BL/10SnPh	CBA/Ph	PH
	B10.K	C57BL/10J	CBA/J	SF
	C57BL/6-$H\text{-}2^k$ ⟨B6-$H\text{-}2^k$⟩	C57BL/6JBoy	AKR/JBoy	BOY
$H\text{-}2^n$	B10.F/Ao	C57BL/HaAo	F/StAo	AO
	B10.F/Eg	C57BL/10SnEg	F/St	EG
	B10.F/Sg	C57BL/10Sn	F/St	SG
	B10.F/Y	C57BL/10Y	F/St	Y
$H\text{-}2^p$	B10.CNB	C57BL/10SnEg	C3H.NB	EG
	B10.NB	C57BL/10SnPh	C3H.NB	PH
	B10.P	C57BL/10Sn	P/J	SG
	C3H.NB	C3H/HeDiSn	NB	SN
$H\text{-}2^{pa}$	B10.Y	C57BL/10Sn	Non-inbred	SN
$H\text{-}2^q$	B10.D1/Ph	C57BL/10SnPh	DBA/1	PH
	B10.D1/Y	C57BL/10Y	DBA/1	Y
	B10.G	C57BL/10SnSg	Gray-lethal linkage testing stock	SG
	B10.Q	C57BL/10SnSg	DBA/1	SG
	C3H.Q	C3H/JSf	STOLI/Lw	SF
$H\text{-}2^r$	B10.RIII	C57BL/10SnSg	RIII/WyJ	SG
	B10.RIII(7NS)	C57BL/10Sn	RIII/WyJ	SN
	LP.RIII	LP/J	RIII/WyJ	SN
$H\text{-}2^s$	A.SW/Kl	A/WySn	Swiss	SN-KL
	A.SW/Sn	A/WySn	Swiss	SN
	B10.ASW	C57BL/10SnPh	A.SW	PH
	B10.S	C57BL/10Sn	A.SW	SG
$H\text{-}2^u$	B10.PL	C57BL/10SnSg	PL/J	SG
	B10.PL(73NS)	C57BL/10SnSg	PL/J	SN
$H\text{-}2^v$	B10.SM/Sg	C57BL/10Sn	SM/J	SG
	B10.SM/Sn	C57BL/10Sn	SM/J	SN

Contributor: Donald C. Shreffler

General References

1. Klein, J. 1973. Transplantation 15:137-153.
2. Shreffler, D. C. Unpublished. Washington Univ. School of Medicine, St. Louis, 1978.

35. COMPOSITION OF RECOMBINANT HAPLOTYPES OF THE *H-2* GENE COMPLEX: MOUSE

Recombinant Haplotype: Haplotype symbols conform to the rules established by Klein, et ai., in 1974. For previous symbols, consult reference 10. **Haplotype of Origin:** Lower case letters indicate haplotype of origin of each region or

continued

subregion; vertical bar indicates crossover position; "?" indicates that origin of region is not determined. **Tla Types:** *a* = Tla.1, 2, 3; *b* = Tla−; *c* = Tla.2. Types in parentheses are inferred from *D*-end parental type. "?" indicates unknown.

<table>
<tr><th>Recombinant Haplotype</th><th>Parental Combination</th><th>Type Strain</th><th>Haplotype of Origin
K A B J E C S G D</th><th>Tla Type</th><th>Reference</th></tr>
<tr><td>a[1]</td><td>d/k</td><td>B10.A, A</td><td>k k k k k | d d d d</td><td>a</td><td>11,22</td></tr>
<tr><td>al</td><td>d/k</td><td>B10.AL, A.AL</td><td>k k k k k k k k | d</td><td>c</td><td>20</td></tr>
<tr><td>o1</td><td>d/k</td><td>B10.OL, C3H.OL</td><td>d d d d d d | k k k</td><td>b</td><td>20</td></tr>
<tr><td>o2</td><td>d/k</td><td>B10.OH, C3H.OH</td><td>d d d d d d d d | k</td><td>b</td><td>21</td></tr>
<tr><td>g</td><td>b/d</td><td>B10.HTG, HTG</td><td>d d d d d d d | ? | b</td><td>b</td><td>7</td></tr>
<tr><td>g1</td><td>b/d</td><td>B10.D2(R101)</td><td>d d d d d d d | ? | b</td><td>(b)</td><td>28</td></tr>
<tr><td>g2</td><td>b/d</td><td>B10.GD, D2.GD</td><td>d d d | b b b b b b</td><td>(b)</td><td>13</td></tr>
<tr><td>g3</td><td>b/d</td><td>B10.D2(R103)</td><td>d d d d d d d | ? | b</td><td>(b)</td><td>28</td></tr>
<tr><td>g4</td><td>b/d</td><td>B10.BDR1</td><td>d d d d d d d | ? | b</td><td>(b)</td><td>16</td></tr>
<tr><td>g5</td><td>b/d</td><td>B10.BDR2</td><td>d d d d d d d | ? | b</td><td>(b)</td><td>16</td></tr>
<tr><td>g6</td><td>b/d</td><td>B6.C-$H\text{-}2^{g6}$</td><td>d d d d d d d | ? | b</td><td>(b)</td><td>2</td></tr>
<tr><td>h</td><td>a/b</td><td>HTH</td><td>k k k k k d d | ? | b</td><td>(b)</td><td>7</td></tr>
<tr><td>h1</td><td>a/b</td><td>B10.A(1R)</td><td>k k k k k d d | ? | b</td><td>b</td><td>25</td></tr>
<tr><td>h2</td><td>a/b</td><td>B10.A(2R)</td><td>k k k k k d d | ? | b</td><td>b</td><td>25</td></tr>
<tr><td>h3</td><td>b/k</td><td>B10.AM</td><td>k k k k k k k k | b</td><td>?</td><td>1</td></tr>
<tr><td>h4</td><td>a/b</td><td>B10.A(4R)</td><td>k k | b b b b b b b</td><td>(b)</td><td>25</td></tr>
<tr><td>h15</td><td>a/b</td><td>B10.A(15R)</td><td>k k k k k d d | ? | b</td><td>(b)</td><td>27</td></tr>
<tr><td>i</td><td>a/b</td><td>HTI</td><td>b b b b b b b | ? | d</td><td>a</td><td>7</td></tr>
<tr><td>i3</td><td>a/b</td><td>B10.A(3R)</td><td>b b b b | k d d d d</td><td>(a)</td><td>25</td></tr>
<tr><td>i5</td><td>a/b</td><td>B10.A(5R)</td><td>b b b | k k d d d d</td><td>a</td><td>25</td></tr>
<tr><td>i7</td><td>b/d</td><td>B10.D2(R107)</td><td>b b b b b b b | ? | d</td><td>(c)</td><td>28</td></tr>
<tr><td>i8</td><td>b/t1</td><td>A.BTR1</td><td>b b b b b b b b | d</td><td>(c)</td><td>17</td></tr>
<tr><td>i9</td><td>b/t1</td><td>A.BTR2</td><td>b b b b b b b | ? | d</td><td>(c)</td><td>17</td></tr>
<tr><td>i18</td><td>a/b</td><td>B10.A(18R)</td><td>b b b b b b b | ? | d</td><td>(a)</td><td>24</td></tr>
<tr><td>i21</td><td>b/t2</td><td>B10.S(21R)</td><td>b b b b b b b b | d</td><td>(a)</td><td>24</td></tr>
<tr><td>m[1]</td><td>k/q</td><td>B10.AKM, AKR.M</td><td>k k k k k k k k | q</td><td>a</td><td>11,12</td></tr>
<tr><td>m1</td><td>a/q</td><td>B10.QAR</td><td>k | ? ? ? ? ? | q q q</td><td>?</td><td>10</td></tr>
<tr><td>t1</td><td>a1/s</td><td>B10.TL, A.TL</td><td>s | k k k k k k k d</td><td>c</td><td>6,20</td></tr>
<tr><td>t2</td><td>a/s</td><td>B10.S(7R), A.TH</td><td>s s s s s s s s | d</td><td>a</td><td>5,6,26</td></tr>
<tr><td>t3</td><td>t1/s</td><td>B10.HTT</td><td>s s s s | k k k k d</td><td>c</td><td>6,15</td></tr>
<tr><td>t4</td><td>a/s</td><td>B10.S(9R)</td><td>s s | ? | k k d d d d</td><td>a</td><td>6,26</td></tr>
<tr><td>t5[1]</td><td>a/s</td><td>B10.BSVS, BSVS</td><td>s s s s s s | d d d</td><td>?</td><td>6,18</td></tr>
<tr><td>t6</td><td>a/s</td><td>B10.S(24R)</td><td>s s s s s s s s | d</td><td>(a)</td><td>24</td></tr>
<tr><td>y1</td><td>a/q</td><td>AQR</td><td>q | k k k k d d d d</td><td>a</td><td>12</td></tr>
<tr><td>y2</td><td>a/q</td><td>B10.T(6R)</td><td>q q q q q q q | ? | d</td><td>a</td><td>26</td></tr>
<tr><td>an1</td><td>t1/f</td><td>A.TFR1</td><td>s k k k k k k | f f</td><td>a</td><td>19</td></tr>
<tr><td>ap1</td><td>a/f</td><td>B10.M(11R)</td><td>f f f f f f f f | d</td><td>(a)</td><td>27</td></tr>
<tr><td>ap2</td><td>t2/f</td><td>A.TFR2</td><td>f f f f f f f | ? | d</td><td>a</td><td>19</td></tr>
<tr><td>ap3</td><td>t2/f</td><td>A. TFR3</td><td>f f | ? ? ? ? | s s d</td><td>a</td><td>19</td></tr>
<tr><td>ap4</td><td>t2/f</td><td>A.TFR4</td><td>f f | ? ? ? ? | s s d</td><td>a</td><td>19</td></tr>
<tr><td>ap5</td><td>t1/f</td><td>A.TFR5</td><td>f f | ? ? | k k k k d</td><td>a</td><td>19</td></tr>
<tr><td>aq1</td><td>a/f</td><td>B10.M(17R)</td><td>k k k k k d d d | f</td><td>(a)</td><td>27</td></tr>
<tr><td>ar1[1]</td><td>d/f/?</td><td>B10.LG, LG</td><td>d f f f f f f f ?</td><td>?</td><td>3,4</td></tr>
<tr><td>as1</td><td>a/s</td><td>B10.S(8R)</td><td>k k | ? ? ? ? | s s s</td><td>(b)</td><td>4,26</td></tr>
<tr><td>at1</td><td>t1/b</td><td>A.TBR1</td><td>s k k k k k k k | b</td><td>(b)</td><td>17</td></tr>
<tr><td>by1</td><td>b/y1</td><td>B10.BYR</td><td>q k k k k d d | ? | b</td><td>?</td><td>10</td></tr>
<tr><td>ia1</td><td>b/da</td><td>B10.D2(R106)</td><td>b b b b b b b | ? | da</td><td>(c)</td><td>28</td></tr>
<tr><td>qp1[1]</td><td>q/s</td><td>B10.DA, DA</td><td>q q q q q q q q | s</td><td>?</td><td>23</td></tr>
</table>

[1] Postulated recombinant type—not directly observed as recombinant.

continued

35. COMPOSITION OF RECOMBINANT HAPLOTYPES OF THE *H-2* GENE COMPLEX: MOUSE

Recombinant Haplotype	Parental Combination	Type Strain	Haplotype of Origin *K*	*A*	*B*	*J*	*E*	*C*	*S*	*G*	*D*	Tla Type	Reference
sq2	*q/s*	B10.QSR2	*s*	*s*	?	?	?	?	*q*	*q*	*q*	?	14
wr7	*w7/k*	B10.WR7	*w*	*w*	*w*	*w*	*w*	*w*	*w*	*w*	*k*	*a*	8
w8	*w/k*	B10.SNA70	*k*	*k*	?	?	?	?	*w*	*w*	*w*	?	9

Contributor: Donald C. Shreffler

References

1. Amos, D. B. Unpublished. Div. of Immunology, Duke Univ. Medical Center, Durham, NC, 1977.
2. Bailey, D. W. Unpublished. Jackson Laboratory, Bar Harbor, ME, 1977.
3. Cherry, M., and G. D. Snell. Unpublished. Jackson Laboratory, Bar Harbor, ME, 1977.
4. David, C. S. Unpublished. Dep. of Genetics, Washington Univ. School of Medicine, St. Louis, MO, 1977.
5. David, C. S., and D. C. Shreffler. 1972. Tissue Antigens 2:241-249.
6. David, C. S., et al. 1975. Ibid. 6:353-365.
7. Gorer, P. A., and Z. B. Mikulska. 1959. Proc. R. Soc. London B151:57-69.
8. Hansen, T. H., and D. C. Shreffler. 1976. J. Immunol. 117:1507-1513.
9. Hauptfeld, M., and J. Klein. 1976. Immunogenetics 3:603-607.
10. Klein, J. 1975. Biology of the Mouse Histocompatibility-2 Complex. Springer-Verlag, New York. pp. 210-211.
11. Klein, J., and D. C. Shreffler. 1971. Transplant. Rev. 6:3-29.
12. Klein, J., et al. 1970. Transplantation 10:309-320.
13. Lilly, F., and J. Klein. 1973. Ibid. 16:530-532.
14. McDevitt, H. O., et al. 1972. J. Exp. Med. 135:1259-1278.
15. Meo, T., et al. 1973. Transplant. Proc. 5:1507-1510.
16. Passmore, H. C. 1970. Ph.D. Thesis. Univ. Michigan, Ann Harbor.
17. Passmore, H. C., and K. W. Beisel. 1977. Immunogenetics 4(4):393-399.
18. Rose, N. R., et al. 1973. Clin. Exp. Immunol. 15:281-287.
19. Shreffler, D. C. Unpublished. Dep. of Genetics, Washington Univ. School of Medicine, St. Louis, MO, 1977.
20. Shreffler, D. C., and C. S. David. 1972. Tissue Antigens 2:232-240.
21. Shreffler, D. C., et al. 1966. Transplantation 4:300-322.
22. Snell, G. D. 1951. J. Natl. Cancer Inst. 11:1299-1305.
23. Snell, G. D. Unpublished. 21 Atlantic Ave., Bar Harbor, ME, 1977.
24. Stimpfling, J. H. Unpublished. McLaughlin Research Institute, Columbus Hospital, Great Falls, MT, 1977.
25. Stimpfling, J. H., and A. Richardson. 1965. Genetics 51:831-846.
26. Stimpfling, J. H., and A. E. Reichert. 1970. Transplant. Proc. 2:39-47.
27. Stimpfling, J. H., et al. 1971. In A. Lengerova and M. Vojtiskova, ed. Immunogenetics of the H-2 System. Karger, Basel. pp. 10-11.
28. Vedernikov, A. A., and I. K. Egorov. 1973. Genetika 9:60-66.

36. *H-2* CONGENIC LINES CARRYING RECOMBINANT HAPLOTYPES: MOUSE

Producer: For full name, *see* list of HOLDERS at front of book.

Differential Locus	Congenic Line	Inbred Partner	Donor Strain	Producer
$H\text{-}2^{al}$	A.AL	A/SnSf	DBA/2J & C3H/J	SF
	B10.AL	C57BL/10SnSf	A.AL	SF
$H\text{-}2^{anl}$	A.TFR1	A/SnSf	A.CA & A.TL	SF
$H\text{-}2^{apl}$	B10.M(11R)	C57BL/10SnSg	B10.A & B10.M	SG
$H\text{-}2^{ap2}$	A.TFR2	A/SnSf	A.CA & A.TH	SF
$H\text{-}2^{ap3}$	A.TFR3	A/SnSf	A.CA & A.TH	SF
$H\text{-}2^{ap4}$	A.TFR4	A/SnSf	A.CA & A.TL	SF
$H\text{-}2^{ap5}$	A.TFR5	A/SnSf	A.CA & A.TL	SF
$H\text{-}2^{aql}$	B10.M(17R)	C57BL/10SnSg	B10.A & B10.M	SG

continued

36. *H-2* CONGENIC LINES CARRYING RECOMBINANT HAPLOTYPES: MOUSE

Differential Locus	Congenic Line	Inbred Partner	Donor Strain	Producer
$H\text{-}2^{ar1}$	B10.LG	C57BL/10SnSf	LG/Ckc	SF
$H\text{-}2^{as1}$	B10.S(8R)	C57BL/10SnSf	A.SW	SG
$H\text{-}2^{at1}$	A.TBR1	A/WySn	A.By & A.TL	PAS
$H\text{-}2^{by1}$	B10.BYR	C57BL/10SnKlj	B10.AQR	KLJ
$H\text{-}2^{g}$	BALB.HTG	BALB/cJ	HTG	LI
	B10.HTG	C57BL/10Sn	HTG	LI
	C3H.HTG(82NS)	C3H/HeDiSn	HTG	SN
$H\text{-}2^{g1}$	B10.D2(R101)	C57BL/10SnEg	B10.D2	EG
$H\text{-}2^{g2}$	D2.GD	DBA/2J	C57BL/6J	LI
	B10.GD	C57BL/10SnSf	D2.GD	SF
$H\text{-}2^{g3}$	B10.D2(R103)	C57BL/SnEg	B10.D2(M504)	EG
$H\text{-}2^{g4}$	B10.BDR1	C57BL/10SnSf	C57BL/6J & DBA/2J	SF
$H\text{-}2^{g5}$	B10.BDR2	C57BL/10SnSf	C57BL/6J & DBA/2J	SF
$H\text{-}2^{g6}$	B6.C-$H\text{-}2^{g6}$	C57BL/6By	BALB/cBy	BY
$H\text{-}2^{h1}$	B10.A(1R)	C57BL/10SnSg	A/WySn	SG
$H\text{-}2^{h2}$	B10.A(2R)	C57BL/10SnSg	A/WySn	SG
$H\text{-}2^{h3}$	B10.AM	C57BL/10SnSg	C57BL/6 & C3H	SG
$H\text{-}2^{h4}$	B10.A(4R)	C57BL/10SnSg	A/WySn	SG
$H\text{-}2^{h15}$	B10.A(15R)	C57BL/10SnSg	B10.A	SG
$H\text{-}2^{ia1}$	B10.D2(R106)	C57BL/10SnEg	B10.D2(M504)	EG
$H\text{-}2^{i3}$	B10.A(3R)	C57BL/10SnSg	A/WySn	SG
$H\text{-}2^{i5}$	B10.A(5R)	C57BL/10SnSg	A/WySn	SG
$H\text{-}2^{i7}$	B10.D2(R107)	C57BL/10SnEg	B10.D2	EG
$H\text{-}2^{i8}$	A.BTR1	A/WySn	A.BY & A.TL	PAS
$H\text{-}2^{i9}$	A.BTR2	A/WySn	A.BY & A.TL	PAS
$H\text{-}2^{i18}$	B10.A(18R)	C57BL/10SnSg	B10.A	SG
$H\text{-}2^{i21}$	B10.S(21R)	C57BL/10SnSg	B10.S(7R)	SG
$H\text{-}2^{m}$	AKR.M	AKR/J	Non-inbred	SN
	B10.AKM	C57BL/10SnSg	AKR.M	SG
$H\text{-}2^{m1}$	B10.QAR	C57BL/10SnKlj	B10.A & B10.G	KLJ
$H\text{-}2^{o1}$	C3H.OL	C3H/JSf	DBA/2J & C3H/J	SF
	B10.OL	C57BL/10SnSf	C3H.OL	SF
$H\text{-}2^{o2}$	C3H.OH/Sn	C3H/DiSn	DBA/2 & C3H/J	SF-SN
	C3H.OH/Sf	C3H/JSf	DBA/2 & C3H/J	SF
	B10.OH	C57BL/10SnSf	C3H.OH	SF
$H\text{-}2^{qp1}$	B10.DA(80NS)	C57BL/10Sn	DA/HuSn	SN
$H\text{-}2^{sq2}$	B10.QSR2	C57BL/10Sn	A.SW & non-inbred	MCD-SF
$H\text{-}2^{t1}$	A.TL	A/SnSf	A.SW & A.AL	SF
	B10.TL	C57BL/10SnSf	A.TL	SF
$H\text{-}2^{t2}$	B10.S(7R)	C57BL/10Sn	A.SW & B10.A	SG
	A.TH	A/SnSf	B10.S(7R)	SF
$H\text{-}2^{t3}$	B10.HTT	C57BL/10SnPh	HTT-non-inbred line carrying $H\text{-}2^{t1}$	PH
$H\text{-}2^{t4}$	B10.S(9R)	C57BL/10SnSg	A.SW & B10.A	SG
$H\text{-}2^{t5}$	B10.BSVS	C57BL/10SnSf	BSVS	SF
$H\text{-}2^{t6}$	B10.S(24R)	C57BL/10SnSg	B10.A & B10.S	SG
$H\text{-}2^{wr7}$	B10.WR7	C57BL/10SnKlj	B10.BR & wild mouse	KLJ-SF
$H\text{-}2^{w8}$	B10.SNA70	C57BL/10SnKlj	B10.BR & wild mouse	KLJ
$H\text{-}2^{y1}$	B10.AQR	C57BL/10SnKlj	B10.A & T138	KLJ
$H\text{-}2^{y2}$	B10.T(6R)	C57BL/10SnSg	Gray-lethal linkage testing stock & strain A	SG
$H\text{-}2^{?}$	B10.P(10R)	C57BL/10SnSg	P/J	SG
	B10.F(13R)	C57BL/10SnSg	F/St	SG
	B10.F(14R)	C57BL/10SnSg	F/St	SG

Contributor: Donald C. Shreffler

General References

1. Klein, J. 1973. Transplantation 15:137-153.
2. Klein, J. 1975. Biology of the Mouse Histocompatibility-2 Complex. Springer-Verlag, New York.
3. Shreffler, D. C. Unpublished. Dep. of Genetics, Washington Univ. School of Medicine, St. Louis, MO, 1977.

Mutant Line & Haplotype: The **New** designations for mutant lines with haplotypes are those adopted at the May 1978 *H-2* Mutant Workshop. The **Old** designations are those previously used for haplotypes. Unofficial designations of mutant haplotypes are given as synonyms. **Type:** GL = gain + loss (new antigenic determinant appears with concomitant loss of normally occurring determinant); L = loss (antigenic determinant is lost without appearance of new determinant). **Location:** Each region is presently defined by marker loci (*D* region = *H-2D* and *H-2L*; *K* region = *H-2K*). The hypothetical *Z1* and *Z2* loci were introduced at a time when there was uncertainty or a difference of opinion as to whether a mutation affected the marker locus or a hypothetical, closely linked locus in the same region. These are indicated where such references exist in the literature. All mutations marked by a single asterisk are non-complementary (allelic) with one another; those with two asterisks are allelic with one another; unmapped mutations are indicated by "·." The apparent conflict between mutations *H-2*^{dm1} and *H-2*^{dm2} which appear to involve different loci, yet are non-complementary, may indicate that one of them involves more than one locus. **Reactions Affected:** GR = graft rejection; MLR = mixed lymphocyte reaction; CML = cell-mediated lympholysis; GVHR = graft vs. host response; HR = humoral response, a qualitative change in antigen detected by serological testing; "+" = affected; "–" = not affected; "·" = not determined; "?" = quantitative changes only or conflicting reports. **Source:** laboratory where mutation was recovered and mutant line developed. For full name of laboratory, *see* list of HOLDERS at front of book.

Mutant Line & Haplotype		Inbred Partner	Donor Strain	Type	Location		Reactions Affected					Source	Reference
New	Old (Synonym)				Region	Locus	GR	MLR	CML	GVHR	HR		
B6.C-*H-2*^{bm1}	*H-2*^{ba} (H(z1))	C57BL/6By	BALB/cBy	GL	*K* or *I-A*	*H-2K* or *Z1**	+	+	+	+	?	BY	3,4, 12
B6.C-*H-2*^{bm2}	*H-2*^{bb} (H(z49))	C57BL/6By	BALB/cBy	GL	·	·	+	·	·	·	·	BY	11
B6-*H-2*^{bm3}	*H-2*^{bd} (M505)	C57BL/6YEg	....	GL	*K* or *I-A*	*H-2K* or *Z1**	+	+	+	+	?	EG	1,11, 12
B6.C-*H-2*^{bm4}	*H-2*^{bf} (H(z170))	C57BL/6By	BALB/cBy	GL	*K* or *I-A*	*H-2K* or *Z1**	+	·	·	·	·	BY	2,24
B6-*H-2*^{bm5}	*H-2*^{bg1}	C57BL/6Kh	....	GL	*K* or *I-A*	*H-2K* or *Z1**	+	+	+	·	–	KH	16, 19
B6-*H-2*^{bm6}	*H-2*^{bg2}	C57BL/6Kh	....	GL	*K* or *I-A*	*H-2K* or *Z1**	+	+	+	·	–	KH	16, 19
B6.C-*H-2*^{bm7}	*H-2*^{bg3}	C57BL/6Kh	BALB/cKh	GL	*K* or *I-A*	*H-2K* or *Z1**	+	·	·	·	·	KH	20
B6-*H-2*^{bm8}	*H-2*^{bh}	C57BL/6Kh	....	GL	*K* or *I-A*	*H-2K* or *Z1**	+	+	+	·	–	KH	16, 19
B6.C-*H-2*^{bm9}	*H-2*^{bi}	C57BL/6Kh	BALB/cKh	GL	*K* or *I-A*	*H-2K* or *Z1**	+	·	·	·	·	KH	20
B6.C-*H-2*^{bm10}	*H-2*^{bj}	C57BL/6Kh	BALB/cKh	GL	*K* or *I-A*	*H-2K* or *Z1**	+	·	·	·	·	KH	20
B6.C-*H-2*^{bm11}	*H-2*^{bk}	C57BL/6Kh	BALB/cKh	GL	*K* or *I-A*	*H-2K* or *Z1**	+	·	·	·	·	KH	20
B6.C-*H-2*^{bm12}	*H-2*^{bm}	C57BL/6Kh	BALB/cKh	GL	·	·	+	·	·	·	·	KH	15, 18
B6.C-*H-2*^{bm13}	*H-2*^{bn}	C57BL/6Kh	BALB/cKh	GL	·	·	+	·	·	·	·	KH	20
B6.C-*H-2*^{bm14}	*H-2*^{bo}	C57BL/6By	BALB/cBy	GL	*D*-end	·	+	–	–	·	–	BY	18
B10-*H-2*^{bm15}	.... (M513)	C57BL/10ScSnEg	....	L	·	·	+	·	·	·	·	EG	7,11
B10.D2-*H-2*^{dm1} 1/	*H-2*^{da} (M504)	C57BL/10SnEg	B10.D2	GL	*D*	*H-2D* or *Z2***	+	+	+	+	+	EG	6,8, 14
C-*H-2*^{dm2}	*H-2*^{db}	BALB/cKh	....	L	*D*	*H-2L* or *Z2***	+	+	+	·	+	KH	9,17, 20
B6.C-*H-2*^{dm3}	*H-2*^{dc}	C57BL/6Kh	BALB/cKh	L	·	·	+	·	·	·	·	KH	20

1/ The *H-2*^{dm1} mutation is also carried by the recombinant line R106 (haplotype *H-2*^{ia}) [ref. 6, 23].

continued

Mutant Line & Haplotype		Inbred Partner	Donor Strain	Type	Location		Reactions Affected					Source	Reference
New	Old (Synonym)				Region	Locus	GR	MLR	CML	GVHR	HR		
C.B6-*H-2^{dm4}*	*H-2^{dd}*	BALB/cKh	C57BL/6Kh	GL	•	•	+	•	•	•	•	KH	20
A.CA-*H-2^{fm1}*	*H-2^{fa}* (M506)	A/SnEg	A.CA	GL	*K*	*H-2K*	+	+	+	+	+	EG	13, 14, 21
B10.M-*H-2^{fm2}*	*H-2^{fb}*	C57BL/10Sn	B10.M	GL	•	•	+	•	•	•	•	MOB & BY	11, 22
CBA-*H-2^{km1}*	*H-2^{ka}* (M523)	CBA/CaLacSto	....	GL	*K*	*H-2K*	+	+	+	+	+	EG	5,10, 14

Contributor: Roger W. Melvold

References

1. Apt, A. S., et al. 1975. Immunogenetics 1:444-451.
2. Bailey, D. W. 1974. Mouse News Lett. 51:9.
3. Bailey, D. W., et al. 1971. Symp. Immunogenet. H-2 Syst. (Liblice, Czechoslovakia), pp. 155-165.

continued

38. WILD-DERIVED HAPLOTYPES, CONGENIC STRAINS,

Part I. B10.W Congenic

Line	Generation	Wild Mouse Origin
B10.BUA1	N12F4	Butcher farm, 1651 Knight Rd., Ann Arbor, Michigan (barn)
B10.CAA2	N10F3	Ribblet house, Menlo Park, California (garage)
B10.CAS2	N10F4	*Mus musculus castaneus* from Michael Potter
B10.CHR51	N12F4	CH farm, Wagner Rd., Ann Arbor, Michigan (granary)
B10.KPA42	N9F7	Kapp farm, 4330 Nixon Rd., Ann Arbor, Michigan (granary)
B10.KPA132	N8F5	Kapp farm, 4330 Nixon Rd., Ann Arbor, Michigan (granary)
B10.KPB128	N9F4	Kapp farm, 4330 Nixon Rd., Ann Arbor, Michigan (corn crib)
B10.LIB18	N8F4	Ks farm, Liberty Rd., Ann Arbor, Michigan (granary)

continued

Part II. *H-2* Chart of B10.W

Symbols: – = antigen absent; • = presence of antigen

Type Strain	*H-2* Haplotype	H-2 Antigenic														
		2	9	19	21	23	26	31	32	102	103	104	105	106	107	108
B10.KPA42	*w1*	–	–	–	21	–	–	–	–	–	–	–	–	–	–	–
B10.KPB68	*w2*	–	–	–	–	–	–	–	–	102	–	–	–	•	•	•
B10.SAA48	*w3*	–	–	–	–	–	–	–	–	–	103	–	–	–	–	–
B10.GAA20	*w4*	–	–	–	–	–	–	–	–	–	–	104	–	•	•	–
B10.KEA5	*w5*	–	–	–	–	–	–	31	–	–	–	–	105	–	107	–
B10.RB7	*w6*	–	–	–	–	–	–	–	–	–	–	–	–	•	•	•
B10.WOA1	*w7*	–	–	–	–	–	–	–	–	–	–	–	–	•	•	•

4. Berke, G., and D. B. Amos. 1973. Nature (London) New Biol. 242:237-238.
5. Blandova, Z., et al. 1975. Immunogenetics 2:291-295.
6. Dishkant, I. P., et al. 1973. Genetika 9(2):83-87.
7. Egorov, I. K., and Z. Blandova. 1972. Genet. Res. 19: 133-143.
8. Forman, J., and J. Klein. 1975. J. Immunol. 115:711-715.
9. Hansen, T. H., and R. B. Levy. 1978. Ibid. 120:1836-1840.
10. Klein, J. 1975. Ibid. 115:716-718.
11. Klein, J. 1975. Biology of the Mouse Histocompatibility-2 Complex. Springer-Verlag, New York.
12. Klein, J., et al. 1974. J. Exp. Med. 140:1127-1132.
13. Klein, J., et al. 1976. Scand. J. Immunol. 5:521-528.
14. Klein, J., et al. 1976. Transplantation 22:572-582.
15. Kohn, H. I. 1975. Mouse News Lett. 52:28.
16. McKenzie, I. F. C., et al. 1976. Immunogenetics 3: 241-251.
17. McKenzie, I. F. C., et al. 1977. Ibid. 4:333-347.
18. McKenzie, I. F. C., et al. 1977. Immunol. Rev. 35: 181-230.
19. Melief, C. J. M., et al. 1975. Immunogenetics 2:337-348.
20. Melvold, R. W., and H. I. Kohn. 1976. Ibid. 3:185-191.
21. Mnatsakanian, Yu.A., and I. K. Egorov. 1975. Genetika 9(10):30-32.
22. Mobraaten, L. E., and D. W. Bailey. 1973. Mouse News Lett. 48:17.
23. Vedernikov, A. A., and I. K. Egorov. 1973. Genetika 9(9):60-66.
24. Zinkernagel, R. M. 1976. J. Exp. Med. 143:437-443.

AND SPECIFICITIES OF THE *H-2* GENE COMPLEX: MOUSE

Line Origins

Line	Generation	Wild Mouse Origin
B10.MOL1	N10F2	*M. musculus molossinus* from Michael Potter
B10.SAA48	N11F5	NB farm, Austin Rd., Saline, Michigan (granary)
B10.SNA57	N8F5	SN farm, Liberty Rd., Ann Arbor, Michigan (corn crib)
B10.SNA70	N9F5	SN farm, Liberty Rd., Ann Arbor, Michigan (corn crib)
B10.STA10	N8F4	Stein farm, 4200 Liberty Rd., Ann Arbor, Michigan (barn)
B10.STA12	N7F5	Stein farm, 4200 Liberty Rd., Ann Arbor, Michigan (barn)
B10.STC77	N9F4	Stein farm, 4200 Liberty Rd., Ann Arbor, Michigan (corn crib)
B10.STC90	N8F6	Stein farm, 4200 Liberty Rd., Ann Arbor, Michigan (corn crib)

Contributor: Jan Klein

and *T/t* Strains of Dallas Group

not tested; ? = presence of antigen questionable.

Specificities																Reference
109	110	111	112	113	114	115	116	117	118	119	120	122	123	124	125	
–	–	–	–	–	–	–	–	–	–	–	120	122	–	–	–	5
•	•	•	•	•	•	•	•	•	•	•	•	•	•	•	•	5
–	–	–	–	–	–	–	–	–	118	119	120?	–	123?	124	–	5
•	–	–	–	–	–	–	–	–	–	–	•	–	–	–	–	5
•	–	–	–	–	–	–	–	–	–	–	–	–	–	–	125	5
•	•	•	•	•	•	•	•	•	•	•	•	•	•	•	•	4
•	•	•	•	•	•	•	•	•	•	•	•	•	•	•	•	6

continued

Part II. *H-2* Chart of B10.W

Type Strain	*H-2* Haplotype															H-2 Antigenic
		2	9	19	21	23	26	31	32	102	103	104	105	106	107	108
B10.SNA70	*w8*	–	–	–	–	23	–	–	–	–	–	–	–	–	–	–
B10.BUA19	*w9*	–	–	–	–	–	–	–	32	•	•	•	•	–	–	•
B10.KEA2	*w10*	–	–	–	–	–	–	–	–	–	–	–	–	–	–	–
B10.CAA2	*w11*	–	9?	–	–	–	26	–	–	–	–	–	–	–	–	•
B10.MOL1	*w12*	–	–	–	–	23	–	–	–	–	–	–	–	–	–	–
B10.STA10	*w13*	–	–	–	–	–	–	–	–	–	–	–	–	–	–	–
B10.STC77	*w14*	–	–	–	–	–	–	31	–	–	–	–	–	–	–	•
B10.STC90	*w15*	–	–	–	–	–	–	–	–	–	–	–	–	–	–	–
B10.BUA1	*w16*	–	–	–	–	–	–	–	–	–	–	–	–	–	–	•
B10.CAS2	*w17*	–	–	–	–	–	–	–	–	–	–	–	–	–	–	•
B10.CHR51	*w18*	–	–	–	–	–	–	31	–	–	–	–	–	–	–	–
B10.KPB128	*w19*	2	–	19	–	–	–	–	–	–	–	–	–	•	•	•
B10.LIB18	*w20*	–	–	–	–	–	–	–	–	–	–	–	–	–	–	–
T/t^{12}	t^{12}	–	–	–	•	–	–	–	–	–	–	–	–	106	107	–
T/t^{w1}	t^{w1}	–	–	–	•	–	–	–	–	–	–	–	–	–	–	108
T/t^{w5}	t^{w5}	–	–	–	•	–	–	–	–	–	–	–	–	–	–	–

Contributor: Jan Klein

References

1. Duncan, W. R., et al. Unpublished. Univ. Texas, Southwestern Medical School, Dallas, 1977.
2. Hammerberg, C., et al. 1976. Transplantation 21:199-212.
3. Hauptfeld, M., and J. Klein. 1976. Immunogenetics 3:603-607.
4. Klein, J. 1971. Proc. Natl. Acad. Sci. USA 68:1594-1597.

Part III. *H-2* Chart of B10.W

The *H-2* haplotypes *w7-w9* and antigens 107-109 of the Prague group were not compared with the series of the Dallas group's B10.W strains. Therefore, it is recommended that the designation of origin ("*p*") be used with the Prague symbols (e.g., "*w7p*"), until direct comparisons and nomenclature rules for *H-2w* haplotypes permit the elimination of discrepancies between the two groups. Data for this Part

Strain	*H-2* Haplotype																							H-2 Antigenic
		1	2	3	4	5	6	7	8	9	11	13	15	16	17	18	19	23	25	28	30	31	32	33
B10.W44	*w7p*	1	–	3	–	5	•	7?	8	–	11?	–	•	–	–	–	–	–	25	•	–	–	•	•
B10.W67	*w8p*	1	–	3	–	5	•	–	8	–	–	13	•	–	–	–	–	–	–	•	–	–	•	•
B10.W625	*w9p*	–	–	–	–	–	•	7?	8	–	–	–	•	–	–	–	–	–	–	•	–	–	•	•

[1] Antigenic specificities are described as H-2.107-109 in reference 1, resulting in the use of the same symbols for different H-2 antigenic specificities (*see* Part II of this Table). [2] Antigenic specificities are described as H-2.107-

Contributor: Milada Micková

References

1. Hammerberg, C., et al. 1976. Transplantation 21:199-212.
2. Micková, M., and P. Iványi. 1976. Folia Biol. (Prague) 22:169-189.

AND SPECIFICITIES OF THE *H-2* GENE COMPLEX: MOUSE

and *T/t* Strains of Dallas Group

Specificities																Reference
109	110	111	112	113	114	115	116	117	118	119	120	122	123	124	125	
–	110	–	–	–	–	–	–	–	–	–	–	–	–	–	–	3
•	–	–	–	–	–	–	•	–	–	–	–	–	–	–	–	1
–	–	–	–	–	–	–	–	–	–	–	–	–	–	–	–	7
•	–	–	–	–	–	–	–	–	–	–	–	–	–	–	–	7
–	–	–	112	–	–	–	–	–	–	–	–	–	–	124	–	7
–	–	–	–	113	–	–	–	–	–	–	120?	122	123	–	125	7
•	–	–	–	–	114	–	–	–	–	119	–	–	–	•	125	7
–	–	–	–	–	–	115	–	117	–	–	–	–	–	–	–	7
•	–	–	–	–	–	–	116	117	–	–	–	–	–	–	–	7
•	–	111	–	–	–	–	–	–	118	119	120	–	123?	124?	–	7
–	–	–	–	–	–	–	–	–	–	–	–	–	–	–	125	7
•	–	–	–	–	–	–	–	–	–	–	–	–	–	–	–	7
–	110	–	–	–	–	–	116	–	–	–	–	–	–	–	–	7
–	–	–	–	–	–	–	–	–	–	–	–	–	–	–	–	2
–	–	–	–	–	–	–	–	–	–	–	–	–	–	–	–	2
109	–	–	–	–	–	–	–	–	–	–	–	–	–	–	–	2

5. Klein, J. 1972. Transplantation 13:291-299.
6. Klein, J. 1975. Immunogenetics 2:297-299.
7. Zalesca-Rutczynska, Z., and J. Klein. 1977. J. Immunol. 119:1903-1911.

Strains of Prague Group

are from reference 2. **H-2 Antigenic Specificities**: Specificities derived from wild mice haplotypes are designated by numbers higher than 100. Private H-2 antigenic specificities are underlined. *Symbols:* – = antigen absent; • = presence of antigen not tested; ? = presence or absence of antigen is uncertain.

Specificities															Ss Allele
41	42	101	102	103	104	105	106	107[1]	108[1]	109[1]	110	107[2]	108[2]	109[2]	
–	42	•	•	•	•	•	•	•	•	•	•	107	–	–	high
–	–	•	•	•	•	•	•	•	•	•	•	–	108	–	high
–	42	•	•	•	•	•	•	•	•	•	•	–	–	109	high

109 in reference 2, resulting in the use of the same symbols for different H-2 antigenic specificities.

39. MAPS OF THE *H-2* GENE COMPLEX: MOUSE

GENE SYMBOLS

T = brachyury
tf = tufted
H-2 = histocompatibility-2
Tla = thymus leukemia antigen
thf = thin fur
H-2K = H-2 antigen K
Ir-1A = immune response locus 1A
Ir-1B = immune response locus 1B
Ia-4 = immune response-associated antigen 4
Ia-5 = immune response-associated antigen 5
Ia-3 = immune response-associated antigen 3
Ss = serum substance
H-2G = H-2 antigen G
H-2D = H-2 antigen D

Part I. Genetic Fine Structure Map of *H-2* Complex

Numbers above the lines indicate map distances.

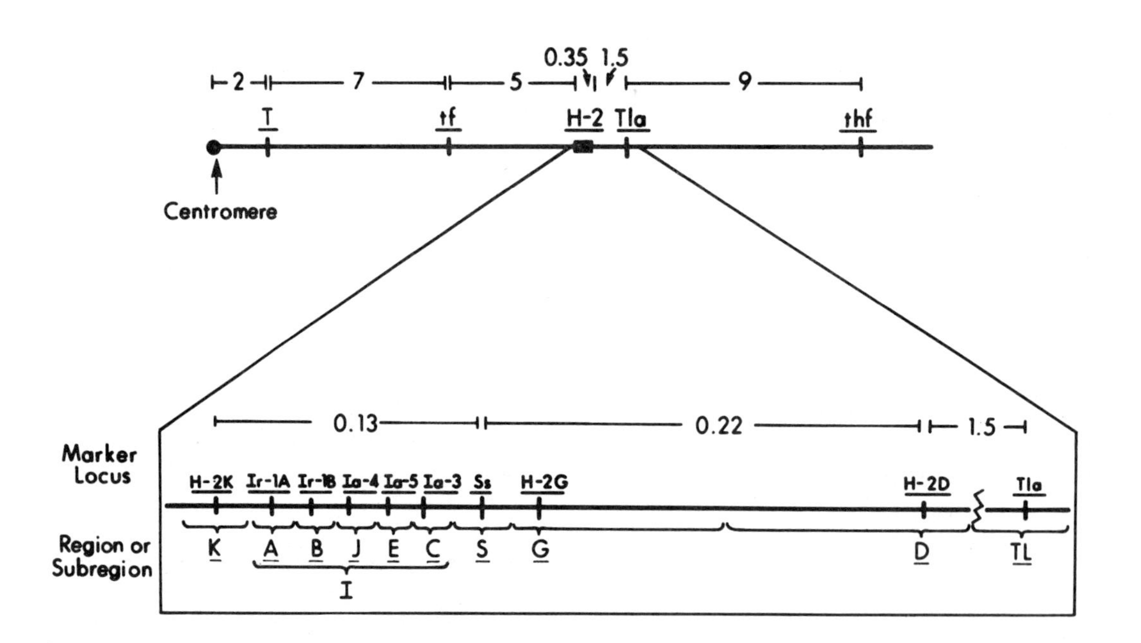

Contributor: Donald C. Shreffler

Reference: Shreffler, D. C. 1977. In E. Sercarz, et al., ed. Immune System: Genetics and Regulation. Academic, New York. pp. 229-240.

continued

Part II. Recombinant Haplotypes Defining Discrete Regions and Subregions of *H-2* Gene Complex

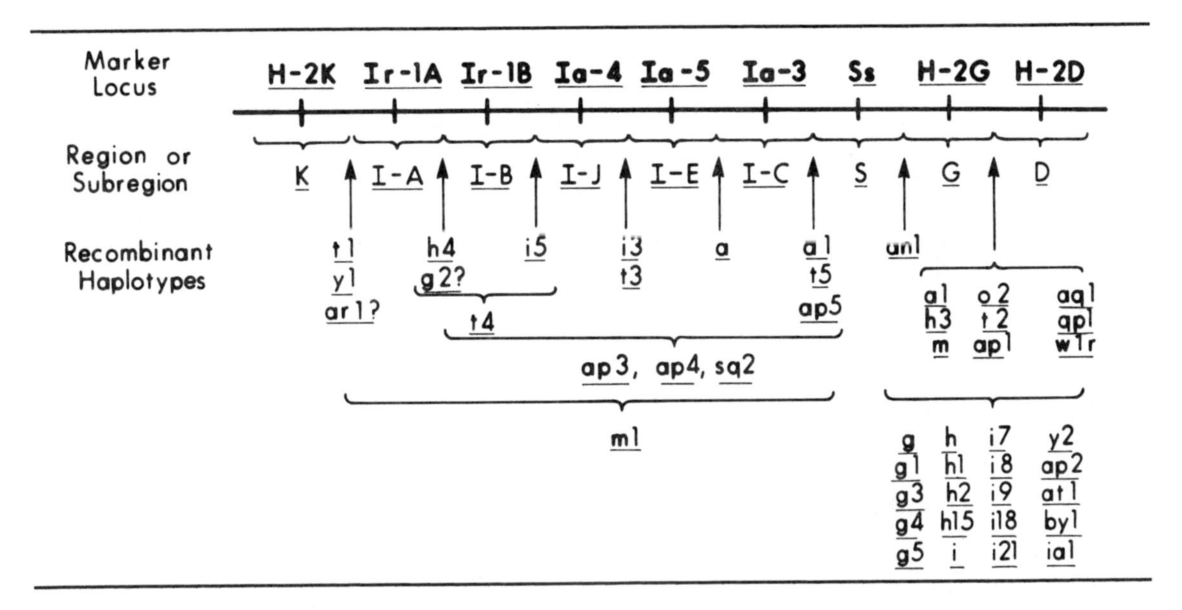

Contributor: Donald C. Shreffler

Reference: Shreffler, D. C., and C. S. David. 1975. Adv. Immunol. 20:125-195.

Part III. Broad Classes of Functions Associated with Various Regions of *H-2* Gene Complex

Region	K	I	S	G	D	TL
Marker Locus	H-2K	Ir-1	Ss	H-2G	H-2D	Tla
Products	Cell Membrane Molecules	Cell Membrane Molecules; Mediators	Serum Proteins	Erythrocyte Membrane Molecules	Cell Membrane Molecules	Cell Membrane Molecules
Histocompatibility Role	Cytotoxic Targets	MLR-GVHR Stimulation	None Defined	None Defined	Cytotoxic Targets	Transplantation Antigens
Function	Marker for Cytotoxicity vs. Deviant Cells	Immune Response; Cell-Cell Interactions	Complement Components	Unknown	Marker for Cytotoxicity vs. Deviant Cells	Unknown

Contributor: Donald C. Shreffler

Reference: Shreffler, D. C. 1977. In E. Sercarz, et al., ed. Immune System: Genetics and Regulation. Academic, New York. pp. 229-240.

40. PRINCIPAL GENETIC TRAITS CONTROLLED BY THE *H-2-Tla* GENE COMPLEX: MOUSE

Trait: CML = cell-mediated lympholysis; CMC = cell-mediated cytotoxicity; MLR = mixed leukocyte reaction; GVHR = graft vs. host reaction; CRL = complement receptor lymphocyte; cAMP = cyclic adenosine 3′,5′-monophosphate.

Trait	Gene Symbol	Controlling Region	Reference
Serologically detected, ubiquitous cellular alloantigens	*H-2K, H-2D, H-2L*	*K, D*	15,28,29
Thymus-leukemia alloantigens	*Tla*	*TL*	3,15,28
I-region-associated lymphocyte alloantigens	*Ia-1, Ia-3, Ia-4*	*I*	5,28
Erythrocyte alloantigen	*H-2G*	*G*	6,18
Lymphocyte alloantigens	*Qa-1, Qa-2*	*TL*	11,30
Transplantation antigens	*H-2K, H-2D*	*K, D*	15,28,29
	H-2A, H-2C	*I*	15,17,20,28
	H-31, H-32	*TL*	12
	H-33	*K*-end	10
CML allogeneic target antigens	*H-2K, H-2D*	*K, D*	15,28
	H-2A, H-2C	*I*	15,23,28
CMC target markers	*H-2K, H-2D*	*K, D*	8,26
MLR-GVHR stimulating antigens	*H-2K, Lad, H-2D*	*K, I, D*	1,16,21,28
Hematopoietic histocompatibility	*Hh*	*K, D, TL*	4,29
Oncogenic virus susceptibility	*Rgv-1*	*K* or *I*	15,19,28,29
In vivo immune response	*Ir-1*	*I*	2,15,28
In vivo immune suppression	*Is*	*I*	2,7,25
Cell-cell interactions	*Ci*	*I*	2,9,24
Complement components	*Ss-Slp*	*S*	27
CRL ontogeny	....	*D-TL*	13
Testosterone levels, organ weights	*Hom-1*	*K-I*	14
Liver cAMP levels	....	*K*-end, *D*-end	22
Mating preference	....	....	31

Contributor: Donald C. Shreffler

References

1. Bach, F. H., et al. 1973. J. Exp. Med. 136:1430-1444.
2. Benacerraf, B., and D. H. Katz. 1975. In B. Benacerraf, ed. Immunogenetics and Immunodeficiency. Univ. Park, Baltimore. pp. 118-177.
3. Boyse, E. A., and L. J. Old. 1969. Annu. Rev. Genet. 3:269-290.
4. Cudkowicz, G., and E. Lotzová. 1973. Transplant. Proc. 5:1399-1405.
5. David, C. S. 1976. Transplant. Rev. 30:299-322.
6. David, C. S., et al. 1975. Immunogenetics 2:131-139.
7. Debré, P., et al. 1975. J. Exp. Med. 142:1447-1454.
8. Doherty, P. C., et al. 1976. Transplant. Rev. 29:89-124.
9. Erb, P., and M. Feldmann. 1975. J. Exp. Med. 142: 460-472.
10. Flaherty, L. 1975. Immunogenetics 2:325-329.
11. Flaherty, L. 1976. Ibid. 3:533-539.
12. Flaherty, L., and S. S. Wachtel. 1975. Ibid. 2:81-85.
13. Gelfand, M. C., et al. 1974. J. Exp. Med. 139:1142-1153.
14. Ivanyi, P., et al. 1972. Folia Biol. (Prague) 18:81-97.
15. Klein, J. 1975. Biology of the Mouse Histocompatibility-2 Complex. Springer-Verlag, New York.
16. Klein, J., and J. M. Park. 1973. J. Exp. Med. 137:1213-1225.
17. Klein, J., et al. 1974. Immunogenetics 1:45-56.
18. Klein, J., et al. 1975. Ibid. 2:141-150.
19. Lilly, F., and T. Pincus. 1973. Adv. Cancer Res. 17: 231-277.
20. McKenzie, I. F. C., and M. Henning. 1976. Immunogenetics 3:253-260.
21. Meo, T., et al. 1973. Transplant. Proc. 5:1507-1510.
22. Meruelo, D., and M. Edidin. 1975. Proc. Natl. Acad. Sci. USA 72:2644-2648.
23. Nabholz, M., et al. 1975. Eur. J. Immunol. 5:594-599.
24. Pierce, C. W., et al. 1976. J. Exp. Med. 144:371-381.
25. Rich, S. S., and R. R. Rich. 1976. Ibid. 143:672-677.
26. Shearer, G. M., et al. 1976. Transplant. Rev. 29:222-248.

continued

27. Shreffler, D. C. 1976. Ibid. 32:140-167.
28. Shreffler, D. C., and C. S. David. 1975. Adv. Immunol. 20:125-195.
29. Snell, G. D., et al. 1976. Histocompatibility. Academic, New York.
30. Stanton, T. H., and E. A. Boyse. 1976. Immunogenetics 3:525-531.
31. Yamazaki, K., et al. 1976. J. Exp. Med. 144:1324-1335.

41. *H-2K, H-2D, H-2L,* AND *H-2G* SPECIFICITIES BY HAPLOTYPE OF THE *H-2* GENE COMPLEX: MOUSE

The *H-2* gene complex produces two types of cell surface antigens: H-2 antigens (molecular weight ∿45,000 and associated with β-2-microglobulin chain) and Ia antigens (two polypeptide chains, molecular weight 33,000 and 26,000, respectively, not associated with β-2-microglobulin). The H-2 antigens are controlled by the *K* region in the *K*-end of *H-2* and by the *D* region in the *D*-end of *H-2*. The *K* region carries the *H-2K* locus which controls the H-2K molecules, and the *D* region contains the *H-2D* and *H-2L* loci which control the H-2D and H-2L molecules, respectively. Another locus, *H-2G*, in the *G* region between *Ss* and *H-2D*, controls a red-cell antigen bearing the antigenic specificity H-2.7. Because of the lack of information on the chemical structure of H-2G antigen(s), it is not known whether they really belong in the same molecular class as the genuine H-2 antigens (H-2K, H-2D, H-2L).

The antigenic characteristics of the H-2 molecules defined by alloimmune sera are "antigenic specificities." *An antigenic specificity is defined by a pattern of reactions on a standard panel of inbred strains of an antibody which behaves as monospecific on the standard panel* [ref. 2]. Hence, an antigenic specificity is a serological phenomenon which need not correspond to a distinct antigenic determinant on the H-2 molecule.

Each *K* or *D* region produces one private specificity, which is unique for each particular allele, and several public specificities. Each public specificity is present in the product of more than one *K* and/or *D* region allele. The public specificities are sometimes serologically more complex than can be indicated in the *H-2* charts. This is particularly true of the specificities H-2.1, H-2.3, and H-2.28 which are actually complexes, or families, of very similar specificities, rather than simple serological entities. They are expressed in some haplotypes in an incomplete or "intermediate" form (for a detailed discussion, consult references 3 and 6. Cells of mice with an intermediate form of H-2.1 react with many, but not necessarily all, anti-H-2.1 sera. The differences in the reactivity of individual anti-H-2.1 sera thus revealed cannot be conveniently expressed in terms of distinct antigenic specificities. The mice with an intermediate form of H-2.1 can, upon immunization with tissues of another H-2.1-positive strain, produce anti-H-2.1 antibodies. An analogous situation exists also with H-2.3 and H-2.28 intermediate haplotypes.

Because the specificity H-2.7 is controlled by the *H-2G* locus in the *G* region [ref. 1 and 4], which is located between *H-2K* and *H-2D*, haplotypes with the same *H-2K* and *H-2D* alleles can nevertheless differ at the central portion of the *H-2* gene complex, including also *H-2G*. Therefore haplotypes of the same type can differ in expression of the specificity H-2.7.

In some instances, mutant *H-2* haplotypes differ from the original haplotype also in the serologically defined specificities. Five such changes were interpreted in terms of distinct antigenic specificities. H-2.40 = a D^d specificity lacking in the mutant $H\text{-}2D^{da}$; since H-2.40 is not present in any allele other than D^d, it represents a subdivision of the private specificity H-2.4. H-2.49 = a public specificity (probably belonging to the H-2.3 family of specificities) present in $H\text{-}2^d$ but absent in $H\text{-}2^{da}$. H-2.50 = a D^{da} specificity present in $H\text{-}2^{da}$ but not in $H\text{-}2^d$, or in any haplotype other than those which have the D^{da} allele. H-2.62 = a K^b specificity present in $H\text{-}2^b$ but not in $H\text{-}2^{bd}$. The specificity H-2.60 is present in *K* region products of $H\text{-}2^k$ but not of $H\text{-}2^{ka}$ mutant.

Symbols: When a number appears, specificity is present. For those specificities which have an intermediate form, the number in roman type indicates that the complete form is present. The intermediate form of a specificity is shown in italics. The symbol of specificity is followed by a period where appropriate tests were not performed to determine whether the form is complete or intermediate. w7 = weak expression of specificity H-2.7; numbers followed by a question mark (?) indicate specificity is likely to be present but conclusive demonstration was not provided. (–) = specificity absent; (·) = no information available.

continued

Data summarize distribution of the H-2 antigenic specificities by haplotypes. This is the most complete and least controversial information, since it involves only interpretation of the serological data in terms of presence, or absence, of

Haplo-																										Antigenic
type	1	2[1]	3	4[1]	5	6	7[2]	8	9[1]	11	12[1]	13	15[3]	16[3]	17[3]	18	19[3]	20[3]	21	22[1]	23[3]	24	25	26[3]	27	28
a	1	–	3	4	5	6	–	8	–	11	–	13	–	–	–	–	–	–	–	–	23	24	25	–	27	28
b	–	2	–	–	5	6	–	–	–	–	–	–	–	–	–	–	–	–	–	–	–	–	–	–	27	28
d	–	–	3	4	–	6	–	8	–	–	–	13	–	–	–	–	–	–	–	–	–	–	–	–	27	28
da	–	–	3	4	–	•	•	8	–	–	•	13	–	–	–	–	–	–	–	–	–	–	–	–	•	28
f	–	–	–	–		6	7	8	9	–	–	–	–	–	–	–	–	–	–	–	–	–	–	26	27	–
g	–	2	–	–	–	6	–	8	–	–	–	–	–	–	–	–	–	–	–	–	–	–	–	–	27	28
h	1	2	3	–	5	6	–, w7	8	–	11	–	–	–	–	–	–	–	–	–	–	23	24	25	–	27	28
i	–	–	3	4	5	6	–	–	–	–	–	13	–	–	–	–	–	–	–	–	–	–	–	–	27	28
j	–	2	–	–	–	6	7	–	–	–	–	–	15	–	–	–	–	–	–	–	–	–	–	–	27	28
k	1	–	*3*	–	5	–	w7	8	–	11	–	–	–	–	–	–	–	–	–	–	23	24	25	–	–	–
m	1	–	*3*	–	5	6	w7	8	–	11	–	13	–	–	–	–	–	–	–	–	23	24	25	–	27	28
o	*1*	–	*3*	–	5	–	–, w7	8	–	–	–	–	–	–	–	–	–	–	–	–	–	–	–	–	–	*28*
p	*1*	–	*3*	–	5	6	7	8	–	–	–	–	–	16	–	–	–	–	–	22	–	24	–	–	–	*28*
q	*1*	–	*3*	–	5	6	–	–	–	11	–	13	–	–	17	–	–	–	–	–	–	–	–	–	27	28
r	*1*	–	*3*	–	5	6	–	8	–	11	–	–	–	–	–	18	–	–	–	–	–	–	25	–	–	*28*
s	*1*	–	*3*	–	5	6	7	–	–	–	12	–	–	–	–	–	19	–	–	–	–	–	–	–	–	28
t	*1*	–	3	4	5	6	–, w7, 7	–	–	–	–	13	–	–	–	–	19	–	–	–	–	–	–	–	27	28
u	*1*	–	3	4	5	6	–	8?	–	–	–	13?	–	–	–	–	–	20	–	–	–	–	–	–	27?	28
v	*1*	–	*3*	–	5?	•	•	–	–	–	–	–	•	–	–	–	–	–	21	–	–	•	–	–	–?	28?
y	•	–	3•	4	5	6	–	–	–	11	•	–	–	–	17	–	–	–	–	–	–	•	–	–	•	•
z	–	–	–	–	5	–	–	–	–	–	•	–	•	–	–	•	–	•	•	•	–	•	–	–	27	–
an	1•	–	?	–	•	6	7	–	9	–	–	–	–	–	–	•	19	•	•	–	–	•	•	–	•	•
ap	–	–	3•	4	–	•	7	8	–	–	–	13	•	•	•	•	•	•	•	–	–	•	•	26	•	28•
aq	1•	–	•	–	5	•	–	8	9	11	•	–	•	•	•	•	•	•	•	–	23	•	25	–	•	•
by	1•	2	3•	–	5	6	–	–	–	11	–	–	–	–	17		–	–	–	–	–	•	–	–	27	28•
qp	•	–	•	–	•	•	•	•	–	11	12	•	–	–	17	–	–	–	–	–	–	•	•	–	•	•
sq	1•	–	3•	–	5	6	–	–	–	–	•	•	•	–	–	•	19	•	•	–	–	•	•	–	27	28•
dx	1•	–	3•	–	5	•	•	8	–	•	–	–	–	–	–	–	–	–	–	–	–	•	•	–	–	*28*

[1] Controlled by the *D* region. [2] Controlled by the *G* region. [3] Controlled by the *K* region.

Contributor: Peter Démant

Part II. *H-2L* Chart

The H-2L molecules express antigenic specificities either of the H-2.28 type or of the H-2.1 type [ref. 1, 2].

Allele	*H-2L* Serotype
b	H-2.28, 64
d	H-2.28, 64, 65
f	*H-2.28*

Allele	*H-2L* Serotype
k	*H-2.1*
q	H-2.28, 64, 65

Allele	*H-2L* Serotype
s	•
dx	H-2.1

Contributor: Peter Démant

References

1. Démant, P., et al. 1975. J. Immunogenet. 2:263-275.
2. Neauport-Sautes, C., et al. 1978. Immunogenetics 6: (in press).

continued

H-2 Chart

an antigenic specificity in a given haplotype without attempting to define the responsible gene. *Symbols:* Private specificities are underlined in the column headings, and their control by *K* or *D* region is indicated whenever known.

Specificities																												
29	30[1]	31[3]	32[1]	33[3]	34	35	36	37	38	39	40[1]	41	42	43	44	45	46	47	49	50[1]	51	52	53	54	55	56	62[3]	63[1]
29	–	–	–	–	–	35	36	–	–	–	40	41	42	43	44	45	–	47	49	–	–	52	–	–	–	–	–	–
29	–	–	–	33	–	35	36	–	–	39	–	–	–	–	–	–	46	–	–	–	–	–	53	54	–	56	62	–
29	–	31	–	–	34	35	36	–	–	–	40	41	42	43	44	–	46	47	49	–	–	–	–	–	–	–	–	–
·	–	31	–	–	·	35	·	–	·	·	–	·	·	·	44	–	·	·	–	50	·	·	·	·	·	·	·	·
–	–	–	–	–	·	–	–	37	–	39	–	–	–	–	–	–	46	–	–	–	–	–	53	–	–	–	–	–
29	–	31	–	–	34	–	–	–	–	–	–	–	–	–	–	–	46	47	–	–	–	–	–	–	–	56	–	–
29	–	–	–	–	·	–	–	–	–	–	–	–	–	–	–	45	·	·	49	–	–	52	–	–	–	56	–	–
29	–	–	–	33	·	35	36	–	–	39	40	41	42	43	44	–	·	·	49	–	–	–	53	54	–	–	62	–
29	–	–	–	–	–	·	·	·	38	·	–	–	–	–	44	45	46?	47?	–	–	51	–	–	–	–	56	–	–
–	–	–	32	–	·	–	–	–	–	–	–	–	–	–	–	45	–	47	49	–	–	52	–	–	–	–	–	–
29	30	–	–	–	·	–	–	–	–	–	–	–	–	–	–	45	–	47	49	–	–	52	–	–	55	56	–	–
29	–	31	32	–	34	–	–	–	–	–	–	–	–	–	–	–	46	47	49	–	–	–	–	–	–	–	–	–
–	–	–	–	–	34?	35	–	37	38	–	–	41	–	–	–	–	46?	·	49	–	–	–	–	–	–	–	–	–
29	30	–	–	–	34	–	–	–	–	–	–	–	–	–	–	45	·	–	49	–	–	52	–	54	55	56	–	–
–	–	–	–	–	–	–	–	–	–	–	–	–	–	–	–	45	·	47	49	–	–	52	–	54	–	–	–	–
–	–	–	–	–	·	–	36	–	–	–	–	–	42	–	–	45	·	–	49	–	51	–	–	–	–	–	–	–
29	–	–	–	–	·	35	36	–	–	–	40	41	42	43	44	–	·	·	49	·	·	·	·	·	·	·	–	–
29	–	–	–	–	·	35	36	–	–	–	40	41	42	43	·	·	·	·	·	·	–	52	53	–	–	–	–	–
·	30?	–	–	–	·	–	–	·	·	·	–	–	–	43	·	45?	·	·	·	·	–	–	–	–	55	–	–	–
·	–	–	–	–	·	·	·	·	·	·	40	·	·	·	·	·	·	·	49	·	·	·	·	·	·	·	·	·
–	–	–	–	–	–	·	·	·	·	·	·	·	·	·	·	·	46	47	·	·	·	·	·	·	·	·	·	·
·	–	–	–	–	·	·	·	–	–	–	–	·	·	·	·	·	·	·	·	–	·	·	·	·	·	·	·	·
·	·	·	·	·	·	·	·	·	·	·	·	·	·	·	·	·	·	·	·	·	·	·	·	·	·	·	·	·
·	·	·	·	·	·	·	·	·	·	·	·	·	·	·	·	·	·	·	·	·	·	·	·	·	·	·	·	·
29	–	–	–	–	–	–	–	–	–	–	–	–	–	–	–	45	–	–	–	–	·	·	·	·	·	·	·	·
·	–	–	–	–	·	·	·	·	·	·	·	·	·	·	·	·	·	·	·	·	–	52	–	54	–	–	·	·
29	30	–	–	–	·	·	·	–	–	–	–	·	·	–	·	45	·	·	·	·	·	·	·	·	·	·	·	·
·	–	31	–	–	·	–	–	–	–	–	–	–	–	–	–	·	·	·	·	–	·	·	·	·	·	·	–	63

References

1. David, C. S., et al. 1975. Immunogenetics 2:131-139.
2. Démant, P. 1973. Transplant. Rev. 15:162-200.
3. Démant, P., et al. 1975. J. Immunogenet. 2:263-275.
4. Klein, J., et al. 1975. Immunogenetics 2:141-150.
5. Neauport-Sautes, C., et al. 1978. Ibid. 6:(in press).
6. Snell, G. D., et al. 1973. Transplant. Rev. 15:3-25.

Part III. *H-2G* Chart

The *H-2G* locus has been shown to control the expression of the antigenic specificity H-2.7 on erythrocytes [ref. 1, 2]. It is not known whether the H-2.7 negative strains express an alternative H-2G product.

Allele	H-2.7
b	–
d	–

Allele	H-2.7
f	7
k	w7

Allele	H-2.7
q	–
s	7

Contributor: Peter Démant

References

1. David, C. S., et al. 1975. Immunogenetics 2:131-139.
2. Klein, J., et al. 1975. Ibid. 2:141-150.

continued

41. *H-2K, H-2D, H-2L,* AND *H-2G* SPECIFICITIES BY HAPLOTYPE OF THE *H-2* GENE COMPLEX: MOUSE

Part IV. Specificities of the *K* and *D* Regions

The chart below incorporates additional information based on analysis of recombinant haplotypes. Consequently, in many instances these data are not complete enough for an unequivocal interpretation. The location of certain specificities in some haplotypes is less certain than often assumed; the chart, therefore, contains a more conservative interpretation than is usually found in the literature. Each *K* or *D* region allele produces one private specificity, which is unique for each particular allele, and several public specificities. Each public specificity is present in the product of more than one *K,* and/or *D* region, allele. Information on the *D* region includes both H-2D and H-2L products.

Allele	Private Specificity	Public Specificities																													
		1	3	5	6	8	11	13	24	25	27	28	29	34	35	36	37	38	39	41	42	43	44	45	49	51	52	53	54	55	56
K^b	33	–	–	5	•	–	–	–	–	–	•	28	•	–	35	36	–	–	39	–	–	–	–	–	–	–	–	53	54	–	–
K^d	31	–	*3*	–	–	8	–	–	–	–	–	28	29	34	–	–	–	–	–	–	–	–	–	–	–	–	–	–	–	–	–
K^k	23	1	*3*	5	–	8	11	–	24	25	–	–	–	–	–	–	–	–	–	–	–	–	–	45	49?	–	52	–	–	–	–
K^q	17	*1?*	*3?*	*5?*	•	–	11	–	–	–	•	•	•	34?	–	–	–	–	–	–	–	–	–	45?	49?	–	52	–	54?	•	•
K^s	19	*1*	3?	5	•	–	–	–	–	–	–	•	–	–	–	36?	–	–	–	–	42?	–	–	45?	•	51?	–	–	–	–	–
K^u	20	*1*	•	5	•	8	–	•	–	–	•	•	•	–	•	•	–	–	–	•	•	•	•	–	•	–	52	53	–	–	–
D^b	2	–	–	–	6	–	–	–	–	–	27	28	29	–	–	–	–	–	–	–	–	–	–	–	–	–	–	–	–	–	56
D^d	4	–	3	–	6	–	–	13	–	–	27	28	29	–	35	36	–	–	–	41	42	43	44	–	49	–	–	–	–	–	–
D^k	32	*1*	•	5	–	–	–	–	–	–	–	–	–	–	–	–	–	–	–	–	–	–	–	–	•	–	–	–	–	–	–
D^q	30	•	3?	•	6	–	•	13	–	–	27	28	29	–	–	–	–	–	–	–	–	–	–	–	49?	–	•	–	?	55	56
D^{dx}	63	*1*	•	5	•	•	•	–	•	•	•	•	•	•	–		–	–	–	–	–	–	•	•	•	•	•	•	•	•	•

Contributor: Peter Démant

References: *See* Part I.

42. *Ss* ALLELES AND PROPERTIES IN THE *H-2* GENE COMPLEX: MOUSE

Quantitative Levels are expressed in arbitrary units relative to a standard reference serum, with B10.D2 male serum set at 1.0. Values are means of 5-25 animals more than 13 weeks old.

Ss Allele	*H-2* Haplotype	Type Strain	Qualitative Type		Quantitative Levels			
			Ss	Slp	Serum Substance (Ss)		Sex-Limited Protein (Slp)	
					Male	Female	Male	Female
Ss^b	$H\text{-}2^b$	B10	Ss^h	Slp–	0.72	0.53	0	0
Ss^d	$H\text{-}2^d$	B10.D2	Ss^h	Slp+	1.00	0.53	1.00	0
Ss^f	$H\text{-}2^f$	B10.M	Ss^h	Slp–	1.01	1.16	0	0
Ss^k	$H\text{-}2^k$	B10.K	Ss^l	Slp–	0.07	0.06	0	0
Ss^p	$H\text{-}2^p$	B10.P	Ss^h	Slp+	0.81	0.50	0.04	0
Ss^q	$H\text{-}2^q$	B10.G	Ss^h	Slp–	0.90	0.58	0	0
Ss^s	$H\text{-}2^s$	B10.S	Ss^h	Slp+	0.93	0.56	0.26	0
Ss^{w7}	$H\text{-}2^{w7}$	B10.WR7	Ss^h	Slp+	1.24	0.74	2.22	1.46

Contributor: Donald C. Shreffler

continued

42. *Ss* ALLELES AND PROPERTIES IN THE *H-2* GENE COMPLEX: MOUSE

General References

1. Hansen, T. H. 1975. Ph.D. Thesis. Univ. Michigan, Ann Arbor.
2. Shreffler, D. C. 1976. Transplant. Rev. 32:140-167.
3. Shreffler, D. C. 1977. Unpublished. Washington Univ. School of Medicine, St. Louis, 1978.

43. TRAITS OF THE *I* REGION OF THE *H-2* GENE COMPLEX: MOUSE

Trait: CML = cell-mediated lympholysis; MLR = mixed leukocyte reaction; GVHR = graft vs. host reaction.

Trait	Gene Symbol	Reference
Susceptibility to oncogenic viruses	*Rgv-1*	6,7,12,21,23
In vivo immune responses[1]	*Ir-1*	2,6,7,21
In vivo immune suppression	*Is*	2,4,6
Histocompatibility antigens	*H-2A, H-2C*	7,10,13,21
CML target antigens	*H-2A, H-2C*	6,7,11,17,21
MLR-stimulating antigens	*Lad*	1,7,14,21
GVHR stimulating antigens	*Lad*	7-9,21
Ia lymphocyte alloantigens[2]	*Ia-1, Ia-3, Ia-4, Ia-5*	3,6,16,21,22,25
Lymphocyte interactions, in vivo	*Ci*	2,6,18,19
Soluble helper factors	....	2,6,15,24
Soluble suppressor factors	....	2,6,20,24
Macrophage factors	....	2,5,6

[1] *See also* Table 44. [2] *See also* Tables 45 and 46.

Contributor: Donald C. Shreffler

References

1. Bach, F. H., et al. 1973. J. Exp. Med. 136:1430-1444.
2. Benacerraf, B., and D. H. Katz. 1975. In B. Benacerraf, ed. Immunogenetics and Immunodeficiency. Univ. Park, Baltimore. pp. 118-177.
3. David, C. S. 1976. Transplant. Rev. 30:299-322.
4. Debre, P., et al. 1975. J. Exp. Med. 142:1447-1454.
5. Erb, P., and M. Feldmann. 1975. Ibid. 142:460-472.
6. Katz, D. 1977. Lymphocyte Differentiation, Recognition and Regulation. Academic, New York.
7. Klein, J. 1975. Biology of the Mouse Histocompatibility-2 Complex. Springer-Verlag, New York.
8. Klein, J., and C. L. Chiang. 1976. J. Immunol. 117: 736-740.
9. Klein, J., and J. M. Park. 1973. J. Exp. Med. 137: 1213-1225.
10. Klein, J., et al. 1974. Immunogenetics 1:45-56.
11. Klein, J., et al. 1977. J. Exp. Med. 145:450-454.
12. Lilly, F., and T. Pincus. 1973. Adv. Cancer Res. 17: 231-277.
13. McKenzie, I. F. C., and M. Henning. 1976. Immunogenetics 3:253-260.
14. Meo, T., et al. 1973. Transplant. Proc. 5:1507-1510.
15. Munro, A. J., and M. J. Taussig. 1975. Nature (London) 256:103-106.
16. Murphy, D. B., et al. 1976. J. Exp. Med. 144:699-712.
17. Nabholz, M., et al. 1975. Eur. J. Immunol. 5:594-599.
18. Paul, W. E., and B. Benacerraf. 1977. Science 195: 1293-1300.
19. Pierce, C. W., et al. 1976. J. Exp. Med. 144:371-381.
20. Rich, S. S., and R. R. Rich. 1976. Ibid. 143:672-677.

continued

21. Shreffler, D. C., and C. S. David. 1975. Adv. Immunol. 20:125-195.
22. Shreffler, D. C., et al. 1977. Cold Spring Harbor Symp. Quant. Biol. 41:477-487.
23. Snell, G. D., et al. 1976. Histocompatibility. Academic, New York.
24. Tada, T. 1977. In E. Sercarz, et al., ed. Regulation of the Immune System. Academic, New York.
25. Tada, T., et al. 1976. J. Exp. Med. 144:713-725.

44. *H-2* GENE COMPLEX-ASSOCIATED IMMUNE RESPONSES: MOUSE

H-2 **Map Position**: The parentheses indicate mapping data do not distinguish between *H-2K* and *I*, or do not distinguish between *I-C* and *S*. However, the assumption is made that the *Ir* gene resides in the *I* region. **Non-*H-2*-Gene**: "Yes" indicates presence of non-*H-2* genes. **IP** = Inheritance pattern; response genes which are expressed in F_1 hybrids as either completely dominant, codominant, or intermediate are listed as "dominant"; D = dominant, R = recessive, NT = not tested. **Response Phenotypes of *H-2* Haplotypes**: H = high response; L = low response or no response; I = intermediate response; V = variable; "•" = not tested or not reported. For information on erythrocyte alloantigen 1, consult references 15-17. For information concerning H-2.2 alloantigen, consult references 28, 29, and 55. Data in brackets refer to the column heading in brackets.

Antigen ⟨Symbol⟩	*H-2* Map Position [Non-*H-2* Gene]	IP	Response Phenotypes of *H-2* Haplotypes																				Reference
			Independent											Recombinant									
			b	*d*	*f*	*j*	*k*	*p*	*q*	*r*	*s*	*u*	*v*	*a*	*g2*	*h2*	*h4*	*i5*	*o1*	*t1*	*t3*	*y1*	
Synthetic Polypeptide Antigens																							
Poly(Glu^{60}Ala^{40}) ⟨GA⟩	?	D	H	H	H	L	H	L	L	H	H	•	•	H	•	•	•	•	•	•	•	•	44
Poly(Glu^{58}Ala^{38}Tyr^{4}) ⟨GAT^{4}⟩	?	NT	H	H	•	•	H	I	L	•	H	•	•	H	•	•	•	•	•	•	•	•	42
Poly(Glu^{60}Ala^{30}Tyr^{10}) ⟨GAT^{10}⟩	*I-A* or *I-B* [Yes]	D	H	H	H	L	H	L	L	H	L	H	H	H	•	H	H	H	H	H	L	H	7,8,12,32, 42,46
Poly(Glu^{57}Lys^{38}Ala^{5}) ⟨GLA^{5}⟩ & Poly(Glu^{54}-Lys^{36}Ala^{10}) ⟨GLA^{10}⟩	? [Yes]	D	L	H	H	H	I	L/I	I	I	H	•	•	I/H	•	I	I	L	H	•	•	•	33,34
Poly(Glu^{55}Lys^{35}Leu^{10}) ⟨GLLeu⟩	Locus *β*: *I-A* or *I-B* Locus *α*: (*I-C*)	D D	L	I	L	H	L	L	L	H	L	I	H	L	L	L	L	I	I	L	H	L	11
Poly(Glu^{58}Lys^{38}Phe^{4}) ⟨GLΦ⟩	Locus *β*: *I-A* or *I-B* Locus *α*: *I-C* or *I-F*	D D	L	H	L	H	L	H	H	H	L	H	L	L	L	L	L	H	H	L	H	L	4,5,9,10, 45,46
Poly(Glu^{57}Lys^{38}Pro^{5}) ⟨GLPro⟩	*I-A* or *I-B*	NT	L	L	L	•	L	L	L	L	H	•	•	L	•	•	•	•	•	L	H	•	8
Poly(Glu^{57}Lys^{38}Tyr^{5}) ⟨GLT^{5}⟩	Similar to GLΦ	D	L	H	L	H	L	L	L	H	L	H	L	L	L	L	L	H	H	•	•	L	5,10,43, 46
Poly(L-His,L-Glu)-poly-(DL-Ala)-poly(L-Lys) ⟨(H,G)-A—L⟩	*I-B*	D	L	I	•	L	H	L	L	•	L	•	•	H	•	•	H	L	L	H	L	H	20,35-39
Poly(L-Phe,L-Glu)-poly-(DL-Ala)-poly(L-Lys) ⟨(Phe,G)-A—L⟩	*I-A* or *I-B*	D	H	H	•	H	H	H	H	•	L	•	•	H	•	H	H	H	H	H	L	H	20,35,37-39
Poly(L-Tyr,L-Glu)-poly-(DL-Ala)-poly(L-Lys) ⟨(T,G)-A—L⟩	(*I-A*)	D	H	I	•	L	L	L	L	L	L	•	•	L	•	L	L	H	H	L	L	L	20,35-39

continued

44. *H-2* GENE COMPLEX-ASSOCIATED IMMUNE RESPONSES: MOUSE

Antigen ⟨Symbol⟩	*H-2* Map Position [Non-*H-2* Gene]	IP	Response Phenotypes of *H-2* Haplotypes																				Reference
			Independent											Recombinant									
			b	*d*	*f*	*j*	*k*	*p*	*q*	*r*	*s*	*u*	*v*	*a*	*g2*	*h2*	*h4*	*i5*	*o1*	*t1*	*t3*	*y1*	
Alloantigens																							
Erythrocyte alloantigen 2.1 ⟨Ea-2⟩	?	D	L	·	·	·	·	·	·	H	·	·	·	·	·	·	·	·	·	·	·	·	56
H-4 alloantigen ⟨H-4⟩	*I-B*	D	H	H	·	·	·	·	·	·	H	·	·	L	·	·	H	H	·	·	·	·	62
IgA myeloma protein: MOPC 467	*(I-A)*	D	L	L	·	·	H	H	L	H	H	·	L	H	·	H	H	L	·	·	·	·	24-27
IgG_{2a} myeloma protein: MOPC 173	*(I-B)*	D	H	L	·	·	L	H	L	H	H	·	H	L	·	L	H	H	·	·	·	·	24-27
Liver-specific F antigen ⟨F-antigen⟩	? [Yes]	R	L	L	·	·	H	·	·	·	·	·	·	·	·	·	·	·	·	·	·	·	54
Male-specific histocompatibility antigen ⟨H-Y⟩	*(I-A)* [Yes]	D	H	I	L	·	I	L	I	I	·	·	·	I	·	I	I	H	·	·	·	·	1,2,17-19
Sex-limited serum protein ⟨Slp⟩	*(I-A)*	D	L	·	I	·	I	·	H	·	·	·	L	·	·	·	V	·	I	I	I	·	31,49
Theta alloantigen 1 ⟨Thy 1.1⟩	Locus A: *(I-A)* Locus B: *I-A* or *I-B* [Yes[2/]]	D D	L	L	·	·	L/H[1/]	·	L	·	·	·	·	L	L	·	L	L	I	·	·	·	63,65-67
Theta alloantigen 2 ⟨Thy 1.2⟩	? [Yes]	R[3/]	·	·	·	·	L	H	H	·	·	H	·	L	·	·	·	·	·	·	·	·	64
Thyroglobulin ⟨TG⟩	*(I-A)* [Yes]	D?	L	L	·	·	H	I	H	·	H	·	·	H	·	H	H	L	L	·	·	·	57,61
Low Doses of Complex Antigens (Hapten-directed Responses)																							
$Benzylpenicilloyl_{25}$-bovine gamma globulin ⟨BPO-BGG⟩	?	NT	L	L	·	·	H	·	L	·	L/H	·	·	H	·	·	·	·	·	·	·	·	58
$Benzylpenicilloyl_{4}$-bovine pancreatic ribonuclease ⟨BPO-RNase⟩	?	NT	L	·	·	·	H	·	I	·	L	·	·	H	·	·	·	·	·	·	·	·	58
Calf type I collagen	*(I-A)* [Yes]	D	H	L	H	L	L	L	L	L	H	L	·	L	·	L	L	H	·	L	·	·	21,47
Calf type I procollagen peptide	*I-A* or *I-B*	D	H	L	·	·	H	·	L	·	·	L	·	H	·	·	·	·	·	·	L	H	48
Dinitrophenyl-bovine gamma globulin ⟨DNP-BGG⟩	*I-A* or *I-B*	D	L	L	·	·	H	L	I	·	L	·	·	H	·	H	V	L	L	H	L	H	14,58
Dinitrophenyl-bovine insulin ⟨DNP-bovine insulin⟩	?	NT	H	H	·	·	L	·	·	·	·	·	·	·	·	·	·	·	·	·	·	·	23
Dinitrophenyl-ovomucoid ⟨DNP-OM⟩	*I-A* [Yes]	D	L	L	L	·	H	L	L	·	L	·	·	H	·	H	H	L	L	H	L	H	13,14,60
Hen albumen lysozyme ⟨HEL⟩	?	D	L	I	·	·	H	·	H	·	·	·	·	H	·	·	·	·	·	·	·	·	22
Leukemia-associated transplantation antigen ⟨X.1⟩	?	NT	H	L	·	·	L	·	·	·	·	·	·	L	·	·	·	·	·	·	·	·	51
Ovalbumin ⟨OA⟩	*I-A* [Yes]	D	H	H	H	H	L	H	H	H	I	·	·	L	·	·	L	H	·	·	·	L	12,59,60
Ovomucoid ⟨OM⟩	?	D	L	L	·	·	H	·	L	·	·	·	·	H	·	·	·	·	·	·	·	·	60

[1/] Major non-*H-2* genes confound the strain distribution arranged according to *H-2* type. [2/] Data suggest two loci in close linkage with the *H-2* complex, plus an additional locus, *Ir-5*, 19 map units to the right of *H-2*. [3/] In most F_1 hybrids tested the response was recessive, but in one F_1 hybrid the response was dominant.

continued

Antigen ⟨Symbol⟩	*H-2* Map Position [Non-*H-2* Gene]	IP	Response Phenotypes of *H-2* Haplotypes																				Reference
			Independent											Recombinant									
			b	*d*	*f*	*j*	*k*	*p*	*q*	*r*	*s*	*u*	*v*	*a*	*g2*	*h2*	*h4*	*i5*	*o1*	*t1*	*t3*	*y1*	
Porcine lactate dehydrogenase-A ⟨LDH_A⟩	?	NT	L	H	H	H	H	H	H	H	H	H	H	H	•	•	•	•	•	•	•	•	41
Porcine lactate dehydrogenase-B ⟨LDH_B⟩	Locus 1: *I-A* or *I-B* Locus 2: *(I-C)*	D D	H	H	H	I	L	H	H	I	H	I	I	L	•	L	H	•	•	•	I	•	40,41
Ragweed pollen extract ⟨RE⟩	*(I-A)*	D	L	H	L	L	H	H	•	•	•	L	•	H	H	H	H	L	•	•	•	•	5,6
Staphylococcal nuclease ⟨Nase⟩	*I-B* [Yes]	D	L	H	•	•	H	•	L	•	H	•	•	H	•	H	L	L	•	•	•	•	3,30
Trinitrophenyl-*H-2D*[d] alloantigen ⟨TNP-*H-2D*[d]⟩	Between *K* & *I-A*	NT	•	H	•	•	L	•	•	•	H	•	•	L	•	•	•	•	•	I	H	•	52,53
2,4,6-Trinitrophenyl-mouse serum albumin ⟨TNP-MSA⟩	?	R	H	•	•	•	•	•	•	•	•	•	•	L	•	•	•	•	•	•	•	•	50

Contributor: Howard C. Passmore

References

1. Bailey, D. W. 1971. Transplantation 11:426-427.
2. Bailey, D. W., and J. Hoste. 1971. Ibid. 11:404-407.
3. Berzofsky, J. A., et al. 1977. J. Exp. Med. 145:123-135.
4. Dorf, M. E., and B. Benacerraf. 1975. Proc. Natl. Acad. Sci. USA 72:3671-3675.
5. Dorf, M. E., et al. 1974. J. Exp Med. 140:859-864.
6. Dorf, M. E., et al. 1974. Eur. J. Immunol. 4:346-349.
7. Dorf, M. E., et al. 1974. J. Immunol. 112:1329-1336.
8. Dorf, M. E., et al. 1975. Ibid. 114:602-605.
9. Dorf, M. E., et al. 1975. J. Exp. Med. 141:1459-1463.
10. Dorf, M. E., et al. 1976. Ibid. 143:889-896.
11. Dorf, M. E., et al. 1976. Eur. J. Immunol. 6:552-556.
12. Dunham, E. K., et al. 1973. J. Immunol. 111:1621-1625.
13. Freed, J. H., et al. 1973. Fed. Proc. Fed. Am. Soc. Exp. Biol. 32:995.
14. Freed, J. H., et al. 1976. J. Immunol. 117:1514-1518.
15. Gasser, D. L. 1969. Ibid. 103:66-70.
16. Gasser, D. L., and D. C. Shreffler. 1972. Nature (London) New Biol. 235:155-156.
17. Gasser, D. L., and D. C. Shreffler. 1974. Immunogenetics 1:133-140.
18. Gasser, D. L., and W. K. Silvers. 1971. J. Immunol. 106:875-876.
19. Gasser, D. L., and W. K. Silvers. 1971. Transplantation 12:412-414.
20. Grumet, F. C., and H. O. McDevitt. 1972. Ibid. 13:171-173.
21. Hahn, E., et al. 1975. Eur. J. Immunol. 5:288-291.
22. Hill, S. W., and E. E. Sercarz. 1975. Ibid. 5:317-324.
23. Keck, K. 1975. Nature (London) 254:78-79.
24. Lieberman, R., and W. Humphrey, Jr. 1971. Proc. Natl. Acad. Sci. USA 68:2510-2513.
25. Lieberman, R., and W. Humphrey, Jr. 1972. J. Exp. Med. 136:1222-1230.
26. Lieberman, R., and W. E. Paul. 1974. Contemp. Top. Immunobiol. 3:117.
27. Lieberman, R., et al. 1972. J. Exp. Med. 136:1231-1240.
28. Lilly, F., et al. 1971. In A. Lengerova and M. Vojiskova, ed. Immunogenetics of the H-2 System. Karger, Basel. pp. 197-199.
29. Lilly, F., et al. 1973. Transplant. Proc. 5:193-196.
30. Lozner, E. C., et al. 1974. J. Exp. Med. 139:1204-1214.
31. Luderer, A. A., and H. C. Passmore. 1976. Immunogenetics 3:277-287.
32. Martin, W. J., et al. 1971. J. Immunol. 107:715-718.
33. Maurer, P. H., and C. F. Merryman. 1974. Immunogenetics 1:174-183.
34. Maurer, P. H., et al. 1974. Ibid. 1:398-406.
35. McDevitt, H. O., and A. Chinitz. 1969. Science 163:1207-1208.
36. McDevitt, H. O., and M. Sela. 1965. J. Exp. Med. 122:517-531.
37. McDevitt, H. O., and M. Sela. 1967. Ibid. 126:969-978.
38. McDevitt, H. O., and M. L. Tyan. 1968. Ibid. 128:1-11.
39. McDevitt, H. O., et al. 1972. Ibid. 135:1259-1278.
40. Melchers, I., and K. Rajewsky. 1975. Eur. J. Immunol. 5:753-759.
41. Melchers, I., et al. 1973. Ibid. 3:754-761.
42. Merryman, C. F., and P. H. Maurer. 1972. J. Immunol. 108:135-141.
43. Merryman, C. F., and P. H. Maurer. 1974. Immunogenetics 1:549-559.
44. Merryman, C. F., and P. H. Maurer. 1976. J. Immunol. 116:739-742.
45. Merryman, C. F., et al. 1972. Ibid. 108:937-940.
46. Merryman, C. F., et al. 1975. Immunogenetics 2:441-448.
47. Nowack, H., et al. 1975. Ibid. 2:331-335.

continued

48. Nowack, H., et al. 1977. Ibid. 4:117-125.
49. Passmore, H. C., and T. H. Hansen. 1975. J. Immunol. 114:1139-1142.
50. Rathbun, W. E., and W. H. Hildemann. 1970. Ibid. 105:98-107.
51. Sato, H., et al. 1973. J. Exp. Med. 138:593-606.
52. Schmitt-Verhulst, A., and G. M. Shearer. 1976. Ibid. 144:1701-1706.
53. Schmitt-Verhulst, A., et al. 1976. Ibid. 143:211-217.
54. Silber, D. M., and D. P. Lane. 1975. Ibid. 142:1455-1461.
55. Stimpfling, J. H., and T. Durham. 1972. J. Immunol. 108:947-951.
56. Stimpfling, J. H., et al. 1976. Ibid. 116:1096-1098.
57. Tomazic, V., et al. 1974. Ibid. 112:965-969.
58. Vaz, N. M., and B. B. Levine. 1970. Science 168:852-854.
59. Vaz, N. M., et al. 1970. J. Immunol. 104:1572-1574.
60. Vaz, N. M., et al. 1971. J. Exp. Med. 134:1335-1348.
61. Vladutiu, A. O., and N. R. Rose. 1971. Science 174: 1137-1139.
62. Wettstein, P. J., and G. Haughton. 1977. Immunogenetics 4:65-78.
63. Zaleski, M. B. 1974. Ibid. 1:226-238.
64. Zaleski, M. B. 1975. Ibid. 2:21-27.
65. Zaleski, M., and J. Klein. 1974. J. Immunol. 113:1170-1177.
66. Zaleski, M. B., and J. Klein. 1976. Ibid. 117:814-817.
67. Zaleski, M., and F. Milgrom. 1973. Ibid. 110:1238-1244.

45. Ia SPECIFICITIES BY HAPLOTYPES OF THE *H-2* GENE COMPLEX: MOUSE

Abbreviations: B10 = C57BL/10.

Part I. Definition of Ia Specificities

Specificity: The first number designates *Ia* marker locus, and the second number, the Ia specificity (e.g., Ia-1.2 designates Ia.2 specificity associated with *Ia-1* (*I-A* subregion) locus [ref. 13]). **Antiserum**: B6 = C57BL/6.

Specificity	Antiserum	Absorption	Target Cell	Reference
Ia-1.1	A.TH anti-A.TL	None	I^f (A.CA)	5
Ia-1.2	(A.CA x B10.HTT)F_1 anti-A.TL	B10	I^k (A.TL)	5
Ia-1.3	(A.TH x B10.HTT)F_1 anti-A.TL	None	I^b (B10)	5
Ia-1.4	(A.TL x A.TFR2)F_1 anti-A.TH	None	I^s (A.TH)	5
Ia-1.5	A.TL anti-A.TH	None	I^f (A.CA)	5
Ia-3.6	B10.A(4R) anti-B10.A(2R)	None	I^d (B10.A(2R))	6
Ia-3.7	(B10 x HTT)F_1, anti-B10.A(5R)	B10.S(8R)	$I\text{-}C^d$ (B10.A(5R))	7
Ia-1.8	(B10.A x A)F_1 anti-B10	None	I^d (B10.D2)	12
Ia-1.9	(B10.D2 x A)F_1 anti-HTI	None	I^s (B10.S)	12
Ia-1.10	B10.AQR anti-B10.T(6R)	None	I^q (B10.T(6R))	10
Ia-1.11	(B10 x A) anti-B10.D2	B10.D2 platelets	I^d (B10.D2)	9
Ia-1.12	(B10.A(2R) x C3H.NB)F_1 anti-B10.RIII	None	I^s (B10.S)	8
Ia-1.13	(B6 x A)F_1 anti-B10.P	None	I^q (B10.G)	8
Ia-1.14	A.TFR1 anti-A.CA	A.CA platelets	I^f (A.CA)	8
Ia-1.15	(A.TH x B10.HTT)F_1 anti-A.TL	None	I^d (B10.D2)	3
Ia-1.16	(B10 x A)F_1 anti-B10.D2	α	I^q (B10.G)	15
Ia-1.17	(C3H.Q x B10.D2)F_1 anti-AQR	None	I^s (B10.S)	1
Ia-1.18	(C3H.Q x B10.D2)F_1 anti-AQR	B10.RIII	I^s (B10.S)	1
Ia-1.19	(C3H.Q x B10.D2)F_1 anti-AQR	B10.S	I^r (B10.RIII)	1
Ia-1.20	(B10.A x A)F_1 anti-B10	B10 platelets	I^b (B10)	11
Ia-3.21	(B6 x A)F_1 anti-B10.P	B10.P platelets	I^p (B10.P)	2
Ia-5.22	(C3H.Q x B10.D2)F_1 anti-AQR	None	B10.A(5R)	14
Ia-5.23	(B10 x A.CA)F_1 anti-B10.D2	D2.GD	B10.HTG	4

Contributor: Chella S. David

continued

45. Ia SPECIFICITIES BY HAPLOTYPES OF THE *H-2* GENE COMPLEX: MOUSE

Part I. Definition of Ia Specificities

References

1. Colombani, J., et al. 1976. Tissue Antigens 7:74-85.
2. Cullen, S. E., et al. 1976. In V. Eisvoogel, ed. Leucocyte Membrane Determinants Regulating Immune Reactivity. Academic, New York. pp. 507-508.
3. David, C. S. 1976. Transplant. Rev. 30:299-322.
4. David, C. S., and J. F. McCormick. 1978. J. Immunol. 120:1659-1662.
5. David, C. S., and D. C. Shreffler. 1974. Transplantation 18:313-323.
6. David, C. S., et al. 1974. Ibid. 17:122-128.
7. David, C. S., et al. 1975. J. Immunol. 114:1205-1214.
8. David, C. S., et al. 1976. Transplantation 21:520-526.
9. Davies, D. A. L., and M. Hess. 1974. Nature (London) 250:228-231.
10. Hauptfeld, V., et al. 1973. Science 181:167-170.
11. Sachs, D. H., and J. L. Cone. 1975. J. Immunol. 114: 165-174.
12. Sachs, D. H., et al. 1975. Transplantation 19:388-397.
13. Shreffler, D. C., and D. B. Murphy. 1978. In H. O. McDevitt, ed. Ir Genes and Ia Antigens. Academic, New York. pp. 609-613.
14. Shreffler, D. C., et al. 1976. In J. Watson, ed. Origins of Lymphocyte Diversity. Cold Spring Harbor Laboratory, Cold Spring Harbor, NY. pp. 477-487.
15. Staines, N. A., et al. 1974. Transplantation 18:192-201.

Part II. Distribution of Ia Specificities by Haplotypes

Specificities: Number indicates specificity is present; "–" indicates antigen is absent.

H-2 Haplotype	Type Strain	Specificities																						
		1	2	3	4	5	6	7	8	9	10	11	12	13	14	15	16	17	18	19	20	21	22	23
Independent Origin																								
b	C57BL/10	–	–	3	–	–	–	–	8	9	–	–	–	–	–	15	–	–	–	–	20	–	–	–
d	B10.D2	–	–	–	–	–	6	7	8	–	–	11	–	–	–	15	16	–	–	–	–	–	–	23
f	B10.M	1	–	–	–	5	–	–	–	–	–	–	–	–	14	–	–	17	18	–	–	–	–	–
k	B10.K	1	2	3	–	–	–	7	–	–	–	–	–	–	–	15	–	17	18	19	–	–	22	–
p	B10.P	–	–	–	–	5	6	7	–	–	–	–	–	13	–	–	–	–	–	–	–	21	–	–
q	B10.G	–	–	3	–	5	–	–	–	9	10	–	–	13	–	–	16	–	–	–	–	–	–	–
r	B10.RIII	1	–	3	–	5	–	7	–	–	–	–	12	–	–	–	–	17	–	19	–	–	–	–
s	B10.S	–	–	–	4	5	–	–	–	9	–	–	12	–	–	–	–	17	18	–	–	–	–	–
Recombinant																								
a	A	1	2	3	–	–	6	7	–	–	–	–	–	–	–	15	–	17	18	19	–	–	22	–
al	A.AL	1	2	3	–	–	–	7	–	–	–	–	–	–	–	15	–	17	18	19	–	–	22	–
g, g1, g3, g4, g5	B10.HTG, B10.BDR, B10.D2(R101), B10.D2(R103)	–	–	–	–	–	6	7	8	–	–	11	–	–	–	15	16	–	–	–	–	–	–	23
g2	D2.GD	–	–	–	–	–	–	–	8	–	–	11	–	–	–	15	16	–	–	–	–	–	–	23
h, h1, h2, h3, h15	HTH, B10.A(1R), B10.A(2R), B10.A(15R), B10.AM	1	2	3	–	–	6	7	–	–	–	–	–	–	–	15	–	17	18	19	–	–	22	–
h2	B10.A(4R)	1	2	3	–	–	–	–	–	–	–	–	–	–	–	15	–	17	18	19	–	–	–	–
i1, i6, i7, i18	HTI, B10.D2(R106), B10.D2(R107), B10.A(18R)	–	–	3	–	–	–	–	8	9	–	–	–	–	–	15	–	–	–	–	20	–	–	–
i3, i5	B10.A(3R), B10.A(5R)	–	–	3	–	–	6	7	8	9	–	–	–	–	–	15	–	–	–	–	20	–	22	–
m	AKR.M	1	2	3	–	–	–	7	–	–	–	–	–	–	–	15	–	17	18	19	–	–	22	–
o1, o2	C3H.OL, C3H.OH	–	–	–	–	–	6	7	8	–	–	11	–	–	–	15	16	–	–	–	–	–	–	23
t1	A.TL	1	2	3	–	–	–	7	–	–	–	–	–	–	–	15	–	17	18	19	–	–	22	–
t2, t5	A.TH, BSVS	–	–	–	4	5	–	–	–	9	–	–	12	–	–	–	–	17	18	–	–	–	–	–
t3	B10.HTT	–	–	–	4	5	–	7	–	9	–	–	12	–	–	–	–	17	18	–	–	–	22	–
t4	B10.S(9R)	–	–	–	4	5	6	7	–	9	–	–	12	–	–	–	–	17	18	–	–	–	22	–

continued

45. Ia SPECIFICITIES BY HAPLOTYPES OF THE *H-2* GENE COMPLEX: MOUSE

Part II. Distribution of Ia Specificities by Haplotypes

H-2 Haplotype	Type Strain	Specificities																						
		1	2	3	4	5	6	7	8	9	10	11	12	13	14	15	16	17	18	19	20	21	22	23
y1	AQR	1	2	3	–	–	6	7	–	–	–	–	–	–	–	15	–	17	18	19	–	–	22	–
y2	B10.T(6R)	–	–	3	–	5	–	–	–	9	10	–	–	13	–	–	16	–	–	–	–	–	–	–
an1	A.TFR1	1	2	3	–	–	–	7	–	–	–	–	–	–	–	15	–	17	18	19	–	–	22	–
ap1, ap2, ap3, ap4	B10.M(11R), A.TFR2, A.TFR3, A.TFR4	1	–	–	–	5	–	–	–	–	–	–	–	–	14	–	–	17	18	–	–	–	–	–
ap5	A.TFR5	1	–	–	–	5	–	7	–	–	–	–	–	–	14	–	–	17	18	–	–	–	–	–
aq1	B10.M(17R)	1	–	–	–	5	–	–	–	–	–	–	–	–	14	–	–	17	18	–	–	–	–	–
ar1	LG/Ckc	1	–	–	–	5	–	–	–	–	–	–	–	–	14	–	–	17	18	–	–	–	–	–
as1	B10.S(8R)	1	2	3	–	–	–	–	–	–	–	–	–	–	–	15	–	17	18	19	–	–	22	–
as2	ASR-1	1	2	3	–	–	–	7	–	–	–	–	–	–	–	15	–	17	18	19	–	–	22	–
at1	A.TBR1	1	2	3	–	–	–	7	–	–	–	–	–	–	–	15	–	17	18	19	–	–	22	–
at2	TBR2	1	2	3	–	–	–	–	–	–	–	–	–	–	–	15	–	17	18	19	–	–	–	–
sq1	A.QSR1	–	–	–	4	5	–	–	–	9	–	–	12	–	–	–	–	17	18	–	–	–	–	–

Contributor: Chella S. David

Reference: David, C. S. 1976. Transplant. Rev. 30:299-322.

46. SUMMARY OF MOLECULAR PROPERTIES OF GENE PRODUCTS OF THE *H-2* COMPLEX: MOUSE

Although it is not normally included in the major histocompatibility complex, the product of the *T/t* locus has been included in this table because the *T/t* genes are mapped in the same linkage group as the major histocompatibility complex genes, and because F9 is a cell surface antigen which appears to have structural homology with the products of the *K, D,* and *Tla* regions. **Method of Isolation**: det sol = detergent solubilization; im prcp = immune precipitation; pap dig = papain digestion; gel filt = gel filtration. **Composition & Molecular Weight**: Except in the case of Ss-Slp gene products, molecular weights were assigned by determining mobility relative to markers of known molecular weight on sodium dodecyl sulfate polyacrylamide gel electrophoresis. The *S* region gene products were assigned molecular weights using either gel filtration or gel electrophoresis. GP = glycoprotein. Data in light brackets refer to the column heading in brackets. Figures in heavy brackets are reference numbers.

Gene Product ⟨Synonym⟩ [Gene]	Tissue Distribution	Method of Isolation [Tissue]	Composition & Molecular Weight	Fine Structure Studies
H-2K & H-2D histocompatibility antigens [*H-2K* & *H-2D*]	Lymphoid cells; erythrocytes; most body tissues; most tumor cells	Det sol, im prcp **[**22**]** Pap dig, im prcp **[**22**]**	GP; 43,000-47,000 **[**22**]**[1] GP; 39,000-44,000; 29,000 **[**22**]**[1]	Tryptic peptide mapping **[**1, 2**]**; *N*-terminal amino acid sequence **[**10,14,27,32**]**; analysis of carbohydrate moiety **[**16,18**]**
Immune-response-gene-associated antigens ⟨Ia⟩ [*Ia-1* through *Ia-3*]	Lymphoid cells; macrophages; some tumor cells; skin; sperm	Det sol, im prcp [Spleen **[**4,5,7,9,29**]**]	GP; 55,000-58,000; subunits of 35,000 & 25,000, each bearing carbohydrate **[**4,5,7**]**. 30,000-35,000 **[**9,29**]**.	Analysis of carbohydrate moiety **[**6,11**]**. Tryptic peptide analysis **[**7**]**.

[1] Associated with β_2-microglobulin which was identified by its own serological or structural characteristics **[**19, 21, 26, 30**]**.

continued

46. SUMMARY OF MOLECULAR PROPERTIES OF GENE PRODUCTS OF THE *H-2* COMPLEX: MOUSE

Gene Product ⟨Synonym⟩ [Gene]	Tissue Distribution	Method of Isolation [Tissue]	Composition & Molecular Weight	Fine Structure Studies
		[Thymocytes or T cells [12,23]]	GP; 35,000 & 25,000 dalton chains [12,23]	
		[Macrophages [9, 24]]	GP; 58,000; subunits of 35,000 & 25,000 daltons [9,24]	
Serum substance ⟨Ss⟩ [*Ss*]	Serum; plasma	Gel filt [3,13,15,25]	Protein; 180,000; subunits or fragments of 95,000, 70,000, & 30,000 daltons [13,15, 25]	Serologically related to human C4 [13,15,25]
			Protein; 120,000; multiple subunits or fragments of 23,000 & 14,000 daltons [3]	
		Affinity chromatography [8]	Protein; 200,000; subunits or fragments of 70,000-80,000, 23,000, 13,000 daltons [8]	
Sex-limited protein ⟨Slp⟩ [*Slp*]	Serum; plasma	Gel filt [13,25]	Protein; 180,000 [13,15, 25]	Slp-positive molecules are a subset of Ss molecules [13, 25]
Thymus leukemia antigen ⟨TL⟩ [*Tla*]	Thymocytes; some leukemias	Det sol, im prcp [28]	GP; 40,000-50,000 [28] 2/	....
		Pap dig, im prcp [17]	GP; 39,000 [17] 2/	Carbohydrate moiety larger than H-2 carbohydrate [17]
F9 antigen [*T/t*]	Sperm; teratocarcinoma	Det sol, im prcp [31]	Protein; 44,000; possibly contains 22,000 daltons subunit [31] 3/	....

2/ Associated with β_2-microglobulin which was identified by its own serological or structural characteristics [20,30].

3/ Probably associated with β_2-microglobulin which was identified by molecular weight only.

Contributor: Susan E. Cullen

References

1. Brown, J. L., and S. G. Nathenson. 1977. J. Immunol. 118:98-102.
2. Brown, J. L., et al. 1974. Biochemistry 13:3174-3178.
3. Capra, J. D., et al. 1975. J. Exp. Med. 142:664-672.
4. Cullen, S. E., and B. D. Schwartz. 1976. J. Immunol. 117:136-142.
5. Cullen, S. E., et al. 1974. Proc. Natl. Acad. Sci. USA 71:648-652.
6. Cullen, S. E., et al. 1975. Transplant. Proc. 7:237-239.
7. Cullen, S. E., et al. 1976. Transplant. Rev. 30:236-270.
8. Curman, B., et al. 1975. Nature (London) 258:243-245.
9. Delovitch, T. L., and H. O. McDevitt. 1974. Immunogenetics 2:39-52.
10. Ewenstein, B. M., et al. 1976. Proc. Natl. Acad. Sci. USA 73:915-918.
11. Freed, J. H., et al. 1976. In J. L. Seligman, et al., ed. Membrane Receptors of Lymphocytes. North Holland, Amsterdam. pp. 241-246.
12. Goding, J. W., et al. 1975. Nature (London) 257:230-231.
13. Hansen, T. H., et al. 1974. Biochem. Genet. 12:281-293.
14. Henning, R., et al. 1976. Proc. Natl. Acad. Sci. USA 73:118-122.
15. Meo, T., et al. 1975. Ibid. 72:4536-4540.
16. Muramatsu, T., and S. G. Nathenson. 1970. Biochemistry 9:4875-4883.
17. Muramatsu, T., et al. 1973. J. Exp. Med. 137:1256-1262.
18. Nathenson, S. G., and T. Muramatsu. 1971. In G. A. Jamieson and T. J. Greenwalt, ed. Glycoproteins of Blood Cells. Lippincott, Philadelphia. pp. 254-262.

continued

19. Natori, T., et al. 1974. Transplantation 18:550-555.
20. Ostberg, L., et al. 1975. Nature (London) 253:735-736.
21. Rask, L., et al. 1974. Ibid. 249:833-834.
22. Schwartz, B. D., et al. 1973. Biochemistry 12:2157-2163.
23. Schwartz, B. D., et al. 1977. Proc. Natl. Acad. Sci. USA 74:1195-1199.
24. Schwartz, R. H., et al. 1976. Scand. J. Immunol. 5:731-743.
25. Shreffler, D. C. 1976. Transplant. Rev. 32:140-167.
26. Silver, J., and L. Hood. 1974. Nature (London) 249:764-765.
27. Silver, J., and L. Hood. 1976. Proc. Natl. Acad. Sci. USA 73:599-603.
28. Vitetta, E., et al. 1972. Cell. Immunol. 4:187-191.
29. Vitetta, E. S., et al. 1974. Immunogenetics 1:82-90.
30. Vitetta, E. S., et al. 1975. J. Immunol. 114:252-254.
31. Vitetta, E. S., et al. 1975. Proc. Natl. Acad. Sci. USA 72:3215-3219.
32. Vitetta, E. S., et al. 1976. Ibid. 73:905-909.

47. NEUROLOGIC MUTANTS: MOUSE

Holder: For full name, *see* list of HOLDERS at front of book; Common = holdings by more than seven laboratories.

Gene Symbol	Gene Name	Chromosome	Clinical Signs	Pathology	Holder	Reference
ac	Absent corpus callosum	....	None	Partial to complete absence of corpus callosum	....	31,42,43,77
ag	Agitans	14	Ataxia, fine tremor	? Atrophy of Purkinje cells & of cells in spinal cord	....	31,38,55,77
asp	Audiogenic seizure prone	4	Susceptibility to audiogenic seizures	None	....	11
av	Ames waltzer	10	Circling; head-shaking; deafness	Defects of membranous labyrinth, like *sh-2*	JAX; KAN; OAK	31,70,77
ax	Ataxia	18	Progressive paralysis with hypertonia; tremor	Reduced spinal cord white matter; abnormal glia & astrocytes; abnormal vertebral column	JAX; MAS; PAS	31,77
bc	Bouncy	18	Severe tremor; bouncing gait	Unknown	JAX	47
bh	Brain hernia	7	Microphthalmia or anophthalmia; hydrocephalus	Cerebral hernia; hydrocephalus; 75% eye defects	JAX	3,31,77
cb	Cerebral degeneration	....	Hydrocephalus at birth	*Ex vacuo* hydrocephalus—cerebral hemispheres, olfactory lobes	LUC	19,31,77
Cd	Crooked	6	Head movements; microphthalmia	Variable dysrhaphic abnormalities; pseudo- to anencephaly	LUC	31,62,77
ch	Congenital hydrocephalus	13	Embryonic hydrocephalus	Abnormal development of cartilaginous skull	JAX	31,34,77
Cm	Coloboma	2	Coloboma; circling; head-shaking	Coloboma	JAX; PAS	69,73
cri	Cribriform degeneration	4	Weakness; ataxia	Cribriform lesions in brainstem & spinal cord	HVD; ILL; JAX	33
ct	Curly tail	....	Spina bifida	Spina bifida; occasionally anencephaly	....	31,35,77
d^l	Dilute lethal	9	Opisthotonus; seizures	Myelin degeneration; like human phenylketonuria	HAR; JAX; MAS; OAK	10,31,77
d^{15}	Dilute-15	9	Behavior like d^l, but some survive	Like d^l	MAS; OAK	31,77
Dc	Dancer	19	Circling; head-shaking; no deafness	Complete absence of macula of utricle; defects of bony labyrinth	HAR; JAX	17,31,77

continued

Gene Symbol	Gene Name	Chromosome	Clinical Signs	Pathology	Holder	Reference
dn	Deafness	....	Deafness	Degeneration of organ of Corti & spiral ganglion	CON; LUC	31,77
Do	Disoriented	....	Abnormal motor coordination	None reported	JAX	25
dr	Dreher	1	Circling; head-shaking; deafness	Hindbrain hydrocephalus; abnormal bony & membranous labyrinths	CIN; HAL; LUC	31,77
dt	Dystonia musculorum	1	Sensory ataxia	Loss of nerve fibers in peripheral nerves, dorsal roots & ganglia, & dorsal & ventrolateral spinal cord columns	JAX; MAS; PAS	31,77
du	Ducky	9	Waddling gait; seizures	None reported	JAX	31,77,79
du^{td}	Torpid	9	Clumsy gait; ? seizures	None reported	....	31,77
dy	Dystrophia muscularis	10	Progressive muscular weakness, particularly hindlimbs	Amyelination of dorsal & ventral roots & peripheral nerves; muscle degeneration	JAX; MCM; NIJ; ORL; YUR	7,31
dy^{2J}	Dystrophia muscularis-2J	10	Much less severe than *dy*	Muscle changes as for *dy*	JAX; MCM	60
eb	Eye-blebs	10	Variable anophthalmia, microphthalmia	None reported	JAX	31,77
El	Epilepsy	....	Tonic-clonic seizures	No brain or visceral lesions	....	77
ey-1	Eyeless-1	....	Variable anophthalmia, microphthalmia	Abnormal development of optic nerve & visual tract	BRN; PAS	9,77
fi	Fidget	2	Circling; head-shaking	Abnormal labyrinth & eye development	CAM; HAR; JAX	31,77,84
Fu	Fused	17	Dysrhaphic disorders	Folding, duplication, overgrowth of neural tube	Common	31,77,82
Fu^{ki}	Kinky	17	Similar to *Fu*; homozygotes nonviable	Extensive CNS & visceral reduplication	ALB; ANN; COR; MIS; PAS	29,31,77
Gy	Gyro	X	Circling; deafness	Inner ear degeneration	HAR; JAX; LUC	31,53,77
ho	Hotfoot	....	Abnormal gait	None reported	JAX	22
hop	Hop-sterile	....	Hopping gait	None reported	LDS	41
hpy	Hydrocephalic polydactyl	6	Hydrocephalus; hopping gait	Hydrocephalus	CIN	39
hy-3	Hydrocephalus-3	8	Hydrocephalus	Non-inflammatory disorder of pia-arachnoid membranes	JAX	4,31,77
jc	Jackson circler	10	Circling behavior	None reported	JAX	80
je	Jerker	4	Circling; head-shaking; deafness	Degeneration of organ of Corti, spiral ganglion, stria vascularis	EDI; HAR; JAX; KAN	12,31,77
ji	Jittery	10	Muscular incoordination; seizures	Polycystic lesions of brain white matter	JAX; MAS	31,37,77
jo	Jolting	....	Tremor; incoordination of hindlimbs	None reported	....	31,77
jp	Jimpy	X	Tremor; convulsions	CNS sudanophilic leukodystrophy	HAR; JAX; PAS	31,76,77
jp^{msd}	Myelin synthesis deficiency	X	Like *jp*; ? more severe	Like *jp*	HVD; JAX	59
js	Jackson shaker	11	Circling; head-shaking; deafness	None reported	JAX	24

continued

Gene Symbol	Gene Name	Chromosome	Clinical Signs	Pathology	Holder	Reference
jv	Jackson waltzer	....	Circling; head-shaking	None reported	JAX	23
kr	Kreisler	2	Circling; head-shaking; deafness	Embryonic malformations of inner ear	HAL; JAX; OAK	15,31,77
Kw	Kinky-waltzer	....	Balance defect	None reported	OAK	30,31,77
la	Leaner	....	Truncal ataxia; hypertonia	Cerebellar degeneration, particularly anterior lobe	....	20,31,77
Lc	Lurcher	6	Swaying hind-quarters; falls to one side	None reported	HAR	31,66,77
lh	Lethargic	2	Ataxia; seizures	None observed in brain or spinal cord	JAX; MCM; TXA	28,31,77
Lp	Loop tail	1	Head wobbling	Failure of neural tube elongation & closure	JAX; NIJ; SHA	31,77,78
ls	Lethal spotting	2	Megacolon	Absence of neural crest-derived ganglion cells in colon	HAR; JAX	31,56,77
lz	Lizard	15	Hindlimb paralysis; tremor	None reported	....	31,63,77
mi	Microphthalmia	6	Small eyes	Abnormal optic cup development, with coloboma	HAL; JAX; LUC; OAK	31,65,77
Mo^{br}	Brindled	X	Slight tremor; ataxia	None reported	Common	31,77
Mp	Micropinna-microphthalmia	....	Microphthalmia	Anophthalmia in homozygotes	JAX; OAK	67
my	Blebs	3	Severe dysrhaphic states	Myelencephalic blebs during embryonic development	JAX	8,31,77
nr	Nervous	8	Ataxia	Purkinje cell loss	HVD; JAX; PAS	44
nv	Nijmegen waltzer	7	Circling; head-shaking	None reported	NIJ	1
oc	Osteosclerotic	19	Circling	Osteopetrosis	JAX	24
oh	Obstructive hydrocephalus	....	Communicating hydrocephalus	Early ventricular dilation; secondary aqueductal stenosis	....	6
opt	Opisthotonus	6	Opisthotonic seizures	None reported	JAX	48
or	Ocular retardation	....	Small eyes	Abnormal retinal layer; no optic nerve in adults	JAX; LUC	83
ot	Oscillator	....	Severe rapid tremor	None reported	....	24
p^s	Pink-eyed sterile	8	Nervous, uncoordinated behavior	? Related to abnormal aromatic amino acid metabolism	....	31,40,77
pa	Pallid	2	Nervousness; ataxia	Absence of otoliths	Common	31,51,77
pi	Pirouette	5	Circling; head-shaking; deafness	Degeneration of neuroepithelia of inner ear	JAX	31,51,77
pr	Porcine tail	....	Circling	Inner-ear abnormality; unequal growth of primitive streak	....	58
Pv	Pivoter	....	Circling; head-shaking; deafness	Like *pi*	LMX	61,77
Q	Quinky	8	Circling; head-shaking	Irregular closure of the neural tube	....	31,71,77
qk	Quaking	17	Tremor	Brain & spinal cord myelin deficiency	ANN; COR; HVD; JAX; ORL; PAS	31,76,77
qv	Quivering	7	Locomoter instability; trembling	Normal brain, spinal cord, & nerve roots	LUC; TXA	31,57,77

continued

Gene Symbol	Gene Name	Chromosome	Clinical Signs	Pathology	Holder	Reference
r	Rodless retina	10	Blindness	Arrested development & degeneration of photoreceptor cells	....	31,77
rd	Retinal degeneration	5	Blindness	Indistinguishable from *r*	HVD; JAX	31,77
rl	Reeler	5	Postural & locomotor ataxia	Abnormal cell distribution in cerebellum & cerebral cortex	HVD; ILL; JAX; PAS	31,36,77
s	Piebald	14	Occasional megacolon	Decreased numbers of neurons in myenteric plexus of GI tract	Common	5,31,77
s^l	Piebald-lethal	14	Megacolon	Lack of ganglion cells in distal colon	CON; JAX	31,45,77
Sey	Small eye	....	Small eyes	Reduced eye size	EDI	68
sg	Staggerer	9	Lurching gait	Small cerebellum, with reduced granule cells	JAX; HVD; MAS; OAK; PAS	31,75,77
sh-1	Shaker-1	7	Circling; head-shaking; deafness	Like *pi*	CAM; HAR; JAX; OAK	13,31,77
sh-2	Shaker-2	11	Circling; head-shaking; deafness	Like *sh-1* & *pi*, but earlier	CAM; HAR	12,31,77
shm	Shambling	11	Wobbly gait; hopping	Phospholipid deposits in lumbar spinal cord	ILL; JAX	31,32,77
Sig	Sightless	6	Blindness	Hydrocephalus	HAR; JAX; LUC	69,72
Sp	Splotch	1	Non-viable homozygotes	Lumbosacral & hindbrain myeloschisis	HAR; JAX; KAN; OAK	2,31,77
Sp^d	Delayed-splotch	1	Lethal postnatally	Like *Sp*, but caudal rachischisis only	JAX	21,31,77
spa	Spastic	12	Rapid tremor; stiffness	None reported	JAX; PAS	31,77
sr	Spinner	....	Circling; head-shaking; deafness	Like *pi*	JAX; KAN	18,31,77
su	Surdescens	....	Deafness	Like *sy*	....	31,77
sv	Snell's waltzer	9	Circling; head-shaking	Like *pi*	JAX; KAN	16,31,77
sw	Swaying	15	Ataxic gait	Malformed anterior vermis of cerebellum	HVD; JAX	74
Swl	Sprawling	....	Abnormal hindlimb behavior	Failure in maturation & myelination of sensory axons	HAR; MCM	27,64
sy	Shaker-with-syndactylism	18	Circling; head-shaking; deafness	Excessive development of mesenchymal tissue in periotic region	HAL; JAX	19,31,77
tb	Tumbler	1	Crab-like walk; side jumps	None reported	JAX	31,69,77
tg	Tottering	8	Wobbly gait; occasional seizures	None reported	ILL; JAX; NAG	31,77
tg^{la}	Leaner	8	Severe ataxia, trunk & limbs	Cerebellar degeneration, particularly anterior lobe	ILL; JAX	31,77
ti	Tipsy	11	Reeling gait, with falling	None reported	....	31,77
tm	Tremulous	....	Marked tremor	None reported	....	31,77
tn	Teetering	11	Progressive hypertonia	None reported	JAX	31,46,77
Tr	Trembler	11	Paresis & seizures when young; head trembling in adults	No lesions of brain or spinal cord	EDI; GLA; ORL; PAS; SYD	31,77
Tw	Twirler	18	Circling; head-shaking	Abnormal or absent semicircular canals	CAM; EDI; HAR; JAX; PAS	31,52,77

continued

Gene Symbol	Gene Name	Chromosome	Clinical Signs	Pathology	Holder	Reference
ub	Unbalanced	....	Head held tilted; no postural reflexes	Otoliths absent	....	31,77
v	Waltzer	10	Circling; head-shaking; deafness	Like *pi*	EDI; HAR; JAX; KAN; NIJ; PAS	13,31,77
v^{df}	Deaf	10	Deafness, with normal behavior; dominant to *v*	Like *pi*	LUC	14,31,77
Va	Varitint-waddler	12	Circling; head-shaking; deafness	Like *pi*	Common	31,49,77
vb	Vibrator	11	Generalized constant tremor	None reported	....	31,50,77
vc	Vacillans	4	Violent tremor; swaying gait	None reported	....	31,77
vl	Vacuolated lens	1	Caudal spina bifida	Spinal bifida	JAX	24,69
wd	Waddler	4	Side-to-side swaying of hindquarters	None reported	....	31
wh	Writher	....	Severe sensory ataxia	None reported	OAK	31,77
wi	Whirler	4	Circling; head-shaking; deafness	Like *sh-2*	JAX; OAK	31,77
wl	Wabbler-lethal	14	Impaired locomotion	Degeneration of vestibulo-spinal & spino-cerebellar tracts	JAX; MAS; PAS	26,31,77
wr	Wobbler	....	Tremor; progressive paralysis	Abnormal lower motor neurons; reduced cord white matter	....	31,77
Wt	Waltzer-type	....	Variable circling; head-shaking	One or more semicircular canals reduced or absent	OAK	31,77,81
wv	Weaver	....	Ataxic gait	Nearly complete absence of cerebellar granule cells; small cerebellum	HVD; ILL; JAX; PAS	31,77
Xt	Extra-toes	13	Cranioschisis	None reported	HAR; JAX; LUC; PAS	31,54

Contributor: Herbert C. Morse

References

1. Abeelen, J. H. F. van, and P. H. W. van der Kroon. 1967. Genet. Res. 10:117-118.
2. Auerbach, R. 1954. J. Exp. Zool. 127:305-329.
3. Bennett, D. 1959. J. Hered. 50:265-268.
4. Berry, R. J. 1961. J. Pathol. Bacteriol. 81:157-167.
5. Bielschowsky, M., and C. C. Schofield. 1962. Aust. J. Exp. Biol. Med. Sci. 40:395-404.
6. Borit, A., and R. L. Sidman. 1972. Acta Neuropathol. 21:316-331.
7. Bradley, W. G., and M. Jenkison. 1973. J. Neurol. Sci. 18:227-247.
8. Carter, T. C. 1959. J. Genet. 56:401-435.
9. Chase, H. B. 1945. J. Comp. Neurol. 83:121-139.
10. Coleman, D. L. 1962. Arch. Biochem. Biophys. 96: 562-568.
11. Collins, R. L. 1970. Behav. Genet. 1:99-109.
12. Deol, M. S. 1954. J. Genet. 52:562-588.
13. Deol, M. S. 1956. Proc. R. Soc. London B145:206-213.
14. Deol, M. S. 1956. J. Embryol. Exp. Morphol. 4:190-195.
15. Deol, M. S. 1964. Ibid. 12:475-490.
16. Deol, M. S., and M. C. Green. 1966. Genet. Res. 8: 339-345.
17. Deol, M. S., and P. W. Lane. 1966. J. Embryol. Exp. Morphol. 16:543-558.
18. Deol, M. S., and M. W. Robins. 1962. J. Hered. 53: 133-136.
19. Deol, M. S., and G. M. Truslove. 1963. Genet. Today Proc. 11th Int. Congr. Genet. (The Hague) 1:183-184 (Abstr. 10.44).
20. Dickie, M. M. 1963. Mouse News Lett. 28:34.
21. Dickie, M. M. 1964. J. Hered. 55:97-101.
22. Dickie, M. M. 1966. Mouse News Lett. 34:29-30.
23. Dickie, M. M. 1966. Ibid. 35:31.
24. Dickie, M. M. 1967. Ibid. 36:39-40.
25. Dickie, M. M. 1968. Ibid. 39:27-28.
26. Dickie, M. M., et al. 1952. J. Hered. 43:283-286.

continued

27. Duchen, L. W. 1975. Neuropathol. Appl. Neurobiol. 1:89-101.
28. Dung, H. C., and R. H. Swigart. 1970. Tex. Rep. Biol. Med. 30:23-39.
29. Gluecksohn-Schoenheimer, S. 1949. J. Exp. Zool. 110:47-76.
30. Gower, J. S., and M. B. Cupp. 1958. Mouse News Lett. 19:37.
31. Green, E. L., ed. 1966. Biology of the Laboratory Mouse. Ed. 2. McGraw-Hill, New York.
32. Green, E. L. 1968. J. Hered. 59:59.
33. Green, M. C., et al. 1972. Science 176:800-803.
34. Grüneberg, H. 1943. J. Genet. 45:1-28.
35. Grüneberg, H. 1954. Ibid. 52:52-67.
36. Hamburgh, M. 1963. Dev. Biol. 8:165-185.
37. Harman, P. J. 1950. Anat. Rec. 106:304(Abstr. D-17).
38. Hoecker, G., et al. 1944. J. Hered. 45:10-14.
39. Hollander, W. F. 1966. Am. Zool. 6:588-589(Abstr. 366).
40. Hollander, W. F., et al. 1960. Genetics 45:413-418.
41. Johnson, D. R., and D. M. Hunt. 1971. J. Embryol. Exp. Morphol. 25:223-236.
42. King, L. S. 1936. J. Comp. Neurol. 64:337-363.
43. King, L. S., and C. E. Keeler. 1932. Proc. Natl. Acad. Sci. USA 18:525-528.
44. Landis, S. C. 1973. J. Cell Biol. 57:782-797.
45. Lane, P. W. 1962. Mouse News Lett. 26:35.
46. Lane, P. W. 1967. Ibid. 37:34.
47. Lane, P. W. 1969. Ibid. 40:30.
48. Lane, P. W. 1972. Ibid. 47:36-37.
49. Lane, P. W. 1972. J. Hered. 63:135-140.
50. Lane, P. W. 1974. Mouse News Lett. 50:43.
51. Lyon, M. F. 1955. J. Embryol. Exp. Morphol. 3:230-241.
52. Lyon, M. F. 1958. Ibid. 6:105-116.
53. Lyon, M. F. 1960. Mouse News Lett. 23:30.
54. Lyon, M. F., et al. 1967. Genet. Res. 9:383-385.
55. Martinez, A., and J. L. Sirlin. 1955. J. Comp. Neurol. 103:131-137.
56. Mayer, T. C., and E. L. Maltby. 1964. Dev. Biol. 9: 269-286.
57. McNutt, W. 1962. Anat. Rec. 142:257.
58. McNutt, W. 1969. Ibid. 163:340.
59. Meier, H., and A. D. MacPike. 1970. Exp. Brain Res. 10:512-525.
60. Meier, H., and J. L. Southard. 1970. Life Sci. 9(2): 137-144.
61. Meredith, R. 1969. Mouse News Lett. 40:35-36.
62. Morgan, W. C., Jr. 1954. J. Genet. 52:354-373.
63. Morris, T. 1966. Mouse News Lett. 35:27.
64. Morris, T. 1967. Ibid. 37:29.
65. Müller, G. 1951. Wiss. Z. Martin-Luther-Univ. Halle-Wittenberg Math. Naturwiss. Reihe 1(4):27-43.
66. Phillips, R. J. S. 1960. J. Genet. 57:35-42.
67. Phipps, E. L. 1965. Mouse News Lett. 33:68.
68. Roberts, R. C. 1967. Genet. Res. 9:121-122.
69. Robinson, R. 1972. Gene Mapping in Laboratory Animals. Plenum, New York. pt. A and B.
70. Schaible, R. H. 1956. Mouse News Lett. 15:29.
71. Schaible, R. H. 1961. Ibid. 24:38-40.
72. Searle, A. G. 1965. Ibid. 33:29.
73. Searle, A. G. 1966. Ibid. 35:27.
74. Sidman, R. L. 1968. In F. R. Carlson, ed. Physiological and Biochemical Aspects of Nervous Integration. Prentice-Hall, Englewood Cliffs, NJ. pp. 163-193.
75. Sidman, R. L., et al. 1962. Science 137:610-612.
76. Sidman, R. L., et al. 1964. Ibid. 144:309-311.
77. Sidman, R. L., et al. 1965. Catalog of the Neurological Mutants of the Mouse. Harvard Univ., Cambridge.
78. Smith, L. J., and K. F. Stein. 1962. J. Embryol. Exp. Morphol. 10:73-87.
79. Snell, G. D. 1955. J. Hered. 46:27-29.
80. Southard, J. L. 1970. Mouse News Lett. 42:30.
81. Stein, K. F., and S. A. Huber. 1960. J. Morphol. 106: 197-203.
82. Theiler, K., and S. Gluecksohn-Waelsch. 1956. Anat. Rec. 125:83-103.
83. Theiler, K., et al. 1976. Anat. Embryol. 150:85-97.
84. Truslove, G. M. 1956. J. Genet. 54:64-86.

48. BEHAVIORAL GENE EFFECTS: MOUSE

Part I. Single Locus Behavior

Gene Symbol	Gene Name	Chromosome	Test Situation	Reference
Aal	Active avoidance learning	9	Shuttle box	7
Aap	Active avoidance performance	....	Jump box	9
asp	Audiogenic seizure prone	4	Bell jar	2,3
Bfo	Bell-flash ovulation	4	Home cage	4
Cpz	Chlorpromazine avoidance	9	Shuttle box	1
Exa	Exploratory activity	4	Open-field	6
Pp	Passive performance	....	Jump box	8

continued

48. BEHAVIORAL GENE EFFECTS: MOUSE

Part I. Single Locus Behavior

Gene Symbol	Gene Name	Chromo-some	Test Situation	Refer-ence
Sac	Saccharin preference	....	Home cage	5
Sco	Scopolamine modification of exploratory activity	17	Open-field	6
Sip	Schedule induced polydipsia	....	Operant chamber	10

Contributor: Richard L. Sprott

References

1. Castellano, C., et al. 1974. Psychopharmacologia 34: 309-316.
2. Collins, R. L. 1970. Behav. Genet. 1:99-109.
3. Collins, R. L., and J. L. Fuller. 1968. Science 162: 1137-1139.
4. Eleftheriou, B. E., and M. B. Kristal. 1974. J. Reprod. Fertil. 38:41-47.
5. Fuller, J. L. 1974. J. Hered. 65:33-36.
6. Oliverio, A., et al. 1973. Physiol. Behav. 10:893-899.
7. Oliverio, A., et al. 1973. Ibid. 11:497-501.
8. Sprott, R. L. 1974. Behav. Biol. 11:231-237.
9. Stavnes, K., and R. L. Sprott. 1975. Psychol. Rep. 36:515-521.
10. Symons, J. P., and R. L. Sprott. 1976. Physiol. Behav. 17:837-839.

Part II. Mutations with Pleiotropic Behavior Effects

Data were adapted from reference 19. **Behavioral Effect:** 1 = avoidance behavior; 2 = circling; 3 = feeding; 4 = motor coordination; 5 = miscellaneous suspected effects. **Holder:** For full name, *see* list of HOLDERS at front of book; Common = holdings by more than seven laboratories.

Gene Symbol	Gene Name	Chromo-some	Behavioral Effect	Holder	Refer-ence
A^{vy}	Viable yellow	2	3	CAL; JAX; LIL; NCI; NCT	7
A^{y}	Yellow	2	3	Common	7
a	Non-agouti	2	5	Common	7
Ad	Adult obesity & diabetes	7	3	CAM	20
ald	Adrenocortical lipid depletion	1	1	JAX	7
av	Ames waltzer	10	2	JAX; KAN; OAK	7
ax	Ataxia	18	4	JAX; MAS; PAS	7
c	Albino	7	5, 1	Common	7
d	Dilute	9	5	Common	7
db	Diabetes	4	3	Common	9
Dc	Dancer	19	2	HAR; JAX	4
Do	Disoriented	....	4	JAX	14
du	Ducky	9	4	JAX	7
dv	Dervish	....	2	OAK	7
ecl	Epistatic circling of C57L/J	....	2	JAX	16
ecs	Epistatic circling of SWR/J	....	2	JAX	16
fi	Fidget	2	4	CAM; HAR; JAX	7
jc	Jackson circler	10	2	JAX	15
je	Jerker	4	4	EDI; HAR; JAX; KAN; LUC; PAS	7
ji	Jittery	10	4	JAX; MAS	7
jp	Jimpy	X	4	HAR; JAX; PAS	7
js	Jackson shaker	11	4	JAX	11
jv	Jackson waltzer	....	2	JAX	[illegible]

Part II. Mutations with Pleiotropic Behavior Effects

Gene Symbol	Gene Name	Chromosome	Behavioral Effect	Holder	Reference
Kw	Kinky waltzer	....	2	OAK	7
Lc	Lurcher	6	4	HAR	7
med	Motor end-plate disease	15	4	HAR; JAX; PAS	5
nr	Nervous	8	4	HVD; JAX; PAS	13
nv	Nijmegen waltzer	7	2	NIJ	1
ob	Obese	6	3	Common	7
pi	Pirouette	5	2	JAX	7
Pv	Pivoter	....	2	LMX	17
qk	Quaking	17	4	ANN; COR; HVD; JAX; ORL; PAS	7
qv	Quivering	7	4	LUC; TXA	7
rd	Retinal degeneration	5	5	HVD; JAX	7
rg	Rotating	....	2	....	2
rl	Reeler	5	4	HVD; ILL; JAX; PAS	7
Rv	Revolving	....	2	....	10
se^{sv}	Short-eared waltzer	9	2	OAK	18
sg	Staggerer	9	4	JAX; HVD; MAS; OAK; PAS	7
sh-1	Shaker-1	7	4	CAM; HAR; JAX; OAK	7
sh-2	Shaker-2	11	4	CAM; JAX	7
shm	Shambling	11	4	ILL; JAX	8
spa	Spastic	12	4	JAX; PAS	7
Spl	Plasma serotonin level	9	5	JAX	6
sr	Spinner	....	2	JAX; KAN	7
sv	Snell's waltzer	9	2	JAX; OAK	3
sw	Swaying	15	4	HVD; JAX	12
tb	Tumbler	1	4	JAX	7
tg	Tottering	8	4	ILL; JAX; NAG	7
tm	Tremulous	....	4	....	7
tn	Teetering	11	4	JAX	7
Tr	Trembler	11	4	EDI; GLA; ORL; PAS; SYD	7
Tw	Twirler	18	2	CAM; EDI; HAR; JAX; PAS	7
v	Waltzer	10	2	EDI; HAR; JAX; KAN; NIJ; PAS	7
Va	Varitint-waddler	12	4	Common	7
vb	Vibrator	11	4	JAX	7
vc	Vacillans	4	4	....	7
wd	Waddler	4	4	....	7
wh	Writher	....	4	OAK	7
wi	Whirler	4	2	JAX; OAK	7
wl	Wabbler-lethal	14	4	JAX; MAS; PAS	7
wr	Wobbler	....	4	....	7
Wt	Waltzer-type	....	2	OAK	7
wv	Weaver	....	4	HVD; ILL; JAX; PAS	7

Contributor: Richard L. Sprott

References

1. Abeelen, J. H. F. van, and P. H. W. van der Kroon. 1967. Genet. Res. 10:117-118.
2. Deol, M., and M. M. Dickie. 1967. J. Hered. 58:69-72.
3. Deol, M., and M. C. Green. 1966. Genet. Res. 8:339-345.
4. Deol, M., and P. W. Lane. 1966. J. Embryol. Exp. Morphol. 16:543-558.
5. Duchen, L. W. 1970. J. Neurol. Neurosurg. Psychiatry 33:238-250.
6. Eleftheriou, B. E., and D. W. Bailey. 1972. J. Endocrinol. 55:225-226.
7. Green, E. L., ed. 1966. Biology of the Laboratory Mouse. Ed. 2. McGraw-Hill, New York.
8. Green, E. L. 1967. J. Hered. 58:65-68.

continued

9. Hummel, K., et al. 1966. Science 153:1127-1128.
10. Lane, P. W., et al. 1966. Mouse News Lett. 35:31-33.
11. Lane, P. W., et al. 1967. Ibid. 36:38-40.
12. Lane, P. W., et al. 1967. Ibid. 36:40-41.
13. Lane, P. W., et al. 1967. Ibid. 37:26-41.
14. Lane, P. W., et al. 1968. Ibid. 39:27-28.
15. Lane, P. W., et al. 1970. Ibid. 42:29-30.
16. Lane, P. W., et al. 1976. Ibid. 55:16-18.
17. Meredith, R. 1969. Ibid. 40:35-36.
18. Russell, L. B. 1971. Mutat. Res. 11:107-123.
19. Searle, A. G., ed. 1977. Mouse News Lett. 56:5-23.
20. Wallace, M. E., et al. 1975. Ibid. 53:19-21.

49. VIRUSES AND INCIDENCE OF MAMMARY TUMORS: MOUSE

Inbred mouse strains are classified as low mammary cancer or high mammary cancer strains. High mammary cancer strains were generally, but not always, obtained by selection combined with inbreeding, e.g., the GR strain. The classical strains were developed in the twenties and thirties by Strong, Little, and others. High mammary cancer strains always have a very high percentage of mammary tumors before one year of age, mostly between 80-100%. They invariably express the mammary tumor virus. Low mammary cancer strains do not necessarily have a low mammary cancer incidence; up to 50%, and sometimes 60%, may develop mammary cancer, but generally the tumors appear after one year of age. Both, $C3H_f$ and BALB/c strains have sublines with quite a high incidence of mammary tumors, but nonetheless are usually referred to as low mammary cancer strains.

With one exception all high mammary cancer strains can be changed into low mammary cancer strains by foster-nursing the newborn females on a female of a low mammary cancer strain. Strains which can be so changed carry the so-called milk agent or milk factor, or mammary tumor virus (MTV). Only the GR strain cannot be changed into a low mammary cancer strain by foster-nursing, since the mammary tumor virus variant of this strain is inherited as a complex of genes.

The group of mammary tumor viruses of the mouse can be classified, according to their mode of transmission, into two distinct groups: (i) endogenous viruses, DNA copies of which are present in the germline and all somatic cells of a mouse strain (the DNA copies are called germinal proviruses); and (ii) exogenous viruses, DNA copies of which are present in a limited number of somatic cells only (these copies are called somatic proviruses). All milk agents have somatic proviruses only. The mode of transmission is mainly through the mother's milk, but can also take place through the father's seminal fluid. For general information on the genetics of resistance and susceptibility to mammary tumorigenesis in the mouse, consult reference 4 in Part I. Figures in heavy brackets are reference numbers.

Abbreviation: MTV = mammary tumor virus.

Part I. Genes Controlling Induction of Endogenous Mammary Tumor Viruses

The table below summarizes present knowledge of the variants of endogenous MTV's of the mouse. The *Mtv-1* gene represents the endogenous MTV strain of the $C3H_f$ mouse. The locus is responsible for relatively early mammary tumorigenesis in BALB/c x (BALB/c x $C3H_f$) backcross females at approximately 12 months of age. The fact that the $C3H_f$ strain is not a high mammary cancer strain may be due to host resistance factors (alleles) that interfere with replication of induced virus. The *Mtv-1* locus is situated on the first linkage group, or chromosome 7, between albino (*c*) and glucose-phosphate-isomerase-1 (*Gpi-1*). The exact location is not yet known. The *Mtv-2* locus represents the highly oncogenic endogenous MTV strain of the GR mouse. The locus is responsible for pregnancy-dependent mammary tumors. It is not allelic with *Mtv-1*, and it has not yet been mapped. A low mammary cancer strain has recently been developed by removing the *Mtv-2* locus from the GR strain in a series of intercrosses and backcrosses [8]. Several

continued

Part I. Genes Controlling Induction of Endogenous Mammary Tumor Viruses

genotypes can inhibit the oncogenic potency of both the *Mtv-1* and *Mtv-2* loci. In case of *Mtv-2*, however, no genotype is known that can prevent carcinogenesis completely.

The BALB/c genotype potentiates the carcinogenic action of *Mtv-1*, probably by enhancing MTV replication [1,9]. The following genotypes inhibit tumorigenesis in F_1 hybrids with $C3H_f$: CBA, O20 [6]; C57BL [3,6]. The C57BL genotype delays mammary carcinogenesis by *Mtv-2* in $((C57BL \times C3H_f)F_1 \times C57BL)$ backcrosses (recessive) [7]. The DBA_f genotype delays mammary carcinogenesis by *Mtv-2* in F_1 hybrids with $C3H_f$ (dominant) [7]. **Holder**: Common = holdings by more than seven laboratories.

Gene Symbol	Gene Name	Chromosome	Strain	Holder	Reference
Mtv-1	Mammary tumor virus inducer 1	7	$C3H_f$	Common	5,9
Mtv-2[1/]	Mammary tumor virus inducer 2	Fn[2/]	GR	Common	1,2,5,7,10,11

[1/] *Mtv-2* was previously called Ms^E [2]. [2/] *Mtv-2* is not located on chromosome 7 and thus is not allelic to *Mtv-1* [7].

Contributor: Jo Hilgers

References

1. Bentvelzen, P., and J. H. Daams. 1969. J. Natl. Cancer Inst. 43:1025-1035.
2. Bentvelzen, P. A. J. 1968. Genetical Control of the Vertical Transmission of the Mühlbock Mammary Tumour Virus in the GR Mouse Strain. Hollandia, Amsterdam.
3. Heston, W. E., and M. K. Deringer. 1952. J. Natl. Cancer Inst. 13:167-175.
4. Hilgers, J., and P. Bentvelzen. 1978. Adv. Cancer Res. 26:143-195.
5. Hilgers, J., et al. 1976. Mouse News Lett. 54:29.
6. Mühlbock, O. 1956. Acta Unio Int. Contra Cancrum 12:665-681.
7. Van Nie, R., and J. Hilgers. 1976. J. Natl. Cancer Inst. 56:27-32.
8. Van Nie, R., and J. de Moes. 1977. Int. J. Cancer 20: 588-594.
9. Van Nie, R., and A. A. Verstraeten. 1975. Ibid. 16: 922-931.
10. Van Nie, R., et al. 1972. In J. Mouriquand, ed. Fundamental Research on Mammary Tumours. INSERM, Paris. pp. 21-29.
11. Van Nie, R., et al. 1977. Int. J. Cancer 19:383-390.

Part II. Genetic Resistance to Exogenous Mammary Tumor Viruses

Data summarize the most important aspects of resistance and susceptibility to exogenous MTV's in the mouse. It is surprising that virtually every mouse strain carries a different MTV. Although an attempt has been made to classify MTV's into different types, according to their biological behavior, host range studies reveal that they all may be different. This is the more surprising since immunological and molecular biological studies have not revealed striking differences so far. One should bear in mind that this table is a key to the literature, and that definitive conclusions can only be drawn from the original papers as experimental conditions vary greatly. In general, foster-nursing is superior to intraperitoneal injection of MTV, even of very large doses (probably due to immunizing effects). Data on MTV strains of low mammary cancer strains are still scarce, and it has yet to be determined whether every mouse strain is capable of producing a full infectious virion. The molecular hybridization studies, which have shown that MTV-DNA copies are present in all strains, suggest that this capability exists.

Tumor incidence is not always an exact function of oncogenicity, since in many strains spontaneous tumors occur to a varying degree. Also, the age at which mammary tumors appear is very often most important. In some cases differences in susceptibility of the same strain (e.g., RIIIf strain for MTV-P) may be due to subline differences of the strain or to the virus variant. Also, the quantity of the virus administered may be an influential factor. Thus, relatively small differences in resistance are not included in this table. **Strain Classification By Tumor Incidence**: Data were generated from (forced) breeders after foster-nursing or intraperitoneal injection of purified MTV or tumor extracts.

continued

Part II. Genetic Resistance to Exogenous Mammary Tumor Viruses

MTV Variant		Strain Classification By Tumor Incidence		
Type	Strain of Origin	≤20%	20-60%	≥60%
MTV-S [5]	A-MTV	CBA [25] C57BL [10,25] DBA$_f$ [25] O20 [23,25,26] RIET [25] WLL$_f$ [25]	....	A$_f$ [23] BALB/c [14,19,22,31] C3H$_f$ [25]
		A$_f$ x C57BL [24] C57BL x DBA$_f$ [23]	A$_f$ x IF [24] O20 x A$_f$ [24]	A$_f$ x CBA [24] A$_f$ x C3H$_f$ [13] A$_f$ x DBA$_f$ [9] A$_f$ x O20 [24]
MTV-S or Bittner virus [7,8]	C3H-MTV	CBA [25] C57BL [1,7,24,25] TSI [3] WLL$_f$ [25] Y [1]	BL [15] CBA [24] C57BL [17,25,26] C57BL/10 [17,27] IF [25] NH [12] O20 [3,7,24,25] OIR [26] STS [26]	BALB/c [1,3,7,19,31] BIMA [17] CE [11] C3H$_f$ [3,7,24,25] IF [24] LIS [26] MA [28] MAS [3,26] O20 [26]
		....	C3H$_f$ x O20 [8] C57BL x BALB/c [8]	C3H$_f$ x A$_f$ [13,24] C3H$_f$ x BALB/c [8] C3H$_f$ x IF [24] C3H$_f$ x WLL$_f$ [24] C57BL x C3H$_f$ [8,24] C57BL x I [1] NH x I [12] O20 x BALB/c [8] O20 x DBA$_f$ [23]
		B10-*H-1*^{b} [17] B10-*H-8*^{b} [17]	B10-*H-1*^{a} [17] B10-*H-2*^{a} [27] B10-*H-2*^{k} [17,27] B10-*H-2*^{m} [27] B10-*H-2*^{pa} [27] B10-*H-2*^{r} [27] B10-*H-3*^{b} [17] B10-*H-9*^{b} [17]	B10-*H-2*^{d} [17,27] B10-*H-2*^{f} [17,27] B10-*H-2*^{i} [27] B10 x B10-*H-2*^{i} [27]
MTV-S [4]	DBA-MTV	A$_f$ [24] C3H$_f$ [25] C57BL [25] RIET [25]	CBA [25] O20 [23-26]	BALB/c [19] DBA$_f$ [23,24] OIR [26] WLL$_f$ [25]
		....	DBA$_f$ x A$_f$ [9] O20 x A$_f$ [24]	C57BL x DBA$_f$ [23] O20 x DBA$_f$ [23,24]
MTV[1]	WLL-MTV	C57BL [26] O20 [26]	....	BALB/c [20,31] WLL$_f$ [23,24]
		....	....	O20 x DBA$_f$ [23] WLL$_f$ x A$_f$ [24] WLL$_f$ x CBA [24] WLL$_f$ x C57BL [24] WLL$_f$ x IF [24] WLL$_f$ x O20 [24]

[1] This MTV variant has not been given a type name.

continued

Part II. Genetic Resistance to Exogenous Mammary Tumor Viruses

MTV Variant		Strain Classification By Tumor Incidence		
Type	**Strain of Origin**	**≤20%**	**20-60%**	**≥60%**
MTV-P [4]	DD-MTV	...	...	BALB/c [20,31] DD$_f$ [31]
MTV-P or Mühlbock virus [7,8]	GR-MTV	STS [26]	O20 [17]	BALB/c [7] BIMA [17] C3H$_f$ [7] C57BL [17,26] LIS [26] MAS [26] OIR [26]
		...	...	C3H$_f$ x BALB/c [7]
		...	B10-*H-1^a* [17] B10-*H-1^b* [17] B10-*H-2^d* [17] B10-*H-2^f* [17] B10-*H-2^k* [17] B10-*H-2^m* [17] B10-*H-3^b* [17] B10-*H-7^b* [17] B10-*H-8^b* [17]	...
MTV-P [4], now called MTV-PS [5]	RIII-MTV	RIII$_f$ [21,22,23]	C3H$_f$ [21,22] C57BL [21,22]	A$_f$ [21,22] BALB/c [19,21,22,29] RIII$_f$ [23]
MTV-ML [5]	BL-MTV	BL [15]	C3H$_f$ [15]	...
MTV-W [5]	"WILD"-MTV	"WILD" [2]	...	BALB/c [2]
MTV-L or nodule-inducing virus [7,8]	C3H$_f$-MTV	...	...	BALB/c [18]
		...	C57BL x A$_f$ [16] C57BL x RIII [16]	...
MTV-L [5]	A$_f$-MTV	...	C57BL x A$_f$ [16]	...
MTV-O [7,8]	BALB-MTV	...	...	BALB/c [6,18]
MTV-X [7,8]	O20-MTV	...	BALB/c [30]	...

Contributor: Jo Hilgers

References

1. Andervont, H. B. 1940. J. Natl. Cancer Inst. 1:147-153.
2. Andervont, H. B., and T. B. Dunn. 1956. Acta Unio Int. Contra Cancrum 12:530-543.
3. Bentvelzen, P. 1972. In P. Emmelot and P. Bentvelzen, ed. RNA Viruses and Host Genome in Oncogenesis. North Holland, Amsterdam. pp. 309-337.
4. Bentvelzen, P. 1972. Int. Rev. Exp. Pathol. 11:259-297.
5. Bentvelzen, P. 1974. Biochim. Biophys. Acta 355: 236-259.
6. Bentvelzen, P. 1975. Cold Spring Harbor Symp. Quant. Biol. 39:1145-1150.
7. Bentvelzen, P., and J. H. Daams. 1969. J. Natl. Cancer Inst. 43:1025-1035.
8. Bentvelzen, P., et al. 1970. Proc. Natl. Acad. Sci. USA 67:377-384.
9. Bittner, J. J. 1936. Proc. Soc. Exp. Biol. Med. 34:42-48.
10. Bittner, J. J. 1940. J. Natl. Cancer Inst. 1:155-168.
11. Bittner, J. J. 1954. Cancer Res. 14:783-789.
12. Bittner, J. J. 1958. Ann. N.Y. Acad. Sci. 71:943-975.
13. Bittner, J. J., and R. A. Huseby. 1946. Cancer Res. 6:235-239.
14. Blair, P. B. 1960. Ibid. 20:635-642.
15. Deringer, M. K. 1970. J. Natl. Cancer Inst. 45:215-218.
16. Dmochowski, L., et al. 1963. Acta Unio Int. Contra Cancrum 19:276-279.
17. Dux, A. 1972. (Loc. cit. ref. 3). pp. 301-308.
18. Hageman, P., et al. 1972. (Loc. cit. ref. 3). pp. 283-300.
19. Hummel, K. P., and C. C. Little. 1959. J. Natl. Cancer Inst. 23:813-821.
20. Matsuzawa, A., et al. 1970. Jpn. J. Exp. Med. 40:159-181.
21. Moore, D. H., et al. 1974. J. Natl. Cancer Inst. 52: 1757-1762.

continued

Part II. Genetic Resistance to Exogenous Mammary Tumor Viruses

22. Moore, D. H., et al. 1976. Ibid. 57:889-896.
23. Mühlbock, O. 1951. Proc. Kon. Ned. Akad. Wet. 54: 386-390.
24. Mühlbock, O. 1956. Acta Unio Int. Contra Cancrum 12:665-681.
25. Mühlbock, O. 1966. Jaarb. Kankeronferz. Kankerbestr. 16:9-16.
26. Mühlbock, O., and A. Dux. 1972. In J. Mouriquand, ed. Fundamental Research on Mammary Tumours. INSERM, Paris. pp. 11-19.
27. Mühlbock, O., and A. Dux. 1974. J. Natl. Cancer Inst. 53:993-996.
28. Murray, W. S. 1963. Ibid. 30:605-610.
29. Squartini, F., et al. 1963. Nature (London) 197:505.
30. Timmermans, A., et al. 1969. J. Gen. Virol. 4:619-621.
31. Vlahakis, G. 1973. J. Natl. Cancer Inst. 51:1711-1712.

Part III. Specific Genes Influencing Mammary Tumorigenesis

This information summarizes the specific genes, except the endogenous viral loci, known to influence mammary tumorigenesis. For the genes *ob, dw,* A^y, and A^{vy}, consult reference 1; for the *H-2* complex, consult reference 4. **Holder**: For full name, *see* list of HOLDERS at front of book; Common = holdings by more than seven laboratories.

Gene Symbol	Gene Name	Effect	Chromosome	Holder	Reference
ob	Obese	Enhances appearance of mammary tumors	6	BRC, EDI, JAX, LIL, NIJ	2,3
dw	Dwarf	Eliminates mammary tumorigenesis	....	Common	3
A^y	Yellow (lethal)	Increases incidence of mammary tumors	2	Common	1,3
A^{vy}	Viable yellow	Increases incidence of mammary tumors	2	JAX, LIL, NCI, NCT	3
*H-2*1/	Major histocompatibility complex	....	17	Common	4

1/ *See also* Part II of this table.

Contributor: Jo Hilgers

References

1. Heston, W. E., and G. Vlahakis. 1961. J. Natl. Cancer Inst. 26:969-983.
2. Heston, W. E., and G. Vlahakis. 1962. Ibid. 29:197-209.
3. Heston, W. E., and G. Vlahakis. 1967. 12th Annu. Symp. Fundam. Cancer Res. (Houston), pp. 347-363.
4. Mühlbock, O., and A. Dux. 1974. J. Natl. Cancer Inst. 53:993-996.

50. PULMONARY ADENOMAS: MOUSE

Strain or Hybrid: In hybrid designations, the strain of the mother is listed first. **Age**: Animals were necropsied at given age, unless otherwise indicated.

Part I. Incidence of Spontaneous Tumors

Strain or Hybrid	Sex	Age, mo	Incidence, %	Reference
A/He	♂♀	12	53	3
		15	79	
		18	90	
(A x L)F_1	♂♀	12	31	3
		15	48	
		18	53	
BALB/cAnDe	♀	17	26	1
BALB/cDeHe	♀	21 1/	28	5

1/ Average age of all animals; mice varied in age at autopsy.

continued

50. PULMONARY ADENOMAS: MOUSE

Part I. Incidence of Spontaneous Tumors

Strain or Hybrid	Sex	Age, mo	Incidence, %	Reference
BL	♂♀	>12	37	6
C3H/He	♂	14	7.8	4
$C3H_eB/De$	♂	14	12.7	4
$C3H_fB/He$	♂	14	10.4	4
	♀	21[1/]	9	5
C57L/He	♂♀	15	0	3
		18	0	
ST	♂	12.6	5	2
	♀	11.3	4	2
SWR	♂♀	>18	80	6

1/ Average age of all animals; mice varied in age at autopsy.

Contributor: Walter E. Heston

References

1. Deringer, M. K. 1965. J. Natl. Cancer Inst. 35:1047-1052.
2. Deringer, M. K., et al. 1953. Ibid. 14:375-380.
3. Heston, W. E. 1942. Ibid. 3:79-82.
4. Heston, W. E., and G. Vlahakis. 1960. Ibid. 24:425-435.
5. Heston, W. E., et al. 1973. Ibid. 51:209-224.
6. Lynch, C. J. 1943. Proc. Soc. Exp. Biol. Med. 52:368-371.

Part II. Effect of Specific Genes or Linkage Groups on Tumor Incidence

Hybrid	Genotype	Sex	Age mo	Incidence %	Reference
$(A \times HR)F_1 \times HR$[1/]	*P/p Hr/hr*	♂	23.9	50	2
		♀	25.3	33	
	p/p Hr/hr	♂	24.8	33	
		♀	25.8	26	
	P/p hr/hr	♂	25.5	8	
		♀	26.3	16	
	p/p hr/hr	♂	23.5	13	
		♀	24.8	11	
(A x V-*ob*[2/])F_2	*ob/+* & *+/+*	♂♀	14	36	3
	ob/ob	♂♀	14	20	
(A x WA[2/])F_1 x WA[1/]	*F/f*	♂♀	18.6	38	4
	f/f	♂♀	18.6	27	
	Sh-2/sh-2	♂♀	18.7	40	
	sh-2/sh-2	♂♀	18.5	25	
(A x Y)F_1	A^y/a	♂♀	15	40	1
	a/a	♂♀	15	21	

1/ First backcross. 2/ Stock, not strain.

Contributor: Walter E. Heston

References

1. Heston, W. E., and M. K. Deringer. 1947. J. Natl. Cancer Inst. 7:463-465.
2. Heston, W. E., and M. K. Deringer. 1951. Ibid. 12:361-367.
3. Heston, W. E., and G. Vlahakis. 1962. Ibid. 29:197-209.
4. Heston, W. E., et al. 1952. Ibid. 12:1141-1157.

Part III. Effect of Diet on Tumor Incidence

Avg No. of Tumors was calculated from the total population studied. Data in brackets refer to the column heading in brackets.

Treatment ⟨Synonym⟩	Strain or Hybrid [Genotype]	Sex	Age [Avg Wt, g]	Incidence, %[1/] [Avg No. of Tumors]	Reference
Full fed	A[2/]	♀	70 wk	52	3
		♂♀	100 wk	58	

1/ Unless otherwise indicated. 2/ Strain A backcross mice.

continued

50. PULMONARY ADENOMAS: MOUSE

Part III. Effect of Diet on Tumor Incidence

Treatment ⟨Synonym⟩	Strain or Hybrid [Genotype]	Sex	Age [Avg Wt, g]	Incidence, %[1/] [Avg No. of Tumors]	Reference
Underfed	A[2/]	♀	70 wk	27	3
		♂♀	100 wk	32	
Fed dog chow	A	♂♀	15 mo	58	2
Low cystine diet ad libitum	A	♂♀	15 mo	29	2
High cystine diet ad libitum	A	♂♀	15 mo	42	2
High cystine diet restricted	A	♂♀	15 mo	32	2
K9 calorie-restricted diet	Swiss	....	....	0.17[3/]	4
K19 calorie-restricted diet	Swiss	....	....	0.21[3/]	4
3-Methylcholanthrene ⟨MCA⟩ only	(A x Y)F_1 [A^y/a]	♂	20 wk [48.5]	98 [7.7]	1
		♀	20 wk [45.6]	100 [13.9]	
	[a/a]	♂	20 wk [34.4]	96 [4.8]	1
			20 wk [35.2]	93 [4.2]	
		♀	20 wk [26.0]	98 [7.4]	
			20 wk [26.1]	98 [6.4]	
3-Methylcholanthrene ⟨MCA⟩ plus food restriction	(A x Y)F_1 [A^y/a]	♂	20 wk [32.1]	92 [4.4]	1
		♀	20 wk [26.1]	100 [6.2]	
3-Methylcholanthrene ⟨MCA⟩ plus aurothioglucose ⟨gold thioglucose⟩	(A x Y)F_1 [A^y/a]	♂	20 wk [50.8]	100 [9.0]	1
		♀	20 wk [50.5]	100 [12.5]	
	[a/a]	♂	20 wk [45.4]	99 [9.9]	1
		♀	20 wk [42.7]	100 [12.6]	

[1/] Unless otherwise indicated. [2/] Strain A backcross mice. [3/] Ratio of percentage of tumors in the restricted group to the percentage in the control group.

Contributor: Walter E. Heston

References

1. Heston, W. E., and G. Vlahakis. 1961. J. Natl. Cancer Inst. 27:1189-1196.
2. Larsen, C. A., and W. E. Heston. 1945. Ibid. 6:31-40.
3. Tannenbaum, A. 1940. Am. J. Cancer 38:335-350.
4. Tannenbaum, A. 1942. Cancer Res. 2:460-467.

51. HEPATOMAS: MOUSE

Age: Animals were necropsied at given age, unless otherwise indicated.

Part I. Incidence of Spontaneous Tumors

Strain	Sex	Age, mo	Incidence, %	Reference
C3H/He	♂	14	85	2
C3H-A^{vy}/He	♂	12	99	3
C3H$_e$B/De	♂	14	79	2
		21	91	1
	♀[1/]	24	59	1
	♀[2/]	21	30	1
	♀[3/]	21	38	1
C3H$_f$B/He	♂	14	72	2
C3H-$A^{vy}{}_f$B/He	♀	21[4/]	75	4
	♀[1/]	16	98	3
	♀[2/]	16	96	3

[1/] Virgins. [2/] Breeders. [3/] Force-bred. [4/] Average age of all animals; mice varied in age at autopsy.

continued

51. HEPATOMAS: MOUSE

Part I. Incidence of Spontaneous Tumors

Contributor: Walter E. Heston

References

1. Deringer, M. K. 1959. J. Natl. Cancer Inst. 22:995-1002.
2. Heston, W. E., and G. Vlahakis. 1960. Ibid. 24:425-435.
3. Heston, W. E., and G. Vlahakis. 1968. Ibid. 40:1161-1166.
4. Heston, W. E., et al. 1973. Ibid. 51:209-224.

Part II. Effect of Genetic Factors on Tumor Incidence

Hybrid: The strain of the mother is listed first. **Avg No. of Tumors** was calculated from the total population studied.

Hybrid	Genotype	Sex	Age, mo	Incidence, %	Avg No. of Tumors	Reference
Effect of A^y Gene						
(C3H x Y)F_1	A/A	♂	16	88	2	2
	A^y/A	♂	16	100	3	
(C3H x YBR)F_1	A/A	♂	16	88	3	1
	A^y/A	♂	16	100	5.6	
Effect of A^{vy} Gene						
(C3H x C3H-A^{vy})F_1 x C3H[1/]	A/A	♂	12	61	1	4
	A^{vy}/A	♂	12	100	8	
Effect of Sex, Allele of A, & Reciprocal Crosses						
(C3H$_f$B x YBR)F_1	A^y/A	♂	14	100	5	3
		♀	14	63	1.2	
	A/a	♂	14	94	2.5	
		♀	14	5	0.05	
(YBR x C3H$_f$B)F_1	A^y/A	♂	14	95	3	3
		♀	14	41	0.5	
	A/a	♂	14	43	0.6	
		♀	14	0	0	

[1/] First backcross.

Contributor: Walter E. Heston

References

1. Heston, W. E. 1963. J. Natl. Cancer Inst. 31:467-474.
2. Heston, W. E., and G. Vlahakis. 1961. Ibid. 26:969-983.
3. Heston, W. E., and G. Vlahakis. 1966. Ibid. 37:839-843.
4. Heston, W. E., and G. Vlahakis. 1968. Ibid. 40:1161-1166.

Part III. Effect of Diet and Bedding on Tumor Incidence

Avg No. of Tumors was calculated from the total population studied. Data in brackets refer to the column heading in brackets.

continued

51. HEPATOMAS: MOUSE

Part III. Effect of Diet and Bedding on Tumor Incidence

Treatment ⟨Synonym⟩	Strain	Sex	Age, mo	Incidence, % [Avg No. of Tumors]	Reference
Effect of Diet					
Fed National Cancer Institute ⟨NCI⟩ Diet	$C3H_fB/He$	♂	14	72	2
Fed Purina chow	$C3H_fB/He$	♂	14	57	2
Effect of Bedding 1/					
Kept in Australia	C3H-A^{vy}	♂	14	28	3
	C3H-$A^{vy}{}_fB$	♂	13.8	18	
		♀	22.9	17	
Kept in United States					
Cedar shavings in bedding	C3H-A^{vy}	♂	12	99	3
	C3H-$A^{vy}{}_fB$	♂	14.7	100	
		♀	16.3	97	
Pine bedding	C3H-A^{vy}	♂	12	95 [2.1]	1
Pine plus cedar bedding	C3H-A^{vy}	♂	12	98 [2.5]	1

1/ Sabine, Horton, and Wicks [ref. 3] concluded that the difference between incidence in Australia and in the United States was probably attributable to the cedar shavings used in bedding in the U.S., but not used in Australia; this conclusion was not confirmed by Heston's work in the U.S. [ref. 1].

Contributor: Walter E. Heston

References

1. Heston, W. E. 1975. J. Natl. Cancer Inst. 54:1011-1014.
2. Heston, W. E., and G. Vlahakis. 1960. Ibid. 24:425-435.
3. Sabine, J. R., et al. 1973. Ibid. 50:1237-1242.

52. TYPE C LEUKEMIA VIRUSES: MOUSE

Part I. Non-defective Murine Leukemia Viruses

Virus Strain: LLV = lymphatic leukemia virus. **Origin:** IdUrd = idoxuridine ⟨IUdR⟩. **Tropism** indicates host range in vitro; A = amphotropic; B = B-tropic; N = N-tropic; NB = NB-tropic; X = xenotropic.

Virus Strain	Origin	Tropism	Reference
AKR-L1 1/	AKR tail extract passaged through NIH mouse cells in vitro	N	3
AKR-V1 1/	SC-1 tissue culture passage of tail extract of a mouse congenic with NIH mice but carrying the *Akv-1* locus	N	3
BALB virus 1	IdUrd induction of BALB/c tissue culture cells	N	1
BALB virus 2	IdUrd induction of BALB/c tissue culture cells	X	1
Friend LLV			
F-SLLV	Extract of Ehrlich ascites tumor passaged through Swiss mice	N	2,5
F-BLLV	F-S virus passaged through BALB/c mice	NB	2,5
Gross passage A	AKR extract passaged through C3H mice	N	2,5
MCF-247	Preleukemic AKR thymus	A	6

1/ These two closely related strains of viruses define a subclass of *Fv-1* alleles; AKR-L1 is not restricted on *Fv-1* n-resistant cells, while AKR-V1 is restricted on these cells [ref. 8].

continued

52. TYPE C LEUKEMIA VIRUSES: MOUSE

Part I. Non-defective Murine Leukemia Viruses

Virus Strain	Origin	Tropism	Reference
Moloney	Extract of sarcoma 37	NB	2,5
NZB-IU	IdUrd induction of NZB tissue culture cells	X	7
RAD LV	Radiation leukemia of C57BL	B	2,5
Rauscher LLV	Swiss mouse tumor passaged through BALB/c mice	NB	2,5
Tennant LLV	Extract of BALB/c leukemia induced by C58 extract	B	2,5
WN1802B	Old BALB/c mouse spleen	B	5
WN1802BNB	Tissue culture passage of WN1802B through NIH cells	NB	5
WN1802N	Old BALB/c mouse spleen	N	5
4070A	Embryo tissue culture isolate from wild mouse (Casitas, California)	A	4

Contributor: Rex Risser

References

1. Aaronson, S. A., and J. R. Stephenson. 1973. Proc. Natl. Acad. Sci. USA 70:2055-2058.
2. Gross, L. 1970. Oncogenic Viruses. Pergamon, New York.
3. Hartley, J. W., and W. P. Rowe. 1975. Virology 65: 128-134.
4. Hartley, J. W., and W. P. Rowe. 1976. J. Virol. 19: 19-25.
5. Hartley, J. W., et al. 1970. Ibid. 5:221-225.
6. Hartley, J. W., et al. 1977. Proc. Natl. Acad. Sci. USA 74:789-792.
7. Levy, J. A. 1973. Science 182:1151-1153.
8. Rowe, W. P., and J. W. Hartley. Unpublished. N.I.A.I.D., National Institutes of Health, Bethesda, MD, 1978.

Part II. Defective Murine Leukemia Viruses

Abbreviation: MuLV = murine leukemia virus.

Virus Strain	Origin	Disease	Reference
Abelson MuLV	Lymphoma of a Moloney MuLV-infected, cortisone-treated BALB/c mouse	B-cell lymphoma	1
Finkel-Biskis-Jenkin virus[1]	Osteosarcoma of CF1 mice	Osteosarcoma	2,3
Friend (SFFV) virus[2]	Extract of Ehrlich ascites carcinoma passaged through Swiss mouse	Erythroleukemia	2
Murine sarcoma viruses			
Harvey	Moloney MuLV passaged through rats	Fibrosarcoma, anaplastic sarcoma	2
Kirsten	MuLV from $C3H_f$ mice passaged through rats	Fibrosarcoma, anaplastic sarcoma	2
Moloney	Rhabdomyosarcoma induced by Moloney MuLV in BALB/c mice	Fibrosarcoma, anaplastic sarcoma	2
Rauscher virus	Swiss mouse tumor passaged through BALB/c mouse	Erythroleukemia	2

[1] Synonym: FBJ virus. [2] Three strains of Friend virus are in common use: F-S, F-B, and B-tropic Friend virus. These have F-SLLV, F-BLLV, and Tennant LLV, respectively, as helper viruses (*see* Part I) [ref. 4].

continued

52. TYPE C LEUKEMIA VIRUSES: MOUSE

Part II. Defective Murine Leukemia Viruses

Contributor: Rex Risser

References

1. Abelson, H. T., and L. S. Rabstein. 1970. Cancer Res. 30:2208-2212.
2. Gross, L. 1970. Oncogenic Viruses. Pergamon, New York.
3. Levy, J. A., et al. 1973. J. Natl. Cancer Inst. 51:525-533.
4. Lilly, F., and R. A. Steeves. 1973. Virology 55:363-370.

Part III. Genetically Transmitted Type C Viruses or Viral Antigens

Mouse Strain	No. of Non-linked Loci for Expression	Locus	Chromosome	Reference
AKR/J	2	*Akv-1*	7	5
		Akv-2	?	
B10.BR/Li	2	......	?	7
C3H/Fg	3	*C3fv-1*	7, ?, ?	6
C58/J	3 or 4	......	?, not 7	7,8
NZB	1 or 2	......	?, not 7	1,2,9
PL/J	1	......	?, not 7	3
RF/J	1	*Rjv-1*	?, not 7	4
129	2	*Gv-1*	....	10,11
		Gv-2	7	

Contributor: Rex Risser

References

1. Chused, T. M., et al. Unpublished. National Institute of Dental Research, National Institutes of Health, Bethesda, MD, 1978.
2. Datta, S. K., and R. S. Schwartz. 1976. Nature (London) 263:412-415.
3. Ishimoto, A., and W. P. Rowe. Unpublished. N.I.A.I.D., National Institutes of Health, Bethesda, MD, 1978.
4. Risser, R., and W. P. Rowe. Unpublished. McArdle Laboratory for Cancer Research, Univ. Wisconsin, Madison, 1978.
5. Rowe, W. P. 1972. J. Exp. Med. 136:1272-1285.
6. Rowe, W. P. 1973. Cancer Res. 33:3061-3068.
7. Rowe, W. P., and J. W. Hartley. Unpublished. N.I.A.I.D., National Institutes of Health, Bethesda, MD, 1978.
8. Stephenson, J. R., and S. A. Aaronson. 1973. Science 180:865-866.
9. Stephenson, J. R., and S. A. Aaronson. 1974. Proc. Natl. Acad. Sci. USA 71:4925-4929.
10. Stockert, E., et al. 1971. J. Exp. Med. 133:1334-1355.
11. Stockert, E., et al. 1976. Proc. Natl. Acad. Sci. USA 73:2077-2081.

Part IV. Genes Affecting Murine Leukemogenesis or Leukemia Antigens

Data do not include endogenous viruses.

Locus	Chromosome	Allele	Strain Distribution	Reference
Fv-1	4	*b*	A/HeJ, A/J, AL/N, BALB/cN, B10.BR/J, C57BL/6, C57BL/10	6,7
		n	AKR, CBA/J, C3H/He, C57BR/cdJ, C57L, C58/J, DBA/2N, NFS/N, NIH, SWR/J	
		n^r	NZB, RF/J, 129/J	

continued

52. TYPE C LEUKEMIA VIRUSES: MOUSE

Part IV. Genes Affecting Murine Leukemogenesis or Leukemia Antigens

Locus	Chromosome	Allele	Strain Distribution	Reference
Fv-2	9	*r*	C57BL/6, C57BL/10, C57BR, C58, I/Sr	3,7
		S	A/J, AKR, BALB/c, CBA, CE/J, DBA/1, DBA/2, NZB, SJL, 129	
Gv-1 [1/], *Gv-2* [1/]	?, 7	1	C3H/An, DBA/2, SJL/J, 101	8
		2	A, AKR, C3H/He, C58, I	
		3	CE/J, 129/J, 129/Sv, 129/Sv-*ter* [2/]	
		....	BALB/c, CBA/J, C3H$_f$/Bi (young), C57BL/6, C57BL/10, C57BR/cdJ, DBA/1J, MA/J, RF/J, SWR/J	
hr	14	*hr/hr* [3/]	....	4
Rfv-1, Rgv-1 [4/]	17	*a* [5/]	....	1-3
		b [5,6/]	....	
		d [7/]	....	
		k [8/]	....	
Sl	10	*Sl/Sl^d* [9,10/]	....	3,5
Tla	17	*a*	A, BDP, C57BR/cdJ, C58, F/St, HTI, NZB, PL/J, SJL/J, SWR	2
		b	AKR, CBA/J, C3H/He, C57BL/6, C57L, DBA/1, HTG, I, MA/J, RF/J	
		c	BALB/c, DBA/2, LP, 129	
W	5	*W/W, W^v/+* [9/]	....	3

[1/] The four classes of G_{IX} antigen expression are quantitatively distinguishable and controlled by two loci, *Gv-1* and *Gv-2*. [2/] Synonym: TER/Sv. [3/] Mice develop an increased incidence of thymic leukemia. [4/] These two loci have been mapped to different regions in the H-2 complex. [5/] Unfavorable to Gross virus leukemogenesis. [6/] Unfavorable to Friend virus leukemogenesis. [7/] Moderately favorable to Gross virus leukemogenesis; favorable to Friend virus leukemogenesis. [8/] Favorable to Gross virus leukemogenesis. [9/] Mice are resistant to Friend virus. [10/] Mice develop an increased incidence of leukemia.

Contributor: Rex Risser

References

1. Chesebro, B., et al. 1974. J. Exp. Med. 140:1457-1467.
2. Klein, J. 1975. Biology of the Mouse Histocompatibility-2 Complex. Springer-Verlag, New York.
3. Lilly, F., and T. Pincus. 1973. Adv. Cancer Res. 17: 231-277.
4. Meier, H., et al. 1969. Proc. Natl. Acad. Sci. USA 63: 759-766.
5. Murphy, E. D. 1977. J. Natl. Cancer Inst. 58:107-110.
6. Pincus, T., et al. 1971. J. Exp. Med. 133:1219-1233.
7. Pincus, T., et al. 1971. Ibid. 133:1234-1241.
8. Stockert, E., et al. 1971. Ibid. 133:1334-1355.

53. STRAINS WITH A HIGH SPONTANEOUS INCIDENCE OF LYMPHOMA AND LEUKEMIA: MOUSE

MLP = mean latent period in days. **Abnormal Cell Types:** L = lymphocytic; R = reticulum cell neoplasm; G = granulocytic; P = plasmacytic; UD = undifferentiated. **Tissue Site of Origin**: T = thymus; LN = lymph nodes; S = spleen. **Thymex** = thymectomized. Data in brackets refer to the column heading in brackets.

Strain [Genotype]	Incidence of Spontaneous Lymphoma, % [MLP]		Abnormal Cell Types	Tissue Site of Origin	Effect of Thymectomy % Leukemias		Reference
	♂	♀			Control	Thymex	
AKR	83.1 [225]	....	L	T	83.1	14.6	4
	79.7 [308]	92.0 [269]	L	T	...	...	5

continued

53. STRAINS WITH A HIGH SPONTANEOUS INCIDENCE OF LYMPHOMA AND LEUKEMIA: MOUSE

Strain [Genotype]	Incidence of Spontaneous Lymphoma, % [MLP]		Abnormal Cell Types	Tissue Site of Origin	Effect of Thymectomy % Leukemias		Reference
	♂	♀			Control	Thymex	
BALB/c_fRIII	43.6 [380]	58.5 [446]	L	T	45.6	5.1	9
C3H/Fg	40.0 [396]	80.9	L, R, G, P	LN, S	...	...	3
C58	88.9	90.2	L	T, S, LN	...	...	6
	90.2 [318]	....	L	....	90.9	34.9	4
HRS [*hr/hr*][1/]	72.0	....	L, G	T, S, LN	...	...	7
HRS [*hr*/+][1/]	21.0	....	L, G	....	...	...	7
PL/J	19.0	50.0	....	....	...	...	10
RF/J	49.0	43.0	....	....	...	...	10
	....	36.0 [449]	....	T, S	...	...	1
SJL/J	....	78 [380]	R, (<2% L)	....	78	60	2
SL	30.8 [443]	68.4 [350]	UD, R	....	57.1	20.7	8

1/ Genotypes of the HRS strain are in effect congenic strains.

Contributor: Michael Potter

References

1. Duran-Reynals, M. L., and C. Cook. 1974. J. Natl. Cancer Inst. 52:1001-1003.
2. Haran-Ghera, N., et al. 1967. Ibid. 39:653-661.
3. Law, L. W. 1957. Ann. N.Y. Acad. Sci. 68:616-635.
4. Law, L. W., and J. H. Miller. 1950. J. Natl. Cancer Inst. 11:253-262.
5. Lynch, C. J. 1954. Ibid. 15:161-176.
6. MacDowell, E. C., and M. N. Richter. 1935. Arch. Pathol. 20:709-724.
7. Meier, H., et al. 1969. Proc. Natl. Acad. Sci. USA 63: 759-766.
8. Nakakuki, K., and Y. Nishizuka. 1961. Mie Med. J. 11:233-250.
9. Squartini, F., et al. 1974. J. Natl. Cancer Inst. 52: 1635-1641.
10. Storer, J. B. 1966. J. Gerontol. 21:404-409.

54. RETICULUM CELL SARCOMAS: MOUSE

Reticulum cell sarcomas are characteristic lymphomas of mice, usually occurring in the second and third years of life. The most widely used classification is that proposed by Dunn **[14]**. The most common type is reticulum cell neoplasm Type B (Hodgkin's-like lesion) which contains a mixture of cell types, including reticulum cells, lymphocytes, plasma cells, and histiocytes **[14, 22]**. Type A is composed almost exclusively of a single cell type which may vary in form from histiocytic to spindle cell. Useful estimates of upper limits of reticulum cell sarcoma incidence can be derived from data reported only as "lymphoma" incidence. The lymphocytic neoplasm, localized and generalized, of Dunn **[14]** accounts for almost all of the lymphomas other than the reticulum cell types that occur spontaneously in mice. Therefore, data on the incidence of lymphocytic neoplasms has been included whenever available for the reported populations. The tumors tabulated are spontaneous tumors in untreated animals, unless treatment is specified.

Strain & Genotype: Consult Staats **[31]** for strain and substrain designations; Green **[16]** for mutant genes; and Cherry **[5]** for congenic strains. **Population: Sex**—V = virgin, B = breeder; **Mean Age**—months are 30-day months. Data in light brackets refer to the column heading in brackets. Figures in heavy brackets are reference numbers. Plus/minus (±) values are standard deviations. Values in parentheses are ranges, estimate "c" (*see* Introduction).

Strain & Genotype	Population			Reticulum Cell Neoplasm				Lymphocytic Neoplasm [Lymphoma, Type Unspecified 1/]		Reference
				Type A		Type B				
	Sex	No.	Mean Age mo 1/	Incidence %	Mean Age mo	Incidence %	Mean Age mo	Incidence %	Mean Age mo	
A 2/	♀	104	....	1	20	4	20	4	7	13
A/WySn	♂V	61	22.6 ± 4.6	...	....	.....	....	[13]	[23.6 ± 5.5]	29
	♀V	60	21.9 ± 5.3	...	....	.....	....	[10]	[28.0 ± 4.4]	29

1/ Unless otherwise indicated. 2/ Material from M. K. Deringer, reported by Dunn.

continued

Strain & Genotype	Population			Reticulum Cell Neoplasm				Lymphocytic Neoplasm [Lymphoma, Type Unspecified 1/]		Reference
				Type A		Type B				
	Sex	No.	Mean Age mo 1/	Incidence %	Mean Age mo	Incidence %	Mean Age mo	Incidence %	Mean Age mo	
(AKR x NH)F_1 3/	♂♀	217	...	2	24	10	19.7	27	16.1	13
BALB/c	♀	108	20(2-30)	2	21(10-28)	17	21(10-28)	6	17	15
BALB/c - +/+	♂♀	650	...	...	...	...	...	[9]	...	34
- *nu/nu*	♂♀	550	...	...	...	...	...	[9]	...	34
BALB/cAnDe	♀B	182	15.4	<1	...	9	20	<1	...	11, 14
CBA/H - +/+	♂♀	830	...	...	...	...	...	[5]	...	34
- *nu/nu*	♂♀	450	...	...	...	...	...	[5]	...	34
CBA/J	♂	17	25.0 ± 3.8	...	...	...	...	[6]	[21.5]	30
	♀	33	21.2 ± 3.9	...	...	...	...	[15]	[21.5(18.2-24.0)]	30
C3H/HeDiSn 4/	♂V	61	22.9 ± 5.5	...	...	...	...	[5]	[25.9 ± 10.2]	29
	♀V	60	21.9 ± 5.1	...	...	...	...	[7]	[26.1 ± 4.1]	29
C3H.K/Sn 4/	♂V	44	17.0 ± 5.1	...	...	...	...	[14]	[17.7 ± 5.4]	29
	♀V	38	17.3 ± 4.7	...	...	...	...	[34]	[18.7 ± 4.5]	29
$C3H_eB$/De	♀V	99	23.3	2	21.5	...	...	1	26.0	9
	♀B	155	19.4	<1	21.0	2	21.7	...	...	9
$C3H_f$ 2/	♀	153	...	<1	26	4	26	4	20	13
($C3H_f$/He x C57BL/He)F_1 5/	♀	100	...	4	29	10	29.2	4	28	13, 19
(C57BL/He x $C3H_f$/He)F_1 5/	♀	99	...	9	27	6	28	4	25	13, 19
C57BL/He	♀B	314	21.1	15	22.3	11	21.3	3	17.1	18
C57BL/HeDe	♀B	59	...	...	...	75 6/	20	...	...	14
(C57BL x C3H/An_f)F_1 7/	♂♀	156	30	49 8/	...	...	...	...	...	6
C57BL/6J	♂V	>45	30.3 9/	...	...	...	...	[32]	[27.7]	32, 33
	♀V	>45	30.0 9/	...	...	...	...	[46]	[27.7]	32, 33
(C57BL/6J x WC/Re)F_1 *Sl*/+, Sl^d/+, +/+	♂♀V	80	27.9 ± 5.4(9.4-39.8)	9	27.6(20.1-32.1)	24	30.6(19.9-36.9)	5	32.2 ± 2.7 (29.0-35.4)	23, 24
Sl/Sl^d	♂♀V	94	14.9 ± 5.8(6.3-30.8)	...	...	...	...	37 10/	12.3 ± 4.9 (6.8-27.6)	23, 24
C57BL/10ScSn 11/	♂V	59	31.3 ± 5.4	...	...	...	...	[29]	[29.6 ± 3.1]	29
	♀V	58	28.0 ± 5.5	...	...	...	...	[36]	[25.2 ± 6.7]	29
B10.AKM/Sn 12/	♂V	60	23.1 ± 5.8	...	...	...	...	[53]	[21.5 ± 5.1]	29
	♀V	57	23.6 ± 5.6	...	...	...	...	[77]	[24.0 ± 5.3]	29

1/ Unless otherwise indicated. 2/ Material from M. K. Deringer, reported by Dunn. 3/ Material from L. W. Law, reported by Dunn. Reciprocal hybrid data combined and recalculated. 4/ Known genetic differences: $H\text{-}1^aC$ vs. $H\text{-}1^bc$ [5]. 5/ Reciprocal hybrid material from W. E. Heston, reported by Dunn. 6/ Authors suggest increased age of population and inclusion of many microscopic lesions as factors contributing to higher than usual incidence of Type B. No data given on Type A or on lymphocytic neoplasms. 7/ Substrains of CUM (for full name, *see* list of HOLDERS at front of book). 8/ Authors interpret all reticulum cell neoplasms in this hybrid as Type A. 9/ Age of 50% survival. 10/ Lymphocytic neoplasms, primary in thymus. Lack of reticulum cell neoplasms in *Sl*/Sl^d may be explained by shortened life-span. 11/ Genotype: $H\text{-}2^b$ Tla^b $H\text{-}1^c$ Hbb^s $Mlv\text{-}1^b$ [5, 35]. 12/ Genotype: $H\text{-}2^m$ Tla^a [5].

continued

Strain & Genotype	Population			Reticulum Cell Neoplasm				Lymphocytic Neoplasm [Lymphoma, Type Unspecified[1]]		Reference
				Type A		Type B				
	Sex	No.	Mean Age mo[1]	Incidence %	Mean Age mo	Incidence %	Mean Age mo	Incidence %	Mean Age mo	
B10.D2(58N)/Sn[13]	♂V	60	28.2 ± 8.5	...	....	.....	....	[8]	[29.2 ± 4.9]	29
	♀V	63	27.3 ± 4.4	...	....	.....	....	[38]	[26.6 ± 3.9]	29
C57L/HeDe	♀B	55	....	...	....	55[6]	20	....	....	14
(C57L x A)F_1[2]	♀	222	....	<1	18	10	23	2	19	13
(C58 x C3H_f)F_1[3]	♂♀	247	....	3	21.5	9	24	36	18.2	13
DBA/2$_e$BDe	♀V	76	22.0	12	21	8	22	3	12	10, 14
	♀B	51	20.9	10	20	18	22	8	13	10, 14
DBA/2J	♂V	67	19.1 ± 7.1	...	....	.....	....	[10]	[20.8(10.0-26.1)]	30
	♀V	51	18.9 ± 5.3	...	....	.....	....	[12]	[22.4(18.0-25.0)]	30
DW/J										
+/+	♂	11	....	...	....	45	14.1 ± 5.7	9	15.2	4
	♀	52	....	...	....	63	13.7 ± 4.8	6	15.4 ± 5.5	4
dw/+	♂	20	....	...	....	60	18.0 ± 4.3	20	19.0 ± 2.1	4
	♀	62	....	...	....	61	16.9 ± 4.1	3	18.6 ± 0.1	4
HR/De										
hr/+	♀V	25	21.6	4	19.0	12	17.3	8	23.0	8,14
	♀B	63	16.2	3	20.5	6	16.0	8	13.0	8,14
hr/*hr*	♀V	24	22.3	4	15.0	8	26.5	4	21.0	8,14
LP/J	♂V	77	24.0 ± 5.3	...	....	.....	....	[1]	[26.1]	30
	♀V	51	23.1 ± 4.5	...	....	.....	....	[8]	[25.4(23.8-28.5)]	30
NZB/Bl	♂	76	....	...	....	.....	....	[5]	....	3
	♀	61	....	...	....	.....	....	[2]	....	3
NZB/BlNJ	♀V	37	15.7 ± 4.1(7.7-24.2)	11	15.3(12.0-20.3)	8	16.3(10.9-24.2)	....	....	24
NZB/BlUmc	♂V	50	9.3[9]	...	....	.....	....	[32[14]]	....	32, 33
	♀V	41	9.0[9]	...	....	.....	....	[25[14]]	....	32, 33
NZO/Bl	♂♀	451	....	2	(16.0-23.7)	5	(12.7-25.2)	<1	(13.7-25.1)	28
RF/Un[15]	♀V	311	....	...	....	50[16]	21.5	10	18.9	7
SJL/J	♂B	33	....	...	....	91	12.5(7.5-20.5)	....	....	21, 24
	♀V	43	....	...	....	91	13.3(8.7-18.6)	....	....	20, 24
	♀V	80	15.7 ± 4.3[17]	...	....	95	15.5 ± 4.3(6.9-24.1)	....	....	24
	♀B	32	....	...	....	88	13.5(8.7-20.0)	....	....	21, 24

[1] Unless otherwise indicated. [2] Material from M. K. Deringer, reported by Dunn. [3] Material from L. W. Law, reported by Dunn. Reciprocal hybrid data combined and recalculated. [6] Authors suggest increased age of population and inclusion of many microscopic lesions as factors contributing to higher than usual incidence of Type B. No data given on Type A or on lymphocytic neoplasms. [9] Age of 50% survival. [13] Genotype: $H\text{-}1^a\ Hbb^d\ Mlv\text{-}1^a$ [5, 35]. [14] Lymphomas, mostly reticulum cell neoplasms, Type B. [15] Non-inbred stock derived from strain RF/Un. [16] Reticulum cell neoplasms not separated into types. [17] Observed until death to determine regressions, gross diagnosis.

continued

Strain & Genotype	Population			Reticulum Cell Neoplasm				Lymphocytic Neoplasm [Lymphoma, Type Unspecified]		Reference
				Type A		Type B				
	Sex	No.	Mean Age mo	Incidence %	Mean Age mo	Incidence %	Mean Age mo	Incidence %	Mean Age mo	
SJL/Wn	♂V	17[18]	...	...	...	71	11.6	...	...	17
	♀V	55	...	...	...	78	12.7(7.5-22.2)	2	7.6	17
		314[19]	...	...	...	85	11.8	...	...	2
SL/Ms	♀V	57	...	...	...	30[20]	11.8(7.5-14.8)	25	11.0(5.6-15.4)	25, 26
SWR/J	♂	68	(9-30)	1[21]	27.0	3[22]	20.5(20-21)	9	12.5(9-19)	27
	♀	243	(5-26)	3[21]	17.9(9-25)	<1[22]	11.0	7	12.6(5-20)	27
(YBR/He x AKR/LwUp)F_1										
+/+	♂V	83	...	1	27.0	8	27.7	5	24.7	12
	♀V	80	...	3	29.0	29	29.4	5	26.2	12
A^y/+	♂V	61	19.9	...	...	10	21.6	3	19.5	12
	♀V	95	...	1	20.0	17	23.0	4	17.7	12
129/RrJ	♂V	41	27.3 ± 7.2	...	...	...	...	[2]	[27.3]	30
	♀V	95	24.3 ± 7.0	...	...	...	...	[7]	[24.3(19.4-28.7)]	30
"Wild"[23]	♂V	54	20.9(2-30)	...	...	6	25.7(23-30)	...	...	1
	♀V	72	24.4(7-33)	...	...	13	25.7(24-33)	...	...	1
	♀B	99	26.6(6-33)	1	28.3(24-30)	11	28.3(24-30)	2	30.0	1

18/ Housed one per cage. 19/ Phosphate-buffered, saline-treated controls. 20/ Includes reticular and anaplastic types as defined by authors; later described as reticulum cell sarcoma in reference 26. Eight additional lymphomas were diagnosed grossly. 21/ Includes authors' "reticulum cell neoplasm, histiocytic." 22/ Includes authors' "mixed lymphocytic-reticulum cell neoplasm." 23/ Wild house mice and their non-inbred descendants maintained under laboratory conditions.

Contributor: Edwin D. Murphy

References

1. Andervont, H. B., and T. B. Dunn. 1962. J. Natl. Cancer Inst. 28:1153-1163.
2. Ben-Yaakov, M., et al. 1975. Ibid. 54:443-448.
3. Bielschowsky, M., and F. Bielschowsky. 1962. Nature (London) 194:692.
4. Chen, H. W., et al. 1972. J. Natl. Cancer Inst. 49:1145-1154.
5. Cherry, M. 1975. Inbred Strains Mice 9:30-34.
6. Chino, F., et al. 1971. J. Gerontol. 26:497-507.
7. Clapp, N. K. 1973. Natl. Tech. Inf. Serv. TID-26373.
8. Deringer, M. K. 1951. J. Natl. Cancer Inst. 12:437-445.
9. Deringer, M. K. 1959. Ibid. 22:995-1002.
10. Deringer, M. K. 1962. Ibid. 28:203-210.
11. Deringer, M. K. 1965. Ibid. 35:1047-1052.
12. Deringer, M. K. 1970. Ibid. 45:1205-1210.
13. Dunn, T. B. 1954. Ibid. 14:1281-1433.
14. Dunn, T. B., and M. K. Deringer. 1968. Ibid. 40:771-821.
15. Ebbesen, P. 1971. Ibid. 47:1241-1245.
16. Green, M. C. 1966. In E. L. Green, ed. Biology of the Laboratory Mouse. Ed. 2. McGraw-Hill, New York. pp. 87-150.
17. Haran-Ghera, N., et al. 1967. J. Natl. Cancer Inst. 39:653-661.
18. Heston, W. E. 1964. Ibid. 32:947-955.
19. Heston, W. E., and M. K. Deringer. 1952. Ibid. 13:167-175.
20. Murphy, E. D. 1963. Proc. Am. Assoc. Cancer Res. 4:46.
21. Murphy, E. D. 1965. 36th Annu. Rep. Jackson Lab., p. 33.
22. Murphy, E. D. 1969. J. Natl. Cancer Inst. 42:797-814.
23. Murphy, E. D. 1977. Ibid. 58:107-110.
24. Murphy, E. D. Unpublished. Jackson Laboratory, Bar Harbor, ME, 1977.
25. Nakakuki, K., and Y. Nishizuka. 1961. Mie Med. J. 11:233-250.
26. Nakakuki, K., and Y. Nishizuka. 1973. Gann 64:161-166.

continued

27. Rabstein, L. S., et al. 1973. J. Natl. Cancer Inst. 50: 751-758.
28. Rappaport, H., et al. 1971. Cancer Res. 31:2047-2053.
29. Smith, G. S., and R. L. Walford. 1978. In D. Bergsma and D. E. Harrison, ed. Genetic Effects on Aging. Liss, New York. v. 14(1), pp. 281-312.
30. Smith, G. S., et al. 1973. J. Natl. Cancer Inst. 50: 1195-1213.
31. Staats, J. 1976. Cancer Res. 36:4333-4377.
32. Stutman, O. 1974. Fed. Proc. Fed. Am. Soc. Exp. Biol. 33:2028-2032.
33. Stutman, O. 1975. Adv. Cancer Res. 22:261-422.
34. Stutman, O. 1978. In J. Fogh and B. Giovanella, ed. The Nude Mouse in Experimental and Clinical Research. Academic, New York.
35. Taylor, B. A., et al. 1973. Nature (London) New Biol. 241:184-186.

55. OCCURRENCE OF HEMANGIOMAS AND HEMANGIOENDOTHELIOMAS: MOUSE

Route: i.p. = intraperitoneal; s.c. = subcutaneous. **Tumor**: **Incidence**—Numerator indicates number of animals developing tumors; denominator indicates total number of animals; **Age at Detection**—Age of animals at time of tumor detection. Data in brackets refer to the column heading in brackets.

Strain or Stock [Phenotype[1]]	Treatment ⟨Synonym⟩ [Route]	Sex	Tumor: Type	Tumor: Incidence	Tumor: Age at Detection[1]	Reference
A	o-Aminoazotoluene [s.c.]	♂	Hemangioendothelioma	1/11	52 wk	4
		♀	Hemangioendothelioma	1/11	52 wk	4
AKR	Urethan [i.p.]	♂	Hemangioma	2/19	...	14
		♀	Hemangioma	2/23	...	14
BALB/c	o-Aminoazotoluene [Diet]	♂	Hemangioendothelioma	4/29	...	5
		♀	Hemangioendothelioma	17/31	...	5
	[s.c.]	♂	Hemangioendothelioma	4/12	47 wk[2]	4
		♀	Hemangioendothelioma	10/12	50 wk[2]	4
	3-Methylcholanthrene ⟨20-Methylcholanthrene⟩ [Diet]	♂♀	Hemangioendothelioma	21/32	...	48
	N-Nitrosodimethylamine ⟨Dimethylnitrosamine⟩ [Diet or s.c.]	♂	Hemangioendothelial sarcoma	15/16	5-9 mo	21
		♀	Angiomatous lesion	20/25	5-9 mo	21
	[Oral]	♂	Hemangioma	9/41	...	42
			Hemangiosarcoma	12/41	...	42
		♀	Hemangioma	15/46	...	42
			Hemangiosarcoma	12/46	...	42
	[s.c.]	♂	Hemangioma	4/22	...	42
			Hemangiosarcoma	1/14	...	42
		♀	Hemangioma	8/16	...	42
			Hemangiosarcoma	2/8	...	42
BALB/cCr	None	♂♀[3]	Hemangioma	12/2065	21.58 mo[2]	22
			Hemangioendothelioma	27/2065	21.89 mo[2]	22
	MSV ⟨Murine sarcoma virus⟩	...	Hemangiomatous tumor	4/5	...	23
BALB/cOs	9,10-Dimethyl-1,2-benzanthracene [Gavage] & splenectomy	♀	Capillary hemangioma, cavernous hemangioma, hemangioendothelioma	7/31	...	1
BALB/cPi	None	♂♀	Angioma	4/67	21 mo	29
CFW	N-Methyl-N′-nitro-N-nitrosoguanidine [i.p.]	♂	Vascular tumor	1/12	14 mo	25

[1] Unless otherwise indicated. [2] Average. [3] 95% female.

continued

Strain or Stock [Phenotype]	Treatment ⟨Synonym⟩ [Route]	Sex	Tumor			Reference
			Type	Incidence	Age at Detection[1]	
CF1	None	♂	Hemangioendothelioma	1/47	....	45
		♀	Hemangioma	1/38	....	45
	Mother treated with 3-methylcholanthrene [Stomach tube]	♀	Angioma	1/41	....	45
			Hemangioma	1/41	....	45
CTM [Albino][4]	None	♂	Angioma	3/88	....	7
		♀	Angioma	1/99	....	7
	Urethan [Drinking water]	♂	Angioma, angiosarcoma	6/293	....	7
		♀	Angioma, angiosarcoma	7/358	....	7
C3H	None	♂	Hemangioendothelioma	1	15 mo	11
	o-Aminoazotoluene [s.c.]	♂	Hemangioendothelioma	1/23	52 wk	4
	3-Methylcholanthrene ⟨20-Methylcholanthrene⟩ [Diet]	♂♀	Hemangioendothelioma	11/50	....	48
	Thorium oxide ⟨Thorium dioxide⟩ [s.c.]	♂	Hemangioma	1/10	18 mo	3
	Urethan [Skin]	♀	Hemangioendothelioma[5]	5%	....	34
C3H/HeOs	9,10-Dimethyl-1,2-benzanthracene [Gavage] & splenectomy	♂	Capillary hemangioma, cavernous hemangioma, hemangioendothelioma	6/39	....	1
C3H$_f$	*N*-Nitrosodimethylamine ⟨Dimethylnitrosamine⟩ [Drinking water]	♂♀	Hemangioma	3/17	58 wk[6]	12
			Capillary hemangioma	14/40	58 wk[6]	12
			Cavernous hemangioma	11/40	58 wk[6]	12
			Hemangioendothelioma	6/40	58 wk[6]	12
			Hemangiosarcoma	3/40	58 wk[6]	12
C57BL	None	♂	Hemangioma	1/30	....	36
		♀	Hemangioma	1/33	....	36
	o-Aminoazotoluene [s.c.]	♂	Hemangioendothelioma	8/34	41 wk[2]	4
		♀	Hemangioendothelioma	4/13	41 wk[2]	4
	9,10-Dimethyl-1,2-benzanthracene [s.c.]	♂	Hemangioma	1/30	....	36
		♀	Hemangioma	3/30	....	36
	Plus trypan blue [Injection]	♂	Hemangioma	3/37	....	36
	Teflon disk [s.c.]	♀	Hemangioma	1/20	....	36
	Plus trypan blue [i.p. or s.c.]	♀	Hemangioma	1/25	....	36
	Trypan blue [i.p.]	♀	Hemangioma	1/19	....	36
C57BL/6Os	9,10-Dimethyl-1,2-benzanthracene [Gavage] & splenectomy	♂	Capillary hemangioma, cavernous hemangioma, hemangioendothelioma	4/31	....	1
		♀	Capillary hemangioma, cavernous hemangioma, hemangioendothelioma	6/27	....	1
C57BR	*p*-Dimethylaminobenzene-1-azo-2-naphthalene [Painting]	♂♀	Hemangioendothelioma	48/50	....	19
C58	Urethan [i.p.]	♂	Hemangioma	11/19	....	14
		♀	Hemangioma	8/18	....	14
DBA	Dibenz[*a,h*]anthracene [Oral]	♂	Hemangioendothelioma	10/14	....	30
		♀	Hemangioma	1/13	....	30
			Hemangioendothelioma	5/13	....	30
	3-Methylcholanthrene ⟨20-Methylcholanthrene⟩ [Cholesterol pellet]	♂	Hemangiomatous lesion	1/10	....	26
		♀	Hemangioma	2/10	....	26
	5,9,10-Trimethyl-1,2-benzanthracene [Painting]	♂♀	Hemangioendothelioma	1/13	....	13

[1] Unless otherwise indicated. [2] Average. [4] Outbred. [5] Malignant mesenchymal tumors. [6] End of experiment.

continued

Strain or Stock [Phenotype]	Treatment ⟨Synonym⟩ [Route]	Sex	Tumor			Reference
			Type	Incidence	Age at Detection[1]	
	Urethan [i.p.]	♀	Hemangioendothelioma	10%	....	34
	[Skin]	♀	Hemangioendothelioma	7%	....	34
DD/N	*N*-Nitrosodimethylamine [Diet]	♂	Hemangioendothelioma	2/11	129[7] & 140[7]	32
			Hemangioendothelial sarcoma	2/21; 4/17	....	33
		♀	Hemangioendothelioma	1/11	166[7]	32
DDD	*N*-Nitrosodimethylamine [Diet or s.c.]	♂	Angiomatous lesion	35/40	4-10 mo[8]	21
			Hemangioendothelial sarcoma	14/40	4-10 mo[8]	21
		♀	Angiomatous lesion	7/40	2-7 mo[8]	21
			Hemangioendothelial sarcoma	1/40	2-7 mo[8]	21
DDI	Mechlorethamine oxide ⟨Nitrogen mustard *N*-oxide⟩ [Injection]	♂♀	Hemangioma	1	259 d[8]	18
Hall	None	♂	Hemangioma	3/79	36-78 wk	24
	Carbamic acid esters [i.p.]	♂	Hemangioma	96/457	36-78 wk	24
HR/De [Haired]	None	♂	Hemangioendothelioma	7/28	20 mo	8
				8/24	23.3 mo	9
		♀	Hemangioendothelioma	6/23	23.7 mo	9
		♀[9]	Hemangioendothelioma	3/25	21.3 mo	8
		♀[10]	Hemangioendothelioma	9/63	19.8 mo	8
		♂♀	Hemangioendothelioma[11]	6/100	22 mo	10
			Hemangioendothelioma[12]	20/100	24 mo	10
	o-Aminoazotoluene ⟨4-*o*-Tolylazo-*o*-toluidine⟩ [Painting]	♂	Hemangioendothelioma	15/23	18.4 mo	9
		♀	Hemangioendothelioma	22/31	18.1 mo	9
	Diethyl ether [Painting]	♂	Hemangioendothelioma	3/15	22.6 mo	8
		♀	Hemangioendothelioma	5/22	24 mo	8
	Ethylene glycol [Painting]	♂♀	Hemangioendothelioma[11]	3/34	22 mo	10
			Hemangioendothelioma[12]	7/34	23 mo	10
	Urethan in ethylene glycol [Painting]	♂♀	Hemangioendothelioma[11]	43/51	14 mo	10
			Hemangioendothelioma[12]	15/51	15 mo	10
[Hairless]	None	♂	Hemangioendothelioma	3/21	23 mo	8
				3/16	18.7 mo	9
		♂[10]	Hemangioendothelioma	2/17	20.5 mo	8
		♀	Hemangioendothelioma	6/25	22.6 mo	9
		♀[9]	Hemangioendothelioma	9/24	22.1 mo	8
		♂♀	Hemangioendothelioma[11]	3/86	23 mo	10
			Hemangioendothelioma[12]	18/86	22 mo	10
	o-Aminoazotoluene ⟨4-*o*-Tolylazo-*o*-toluidine⟩ [Painting]	♂	Hemangioendothelioma	12/22	18.3 mo	9
		♀	Hemangioendothelioma	19/25	20.2 mo	9
	Diethyl ether [Painting]	♀	Hemangioendothelioma	4/22	21.5 mo	8
	Ethylene glycol [Painting]	♂♀	Hemangioendothelioma[11]	2/37	24 mo	10
			Hemangioendothelioma[12]	6/37	23 mo	10
	3-Methylcholanthrene ⟨20-Methylcholanthrene⟩ [Painting]	♂	Hemangioendothelioma	1/15	12 mo	8
	Urethan in ethylene glycol [Painting]	♂♀	Hemangioendothelioma[11]	34/48	14 mo	10
			Hemangioendothelioma[12]	9/48	16 mo	10

[1] Unless otherwise indicated. [7] Experimental days. [8] Average survival. [9] Virgin. [10] Breeder. [11] Liver. [12] Other sites.

continued

Strain or Stock [Phenotype[1/]]	Treatment ⟨Synonym⟩ [Route]	Sex	Tumor			Reference
			Type	Incidence	Age at Detection[1/]	
ICR	*N*-Nitrosodiethylamine [Diet]	♂	Hemangioendothelial sarcoma	1/11	...	33
	N-Nitrosodimethylamine [Diet]	♂	Hemangioendothelial sarcoma	7/34	...	33
[Nude]						
[nu/nu][13/]	Polyoma virus [i.p.]	...	Hemangioma	3/20	...	46
[nu/+][13/]	Polyoma virus [i.p.]	...	Hemangioma	1/20	...	46
RF	*N*-Nitrosodimethylamine ⟨Dimethylnitrosamine⟩ [Oral]	♂	Hemangioendothelioma	5/90	...	6
			Hemangiosarcoma	84/90	...	6
Swiss	None	♂	Blood vessel tumor	6/99	...	28
			Angiosarcoma	6/99	88 wk	44
			Hemangioma	2/110	...	40
		♀	Blood vessel tumor	5/99	...	28
			Angioma	2/17	85-90 wk	35
			Angiosarcoma	5/99	113 wk	44
			Hemangioma	4/100	...	40
	n-Amylhydrazine hydrochloride [Drinking water]	♂	Angioma, angiosarcoma	7/48	47-100 wk	28
		♀	Angioma, angiosarcoma	11/50	62-100 wk	28
	9,10-Dimethyl-1,2-benzanthracene [s.c.]	♂	Angioma	1/55	...	36
		♀	Angioma	3/60	...	36
	Plus trypan blue [i.p.]	♂	Angioma	1/30	...	36
		♀	Angioma	2/30	...	36
	[i.p. & s.c.]	♀	Angioma	1/87	...	36
	1,2-Dimethylhydrazine dihydrochloride					
	[Drinking water]	♂	Angiosarcoma	46/50	42 wk[14/]	40
		♀	Angiosarcoma	49/50	45 wk[14/]	40
	[Injection, single]	♂	Angioma, angiosarcoma	12/50	75 wk	43
		♀	Angioma, angiosarcoma	10/50	88 wk	43
	[Injection, 10]	♂	Angioma, angiosarcoma	25/50	54 wk	43
		♀	Angioma, angiosarcoma	23/50	61 wk	43
	Ethylhydrazine hydrochloride [Drinking water]	♂	Angioma, angiosarcoma	8/50	68 wk	27
		♀	Angioma, angiosarcoma, angiopericytoma	30/50	71 wk	27
	N-Nitrosodimethylamine ⟨Dimethylnitrosamine⟩ [Drinking water]					
	25 µg	♀	Angioma	1/8	63 wk	35
	0.0005%	♂	Angioma	9/15	37-49 wk	35
			Angiosarcoma	1/15	...	35
	0.0025%	♀	Angioma	5/12	29-38 wk	35
	0.005%	♀	Angioma	2/4	63-69 wk	35
	Tetramethylhydrazine hydrochloride [Drinking water]	♂	Angiosarcoma	44/49	37 wk	44
		♀	Angioma	1/50	38 wk	44
			Angiosarcoma	47/50	38 wk	44
	Trimethylhydrazine hydrochloride [Drinking water]	♂	Angiosarcoma	39/49	39 wk	20
		♀	Angioma	1/49	37 wk	20
			Angiosarcoma	44/49	37 wk	20
	Trypan blue [s.c.]	♀	Angioma	2/50	...	36
	Ultraviolet light	♂♀	Angioma	2/40	...	31
	Urethan [Drinking water]	♂♀	Hemangioma, small[11/]	Multiple	...	41
			Hemangioma	4	...	41
	Vinyl chloride [Inhalation]	♂♀	Angioma	9/510	...	17
			Fibroangioma	5/510	...	17
			Angiosarcoma	18/510	...	17

[1/] Unless otherwise indicated. [11/] Liver. [13/] Genotype. [14/] Average latent period.

continued

55. OCCURRENCE OF HEMANGIOMAS AND HEMANGIOENDOTHELIOMAS: MOUSE

Strain or Stock [Phenotype]	Treatment ⟨Synonym⟩ [Route]	Sex	Tumor			Reference
			Type	Incidence	Age at Detection[1]	
Swiss albino	None	♂	Angioma	3/99	....	39
			Angiosarcoma	3/99	....	39
		♀	Angioma	2/99	....	39
			Angiosarcoma	3/99	....	39
				4/110	....	37
	1,1-Dimethylhydrazine [Drinking water]	♂	Angiosarcoma	42/50	....	37
		♀	Angiosarcoma	37/50	....	37
	Phenicarbazide ⟨1-Carbamyl-2-phenylhydrazine⟩ [Drinking water]	♂	Angioma	2/49	....	39
			Angiosarcoma	4/49	....	39
		♀	Angioma	3/50	....	39
			Angiosarcoma	4/50	....	39
	β-Phenylhydrazine sulfate [Drinking water]	♂	Angiosarcoma	8% (4 tumors)	50 wk	38
		♀	Angioma, angiosarcoma	44% (22 tumors)	83 wk	38
Swiss NIH	Methylazoxyoctane [i.p.]	♂	Hemangiosarcoma	4/16	....	47
Wild	None	♀[9]	Hemangioendothelioma	2/12	25 mo[8]	2
		♀[10]	Hemangioendothelioma	8/99	23 mo[8]	2
(C57BL x C3H)F_1	None	♂	Hemangioma	3/150	2 yr	34
			Hemangioendothelioma	1/150	2 yr	34
		♀	Hemangioma	2/150	2 yr	34
	MSV ⟨Murine sarcoma virus⟩	♀	Hemangiosarcoma	....	....	15
	Urethan [i.p.]	♀	Hemangioendothelioma[5]	23%	....	34
	[Skin]	♂♀	Hemangioendothelioma[5]	12%	....	34
(DBA x C57BL)F_1 & (C57BL x DBA)F_1	5,9,10-Trimethyl-1,2-benzanthracene [Painting]	♂♀	Hemangioendothelioma	3/48	....	13
ABC, C3H, DBA	1,2:5,6-Dibenzanthracene [Oral]	♂♀	Hemangioendothelioma	8/70	....	16
	3-Methylcholanthrene ⟨20-Methylcholanthrene⟩ [Oral]	♂♀	Hemangioendothelioma	1/30	....	16

[1] Unless otherwise indicated. [5] Malignant mesenchymal tumors. [8] Average survival. [9] Virgin. [10] Breeder.

Contributor: Margaret K. Deringer

References

1. Akamatsu, Y. 1975. J. Natl. Cancer Inst. 55:893-897.
2. Andervont, H. B., and T. B. Dunn. 1962. Ibid. 28: 1153-1163.
3. Andervont, H. B., and M. B. Shimkin. 1940. Ibid. 1: 349-353.
4. Andervont, H. B., et al. 1942. Ibid. 3:131-153.
5. Andervont, H. B., et al. 1944. Ibid. 4:583-586.
6. Clapp, N. K., et al. 1968. Ibid. 41:1213-1227.
7. Della Porta, G., et al. 1963. Tumori 49:413-428.
8. Deringer, M. K. 1951. J. Natl. Cancer Inst. 12:437-445.
9. Deringer, M. K. 1956. Ibid. 17:533-539.
10. Deringer, M. K. 1962. Ibid. 29:1107-1121.
11. Edwards, J. E., et al. 1942. Ibid. 2:479-490.
12. Engelse, L. den, et al. 1974. Eur. J. Cancer 10:129-135.
13. Hartwell, J. L., and H. L. Stewart. 1942. J. Natl. Cancer Inst. 3:277-285.
14. Kawamoto, S., et al. 1961. Cancer Res. 21:71-74.
15. Law, L. W., and R. C. Ting. 1970. J. Natl. Cancer Inst. 44:615-621.
16. Lorenz, E., and H. L. Stewart. 1947. Ibid. 7:227-238.
17. Maltoni, C., and G. Lefemine. 1975. Ann. N.Y. Acad. Sci. 246:195-218.
18. Matsuyama, M., et al. 1969. Br. J. Cancer 23:167-171.
19. Mulay, A. S., and E. A. Saxén. 1953. J. Natl. Cancer Inst. 13:1259-1273.
20. Nagel, D., et al. 1976. Ibid. 57:187-189.
21. Otsuka, H., and A. Kuwahara. 1971. Gann 62:147-156.
22. Peters, R. L., et al. 1972. Int. J. Cancer 10:273-282.
23. Peters, R. L., et al. 1974. J. Natl. Cancer Inst. 53: 1725-1729.
24. Pound, A. W., and T. A. Lawson. 1976. Cancer Res. 36:1101-1107.

continued

25. Schoental, R., and J. P. M. Bensted. 1969. Br. J. Cancer 23:757-764.
26. Shear, M. J., and F. W. Ilfeld. 1940. Am. J. Pathol. 16: 287-293.
27. Shimizu, H., et al. 1974. Int. J. Cancer 13:500-505.
28. Shimizu, H., et al. 1975. Br. J. Cancer 31:492-496.
29. Smith, C. S., and H. I. Pilgrim. 1971. Proc. Soc. Exp. Biol. Med. 138:542-544.
30. Snell, K. C., and H. L. Stewart. 1962. J. Natl. Cancer Inst. 28:1043-1051.
31. Stenbäck, F. 1975. Oncology 31:61-75.
32. Takayama, S., and K. Oota. 1963. Gann 54:465-472.
33. Takayama, S., and K. Oota. Ibid. 56:189-199.
34. Tannenbaum, A. 1961. Acta Unio Int. Contra Cancrum 17:72-87.
35. Terracini, B., et al. 1966. Br. J. Cancer 20:871-876.
36. Tomatis, L. 1966. Tumori 52:1-16.
37. Toth, B. 1973. J. Natl. Cancer Inst. 50:181-194.
38. Toth, B. 1976. Cancer Res. 36:917-921.
39. Toth, B., and H. Shimizu. 1974. J. Natl. Cancer Inst. 52:241-251.
40. Toth, B., and R. B. Wilson. 1971. Am. J. Pathol. 64: 585-600.
41. Toth, B., et al. 1961. Br. J. Cancer 15:322-326.
42. Toth, B., et al. 1964. Cancer Res. 24:1712-1721.
43. Toth, B., et al. 1976. Am. J. Pathol. 84:69-86.
44. Toth, B., et al. 1976. J. Natl. Cancer Inst. 57:1179-1183.
45. Turusov, V., et al. 1973. Int. Agency Res. Cancer Sci. Publ. 4:84-91.
46. Vandeputte, M., et al. 1974. Int. J. Cancer 14:445-450.
47. Ward, J. M., et al. 1974. J. Natl. Cancer Inst. 53:1181-1185.
48. White, J., and H. L. Stewart. 1942. Ibid. 3:331-347.

56. OCCURRENCE OF ADRENAL CORTICAL TUMORS: MOUSE

Route: i.p. = intraperitoneal. **Tumor**: **Incidence**—Numerator indicates number of animals developing tumors; denominator indicates total number of animals; **Age at Detection**—Age of animals at time of tumor detection. Data in brackets refer to the column heading in brackets.

Strain or Stock ⟨Synonym⟩	Treatment [Route]	Sex	Tumor			Reference
			Type	Incidence	Age at Detection[1/]	
A	Castration	♀	Adenoma	4/33	....	16
	Ovariectomy plus adrenocorticotropic hormone	♀	Carcinoma	1	537 d	6
	Ovariectomy plus transplanted seminal vesicle	♀	Adenoma	3/3	14-16 mo	7
BALB/c ⟨Bagg albino⟩	None	♂	Adenoma	3/8	....	7
			Carcinoma	1/8	....	7
		♀	....	1	24.5 mo	2
	Castration	♂	Adenoma	9/10	2 yr	7
			Carcinoma	1/10	2 yr	7
	Ovariectomy plus transplanted seminal vesicle	♀	Adenoma	1/9	>10 mo	7
			Carcinoma	8/9	>10 mo	7
CBA	Ovariectomy plus transplanted seminal vesicle	♀	Adenoma	6/6	12 mo	7
CE	Ovariectomy	♀	Carcinoma	14/24	18 mo[2/]	18
CE[3/]	None	♂	....	0/15	....	24
		♀	....	0/11	....	23
	Castration	♂	Carcinoma	37/51	3-28 mo	5
	At 1-3 days old	♂	Carcinoma	15/19	7-12 mo	24
	Ovariectomy	♀	Carcinoma	58/63	3-22 mo	5
	At 1-3 days old	♀	Carcinoma	21/21	6-12 mo	23
C3H/Bcr	Gonadectomy	♂♀	Carcinoma	2/36	89, 90 wk	12
	Plus 9,10-dimethyl-1,2-benzanthracene[4/]	♂♀	Carcinoma	15/62	43-82 wk	12

[1/] Unless otherwise indicated. [2/] Age at death. [3/] Genotype: Extreme dilution. [4/] Synonym: 7,12-Dimethylbenzanthracene.

continued

Strain or Stock ⟨Synonym⟩	Treatment [Route]	Sex	Tumor			Reference
			Type	Incidence	Age at Detection[1]	
C3H/Bi ⟨Z⟩	Castration	♀	Adenoma	20/30	....	16
			Adenocarcinoma	3/33	....	16
	9,10-Dimethyl-1,2-benzanthracene					
	[i.p.]	♀	....	1/10	9 mo	13
	[Skin painting] plus mammectomy	♀	....	4/12	>10 mo	13
	Caged with vasectomized ♂	♀	....	3/10	9-13 mo	13
	Ovariectomy at birth	♀	Nodular hyperplasia[5]	20	....	21
	Ovariectomy plus transplanted seminal vesicle	♀	Adenoma	6/6	12 mo	7
C3H/HeCrgl	Parabiosis					
	Prespayed with intact	♀	Nodular hyperplasia	13/15	....	14
	Spayed with intact	♀	Nodular hyperplasia	2/5	....	14
	Spayed with spayed	♀	Nodular hyperplasia	2/12	....	14
C3H/Tw	Orchiectomy	♂	Carcinoma	1	....	17
C57BL	Ovariectomy at birth	♀	Nodular hyperplasia[6]	30	....	21
DBA	Castration, some at birth, some later	♂	Nodular hyperplasia	13/13	>10 mo	22
	Ovariectomy	♀	Malignant	3/82	....	20
	At birth	♀	Nodular hyperplasia[6]	125	....	21
NH	None	♂	....	1/8	>12 mo	11
		♀	Adenoma	13/14	>12 mo	11
	Castration	♂	Adenoma	57/57	12 mo	7
	Ovariectomy at 43-65 days old	♀	....	13/15	(495-726 d)[2]	8
	Ovariectomy plus transplanted seminal vesicle	♀	Adenoma	15/15	12 mo	7
Slye stock	None	♂♀	Adenoma	1/33,000	....	15
(A x C3H)F_1 ⟨AZF_1⟩	Castration	♀	Adenoma	6/14	....	16
			Adenocarcinoma	5/14	....	16
	Gonadectomy at 1-3 days old	♂	Carcinoma	8/13	>14 mo	4
		♀	Carcinoma	14/18	>14 mo	4
(C3H x A)F_1 ⟨ZAF_1⟩	Castration	♀	Adenoma	4/14	....	16
			Adenocarcinoma	3/14	....	16
	Gonadectomy at 1-3 days old	♂	Carcinoma	11/11	>14 mo	4
		♀	Carcinoma	10/10	>14 mo	4
((A x C3H)F_1 x C3H) & ((C3H x A)F_1 x C3H) ⟨ZBC⟩	Castration	♀	Adenoma	21/55	....	16
			Adenocarcinoma	25/55	....	16
((A_f x $C3H_f$)F_1 x C3H) & (($C3H_f$ x A_f)F_1 x C3H) ⟨ZBC, fostered⟩	Castration	♀	Adenoma	5/24	....	16
			Adenocarcinoma	12/24	....	16
(CE x C57BL)F_1	Gonadectomy at 1-3 days old	♂	Carcinoma	9/15	>14 mo	4
		♀	Carcinoma	14/15	>14 mo	4
(C57BL x CE)F_1	Gonadectomy at 1-3 days old	♂	Carcinoma	11/13	>14 mo	4
		♀	Carcinoma	20/20	>14 mo	4
(C57BL x NH)F_1	Ovariectomy & parabiosed with intact ♀	♀	Adenoma	9/9	....	18
(C57L x A)F_1 ⟨LAF_1⟩	None	♂♀	Adenoma	0.16%	....	19
	Irradiation					
	Atomic bomb test	♂	....	1	....	1
			Adenocarcinoma	0.05%	....	19
		♀	Adenoma	0.60%	....	19

[1] Unless otherwise indicated. [2] Age at death. [5] Extensive. [6] Slight.

continued

56. OCCURRENCE OF ADRENAL CORTICAL TUMORS: MOUSE

Strain or Stock (Synonym)	Treatment [Route]	Sex	Tumor			Ref- er- ence
			Type	Inci- dence	Age at Detection[1]	
	Neutrons	♀	....	3/109	84 wk[7]	9
				4/121	86 wk[7]	9
	X rays	♀	....	1/127	89 wk[2]	9
				1/117	73.5 wk[2]	9
	Plus adrenalectomy	♀	....	3/118	66 wk[7]	9
				6/120	85.5 wk[7]	9
(DBA x CE)F_1	Gonadectomy at 1-3 days old	♂	Carcinoma	24/24	>14 mo	4
		♀	Carcinoma	19/19	>14 mo	4
(CE x DBA)F_1	Gonadectomy at 1-3 days old	♂	Carcinoma	16/16	>14 mo	4
		♀	Carcinoma	21/21	>14 mo	4
(DBA x C3H)F_1	DBA ovaries grafted to spleens in gonadectomized animals and transferred to F_1 hybrids	♀	....	42/46	....	10
(DBA x C57BL)F_1	Gonadectomy at 1-3 days old	♂	Carcinoma	4/16	>14 mo	4
		♀	Carcinoma	9/11	>14 mo	4
(C57BL x DBA)F_1	Gonadectomy at 1-3 days old	♂	Carcinoma	3/14	>14 mo	4
		♀	Carcinoma	10/16	>14 mo	4
(DE x DBA)F_1	None	♂	"A" cell tumor	0/20	27 mo	3
		♀	"A" cell tumor	1/25	27 mo	3
	Gonadectomy	♂	Carcinoma	13/13	22 mo	3
		♀	Carcinoma	23/23	12 mo	3
(NH x A)F_1	X ray (1000 rads)	♂	Carcinoma	1	....	11

[1] Unless otherwise indicated. [2] Age at death. [7] Mean survival.

Contributor: Margaret K. Deringer

Specific References

1. Cohen, A. I., et al. 1957. Am. J. Pathol. 33:631-651.
2. Dalton, A. J., et al. 1943. J. Natl. Cancer Inst. 4:329-338.
3. Dickie, M. M., and P. W. Lane. 1956. Cancer Res. 16: 48-52.
4. Dickie, M. M., and G. W. Woolley. 1949. Ibid. 9:372-384.
5. Fekete, E., and C. C. Little. 1945. Ibid. 5:220-226.
6. Flaks, J. 1949. J. Pathol. Bacteriol. 61:266-269.
7. Frantz, M. J., and A. Kirschbaum. 1949. Cancer Res. 9:257-276.
8. Gardner, W. U. 1941. Ibid. 1:632-637.
9. Haran-Ghera, N., et al. 1959. Ibid. 19:1181-1187.
10. Hummel, K. P. 1954. J. Natl. Cancer Inst. 15:711-715.
11. Kirschbaum, A., et al. 1946. Cancer Res. 6:707-711.
12. Marchant, J. 1967. Br. J. Cancer 21:750-754.
13. Mody, J. K. 1965. Indian J. Pathol. Bacteriol. 8:324-326.
14. Pilgrim, H. J. 1960. Cancer Res. 20:1555-1560.
15. Slye, M., et al. 1921. J. Cancer Res. 6:305-336.
16. Smith, F. W. 1948. Cancer Res. 8:641-651.
17. Takasugi, N., et al. 1975. Gann 66:57-67.
18. Tullos, H. S., et al. 1961. Cancer Res. 21:730-734.
19. Upton, A. C., et al. 1960. Ibid. 20(8-2):1-60.
20. Woolley, G., et al. 1939. Proc. Natl. Acad. Sci. USA 25:277-279.
21. Woolley, G., et al. 1940. Proc. Soc. Exp. Biol. Med. 45:796-798.
22. Woolley, G., et al. 1941. Endocrinology 28(2):341-343.
23. Woolley, G. W., and C. C. Little. 1945. Cancer Res. 5:193-202.
24. Woolley, G. W., and C. C. Little. 1945. Ibid. 5:211-219.

General References

25. Dunn, T. B. 1970. J. Natl. Cancer Inst. 44:1323-1389.
26. Gardner, W. U., et al. 1959. In F. Homburger, ed. Physiopathology of Cancer. Ed. 2. Hoeber-Harper, New York. pp. 190-192.
27. Russfield, A. B. 1966. Public Health Serv. Publ. 1332.
28. Woolley, G. 1950. Rec. Prog. Horm. Res. 5:383-405.
29. Woolley, G. W. 1958. In G. E. W. Wolstenholme and M. O'Connor, ed. Ciba Found. Colloq. Endocrinol. Proc. 12:122-136.

57. PITUITARY TUMORS: MOUSE

For reviews, consult **General References** in Part I.

Part I. Spontaneous

All tumors are in the anterior lobe unless otherwise indicated. Data in brackets refer to the column heading in brackets.

Strain or Hybrid	Sex	Tumor Incidence [No. of Tumors]	Age	Remarks	Reference
C57BR/cd	♀1/	33.0%	Old	....	6
C57L	♀1/	33.0%	Old	....	6
EI	♀	[1]	695 d	Associated with bilateral ovarian tumors & mammary tumors	4
NZY	♀	~25%	12-18 mo	Associated with high incidence of mammary cancer	1
RF/Un	♀	7/526; 1.3%	677 d	Based on gross diagnosis	2
		6/427; 1.4%	676 d	Based on microscopic diagnosis	2
Slye stock	....	1/11,188	....	....	8
($C3H_eB$ x SWR x C57L x 129/Rr)	♀2/	19/175	896 d3/	2 tumors in pars intermedia	7
(C57BL x C57BR)	....	[2]	....	....	3
$(C57L \times A)F_1$	♀	8.0%	Old	....	5
$(O20 \times DBA_f)F_1$	♀	~10%	Old	Genetic factors suggested	5

1/ Breeder. 2/ Virgins. 3/ Average.

Contributor: Annabel G. Liebelt

Specific References

1. Bielschowski, M., et al. 1956. Br. J. Cancer 10:688-699.
2. Clapp, N. K., et al. 1974. Radiat. Res. 57:158-186.
3. Cloudman, A. M. 1941. In G. Snell, ed. Biology of the Laboratory Mouse. Dover-Blakiston, Philadelphia. pp. 168-233.
4. Gardner, W. U., et al. 1936. Am. J. Cancer 26:541-546.
5. Kwa, H. G. 1961. An Experimental Study of Pituitary Tumours; Genesis, Cytology and Hormone Content. Springer-Verlag, Berlin. p. 4.
6. Murphy, E. D. 1966. In E. L. Green, ed. Biology of the Laboratory Mouse. Ed. 2. McGraw-Hill, New York. pp. 521-570.
7. Poel, W. E., et al. 1969. Proc. Am. Assoc. Cancer Res. 10:69(Abstr. 273).
8. Slye, M., et al. 1931. Am. J. Cancer 15:1387-1400.

General References

9. Clifton, K. H. 1959. Cancer Res. 19:2-22.
10. Clifton, K. H., and B. N. Sridharan. 1975. In F. F. Becker, ed. Cancer, A Comprehensive Treatise. Plenum. New York. v. 3, pp. 249-285.
11. Furth, J. 1953. Cancer Res. 13:477-492.
12. Furth, J., et al. 1973. Methods Cancer Res. 10:201-277.
13. Gardner, W. U., et al. 1959. In F. Homburger, ed. The Physiopathology of Cancer. Hoeber-Harper, New York. pp. 152-237.
14. Gorbman, A. 1956. Cancer Res. 16:99-105.
15. Kirschbaum, A. 1960. In W. W. Nowinski, ed. Fundamental Aspects of Normal and Malignant Growth. Elsevier, New York. pp. 823-876.
16. Liebelt, A. G., and R. A. Liebelt. 1978. IARC Sci. Publ. (in press).

Part II. Induced

All tumors are in the anterior lobe unless otherwise specified. **Age** = age at time of tumor detection or, if no tumors present, age at which animals were examined.

Induction Method ⟨Synonym⟩	Strain or Hybrid ⟨Synonym⟩	Sex	Tumor Incidence [No. of Tumors1/]	Age1/	Remarks	Reference
Hypophyseal Isografts						
Subcutaneous transplant						
♂ or ♀ donor, 1-4 grafts	BC	♀	14/44	673-750 d	Chromophobe adenomas. Tendency for grafts to become	22
	C57	♀	2/5	739-760 d		
	(C57 x BC)	♀	6/6	576-839 d		

1/ Unless otherwise indicated.

continued

Part II. Induced

Induction Method ⟨Synonym⟩	Strain or Hybrid ⟨Synonym⟩	Sex	Tumor Incidence [No. of Tumors]	Age	Remarks	Reference
	(C3H x BC)	♀	4/8	689-870 d	tumorous may be strain-limited, and may parallel the susceptibility after estrogen.	
C57BL ♂ origin	(C57BL x DBA_f)F_1	♀	5/6	22-29 mo	Mammotropic	3
Single donor	(C57BL x $DBA/2_f$)F_1	♂	7/7	>600 d	Tumor size at autopsy, 0.5 X 0.5 cm. Mammotropic; 3/7 had mammary cancer.	30
Subcutaneous or intramuscular, 2-4 grafts	A	♂♀	0/1	....	Grafts were somatotropic & mammotropic. C57 susceptible; BC may also be.	36
	BALB/c	♂♀	0/1	....		
	CBA	♂♀	0/3	....		
	C3H	♂♀	0/1	....		
	C57	♂♀	2/8	....		
	(A x BALB/c)	♂♀	0/1	....		
	(A x C3H)	♂♀	0/6	....		
	(A x C57)	♂♀	7/10	....		
	(BALB/c x C57)	♂♀	7/17	....		
	(BC x C57)	♂♀	1/1	....		
	(C3H x BC)	♂♀	12/17	....		
Eye	A	♂	1/16	815 d	....	4
		♀	0/16	287 d	Early mammary cancer due to mammotropic effect of graft	4
	CBA	♀	5/14	543 d	....	4
	CE	♀	2/15	654 d	....	4
	C3H	♀	0/18	290 d	Early mammary cancer due to mammotropic effect of graft	4
	$C3H_f$⟨Zb⟩	♂	0/9	555 d	....	4
	C57BL	♀	1/11	501 d	....	4
	(CBA x RIII)F_1	♀	13/14	393 d	Hypophysectomy prevented tumor; ovariectomy decreased tumor	4
	(RIII x CBA)F_1	♂	0/9	546 d	....	4
		♀	8/16	369 d	....	4
		♀[2]	High	240 d	Mammotropic	5
Single graft into anterior chamber						
RIII ♂ origin	(C57BL x RIII)F_1	♀[2]	10/10	15-22 mo	....	3
		♀[3]	2/2	9-11 mo	Resemble chromophobe adenomas	3
(RIII x CBA) ♀ origin	(RIII x CBA)F_1	♀	High	8 mo	....	3
Kidney, single graft					Chromophobe adenomas; mammotropic & somatotropic	24
60-d donor	(C57L x A/He)F_1	♂	2/6; 33%	665 d		
		♀	24/30; 80%	600 d		
300-d donor	(C57L x A/He)F_1	♀	13/22; 60%	420 d		
Endogenous Hormones						
Gonadectomy						
Neonatal	(CE x DBA)	♀	(50-60%)	14 mo	....	26
	(DBA x CE)	♂	(75-85%)	14 mo		

[2] Virgin. [3] Breeder.

continued

Part II. Induced

Induction Method (Synonym)	Strain or Hybrid (Synonym)	Sex	Tumor Incidence [No. of Tumors[1]]	Age	Remarks	Reference
At 1-3 d	$(A \times C3H)F_1$	♂	2/13	14 mo	No tumors at 6-13 mo; basophilic adenomas; adrenal cortical carcinomas. Mammary glands overdeveloped. F_1 mice from CE or DBA have most tumors.	15
		♀	4/18	14 mo		
	$(C3H \times A)F_1$	♂	1/11	14 mo		
		♀	1/10	14 mo		
	$(CE \times DBA)F_1$	♂	14/16	14 mo		
		♀	12/21	14 mo		
	$(DBA \times CE)F_1$	♂	18/24	14 mo		
		♀	9/19	14 mo		
	$(CE \times C57)F_1$	♂	0/15	14 mo		
		♀	4/15	14 mo		
	$(C57 \times CE)F_1$	♂	2/13	14 mo		
		♀	7/20	14 mo		
	$(C57 \times DBA)F_1$	♂	0/14	14 mo		
		♀	4/16	14 mo		
	$(DBA \times C57)F_1$	♂	0/16	14 mo		
		♀	1/11	14 mo		
	$(CE \times DBA)F_1$	♂[4]	0/28	....	Basophilic adenomas; gonadectomized have adrenal dysplasia	16
		♂[5]	24/27; 88%	>18 mo		
		♀[4]	0/28	....		
		♀[5]	(50-60%) [25[6]]	>15 mo		
	$(DE \times DBA)F_1$[7]	♂[4]	0/20	....		
		♂[5]	13/13	20 mo		
		♀[4]	0/25	....		
		♀[5]	23/23	18 mo		
Exogenous Hormones						
Estrogen	A	...	None	....	Only tumors >12 mg were counted. C57BL most susceptible, O20 next most susceptible; CBA has low susceptibility; A, C3H, DBA, & RIET not susceptible.	34
	CBA	...	<70%	....		
	C3H	...	None	....		
	$C3H_f$	...	None	....		
	C57BL	...	>90%[8]	....		
	DBA	...	None	....		
	O20	...	(70-90%)	....		
	RIET	...	None	....		
	$(C57BL \times DBA)F_1$	...	(70-90%)[8]	....		
	$(O20 \times DBA_f)F_1$	...	(70-90%)	....		
Cutaneous	....	♂	3/12	44 wk	Chromophobe adenomas	13
Subcutaneous	C57	...	15/106	....	Chromophobic (only tumors >12 mg were counted). C57 susceptible.	23
		♂♀	14/68	....		
	A; CBA; C3H; C121; JK; N	...	None	....	....	23
Plus corn oil in digestive tract	TM	♀	9/230	....	2 tumors in pars intermedia[9]	7
Control: corn oil and fatty acid only	TM	♀	2/285	....	1 tumor in pars intermedia[9]	7
Diethylstilbestrol (DES; stilbestrol)						
Pellet	$(C57L \times A)F_1$ (LAF_1)	♀	51.0%	....	....	19
Subcutaneous	C57BL	♂	1/31	19 mo	....	2
			5/29	17 mo		

[1] Unless otherwise indicated. [4] Intact animals. [5] Gonadectomized animals. [6] Total number of animals. [7] DE has CE background, selected for extreme dilution. [8] >60% incidence in 500 d. [9] Synonym: Intermediary lobe.

continued

57. PITUITARY TUMORS: MOUSE

Part II. Induced

Induction Method ⟨Synonym⟩	Strain or Hybrid ⟨Synonym⟩	Sex	Tumor Incidence [No. of Tumors]	Age	Remarks	Reference
	$(BALB/c \times C57BL)F_1$	♂	2/38	14 mo		
	$(BALB/c \times RIII)F_1$	♂	3/37	14 mo		
2.5% or 5%	BALB/c	♀	[0]	....	C57BL most susceptible; BALB/c & I appear next in susceptibility	1
	C3H	♀	[0]	....		
	C57BL	♀	[0]	....		
	DBA/2	♀	[0]	....		
	I	♀	[0]	....		
	$(BALB/c \times C3H)F_1$	♀	[0]	....		
	$(BALB/c \times C57BL)F_1$	♀	[0]	....		
	$(BALB/c \times DBA/2)F_1$	♀	[0]	....		
	$(BALB/c \times I)F_1$	♀	1/20	>17 mo		
	$(BALB/c \times RIII)F_1$	♀	1/56	>17 mo		
	$(C57BL \times C3H)F_1$	♀	[0]	....		
	$(DBA/2 \times C3H)F_1$	♀	[0]	....		
	$(DBA/2 \times C57BL)F_1$	♀	[0]	....		
	$(DBA/2 \times I)F_1$	♀	[0]	....		
	$(I \times C3H)F_1$	♀	[0]	....		
	$(I \times C57BL)F_1$	♀	[0]	....		
	$(RIII \times C3H)F_1$	♀	[0]	....		
10% or 20%, plus castration	BALB/c	♂	[0]	....	C57BL most susceptible; BALB/c & I appear next in susceptibility	1
	C3H	♂	[0]	....		
	C57BL	♂	3/45	>17 mo		
	DBA/2	♂	[0]	....		
	I	♂	[0]	....		
	$(BALB/c \times C3H)F_1$	♂	1/97	>17 mo		
	$(BALB/c \times C57BL)F_1$	♂	[0]	....		
	$(BALB/c \times DBA/2)F_1$	♂	1/49	>17 mo		
	$(BALB/c \times I)F_1$	♂	[0]	....		
	$(BALB/c \times RIII)F_1$	♂	1/15	>17 mo		
	$(C57BL \times C3H)F_1$	♂	2/30	>17 mo		
	$(DBA/2 \times C3H)F_1$	♂	[0]	....		
	$(DBA/2 \times C57BL)F_1$	♂	1/26	>17 mo		
	$(DBA/2 \times I)F_1$	♂	3/24	>17 mo		
	$(I \times C3H)F_1$	♂	1/26	>17 mo		
	$(I \times C57BL)F_1$	♂	8/51	>17 mo		
	$(RIII \times C3H)F_1$	♂	[0]	....		
10%, at sea level	C3H	...	11/500	....	....	33
at high altitude	C3H	...	0/500	....		
20%	(C57L x RIII)	♂	48/90; 53.3%	16.3 mo	Chromophobe adenomas; gonadotropin assay of tumors low	39
		♀	61/91; 67%	17.7 mo		
25 mg, weekly	(CBA x C57BL)	...	[1]	....	Continuous treatment until death. Tumor wt, 59.8 mg (only tumors >12 mg were counted). C57BL susceptible.	21
	(C57BL x CBA)	...	[3]	....	Continuous treatment until death. Tumor wt, 16.5-120.5 mg (only tumors >12 mg were counted). C57BL susceptible.	21

continued

Part II. Induced

Induction Method ⟨Synonym⟩	Strain or Hybrid ⟨Synonym⟩	Sex	Tumor Incidence [No. of Tumors]	Age	Remarks	Reference
			[8]	....	Treatment period, 32-130 days of age; treatment stopped before autopsy. Tumor wt, 16.5-111.3 mg (only tumors >12 mg were counted). C57BL susceptible.	21
	(C57BL x C3H)	...	[8]	....	Continuous treatment until death. Tumor wt, 21.0-138.0 mg (only tumors >12 mg were counted). C57BL susceptible.	21
			[3]	....	Treatment period, 62-138 days of age; treatment stopped before autopsy. Tumor wt, 51.5-53.5 mg (only tumors >12 mg were counted). C57BL susceptible.	21
Estradiol benzoate 16.6 μg or 50 μg, 2 ml weekly	$(CBA \times C57)F_1$	♂	18/24; 75%	480 d	Only tumors >12 mg were counted. Brown degeneration, and adenomas in adrenal.	20
		♀	9/28; 32%	562 d		
	$(C57 \times CBA)F_1$	♂	19/23; 83%	513 d		
		♀	16/30; 53%	599 d		
16.6 μg in sesame oil, weekly	(C57BL x C3H)	...	[9]	....	Continuous treatment until death. Tumor wt, 9.0-118.5 mg (only tumors >12 mg were counted). C57BL susceptible.	21
			[4]	....	Treatment period, 32-61 days of age; treatment stopped before autopsy. Tumor wt, 16.5-111.3 mg (only tumors >12 mg were counted). C57BL susceptible.	21
	[(CBA x C57BL) x CBA]	...	[2]	....	Continuous treatment until death. Tumor wt, 28.5-44.3 mg (only tumors >12 mg were counted). C57BL susceptible.	21
			[2]	....	Treatment period, 90-136 days of age; treatment stopped before autopsy. Tumor wt, 17.3-53.3 mg (only tumors >12 mg were counted). C57BL susceptible.	21
	[(CBA x C57BL) x C57BL]	...	[10]	....	Continuous treatment until death. Tumor wt, 14.0-132.5 mg (only tumors >12 mg were counted). C57BL susceptible.	21
25 μg in sesame oil, weekly	(A x C57BL)	...	[1]	....	Treatment at 24 days of age; treatment stopped before autopsy. Tumor wt, 18.3 mg (only tumors >12 mg were counted). C57BL susceptible.	21
	(C57BL x C3H)	...	[10]	....	Continuous treatment until death. Tumor wt, 15.0-117.5 mg (only tumors >12 mg were counted). C57BL susceptible.	21
			[1]	....	Treatment at 61 days of age; treatment stopped before autopsy. Tumor wt, 15.0 mg (only tumors >12 mg were counted). C57BL susceptible.	21

continued

Part II. Induced

Induction Method ⟨Synonym⟩	Strain or Hybrid ⟨Synonym⟩	Sex	Tumor Incidence [No. of Tumors]	Age	Remarks	Reference
	[(CBA x C57BL) x CBA]	...	[2]	....	Continuous treatment until death. Tumor wt, 22.3-47.3 mg (only tumors >12 mg were counted). C57BL susceptible.	21
	[(CBA x C57BL) x C57BL]	...	[3]	....	Continuous treatment until death. Tumor wt, 12.5-44.0 mg (only tumors >12 mg were counted). C57BL susceptible.	21
Estradiol dipropionate 1 μg plus corn oil, daily	TM	...	[2] 15%	....	1 chromophobe; 1 tumor in pars intermedia[9/]. Tumor wt, 15-20 mg.	41, 45
Corn oil supplement only	TM	...	None	....	....	41, 45
Control	TM	...	None	....	....	41, 45
25 μg in sesame oil, weekly	(CBA x C57BL)	...	[4]	....	Continuous treatment until death. Tumor wt, 20.0-79.8 mg (only tumors >12 mg were counted). C57BL susceptible (*see* backcross for estradiol benzoate).	21
			[3]	....	Treatment at 41-49 days of age; treatment stopped before autopsy. Tumor wt, 47.5-215 mg (only tumors >12 mg were counted). C57BL susceptible (*see* backcross for estradiol benzoate).	21
	(C57BL x CBA)	...	[5]	....	Continuous treatment until death. Tumor wt, 12.0-42.3 mg (only tumors >12 mg were counted). C57BL susceptible (*see* backcross for estradiol benzoate).	21
Estradiol 17-valerate ⟨Delestrogen⟩ 10 μg, subcutaneously, once every 2 wk	C57BL	...	19/62; 30.7%	573 d	9 anterior lobe tumors; 6 tumors in pars intermedia[9/]	42
Estrone	Stock	...	1/567	....	Chromophobe adenoma	8
Pellets administered subcutaneously	(CBA x C57BL)	...	[1]	....	Continuous treatment until death. Tumor wt, 80 mg (only tumors >12 mg were counted). C57BL susceptible.	21
	(C57BL x CBA)	...	[1]	....	Continuous treatment until death. Tumor wt, 20.3 mg (only tumors >12 mg were counted). C57BL susceptible.	21
	(C57BL x C3H)	...	[3]	....	Continuous treatment until death. Tumor wt, 27.3-281.5 mg (only tumors >12 mg were counted). C57BL susceptible.	21

[9/] Synonym: Intermediary lobe.

continued

Part II. Induced

Induction Method ⟨Synonym⟩	Strain or Hybrid ⟨Synonym⟩	Sex	Tumor Incidence [No. of Tumors]	Age	Remarks	Reference
Plus estradiol & diethylstilbestrol ⟨DES; stilbestrol⟩	(CBA x C57BL)	♂	18/24; 75%	480 d	Only tumors >12 mg were counted. C57BL susceptible (*see* backcross).	21
	(C57BL x CBA)	♂	19/23; 83%	513 d		
	[(CBA x C57BL) x CBA]	♂	8/38; 21%	517 d		
	[(CBA x C57BL) x C57BL]	♂	26/41; 63%	419 d		
	[(C57BL x CBA) x CBA]	♂	7/34; 21%	525 d		
	(C3H x C57BL)	♂	0/38	....		
	(C57BL x C3H)	♂	22/33; 67%	514 d		
	($C3H_eB$ x C57BL)	♂	11/22; 50%	508 d		
Ethynylestradiol plus mestranol						
Untreated	BDH-SPF	♂	1.8%	....	Total number of mice, untreated & treated, 7000	29
		♀	0.0%	....		
	CF-LP	♂	1.7%; 1.7%	....		
		♀	3.4%; 6.8%	....		
	Swiss random	♂	0.0%	....		
		♀	0.0%	....		
Treated	BDH-SPF; CF-LP; Swiss random	♂	10-fold increase over untreated	....	Steep dose-response curve; low dose similar to control	29
		♀	4-fold increase over untreated	....		
Progestins ⟨Progestogens⟩: delmadinone acetate ⟨chlormadinone acetate⟩, & megestrol compounds;[10] ethynodiol diacetate, lynestrenol compounds, & norethindrone acetate ⟨norethisterone acetate⟩[11]	BDH-SPF & Swiss	♂♀	No increase in incidence over untreated (*see* above)	....	....	29
Progestins plus estrogens, in peanut oil	C3H	...	Low	>84 wk[12]	Mammotropic	37, 38
Norethynodrel & mestranol (50:1)						
7 µg	C57L	...	6/11	>84 wk[12]	Mammotropic; tumor size, 1-5 µl	37, 38
70 µg	C57L	...	7/7	>84 wk[12]	Mammotropic; tumor size, 2-7 µl	37, 38
Norethindrone & ethynylestradiol (50:1)						
7 µg	C57L	...	7/15	>84 wk[12]	Mammotropic; tumor size, 2-6 µl	37, 38
70 µg	C57L	...	5/8	>84 wk[12]	Mammotropic; tumor size, 2-4 µl	37, 38
Peanut oil only	C57L	...	1/8	>84 wk[12]	Mammotropic	37, 38
Saline only	C57L	...	1/7	>84 wk[12]	Mammotropic	37, 38
Thyroid Deficiency						
Thyroidectomy						
Controls	C57BL	♂♀	0/11, 0/100	....	....	14
Surgical	C57BL	♂♀	10/15	290-598 d	Thyrotropic; few with thyroid regeneration	14
	$(C57L \times A)F_1$	♂♀	2/6	12-18 mo	....	31
Iodine-131 ⟨^{131}I⟩	CBA	...	<70%	....	Only tumors >12 mg were counted.	34
	$C3H_f$	...	<70%[8]	....		

[8] >60% incidence in 500 d. [10] Related to 17α-hydroxyprogesterone. [11] Related to nandrolone ⟨19-nortestosterone⟩. [12] Age at which animals were killed.

continued

Part II. Induced

Induction Method ⟨Synonym⟩	Strain or Hybrid ⟨Synonym⟩	Sex	Tumor Incidence [No. of Tumors]	Age	Remarks	Reference
	C57BL	...	>90%[8]	....		
	O20	...	(70-90%)	....		
	(C57BL x DBA_f)F_1	...	(70-90%)[8]	....		
	(O20 x DBA_f)F_1	...	(70-90%)	....		
180 μCi	C57	...	....	12-18 mo	Thyrotropic; tumors >20 mg used for purification	12
200 μCi, intraperitoneally	(C57L x A)F_1	♂♀	5/7	12-18 mo	....	31
Thyroid blocking agent[13]						
Propylthiouracil ⟨6-Propyl-2-thiouracil⟩					All control pituitaries weighed ⩽1.5 mg; only tumors >4 mg were counted	27
10 g/kg in diet	C57BL	...	15/24; 62%	⩽18 mo		
12 g/kg in diet	C57BL	...	21/29; 72%	⩽18 mo		
Plus 2,4-dinitrophenol						
10 g/kg in diet	C57BL	...	3/20; 15%	⩽18 mo		
12 g/kg in diet	C57BL	...	3/29; 10%	⩽18 mo		
Diets						
Diet supplemented with egg						
Extract of whole egg	TM	...	[1[14]]	300-950 d	4 tumors out of total of 100 mice	43
Egg yolk	TM	...	[2[14]]; 12%			
Alcohol extract of egg yolk	TM	...	[1[14]]			
Raw egg yolk	C57BL	...	15/26	....	10 diagnosed: 2 acidophilic; 2 chromophobic; 6 pars intermedia[9]	40
Diet supplemented with liquids	TM	...	....	>300 d	5 tumors out of total of 1455 mice	44
Corn oil and cholesterol	TM	...	[1]	614 d	Pars intermedia[9]	44
			[1]	614 d	Fibrosarcoma	44
Lecithin & cholesterol	TM	...	[1]	334 d	Pars intermedia[9]	44
Glyceryl monostearate ⟨Monostearin⟩	TM	...	[2]	482 d	Chromophobe adenoma	44
Irradiation						
$^{90}Sr(NO_3)_2$, 0.7 μCi/g body wt, intraperitoneally	CBA	♀[15]	0/97	15-16 mo	....	35
		♀[16]	2/97	15-16 mo		
Protons or X rays, whole body, (47-400) rads[17]	RF/Un	♀	65/2139; 3.0%	....	Not dose-dependent	9
X rays						
Control	(C57BL/6 x C3H)F_1	♂	0/100	56 wk[18]	3 groups of animals: ages at start of experiment were 1 d, 15 d, & 42 d. Younger animals more susceptible.	49
			0/100	86 wk[18]		
			0/190	166 wk[18]		
		♀	0/98	56 wk[18]		
			0/99	86 wk[18]		
			1/182	109 wk[18]		

[8] >60% incidence in 500 d. [9] Synonym: intermediary lobe. [13] For additional information on thyroid blocking agent or radiothyroidectomy, consult references 17 and 32. [14] 20-30 animals per group. [15] Unmated animals. [16] Mated animals. [17] For controls, *see* Part I. [18] Average age.

continued

Part II. Induced

Induction Method (Synonym)	Strain or Hybrid (Synonym)	Sex	Tumor Incidence [No. of Tumors[1]]	Age[1]	Remarks	Reference
Exposed: 1 exposure to 320R; or 2 exposures to 160R at 4-d intervals; or 4 exposures to 80R at 4-d intervals; or 4 exposures to 80R at 7-d intervals; or 4 exposures to 120R at 7-d intervals	(C57BL/6 x C3H)F_1	♂	15/561; 2.67%	73 wk[18]		
		♀	27/554; 4.87%	73 wk[18]		
Control: 0 rads	A	♀	None	....	100-150 animals/group. Strain A has low susceptibility, and C57L has high susceptibility, to spontaneous & induced tumors.	18
	C57L ⟨L⟩	♀	3.5%	....		
	(C57L x A)F_1 ⟨LAF_1⟩	♀	1.0%	....		
Exposed: 475 rads	A	♀	1.8%	....		
	C57L ⟨L⟩	♀	4.7%	....		
	(C57L x A)F_1 ⟨LAF_1⟩	♀	4.8%	....		
Control	(C57L x A)F_1 ⟨LAF_1⟩	♀	1.0%	....	100-150 animals per group	18
Exposed						
Total body						
400 rads	(C57L x A)F_1 ⟨LAF_1⟩	♀	2.4%	....		
475 rads	(C57L x A)F_1 ⟨LAF_1⟩	♀	4.8%	....		
Head						
500 rads	(C57L x A)F_1 ⟨LAF_1⟩	♀	7.4%	....		
1300 rads	(C57L x A)F_1 ⟨LAF_1⟩	♀	1.5%	....		
Abdomen						
400 rads	(C57L x A)F_1 ⟨LAF_1⟩	♀	None	....		
750 rads	(C57L x A)F_1 ⟨LAF_1⟩	♀	None	....		
Exposed at 8-10 wk						
0 rads	(C57L/J x A/HeJ)F_1	♀	0/98	70-115 wk[19]	....	50
550 rads	(C57L/J x A/HeJ)F_1	♀	6/78; 7.7%	70-115 wk[19]	4 thyrotropic; 4 mammotropic	50
775 rads	(C57L/J x A/HeJ)F_1	♀	4/83; 4.8%	70-115 wk[19]		
1000 rads	(C57L/J x A/HeJ)F_1	♀	4/84; 4.8%	70-115 wk[19]		
γ rays, exposed at 6-12 wk					Most tumors chromophobic & monomorphous; most mammotropic	47, 48
Control	(C57L x A/He)F_1	♂	0.98% [306[6]]	167.7 wk		
		♀	3.57% [356[6]]	139.9 wk		
Exposed						
223 rads	(C57L x A/He)F_1	♂	1.43% [210[6]]	140.5 wk		
		♀	4.25% [212[6]]	114.5 wk		
368 rads	(C57L x A/He)F_1	♂	0.95% [316[6]]	128.7 wk		
		♀	7.59% [290[6]]	109.5 wk		
578 rads	(C57L x A/He)F_1	♂	1.96% [306[6]]	107.4 wk		
		♀	12.42% [298[6]]	100.1 wk		
697 rads	(C57L x A/He)F_1	♂	0.74% [270[6]]	94.5 wk		
		♀	3.38% [266[6]]	90.9 wk		
γ rays or neutrons; exposed at 6-12 wk					16,000 animals in experiment. Neutrons more effective in ♀.	46, 48
Control	(C57L x A/He)F_1	♂	0.98%	....		
		♀	4.22%	....		
γ rays	(C57L x A/He)F_1	♂	1.62%	....		
		♀	7.88%	....		
Neutron	(C57L x A/He)F_1	♂	1.37%	....		
		♀	14.82%	....		

[1] Unless otherwise indicated. [6] Total number of animals. [18] Average age. [19] Time post-irradiation.

continued

Part II. Induced

Induction Method ⟨Synonym⟩	Strain or Hybrid ⟨Synonym⟩	Sex	Tumor Incidence [No. of Tumors]	Age[1]	Remarks	Reference
Neutron: total body[20]					100-150 animals per group	18
400 rads	$(C57L \times A)F_1$ ⟨LAF_1⟩	♀	4.1%	....		
475 rads	$(C57L \times A)F_1$ ⟨LAF_1⟩	♀	9.2%	....		
Unspecified						
Total body, 475 rads	$(C57L/J \times A/HeJ)F_1$	♀	4.9%	....	....	50
Head & neck						
500 rads	$(C57L/J \times A/HeJ)F_1$	♀	7.4%	....	....	50
550 rads	$(C57L \times A)F_1$ ⟨LAF_1⟩	♀	6.4%	....	720 animals. For 550 rads, combined adrenalectomy & gonadectomy decreased tumor incidence to 1.1%.	19
775 rads	$(C57L \times A)F_1$ ⟨LAF_1⟩	♀	3.5%	....		
1000 rads	$(C57L \times A)F_1$ ⟨LAF_1⟩	♀	3.6%	....		
Combination Treatments						
Hypophyseal isograft to kidney						
Orchiectomized	$(C57BL \times CBA)F_1$	...	High	400 d[21]	Progesterone decreases tumor; estrogen increases tumor	6
1 isograft	C57BL/He	♀[2]	10/12	23 mo	Resembled chromophobe adenomas. Mammary cancers.	25
5 isografts	C57BL/He	♀[2]	17/21	21 mo		
Plus DES pellet					Resembled chromophobe adenomas	25
1 isograft	C57BL/He	♂[5]	10/12	21 mo		
5 isografts	C57BL/He	♂[5]	9/13	17 mo		
Hypophyseal isograft to kidney plus radiothyroidectomy						
Intact host						
Controls—no graft	$(C57L \times A)F_1$ ⟨LAF_1⟩	♀	None	....	....	10
Grafted						
Intact donor	$(C57L \times A)F_1$ ⟨LAF_1⟩	♀	6/11	....	Grafting significant	10
Thyroidectomized donor	$(C57L \times A)F_1$ ⟨LAF_1⟩	♀	5/7			
Thyroidectomized host						
Controls—no graft	$(C57L \times A)F_1$ ⟨LAF_1⟩	♀	19/22	420-600 d	Host susceptible to radiothyroidectomy	10
Grafted						
Intact donor	$(C57L \times A)F_1$ ⟨LAF_1⟩	♀	12/12	420-600 d	Graft susceptible	10
Thyroidectomized donor	$(C57L \times A)F_1$ ⟨LAF_1⟩	♀	10/10			
Iodine-131 ⟨^{131}I⟩, 60-65 μCi						
Intact host						
Controls—no graft	$(C57L \times A)F_1$ ⟨LAF_1⟩	♀	None	....	....	11
Grafted						
Intact donor	C57BL	♂	0/30	....	Graft not susceptible	11
	$(C57L \times A)F_1$ ⟨LAF_1⟩	♀	10/14	....	Only tumors >8 mg were counted. Graft susceptible; mammotropic.	11
Thyroidectomized donor	C57BL	♂	0/30	....	Graft not susceptible	11
	$(C57L \times A)F_1$ ⟨LAF_1⟩	♀	7/13	....	Only tumors >8 mg were counted. Graft susceptible; mammotropic.	11

[1] Unless otherwise indicated. [2] Virgin. [5] Gonadectomized animals. [20] For controls, *see* above data also from reference 18. [21] End-point of experiment.

continued

Part II. Induced

Induction Method (Synonym)	Strain or Hybrid (Synonym)	Sex	Tumor Incidence [No. of Tumors]	Age	Remarks	Reference
Thyroidectomized host						
Controls—no graft	C57BL	♂	33/33	390 d	Thyrotropic in situ; host susceptible to radiothyroidectomy	11
	(C57L x A)F_1 (LAF$_1$)	♀	16/19	....		
Grafted						
Intact donor	C57BL	♂	0/33	....	Graft not susceptible	11
	(C57L x A)F_1 (LAF$_1$)	♀	9/15	500 d	Only tumors >8 mg were counted. Graft susceptible; mammotropic.	11
Thyroidectomized donor	C57BL	♂	0/33	....	Graft not susceptible	11
	(C57L x A)F_1 (LAF$_1$)	♀	10/14	500 d	Only tumors >8 mg were counted. Graft susceptible; mammotropic.	11
Hypophyseal isograft to kidney plus radiothyroidectomy, plus gonadectomy	C57BL	♂	0/44	....	....	11
Endogenous hormone plus thyroid deficiency: gonadectomized plus radiothyroidectomy—iodine-131 (^{131}I), 200 μCi, intraperitoneally	CBA	♂	1	....	Only tumors >12 mg were counted. Earliest tumor appearance in C57BL; next in F_1 hybrids—(C57BL x DBA$_f$)F_1, followed by (O20 x DBA$_f$)F_1.	28
	C3H$_f$	♂	Few	....		
	C57BL	♂	All	....		
	O20	♂	Half	....		
	(C57BL x DBA$_f$)F_1	♂	High	....		
	(O20 x DBA$_f$)F_1	♂	Some	....		
Exogenous hormone plus thyroid deficiency						
Estrone, 2000 μg/liter drinking H_2O	CBA	♂[5]	Some	....	Only tumors >12 mg were counted. Earliest tumor appearance in C57BL; next earliest in hybrids—(C57BL x DBA$_f$)F_1, followed by (O20 x DBA$_f$)F_1.	28
	C3H$_f$	♂[5]	None	....		
	C57BL	♂[5]	High	....		
	O20	♂[5]	High	....		
	(C57BL x DBA$_f$)F_1	♂[5]	High	....		
	(O20 x DBA$_f$)F_1	♂[5]	High	....		
Plus methylthiouracil, 0.4% in food pellets	C57BL	♂[5]	Some	....	Only tumors >12 mg were counted. Earliest tumor appearance in C57BL; next earliest in hybrids—(C57BL x DBA$_f$)F_1, followed by (O20 x DBA$_f$)F_1.	28
	(C57BL x DBA$_f$)F_1	♂[5]	Some	....		
	(O20 x DBA$_f$)F_1	♂[5]	Few	....		
Methylthiouracil only, 0.4% in food pellets	C57BL	♂[5]	Some	....	Only tumors >12 mg were counted. Earliest tumor appearance in C57BL; next earliest in hybrids—(C57BL x DBA$_f$)F_1, followed by (O20 x DBA$_f$)F_1.	28
	(C57BL x DBA$_f$)F_1	♂[5]	Few	....		
	(O20 x DBA$_f$)F_1	♂[5]	Some	....		
Exogenous hormone plus irradiation						
DES pellet, 0.05 mg, plus irradiation, 550 rads	(C57L x A)F_1 (LAF$_1$)	♀	68.6%	....	....	19

[5] Gonadectomized animals.

continued

57. PITUITARY TUMORS: MOUSE

Part II. Induced

Induction Method (Synonym)	Strain or Hybrid (Synonym)	Sex	Tumor Incidence [No. of Tumors]	Age	Remarks	Reference
DES, 0.05 mg; X ray at 8-10 wk					....	50
Control	(C57L/J x A/HeJ)F_1	♀	0/98	<800 d		
Ovariectomized-adrenalectomized	(C57L/J x A/HeJ)F_1	♀	0/93	<800 d		
Plus 550 rads	(C57L/J x A/HeJ)F_1	♀	1/90; 0.9%	<800 d		
550 rads only	(C57L/J x A/HeJ)F_1	♀	6/78; 7.7%	<800 d		
DES only	(C57L/J x A/HeJ)F_1	♀	53/103; 51%	<800 d		
Plus 550 rads	(C57L/J x A/HeJ)F_1	♀	57/83; 69%	<800 d		

Contributor: Annabel G. Liebelt

References

1. Andervont, H. B., et al. 1958. J. Natl. Cancer Inst. 21:783-811.
2. Andervont, H. B., et al. 1960. Ibid. 25:1069-1081.
3. Bardin, C. W., and A. G. Liebelt. 1960. Proc. Am. Assoc. Cancer Res. 3:93(Abstr. 7).
4. Bardin, C. W., et al. 1962. Ibid. 3:302(Abstr. 14).
5. Bardin, C. W., et al. 1966. J. Natl. Cancer Inst. 36: 259-275.
6. Boot, L. M., et al. 1973. Eur. J. Cancer 9:185-193.
7. Boschetti, N. V. 1974. Anat. Rec. 178:315.
8. Burrows, H. 1936. Am. J. Cancer 28:741-745.
9. Clapp, N. K., et al. 1974. Radiat. Res. 57:158-186.
10. Clifton, K. H. 1962. Ibid. 16:601-602.
11. Clifton, K. H. 1963. Proc. Soc. Exp. Biol. Med. 114: 559-565.
12. Condliffe, P. G., et al. 1969. Endocrinology 85:453-464.
13. Cramer, W., and E. S. Horning. 1936. Lancet 230:247-249.
14. Dent, J. W., et al. 1956. Cancer Res. 16:171-174.
15. Dickie, M. M., and G. W. Woolley. 1949. Ibid. 9:372-384.
16. Dickie, M. M., and P. W. Lane. 1956. Ibid. 16:48-52.
17. Dingemans, K. P. 1973. Virchows Arch. B12:338-359.
18. Furth, J., et al. 1960. Acta Unio Int. Contra Cancrum 16:138-142.
19. Furth, J., et al. 1960. Radiat. Res. 12:435-436.
20. Gardner, W. U. 1941. Cancer Res. 1:345-358.
21. Gardner, W. U. 1954. J. Natl. Cancer Inst. 15:693-709.
22. Gardner, W. U. 1962. In A. Haddow, et al. On Cancer and Hormones. Univ. Chicago, Chicago. pp. 89-106.
23. Gardner, W. U., and L. C. Strong. 1940. Yale J. Biol. Med. 112:365-366.
24. Haran-Ghera, N. 1965. Proc. Soc. Exp. Biol. Med. 120:228-230.
25. Heston, W. E. 1964. J. Natl. Cancer Inst. 32:947-955.
26. Hilton, F. K., et al. 1962. Anat. Rec. 142:241.
27. King, D. W., et al. 1963. Proc. Soc. Exp. Biol. Med. 112:365-366.
28. Kwa, H. G. 1961. An Experimental Study of Pituitary Tumours; Genesis, Cytology and Hormone Content. Springer-Verlag, Berlin.
29. Leonard, B. J. 1974. Acta Endocrinol. (Copenhagen), Suppl. 185:34-73.
30. Liebelt, A. G., and R. A. Liebelt. 1961. Cancer Res. 21:86-91.
31. Lundin, P. M., and U. Schelin. 1974. Lab. Invest. 13: 62-68.
32. Moore, G. E., et al. 1953. Proc. Soc. Exp. Biol. Med. 82:643-645.
33. Mori-Chavez, P., and M. Salazar. 1961. Proc. Am. Assoc. Cancer Res. 3:253(Abstr. 200).
34. Mühlbock, O., and H. G. Kwa. 1962. Acta Unio Int. Contra Cancrum 18:275-279.
35. Nilsson, A. 1967. Acta Radiol. 6:33-52.
36. Petrea, I., and W. U. Gardner. 1973. Int. J. Cancer 11:40-57.
37. Poel, W. E. 1966. Science 154:402-403.
38. Poel, W. E. 1968. Proc. Am. Assoc. Cancer Res. 9: 58(Abstr. 230).
39. Richardson, F. L. 1957. J. Natl. Cancer Inst. 18:813-829.
40. Szepsenwol, J. 1966. Anat. Rec. 154:430.
41. Szepsenwol, J., and N. V. Boschetti. 1973. Fed. Proc. Fed. Am. Soc. Exp. Biol. 32:227(Abstr. 91).
42. Szepsenwol, J., and A. A. Ydrach. 1969. Cellule 68: 57-66.
43. Szepsenwol, J., et al. 1966. Proc. Soc. Exp. Biol. Med. 122:1284-1288.
44. Szepsenwol, J., et al. 1969. Cellule 68:43-55.
45. Szepsenwol, J., et al. 1975. Proc. Am. Assoc. Cancer Res. 16:76(Abstr. 302).
46. Upton, A. C., and J. Furth. 1955. J. Natl. Cancer Inst. 15:1005-1021.
47. Upton, A. C., and A. W. Kimball. 1960. Radiat. Res. 12:482(Abstr. 177).
48. Upton, A. C., et al. 1960. Cancer Res. 20(8-2):1-60.
49. Vesselinovich, S. D., et al. 1971. Ibid. 31:2133-2142.
50. Yokoro, K., et al. 1961. Ibid. 21:178-186.

Spontaneous ovarian tumors appear infrequently in most inbred strains of mice, but they are more common in a few selected strains—CE, $C3H_fB$, $C3H_eB$/De, $C3H_eB$/Fe, NZC/Bl, RIII, LT—and in certain F_1 hybrids—(DBA x CE)F_1, (CE x DBA)F_1. Under various experimental conditions, ovarian tumors are inducible depending on genetic background. The more successful etiologic or enhancing factors are various types of radiation, certain chemical carcinogens, such as 9,10-dimethyl-1,2-benzanthracene and 3-methylcholanthrene, or selected hormonal treatment/manipulations including ovarian grafting to the spleen of gonadectomized hosts. Also, specific gene mutations (W^x/W^v) have been shown to be important. Viruses have not been implicated.

Data in this table are not meant to be all-inclusive. For the most part, the experiments enumerated below used standard inbred strains, had adequate numbers of animals per group, gave age at death, described morphology or other characteristics, and had a clearly defined protocol. The table does not include any data on, or references to, transplantable tumor lines.

The morphology generally suggested that granulosa cell tumors (G.C.T.) were the most common (many authors noted that G.C.T.-bearing animals often had hyperplasia of the uterus and breast). Tubular adenomas also were frequent, and some luteomas and papillary cystadenomas were present; many tumors were of mixed types. Teratomas and teratocarcinomas were rare except under certain conditions.

An underlying factor in the development of a high percentage of ovarian tumors appears to be an "aging" factor, i.e., sterility as in W^x/W^v or damage (atrophy) caused by radiation or certain chemicals. In transplantation experiments, the site of transplant is of great importance; i.e., those ovaries in the spleen of gonadectomized hosts of either sex have the highest incidence of tumor development. Altered homeostasis and feedback within the endocrine system is important in the progression of normal ovarian cells into tumors.

The relationship of ovarian tumors in mice to those in the human is not clear at the present time, although several of the experiments referenced suggest appropriate animal-human correlates with respect to possible diagnostic similarities. These experiments, however, have provided useful models of the hormone dependency of tumors, and of autoregulation and the aging phenomena.

Reviews

1. Charyulu, K. K. N. 1976. In K. K. N. Charyulu and A. Sudarsanam, ed. Hormones and Cancer. Academic, New York. pp. 9-20.
2. Jull, J. W. 1973. Methods Cancer Res. 7:131-186.
3. Lipschutz, A. 1963. Gynaecologia 156:93-115.
4. Magee, P. N. 1969. Environ. Res. 2:380-396.
5. Sherman, M. I., and D. Solter, ed. 1975. Teratomas and Differentiation. Academic, New York.
6. Stevens, L. C. 1975. Symp. Soc. Dev. Biol. (33):93-106.
7. Thung, P. J. 1961. In G. H. Bourne, ed. Structural Aspects of Ageing. Hafner, New York. pp. 109-142.

Miscellaneous References

8. Ageenko, A. I. 1962. Folia Biol. (Prague) 8:7-11.
9. Alexandrov, S. N., et al. 1968. Vopr. Onkol. 14(3): 47-52.
10. Artzt, K., and D. Bennett. 1972. J. Natl. Cancer Inst. 48:141-158.
11. Bailliff, R. N., et al. 1970. Anat. Rec. 166:274.
12. Borovkov, A. I. 1968. Arkh. Patol. 30(12):31-36.
13. Bruzzone, S. 1968. Cancer Res. 28:159-165.
14. Bruzzone, S. 1971. Br. J. Cancer 25:158-165.
15. Burdett, W. J. 1962. Surg. Clin. North Am. 42:289-303.
16. Cutler, M. G., and R. Schneider. 1973. Food Cosmet. Toxicol. 11:443-457.
17. Damjanov, I., et al. 1975. Z. Krebsforsch. 83:261-267.
18. Dixon, M. S., et al. 1961. J. Food Sci. 26:611-617.
19. Erickson, R. P., and B. Gondos. 1976. Lancet 1:407-409.
20. Grahn, D., and K. F. Hamilton. 1964. Radiat. Res. 22:191(Abstr.).
21. Ird, E. A. 1975. Vopr. Onkol. 21(12):72-74.
22. Ivanov-Golitsyn, M. N. 1974. Ibid. 20(8):65-68.
23. Kirschbaum, A. 1957. Cancer Res. 17:432-453.
24. Koprowska, I., and H.-Y. Park. 1969. Proc. Am. Assoc. Cancer Res. 10:47(Abstr. 185).
25. Lipschutz, A., et al. 1960. Acta Unio Int. Contra Cancrum 16:206-210.
26. McBurney, M. W., and E. D. Adamson. 1976. Cell 9:57-70.
27. Murphy, E. D. 1972. J. Natl. Cancer Inst. 48:1283-1295.
28. Murphy, E. D., and W. G. Beamer. 1973. Cancer Res. 33:721-723.
29. Murthy, A. S. K., and A. B. Russfield. 1970. Endocrinology 86:914-917.
30. Smith, G. S., et al. 1973. J. Natl. Cancer Inst. 50: 1195-1213.
31. Squartini, F., et al. 1967. Lav. Ist. Anat. Istol. Patol. Univ. Studi Perugia 27:137-156.
32. Upton, A. C. 1967. Symp. Fundam. Cancer Res. 20: 631-675.

continued

58. OVARIAN TUMORS: MOUSE

Part I. Spontaneous

Strain or Stock: Data are for females. **Tumor Type:** G.C.T. = granulosa cell tumor; T.A. = tubular adenoma. Data in brackets refer to the column heading in brackets. For additional information, consult references 3, 12, 25-27, 31, 49, 50, 62, and 81.

Strain or Stock ⟨Synonym⟩	Tumor Type	Incidence, %	Age at Death mo	Reference
Low-Incidence Strains				
A	....	0	....	40
	G.C.T.; T.A.	0-1.6	....	34
AKR	....	0	....	30
AKR/J	....	0	....	24
ASW/Sn	....	0	....	40
	Adenoma; carcinoma	1.9	....	72
BALB/c ⟨Bagg⟩	....	0	....	4,37
	G.C.T.; T.A.	<1.0	....	9
		1.8-5.5	....	34
		3.0	....	42-44
BALB/cCr	G.C.T.	<1.0[1]	≥18	45
BALB/cOs	G.C.T.	1.6	....	1
		4.0	....	2
BALB/cPi	Carcinoma	1.5	....	65
BALB/c_fRIII	G.C.T.; luteoma	<1.0[1]	12 to ≥18	66
BTOs	....	0	....	1
	G.C.T.	2.0	....	2
CBA	G.C.T.	<1.0	....	9
CF1	....	0	....	47
	G.C.T.; luteoma; papillary cystadenoma	0-1.0	....	74
		17.0	....	8
C3H	G.C.T.; T.A.	1.8	....	34
		1.9	≥18	77
		20.0	....	30
C3H-A^{vy}	....	0	....	30
C3H/HeOs	G.C.T.	2.4	....	2
		8.6	....	1
C3H/Tw	....	0	....	36
$C3H_e$/A	....	6[2]	≥18	59,60
		9[3]	≥18	
$C3H_f$⟨C3Hb⟩	G.C.T.; luteoma; T.A.	20	....	6,30
$C3H_f$/Dp	G.C.T.; luteoma	12.0	....	70
$C3H_f$/He	G.C.T.; T.A.	6.0-8.0	....	23,53
C57BL	....	0	....	30
	G.C.T.; T.A.	0-2	....	34
C57BL/He	G.C.T.; T.A.	<1.0[1]	....	53
C57BL/Icrf-a^t	G.C.T.; T.A.	0-4.2	....	61
C57BL/6J	....	0	....	24
C57BL/6Os	....	0	....	2
	G.C.T.	0-1.6	....	1
C57L/J	....	0	....	24
DBA	T.A.	7.7	<6	35
DBA/2J	....	0	....	24
ddY/F	....	0	....	36

[1] Extremely low incidence appears to be related to shortened life-span due to development of other disease. [2] Young. [3] Adult.

continued

58. OVARIAN TUMORS: MOUSE

Part I. Spontaneous

Strain or Stock ⟨Synonym⟩	Tumor Type	Incidence, %	Age at Death mo	Reference
HR/De	G.C.T.; T.A.	4.0	....	22
		4.0-6.0	....	53
IC	....	0[2]	≥18	59,60
		0[3]	12-18	
ICI-SPF	G.C.T.; T.A.	<1.0	....	9
ICR/Jcl	Teratoma; T.A.	0-1.0	....	56,57
NZB/Bl	....	0[1]	....	7
NZO/Bl	G.C.T.	4.5-15.6	....	7
NZY/Bl	G.C.T.	2.2-3.7	....	7
RF	....	1.0	....	75
		11.0	....	15
		20.0	....	14
RF/Un	G.C.T.; luteoma; T.A.	1.0-2.9	....	52
		3.3	....	51
		3.8	....	10,11
		9.0	....	76
		15.0	....	16,17
RF/Up	....	7.0	....	13
Stock A	G.C.T.; luteoma; T.A.	1.8	....	28
Stock R	G.C.T.; T.A.	1.2	....	28
Stock S	....	0	....	28
Swiss	Stromal; adenocarcinoma; papillary adenoma; G.C.T.; T.A.	<1.0	....	9
		3.0	....	63,64
		4.59	....	40
		5.0	≥18	77
SWR/J	Adenoma; carcinoma	1.0	....	24
XLII	....	0[2]	≥18	59,60
		0[3]	12-18	
101	Fibrosarcoma	<1.0	....	9
(CBA/CbSe x BALB/c_f/CbSe)	G.C.T.; luteoma	16.1	≥18	5
(C3H x A)F_1 ⟨C3AF_1⟩	....	0	....	80
(C3H x ICRC)F_1	G.C.T.	14.28	....	58
(ICRC x C3H)F_1	G.C.T.	3.7	....	58
(C3H x 129)F_1	....	0	....	55
(C57BL x C3H)F_1	....	2.5	12-18	78
		3.9	≥18	
(C57BL/6 x C3H)F_1 ⟨B6C3F_1⟩	....	0	....	80
	G.C.T.; T.A.	1.0-2.0	....	79
(C57L x A)F_1 ⟨LAF_1⟩	....	0	....	73
	G.C.T.; luteoma; T.A.	2.0	....	29
		15.6	....	38,39
(C57L x A/He)F_1J ⟨LAF_1J⟩	Adenoma	3.0	....	20
Wild mice	G.C.T.; T.A.	3.0-6.0	....	53
High-Incidence Strains				
CE	G.C.T.; T.A.	34.0	....	53
C3H_eB/De	G.C.T.; T.A.	29.0-47.0	....	33,53
C3H_eB/Fe	G.C.T.; T.A.	22.0-64.0	....	21,53
C3H_fB	G.C.T.; T.A.	23.1-36.0	....	34
DBA	T.A.	55.5	12 to >18	35

1/ Extremely low incidence appears to be related to shortened life-span due to development of other disease. 2/ Young. 3/ Adult.

continued

58. OVARIAN TUMORS: MOUSE

Part I. Spontaneous

Strain or Stock ⟨Synonym⟩	Tumor Type	Incidence, %	Age at Death mo	Reference
NZC/Bl	G.C.T.	20.0-33.0	....	7
(101 x C3H)F_1 ⟨101C3F_1⟩	....	21.0	....	15
		Highest-Incidence Strains		
BALB/c	G.C.T.; luteoma	75.8	....	41
LT	Teratoma[4]	40.0	<6	67-69
		50.0	....	32
		Miscellaneous		
C3H/N (uterus painted with carcinogen)	Teratoma	<1.0	....	71
		Effect of Genetics on Incidence		
CE/Wy	G.C.T.; T.A.	34.8	>18	23
DBA/2Wy	....	0	....	23
(CE x DBA)F_1	G.C.T.; T.A.	19.0	....	23
(DBA x CE)F_1	G.C.T.; T.A.	33.3	....	23
F_1 backcross to CE	G.C.T.; T.A.	52.0[5,6]	>18	23
CE x (CE x DBA)	G.C.T.; T.A.	46.2	....	23
CE x (DBA x CE)	G.C.T.; T.A.	41.7	....	23
(CE x DBA) x CE	G.C.T.; T.A.	53.8	....	23
(DBA x CE) x CE	G.C.T.; T.A.	66.7	....	23
F_1 backcross to DBA	Papilliferous cystadenoma	11.5[5]	....	23
DBA x (CE x DBA) & DBA x (DBA x CE)	Papilliferous cystadenoma	20.0	....	23
(CE x DBA) x DBA & (DBA x CE) x DBA	Papilliferous cystadenoma	9.5	....	23
(C3H/He x C57BL/6); (C57BL/6 x C3H/He)				
Controls: $W^x/+$, $W^v/+$, $+/+$	Teratoma	1.4	....	18,19,46,48
Experimentals				
W^j/W^v	Teratoma	37.5	12-18	54
W^x/W^y; W^x/W^v[7]	T.A.	100; 78.0[8]	....	18,19,46

[4] Tumors developed very early. [5] Overall incidence. [6] Incidence in extreme dilution mice, 58.6%; in black agouti mice, 42.9%. [7] Black-eyed whites are sterile. [8] Diagnosed from peritoneal fluid.

Contributor: Annabel G. Liebelt

References

1. Akamatsu, Y. 1975. J. Natl. Cancer Inst. 55:893-897.
2. Akamatsu, Y., and B. P. Barton. 1974. Ibid. 52:377-385.
3. Bayreuther, K. 1960. Nature (London) 186:6-9.
4. Biancifiori, C., and F. Caschera. 1962. Br. J. Cancer 16:722-730.
5. Biancifiori, C., and R. Ribacchi. 1961. Lav. Ist. Anat. Istol. Patol. Univ. Studi Perugia 21:23-28.
6. Biancifiori, C., et al. 1961. Br. J. Cancer 15:270-283.
7. Bielschowsky, M., and E. F. D'Ath. 1973. Pathology 5:303-310.
8. Breslow, N. E., et al. 1974. J. Natl. Cancer Inst. 52: 233-239.
9. Carter, R. L. 1968. Eur. J. Cancer 3:537-543.
10. Clapp, N. K. 1974. Radiat. Res. 59:18(Abstr. A9-4).
11. Clapp, N. K., et al. 1974. Ibid. 57:158-186.
12. Cloudman, A. M. 1941. In G. D. Snell, ed. Biology of the Laboratory Mouse. Blakiston, Philadelphia. pp. 222-225.
13. Conklin, J. W., et al. 1963. Radiat. Res. 19:156-168.
14. Conklin, J. W., et al. 1965. Cancer Res. 25(1):20-28.
15. Cosgrove, G. E., et al. 1965. Ibid. 25(1):938-945.
16. Darden, E. B., Jr., et al. 1967. Int. J. Radiat. Biol. 12: 435-452.
17. Darden, E. B., Jr., et al. 1969. Radiat. Res. 39:537-538(Abstr. Fd-4).
18. Davis, R. H., et al. 1970. Proc. Soc. Exp. Biol. Med. 134:434-436.
19. Davis, R. H., et al. 1972. Experientia 28:93-94.
20. Davis, W. E., Jr., et al. 1970. Radiat. Res. 41:400-408.
21. Deringer, M. K. 1959. J. Natl. Cancer Inst. 22:995-1002.

continued

Part I. Spontaneous

22. Deringer, M. K. 1962. Ibid. 29:1107-1121.
23. Dickie, M. M. 1954. Ibid. 15:791-799.
24. Diwan, B. A., and H. Meier. 1974. Cancer Res. 34: 764-770.
25. Dominguez, O. V., and R. A. Huseby. 1969. Endocrinology 84:1039-1047.
26. Fathalla, M. F. 1972. Obstet. Gynecol. Surv. 27:751-768.
27. Fekete, E., and M. A. Ferrigno. 1952. Cancer Res. 12: 438-440.
28. Furth, J., and J. S. Butterworth. 1936. Am. J. Cancer 28:66-95.
29. Furth, J., et al. 1960. Acta Unio Int. Contra Cancrum 16:138-142.
30. Gabrielides, C. G., et al. 1972. Pathol. Eur. 7:155-159.
31. Gardner, W. U., and S. C. Pan. 1948. Cancer Res. 8: 241-256.
32. Hecht, F., et al. 1976. Lancet 2:1311.
33. Heston, W. E. 1963. In W. J. Burdette, ed. Methodology in Mammalian Genetics. Holden-Day, San Francisco. pp. 247-268.
34. Heston, W. E., et al. 1973. J. Natl. Cancer Inst. 51: 209-224.
35. Hummel, K. P. 1954. Ibid. 15:711-715.
36. Komuro, M. 1976. J. Radiat. Res. 17:99-105.
37. Krarup, T. 1970. Br. J. Cancer 24:168-186.
38. Lesher, S., et al. 1960. Radiat. Res. 12:451(Abstr. 93).
39. Lesher, S., et al. 1965. Ibid. 24:239-277.
40. Li, Y., et al. 1976. Z. Krebsforsch. Klin. Onkol. 86: 165-170.
41. Lipschutz, A., et al. 1962. Nature (London) 196:946-948.
42. Lipschutz, A., et al. 1966. Ibid. 212:686-688.
43. Lipschutz, A., et al. 1967. Br. J. Cancer 21:144-152.
44. Lipschutz, A., et al. 1967. Ibid. 21:153-159.
45. Madison, R. M., et al. 1968. J. Natl. Cancer Inst. 40: 683-685.
46. McGowan, L., and R. H. Davis. 1970. Obstet. Gynecol. 35:878-890.
47. McGowan, L., and R. H. Davis. 1971. Ibid. 38:125-135.
48. McGowan, L., et al. 1971. Acta Cytol. 15:306-309.
49. Meier, H., et al. 1970. Cancer Res. 30:30-34.
50. Mody, J. K. 1967. Indian J. Pathol. Bacteriol. 10: 193-196.
51. Mori-Chavez, P., et al. 1970. Cancer Res. 30:913-928.
52. Mori-Chavez, P., et al. 1974. Ibid. 34:328-336.
53. Murphy, E. D. 1966. In E. L. Green, ed. Biology of the Laboratory Mouse. Ed. 2. McGraw-Hill, N.Y. pp. 521-562.
54. Murphy, E. D., and E. S. Russell. 1963. Acta Unio Int. Contra Cancrum 19:779-782.
55. Nishizuka, Y., et al. 1973. Excerpta Med. Int. Congr. Ser. 267:171-180.
56. Nomura, T. 1976. Br. J. Cancer 33:521-534.
57. Nomura, T., et al. 1972. Med. J. Osaka Univ. 23:121-131.
58. Ranadive, K. J., et al. 1972. Indian J. Cancer 9:14-26.
59. Riviere, M. R., et al. 1962. Bull. Cancer 49:206-216.
60. Riviere, M. R., et al. 1962. C. R. Soc. Biol. 156:1605-1607.
61. Rowlatt, C., et al. 1976. Lab. Anim. 10:419-442.
62. Schueler, R. L., and R. Ediger. 1975. Am. J. Vet. Res. 36:341-342.
63. Shimizu, H., et al. 1974. Int. J. Cancer 13:500-505.
64. Shimizu, H., et al. 1975. Br. J. Cancer 31:492-496.
65. Smith, C. S., and H. I. Pilgrim. 1971. Proc. Soc. Exp. Biol. Med. 138:542-544.
66. Squartini, F., et al. 1974. J. Natl. Cancer Inst. 52: 1635-1641.
67. Stevens, L. C. 1975. In M. I. Sherman and D. Solter, ed. Teratomas and Differentiation. Academic, New York. pp. 17-32.
68. Stevens, L. C. 1976. Am. J. Pathol. 85:809-811.
69. Stevens, L. C., and D. S. Varnum. 1974. Dev. Biol. 37:369-380.
70. Terracini, B., et al. 1976. Br. J. Cancer 33:427-439.
71. Thiery, M. 1963. Ibid. 17:231-234.
72. Toth, B., and T. Toth. 1970. Tumori 56:315-324.
73. Ueda, G., et al. 1974. Acta Obstet. Gynaecol. Jpn. 21: 135-142.
74. Uematsu, K., and C. Huggins. 1968. Mol. Pharmacol. 4:427-434.
75. Upton, A. C., et al. 1960. Proc. Soc. Exp. Biol. Med. 104:769-772.
76. Upton, A. C., et al. 1970. Radiat. Res. 41:467-491.
77. Vesselinovitch, S. D., et al. 1967. Cancer Res. 27(1): 2333-2337.
78. Vesselinovitch, S. D., et al. 1970. Ibid. 30:2548-2551.
79. Vesselinovitch, S. D., et al. 1971. Ibid. 31:2133-2142.
80. Vesselinovitch, S. D., et al. 1974. Ibid. 34:2530-2538.
81. Whiteley, H. J., and D. L. Horton. 1964. Br. J. Cancer 18:252-254.

Part II. Hormonal Effects

Strain or Stock: Data are for females unless otherwise indicated. **Tumor Type:** G.C.T. = granulosa cell tumor; T.A. = tubular adenoma. For additional information, consult references 5-8, 13-17, 23, 28-30, 33, and 34.

Procedure or Agent	Strain or Stock	Tumor Type	Incidence, %	Age at Death mo	Reference
Endogenous Hormones					
Castration, subtotal	BALB-A	Luteoma	18.8	....	18
	BALB/c	Luteoma	55.6	....	19

continued

58. OVARIAN TUMORS: MOUSE

Part II. Hormonal Effects

Procedure or Agent	Strain or Stock	Tumor Type	Incidence, %	Age at Death mo	Reference
Ova transfer (C3H ova in C57BL)	$C3H_eB$	G.C.T; luteoma; T.A.	(29.3-47.5)	>18	4
Transplantation of ovary					
To testis, with unilateral gonadectomy	BALB-A ♂	G.C.T.; luteoma	(94-100)	....	21
	C-I ♂	Luteoma[1]	56.9	(6-12)	12
	C57 ♂	Luteoma	50.0	(6-12)	12
To spleen					
Gonadectomized	BALB-A ♂	G.C.T.; luteoma	(80-100)	....	21
	ddO	G.C.T.; luteoma (very small)	100	....	35
	$(C57L \times A)F_1$[2]	G.C.T.; luteoma	91.7	....	36
Ovariectomized	MA_f/Sp	G.C.T.; luteoma	81.8	<6	9
To spleen or ovarian bursa following short span in spleen of gonadectomized DBA	DBA x C3H	T.A.	61.0	(12-18)	11
To spleen & kidney	BALB-A ♂	G.C.T.; luteoma[3]	86.1	....	20,21,24
To pancreas					
Bilaterally ovariectomized	KM stock	G.C.T.; luteoma	84.9	(0-12)	12
With estrogen	KM stock	G.C.T.; luteoma	(14.3-35.5)	....	12
Transplantation of egg cylinder to renal capsule	A; CBA; C3H/H; C57BL	Teratoma, teratocarcinoma	100	<6	1-3
Thymectomy	$(C3H \times 129)F_1$	G.C.T.; T.A.	35	(12 to >18)	32
	$(C57BL \times A/J)F_1$	G.C.T.; T.A.	(33.0-44.4)	....	31,32
Exogenous Hormones					
Progestins					
Norethindrone, 7.7 µg/d	BALB/c	G.C.T.	52.0	....	25
Norethynodrel, 5.5 µg/d	BALB/c	G.C.T.	8.3	....	25
19-Norprogesterone, >15 µg/d	BALB/c	G.C.T.; luteoma	23.5	....	22,25
Progesterone					
30 µg pellet, subcutaneous treatment	BALB/c	G.C.T.	32.5	....	25
59-900 µg/d, treatment for 13 or 18 mo	BALB/c	G.C.T.	(0-44.8)[4]	....	25-27
Progestins plus estrogens: norethynodrel + mestranol[5], 5-20 µg/g food	A	....	(0-1.9)	....	10
	BALB/c	....	(3.9-9.1)	....	10
	C3H	....	1.8	....	10
	$C3H_fB$	G.C.T.; T.A.	(5.7-26.8)[4]	....	10
	C57BL	....	(2.0-4.3)	....	10

[1] In addition, 3 interstitial cell tumors. [2] Synonym: LAF_1. [3] Second group all microscopic. [4] Correlation between increasing dose and tumor incidence. [5] Synonym: Enovid.

Contributor: Annabel G. Liebelt

References

1. Damjanov, I., and D. Solter. 1974. Z. Krebsforsch. 81:63-69.
2. Damjanov, I., and D. Solter. 1975. In M. Sherman and D. Solter, ed. Teratomas and Differentiation. Academic, New York. pp. 209-220.
3. Damjanov, I., and D. Solter. 1976. Am. J. Pathol. 83: 241-244.
4. Deringer, M. K. 1959. J. Natl. Cancer Inst. 22:995-1002.
5. Dunn, T. B., and A. W. Green. 1963. Ibid. 31:425-455.
6. Gardner, W. U. 1955. Cancer Res. 15:109-117.
7. Gardner, W. U. 1961. J. Natl. Cancer Inst. 26:829-853.

continued

Part II. Hormonal Effects

8. Gardner, W. U., et al. 1959. In F. Homburger, ed. The Physiopathology of Cancer. Ed. 2. Hoeber-Harper, New York. pp. 152-237.
9. Guthrie, M. J. 1957. Cancer 10:190-203.
10. Heston, W. E., et al. 1973. J. Natl. Cancer Inst. 51: 209-224.
11. Hummel, K. P. 1954. Ibid. 15:711-715.
12. Institute of Experimental Medicine, Chinese Academy of Medical Sciences. 1968. China's Med. 3:174-187.
13. Jett, J. D., et al. 1961. Proc. Am. Assoc. Cancer Res. 3:238(Abstr. 142).
14. Jull, J. W. 1969. J. Natl. Cancer Inst. 42:967-972.
15. Jull, J. W., et al. 1966. Ibid. 37:409-420.
16. Lancaster, M. C. 1974. 2nd Int. Norgestrel Symp. Amsterdam, pp. 16-27.
17. Li, M. H., and W. U. Gardner. 1949. Cancer Res. 9: 35-41.
18. Lipschutz, A. 1960. Acta Unio Int. Contra Cancrum 16:149-156.
19. Lipschutz, A. 1968. Perspect. Biol. Med. 11:461-474.
20. Lipschutz, A., and H. Cerisola. 1962. Nature (London) 193:145-147.
21. Lipschutz, A., and V. I. Panasevich. 1972. Perspect. Biol. Med. 16:22-35.
22. Lipschutz, A., et al. 1962. Nature (London) 196:946-948.
23. Lipschutz, A., et al. 1964. Ibid. 202:503-504.
24. Lipschutz, A., et al. 1964. Acta Unio Int. Contra Cancrum 20:1412-1416.
25. Lipschutz, A., et al. 1966. Nature (London) 212:686-688.
26. Lipschutz, A., et al. 1967. Br. J. Cancer 21:144-152.
27. Lipschutz, A., et al. 1967. Ibid. 21:153-159.
28. Marchant, J. 1961. Ibid. 15:821-827.
29. Miller, O. J., and W. U. Gardner. 1954. Cancer Res. 14:220-226.
30. Myhre, E. 1972. Acta Pathol. Microbiol. Scand. A80 (Suppl. 233):67-75.
31. Nishizuka, Y., et al. 1972. Gann 63:139-140.
32. Nishizuka, Y., et al. 1973. Excerpta Med. Int. Congr. Ser. 267:171-180.
33. Poel, W. E., et al. 1968. Cancer Res. 28:845-859.
34. Trentin, J. J., and B. T. Morris. 1961. Proc. Am. Assoc. Cancer Res. 3:273(Abstr. 281).
35. Ueda, G., et al. 1972. Acta Obstet. Gynaecol. Jpn. 19:16-24.
36. Ueda, G., et al. 1974. Ibid. 21:135-142.

Part III. Other Induction Methods

Dose: r = rad. **Route**: i.p. = intraperitoneal; i.v. = intravenous; s.c. = subcutaneous. **Strain & Condition**: Data are for females unless otherwise indicated. **Tumor Type**: G.C.T. = granulosa cell tumor; T.A. = tubular adenoma. Data in brackets refer to the column heading in brackets. For additional information, consult references 1, 4, 5, 9, 11-13, 25, 26, 31-43, 45-50, 53-55, 57, 64-66, 70, 71, 74, 75, 79, 80, 86, 88, 90, 98, and 107-109.

Agent (Synonym)	Dose	Route	Strain & Condition (Synonym)	Tumor Type	Incidence %	Age at Death mo	Reference
Radiation							
Protons	47-372 MeV	Total body	RF/Un	G.C.T.; luteoma; adenoma	38.0-55.2	12-18	14, 15
X rays	0-780 r	Total body	A	G.C.T.; benign cystadenoma	≤20.2[1]	≥18	62, 63
	50-400 r, all at one time	Total body; received in utero or following birth up to 180 d of age	RF	G.C.T.; luteoma; adenoma	7.0-66.0[1]	12-18	14, 15, 97
	80-160 r, single or multiple doses	Total body	(C57BL/6 x C3H)F_1				
			1 d old	G.C.T.; T.A.	31.0-73.8	......	103
			15 d old	G.C.T.; T.A.	61.0-90.0	......	103
			42 d old	G.C.T.; T.A.	42.0-84.0	......	103

[1] Lack of definite correlation between incidence and dose.

continued

Part III. Other Induction Methods

Agent ⟨Synonym⟩	Dose	Route	Strain & Condition ⟨Synonym⟩	Tumor Type	Incidence %	Age at Death mo	Reference
	85-90 r	Total body	RF/Up	....	37.0	12-18	17
	130 or 260 r	Total body	C3H/Tw	G.C.T.; luteoma; T.A.	72.4	6-18	52
			ddY/F	Luteoma; G.C.T.; T.A.	39.4	6-18	52
	200 r	Total body	BALB/c	G.C.T.; luteoma; T.A.	90.9	12-18	51
			Stock A	G.C.T.; luteoma; T.A.	27.9	......	28
			Stock R	G.C.T.; T.A.	18.5	......	28
			Stock S	G.C.T.; T.A.	4.1	......	28
		Local to one ovary	BALB/c	G.C.T.; luteoma; T.A.	8.1	......	51
	300 r	Total body	$C3H_e/A$	....	92[2] 67[3]	⩾18 12-18	82, 83
			IC	....	14[2] 10[3]	⩾18 12-18	82, 83
			RF	....	39.0	12-18	18
			XLII	....	43[2] 26[3]	⩾18 12-18	82, 83
	300-500 r	Total body	CF1	G.C.T.; luteoma; papillary cystadenoma[4]	38.0	......	68
	320 r	Total body	$(C57BL/6 \times C3H)F_1$, at 1, 15, & 42 d old	T.A.; G.C.T.	>50.0	......	104
	400 r	Abdomen	$(C57L \times A)F_1$ ⟨LAF_1⟩	....	74.0	......	29
	400, 475 r	Total body	$(C57L \times A)F_1$ ⟨LAF_1⟩	....	69.0	......	29
	500 r	Head	$(C57L \times A)F_1$ ⟨LAF_1⟩	....	6.0	......	29
At sea level	150 r	Total body	RF/Un	G.C.T.; luteoma; T.A.; angioma	31.9	......	72
	300 r	Total body	RF/Un	G.C.T.; luteoma; T.A.; angioma	45.6	......	72
At low altitude	150 r	Total body	RF/Un	G.C.T.; luteoma; T.A.	47.8; 51.2	......	73
At high altitude	150 r	Total body	RF/Un	G.C.T.; luteoma; T.A.; angioma	37.1	......	72, 73
				G.C.T.; luteoma; T.A.	60.9	......	73
	300 r	Total body	RF/Un	G.C.T.; luteoma; T.A.; angioma	30.5	......	72
Atomic bomb source[5]	....	Total body	RF, $(C57L \times A/He)F_1$	G.C.T.; luteoma	35.5	......	96
γ rays	0-300 r	Total body	RFM_f/Un	....	2.4-47.8	......	94
	618-2408 r	Total body	$(C57L \times A)F_1$ ⟨LAF_1⟩	....	50.0-86.7	⩾18	78
^{60}Co source	....	Total body	$(C57L \times A)F_1$ ⟨LAF_1⟩	G.C.T.; luteoma; T.A.	0.67-29.2	......	59, 60
	40 different doses	Total body	RF/Un	G.C.T.; luteoma; T.A.	35.0[6]	⩾18	99
	100, 300, 600 r						
	.0035 r/min	Total body	RF	....	∿5.0-21.0	......	95
	7 r/min	Total body	RF	....	∿26.0-45.0	......	95

[2] Young. [3] Adult. [4] Peritoneal aspirates, morphological study. [5] Easy Shot, George Shot. [6] Dose-incidence relationship follows sigmoid curve.

continued

Part III. Other Induction Methods

Agent ⟨Synonym⟩	Dose	Route	Strain & Condition ⟨Synonym⟩	Tumor Type	Incidence %	Age at Death mo	Reference
Atomic bomb source[5]	223-697 r	Total body	RF, (C57L x A/He)F_1	G.C.T.; luteoma	26.7-40.3	......	96
Neutrons	0-188 r	Total body	RFM$_f$/Un	....	0-51.6	......	94
	0-930 r; 28 different doses	Total body	RF/Un	G.C.T.; luteoma; T.A.	$\geq$19.0	$\geq$18	99
	14 MeV	Total body	RF/Un	Non-reticular neoplasm	43.0-62.0	......	20, 21
Atomic bomb source[5]	....	Total body	RF, (C57L x A/He)F_1	G.C.T.; luteoma	29.7	......	96
Radiomimetic: 4-(*p*-dimethylaminostyryl)-quinoline ⟨4M20⟩	....	....	RF	....	18.0-24.0	$\geq$18	19
			(101 x C3H)F_1 ⟨101C3F_1⟩	....	35.0	$\geq$18	19
Chemical Carcinogens							
n-Amylhydrazine hydrochloride ⟨AH⟩	....	....	Swiss	Adenocarcinoma; papillary adenoma	2.0	$<$6	84
Benzo[*a*]pyrene ⟨3,4-benzopyrene; BP⟩	....	....	BALB/c				
			Virgin	....	0	6-12	6
			Pseudopregnant	....	0	6-12	6
	Single & massive pulse dose	i.v.	CF1	....	0	......	93
In almond oil	....	Stomach tube, skin	C3H$_f$⟨C3Hb⟩	Luteoma	4.6	6-12	8
In peanut ⟨arachis⟩ oil	....	Skin	IF	G.C.T.; luteoma	7.5	......	69
Busulfan ⟨Myleran⟩	....	i.v.	RF	....	45.0	$\geq$18	18
DDT	2-250 ppm	Oral	CF1	....	14.8	$\geq$18	10
1,2:5,6-Dibenzanthracene ⟨DBA⟩	....	....	BALB/c				
			Virgin	....	0	12-18	6
			Pseudopregnant	....	0	12-18	6
		Skin	C3H$_f$⟨C3Hb⟩	G.C.T.; luteoma	38.5	6-12	8
In peanut ⟨arachis⟩ oil	....	Skin	IF	....	0	......	69
1-(4-Dimethylaminobenzylidene)indene ⟨DABI⟩	....	....	AKR	G.C.T.; T.A.	33.3	6-12	30
			C3H	G.C.T.; T.A.	71.4	6-18	30
			C3H$_f$⟨C3Hb⟩	G.C.T.; T.A.	94.4	12-18	30
			C3H-A^{vy}	G.C.T.; T.A.	53.3	6-12	30
			C57BL	....	0	6-12	30
9,10-Dimethyl-1,2-benzanthracene ⟨DMBA⟩	....	....	BALB/cOs				
			Normal	G.C.T.	10.3	......	2
			Splenectomized	G.C.T.	9.1	......	2
			BTOs				
			Normal	G.C.T.	4.6	......	2
			Splenectomized	G.C.T.	4.4	......	2
			C3H/HeOs				
			Normal	....	0	......	2
			Splenectomized	....	0	......	2
			C57BL/6Os				
			Normal	G.C.T.	9.7	......	2
			Splenectomized	G.C.T.	2.6	......	2

[5] Easy Shot, George Shot.

continued

Part III. Other Induction Methods

Agent ⟨Synonym⟩	Dose	Route	Strain & Condition ⟨Synonym⟩	Tumor Type	Incidence %	Age at Death mo	Reference
		Oral	BALB/c				
			Virgin	G.C.T.	63.9	6-12	6
			Pseudopregnant	G.C.T.	44.1	6-12	6
		or i.p.	BALB/c ⟨Bagg⟩	G.C.T.; luteoma	100	<18	56
		i.p.	C3H	G.C.T.	0-57.0	⩽6	58
		i.v.	C3H	G.C.T.	29-100	6-12	58
		Single feeding	C3H	....	59	......	58
		Single feeding, i.p., or i.v.	C57BL	....	0	6-12	58
	Pulse dose						
	Single massive	....	CF1	....	32.0	......	93
	Multiple	....	CF1	....	28.0-42.0	0-6	93
	50 µg	s.c.	BALB/cCbSe	G.C.T.; luteoma	25.0	......	81
	100 µg	s.c.	BALB/cCbSe	G.C.T.; luteoma	57.9	......	81
In oil	....	Skin	A	....	0	......	8
			C3H	....	0	......	8
			C57BL	....	11.0	......	8
			(IF x A)	....	40.0	......	8
			(IF x C57BL)	....	70.0	......	8
Almond	....	Skin plus stomach tube	$C3H_f$⟨C3Hb⟩	G.C.T.; luteoma	78.0	0-12	8
Olive	....	Skin	IF/Or; (IF x A); (IF x C57)	G.C.T.	60	6-12[7]	44
Peanut ⟨arachis⟩	....	Skin	IF	G.C.T.; luteoma	20	<18	69
Ethylhydrazine hydrochloride ⟨EH⟩	....	....	Swiss	Adenocarcinoma; papillary adenoma	3.0	0-6	84
1-Ethyl-1-nitrosourea ⟨ENU⟩	....	....	AKR/J	....	0	......	24
			C57BL/6J	....	0	......	24
			C57L/J	G.C.T.; carcinoma	4.0	......	24
			DBA/2J	....	0	......	24
			SWR/J	G.C.T.; carcinoma	4.6	......	24
			(C3H x A)F_1 ⟨C3AF_1⟩; (C57BL x C3H)F_1 ⟨B6C3F_1⟩				
			Newborn	G.C.T.; T.A.	42.7	⩾18	106
			Young	G.C.T.; T.A.	26.9	⩾18	106
Isoniazid ⟨INH⟩	....	....	ASW/Sn	....	0[8]	0-6	89
Mechlorethamine ⟨Methylbis(β-chloroethyl)-amine; nitrogen mustard; HN_2⟩	....	i.v.	RF	....	26.0	12-18	18
			RF/Up	....	13.0	⩾18	17
3-Methylcholanthrene ⟨MC⟩	....	....	$C3H_f$⟨C3Hb⟩				
			Virgin	G.C.T.	17.0	6-12	7
			Pseudopregnant	G.C.T.	19.0	6-12	7
			Olfactory lobe removed	G.C.T.	12.0	6-12	7
			NZY/Bcr				
			Pseudopregnant	G.C.T.	7.9	6-12	67
			Lactating	G.C.T.	3.8	12-18	67

[7] Average. [8] Extremely low incidence appears to be related to shortened life-span due to development of other disease.

continued

Part III. Other Induction Methods

Agent ⟨Synonym⟩	Dose	Route	Strain & Condition ⟨Synonym⟩	Tumor Type	Incidence %	Age at Death mo	Reference
		Gavage	BALB/cOs	....	5.0	......	3
			BTOs	....	2.0	......	3
			C3H/HeOs	....	0	......	3
			$C3H_eB$/Os	....	11.0	......	3
			C57BL/6Os	....	0	......	3
		Oral	BALB/c				
			Virgin	G.C.T.	46.2	......	6
			Pseudopregnant	G.C.T.	7.1	......	6
		Skin	BALB/c, olfactory lobe removed	....	0	......	6
	Single & massive pulse dose	....	CF1	....	0	......	93
In oil	....	Oral	$DBA/2_f$	....	1.0	......	8
		Skin	BALB/c	....	0	......	8
			IF	....	0-5.0	......	8
			(IF x A)	....	5.0	......	8
			(IF x C57BL)	....	50.0	......	8
Peanut ⟨arachis⟩	....	....	$C3H_f$⟨C3Hb⟩	G.C.T.	46.0	6-18	8
N-Nitrosodimethylamine ⟨Dimethylnitrosamine; DMN⟩	....	....	A	....	0	......	61
			ASW/Sn	....	0	......	61
			Swiss	Stromal	1.1	......	61
N-Nitroso-*N*-methylurea ⟨NMU⟩[9]	Lower	....	$C3H_f$/Dp				
			1 d old	G.C.T.	11.0	<18	87
			70 d old	G.C.T.	20.0	<18	87
	Higher	....	$C3H_f$/Dp				
			1 d old	....	0	<18	87
			21 d old	G.C.T.	47.0	<18	87
			70 d old	G.C.T.	73.0	<18	87
Triethylenemelamine ⟨TEM⟩	....	i.p.	RF	....	52.0	12-18	18
			RF/Up	....	58.0	⩾18	16, 17
5,9,10-Trimethylbenzanthracene ⟨7,8,12-Trimethylbenzanthracene; TMBA⟩	....	....	CF1	....	50.0	......	93
Urethan	....	Drinking water	CF1	....	25.0	......	10
		Injection	ICR/Jcl				
			Fetus				76, 77
			11-16 d	Teratoma; T.A.	5.9	......	
			14-19 d	....	0	......	
			Neonate	....	0	......	76, 77
			Weanlings	T.A.; cystadenoma	30.0	......	76, 77
			Adult virgin	T.A.; cystadenoma	10.0	......	76, 77
			Pregnant	T.A.; cystadenoma	9.1	......	76, 77

[9] One luteoma occurred in entire group.

continued

Part III. Other Induction Methods

Agent ⟨Synonym⟩	Dose	Route	Strain & Condition ⟨Synonym⟩	Tumor Type	Incidence %	Age at Death mo	Reference
		i.p.	$(C57BL \times C3H)F_1$				
			Prenatal	G.C.T.; T.A.	<30.6	......	105
			Postnatal	G.C.T.; T.A.	<40.4[10]	......	105
			Newborn	....	<31.0[10]	......	100
		s.c.	ICR/Jcl, pregnant	T.A.; cystadenoma	3.0	......	76, 77
			Swiss	G.C.T.; T.A.	70.9	12-18	101
	None	i.p.	$(C57BL \times C3H)F_1$	....	3.9	⩾18	102
	1.5 mg/g	i.p.	$(C57BL \times C3H)F_1$, at 1 d old	....	10.0	6-12	102
	3.0 mg/g	i.p.	$(C57BL \times C3H)F_1$				
			At 4 d old	....	24.3	6-12	102
			At 25 wk old	....	12.2	12-18	102
	6.0 mg/g	i.p.	$(C57BL \times C3H)F_1$, at 25 wk old	....	29.4	12-18	102
	12.0 mg/g	i.p.	$(C57BL \times C3H)F_1$, at 25 wk old	....	57.8[11]	12-18	102
In 40% ethylene glycol	....	Skin painted	HR/De				
			hr/+, haired	Rare carcinoma	4.0	12-18	23
			hr/*hr*, hairless	Rare carcinoma	8.0	12-18	23
Ethylene glycol alone	....	Skin painted	HR/De				
			hr/+, haired	Rare carcinoma	21.1	⩾18	23
			hr/*hr*, hairless	Rare carcinoma	13.6	⩾18	23
Combined Treatment							
X-rayed ovary, plus immediate removal of untreated ovary	200 r	Local to one ovary	BALB/c	G.C.T.; luteoma; T.A.	87.9	......	51
X ray, plus immediate ovarian grafts	200 r	Total body	BALB/c	....	0	......	51
X-rayed ovary grafted subcutaneously	200 r	Total body of donor	BALB/c				
			Intact ♂	G.C.T.; luteoma; T.A.	54.0	6-18	51
			Gonadectomized ♂	G.C.T.; luteoma; T.A.	85.0	......	51
DMBA	....	i.v.	$(C57L \times A)F_1$ ⟨LAF_1⟩	G.C.T.; luteoma	26.7	......	92
Plus ovariectomy & ovarian transplant	....	i.v.	$(C57L \times A)F_1$ ⟨LAF_1⟩	G.C.T.; luteoma	90.3	......	92
		....	ddO	G.C.T.[12]; luteoma	100	......	91
X rays, AET ⟨2-aminoethylisothiuronium bromide hydrobromide⟩, or X rays plus AET	....	....	$(C57L \times A/He)F_1J$ ⟨LAF_1J⟩	Adenoma	4.0-85.0	......	22
X rays plus 3-methylcholanthrene ⟨MC⟩	87-175 r or 350 r	X ray, total body; 3-methylcholanthrene, skin	$(RF \times AK)F_1$	....	3.8-94.1	>18	27

[10] Multiple injections induce more tumors. [11] Younger age and higher dose induce more tumors. [12] More common.

Contributor: Annabel G. Liebelt

continued

Part III. Other Induction Methods

References

1. Akamatsu, Y. 1971. Proc. Am. Assoc. Cancer Res. 12:83(Abstr. 330).
2. Akamatsu, Y. 1975. J. Natl. Cancer Inst. 55:893-897.
3. Akamatsu, Y., and B. P. Barton. 1974. Ibid. 52:377-385.
4. Alexandrov, S. N., et al. 1971. Vopr. Onkol. 17(1): 51-54.
5. Bennette, J. G. 1969. Annu. Rep. Br. Campaign Res., pp 13-14.
6. Biancifiori, C., and F. Caschera. 1962. Br. J. Cancer 16:722-730.
7. Biancifiori, C., and F. Caschera. 1963. Ibid. 17:116-118.
8. Biancifiori, C., et al. 1961. Ibid. 15:270-283.
9. Boone, I. U. 1960. In B. B. Watson, ed. Delayed Effects of Whole-Body Radiation. Johns Hopkins, Baltimore. pp. 19-37.
10. Breslow, N. E., et al. 1974. J. Natl. Cancer Inst. 52: 233-239.
11. Bulay, O. M., and L. W. Wattenberg. 1971. Ibid. 46: 397-402.
12. Burki, H. R., and G. T. Okita. 1969. Ibid. 43:643-651.
13. Burki, H. R., and G. T. Okita. 1969. Br. J. Cancer 23: 591-596.
14. Clapp, N. K. 1974. Radiat. Res. 59:18(Abstr. A9-4).
15. Clapp, N. K., et al. 1974. Ibid. 57:158-186.
16. Conklin, J. W., and A. C. Upton. 1962. Proc. Am. Assoc. Cancer Res. 3:311(Abstr. 50).
17. Conklin, J. W., et al. 1963. Radiat. Res. 19:156-168.
18. Conklin, J. W., et al. 1965. Cancer Res. 25(1):20-28.
19. Cosgrove, G. E., et al. 1965. Ibid. 25(1):938-945.
20. Darden, E. B., Jr., et al. 1967. Int. J. Radiat. Biol. 12: 435-452.
21. Darden, E. B., Jr., et al. 1969. Radiat. Res. 39:537-538(Abstr. Fd-4).
22. Davis, W. E., Jr., et al. 1970. Ibid. 41:400-408.
23. Deringer, M. K. 1962. J. Natl. Cancer Inst. 29:1107-1121.
24. Diwan, B. A., and H. Meier. 1974. Cancer Res. 34: 764-770.
25. Duhig, J. T. 1965. Arch. Pathol. 79:177-184.
26. Flaks, A. 1968. Eur. J. Cancer 4:579-585.
27. Furth, J., and M. C. Boon. 1947. Cancer Res. 7:241-245.
28. Furth, J., and J. S. Butterworth. 1936. Am. J. Cancer 28:66-95.
29. Furth, J., et al. 1960. Acta Unio Int. Contra Cancrum 16:138-142.
30. Gabrielides, C. G., et al. 1972. Pathol. Eur. 7:155-159.
31. Gardner, W. U. 1961. J. Natl. Cancer Inst. 26:829-853.
32. Graham, J., and R. Graham. 1967. Environ. Res. 1: 115-128.
33. Gubareva, A. V. 1968. Biull. Eksp. Biol. Med. 65(4): 83-84.
34. Gubareva, A. V. 1968. Vopr. Onkol. 14(9):45-52.
35. Gubareva, A. V. 1969. Ibid. 15(10):89-93.
36. Gubareva, A. V. 1969. Akush. Ginekol. (Moscow) 45(3):66-67.
37. Gubareva, A. V. 1969. Biull. Eksp. Biol. Med. 67(5): 61-63.
38. Gubareva, A. V. 1972. Arkh. Patol. 34(5):17-21.
39. Gubareva, A. V., and K. F. Galkovskaya. 1968. Med. Radiol. 13(4):42-45.
40. Gubareva, A. V., and K. F. Galkovskaya. 1971. Vopr. Onkol. 17(7):48-52.
41. Hamlett, J. D., et al. 1971. J. Pathol. 105:111-124.
42. Hilfrich, J. 1975. Arch. Gynaekol. 219:188-189.
43. Hilfrich, J. 1975. Br. J. Cancer 32:588-595.
44. Howell, J. S., et al. 1954. Ibid. 8:635-646.
45. Jull, J. W. 1969. J. Natl. Cancer Inst. 42:961-966.
46. Jull, J. W. 1969. Ibid. 42:967-972.
47. Jull, J. W., and A. Russell. 1970. Ibid. 44:841-844.
48. Jull, J. W., et al. 1966. Ibid. 37:409-420.
49. Jull, J. W., et al. 1966. Ibid. 37:421-424.
50. Jull, J. W., et al. 1968. Ibid. 40:687-706.
51. Kirschbaum, A., et al. 1956. Proc. Soc. Exp. Biol. Med. 92:221-222.
52. Komuro, M. 1976. J. Radiat. Res. 17:99-105.
53. Krarup, T. 1967. Acta Pathol. Microbiol. Scand. 70: 241-248.
54. Krarup, T. 1969. Int. J. Cancer 4:61-75.
55. Krarup, T. 1970. Acta Endocrinol. (Copenhagen) 64: 489-507.
56. Krarup, T. 1970. Br. J. Cancer 24:168-186.
57. Krarup, T., and H. Loft. 1971. Acta Pathol. Microbiol. Scand. 79A:139-149.
58. Kuwahara, I. 1967. Gann 58:253-266.
59. Lesher, S., et al. 1960. Radiat. Res. 12:451(Abstr. 93).
60. Lesher, S., et al. 1965. Ibid. 24:239-277.
61. Li, Y., et al. 1976. Z. Krebsforsch. Klin. Onkol. 86: 165-170.
62. Lindop, P. J., and J. Rotblat. 1961. Nature (London) 189:645-648.
63. Lindop, P. J., and J. Rotblat. 1961. Proc. R. Soc. London B154:332-349, 350-368.
64. Marchant, J. 1960. Br. J. Cancer 14:514-518.
65. Marchant, J. 1960. Ibid. 14:519-523.
66. Marchant, J. 1961. Ibid. 15:821-827.
67. Marchant, J. 1966. Ibid. 20:210-215.
68. McGowan, L., and R. H. Davis. 1971. Obstet. Gynecol. 38:125-135.
69. Mody, J. K. 1960. Br. J. Cancer 14:256-266.
70. Mody, J. K. 1966. Indian J. Med. Sci. 20:941-944.
71. Mody, J. K. 1967. Gann 58:291-295.
72. Mori-Chavez, P., et al. 1970. Cancer Res. 30:913-928.
73. Mori-Chavez, P., et al. 1974. Ibid. 34:328-336.
74. Nilsson, A. 1967. Acta Radiol. Ther. Physics Biol. 6: 33-52.

continued

58. OVARIAN TUMORS: MOUSE

Part III. Other Induction Methods

75. Nishizuka, Y., et al. 1962. Mie Med. J. 12:77-93.
76. Nomura, T. 1976. Br. J. Cancer 33:521-534.
77. Nomura, T., et al. 1972. Med. J. Osaka Univ. 23:121-131.
78. Nowell, P. C., and L. J. Cole. 1965. Science 148:96-97.
79. Patricio, M. B., et al. 1975. J. Surg. Oncol. 7:57-61.
80. Poel, W. E. 1962. Proc. Am. Assoc. Cancer Res. 3: 351(Abstr. 209).
81. Ribacchi, R., and G. Giraldo. 1964. Lav. Ist. Anat. Istol. Patol. Univ. Studi Perugia 24:195-202.
82. Riviere, M. R., et al. 1962. Bull. Cancer 49:206-216.
83. Riviere, M. R., et al. 1962. C. R. Soc. Biol. 156:1605-1607.
84. Shimizu, H., et al. 1974. Int. J. Cancer 13:500-505.
85. Shimizu, H., et al. 1975. Br. J. Cancer 31:492-496.
86. Shisa, H., and Y. Nishizuka. 1968. Ibid. 22:70-76.
87. Terracini, B., et al. 1976. Ibid. 33:427-439.
88. Tomatis, L., and C. M. Goodall. 1969. Int. J. Cancer 4:219-225.
89. Toth, B., and T. Toth. 1970. Tumori 56:315-324.
90. Trentin, J. J., and B. T. Morris. 1961. Proc. Am. Assoc. Cancer Res. 3:273(Abstr. 281).
91. Ueda, G., et al. 1972. Acta Obstet. Gynaecol. Jpn. 19:16-24.
92. Ueda, G., et al. 1974. Ibid. 21:135-142.
93. Uematsu, K., and C. Huggins. 1968. Mol. Pharmacol. 4:427-434.
94. Ullrich, R. L., et al. 1976. Radiat. Res. 68:115-131.
95. Upton, A. C. 1961. Cancer Res. 21:717-729.
96. Upton, A. C., and A. W. Kimball. 1960. Radiat. Res. 12:482(Abstr. 177).
97. Upton, A. C., et al. 1960. Proc. Soc. Exp. Biol. Med. 104:769-772.
98. Upton, A. C., et al. 1960. Cancer Res. 20(8-2):1-60.
99. Upton, A. C., et al. 1970. Radiat. Res. 41:467-491.
100. Vesselinovitch, S. D., and N. Mihailovich. 1967. Cancer Res. 27(1):1422-1429.
101. Vesselinovitch, S. D., et al. 1967. Ibid. 27(1):2333-2337.
102. Vesselinovitch, S. D., et al. 1970. Ibid. 30:2548-2551.
103. Vesselinovitch, S. D., et al. 1971. Ibid. 31:2133-2142.
104. Vesselinovitch, S. D., et al. 1971. Radiat. Res. 47: 253(Abstr. Ce-9).
105. Vesselinovitch, S. D., et al. 1971. Cancer Res. 31: 2143-2147.
106. Vesselinovitch, S. D., et al. 1974. Ibid. 34:2530-2538.
107. Vol'fson, N. I. 1976. Neoplasma 23:151-160.
108. Vol'fson, N. I. 1976. Vopr. Onkol. 22(3):68-75.
109. Yuhas, J. M. 1974. Radiat. Res. 60:321-332.

59. INTERSTITIAL CELL TESTICULAR TUMORS: MOUSE

Interstitial cell tumors of the testis ⟨Ict⟩ are induced with estrogens within one year in >80% of strains BALB/c and A. Induction is related to dose and length of estrogen exposure; the most convenient source is a 5-10 mg fused pellet of cholesterol, containing 10-20% diethylstilbestrol, implanted subcutaneously. IFS, WLL, JK, and DBA/2 strains are intermediate in susceptibility to estrogen-induced Ict. C57BL, I, Y, and CBA strains are more resistant. Data on strains RIII and C3H are not consistent and may reflect substrain differences. Susceptibility of F_1 mice lies between parent strains, but may resemble the susceptible (C x Y) or the resistant (C x I) parent. Ict induction is not affected by reciprocal crossing, nor by foster-nursing. Ict are transplantable to estrogenized mice, but may be estrogen-independent or acclimated to such independence. Spontaneous Ict are encountered rarely in mice two years old or older. **Estrogen: Name**—EB = estradiol benzoate; DES = diethylstilbestrol; TPE = 1,2-triphenylethylene; TACE = chlorotrianisene ⟨tri-*p*-anisylchloroethylene⟩; **Dosage**—administered subcutaneously. **No. of Mice: Effective**—mice alive at 6 months after beginning of experiment. **Tumors: Mean Time**—mean time to detection of gross tumor after treatment.

Strain ⟨Synonym⟩	Estrogen		No. of Mice		Tumors			Reference
	Name	Dosage	Original	Effective	No.	%	Mean Time mo	
A	EB	16-50 µg in oil/wk	49	24	10	42	8+	6
	DES	0.25 mg in oil/wk	47	32	29	94	8+	
A	TPE	2.5-5 mg in oil/wk	?	17	7	41	8+	3
C3H	TPE	2.5-5 mg in oil/wk	?	14	1?	7?	8+	
JK	TPE	2.5-5 mg in oil/wk	?	13	7	54	8+	
C57BL x CBA	TACE	50 µg in oil/wk	92	?	46	50	25	5
A	......	....	?	24	0	0	....	4
C3H x A	......	....	?	25	1	4	32	
C57BL x A	......	....	?	26	2	6	27	

continued

59. INTERSTITIAL CELL TESTICULAR TUMORS: MOUSE

Strain (Synonym)	Estrogen		No. of Mice		Tumors			Reference
	Name	Dosage	Original	Effective	No.	%	Mean Time mo	
A	TPE	3 mg in oil/wk	40	31	29	94	15	2
RIII	TPE	3 mg in oil/wk	87	67	21	31	17	
IFS	TPE	3 mg in oil/wk	20	18	4	22	19	
WLL	TPE	3 mg in oil/wk	39	35	3	9	19	
CBA	TPE	3 mg in oil/wk	62	38	0	0	.....	
BALB/c	......	....	154	137	2	1	24	1
BALB/c ⟨C⟩	DES	10-20% in cholesterol pellet	76	75	61	81	10	
DBA/2	DES	10-20% in cholesterol pellet	50	47	5	11	16	
C3H	DES	10-20% in cholesterol pellet	32	32	2	6	17	
C57BL	DES	10-20% in cholesterol pellet	60	60	1	2	23	
I	DES	10-20% in cholesterol pellet	15	14	0	0	.....	
RIII	DES	10-20% in cholesterol pellet	45	32	0	0	.....	
Y	DES	10-20% in cholesterol pellet	31	30	0	0	.....	
C x DBA/2	DES	10-20% in cholesterol pellet	24	24	16	67	11	
C x Y	DES	10-20% in cholesterol pellet	26	26	17	65	11	
C x C3H	DES	10-20% in cholesterol pellet	27	27	12	44	12	
C x C57BL	DES	10-20% in cholesterol pellet	38	38	10	26	13	
C x I	DES	10-20% in cholesterol pellet	37	37	1	3	11	
C x RIII	DES	10-20% in cholesterol pellet	37	28	0	0	.....	

Contributor: Michael B. Shimkin

References

1. Andervont, H. B., et al. 1960. J. Natl. Cancer Inst. 25: 1069-1081.
2. Bonser, G. M. 1944. J. Pathol. Bacteriol. 56:15-26.
3. Gardner, W. U. 1943. Cancer Res. 3:92-99.
4. Gardner, W. U. 1943. Ibid. 3:757-761.
5. Gardner, W. U., and J. Boddaert. 1950. Arch. Pathol. 50:750-764.
6. Hooker, C. W., and C. A. Pfeiffer. 1942. Cancer Res. 2:759-769.

60. SPONTANEOUS TESTICULAR TERATOMAS: MOUSE

Strain	Genotype	No. of Mice	Incidence of Males with Teratomas, %	Laterality, %		
				Left	Right	Bilateral
129/J	+/+	588	1.70	...	...	...
129/Re	+/+	382	0.26	...	...	...
129/Rr	+/+	1452	0.55	...	...	...
129/Sv	....	662	2.1	1.1	0.9	0.1
129/Sv-*A^yCP*	*A^y*/+	288	0.0	...	...	...
	+/+	294	0.0	...	...	...
129/Sv-*CP*	....	1020	6.5	3.8	1.9	0.8
129/Sv-*Sl^dCP*	*Sl^d*/+	178	16.5	11.8	2.8	1.9
	+/+	315	9.5	5.0	1.3	3.2
129/Sv-*Sl^JCP*	*Sl^J*/+	918	17.8	11.4	5.0	1.4
	+/+	667	10.0	6.1	2.9	1.0
129/Sv-*ter* 1/	....	474	33.3	12.0	8.0	13.3

1/ Formerly referred to as 129/*ter*Sv.

continued

60. SPONTANEOUS TESTICULAR TERATOMAS: MOUSE

Strain	Genotype	No. of Mice	Incidence of Males with Teratomas, %	Laterality, %		
				Left	Right	Bilateral
129/Sv-*ter iv*	*iv/iv*[2]	244	7.8	0.8	3.3	3.7
	iv/iv[3]	186	4.2	1.1	1.1	2.2
	iv/+[3]	112	11.7	6.3	3.6	1.8
129/Sv-W^vCP	$W^v/+$	314	5.4	1.6	2.5	1.3
	+/+	401	9.9	4.7	2.7	2.5
All other strains except A/He & DBA/2J[4]	+/+	Millions	0.00	...	...	...

[2] Reversed phenotypically. [3] Not reversed phenotypically. [4] Two teratomas have been observed in strain A/He and one in DBA/2J.

Contributor: Leroy C. Stevens

General References

1. Meier, H., et al. 1970. Cancer Res. 30:30-34.
2. Stevens, L. C. 1973. J. Natl. Cancer Inst. 50:235-242.
3. Stevens, L. C., and J. A. Mackensen. 1961. Ibid. 27: 443-453.

61. INTESTINAL TUMORS: MOUSE

Data include intestinal tumors other than hematopoietic. Naturally occurring intestinal tumors are rare in most strains of mice. In order to ascertain the presence or absence of such tumors, the entire intestinal tract must be carefully opened at the time of necropsy. The naturally occurring tumors reported below were mostly polyps. The vast majority of induced tumors are carcinomas; few adenomas (polyps) were reported. Tumor incidence after administration of an intestinal carcinogen depends on the dose, dosage schedule, route of administration, and, to a lesser extent, on the sex of the mouse. There are few experimental comparisons of these variables, much less comparisons of the effect of genes.

Part I. Naturally Occurring Tumors

Strain	Small Intestine		Large Intestine		Reference
	Type	Incidence %[1]	Type	Incidence %	
BALB/c	Carcinoma	Rare	....	....	4
C3H, C57, SWR (random-bred)	Polyps	1	....	....	7
C57BL	....	....	Polyps	10	4
	Polyps, carcinomas	10	Polyps	1	8
C57BL/6N	Polyps	12.5	....	0	10
DBA	Polyps	0-86	....	0	5
Stock albino	Carcinoma	1	Polyps	1	1
CN hybrids	Carcinoma	1	Polyps	1	6
$(C57BL \times A/He)F_1$	....	....	Leiomyoma, rectal squamous cell carcinoma	1	9

[1] Unless otherwise specified.

continued

61. INTESTINAL TUMORS: MOUSE

Part I. Naturally Occurring Tumors

Contributors: Jerrold M. Ward and Myra E. Grabin

References

1. Bonser, G. M., et al. 1956. Br. J. Cancer 10:653-667.
2. Corbett, T. H., et al. 1975. Cancer Res. 35:2434-2439.
3. Double, J. A., et al. 1975. J. Natl. Cancer Inst. 54: 271-275.
4. Dunn, T. 1965. In W. J. Burdette, ed. Carcinoma of the Alimentary Tract. Univ. Utah, Salt Lake City. pp. 45-54.
5. Hare, W. V., and H. L. Stewart. 1956. J. Natl. Cancer Inst. 16:889-911.
6. Miller, E., and F. Pybus. 1956. Br. J. Cancer 10:89-109.
7. Poel, W. E., et al. 1968. Cancer Res. 28:845-859.
8. Rowlett, C., et al. 1969. J. Natl. Cancer Inst. 43:1353-1364.
9. Upton, A. C., et al. 1960. Cancer Res. 20(8-2):1-60.
10. Ward, J. M., and E. K. Weisburger. 1975. Ibid. 35: 1938-1943.

Part II. Experimentally Induced Tumors

For information on experimentally induced, transplantable colon carcinomas, consult reference 2 (Part I above) for strains BALB/c and C57, and reference 3 (Part I above) for strain NMRI.

Chemical Carcinogen	Strain	Tumor Incidence, %		Reference
		Small Intestine	Large Intestine	
Azoxymethane	BALB/c	0	≤100	11
	CFW/N	0	≤100	11
	C3H/HaN	0	≤100	11
	C57BL/6J	0	≤100	11
	NIH Swiss	0	≤100	11
Butylnitrosourea	C57BL/6	35-100	2-15	10
1,2:5,6-Dibenzanthracene & 3-methylcholanthrene[1]	....	....	1[2]	7
1,2-Dimethylhydrazine	BALB/c	....	10.0	1
	CF1	....	90	8
	C3H	....	12.3	1
	C57BL/6	....	16.9	1
	C57BL/Ha[3]	....	0	2,3
	DBA	....	13.1	1
	DBA/2	....	0	2
	ICR/Ha[3]			
	Inbred	....	100	2,3
	Random-bred	....	67	2
	NMRI	5	90	12
		....	100	4
	Swiss	....	8.8	1
		....	80-90	9
	(ICR x C57BL)	....	32-100	3
Ethylnitrosoguanidine	CBA/H	60	....	5
	CBA/H-T6	72	....	5

[1] Synonym: 20-Methylcholanthrene. [2] Tumor type: polyps. [3] The difference in the response of ICR/Ha and C57BL/Ha appears to be controlled by a single dominant gene which so far is not linked to seven studied loci [ref. 3].

continued

61. INTESTINAL TUMORS: MOUSE

Part II. Experimentally Induced Tumors

Chemical Carcinogen	Strain	Tumor Incidence, %		Reference
		Small Intestine	Large Intestine	
	C3H/He	78	....	5
	C57BL/6	78	....	5
	DD/I	54	....	5
Methylnitrosoguanidine				
Intrarectal	BALB/c	....	13.1	1
	C3H	....	4.5	1
	C57BL/6	....	12.5	1
	DBA	....	8.8	1
Oral	CBA/H-T6	38	....	5
	C3H/He	29	....	5
	DD/I	24	....	5
Methylnitrosourea	BALB/c	....	34.4	1
	C3H	....	0-15.4	1
	C57BL/6	....	30-53	1
	DBA	....	6.8-11.1	1
	ICR/Ha	0	4-65	6

Contributors: Jerrold M. Ward and Myra E. Grabin

References

1. Corbett, T. H., et al. 1975. Cancer Res. 35:2434-2439.
2. Evans, J. T., et al. 1974. J. Natl. Cancer Inst. 52:999-1000.
3. Evans, J. T., et al. 1977. Cancer Res. 37:134-136.
4. Haase, P., et al. 1973. Br. J. Cancer 28:530-543.
5. Matsuyama, M., et al. 1975. Gann Monogr. 17:269-281.
6. Narisawa, T., and J. H. Weisburger. 1975. Proc. Soc. Exp. Biol. Med. 148:166-169.
7. Stewart, H. L., and E. Lorenzo. 1947. J. Natl. Cancer Inst. 7:239-269.
8. Thurnherr, N., et al. 1973. Cancer Res. 33:940-945.
9. Toth, B., et al. 1976. Am. J. Pathol. 84:69-86.
10. Ward, J. M., and E. K. Weisburger. 1975. Cancer Res. 35:1938-1943.
11. Ward, J. M., et al. 1974. J. Am. Vet. Med. Assoc. 164: 729-732.
12. Wiebecke, B., et al. 1973. Virchows Arch. A360:179-193.

62. STRAIN INCIDENCE OF SPONTANEOUS FIBROSARCOMA: MOUSE

Average Age: Mean life-span unless otherwise indicated.

Strain or Hybrid (Synonym)	Subjects		Average Age, mo	Incidence, %	Reference
	Sex	No.			
BALB/cCr	♂♀	4500	22.89[1]	0.5	5
BALB/cKh	♂♀	37	26.2	3.0	8

[1] Age at tumor development.

continued

62. STRAIN INCIDENCE OF SPONTANEOUS FIBROSARCOMA: MOUSE

Strain or Hybrid ⟨Synonym⟩	Subjects		Average Age, mo	Incidence, %	Reference
	Sex	No.			
BALB/cPi[2]	♂♀	67	20.00	0.0	9
CBA/J	♂	17	24.88	0[3]	10
	♀	33	21.16	3[3]	10
C3H	♂	29	26.27	0[3]	10
	♀	1774	25	3.2	2,4
C3H.K	♂	25	24.18	4[3]	10
C57BL/Icrf-a^t	♂	497	27[1]	0.6	6
	♀	293	30[1]	1	6
C57BL/6NIcr	♂	93	....	2.7	1
C57BL/10ScSn	♂	39	27.44	0[3]	10
	♀	35	23.02	0[3]	10
DBA/2J	♂	67	19.06	3[3]	10
	♀	51	18.83	0[3]	10
LP/J	♂	77	23.95	6.5[3]	10
	♀	51	23.02	5.9[3]	10
NZO/Bl	♂	194	14.30[1]	4.6	3
	♀	217	18.30[1]	4.6	3
129/J	♂	41	27.20	2.4[3]	10
	♀	95	24.18	1.1[3]	10
(AKR x DBA/2)F_1 ⟨AKD2F_1⟩	♂	101	17.23	0.99	7
	♀	102	18.91	2.9	7
(BALB/c x A)F_1 ⟨CAF_1⟩	♂	138	22.79	0	7
	♀	139	23.19	0	7
(BALB/c x C57BL/6)F_1 ⟨CB6F_1⟩	♂	38	28.42	0	7
	♀	34	28.42	2.94	7
(C57BL/6 x A)F_1 ⟨B6AF_1⟩	♂	122	21.61	1.63	7
	♀	112	24.01	0.89	7
(C57BL/6 x C57BR)F_1 ⟨B6BRF_1⟩	♂	94	24.47	0	7
	♀	84	25.06	0	7
(C57BL/6 x DBA/2)F_1 ⟨B6D2F_1⟩	♂	48	24.47	0	7
	♀	60	25.36	1.66	7
(C57L x A)F_1 ⟨LAF_1⟩	♂	64	22.26	0	7
	♀	58	22.79	1.72	7

[1] Age at tumor development. [2] Germfree. [3] Includes fibromas and tumors possibly originating from muscles.

Contributor: David D. Myers

References

1. Diller, I. C. 1974. Growth 38:507-517.
2. Dunn, T. B., et al. 1956. J. Natl. Cancer Inst. 17:639-656.
3. Goodall, C. M., et al. 1973. Lab. Anim. 7:65-71.
4. Heston, W. E. 1963. In W. J. Burdette, ed. Methodology in Mammalian Genetics. Holden-Day, San Francisco. pp. 247-268.
5. Peters, R. L., et al. 1972. Int. J. Cancer 10:273-282.
6. Rowlatt, C., et al. 1976. Lab. Anim. 10:419-442.
7. Russell, E. S. Unpublished. Jackson Laboratory, Bar Harbor, ME, 1978.
8. Sanford, B. H., et al. 1973. J. Immunol. 110:1437-1443.
9. Smith, C. S., et al. 1971. Proc. Soc. Exp. Biol. Med. 138:542-544.
10. Smith, G. S., et al. 1973. J. Natl. Cancer Inst. 50:1195-1213.

63. CLASSIFICATION OF GENUS *MUS*

Part I. List of Better-known Species and Principal Subspecies of the House Mouse and Allies

Subgenus *Pyromys* [Spiny mice]
- *Mus shortridgei*
- *Mus saxicola*—including subspecies *gurkha, sadhu,* and *saxicola*
- *Mus platythrix*
- *Mus phillipsi*
- *Mus fernandoni*

Subgenus *Coelomys* [Shrew-like mice]
- *Mus mayori*—including subspecies *pococki* and *mayori*
- *Mus pahari*—including subspecies *gairdneri* and *pahari*
- *Mus famulus*
- *Mus crociduroides*
- *Mus vulcani*

Subgenus *Mus* [House and ricefield mice]
- *Mus caroli*—including subspecies *caroli* and *ouwensi*
- *Mus cervicolor*—including subspecies *cervicolor, popaeus,* and *annamensis*
- *Mus cookii*—including subspecies *cookii, nagarum,* and *palnica*
- *Mus booduga*—including subspecies *lepidoides, booduga,* and *fulvidiventris*
- *Mus dunni*
- *Mus musculus*
  - Outdoor subspecies include *bactrianus, gansuensis, homourus, hortulanus, manchu, molossinus, mongolium, musculus, praetextus, spicilegus, spretus, tantillus, wagneri, yamashinai,* and *yesonis*
  - Indoor subspecies include *brevirostris, castaneus, domesticus, muralis, poschiavinus, tytleri,* and *urbanus*

Subgenus *Nannomys* [African pygmy mice]
- *Mus bufo, callewaerti, goundae, gratus, haussa, indutus, mattheyi, minutoides, neavei, oubanguii, pasha, setulosus, sorella, tenellus,* and *triton* are species of this karyologically unique group being studied by R. Matthey and F. Petter.

Contributor: Joe T. Marshall, Jr.

Part II. Wild Mice of Eurasia

VOUCHER SPECIMENS

Because of the importance of the flood of cytologic and virologic discoveries accruing from studies of wild mice, it is necessary that the sources be verifiable. This means preparation of a voucher specimen—at least one for each discovery—which is deposited in a museum such as the Smithsonian Institution, and listed with its specimen number *in your publication.* The identification and locality of origin of each mouse so selected *can then be verified.*

METHODS

Sherman live-traps, baited with peanut butter, were set at 20-step intervals. Measurements of the specimens were taken before skinning and included total length, tail measured from rump while held at right angle to the back, hind foot from heel to tip of claw, ear from notch, and body weight. (The tail measurement was subtracted from the total length to give head plus body length.) These measurements together with mammary formula, habitat (including the indispensable "caught in house" when true), date, locality, and field number were written on a museum label attached to the foot. After skinning, by turning the skin inside out from a slit in the belly, the skin was righted around a cotton filler. The brain was removed through the intact foramen magnum by a jet of water from a blunt-needled syringe. The skull was dried and then cleaned by beetles, *Dermestes maculatus* ⟨*D. vulpinus*⟩, obtained from a mounted deer head and a dried bat carcass fallen from a temple roof. Any other treatment of these delicate skulls is disastrous.

SCALE

All drawings of the same aspect are comparable among themselves. The original drawings were prepared according to the following scales: external characteristics, life size; entire skull X3; foot X4; side view of rostrum X5; palate X10; zygomatic plate X10; molars X20. All reductions were the same (∿25%), so the above ratios are valid. A scale of 10 millimeters is included below fig. 3.

KEY TO ASIAN SPECIES OF THE GENUS *MUS*

Genus *Mus.* Length of first upper molar more than half the toothrow; head plus body 50 to 120 mm.; skull 17 to 30 mm.; postero-internal cusp (fig. 1) absent; two plantar and four interdigital foot pads round, small, and smooth (fig. 2)

KEY TO ASIAN SUBGENERA

A. Supraorbital ridge present (fig. 3), like that of *Rattus*; fur spiny (fig. 4 shows dorsal spine and cross section) except in the subspecies *Mus saxicola gurkha*; lives in grass beneath woodland . . . . Subgenus *Pyromys*, C
No supraorbital ridge . . . . .B

B. Interorbital width (fig. 5) more than 4 mm.; incisive foramina (fig. 6A) broad and short; interpterygoid space broad (fig. 6B); fur velvety or spiny; eye small, opening in skin usually less than 3 mm.; zygomatic plate narrow and leaning backward (fig. 7) except in old individuals; like a shrew, lives on the evergreen forest floor . . . . Subgenus *Coelomys*, G

continued

Part II. Wild Mice of Eurasia

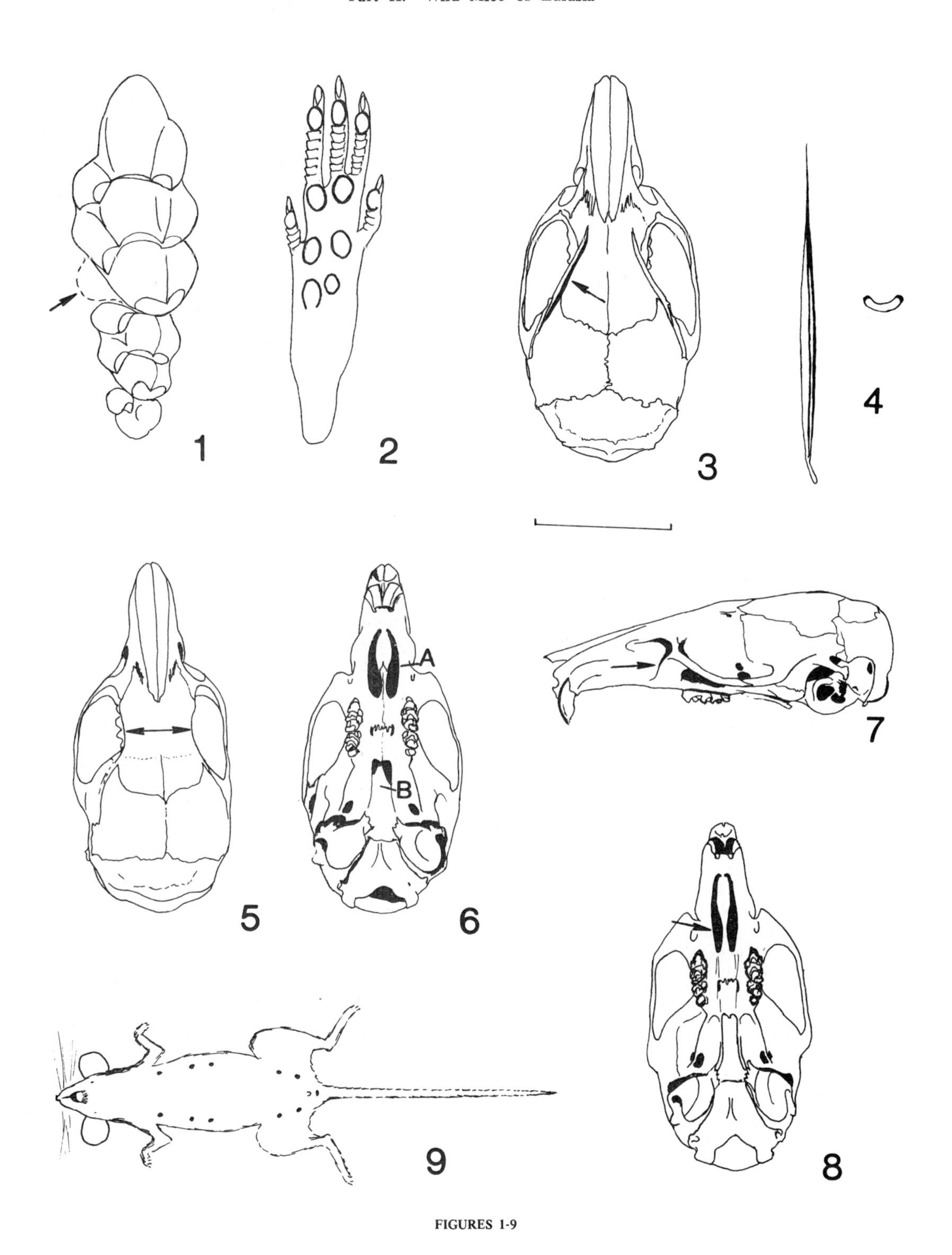

FIGURES 1-9

continued

Interorbitum narrower than 4 mm.; fur not spiny; incisive foramina (fig. 8) slender and long; eye large compared with *Coelomys*; mammae 3+2 (fig. 9); lives in grass and capable of occupying ricefield as a commensal with man . . . . Subgenus *Mus*, K

KEY TO SPECIES OF THE SUBGENUS *PYROMYS*

C. The only southeast Asian species; size large, skull averages 28 mm. and molar crowns 4.8 mm.; fur spiny, dorsal color light grayish brown, ventral fur white with gray bases; anterior border of zygomatic plate curved backward (fig. 10A); upper incisors notched (fig. 10B); incisive foramina long, averaging 6.7 mm. and usually penetrating deep between first molars (fig. 11); molars broad, lacking accessory cusp; mammae 3+2; ears large, averaging 18 mm.; inhabit grass beneath deciduous forest in Burma, Thailand, Cambodia . . . . *Mus shortridgei*

Indian species; ventral fur pure white, including bases; incisive foramina not more than 6.5 mm. . . . . D

D. Mammae 4+2 (fig. 12); anterior border of zygomatic plate arched forward convexly (fig. 13A); rostrum narrow; incisors without a notch (fig. 13B); anterior accessory cusp on elongated anterior cusp of first upper molar (fig. 14A); incisive foramina cut deep between anterior molars (fig. 14B); size medium, skull averaging 25 mm.; fur soft or spiny depending on geographic population, these spines narrower than those of other spiny mice; dorsal color varying from pale sandy brown to darker grayish brown depending on location; interpterygoid space narrow (fig. 14C); common and widespread from Pakistan and Himalayan foothills to southern India . . . . *Mus saxicola*

Mammae 3+2; anterior border of zygomatic plate (fig. 15) rises vertically then is swept back in an arc of a quarter circle at the zygomatic notch; incisive foramina usually reaching only to level of anterior cusp of first molar (fig. 16A); no accessory cusp; fur spiny above and below; dorsal pelage dusky brown or purplish brown; rostrum and interpterygoid space (fig. 16B) relatively broad; incisors sometimes notched. . . . . .E

E. Large size as with *M. shortridgei* but incisive foramina shorter, averaging 5.9 mm., skull 26 to 30 mm., averaging 27; India from Bihar and Maharashtra states to the south . . . .*Mus platythrix*

Skull less than 26 mm. . . . . .F

F. Size small, skull averaging 22 mm. (21 to 23), head plus body less than 90 mm., tail less than 70 mm.; India from southern Rajasthan State to the south. . . . . *Mus phillipsi*

Size medium, equal to *Mus saxicola*, various skull dimensions halfway between those of *M. platythrix* and *M. phillipsi*; skull length 24 mm.; upper incisors very recurved, thick, short and unnotched (fig. 17); Sri Lanka . . . . *Mus fernandoni*

KEY TO SPECIES OF THE SUBGENUS *COELOMYS*

G. Pelage spiny or stiff, not woolly; ear small; mammae 3+2 . . . . H

Pelage dense, woolly, velvety; ear large; three pairs of mammae I

H. Size large, skull length greater than 27 mm.; molars (fig. 18) broad and squarish; Sri Lanka . . . .*Mus mayori*

Size medium, skull length under 27 mm., molars narrow (fig. 19), Sikkim to Vietnam . . . . *Mus pahari*

I. Dorsum rich chocolate brown, sharply demarcated in a precise line along the flanks from the bright ochraceous-buff underparts; parietal bones with anterior spine as is usual in the genus *Mus* (fig. 20); mammae 1+2; Nilgiri Hills, Tamil Nadu State, India . . . . *Mus famulus*

Dorsum deep blackish brown or purplish brown, becoming gradually paler beneath, where fur is slate with silvery or bronzy tips; parietal bones lack the spine thus imparting a squarish appearance to the cranial roof (fig. 21); mountains of Sumatra and Java . . . . .J

J. Dorsal hairs steely in color, ventral hairs with silvery tips, feet white, tail bicolored; tail long, averaging 120 mm.; incisive foramina short (fig. 22), 4.6 mm.; molars large, 4.0 mm.; mammae 2+?; Sumatra . . . . *Mus crociduroides*

Dorsal hairs blackish brown, ventral hairs bronzy or silvery tipped, feet and short

continued

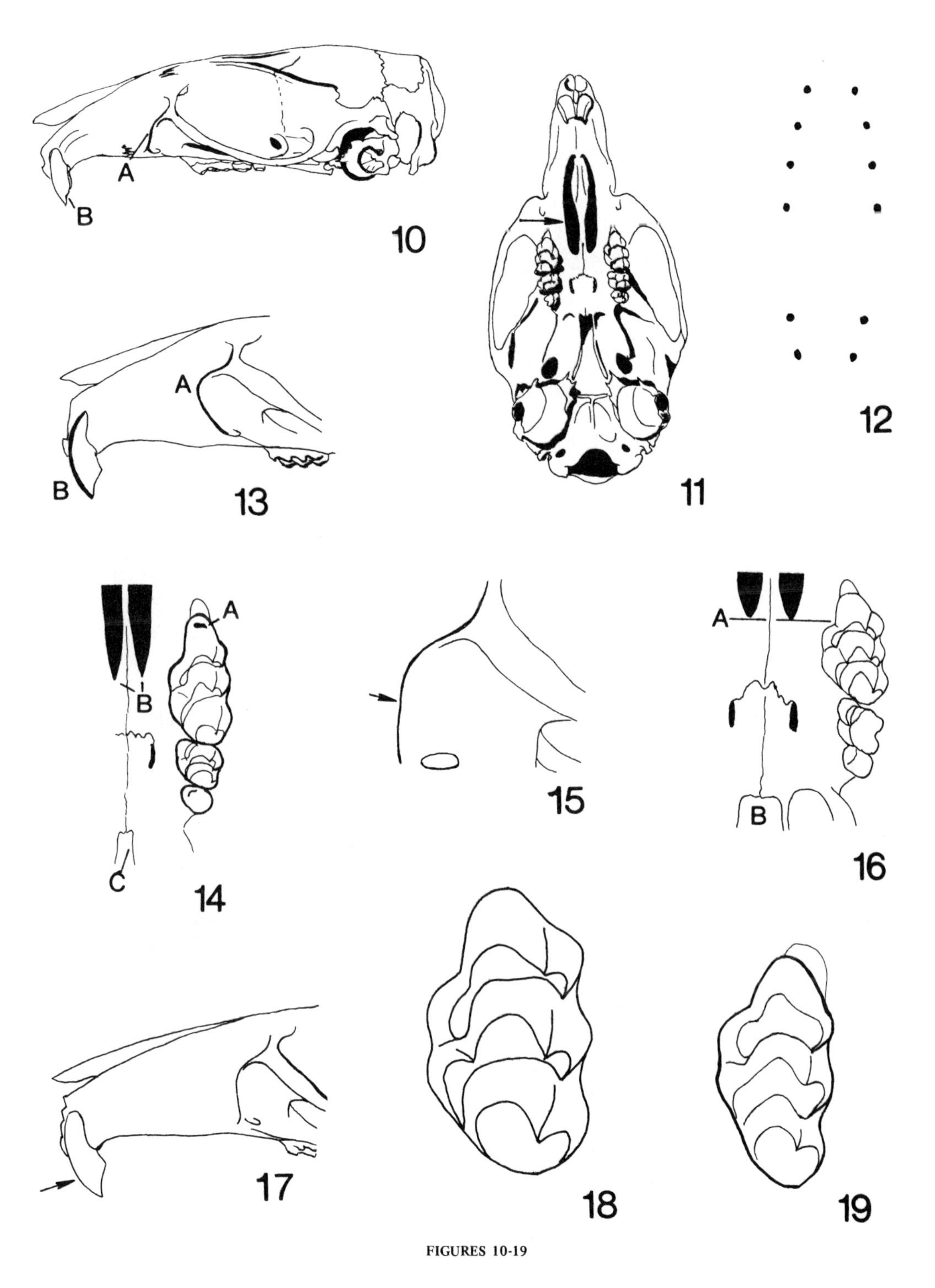

FIGURES 10-19

continued

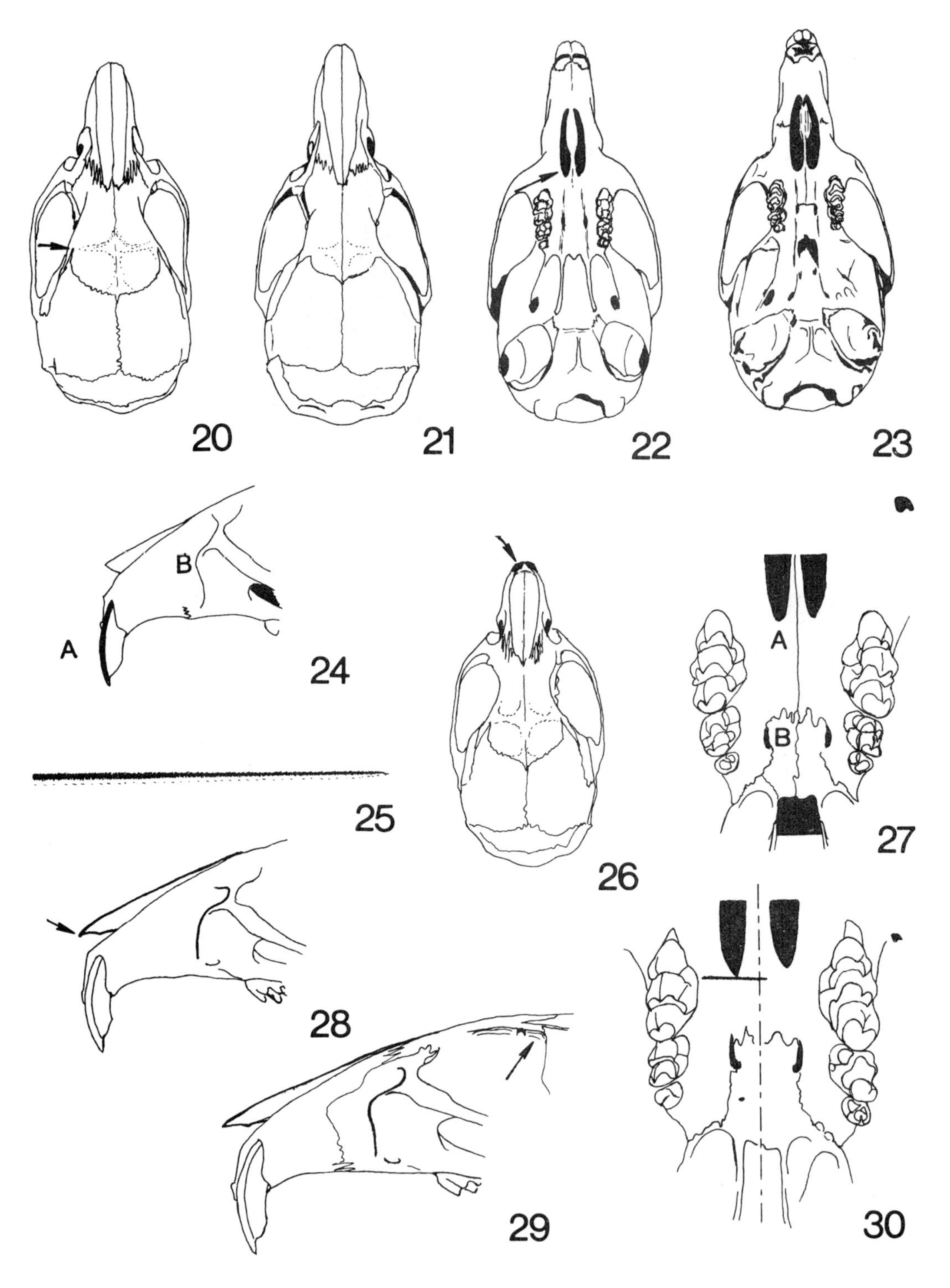

FIGURES 20-30

continued

tail blackish; tail averaging 87 mm.; incisive foramina longer (fig. 23), 5.1 mm.; molars small, 3.5 mm.; mammae 2+1 or 1+2; Java . . . .*Mus vulcani*

KEY TO SPECIES OF THE SUBGENUS *MUS*

K. Upper incisors recurved (opisthodont, figs. 31, 32); ventral color whitish or dark . M
Upper incisors (fig. 24A) curve forward and downward perpendicular to palate (pro-odont); narrow interpterygoid space; ventral fur whitish, tail bicolored (fig. 25, 75 mm.) L

L. Nasals short, exposing to dorsal view the dark tan or brownish orange anterior surface of upper incisors (fig. 26); narrow zygomatic plate with S-shaped anterior border (fig. 24B); tail bicolored, as long as head plus body, and blackish on top (fig. 25); incisive foramina (fig. 27A) extending only to the anterior cusp of first molar; posterior palatine foramina (fig. 27B) at rear of palatal bridge unlike those of all other mice; ricefields and grassy habitats in Ryukyu Is., Taiwan, southeastern China to Vietnam and Thailand, reappearing in Kedah (West Malaysia) and Indonesia . . . . .*Mus caroli*
Nasals long (fig. 28), overhanging the buff-colored upper incisors; old individuals develop a supraorbital shelf (fig. 29); bicolored tail paler gray on top than in *caroli* and shorter than head plus body; feet white; incisive foramina penetrating deep between anterior molars (fig. 30, on left *cervicolor*, on right *popaeus*); posterior palatine foramina in middle of palatal bridge; Nepal, Manipur, Burma, Thailand, Laos, and Vietnam, reappearing in Sumatra and Java. . . . .*Mus cervicolor*

M. Rostrum long and shallow, its least depth (fig. 31A) only one-half of rostral length (fig. 31B); ventral fur white, pale gray, or white with gray bases; tail bicolored . . . . N
Rostrum short and deep (fig. 32), its least depth two-thirds of length; interpterygoid space narrow; depending on location and habitat ventral color white, white with gray bases, golden buff, salmon buff, ochraceous, gray, or slate, and tail is bicolored or unicolored; suture between palatal foramina (fig. 33) more posterior than in other mice; Eurasia and northern Africa, introduced to Americas, southern Africa, Australia and Oceania . . . . *Mus musculus*, with various groups of subspecies. . . . . Q

N. Larger mice, breadth of cranium 9 to 11 mm., hind foot 16 to 21 mm., skull elongated (fig. 34), evenly tapered; zygomatic plate square; incisors thick and curved in a short radius (fig. 31); incisive foramina extend only to anterior cusp of first molar (as in fig. 27, *Mus caroli*); first upper molar moderately slender, anterior cusp long (fig. 35); interpterygoid space broad; mostly in mountains and forests, Vietnam, Yunnan, Laos, northern Thailand, northern Burma (*cookii*); Nepal and India (*nagarum*) . . . . *Mus cookii* (provisionally including as subspecies the taxa *nagarum* and *palnica*).
Smallest Eurasian mice, breadth of braincase 8.0 to 9.2 mm., hind foot 14 to 16 mm., live in grass, wheat fields, ricefields; Burma, Nepal, Pakistan, India, Sri Lanka . . . . *booduga/dunni* Complex, O

O. Underparts gray; upper incisors straighter (fig. 36); incisive foramina shorter . . . .P
Underparts white; upper incisors short and markedly recurved (fig. 39A); incisive foramina longer; upper parts brown; head and body length 65 to 70 mm., first upper molar broad, anteroexternal cusp prominent (fig. 37) . . . .
Superspecies or species *Mus booduga* consisting of taxa from:
(a) Burma: back light brown, brown chest spot, some individuals with gray bases of ventral fur; tail average only 50 mm.; skull short and broad with relatively large bullae (fig. 38) and teeth, zygomatic plate rounded and set out laterally, large masseteric knob (fig. 39B); ricefields at Myingyan and bamboo slopes of Mt. Popa . . . .*Mus (?booduga) lepidoides*
(b) Sri Lanka: back dark brown; tail 60 to 65 mm., skull measurements slightly less than those of *Mus booduga* (below) except for interorbital breadth which is greater (average 3.4 mm.); zygomatic plate (fig. 40) square and interorbital space broad as in *Mus cookii* but molar shape agrees with *M. booduga* . . . .*Mus (?booduga) fulvidiventris*
(c) India: back light brown with yellowish highlights, venter pure white; tail 55 mm. average; zygomatic plate similar to that

continued

of *Mus lepidoides* (fig. 39), otherwise skull more elongate and similar to that of *M. fulvidiventris* except for slightly longer incisive foramina, averaging 4.7

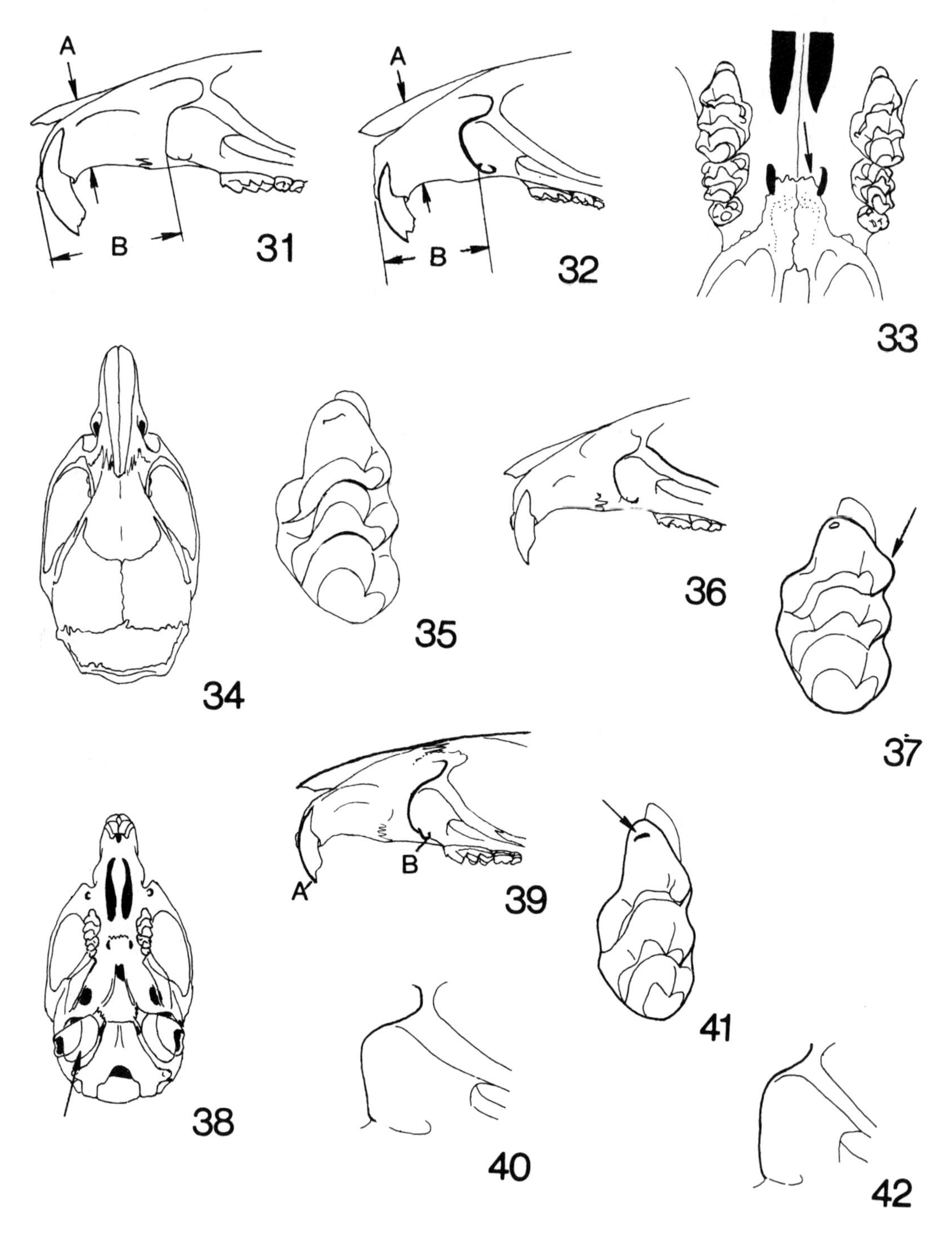

FIGURES 31-42

continued

mm.; X-chromosome telocentric as are autosomes. Description based on specimens of known karyotype from Varanasi, Khajuraho, and Madras. . . .*Mus booduga*

P. Back dull brownish gray; head plus body 65 to 70 mm.; first upper molar long and slender with long anterior cusp surmounted by an accessory cusp (fig. 41), antero-external cusp inconspicuous; X-chromosome a huge metacentric; tail about 65 mm.; incisive foramina averaging 4.2 mm.; zygomatic plate (fig. 42) usually narrow. This description based on samples of known karyotype from fields at Varanasi, Khajuraho, Poona, and Madras. [Specimens from Pakistan and Gujarat approach *Mus dunni* in most measurements but have white underparts and variable molar shape.] . . *Mus dunni*

Back brown; head and body length about 50 mm.; tail 55 mm.; skull length average 17 mm. (Status uncertain, only 10 tiny specimens are known; molars either worn or worn flat; from India, Nepal, and Pakistan; too small and too gray beneath to fit anywhere else.) . . . . . . *Mus terricolor*

Q. Zygomatic plate usually narrow, with straight anterior border (fig. 43) and small masseteric knob; weight about 16 grams; Europe, northern Africa, and western Asia thence extending narrowly eastward along the Himalayas to Sikkim; absent from India (south of the Himalayas and Punjab) and absent from southeast Asia . . *musculus* Group of subspecies (this group includes *Mus musculus musculus* of northern and eastern Europe and *Mus musculus domesticus* west of the Elbe; the latter is progenitor of the ordinary albino laboratory mouse).

(a) Afghanistan to Pakistan in fields and houses of lowlands: bright pale buffy brown on back, pure white beneath, white feet, bicolored tail; zygomatic plate tending toward a rounded outline (fig. 44A,B) . . . *Mus musculus bactrianus*

(b) Himalayas at high elevations from Pakistan to Sikkim in houses and outdoor areas disturbed by man: dark brown back with peculiar singed brown along flanks, venter gray with white tips, feet white, tail bicolored; zygomatic plate of variable outline (fig. 45A, B) but not rounded (some individuals have a fairly elongated rostrum) . . . . . *Mus musculus homourus*

Zygomatic plate always broad with anterior border arched convexly forward in a semicircle (fig. 46A), with large masseteric knob (fig. 46B); weight about 13 grams; Asia eastward from India and Mongolia, the outdoor races occurring only in the north. . . . . *castaneus* Group of subspecies

(a) Outdoor pale desert races; fur long and silky; pale brown above, pure white beneath; feet white; tail short and bicolored:
Mongolia . . . . *Mus musculus mongolium*
China . . . . *Mus musculus gansuensis*

(b) Outdoor dark races; dorsum dark brownish gray, ventral hairs white with gray bases; feet white; tail short and bicolored:
Manchuria . . . . . *Mus musculus manchu*
Korea . . . . . . *Mus musculus yamashinai*
Japan, Quelparte Is., and Okinawa . . . . *Mus musculus molossinus*
Szechwan to Shensi . . . . *Mus musculus tantillus*

(c) Urban, indoor mice, underparts dark, ochraceous, and same color as or only slightly paler than the back, feet, and unicolored, long tail; no white except on tips of toes; India to southeast Asia, the precise area which lacks outdoor subspecies of *Mus musculus*:
Cities of north-central India; golden buff . . . .*Mus musculus tytleri*
Kathmandu, Darjeeling, and Myingyan, Burma; coastal and island cities throughout southeast Asia, India, Burma, and southern China. Grayish brown with ochraceous buff or dark gray tips on ventral fur . . . . . . *Mus musculus castaneus*

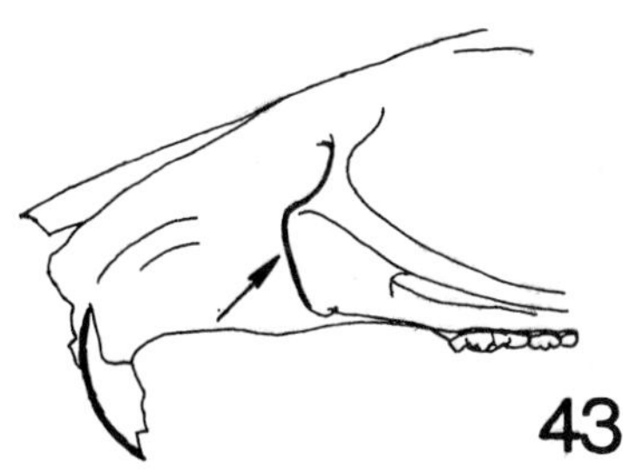

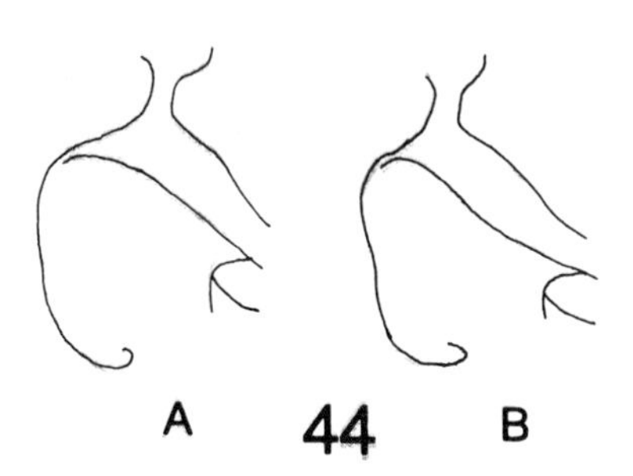

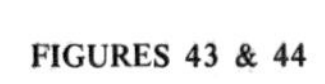

FIGURES 43 & 44

continued

63. CLASSIFICATION OF GENUS *MUS*

Part II. Wild Mice of Eurasia

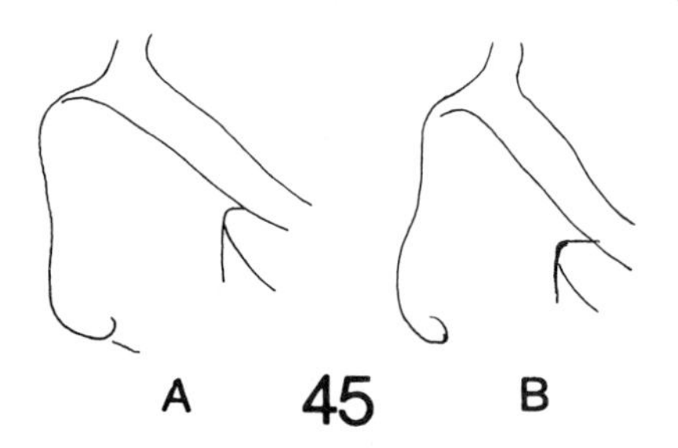

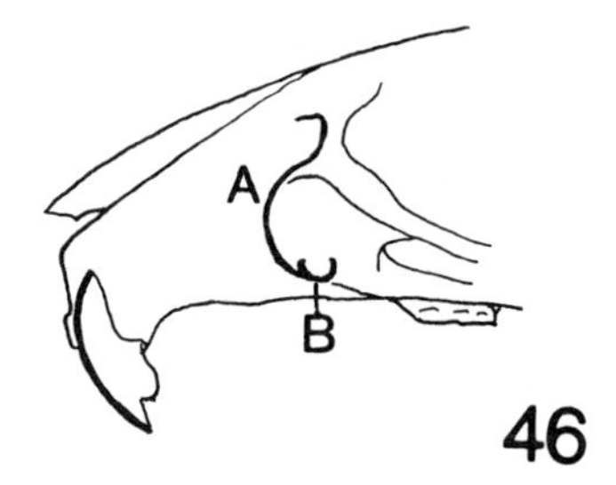

FIGURES 45 & 46

Contributor: Joe T. Marshall, Jr.

Reference: Marshall, J. T., Jr. 1977. Bull. Am. Mus. Nat. Hist. 158:177-220.

64. POLYMORPHISMS: WILD MOUSE

Part I. Electrophoretic Polymorphisms in Feral Populations of *Mus musculus*

For additional data on enzymes associated with locus designations, *see* Table 19. Additional enzymes and proteins scored as monomorphic: xanthine dehydrogenase (XDH); fumarate hydratase ⟨fumarase; FUM⟩; fructose-bisphosphate aldolase—liver ⟨aldolase—liver; ALD-1, ALD-2⟩; RBC esterases Es-A, Es-B, Es-C, Es-D; plasma proteins prealbumin-3, Pp-A, Pp-B, Pp-C; RBC protein Ec-1. **Gene Symbol & Gene Name**: Nomenclature for enzyme loci in cited literature may differ from standardized nomenclature used in this table. **Average Heterozygosity/Locus**: Heterozygosity per locus was derived from the Hardy-Weinberg formula in which allele frequencies of p and q are multiplied to determine genotype frequencies—$(p + q)^2 = p^2 + 2pq + q^2$. The heterozygosity of a locus is estimated as 2pq. Average heterozygosity is the unweighted average over all the populations scored for a particular locus. **Populations Polymorphic**: the proportion of populations tested which are polymorphic for a given locus. Data in brackets refer to the column heading in brackets. Values in parentheses are ranges, estimate "c" (*see* Introduction).

Gene Symbol	Gene Name	Trapping Region	Average Heterozygosity/Locus [Populations Polymorphic]	Reference
Adh-1	Alcohol dehydrogenase	Continental Europe		
		North Denmark	0.31	2,6
		South Denmark	0.06	2,6
		North America: California	0	5
		Summary of all regions	0.12(0-0.31) [2/3]	
Alb-1	Albumin	Continental Europe		
		North Denmark	0	2,6
		South Denmark	0	2,6
		North America: California	0	5
		Summary of all regions	0 [0/3]	
Ap-1[1/]	....	Continental Europe		
		North Denmark	0	2,6
		South Denmark	0	2,6
		North America: California	0.10	5
		Summary of all regions	0.03(0-0.10) [1/3]	

[1/] So designated because the staining technique uses the substrate for acid phosphatase.

continued

64. POLYMORPHISMS: WILD MOUSE

Part I. Electrophoretic Polymorphisms in Feral Populations of *Mus musculus*

Gene Symbol	Gene Name	Trapping Region	Average Heterozygosity/Locus [Populations Polymorphic]	Reference
Dip-1	Dipeptidase-1	United Kingdom		
		Pembrokeshire	0.59	1
		Skokholm	0.34	1
		Somerset	0.64	1
		Summary of all regions	0.52(0.34-0.64) [3/3]	
Es-1	Esterase-1	Continental Europe		
		North Denmark	0	2,6
		South Denmark	0	2,6
		North America		
		Alberta	0	4
		Ontario: Windsor	0	3
		California	0.28	5
		North Carolina	0	4
		Vermont	0	4
		Pacific: Hawaii	0.50	7
		Summary of all regions	0.10(0-0.5) [2/8]	
Es-2	Esterase-2	Continental Europe		
		North Denmark	0.15	2,6
		South Denmark	0	2,6
		United Kingdom		
		Pembrokeshire	0.25	1
		Skokholm	0.43	1
		Somerset	0.46	1
		North America		
		Alberta	0.48	4
		Ontario: Windsor	0.39	3
		California	0.53	5
		North Carolina	0.38	4
		Vermont	0.0	4
		Pacific: Hawaii	0.50	7
		Summary of all regions	0.32(0-0.53) [9/11]	
Es-3	Esterase-3	Continental Europe		
		North Denmark	0.50	2,6
		South Denmark	0.46	2,6
		North America		
		Alberta	0.56	4
		Ontario: Windsor	0.13	3
		California	0.25	5
		North Carolina	0.25	4
		Vermont	0.49	4
		Pacific: Hawaii	0.23	7
		Summary of all regions	0.36(0.13-0.56) [8/8]	
Es-5	Esterase-5	Continental Europe		
		North Denmark	0.10	2,6
		South Denmark	0.39	2,6
		United Kingdom		
		Pembrokeshire	0.15	1
		Skokholm	0.01	1
		Somerset	0.04	1
		North America		
		Ontario: Windsor	0.41	3
		California	0.11	5

continued

64. POLYMORPHISMS: WILD MOUSE

Part I. Electrophoretic Polymorphisms in Feral Populations of *Mus musculus*

Gene Symbol	Gene Name	Trapping Region	Average Heterozygosity/Locus [Populations Polymorphic]	Reference
		Pacific: Hawaii	0.17	7
		Summary of all regions	0.17(0.01-0.41) [8/8]	
Gdc-1	NAD α-glycerol phosphate dehydrogenase	Continental Europe		
		North Denmark	0	2,6
		South Denmark	0	2,6
		Summary of all regions	0 [0/2]	
Gpd-1	Glucose-6-phosphate dehydrogenase, autosomal	Continental Europe		
		North Denmark	0	2,6
		South Denmark	0.50	2,6
		North America		
		Alberta	0.36	4
		California	0.50	5
		North Carolina	0.42	4
		Vermont	0.34	4
		Summary of all regions	0.35(0-0.50) [6/6]	
Gpd-2	Glucose-6-phosphate dehydrogenase, sex-linked	Continental Europe		
		North Denmark	0	2,6
		South Denmark	0	2,6
		North America		
		Alberta	0	4
		California	0	5
		North Carolina	0	4
		Vermont	0	4
		Summary of all regions	0 [0/6]	
Gpi-1	Glucose phosphate isomerase	Continental Europe		
		North Denmark	0	2,6
		South Denmark	0.13	2,6
		North America: California	0	5
		Summary of all regions	0.04(0-0.13) [1/3]	
Hba	Hemoglobin α-chain	Continental Europe		
		North Denmark	0	2,6
		South Denmark	0	2,6
		United Kingdom		
		Pembrokeshire	0	1
		Skokholm	0	1
		Somerset	0	1
		North America: California	0	5
		Summary of all regions	0 [0/6]	
Hbb	Hemoglobin β-chain	Continental Europe		
		North Denmark	0.40	2,6
		South Denmark	0.28	2,6
		United Kingdom		
		Pembrokeshire	0.42	1
		Skokholm	0.48	1
		Somerset	0.04	1
		North America		
		Ontario: Windsor	0.39	3
		California	0.49	5
		Pacific: Hawaii	0.45	7
		Summary of all regions	0.37(0.04-0.49) [8/8]	

continued

64. POLYMORPHISMS: WILD MOUSE

Part I. Electrophoretic Polymorphisms in Feral Populations of *Mus musculus*

Gene Symbol	Gene Name	Trapping Region	Average Heterozygosity/Locus [Populations Polymorphic]	Reference
Id-1	Isocitrate dehydrogenase	Continental Europe		
		North Denmark	0.23	2,6
		South Denmark	0.13	2,6
		North America		
		Alberta	0.44	4
		California	0.49	5
		North Carolina	0.50	4
		Vermont	0.06	4
		Summary of all regions	0.31(0.06-0.50) [6/6]	
Id-2	Isocitrate dehydrogenase, mitochondrial	Continental Europe		
		North Denmark	0	2,6
		South Denmark	0	2,6
		North America: California	0	5
		Summary of all regions	0 [0/3]	
Ipo-1	Indophenol oxidase-1 [2/]	Continental Europe		
		North Denmark	0.10	2,6
		South Denmark	0	2,6
		North America: California	0	5
		Summary of all regions	0.03(0-0.10) [1/3]	
Ld-A	Lactate dehydrogenase, A form	Continental Europe		
		North Denmark	0	2,6
		South Denmark	0	2,6
		North America		
		Alberta	0	4
		North Carolina	0	4
		Vermont	0	4
		Summary of all regions	0 [0/5]	
Ld-B	Lactate dehydrogenase, B form	Continental Europe		
		North Denmark	0	2,6
		South Denmark	0	2,6
		North America		
		Alberta	0	4
		North Carolina	0	4
		Vermont	0	4
		Summary of all regions	0 [0/5]	
Ldr-1	Lactate dehydrogenase regulator	North America		
		Alberta	0.32	4
		Ontario: Windsor	0.50	3
		California	0.47	5
		North Carolina	0.47	4
		Vermont	0.36	4
		Summary of all regions	0.42(0.32-0.50) [5/5]	
Mod-1	Malic enzyme, supernatant	Continental Europe		
		North Denmark	0.47	2,6
		South Denmark	0	2,6
		North America		
		Alberta	0.44	4
		California	0	5
		North Carolina	0.13	4
		Vermont	0.46	4

[2/] Probably homologous to human superoxide dismutase.

continued

Part I. Electrophoretic Polymorphisms in Feral Populations of *Mus musculus*

Gene Symbol	Gene Name	Trapping Region	Average Heterozygosity/Locus [Populations Polymorphic]	Reference
		Summary of all regions	0.25(0-0.47) [4/6]	
Mod-2	Malic enzyme, mitochondrial	Continental Europe		
		North Denmark	0	2,6
		South Denmark	0	2,6
		North America: California	0	5
		Summary of all regions	0 [0/3]	
Mor-1	Malate dehydrogenase, mitochondrial	Continental Europe		
		North Denmark	0.18	2,6
		South Denmark	0.0	2,6
		North America: California	0	5
		Summary of all regions	0.06(0-0.18) [1/3]	
Mor-2	NAD malate dehydrogenase, cytoplasmic	Continental Europe		
		North Denmark	0	2,6
		South Denmark	0	2,6
		North America: California	0	5
		Summary of all regions	0 [0/3]	
Pgd	6-Phosphogluconate dehydrogenase	Continental Europe		
		North Denmark	0.42	2,6
		South Denmark	0	2,6
		North America		
		Alberta	0	4
		California	0.39	5
		North Carolina	0	4
		Vermont	0	4
		Summary of all regions	0.14(0-0.42) [2/6]	
Pgm-1	Phosphoglucomutase-1	Continental Europe		
		North Denmark	0.16	2,6
		South Denmark	0.0	2,6
		North America: California	0	5
		Summary of all regions	0.05(0-0.16) [1/3]	
Pgm-2	Phosphoglucomutase-2	Continental Europe		
		North Denmark	0.50	2,6
		South Denmark	0.0	2,6
		North America: California	0.13	5
		Summary of all regions	0.21(0-0.50) [2/3]	
Pre-1	Prealbumin protein-1	Continental Europe		
		North Denmark	0.06	2,6
		South Denmark	0.48	2,6
		North America: California	0.50	5
		Summary of all regions	0.35(0.06-0.50) [3/3]	
Trf	Transferrin	Continental Europe		
		North Denmark	0	2,6
		South Denmark	0	2,6
		United Kingdom		
		Pembrokeshire	0.09	1
		Skokholm	0.02	1
		Somerset	0.17	1
		North America		
		Ontario: Windsor	0.0	3
		California	0	5
		Summary of all regions	0.04(0-0.17) [3/7]	

continued

64. POLYMORPHISMS: WILD MOUSE

Part I. Electrophoretic Polymorphisms in Feral Populations of *Mus musculus*

Gene Symbol	Gene Name	Trapping Region	Average Heterozygosity/Locus [Populations Polymorphic]	Reference
Trp-1	Tripeptidase-1[3]	United Kingdom		
		Pembrokeshire	0.27	1
		Skokholm	0.02	1
		Somerset	0.05	1
		Summary of all regions	0.11(0.02-0.27) [3/3]	

[3] Homologous to human peptidase-B.

Contributors: Verne M. Chapman and Robert K. Selander

References

1. Berry, R. J., and H. M. Murphy. 1970. Proc. R. Soc. London B176:87-103.
2. Hunt, W. G., and R. K. Selander. 1973. Heredity 31: 11-33.
3. Petras, M. L., et al. 1969. Can. J. Genet. Cytol. 11: 497-573.
4. Ruddle, F. H., et al. 1969. J. Hered. 60:321-322.
5. Selander, R. K., and S. Y. Yang. 1969. Genetics 63: 653-667.
6. Selander, R. K., et al. 1969. Evolution 23:379-390.
7. Wheeler, L. L., and R. K. Selander. 1972. Stud. Genet. Univ. Texas 7:269-296.

Part II. Loci With Unusual or Rare Electrophoretic Alleles in Subspecies and Feral *Mus musculus*

Data in brackets refer to the column heading in brackets.

Species [Region]	Gene Symbol	Gene Name	Reference
Mus musculus [North America]	*Adh-1*	Alcohol dehydrogenase	13
	Alb-1	Albumin	12
	Es-2	Esterase-2	13
	Id-1	Isocitrate dehydrogenase	13
	Mor-1	Malate dehydrogenase, mitochondrial	15
	Pgd	6-Phosphogluconate dehydrogenase	13
M. musculus castaneus [Thailand]	*Amy-2*	Pancreatic α-amylase	7,9
	Es-2	Esterase-2	2
	Es-7	Esterase-7	2
	Es-8	Esterase-8	2
	Gdc-1	NAD α-glycerol phosphate dehydrogenase	2
	Got-1	Glutamate oxaloacetate transaminase-1	5
	Gpd-1	Glucose-6-phosphate dehydrogenase, autosomal	2
	Gpt-1	Glutamate-pyruvate-transaminase	6
	Gus	β-Glucuronidase	4
	Id-1	Isocitrate dehydrogenase	2
	Np-1	Nucleoside phosphorylase-1	2
M. musculus domesticus [United Kingdom]	*Es-10*	Esterase-10	11
	Trp-1[1]	Tripeptidase-1	1

[1] Homologous to human peptidase-B.

continued

Part II. Loci With Unusual or Rare Electrophoretic Alleles in Subspecies and Feral *Mus musculus*

Species [Region]	Gene Symbol	Gene Name	Reference
M. musculus molossinus [Japan]	*Ags*	α-Galactosidase	8
	Apk	Acid phosphatase—kidney	16
	Gpt-1	Glutamate-pyruvate-transaminase	6
	Id-1	Isocitrate dehydrogenase	16
	Np-1	Nucleoside phosphorylase-1	16
M. musculus musculus [Denmark]	*Adh-1*	Alcohol dehydrogenase	14
	Alb-1	Albumin	14
	Es-2	Esterase-2	14
	Es-10	Esterase-10	3
	Mod-1	Malic enzyme, supernatant	14
	Mor-1	Malate dehydrogenase, mitochondrial	14
	Pgd	6-Phosphogluconate dehydrogenase	3
	Pgk-1	Phosphoglycerate kinase-1	10

Contributor: Verne M. Chapman and Robert K. Selander

References

1. Berry, R. J., and H. M. Murphy. 1970. Proc. R. Soc. London B176:87-103.
2. Chapman, V. M. 1973. Mouse News Lett. 48:44-45.
3. Chapman, V. M. 1975. Biochem. Genet. 13:849-856.
4. Chapman, V. M. Unpublished. Roswell Park Memorial Institute, Buffalo, NY, 1978.
5. Chapman, V. M., and F. H. Ruddle. 1972. Genetics 70:299-305.
6. Eicher, E. M., and J. E. Womack. 1977. Biochem. Genet. 15:1-8.
7. Kaplan, R. D., et al. 1973. J. Hered. 64:155-157.
8. Lusis, A. J., and J. D. West. 1976. Biochem. Genet. 14:849-856.
9. Nielsen, J. T. Unpublished. Univ. Aarhus, Dep. Genetics, Denmark, 1978.
10. Nielsen, J. T., and V. M. Chapman. 1977. Genetics 87: 319-325.
11. Peters, J., and H. R. Nash. 1976. Biochem. Genet. 14:119-125.
12. Petras, N. L. 1972. Ibid. 7:273.
13. Selander, R. K., and S. Y. Yang. 1969. Genetics 63: 653-667.
14. Selander, R. K., et al. 1969. Evolution 23:379-390.
15. Shows, T. B., et al. 1970. Biochem. Genet. 4:707-718.
16. Womack, J. E. Unpublished. Jackson Laboratory, Bar Harbor, ME, 1978.

Part III. Electrophoretic Differences Between *Mus musculus* and Other *Mus* Species

Species	Phenotype	Enzyme or Protein Name ⟨Synonym⟩	Reference
Mus caroli[1]	ADA	Adenosine deaminase	1
	DIP-2	Dipeptidase	1
	ENO	Enolase	2
	Fn[2]	Esterase	1
	GOT-1	Aspartate aminotransferase ⟨Glutamate oxaloacetate transaminase⟩	2
	G6PD	Glucose-6-phosphate dehydrogenase	2
	HBB	Hemoglobin β-chain	1
	HPRT	Hypoxanthine phosphoribosyltransferase	2
	ID-1	Isocitrate dehydrogenase	1

[1] Region: Thailand. [2] Homologies not established.

continued

64. POLYMORPHISMS: WILD MOUSE

Part III. Electrophoretic Differences Between *Mus musculus* and Other *Mus* Species

Species	Phenotype	Enzyme or Protein Name ⟨Synonym⟩	Reference
	NP	Nucleoside phosphorylase	2
	PGK-1	Phosphoglycerate kinase	2
	TRP-1	Tripeptidase	1
M. cervicolor cervicolor	ADA	Adenosine deaminase	1
	DIP-2	Dipeptidase	1
	HBB	Hemoglobin β-chain	1
	ID-1	Isocitrate dehydrogenase	1
	MOD-2	Malate dehydrogenase, decarboxylating ⟨Malic enzyme⟩, mitochondrial	1
	PGD	Phosphogluconate dehydrogenase	1,3
	TRP-1	Tripeptidase	1
M. dunni	G6PD	Glucose-6-phosphate dehydrogenase	1,3
	GPI	Glucosephosphate isomerase	1,3
	HPRT	Hypoxanthine phosphoribosyltransferase	1
	PGK-1	Phosphoglycerate kinase	1
M. shortridgei	GDC-1	Glycerol-3-phosphate dehydrogenase (NAD^+) ⟨NAD-α-glycerol phosphate dehydrogenase⟩	1
	G6PD	Glucose-6-phosphate dehydrogenase	1
	GPI	Glucosephosphate isomerase	1
	MOD-1	Malate dehydrogenase, decarboxylating ⟨Malic enzyme⟩, supernatant	1
	MOR-2	Malate dehydrogenase, cytoplasmic	1
	PGD	Phosphogluconate dehydrogenase	1

Contributors: Verne M. Chapman and Robert K. Selander

References

1. Chapman, V. M. Unpublished. Roswell Park Memorial Institute, Buffalo, NY, 1978.
2. Chapman, V. M., and T. B. Shows. 1976. Nature (London) 259:665-667.
3. Selander, R. K. Unpublished. Univ. Rochester, Dep. Biology, Rochester, NY, 1978.

65. TYPE C RNA VIRUSES: WILD MOUSE

Part I. Biological Characteristics

All trapping areas are located in or near Los Angeles County, California. The indigenous C type virus activity and spontaneous tumor occurrence in wild mice is high at the La Puente and Lake Casitas trapping sites [ref. 10, 11, 14, 15, 21], and low at the Munneke, Bouquet Canyon, and Hartz Mountain sites [ref. 7, 9]. Other latent indigenous viruses include polyoma and lymphocytic choriomeningitis in Bouquet Canyon mice [ref. 12], and cytomegalovirus in wild mice from all areas [ref. 13]. **Virus: Nomenclature**—FI = naturally occurring field isolate; A-clone = amphotropic virus clone derived from FI by end-point titration in vitro; E-clone = ecotropic virus clone derived from field isolate by end-point dilution in vitro. Amphotropic murine leukemia viruses ⟨MuLV⟩ grow in both mouse and heterologous species cells; ecotropic MuLV grow only in mouse and rat cells. Based upon cross-neutralization and interference, the amphotropic and ecotropic viruses of wild mice are separate classes of MuLV [ref. 5, 17, 25]. By envelope antigenicity [ref. 5, 17, 25] and gp70 peptide profile [ref. 6], ecotropic viruses are closest to AKR virus, and amphotropic viruses most resemble the xenotropic viruses of laboratory mice. **In Vitro Host Range:** (+) = virus will grow, (−) = virus will not grow after addition to the respective mouse or heterologous species (non-mouse or rat) cells; **Mouse Cell** lines included SC-1 [ref. 16], NIH 3T3 [ref. 19], BALB 3T3 [ref. 1]; **Heterologous Species Cells** [ref. 24] included rabbit SIRC [ref. 3], human RD [ref. 22], and mink [ref. 18]. Both ecotropic and amphotropic virus clones are N-tropic

continued

for mouse cells [ref. 5, 17, 25], and thus are subject to *Fv-1* gene restriction. **XC Test on Mouse Cells:** XC test is an assay for ecotropic MuLV [ref. 20, 26, 27]; (+) = virus gives a positive XC test; (−) = virus gives a negative XC test. **In Vivo Pathogenesis:** In vivo transmission studies with purified virus clones are still in progress and, thus, these pathogenesis results are only preliminary. (+) = induced by virus on experimental transmission to newborn NIH Swiss mice; **Lymphoma**—non-thymic B or null cell origin [ref. 4]; latent period for experimental transmission, 7-15 months; **Paralysis**—spongiosis and gliosis of lower spinal cord anterior lateral horns [ref. 2, 11, 23]; latent period, 2-10 months.

Virus			In Vitro Host Range		XC Test on Mouse Cells	In Vivo Pathogenesis		Remarks	Reference
Name	Tissue of Origin	Nomenclature	Mouse Cells	Heterologous Species Cells		Lymphoma	Paralysis		
Mice Trapped at Bouquet Canyon Squab Farm									
2221	Spontaneous liposarcoma culture	FI	+	+	−	...	...	FI not cloned. Culture also contains lymphocytic choriomeningitis ⟨LCM⟩ virus.	24
4996	Spontaneous lymphoma; spleen extract on SC-1 cells	FI	+	−	+	+	−	FI consists solely of ecotropic virus	25
Mice Trapped at Hartz Mountain Bird Seed Plant									
1313	Spontaneous lymphoma; tumor Moloney concentrate on SC-1 and RD cells	FI	+	+	−	+	−	FI consists solely of amphotropic virus	25
Mice Trapped at Lake Casitas Squab Farm									
292	NIH Swiss embryo cells infected with pooled spinal cord extracts of 3 naturally paralyzed wild mice	FI	+	+	+	+	+	FI now at cell passage 150 with undiminished pathogenicity	11, 24
		A-clone	+	+	−	+	−		
		E-clone	+	−	+	+	+		
4007	Whole embryo-derived tissue culture line	FI	+	+	+	+	−	Based upon positive XC test, an ecotropic virus is present in this FI. It has not yet been cloned and may not exist in high enough titer to induce paralysis.	5
		A-clone	+	+	−	+	−		
4070	Whole embryo-derived tissue culture line	FI	+	+	+	+	+	E-clone stable on repeated passage in mouse cell lines; A-clone stable in mouse & heterologous species cells. This also applies to A- and E-clones from other FI's.	5,24
		A-clone	+	+	−	+	−		
		E-clone	+	−	+	+	+		
7926	Kidney fibroblastic culture from normal 1-d-old mouse	FI	+	+	−	...	...	FI not cloned	24
11235	Spontaneous lymphoma-derived lymphoid suspension culture	FI	+	+	−	+	−	FI consists solely of amphotropic virus	5,24

continued

65. TYPE C RNA VIRUSES: WILD MOUSE

Part I. Biological Characteristics

Virus			In Vitro Host Range		XC Test on Mouse Cells	In Vivo Pathogenesis		Remarks	Reference
Name	Tissue of Origin	Nomenclature	Mouse Cells	Heterologous Species Cells		Lymphoma	Paralysis		
Mice Trapped at LaPuente Duck Farm									
1504	Whole embryo-derived tissue culture line	FI	+	+	+	+	+	Original wild mouse C type virus isolate. FI now at cell passage 185 with undiminished pathogenicity.	5,8, 17, 23, 24
		A-clone	+	+	−	+	−		
		E-clone	+	−	+	+	+		
1740	NIH Swiss embryo cells infected with brain extract of naturally paralyzed wild mouse	FI	+	+	+	+	+	....	5
		A-clone	+	+	−	+	−		
		E-clone	+	−	+	+	+		
Mice Trapped at Munneke Egg Ranch									
672	Tissue culture line derived from 3-methylcholanthrene-induced sarcoma	FI	+	+	+	+	+	Amphotropic virus present but not yet cloned	5,24
		E-clone	+	−	+	+	+		

Contributor: Murray B. Gardner

References

1. Aaronson, S. A., and G. J. Todaro. 1968. J. Cell Physiol. 72:141-148.
2. Andrew, J. M., and M. B. Gardner. 1974. J. Neuropathol. Exp. Neurol. 33:285-307.
3. Benveniste, R. E., et al. 1974. Proc. Natl. Acad. Sci. USA 71:602-606.
4. Blankenhorn, E. P., et al. 1975. J. Natl. Cancer Inst. 54:665-672.
5. Bryant, M. L., and V. Klement. 1976. Virology 73: 532-536.
6. Elder, J. H., et al. 1977. Nature (London) 267:23-28.
7. Gardner, M. B., et al. 1971. Ibid. 232:617-620.
8. Gardner, M. B., et al. 1973. Bibl. Haematol. (Basel) 39:335-344.
9. Gardner, M. B., et al. 1973. J. Natl. Cancer Inst. 50: 719-734.
10. Gardner, M. B., et al. 1973. Ibid. 50:1571-1579.
11. Gardner, M. B., et al. 1973. Ibid. 51:1243-1254.
12. Gardner, M. B., et al. 1974. Ibid. 52:979-981.
13. Gardner, M. B., et al. 1974. Infect. Immun. 10:966-969.
14. Gardner, M. B., et al. 1976. J. Natl. Cancer Inst. 57: 585-590.
15. Gardner, M. B., et al. 1976. Cancer Res. 36:574-581.
16. Hartley, J. W., and W. P. Rowe. 1975. Virology 65: 128-134.
17. Hartley, J. W., and W. P. Rowe. 1976. J. Virol. 19: 19-25.
18. Henderson, I. C., et al. 1974. Virology 60:282-287.
19. Jainchill, J. L., et al. 1969. J. Virol. 4:549-553.
20. Klement, V. 1969. Proc. Natl. Acad. Sci. USA 63:753.
21. Klement, V., et al. 1976. J. Natl. Cancer Inst. 57: 1169-1173.
22. McAllister, R. M., et al. 1969. Cancer 24:520-526.
23. Officer, J. E., et al. 1973. Science 181:945-947.
24. Rasheed, S., et al. 1976. J. Virol. 19:13-18.
25. Rasheed, S., et al. 1977. Intervirology 8:323-325.
26. Rowe, W. P., et al. 1970. Virology 42:1136-1139.
27. Svoboda, J. 1960. Nature (London) 186:980.

Part II. Relationship to Other Retroviruses

Type C viruses were isolated from inbred mouse strains (*Mus musculus musculus*) and wild mice trapped in the islands of Japan (*M. musculus molossinus*), Thailand (*M. musculus castaneus, M. caroli,* and *M. cervicolor*), and Lake Casitas and Lake Puente in California (*M. musculus domesticus*). The relationship of these viruses to other retroviruses is based on antigenic cross-reactivity or nucleic acid hybridization. **Host Range**: Ampho = amphotropic; replicates in cell lines derived from *Mus* and other genera; Eco = ecotropic; replicates in cell lines derived from *Mus* but not

continued

Part II. Relationship to Other Retroviruses

lines from other genera; Xeno = xenotropic; replicates in cell lines from genera other than *Mus* but not *Mus* cell lines. **Type of Assay:** H = hybridization of ^{3}H-DNA probes of **Representative Virus** with RNA from cells producing prototype viruses (data represent percent hybridization of the ^{3}H-DNA probe); N = virus neutralization by focus-reduction assays with antisera prepared against the **Representative Viruses**; p30 = reactivity of p30 proteins of prototype viruses in homologous assays for the p30 protein of the **Representative Viruses**; RT = inhibition of RNA-directed DNA polymerase by antibody to the enzyme of the **Representative Virus**; N, p30, and RT values given on an arbitrary scale of 0 = no reactivity to 4 = maximal reactivity.

Prototype Virus	Host Range	Test Species	Representative Virus	Type of Assay	Value	Reference
Isolated from *Mus musculus musculus* [1]						
AKR-LI	Eco	*M. musculus musculus*	Eco	N	4	5,10
				p30	4	
				RT	4	
				H	100	
			Xeno	N	0	
				p30	1	
				H	50	
		M. musculus [2]	1504-A	N	0	
			1504-M	N	0	
		M. caroli	Caroli CI	H	9	
		M. cervicolor	Cerv CI	H	4	
			Cerv CII	H	51	
		Gibbon	SSAV/GALV [3]	p30	1	
				RT	0	
				H	21	
BALB virus 2	Xeno	*M. musculus musculus*	Eco	H	57	1,4
			Xeno	H	100	
		M. caroli	Caroli CI	H	2	
		M. cervicolor	Cerv CI	H	3	
			Cerv CII	H	27	
		Gibbon	SSAV/GALV [3]	H	12	
MCF	Ampho	*M. musculus musculus*	Eco	N	2	8
			Xeno	N	1	
		M. musculus [2]	1504-A	N	0	
			1504-M	N	0	
NZB-IU-1	Xeno	*M. musculus musculus*	Eco	N	0	4,9,10,13
				p30	4	
				RT	4	
				H	39	
			Xeno	N	4	
				H	66	
		Gibbon	SSAV/GALV [3]	RT	0	
Isolated from *M. musculus* [4]						
CAS-IU-1	Xeno	*M. musculus musculus*	Eco	N	0	6
			Xeno	N	4	
1504-A	Ampho	*M. musculus musculus*	Eco	N	1	7
			Xeno	N	0	
		M. musculus [2]	1504-A	N	4	
			1504-M	N	0	

[1] Inbred strains. [2] Trapped at Lake Puente. [3] Antigenic similarity of Caroli CI P30 & RT and Cerv CI P30 & RT to SSAV/GALV, and their nucleic acid homology, suggests that the gibbon ape endogenous virus was acquired from the mouse. [4] Trapped at Lake Casitas, CA.

continued

65. TYPE C RNA VIRUSES: WILD MOUSE

Part II. Relationship to Other Retroviruses

Prototype Virus	Host Range	Test Species	Representative Virus	Type of Assay	Value	Reference
1504-M	Eco	*M. musculus musculus*	Eco	N	2	7
			Xeno	N	0	
		M. musculus[2]	1504-A	N	0	
			1504-M	N	4	
Isolated from *M. musculus molossinus*						
MOL-NIH	Eco	*M. musculus musculus*	Eco	p30	4	3,11
				RT	2	
				H	95	
			Xeno	H	65	
MOL-8155	Xeno	*M. musculus musculus*	Eco	p30	3	3,10,11
				RT	4	
				H	60	
			Xeno	H	90	
		Gibbon	SSAV/GALV[3]	RT	0	
Isolated from *M. musculus castaneus*						
CAST-8155	Xeno	*M. musculus musculus*	Eco	H	60	3,12
			Xeno	H	85	
Isolated from *M. caroli*						
Caroli CI	Xeno	*M. musculus musculus*	Eco	p30	0	12
				RT	0	
			Xeno	H	1	
		M. caroli	Caroli CI	H	100	
		M. cervicolor	Cerv CI	H	54	
			Cerv CII	H	14	
		Gibbon	SSAV/GALV[3]	p30	3	
				RT	3	
				H	6	
Isolated from *M. cervicolor*						
Cerv CI	Xeno	*M. musculus musculus*	Eco	p30	2	2
				RT	1	
			Xeno	H	3	
		M. caroli	Caroli CI	H	60	
		M. cervicolor	Cerv CI	H	100	
			Cerv CII	H	9	
		Gibbon	SSAV/GALV[3]	p30	4	
				RT	3	
				H	7	
Cerv CII	Xeno	*M. musculus musculus*	Eco	p30	2	2
				RT	1	
			Xeno	H	17	
		M. caroli	Caroli CI	H	11	
		M. cervicolor	Cerv CI	H	10	
			Cerv CII	H	100	
		Gibbon	SSAV/GALV[3]	p30	0	
				RT	0	
				H	0	

[2] Trapped at Lake Puente. [3] Antigenic similarity of Caroli CI P30 & RT and Cerv CI P30 & RT to SSAV/GALV, and their nucleic acid homology, suggests that the gibbon ape endogenous virus was acquired from the mouse.

continued

65. TYPE C RNA VIRUSES: WILD MOUSE

Part II. Relationship to Other Retroviruses

Contributor: Herbert C. Morse and Thomas M. Chused

References

1. Benveniste, R. E., et al. 1974. Proc. Natl. Acad. Sci. USA 71:602-606.
2. Benveniste, R. E., et al. 1977. J. Virol. 21:845-862.
3. Callahan, R., and G. J. Todaro. 1978. In H. C. Morse III, ed. Origins of Inbred Mice. Academic, New York. (In press).
4. Callahan, R., et al. 1974. J. Virol. 14:1394-1403.
5. Gross, L. 1951. Proc. Soc. Exp. Biol. Med. 76:27.
6. Hartley, J. W. Unpublished. N.I.A.I.D., National Institutes of Health, Bethesda, MD, 1978.
7. Hartley, J. W., and W. P. Rowe. 1976. J. Virol. 19:19.
8. Hartley, J. W., et al. 1977. Proc. Natl. Acad. Sci. USA 74:789-792.
9. Levy, J. A. 1973. Science 182:1151.
10. Lewis, H. A., et al. 1975. J. Virol. 15:1378-1384.
11. Lieber, M. M., et al. 1975. Int. J. Cancer 15:211-220.
12. Lieber, M. M., et al. 1975. Proc. Natl. Acad. Sci. USA 72:2315-2319.
13. Todaro, G. J., et al. 1973. Ibid. 70:859-862.

II. RAT

RAT TAXONOMY

The genus *Rattus* consists of some 137 valid species [ref. 5], seemingly a conservative estimate. Ellerman [ref. 2] lists some 500 named forms, a figure which probably contains many subspecies or races of doubtful standing. Missonne [ref. 4] has reconsidered the classification of the Muridae from an evolutionary viewpoint. Specifically, he has attempted to revise the large *Rattus* genus. Only future studies can decide on the wisdom of this work. It could be argued that systematic principles alone will be incapable of providing a final solution.

The two most relevant *Rattus* species are *R. rattus* (Alexandrine rat, black rat, roof rat, ship rat, etc.) and *R. norvegicus* (brown rat, Norway rat). *R. rattus* has been credited with a large number of subspecies and races [ref. 2]. This seems legitimate because the species inhabits innumerable islands and land masses in Southeast Asia and appears to be subject to divergent evolution. This is the presumed origin of *R. rattus,* although the species has a world-wide distribution, largely abetted inadvertently by man. The original diploid chromosome number for the species complex would seem to be 42 (East and Southeast Asia), but animals with 40 (Sri Lanka) and 38 (Europe, Americas, Africa, Australia and New Guinea) chromosomes have been reported. The reduction in number is due to successive fusions of acrocentric chromosomes to form metacentrics [ref. 1, 8, 9]. These fusions are depicted as occurring during the migration of *R. rattus* from Asia to Europe, thence to Australia and neighboring islands. The four acrocentrics involved have been identified [ref. 7].

Ellerman [ref. 2] recognizes five subspecies of *R. norvegicus: caraco, norvegicus, praestans, primarius* and *socer.* The existence of these forms may be taxonomically valid but, beyond this, little information seems to be available. At this time, *R. norvegicus* is truly cosmopolitan. Specimens from any part of the world will freely interbreed and are generally regarded as constituting a single species. The wide distribution is due almost entirely to human agency. Rats, like house mice, are highly successful commensals with man. If aboriginal populations do exist, these must be as isolates in Central Asia or Northern China, the presumed centers of dispersal of the species [ref. 3].

The early history of both rats is obscure [ref. 3, 6]. The westward migration of each species began with the development of trade routes with the East. The first to arrive was *R. rattus,* reaching Europe sometime in the twelfth century and becoming a prime vector in the great plagues. It was subsequently carried to the new world by the ships of the early explorers. *R. norvegicus* was not seen in appreciable numbers in Europe until the early part of the 18th century, whence it spread rapidly. It reached North America by the latter part of that century. An interesting aspect of these successive invasions is the ease by which *R. norvegicus* can supplant *R. rattus.* Exceptions occur in the warmer climatic regions. Although *R. norvegicus* is a larger, more robust animal than *R. rattus,* it evidently is less adaptable to tropical or even subtropical conditions. *R. norvegicus* is reputedly a less agile climber, preferring more subterranean environments.

Contributor: Roy Robinson

References

1. Capanna, E., and M. V. Civitelli. 1971. Boll. Zool. (Naples) 38:151-157.
2. Ellerman, J. R. 1941. The Families and Genera of Living Rodents. British Museum (Natural History), London. v. 2.
3. Hinton, M. A. C. 1920. Br. Mus. (Nat. Hist.) Econ. Ser. 8.
4. Missonne, X. 1969. Ann. Mus. R. Afr. Cent. Zool. 172:1-219.
5. Morris, D. 1965. The Mammals. Hodder and Stoughton, London.
6. Robinson, R. 1965. Genetics of the Norway Rat. Pergamon, London.
7. Yosida, T. H., and T. Sagai. 1972. Chromosoma 37: 387-394.
8. Yosida, T. H., et al. 1971. Ibid. 33:252-267.
9. Yosida, T. H., et al. 1972. Jpn. J. Genet. 47:451-454.

ORIGIN OF THE LABORATORY RAT

The rat—unlike the mouse, guinea pig, and rabbit—has only recently been domesticated, and some of the common laboratory stocks, e.g., the Long Evans, are known to have been developed after crosses involving wild rats. For many stocks the exact ancestry is obscure. The two main stocks, the so-called "Wistar" and "Sprague-Dawley," have become widely dispersed. Sublines have been developed through genetic sampling, selection, and possibly genetic contamination, so that there is now substantial evidence that stocks with the same name are not necessarily genetically identical. Care, therefore, should always be taken to describe the origin and history of stocks used in research, and investigators are strongly urged to consider the use of inbred strains rather than outbred stocks in any studies where the genotype of the animal needs to be specified.

Some of the main events in the history of the laboratory rat, with special emphasis on the origin of ancestral stocks and strains, are given below:

Date	Event	Reference
1551	Possible first mention of *Rattus norvegicus* by Gesner, although apparently he himself had not seen the animal	4
1727-1730	*R. norvegicus* reached Europe from Asia and began to displace *R. rattus*	4
1775	*R. norvegicus* reached the Atlantic seaboard of the United States by boat	4
1800-1870	Rat baiting developed into a popular sport (1-200 rats placed in arena; time recorded for a terrier to kill all). Large numbers of wild rats trapped; albinos kept for show and breeding.	4
1856	First recorded use of rats for experimental work. Waller and Philipeaux studied effects of adrenalectomy in rats. Unknown whether these animals came from an established colony.	10
1877-1885	First recorded breeding experiments by Crampe. Studied crosses between wild and albino rats that differed at the albino, agouti, and piebald loci.	2
1898-1900	Rat first used in psychological research by Steward, Kline, and Small	11
1900	Albino rats of European origin brought to America by Dr. H. H. Donaldson and others. Colony established at the University of Chicago.	2
1906-1907	Donaldson's colony transferred to the Wistar Institute	2,10
1907	First rat colony established for nutritional research by McCollum at Wisconsin from albino rats purchased from a pet shop. Later work (in 1913) with this colony led to the discovery of vitamin A by McCollum and Davis.	7,13
1909	First inbred strain of rats started from an albino Wistar stock by Dr. Helen King; sublines now known as PA and WKA	5
1909	Osborne-Mendel stock established at Yale University for nutritional studies	7
1915	Drs. J. A. Long and H. N. Evans established the "hybrid" Long-Evans black-hooded strain from crosses of a captured wild male with several albino females	12
1919	Wild rat colony designated "captive grey" established at the Wistar Institute to study the genetics of domestication. By 1932, records available on 45,000 individuals. Following mutants discovered: piebald, non-agouti, pink-eyed dilution, albino, curly coat, brown, and a possible "silver." Brown mutation maintained in the BN inbred strain developed by Billingham and Silvers.	1,6
1919	Large colony of rats established at the Institute for Cancer Research, Columbia University, by Drs. Curtis, Dunning, and Bullock. Initial stocks purchased from seven different breeders, includng Fischer, August, Marshall, Zimmerman, and a stock from Copenhagen, Denmark. Subsequent inbreeding led to the development of inbred strains: F344; various sublines of August rats, including AUG, A990, A7322, A28807, and A35322; M520; Z61; and COP. Crosses also led to the development of the ACI and ACH strains.	3
1923	"Hooded Lister" strain established and maintained by "inbreeding" at the Lister Institute, United Kingdom, from a stock supplied by Dr. Gladys Hartwell, Kings College for Women, London. Strain PVG derived from this stock.	8
1924	Sprague-Dawley colony established by the Sprague-Dawley Company as a closed colony. Stock known to include rats from the Wistar colony.	9
1947	A total of 23 mutants recorded by Castle	2
1959	First list of inbred strains of rats published by Billingham and Silvers	1

Contributor: Michael F. W. Festing

continued

References

1. Billingham, R. E., and W. K. Silvers. 1959. Transplant. Bull. 6:399-406.
2. Castle, W. E. 1947. Proc. Natl. Acad. Sci. USA 33: 109-117.
3. Curtis, M. R., et al. 1931. Am. J. Cancer 15:67-121.
4. Donaldson, H. H. 1924. Mem. Wistar Inst. Anat. Biol. 6.
5. King, H. D. 1918. J. Exp. Zool. 26:1-54.
6. King, H. D. 1932. Proc. 6th Int. Congr. Genet., pp. 108-110.
7. McCollum, E. V. 1953. Annu. Rev. Biochem. 22:1-16.
8. McGaughey, C. A. 1947. In A. N. Warden, ed. The UFAW Handbook on the Care and Management of laboratory Animals. Baillière, Tindall and Cox, London. p. 110.
9. Palm, J., and G. Black. 1971. Transplantation 11: 184-189.
10. Richter, C. P. 1954. J. Natl. Cancer Inst. 15:727-738.
11. Robinson, R. 1965. Genetics of the Norway Rat. Pergamon, New York.
12. Weisbroth, S. H. 1969. Lab. Anim. Care 19:733-737.
13. Yang, M. G., and O. Mickelsen. 1974. In W. I. Gay, ed. Methods of Animal Experimentation. Academic, New York. v. 5, pp. 1-40.

SUITABILITY OF THE RAT FOR DIFFERENT INVESTIGATIONS

With the exception of the mouse, the rat is the most-widely used animal for research, accounting for approximately 20-25% of all laboratory animals [ref. 5, 6], and there are indications that its relative popularity is increasing [ref. 6].

Interest in the rat as a research animal varies among different disciplines. A 1975 sampling of published papers, on studies in which the rat was used as the experimental animal of choice, showed that the rat is apparently used less than would be expected in immunology, considerably more than expected in toxicology and physiology/behavior, and about as expected in cancer research—if expectation is based on the relative numbers of rats used in scientific studies.

The choice of the rat for any particular investigation is probably based partly on historical accident, partly on questions of economics and convenience, and partly on scientific considerations. The importance of historical accident should not be overlooked. If an important discovery happens to have been made by work on the rat, this is likely to reinforce use of the rat in that particular discipline, even though in principle the work could have been done equally well on the mouse. Once an area of investigation has been opened up, the scientific investment in terms of background data and knowledge becomes an important factor in the continued use of the original species.

It seems probable that the rat could be used in most biomedical research disciplines. However, in specific investigations other species may be chosen because of their particular anatomical, physiological, biochemical, behavioral, genetic, or other features. For example, although the rat may be suitable for a wide range of cancer studies involving transplantable tumors, some specific projects might be carried out more conveniently by using an anatomical feature such as the hamster cheek pouch. Similarly, specific genetic models may be available only in the mouse even though other considerations would indicate use of the rat. Some of the characteristics of the rat which may influence its choice as an experimental subject are given in the Table below.

Characteristic	Comments
Size	Body weight at birth, ∼5 g; at maturity—♀, ∼350 g; ♂, ≳500 g. Mature weight strongly influenced by both genetic & environmental factors. Advantages because of size: Animals economical to maintain—large numbers may be housed in a small area; diet, bedding, & caging costs low. Small quantities of test substances needed. Animals large enough for surgical procedures, such as kidney transplantation. Dissection somewhat easier than in the mouse. Size of body organs convenient for histological work.
Extended growth period	Rapid from birth to ∼10-12 wk. Body weight increases for several months, thus growth is convenient experimental end point.
Prolificity	One breeding ♀ will usually produce 25-50 young in a 6-mo period (considerably more prolific than the guinea pig which will only produce 6-9 young in same period). Under laboratory conditions, virtually no seasonal fluctuation in production; rat therefore is economical to breed. The relatively large litter size (6-18 young) useful in reproductive studies & particularly in teratology.
Behavior	Usually gentle & non-aggressive with both humans & each other. In contrast with the mouse, ♂ housed together rarely damage each other by fighting. Animals of all ages & both sexes usually easy to handle. Readily adapt to wide range of experimental conditions, making them more amenable to experimental use than some other species. Show wide range of different types of behavior which have been studied intensively by psychologists as a model of mammalian behavior [ref. 1].
Genetically defined stocks	Now more than 100 inbred strains. Some more widely used strains reasonably well characterized. Approximately 60 Mendelian genetic loci described [ref. 7]. Some mutants of biomedical interest. Following strains, mutants, and stocks are of particular interest: Inbred strains (*see also* Table 66) ACI (congenital urogenital abnormalities) BN (congenital hydronephrosis) CAR, CAS (dental caries) GH, SHR (hypertension) LOU (plasmacytomas) MR, MNR ("emotionality") TMB, TMD (maze learning)

continued

Characteristic	Comments
	Selected stocks AA, ANA (alcohol acceptance) [ref. 4] RHA, RLA, RCA (avoidance learning) [ref. 2] Mutants *di* (diabetes insipidus, the Brattleboro rat) *fa* (fatty, the Zucker rat) *j* (jaundice, the Gunn rat) *Ra* (renal adenoma) [ref. 3] *rdy* (retinal dystrophy)
Disease-free stocks	Specific pathogen-free (SPF) & gnotobiotic stocks readily available; breed well under germ-free conditions (in contrast with guinea pig). Respiratory infections do not hinder use to same extent as in the past. Also relatively resistant to wound infections, which is advantageous in experimental surgery.
Omnivorousness	Normal diets relatively economical. Rat used extensively in nutritional research after discovery of vitamin A in experiments in which the animal was involved; however, the fact that it is coprophagous may be a disadvantage in some nutritional studies.
Background data	Substantial amount of information available on biology of the rat. Such background data substantially increases the animal's value in research.

Contributor: Michael F. W. Festing

References

1. Barnett, S. A. 1975. The Rat: A Study in Behavior. Univ. Chicago Press, Chicago and London.
2. Bignami, G. 1965. Anim. Behav. 13:221-227.
3. Eker, R., and J. Mossige. 1961. Nature (London) 189:858-859.
4. Eriksson, K., and M. Närhi. 1972. In A. Spiegel, ed. The Laboratory Animal in Drug Testing. Fischer, Stuttgart. pp. 163-171.
5. Lane-Petter, W., and A. E. G. Pearson. 1971. The Laboratory Animal: Principles and Practice. Academic, London and New York. p. 36.
6. Medical Research Council of Great Britain. 1973. Med. Res. Counc. Lab. Anim. Cent. (Carshalton, Surrey) Man. Ser. 3.
7. Palm, J. E. 1975. In R. C. King, ed. Handbook of Genetics. Plenum, New York and London. v. 4, pp. 243-254.

The first listing of inbred strains of rats was by Billingham and Silvers [9]. The fifth listing, given below, is a substantial revision although it is heavily dependent on the previous compilations [31]. In contrast to the earlier listings, an attempt has been made to cite the original reference for a strain's characteristics. Particular attention has also been given to studies involving several inbred strains; where possible, the strains have been ranked according to characteristics, with the approximate top quartile classified as "high" and the approximate bottom quartile as "low". It should be strongly emphasized that the ranking may be dependent on environmental as well as genetic factors. In a different environment strains might rank differently.

It is clear from this listing that there are already a very large number of inbred strains of rats, but virtually nothing is known about the characteristics of many of them. Future priority should go to the study and characterization of those strains that are already available, rather than the development of new, general-purpose, inbred strains. Comparative studies, involving six or more inbred strains are particularly valuable, and, if possible, such studies should include F344 and/or LEW, which seem likely to emerge as the most widely used of all inbred rat strains. New inbred strains with diseases or conditions of biomedical interest may well be worth developing, and there may be a need for a much wider range of congenic lines.

Generations Inbred: Where more than one number if given, the first F number is the generation at which the animals were received, and the second F represents the number of generations carried by the current holder; Fx indicates no definite information available, but inbred more than 20 generations, unless otherwise indicated. **Source** gives only the name of the holder immediately preceding the current one. **Holder**: For full name, see list of HOLDERS at front of book. This list is not exhaustive; a maximum of three locations of any one stock has been given. Additional information may be obtained from current issues of *Rat News Letter* and the *International Index of Laboratory Animals*, both of which are obtainable from Medical Research Council, Laboratory Animals Centre, Woodmansterne Road, Carshalton, Surrey, SM5 4EF, United Kingdom. **Characteristics**: GVH = graft vs. host; SPF = specific-pathogen-free; IgA = immunoglobulin A; IgE = immunoglobulin E; IgG_1 = immunoglobulin G_1; DNP = 2,4-dinitrophenol; GABA = γ-aminobutyric acid. **Rank**: The first number gives the position and the second gives the total number of strains tested; e.g., "low litter size [10/12]" indicates that the strain in question ranked 10th out of 12 strains examined for litter size. Data in light brackets refer to the column heading in brackets. Figures in heavy brackets are reference numbers. Plus/minus (±) values are standard deviations, unless otherwise indicated.

Strain ⟨Synonym⟩	Coat Color [Relevant Genes]	Generations Inbred	Origin [Source]	Holder	Characteristics [Rank]
ACH ⟨AXC 9935 piebald⟩	Black hooded [*a*, *h*]	F73	Curtis & Dunning, at Columbia Univ. Institute for Cancer Research, 1926	DU	High incidence of spontaneous lymphosarcoma of ileocecal mesentery; will grow transplantable tumors: R-2788, IRS 6820, B-P839 [30]
ACI ⟨AXC 9935 Irish⟩	Black with white belly & feet [*a*, h^i]	>F100	Curtis & Dunning at Columbia Univ. Institute for Cancer Research, 1926; to Heston, 1945, at F30; to NIH, 1950, at F41. Subsequent sublines from either Dunning or NIH.	N	Poor reproductive performance [9/12] & low litter size [11/12] [42]. High in-utero embryo mortality which depends on maternal genotype in various crosses [23]. High early prenatal mortality (11%) [2/8] & high incidence of congenital malformations (10%) [1/8] [91]. These abnormalities have a polygenic mode of inheritance [23]. Some conditions include absent, hypoplastic, or cystic kidneys on one side (♂, 28%; ♀, 20%), sometimes associated with an absent or defective uterine horn or atrophic testis on the same side [64].
		F113	As above. From N to TBI; to HOK, 1967, at F87; to HKM, 1974, at F109.	HKM	
		Fx + F10	As above. [MAI, 1971]	MAX	
		Fx + F18	As above. [CR, 1971]	PIT	

continued

Strain (Synonym)	Coat Color [Relevant Genes]	Generations Inbred	Origin [Source]	Holder	Characteristics [Rank]
					Long latency to emerge into familiar [♂, 11/12; ♀, 9/12] & novel environment [♂, 12/12; ♀, 11/12] [43]. Low systolic blood pressure [17/17] [42]. Low serum thyroxine [5/5] [28]. High hepatic metabolism of aniline in ♀ [2/10] [75]. Absorbs diethylstilbestrol at intermediate rate, leading to a high incidence of mammary tumors [27]. High dose of pentobarbital sodium required for LD_{50} (120 mg/kg) [1/7] [90]. Median survival: ♂, 113 wk; ♀, 108 wk [64]. Spontaneous tumors: ♂—testis 46%, adrenal 16%, pituitary 5%, skin & ear duct 6%, fewer of other types; ♀—pituitary 21%, uterus 13%, mammary gland 11%, adrenal 6%, fewer of other types [64]. These abnormalities have a polygenic mode of inheritance [23]. Spontaneous adenocarcinomas of ventral prostate seen in 7/41 untreated ♂ at 34-37 mo [89]. Will grow transplantable tumors: M-C961, 970, R-3234, R-3559 [31].
ACP (ACP 9935 Irish piebald)	Black with white belly & feet [h^i]	Fx + F11	Dunning; to NCI, 1967, at F54. [CR, 1971]	PIT	....
AGA	Black [*a*]	F20	Nakic, Zagreb [100]	....	Used for immunological studies. No further information.
AGUS	Albino [*a*, *c*, *H*]	F35	Germ-free strain developed by Gustafsson from stock (Sprague-Dawley?) hysterectomized in 1948 at F10. To LAC, 1968, at F26.	LAC	Good breeding performance, though sensitive to environmental influences; about 1-4% tailless young, which are infertile [29]. Susceptible to experimental allergic encephalomyelitis [52]. Relatively resistant to infection by *Entamoeba histolytica* [71].
ALB (Albany)	Dilute brown [*a*, *b*, *d*?]	F41	Wolf & Wright, Albany Medical College; to NIH, 1950. No inbreeding records prior to transfer.	N, HOK	Poor reproductive performance [11/12] & small litter size [10/12]; variable frequency of fetal resorption which affects reproductive performance [42]. Docile behavior [31]. Low hepatic metabolism of aniline in ♀ [10/10] [75]. Some mammary fibroadenoma [31].
		Fx + F11	As above. [CR, 1973]	PIT	
		F68	As above. [N, 1974 at F62]	WIS	
AM	Yellow	F26	Torres, Rio de Janeiro, from outbred stock	TOR	....
AMDIL	Dilute yellow	F25	Torres, as above	TOR	....
AO	Albino [*A*, *c*, *h*] 1/	F? + F48	From ARC Compton, probably as "WAG," to Gowans, Oxford, 1957	IAP	....
AS	Albino [*c*]	F60+	Univ. of Otago, from Wistar rats, imported	OSU	Good reproductive performance [31]. Hypertensive, though not to

1/ Appears to differ from some other WAG sublines in having *A* at the agouti locus.

continued

Strain ⟨Synonym⟩	Coat Color [Relevant Genes]	Generations Inbred	Origin [Source]	Holder	Characteristics [Rank]
			from England in 1930. May be a subline of GH, with which it is histocompatible [48].		such an extent as GH [48]. Susceptible to development of experimental allergic encephalomyelitis [52].
		Fx + F22	As above. [CSR, 1967]	MAX	
AS2	Albino [*c*]	F60+	Outbred rats at Univ. of Otago Medical School, to Dept. of Surgery, 1963, at F22-24	OSU	....
		Fx + F23	As above. [CSR, 1967]	MAX	
AUG ⟨August⟩	Dilute hooded [*h*, *p*]	F? + F34	Derived from one of the U.S. "August" sublines in 1951 and distributed by CBI	ORL, IAP	Susceptible to experimental allergic encephalomyelitis [52]. Resistant to induction of autoimmune thyroiditis [83].
AVN	Albino [*c*]	Fx + F23	[Stark, CUB, 1971]	MAX	Will grow ferridextron-induced sarcoma FEDEX-AVN [57]
A990 ⟨August 990⟩	Agouti or nonagouti hooded [*A*, *C*, *h*, or *a*, *C*, *h*]	F87	Curtis at Columbia Univ. Institute for Cancer Research, 1921	DU	Good reproduction [31]. High open-field defecation [♂, 2/12; ♀, 3/12] & low ambulation [11/12] [45]. Low wheel activity [12/12] [44]. Life-span, 14 ± 1 mo. Resistant to cysticercus. Susceptible to estrogen-induced mammary & adrenal tumors. Will grow transplantable tumors IRC-855 & R-3409. [31]
A7322 ⟨August 7322⟩	Pink-eyed hooded, dilute [*h*, *p*]	F64	Curtis at Columbia Univ. Institute for Cancer Research, 1925	DU	Life-span, 14 ± 1 mo. Resistant to cysticercus. Spontaneous mammary tumors frequent. Will grow transplantable tumors R-2426, R-2737, R-2857, R-3442. [31]
A28807 ⟨AUG-28807; August 28807⟩	Pink-eyed hooded, dilute [*h*, ?]	F53	Subline of A7322 derived from a half-brother x sister mating at F15	DU	Similar to A7322 [31]
	Pink-eyed agouti, piebald	F60 + F11	As above. [CR, 1972]	PIT	
		F70	As above. [CR, 1972, at F57]	WIS	
A35322	Black hooded [*a*, *h*]	F50	Curtis & Dunning, 1942, from a mutation originating in an aunt x nephew mating at F27 of animals of strain A990	DU, CR	Short latency to emerge from cage into familiar [2/12] & into novel [4/12] environment [43]. High open-field defecation [♂, 3/12; ♀, 1/12] [45]. High wheel activity [♂, 4/12; ♀, 3/12] [44]. Vaginal prolapse frequent. Will grow transplantable tumors R-3280 (bronchiogenic carcinoma) & R-3371. [31]

continued

Strain ⟨Synonym⟩	Coat Color [Relevant Genes]	Generations Inbred	Origin [Source]	Holder	Characteristics [Rank]
B	Albino [*c*]	F69	P. Swanson from Wistar stock; to E. Dempster at F43	Har	Large body size—68 g at 28 d. Poor maternal instincts; fertile. [31]
BDE	Black hooded [*a, h*]	F25	Zentralinstitut für Versuchstierzucht Hannover, from a cross between BD VII & E3	HAN	Low incidence of megaesophagus/ aperistalsis & microphthalmia [31]
BD I	Yellow pink-eyed [*A, C, H, p*]	F? + F50	Druckrey, 1937, from a yellow, pink-eyed strain. Inbred & reduced to 1 pair after World War II. Crosses with Wistar stock and subsequent inbreeding led to development of BD II. According to Druckrey [26], strains BD III through BD X were then developed from a cross of a single BD I x BD II mating pair, with subsequent selection for coat color alleles[2]	BU	All strains (BD I-BD X) have a median life-span of 700-950 d, depending on strain, & a low tumor incidence [26]
BD II	Albino [*a, c, h, P*]	F? + F50	*See* BD I	BU, HAN	Frequent microphthalmia [26]. *See also* BD I.
BD III	Pink-eyed, yellow, hooded [*A, C, h, p*]	F? + F50	*See* BD I	BU	*See* BD I
BD IV	Black hooded [*a, C, h, P*]	F? + F50	*See* BD I	BU	*See* BD I
BD V	Pink-eyed, non-agouti, hooded [*a, C, h, p*]	F? + F50	*See* BD I	BU, KIEL	*See* BD I
	Sandy hooded	Fx + F13	*See* BD I. [Druckrey, Freiburg, W. Germany, 1971]	MAX	
BD VI	Black [*a, C, H, P*]	F? + F50	*See* BD I	BU	*See* BD I
BD VII	Pink-eyed sandy [*a, C, H, p*]	F? + F50	*See* BD I	BU	*See* BD I
		Fx + F2	*See* BD I. [Druckrey, Freiburg, W. Germany, 1976]	MAX	
BD VIII	Agouti hooded [*A, C, h, P*]	F? + F50	*See* BD I	BU	Occasional vaginal atresia [26]. *See also* BD I.
BD IX	Agouti [*A, C, H, P*]	F? + F50	*See* BD I	BU	*See* BD I
BD X	Albino [*a, c, h, p*]	F? + F50	*See* BD I	BU	*See* BD I
BIRMA ⟨Birmingham A⟩	Albino [*c*]	F27	A. M. Mandl, 1952, from rats purchased at Birmingham market	BUA	....
BIRMB ⟨Birmingham B⟩	Albino [*c*]	F25	As for BIRMA	BUA	....

2/ However, the strains have four different *H-1* (*Ag-B*) haplotypes ($H\text{-}1^d$, $H\text{-}1^w$, $H\text{-}1^l$, & $H\text{-}1^e$) rather than the two that would be expected from such a cross [98]. The strains cannot be regarded as a set of recombinant strains as defined by Bailey [2], though their definition by coat color alleles makes the set easily identifiable and should help to insure authenticity.

continued

Strain ⟨Synonym⟩	Coat Color [Relevant Genes]	Generations Inbred	Origin [Source]	Holder	Characteristics [Rank]
BN	Brown [*a*, *b*, *h^i*]	F35	Silvers & Billingham, 1958, from a brown mutation maintained by D. H. King & P. Aptekman in a pen-bred colony	N, HAN, ORL	Low fertility [31]. Congenital hydronephrosis, 30% [21]. Somewhat vicious [31]. Intermediate susceptibility to pentobarbital sodium [3/7] with LD_{50} of 90 mg/kg [90]. Resistant to experimental allergic encephalomyelitis [1/7] [35,65]. Endocardial disease 7% at avg age of 31 mo [13]. Estimated median life-span >24 mo in ♂ and >25 mo in ♀ [11]. Survival curves based on 74 ♂ & 236 ♀ show a median life-span of 29 mo in ♂ & 31 mo in ♀ [17]. Most common neoplastic lesions in ♂: urinary bladder carcinoma, 35%; pancreatic islet adenoma, 15%; pituitary adenoma, 14%; lymphoreticular sarcomas, 14%; adrenal cortical adenoma, 12%; thyroid medullary carcinoma, 9%; adrenal pheochromocytoma, 8%; 4 other types of tumors observed. In ♀: pituitary adenoma, 26%; ureter carcinoma, 22%; adrenal cortical adenoma, 19%; cervix sarcoma, 15%; mammary gland fibroadenoma, 11%; pancreas islet adenoma, 11%; 12 other tumor types observed. [17] Tumors of epithelium—28% in ♂, 2% in ♀; tumors of ureter—6% in ♂, 20% in ♀ [11]. Vaginal & cervical tumors, mostly sarcomas but also 7 squamous cell carcinomas & 4 leiomyomas, seen in 20% of animals that died naturally. Cervical & vaginal tumors studied in more detail by Burek, et al. [20] Chance of death from metastases reaches peak at 25-30 mo in ♂, but increases with age in ♀ [18]. Further details of the aging colony are given by Hollander [50] & by Burek & Hollander [19].
		Fx + F15	As above. [Stark, CUB, 1970]	MAX	
		Fx + F16	As above. [CR, 1971]	PIT	
		F60	As above. [Silvers, 1976, at F57]	WIS	
BP ⟨Black Praha⟩	Black Irish hooded [*a*, *h^i*]	F25	Sekla. Strain selected for resistance to Walker 256 tumor.	CUB	Low fertility. Adult spleen cells highly effective in induction of GVH reaction in allogeneic newborn, but BP newborn resistant to GVH reaction. [31]
BROFO	Albino [*c*]	F28	Medical Biological Lab. Defense Res. Org. of Netherlands	RIJ	Large Wistar type of rat maintained SPF & germfree [31].
BS	Black [*a*]	F41	Univ. of Otago Medical School, from a cross of wild rats with Wistar stock, with the black phenotype backcrossed to the Wistar [106]	OSU	Fair reproduction. Low incidence of hydrocephalus. Docile. [31]
		Fx + F19	As above. [Heslop, Dunedin, New Zealand, 1967]	MAX	
BUF ⟨Buffalo⟩	Albino [*c*]	F58	Heston, 1946, from Buffalo stock of H. Morris; to NIH in 1950, at F10	N, MAX	Intermediate breeding performance [5/12] & litter size [8/12] [42]. High incidence of prenatal mortality early (18%) [1/8] & late (11%) [2/8], but no congenital abnormalities detected [91]. Low hepatic metabolism of ethylmorphine [♂,
		F71	As above. To HOK, 1956; to HKM, 1974, at F64	HKM	

continued

Strain ⟨Synonym⟩	Coat Color [Relevant Genes]	Generations Inbred	Origin [Source]	Holder	Characteristics [Rank]
		Fx + F16	As above. [CR, 1971]	PIT	7/10; ♀, 10/10], but high metabolism of aniline [♂, 3/10; ♀, 1/10] [75]. Autoimmune thyroiditis seen in 6/11 ♂ at 36 wk, but not seen in 5 other strains [40]. Spontaneous autoimmune thyroiditis with mononuclear cell infiltration of the thyroid in 26% of animals >1 yr old [72]. Autoimmune thyroiditis develops spontaneously and after ingestion of 3-methylcholanthrene, but reaches nearly 100% incidence after neonatal thymectomy [94]. Poor immune response to sheep erythrocytes [6/7] [101]. Low incidence of dental caries [31]. Survival 58% at 2 yr [60]. Spontaneous tumors of anterior pituitary 30% and of adrenal cortex 25% in older animals [31]. Incidence of thyroid carcinoma at 2 yr of age 25% [60]. Will grow hepatoma 5123 (which has some enzymes similar to normal liver), & Yoshida ascites sarcoma (16%); also Morris hepatoma & pituitary tumor [31].
B3	Pink-eyed yellow, non-agouti, brown hooded [*a, C, h, p*]	F10[3]	From stock bred for homozygosity of coat color genes, as a natural recombinant of the major histocompatibility complex [36]	PIT	....
CAP	Albino [*a, h*]	F30+	Polish Academy of Sciences, Krakow [99]	KIEL	....
CAR ⟨Hunt's caries resistant⟩	Albino [*c*]	F77	Hunt, 1937; developed for resistance to dental caries	N	Poor reproductive performance [10/12] & low litter size [9/12] [42]
CAS ⟨Hunt's caries susceptible⟩	Albino [*c*]	F38	Hunt, 1937; developed for high incidence of dental caries	N	Poor reproductive performance [12/12] & low litter size [12/12] [42]
COP ⟨COP-2331; Copenhagen 2331⟩	Agouti hooded	Fx + F9	[NIH, Bethesda, USA, 1971]	MAX	Small pituitaries [27]. Resistant to cysticercus [31]. Mean life-span 20 ± 0.2 mo. Slow absorption of diethylstilbestrol leading to death from bladder calculi & papillomas. Spontaneous tumors of thymus. Resistant to mammary tumor induction. [27] Will grow transplantable tumors IRS 4337 & R-3327 prostate adenocarcinoma, which is a model of human prostate cancer [61].
CPBB ⟨CPB-B⟩	Agouti hooded [*A, h*]	F65	Hagedoorn, Holland; to CPB, 1949, at F15.	CPB	Susceptible to audiogenic seizures [102]
DA	Agouti [*A, B, C*]	F? + F50	Not known, but Palm & Black [76] suggest that it may be related to COP	HAN, IAP	Susceptible to induction of autoimmune thyroiditis [83]
		Fx + F16	As above. [Ramseier, Zürich, Switzerland, 1967]	MAX	
		F9 + F15	As above. [PM, 1972, Wistar Institute, Philadelphia]	PIT	
		F46	As above. [D. B. Wilson, Univ. of Pennsylvania, 1967, at F25]	WIS	

[3] Only part-inbred.

continued

Strain (Synonym)	Coat Color [Relevant Genes]	Generations Inbred	Origin [Source]	Holder	Characteristics [Rank]
DONRYU	Albino [*c*]	F59	R. Sato, 1950, by inbreeding Japanese albino rats	SATO	Relatively small size; mild nature. Will accept 100% of transplantable Yoshida sarcoma & ascites hepatoma. [87]
E3	Non-agouti, yellow-brown hooded [*a*?, *C*, *h*]	F55	Kröning, Göttingen, 1949, from rats of unknown origin; to HAN, 1957	HAN	....
F344 (Fischer; Fischer 344)	Albino [*a*, *c*, *h*]	F86	Curtis at Columbia Univ. Institute for Cancer Research, 1920; to Heston, 1949; then to NIH, Bethesda, 1950, at F51. Subsequent sublines from either Dunning or the NIH colonies.	N, LAC, HAN	Good breeding performance [1/12] & large litter size [1/12] [42]. Large pituitaries [27]. Low wheel activity [♂, 8/12; ♀, 11/12] [44]; low open-field defecation [♂, 10/12] [45]. Low serum insulin [5/5] [28]. High specific activity but low inducibility of NADPH—cytochrome reductase compared with outbred Sprague-Dawley rats [37]. Resistant to development of salt hypertension [41]. Incidence of nephritis, 21-30% [31]. High hepatic metabolism of ethylmorphine & aniline in ♂ [2/10 in each case] [75]. Rapidly absorbs implanted diethylstilbestrol pellets, leading to death; fatty liver most common finding at autopsy [27]. Low LD_{50} of pentobarbital sodium (70 mg/kg) [5/7] [90]. Low primary & secondary immune response to sheep erythrocytes [7/7] [101]. Susceptible to cysticercus infection [27]. Median life-span: ♂, ∿31 mo; ♀, 29 mo; ∿87% survival to 24 mo in both sexes [86]. Mean life-span: ♂, 725 d; ♀, 675 d [54]. Median life-span in SPF ♀, 25 mo [68]. Mean life-span in presence of severe pulmonary infection: ♂♀, 24 mo [25]. Incidence of tumors in ♂: testicular interstitial cell tumors—85% [86], 68% [54], 65% [25]; pituitary adenomas, 24% [86]; mammary tumors, 23% [86]; other tumor types less common [86]. Incidence of tumors in ♀: mammary tumors, 41% [86]; pituitary adenomas, 36% [86]; mononuclear cell leukemia, 24% [25]; uterine polyploid tumors of endometrial origin, 21% [54]; subcutaneous fibroadenoma, 9% [25]. Incidence in ♂♀: thyroid carcinoma, 22% [60]; nodular hyperplasia of liver, 5% [25]. In SPF ♀, 21/86 animals developed a unique mononuclear cell leukemia with uniform involvement of liver & spleen [68]. Incidence of tumors under germfree conditions in ♂: leukemia, 26%; mammary tumors, 12%; all others, 9% [85]. Incidence of tumors under germfree conditions in ♀: leukemia, 36%; mammary tumors, 20%; all others, 5% [85]. Will grow following transplantable tumors: Dunning hepatoma, hepatoma LC-18, Novikoff hepatoma; Dunning leukemia, leukemias HLF1, IRC-741, R-3149, R-3323, R-3330, R-3399, & R-3432; lymphosarcoma R-3251; uterine sarcoma F-529; mammary carcinomas HMC & R-3230, mammary fibroma F-609; fibrosarcoma R-3244; sarcomas IRS 9802 & R-3259; Walker 256 carcinosarcoma; & pituitary tumors MtT and MtTf4 [42].
		F113	Curtis at Columbia Univ. Institute for Cancer Research, 1920; to Heston, 1948; to N, 1950, at F55; to HOK, 1956; to HKM, 1974, at F107	HKM	
		(Fx + F18) + F15[4/]	[CR, 1971]	PIT	

[4/] Fx + F18 when received.

continued

Strain ⟨Synonym⟩	Coat Color [Relevant Genes]	Generations Inbred	Origin [Source]	Holder	Characteristics [Rank]
G/Cpb ⟨CPB-G⟩	Albino [*c*]	F86	Gorter, Holland; to Hagedoorn; to CPB at F35 [102]	CPB	....
GH ⟨Genetic hypertension⟩	Albino [*c*]	F31	Univ. of Otago Medical School, from rats of Wistar origin imported from England in 1930. Selection for high blood pressure started by Smirk, 1955. A number of sublines have been developed. Closely related to strain AS. [48]	OMR	Hypertension, cardiac hypertrophy, & vascular disease [78]. Heart rate ∿20% greater, heart wt ∿50% greater, & body fat lower than in normotensive strains [78]. Genetic hypertension in GH (but not in SHR) may be associated with a defect in renal prostaglandin catabolism [1]. Strain characteristics compared to SHR reviewed by Simpson et al. [95].
GHA	Ginger hooded [?, *h*]	F20	QEH, from mixed Wistar, Lewis, & colored stock	QEH	....
HCS ⟨Harvard caries susceptible⟩	Albino [*c*]	F?	Harvard; to Liverpool, U.K., 1960. May be a subline of CAS (?).	LHR	....
HO (*see also* PVG)	Black hooded	Fx + F11	[OX, 1972]	PIT	*See* PVG
HS	Black hooded [*a*, *h*]	F20+	Probably from same cross as BS [106]	OSU	Fair reproduction. ∿12% hydrocephalus. Docile. [31]
INR ⟨Iowa non-reactive⟩	Black hooded [*a*, *h*]	F21	Har, 1962, from a stock selected by Hall for low open-field defecation	Har	Short latency for emergence from cage into both familiar & novel environment [1/12] [43]. Low open-field defecation in ♀ [12/12], & high open-field ambulation in ♂♀ [1/12] [45].
IR ⟨Iowa reactive⟩	Pale cinnamon beige hooded [*a*, c^d?, *h*]	F20	Har, 1962, from mutation of a Michigan stock	Har	Vigorous & healthy, but high incidence of physical anomalies [31]. Long latency to emerge into familiar environment [♂, 12/12; ♀, 10/12] [43]; high open-field exploration in ♀ [3/12] [45]; low wheel activity [♂, 11/12; ♀, 10/12] [44].
IS	Agouti [*A*]	F23	From a cross between a wild ♂ and a Wistar ♀, with sib mating since 1968 [53]	IS	Malformations of the thoracolumbar vertebrae, leading to kyphoscoliosis with restricted spinal canals & compressed spinal cords. These malformations occur from the 12th thoracic to 6th lumbar vertebra; usually 2-4, but sometimes as many as 7, vertebrae are affected. Some degree of abnormality occurs in ∿90% of individuals. [53]
K	Albino [*c*]	F40	E. Matthies, Halle-Wittenberg, 1958, from outbred Wistar stock	HALLE	Good breeding performance. Wt at 100 d is 290 g in ♂, 200 g in ♀. Low spontaneous tumor incidence (<0.5%). Developed by selection for resistance to a range of transplantable tumors. [63]

continued

Strain ⟨Synonym⟩	Coat Color [Relevant Genes]	Generations Inbred	Origin [Source]	Holder	Characteristics [Rank]
KGH	Albino	F39 + F17	[HVD, 1968 **[**58**]**]	PIT	Naturally occurring recombinant between loci *Ag-B7* & *MLR-1* **[**24, 58**]**
KX ⟨NEDH; Slonaker⟩	Ablino[5]	F45	From Slonaker colony at Univ. of Chicago, ca. 1928.	KNOX	....
KYN ⟨Kyoto notched⟩	[h^n]	F?	Makino, Hokkaido Univ., 1960, from stock carrying the "notched" character isolated by Nakata from wild rats in Kyoto, and from a cross involving WKA **[**91**]**	HOK	Abnormalities of the urogenital organs, 11% **[**92**]**
LA/N	Non-agouti, brown [*a, b, H*]	F23	From a cross between ALB/N and a hooded stock of unknown origin **[**42**]**	N	....
LE	Black hooded [*a, h*]	F34	M. Sabourdy, ∿1960; to Amsterdam	HAN, ORL	....
LEJ ⟨Long-Evans⟩	Black hooded [*a, h*]	F39	From Pacific Farms to MS, 1956	MS, HOK	....
		F27	From Pacific Farms to MS and to HOK, 1956; to HKM, 1974, at F22	HKM	
LEP ⟨Long-Evans Praha⟩	Dark agouti	Fx + F37	[Štark, CUB, 1973]	MAX	....
LEW ⟨Lewis⟩	Albino [*a, c, h*]	F67	Lewis from Wistar stock; to Aptekman & Bogden, 1954, at F20; to Silvers, 1958, at F31. Subsequently distributed by Silvers. Used as the inbred partner for a number of congenic strains at the major histocompatibility complex **[**97**]**.	HAN, LAC, N	High fertility. Docile. **[**31**]** High serum growth hormone [1/5], high serum thyroxine [1/5], & high serum insulin [1/5] **[**28**]**. Becomes obese on high fat diet [2/7] **[**88**]**. High hepatic metabolism of ethylmorphine in ♀ [3/10] **[**75**]**. Susceptible to induction of experimental allergic encephalitis & adjuvant-induced arthritis **[**77**]**. Susceptible to development of experimental allergic encephalomyelitis after challenge with guinea pig myelin basic protein (in contrast with strain BN) **[**52,65**]**. Highly susceptible to development of induced autoimmune myocarditis **[**34**]**. Survival 26% at 2 yr of age **[**60**]**. Host for lymphoma 8; kidney sarcoma; sarcoma No. 3 Lewis; fibrosarcomas MC-39, ML-1, ML-7; & carcinoma No. 10 Lewis **[**42**]**.
		Fx + F22	As above. [Štark, CUB, 1969]	MAX	
		F36 + F7	As above. [N, 1972]	PIT	
LGE	Black hooded [*a, h*]	Fx + F15	Long Evans stock from CR to PIT, 1971	PIT	....
LOU/C ⟨Louvain⟩	?	F20 + F?	Bazin & Beckers, from rats of presumed Wistar origin kept at the Univ. of	WSL	Develop spontaneous plasmacytomas after ∿8 mo of age; incidence: ♂, ∿30%; ♀, ∿16%. Tumors develop rapidly, and may be detected by

5/ Sublines carrying color genes *C* & *H*, and carrying *ic* ⟨infantile ichthyosis⟩, are also kept **[**56**]**.

continued

Strain ⟨Synonym⟩	Coat Color [Relevant Genes]	Generations Inbred	Origin [Source]	Holder	Characteristics [Rank]
			Louvain. From 28 parallel sublines LOU/C was selected for its high incidence of plasmacytomas, and LOU/M for its low incidence. The two are histocompatible, and this is maintained by selection of LOU/M ♂ on basis of acceptance of skin grafts from LOU/C animals. [4]		palpation. They usually develop in ileocecal lymph nodes, and ∿60% synthesize monoclonal immunoglobulins ⟨Bence-Jones proteins⟩ of IgG_1 (35%), IgE (36%), or IgA classes. [3,6-8] (Amino-acid sequence of a kappa Bence-Jones protein has been studied by Starace & Querinjean [96].) Tumors are transplantable in solid or ascites form, and retain their secretory properties through successive passages [3,6-8]. More than 800 transplantable immunocytomas are now available.
LOU/M ⟨Louvain⟩	?	F?	*See* LOU/C	WSL	Good IgE antibody response to ovalbumin & DNP hapten [5] (*see also* LOU/C)
MAXX	Black hooded [*a*, *h*]	F20 + F10	From a cross of BN with LEW, with subsequent inbreeding. [MAI, 1973]	PIT	....
MNR ⟨Maudsley non-reactive⟩	Albino [*A* or *a*, *c*, *h*]	F44+	P. L. Broadhurst, 1954, from a commercial Wistar stock, with selection for low defecation response in an open field. A number of parallel sublines are in existence; these differ at least at the agouti ⟨*A*⟩ and the major histocompatibility ⟨*Ag-B* or *H-1*⟩ loci.	BRH, N	Good breeding performance [3/12] & large litter size [3/12] [42]. Long latency to emerge into familiar [11/12] & novel [12/12] environment in ♀ [43]. Low open-field defecation [♂, 12/12; ♀, 11/12 to 9/12, depending on subline] [45]. Low open-field defecation [8/8] in hybrid offspring [79]. High wheel activity [1/12] [44]. *See also* strain MR. An extensive review of the difference between MNR & MR is given by Broadhurst [16].
MNRA	Albino [*a*, *c*, *h*]	?	? Subline of MNR	Har	Differs from MNR in a number of characteristics and carries the *a* allele at the agouti locus [46]
MR ⟨Maudsley reactive⟩	Albino [*a*, *c*, *h*]	F40	P. L. Broadhurst, 1954, from commercial Wistar stock, with selection for high defecation response in open field (*see also* MNR)	BRH, N	Good breeding performance [4/12] & large litter size [2/12] [42]. Long latency to emerge into familiar environment in ♀ [12/12], & into novel environment in ♂ [11/12] [43]. High open-field defecation [♂, 1/12; ♀, 2/12] [45]. Compared with MNR the strain has a low rearing, low ambulation, high open-field defecation, low shock-avoidance conditionability, and tends to be more "emotional" in a wide range of behavior [16]. *See also* MNR.
		F59 + F5	As above. [N, 1975]	PIT	
MSUBL/Icgn	Black [*a*]	F23	Stroeva, Institute of Developmental Biology, Moscow, from a cross of wild rats	NOVO	High incidence of microphthalmia & anophthalmia. High reactivity of adrenals to ACTH [1/4]. [14] Immunoglobulin (kappa chain) allo-

continued

Strain ⟨Synonym⟩	Coat Color [Relevant Genes]	Generations Inbred	Origin [Source]	Holder	Characteristics [Rank]
			with MSU microphthalmic rats obtained from Brouman at Montana State Univ.		type reported by Rokhlin & Nezlin [82].
MW	Albino [*c*]	?[6]	....	MGM	Has renal abnormality in which some glomeruli are near kidney surface. Also suffers from retarded growth associated with a deficiency of plasma growth hormone. [62]
M14	Albino [*c*]	F40+	A. B. Chapman, 1940, from Sprague-Dawley stock, with selection for low ovarian response to pregnant mare's serum	CP	....
M17	Albino [*c*]	F40+	A. B. Chapman, 1940, from Sprague-Dawley stock, with selection for high ovarian response to pregnant mare's serum	CP	....
M520 ⟨Marshall 520⟩	Albino [*a, c, h*]	F86	Curtis at Columbia Univ. Institute for Cancer Research, 1920; to Heston, 1949, at F49; to NIH, 1950, at F51	N	Good breeding performance [2/12] & intermediate litter size [6/12] [42]. Low systolic blood pressure [16/17] [42]. Low hepatic metabolism of aniline [♂, 9/10; ♀, 7/10], but high metabolism of ethylmorphine [1/10] [75]. Highly susceptible to nephritis [42]. Susceptible to cysticercus infection [31].
		F123 + F4	As above. [N, 1976]	PIT	NIH subline has 21-25% incidence of adrenal medullary tumors. Low frequency of pancreatic exocrine tumors. Interstitial cell tumors & tumors of uterus, anterior pituitary, adrenal cortex & medulla range from 0-10% in animals <18 mo old; but after 18 mo, tumor incidence is as follows: adrenal medulla, 60-85%; adrenal cortex, 20-45%; anterior pituitary, 20-40%; interstitial cell, 35% of virgin ♂; uterus, 12-50% in ♀. Susceptible to induction of tumors by *N*-2-fluorenylacetamide ⟨2-acetylaminofluorene⟩. Will grow Jensen sarcoma & Yoshida ascites tumors, and 75% will grow hepatomas 7974 & 130. Host for mammary tumor BICR/MI, Harderian gland carcinoma 2226, carcinomas 338 & 343, osteogenic sarcoma 344, & sarcoma E-2730. [42]
NBR ⟨NIH black⟩	Black selfed [*a*]	Fx + F2	Poiley, 1966, from heterogeneous stock; to PIT, 1977	PIT	Good reproduction. Small; active. S5B may be closely related (*see* S5B). [76]
NIG-III	[*a, B, C, H,* p^m]	....	From a mating between a wild rat trapped in Misima City and Castle's black rat	MS	....
NSD ⟨SD/N⟩	Albino [*c*]	F50	NIH, Bethesda, 1964, from non-inbred Sprague-Dawley stock	N	Intermediate breeding performance [6/12] & litter size [5/12]. High systolic blood pressure [4/17]. [42] Used for studies of mammary carcinomas.

[6] It is not clear whether this is an inbred strain.

continued

Strain ⟨Synonym⟩	Coat Color [Relevant Genes]	Generations Inbred	Origin [Source]	Holder	Characteristics [Rank]
NZR/Gd	Albino [*c*]	F32+	Developed as a subline of AS2 at F32	GD	Tumors—atriocaval epithelial mesotheliomas—of right atrium or inferior vena cava occur in ∿20% of ♂♀ animals >1 yr of age. These tumors are slow growing, but apparently malignant, and closely resemble tumors found in humans. [38]
OKA/Wsl ⟨Okamoto⟩[7]	Albino [*c*]	Fx + F5	Faculty of Medicine, Kyoto, Japan; to J. Roba, Machelen, Belgium, 1970; to H. Bazin, 1971; to PIT, 1975	PIT	....
OM ⟨Osborne-Mendel⟩	Albino [*c*]	F30+	Heston, 1946, from non-inbred Osborne-Mendel stock obtained from J. White; to NIH at F10	N, HAN	Intermediate breeding performance [8/12] & good litter size [4/12] [42]. High incidence of retinal degeneration in animals >10 mo old [104]. High systolic blood pressure [2/17] [42]. Becomes obese on a high fat diet [1/7] [88]. Adrenal cortical tumors, 94%, in rats >18 mo old [42]. Spontaneous thyroid carcinoma, 33% [60]. Mammary tumors 26-30% [31]. Pituitary tumors, 8-18%, in rats 18 mo old [42].
PA ⟨P. A. King albino⟩[8]	Albino [*c*]	F180	King, 1909, from Wistar Institute stock; to Aptekman, 1946, at F135; to Bogden, 1958, at F155. The oldest inbred strain of rats.	CR	Good reproduction. Healthy, vigorous, & vicious. Will grow ascites tumor 9A, carcinoma 5, leukemia LK2, & lymphoma 6. [31]
PETH[9]	Pink-eyed hooded [*a, h, p*]	F39	Bourne, 1938; to Sidman; to NIH, 1966, at F9N1F18.	N	Carries gene for retinal dystrophy ⟨*rdy*, formerly *re*⟩, causing cataracts at 3 mo of age. Secondary effects include unusual lens reflection & persistent hyaloid artery. Low systolic blood pressure [13/17]. [42]
PVG ⟨B(black hooded); HO⟩	Black hooded [*a, h*]	F70+	Kings College of Household Science; to Lister Institute; to Virol; to Glaxo, 1946. Inbred by Glaxo.	IAP, LAC	Good breeding performance [31]. Lower incidence of polyspermia than in WAG [15]. Docile; low defecation & activity in open field in Broadhurst's subline [31]. Resistant to experimental allergic encephalomyelitis [52]. Resistant to induction of autoimmune thyroiditis [83]. Susceptible to infection by *Entamoeba histolytica* [71]. A subline with a chromosome marker analogous to mouse strain CBA/H-T6 has been developed by Howard [51]; PVG is also being used as the background strain in the development of a number of congenic lines [32].
		Fx + F10	As above. [G, 1973]	PIT	

[7] This should probably be regarded as a subline of SHR, although skin grafts between OKA & SHR are rejected after ∿30-45 d. [8] WKA is probably a subline of this strain. [9] This should probably be regarded as a subline of RCS.

continued

Strain ⟨Synonym⟩	Coat Color [Relevant Genes]	Generations Inbred	Origin [Source]	Holder	Characteristics [Rank]
R	Albino [*c*]	F78	Mühlbock, 1947, from a Wistar stock	LN	....
RCS[10]	Pink-eyed hooded [*a, h, p*]	F20+	Developed prior to 1965 by Sidman, from a stock obtained from Sorsby of the Royal College of Surgeons, London [93].	HV	Mean litter size 6.8 ± 0.1[11]; body wt plateaus at 5-6 mo of age at ∿275 g in ♂ & ∿185 g in ♀ [59]. Carries gene for retinal dystrophy; cataracts develop in 24%, & microphthalmia in 4% [93]. A congenic strain RCS-*p*/+ has been developed in which the retinal dystrophy develops more slowly [93].
RHA/N	Albino [*c*]	F23	Bignami; selected for high avoidance conditioning with light as a conditioned stimulus and electric shock as an unconditioned stimulus. This outbred stock to NIH in 1968, where b x s mating was initiated. [12]	N	....
RLA/N	Albino [*c*]	F33	As for RHA, except original stock selected for low avoidance conditioning[12]	N	....
SD	Albino [*c*]	F25	Halberg, Univ. of Minnesota; to NIH at F25	N?	Grows Walker 256 tumor [31]
⟨Sprague-Dawley⟩	Albino	F24	From laboratory of Takeda Pharmaceutical Co. to HOK, 1966, inbreeding started by Makino; to HKM, 1976, at F22	HKM	....
SEL ⟨Selfed 36670⟩	Black [*a*]	F28	Dunning, 1948	DU	Will grow transplantable tumors R-3401 & R-3478 [31]
SHR ⟨Spontaneously hypertensive rat⟩	Albino [*c*]	F23+	Okamoto, 1963, from outbred Wistar Kyoto rats. From a colony of many animals, one ♂ with spontaneous hypertension was mated with a ♀ with elevated blood pressure. B x s mating with continued selection for spontaneous hypertension was then started. [73] A number of sublines with a ten-	HAN, N, ORL	Incidence of hypertension is high and there are no obvious primary organic lesions in kidneys or adrenal glands; hypertension is very severe, with blood pressure frequently >200 mm Hg; there is a high incidence of cardiovascular disease [74]. Genetic analysis indicates that the condition is controlled by 3-4 genetic loci, one of which may be a "major" locus [105]. It is suggested that the blood pressure maintenance in the hypothalamus is deranged; alloxan diabetes further increases blood pressure, but the

[10] Presumed to be very similar to PETH. [11] Plus/minus (±) value is standard error. [12] The original outbred stock and, possibly, some independently derived inbred strains are still in existence so it is important to retain the "N" subline notation in describing this strain.

continued

Strain ⟨Synonym⟩	Coat Color [Relevant Genes]	Generations Inbred	Origin [Source]	Holder	Characteristics [Rank]
			dency to develop cerebrovascular lesions & stroke have also been developed [70].		animals respond to anti-hypertensive drugs [73]. Roba [81] has concluded that the strain is a suitable model for screening drugs for anti-hypertensive action. In young SHR rats the plasma levels of both norepinephrine ⟨noradrenaline⟩ & dopamine β-monooxygenase ⟨dopamine β-hydroxylase⟩ were increased over levels in control WKA rats, but total catecholamines were not significantly different; catecholamine content of the adrenals was reduced [39]. Circulating thyrotropin levels were markedly elevated over those of 2 control strains [103]. There was a reduced ^{131}I metabolism & increased thyroid weight relative to Wistar controls [33]. The "Committee on Care and Use of Spontaneously Hypertensive ⟨SHR⟩ Rats" has issued guidelines for the breeding, care, & use of this strain [22].
S5B	Albino [*a*, *c*]	F38	Poiley, 1955, from a cross of outbred NBR rats with Sprague-Dawley, with 5 generations of backcrossing of the albino gene, followed by sib mating [76]	CR	Good learning [31]. Resistant to dietary induction of obesity [1/7] [88]. High tolerance of toxic compounds (?) [31].
TMB ⟨Tryon Maze Bright; Berkeley S1; S1; TS1⟩	Agouti, Irish? [h^i?]	F20+	Tryon, 1929, from a Berkeley stock selectively bred for good maze learning performance	BRH	Low open-field activity [8/8] [79], [12/12] [45]; low wheel activity in ♂ [10/12] [44]. High voluntary alcohol acceptance concentration [1/4] [84].
TMD ⟨Tryon Maze Dull; Berkeley S3; S3; TS3⟩	Black [*a*]	F20+	*See* TMB, but selected for poor maze performance	BRH	Short latency to emerge into familiar [♂, 3/12; ♀, 4/12] & novel [♂, 3/12; ♀, 2/12] environment [43]; high open-field ambulation [4/12] & low defecation [♂, 9/12; ♀, 10/12] [45]. Low shock avoidance [7/8] [79]. Low voluntary alcohol final acceptance level [4/4] [84]. High brain GABA [3/8] [80].
TO ⟨Tokyo⟩	Albino [*c*]	F38	A breeder in Tokyo; to Hokkaido Univ., 1952	HOK	Will grow Yoshida sarcoma [31]
		F52	As above. Inbreeding started by Makino; to HKM, 1974, at F47.	HKM	
U	....	F48	Zootechnical Inst. Utrecht, 1958	TOR	....
W	Albino	F65	Wistar Institute to Tokyo Univ., 1938; to Hokkaido Univ., 1944; inbreeding started by Makino	HOK, MS	Congenital cleft palate in 0.5% [91]
⟨Wistar/Mk⟩	Albino	F77	As above; to HKM, 1974, at F70	HKM	

continued

Strain ⟨Synonym⟩	Coat Color [Relevant Genes]	Generations Inbred	Origin [Source]	Holder	Characteristics [Rank]
WA	Albino [*a*, *c*, *h*]	F66	Wistar stock, to St. Thomas' Hospital; to LAC, 1964, at F43	LAC	High incidence of polycystic nephrosis. Mean life-span: ♂, 645 ± 30 d; ♀, 749 ± 40 d (based on 22 ♂ & 25 ♀ maintained in SPF conditions). ⟦30⟧
WAB ⟨Wistar albino Boots⟩	Albino	F81	From same stock as WAG, but separated in 1926, prior to inbreeding	BELL	Benign thymoma in 23% of individuals >2 yr of age, with incidence in castrated ♂ of 50%, in spayed ♀ of 57% ⟦49⟧
WAG ⟨Wistar albino Glaxo⟩	Albino [*a*, *c*, *h*; *A*, *c*, *h*; or *A*, *c*, *H*, depending on subline] 13/	F101	A. L. Bacharach, 1924, from Wistar stock	LAC, ORL	Higher incidence of polyspermia than in PVG ⟦15⟧. Short latency to emerge from cage into familiar environment [♂, 4/12; ♀, 3/12] & novel environment [♂, 2/12; ♀, 3/12] ⟦43⟧; high open-field ambulation [2/12] ⟦45⟧; [1/8] ⟦79⟧; low open-field defecation [7/8] ⟦79⟧; high wheel activity [2/12] ⟦44⟧; high shock-avoidance learning [1/8] ⟦79⟧. Low brain GABA [8/8] ⟦80⟧. Susceptible to iron deficiency ⟦31⟧. Endocardial disease 4% at avg age of 31 mo ⟦13⟧. Resistant to experimental allergic encephalomyelitis ⟦52⟧. Susceptible to induction of autoimmune thyroiditis ⟦83⟧. Some sublines may carry recessive gene *dx*, preventing anaphylactoid reaction to dextran ⟦47⟧. Mean life-span: ♂, >22 mo ⟦12⟧; ♀, 31 mo ⟦10⟧, >31 mo ⟦12⟧. Incidence of tumors: pituitary adenoma, 69% of ♀ ⟦10⟧; chromophobe adenoma of pituitary, 68% ⟦12⟧; thyroid medullary carcinoma, 40% ⟦10⟧, 27% ⟦12⟧; adrenal cortical adenoma, 29% ⟦10, 12⟧; fibroadenoma of breast, 21% ⟦10⟧; 19 other types of tumors found in 290 animals ⟦10⟧; pheochromocytoma, 2% ⟦12⟧; pancreatic islet cell adenoma, 1% ⟦12⟧. Further details of aging colony given by Hollander ⟦50⟧.
WAG/Rij	Albino	Fx + F12	[Štark, CUB, 1970]	MAX	
WE/Cpb ⟨CPB-WE⟩	Light beige	F38	Outcross involving strains B, WAG, & others, 1956, followed by inbreeding	CPB	Relatively aggressive ⟦102⟧
WF ⟨Wistar Furth⟩	Albino [*c*]	Fx + F12	J. Furth, 1945, from a commercial Wistar stock, in an attempt to develop a rat strain with high incidence of leukemia. CR to PIT, 1971.	PIT	Strain carries a distinctive heteropyknotic Y chromosome, which may be used as a cellular marker ⟦107⟧. Low serum growth hormone [5/5] ⟦28⟧. Resistant to adrenal regeneration hypertension ⟦69⟧. Mean life-span: ♂, 23 mo; ♀, 21 mo ⟦67⟧. Incidence of leukemia 15-22%, characterized by unusual type of mononuclear cell containing reddish granules; administration of 3-methylcholanthrene to young increased leukemia, but X rays decreased incidence ⟦67⟧. Carcinoma of colon

13/ The presence of different coat color alleles implies that this strain, or some sublines, may have become genetically contaminated at some point in the past. It is therefore important that the subline be stated carefully in published work. Most common sublines are *a*, *c*, *h*.

continued

Strain ⟨Synonym⟩	Coat Color [Relevant Genes]	Generations Inbred	Origin [Source]	Holder	Characteristics [Rank]
					found in 8/12 ♂, 8/29 ♀ ⟦66⟧. Tumors in ♀ include: pituitary, 27%; mammary, 21%; leukemia (generalized), 9%; malignant lymphoma, 7%; adrenal, 3%; lipomas, 3%; and unclassified, 4% ⟦55⟧.
WKA[14] ⟨Wistar King A⟩	Albino	F201	King, 1909, from Wistar Institute stock; to Aptekman, 1946, at F135; to HOK, 1953, at F148	HOK, MS	Congenital clubfoot & polydactyly, 2% ⟦91⟧
		F211	As above; to HKM, 1974, at F206	HKM	
		F205 + F9	As above. [HOK, 1973]	PIT	
WKY/N	Albino [*c*]	F15[3]	Outbred Wistar stock from Kyoto School of Medicine; to NIH, Bethesda, 1971. Inbred as a normotensive control strain for SHR ⟦42⟧.	N	....
WM	Albino	F60	From WIS to Tokyo Univ., 1938; to HOK, 1944; to MS, 1951	MS	....
WN ⟨Inbred Wistar[15]⟩	Albino [*c*]	F34	Heston, 1942, from Wistar stock of Nettleship; to NIH, 1950, at F15	N	Intermediate breeding performance [7/12] & litter size [7/12] ⟦42⟧. Low systolic blood pressure [13/17] ⟦42⟧. Squamous cell hyperplasia & metaplasia of the thyroid has been found ⟦31⟧. Incidence of spontaneous benign & malignant mammary tumors, 30-50%; anterior pituitary tumors, 0-35% at <18 mo of age, but 40-93% at >18 mo old; tumors of adrenal cortex & medulla, 0-10% at <18 mo old, but adrenal cortical tumors, 25-50% at >18 mo old; no tumors of uterus or interstitial cells found at any age ⟦31⟧.
WR	....	F30+	Sykora, Rosice ⟦100⟧. No further information.	....	....
YO	Albino	F16+	CR, 1972; to PIT	PIT	....
Y59	....	....	Developed in Zagreb, Yugoslavia	?	....
Z61	Albino [*a*, *c*, *h*]	F70	Curtis at Columbia Univ. Institute for Cancer Research, 1920	DU	Susceptible to cysticercus infection. Susceptible to estrogen-induced tumors, & tumors induced by *N*-2-fluorenylacetamide ⟨2-acetylaminofluorene⟩. Will grow Jensen sarcoma, R-3449, & R-92. ⟦31⟧

[3] Only part-inbred. [14] This should probably be considered a subline of PA. [15] There may be a number of so-called "Inbred Wistar" stocks which are not WN.

continued

66. INBRED STRAINS: RAT

Contributors: Michael F. W. Festing; Thomas J. Gill III and H. W. Kunz; Eberhard Günther; M. Aizawa; Dietrich Götze

References

1. Armstrong, J. M., et al. 1976. Nature (London) 260: 582-586.
2. Bailey, D. W. 1971. Transplantation 11:325-327.
3. Bazin, H. 1974. Ann. Immunol. (Inst. Pasteur) 125C: 277-279.
4. Bazin, H. 1977. Rat News Lett. 1:27.
5. Bazin, H., and B. Platteau. 1976. Immunology 30: 679-683.
6. Bazin, H., et al. 1972. Int. J. Cancer 10:568-580.
7. Bazin, H., et al. 1973. J. Natl. Cancer Inst. 51:1359-1361.
8. Bazin, H., et al. 1974. Immunology 26:713-723.
9. Billingham, R. E., and W. K. Silvers. 1959. Transplant. Bull. 6:399-406.
10. Boorman, G. A., and C. F. Hollander. 1973. J. Geron-tol. 28:152-159.
11. Boorman, G. A., and C. F. Hollander. 1974. J. Natl. Cancer Inst. 52:1005-1008.
12. Boorman, G. A., et al. 1972. Arch. Pathol. 94:35-41.
13. Boorman, G. A., et al. 1973. Ibid. 96:39-45.
14. Borodin, P. M. Unpublished. Institute of Cytology and Genetics, Academy of Sciences of the USSR, Siberian Branch, Novosibirsk, 1978.
15. Braden, A. W. H. 1958. Fertil. Steril. 9:243-246.
16. Broadhurst, P. L. 1975. Behav. Genet. 5:299-319.
17. Burek, J. D., and C. F. Hollander. 1975. Annu. Rep. 1975. Organ. Health Res., pp. 235-237.
18. Burek, J. D., and C. F. Hollander. 1975. Ibid., pp. 238-241.
19. Burek, J. D., and C. F. Hollander. 1977. J. Natl. Cancer Inst. 58:99-105.
20. Burek, J. D., et al. 1976. Ibid. 57:549-554.
21. Cohen, B. J., et al. 1970. Lab. Anim. Care 20:489-493.
22. Committee on Care and Use of Spontaneously Hypertensive (SHR) Rats. 1976. ILAR News 19:G1-G20.
23. Cramer, D. V., and T. J. Gill, III. 1975. Teratology 12:27-32.
24. Cramer, D. V., et al. 1974. J. Immunogenet. 1:421-428.
25. Davey, F. R., and W. C. Moloney. 1970. Lab. Invest. 23:327-334.
26. Druckrey, H. 1971. Arzneim. Forsch. 21:1274-1278.
27. Dunning, W. F., et al. 1947. Cancer Res. 7:511-521.
28. Esber, H. J., et al. 1974. Proc. Soc. Exp. Biol. Med. 146:1050-1053.
29. Festing, M. F. W. Unpublished. MRC Laboratory Animals Centre, Carshalton, Surrey, England, 1978.
30. Festing, M. F. W., and D. K. Blackmore. 1971. Lab. Anim. 5:179-192.
31. Festing, M. F. W., and J. Staats. 1973. Transplantation 16:221-245.
32. Ford, W. L. Unpublished. Univ. Manchester, Dep. Pathology, England, 1978.
33. Fregly, M. J. 1975. Proc. Soc. Exp. Biol. Med. 149: 124-132.
34. Friedman, I., et al. 1970. Experientia 26:1143-1145.
35. Gasser, D. L., et al. 1975. J. Immunol. 115:431-433.
36. Gill, T. J., III, et al. 1977. Transplant. Proc. 9:567-569.
37. Gold, G., and C. C. Widnell. 1975. Biochem. Pharmacol. 24:2105-2106.
38. Goodall, C. M., et al. 1975. J. Pathol. 116:239-252.
39. Grobecker, H., et al. 1975. Nature (London) 258: 267.
40. Hajdu, A., and G. Rona. 1969. Experientia 25:1325-1327.
41. Hall, C. E., et al. 1976. Life Sci. 18:1001-1008.
42. Hansen, C. T., et al. 1973. U.S. Dep. HEW Publ. (NIH) 74-606.
43. Harrington, G. M. 1971. Psychonom. Sci. 23:348-349.
44. Harrington, G. M. 1971. Ibid. 23:363-364.
45. Harrington, G. M. 1972. Ibid. 27:51-53.
46. Harrington, G. M. Unpublished. Univ. Northern Iowa, Dep. Psychology, Cedar Falls, 1978.
47. Harris, J. M., et al. 1963. Genet. Res. 4:346-355.
48. Heslop, B. F., and E. L. Phelan. 1973. Lab. Anim. 7: 41-46.
49. Hinsull, S. M., and D. Bellamy. 1977. J. Natl. Cancer Inst. 58:1609-1614.
50. Hollander, C. F. 1976. Lab. Anim. Sci. 26:320-328.
51. Howard, J. C. 1971. Transplantation 12:95-97.
52. Hughes, R. A. C., and J. Stedronska. 1972. Immunology 24:879-884.
53. Ishibashi, M. 1976. Teratology 14:242.
54. Jacobs, B. B., and R. A. Huseby. 1967. J. Natl. Cancer Inst. 39:303-307.
55. Kim, U., et al. 1960. Ibid. 24:1031-1055.
56. Knox, W. E. 1977. Rat News Lett. 1:26.
57. Kren, V., et al. 1970. Neoplasma 17:329-337.
58. Kunz, H. W., and T. J. Gill, III. 1974. J. Immunogenet. 1:413-420.
59. La Vail, M. M., et al. 1975. J. Hered. 66:242-244.
60. Lindsay, S., et al. 1968. Arch. Pathol. 86:353-364.
61. Lubaroff, D. M., et al. 1977. J. Natl. Cancer Inst. 58: 1677-1689.
62. Martin, J. B., et al. 1974. Endocrinology 94:1359-1363.
63. Matthies, E., and W. Ponsold. 1973. Z. Krebsforsch. 80:27-30.
64. Maekawa, A., and S. Odashima. 1975. J. Natl. Cancer Inst. 55:1437-1445.

continued

65. McFarlin, D. E., et al. 1975. J. Immunol. 115:1456-1458.
66. Miyamoto, M., and S. Takizawa. 1975. J. Natl. Cancer Inst. 55:1471-1472.
67. Moloney, W. C., et al. 1969. Cancer Res. 29:938.
68. Moloney, W. C., et al. 1970. Ibid. 30:41-43.
69. Molteni, A., et al. 1975. Proc. Soc. Exp. Biol. Med. 150:80-84.
70. Nagaoka, A., et al. 1976. Am. J. Physiol. 230:1354-1359.
71. Neal, R. A., and W. G. Harris. 1975. Trans. R. Soc. Trop. Med. Hyg. 69:429-430.
72. Noble, B., et al. 1976. J. Immunol. 117:1447-1455.
73. Okamoto, K. 1969. Int. Rev. Exp. Pathol. 7:227-270.
74. Okamoto, K., et al. 1973. Clin. Sci. Mol. Med. 45: 11S-14S.
75. Page, J. G., and E. S. Vesell. 1969. Proc. Soc. Exp. Biol. Med. 131:256-261.
76. Palm, J., and G. Black. 1971. Transplantation 11: 184-189.
77. Perlik, F., and Z. Zidek. 1974. Z. Immunitaetsforsch. Allerg. Klin. Immunol. 147:191-193.
78. Phelan, E. L. 1968. N. Z. Med. J. 67:334.
79. Rick, J. T., and D. W. Fulker. 1972. Prog. Brain Res. 36:105-112.
80. Rick, J. T., et al. 1971. Brain Res. 32:234-238.
81. Roba, J. L. 1976. Lab. Anim. Sci. 26:305-319.
82. Rokhlin, O. V., and R. S. Nezlin. 1973. Eur. J. Immunol. 3:732-738.
83. Rose, N. R. 1975. Cell. Immunol. 18:360-364.
84. Russell, K. E., and M. H. Stern. 1973. Physiol Behav. 10:641-642.
85. Sacksteder, M. R. 1976. J. Natl. Cancer Inst. 57:1371-1373.
86. Sass, B., et al. 1975. Ibid. 54:1449-1456.
87. Sato, S. Unpublished. Nippon Rat Co., Ltd., Saitama, Japan, 1978.
88. Schemmel, R., et al. 1970. J. Nutr. 100:1041-1048.
89. Shain, S. A., et al. 1975. J. Natl. Cancer Inst. 55: 177-180.
90. Shearer, D., et al. 1973. Lab. Anim. Sci. 23:662-664.
91. Shoji, R. 1977. Proc. Jpn. Acad. 53:54-67.
92. Shoji, R., and M. Harata. 1977. Lab. Anim. 11:247-249.
93. Sidman, R. L., and R. Pearlstein. 1965. Dev. Biol. 12.93-116.
94. Silverman, D. A., and N. R. Rose. 1975. J. Immunol. 114:145-147.
95. Simpson, F. O., et al. 1973. Clin. Sci. Mol. Med. 45: 155-215.
96. Starace, V., and P. Querinjean. 1975. J. Immunol. 115:59-62.
97. Štark, O., and V. Kren. 1969. Transplantation 8: 200-203.
98. Štark, O., and I. Zeiss. 1970. Z. Versuchstierk. 125: 27-40.
99. Štark, O., et al. 1968. Folia Biol. (Prague) 14:169-175.
100. Štark, O., et al. 1968. Ibid. 14:425-432.
101. Tada, N., et al. 1974. J. Immunogenet. 1:265-275.
102. Van Vliet, J. C. J. 1977. Rat News Lett. 1:63.
103. Werner, S. E., et al. 1975. Proc. Soc. Exp. Biol. Med. 148:1013-1017.
104. Von Sallmann, L., and P. Grimes. 1972. Arch. Ophthalmol. 88:404-411.
105. Yen, T. T., et al. 1974. Heredity 33:309-316.
106. Zeiss, I. M. 1966. Transplantation 4:48-55.
107. Zieverink, W. D., and W. C. Moloney. 1965. Proc. Soc. Exp. Biol. Med. 119:370-373.

67. CONGENIC STRAINS: RAT

For full name of holders, *see* list of HOLDERS at front of book.

Differential Locus ⟨Synonym⟩	Congenic Strain	Inbred Partner ⟨Synonym⟩	Donor Strain ⟨Synonym⟩	Generations
Maintained by CUB, MAX, PIT[1]				
Ag-B2 ⟨*H-1*^{w}⟩	LEW.1W[3]	LEW	WP	N6F22
Ag-B3 ⟨*H-1*^{n}⟩	BP.1N[2]	BP	BN	N10F5
	LEW.1N[3]	LEW	BN	N8F30
Ag-B4 ⟨*H-1*^{a}⟩	LEW.1A[3]	LEW	AVN	N4F30
	LEW.1A[3]	LEW	AVN	N4F12N4F18
Ag-B5 ⟨*H-1*^{c}⟩	LEW.1C[2]	LEW	AUG	N11F5
Ag-B6 ⟨*H-1*^{b}⟩	BN.1B[2]	BN	BP	N9F4
	LEW.1B[3]	LEW	BP	N5F25

[1] Unless otherwise indicated. [2] Not maintained by MAX or PIT. [3] Not maintained by PIT.

continued

Differential Locus ⟨Synonym⟩	Congenic Strain	Inbred Partner ⟨Synonym⟩	Donor Strain ⟨Synonym⟩	Generations
Ag-B9 ⟨*H-1*^{d}⟩	LEW.1D	LEW	BD V	N7F31
Ag-B10 ⟨*H-1*^{f}⟩	LEW.1F	LEW	AS2	N9F12
Maintained by OSU				
Ag-B1 ⟨*H-1*^{l}⟩	AS2.AS	AS2	AS	G13F2
Ag-B10 ⟨*H-1*^{f}⟩	AS.AS2	AS	AS2	G13F2
Maintained by PIT				
Ag-B1 ⟨*H-1*^{l}⟩	BN.LEW	BN	LEW	N12F2
Ag-B2 ⟨*H-1*^{w}⟩	BN.WF	BN	WF	N12F1
	BN.YO	BN	YO38366	N11
Ag-B3 ⟨*H-1*^{n}⟩	AUG.BN	A28807 ⟨AUG28807⟩	BN	N12F2
	DA.B3	DA	B3	N9
	LEW.BN	LEW	BN	N12F2
Ag-B4 ⟨*H-1*^{a}⟩	BN.DA	BN	DA	N12F2
Ag-B5 ⟨*H-1*^{c}⟩	BN.AUG	BN	A28807 ⟨AUG28807⟩	N12F5
Ag-B6 ⟨*H-1*^{b}⟩	BN.BUF	BN	BUF	N12F1
Ag-B7 ⟨*H-1*^{g}⟩	BN.KGH	BN	KGH	N13F1
	PVG.KGH	PVG	KGH	N12F3
Ag-B8 ⟨*H-1*^{k}⟩	BN.WKA	BN	WKA	N12F1
	PVG.WKA	PVG	WKA	N12F3
Ag-B13 ⟨*H-1*^{m}⟩	PVG.MNR	PVG	MNR	N9
Ag-C1	PVG.AUG	PVG	A28807 ⟨AUG28807⟩	N7
Ag-C2	AUG.PVG	A28807 ⟨AUG28807⟩	PVG	N7
MLR-4	BN.B3	BN	B3	N6
Maintained by WIS				
Ag-B1 ⟨*H-1*^{l}⟩	AUG.B1F	A28807 ⟨August-28807⟩	F344/Mai ⟨Fischer-344/Mai⟩	N5
	BN.B1L	BN	LEW/Mai	N6
Ag-B2 ⟨*H-1*^{w}⟩	BN.B2/P	BN	WF ⟨WIF⟩	N11F7
Ag-B3 ⟨*H-1*^{n}⟩	AUG.B3/P	A28807 ⟨August-28807⟩	BN	N11F3
	F.B3	F344	AUG.B3	N4
Ag-B4 ⟨*H-1*^{a}⟩	BN.B4/P	BN	DA	N12F3
	F.B4	F344	DA	N6
Ag-B5 ⟨*H-1*^{c}⟩	BN.B5	BN	A28807 ⟨August-28807⟩	N3
	F.B5	F344	A28807 ⟨August-28807⟩	N5
Ag-B6 ⟨*H-1*^{b}⟩	BN.B6	BN	ALB ⟨Albany⟩	N3
Ag-C2	BN.B2-C2	BN.B2/P	DA	N4
Maintained by WSL				
Ag-B8 ⟨*H-1*^{k}⟩	LOU/C.Ag-B8 (OKA)	LOU/C	OKA	N4
Iκ(1b) ⟨Locus for kappa light chain⟩	LOU/C.Iκ(OKA)	LOU/C	OKA	N7
Iα(1.) ⟨Locus for Ig heavy chain⟩	LOU/C.IH(AUG)	LOU/C	AUG	N7
	LOU/C.IH(AxC)	LOU/C	AxC	N7
Iα(1.) and Iγ2b(1b) ⟨Loci for Ig heavy chains⟩	LOU/C.IH(OKA)	LOU/C	OKA	N10

Contributors: H. W. Kunz and Thomas J. Gill III; Eberhard Günther; Dietrich Götze; Michael F. W. Festing

Gene Symbol	Gene Name	Main Effect or Area of Action	Reference
Ag-A	Agglutinogen-A	Antigen production	30
Ag-C	Agglutinogen-C	Antigen production	30
Ag-D	Agglutinogen-D	Antigen production	30
Ag-E	Histocompatibility	Antigen production	28
Ag-F	Lymphocytotoxin	Antigen production	5
Alp	Alkaline phosphatase	Electrophoretic variation	13,33
am	Anemia-2	Hematopoiesis	35
an	Anemia-1	Hematopoiesis	32
at	Atrichosis	Hypotrichosis	20
Ca	Cataract	Eye anomaly	32
cat	Cataract lens	Eye anomaly	21
cb	Chubby	Skeleton	38
cp	Corpulent	Physiology	16
cr	Crawler	Behavior	38
ct	Cataractous	Eye anomaly	36
Cu-1	Curly-1	Hair anomaly	32
Cu-2	Curly-2	Hair anomaly	32
cw	Cowlick	Trichonosis	32
di	Diabetes insipidus	Physiology	34
dw-1	Dwarf-1	Body size	32
dw-2	Dwarf-2	Body size	32
dx	Anaphylactoid reaction	Physiology	32
Es-1	Esterase-1	Electrophoretic variation	25
Es-2	Esterase-2	Electrophoretic variation	8
Es-3	Esterase-3	Electrophoretic variation	25
Es-4	Esterase-4	Electrophoretic variation	25
Es-5	Esterase-5	Electrophoretic variation	25
Es-12	Esterase-12	Electrophoretic variation	13
Esc	Esterase concentration	Electrophoretic variation	41
fa	Fatty	Physiology	32
fz	Fuzzy	Hypotrichosis	29,33
Gl-1	Plasma protein	Electrophoretic variation	11
H-1	Histocompatibility-1	Antigen production	18
H-2	Histocompatibility-2	Antigen production	18
H-3	Histocompatibility-3	Antigen production	18
H-4	Histocompatibility-4	Antigen production	18
H-5	Histocompatibility-5	Antigen production	18
H-Y	Histocompatibility-Y	Antigen production	12
Hbb	Hemoglobin β-chain	Hematopoiesis	6
hd	Hypodactyly	Skeleton	26
he	Hematoma	Vascular system	32
hr	Hairless	Hypotrichosis	32
hy	Hypotrichosis	Hypotrichosis	24
Hyp-1	Hypertension-1	Physiology	31
in	Incisorless	Skeleton	32
Ir-1	Immune response-1	Immunology	33
Ir-2	Immune response-2	Immunology	33
Ir-BGG	Immune response to bovine gamma globulin	Immunology	37
Ir-EAE	Immune response to experimental allergic encephalomyelitis	Immunology	9
Ir-GA	Immune response to poly(Glu, Ala)	Immunology	2
Ir-GT	Immune response to poly(Glu, Tyr)	Immunology	2
Ir-LDHA	Immune response to lactate dehydrogenase	Immunology	40
Ir-SRBC	Immune response to sheep erythrocytes	Immunology	37
Ix(1a)	Immunoglobulin-x	Immunology	3

continued

68. MUTANT GENES: RAT

Gene Symbol	Gene Name	Main Effect or Area of Action	Reference
j	Jaundice	Physiology	32
k	Kinky	Trichonosis	32
lg	Grüneberg's lethal	Skeleton	32
lk	Kon's lethal	Growth	32
lx	Polydactyly-luxate	Skeleton	17
Mi	Microphthalmia	Eye anomaly	39
mk	Masked	Trichonosis	14
n	Naked	Hypotrichosis	32
Ne	Hydronephrosis	Physiology	22
op	Osteopetrosis	Skeleton	27
Pgd	Phosphogluconate dehydrogenase	Electrophoretic variation	15
rdy	Retinal dystrophy	Eye anomaly	19,32
Re	Rex	Trichonosis	33
RI-1	Immunoglobulin-1	Immunology	11
rt	Runt	Body size	38
Rw	Warfarin resistance	Physiology	11
sb	Stubby	Skeleton	38
Sh	Shaggy	Trichonosis	32
sk	Skinny	Body size	38
sn	Spinner	Behavior	33
sp	Spastic	Behavior	38
sr	Shaker	Behavior	32
st	Stub	Skeleton	32
Svp-1	Seminal vesicle protein	Electrophoretic variation	7
Tfm	Testicular feminization	Physiology	1
tl	Toothless	Skeleton	4
tr	Trembler	Behavior	33
Tu	Renal tumor	Physiology	32
vb	Vibrissaeless	Trichonosis	23
w	Waltzing	Behavior	32
wo	Wobbly	Behavior	32

Contributor: Roy Robinson

References

1. Allison, J. E., et al. 1965. Anat. Rec. 153:85-92.
2. Armerding, D., et al. 1974. Immunogenetics 1:340-351.
3. Bazim, H., et al. 1974. J. Immunol. 112:1035-1041.
4. Cotton, W. R., and J. F. Gaines. 1974. Proc. Soc. Exp. Biol. Med. 146:554-561.
5. DeWitt, C. E., and M. McCullough. 1975. Transplantation 19:310-317.
6. French, C. A., et al. 1971. Biochem. Genet. 5:397-404.
7. Gasser, D. L. 1972. Ibid. 6:61-63.
8. Gasser, D. L., et al. 1973. Ibid. 10:203-217.
9. Gasser, D. L., et al. 1975. J. Immunol. 115:431-433.
10. Greaves, J. H., and P. Ayers. 1967. Nature (London) 215:877-878.
11. Gutman, G. A., and I. L. Weissman. 1971. J. Immunol. 107:1390-1393.
12. Heslop, B. F. 1973. Transplantation 15:31-35.
13. Jimenez-Marin, D. 1974. J. Hered 65:235-237.
14. Kent, R. L., et al. 1976. Ibid. 67:3-5.
15. Koga, A., et al. 1972. Jpn. J. Genet. 47:335-338.
16. Koletsky, S. 1973. Exp. Mol. Pathol. 19:53-60.
17. Kren, V. 1975. Acta Univ. Carol. Monogr. 68.
18. Kren, V., et al. 1973. Transplant. Proc. 5:1463-1466.
19. Lavail, M. M., et al. 1975. J. Hered. 66:242-244.
20. Lefebvres-Boisselot, J. 1968. C. R. Acad. Sci. D267: 1636-1638.
21. Leonard, A., and J. R. Maisin. 1965. Nature (London) 205:615-616.
22. Lozzio, B. B., et al. 1967. Science 156:1742-1744.
23. Lutzner, M. A., and C. T. Hansen. 1975. J. Invest. Dermatol. 65:212-216.
24. McGregor, J. F. 1974. Can. J. Genet. Cytol. 16:341-348.
25. Moutier, R., et al. 1973. Biochem Genet. 8:321-328.
26. Moutier, R., et al. 1973. J. Hered. 64:99-100.
27. Moutier, R., et al. 1974. Ibid. 65:373-375.
28. Palm, J. 1971. Transplant. Proc. 3:169-181.

continued

29. Palm, J. 1975. In R. C. King, ed. Handbook of Genetics. Plenum, New York. v. 4.
30. Palm, J., and G. Black. 1971. Transplantation 11: 184-189.
31. Rapp, J. R., and L. K. Dahl. 1976. Biochemistry 15: 1235-1242.
32. Robinson, R. 1965. Genetics of the Norway Rat. Pergamon, London.
33. Robinson, R. 1978. In H. J. Baker, et al., ed. Biology of the Laboratory Rat. Academic, New York.
34. Saul, G. B., et al. 1968. J. Hered. 59:113-117.
35. Sladic-Simic, D., et al. 1966. Genetics 53:1079-1089.
36. Smith, R. S., et al. 1969. Arch. Ophthalmol. 81:259-264.
37. Tada, N., et al. 1974. J. Immunogenet. 1:265-275.
38. Taylor, B. A. 1968. Genetics 60:559-565.
39. Vasenius, L. 1971. Acta Vet. Scand. 12:109-110.
40. Wurzburg, U. 1971. Eur. J. Immunol. 1:496-497.
41. Zemaitis, M. A., et al. 1974. Biochem. Genet. 12: 295-308.

69. LINKAGE GROUPS: RAT

The *Ag-C* locus is linked to an unsymbolized sex dependent esterase [ref. 5]. The esterases determined by the loci *Es-1*, *Es-3*, *Es-4*, and *Es-5* are closely linked [ref. 8]. The major histocompatibility complex (*H-1*) is closely linked to an immune response complex (*Ir-1*) [ref. 11]. **Crossover Percentage**: All mean values found by weighting by the reciprocal of variance. Plus/minus (±) values are standard errors.

Linkage Group	Gene Loci	Crossover Percentage	Reference
I	*Ag-F—c*	4.4 ± 3.0	2
	c—fz	17.7 ± 3.2	10
	c—Hbb	8.2 ± 1.3	1,3,7
	c—lg	10.6 ± 5.2	10
	c—p	18.4 ± 0.4	10
	c—r	0.3 ± 0.1	10
	c—Rw	21.8 ± 2.5	10
	c—w	42.5 ± 1.5	10
	Hbb—p	24.0 ± 2.6	1,3,7
	he—p	29.2 ± 5.6	10
	lg—p	22.1 ± 1.6	10
	p—r	19.0 ± 1.1	10
	p—Rw	26.7 ± 3.6	10
	p—w	52.0 ± 6.2	10
	r—w	36.8 ± 5.9	10
II	*an—b*	45.2 ± 2.3	10
	an—Cu-1	2.3 ± 0.8	10
	an—in	13.7 ± 2.6	10
	b—Cu-1	45.2 ± 1.0	10
	b—s	7.6 ± 1.2	10
	b—Sh	51.2 ± 4.5	10
	Cu-1—in	14.9 ± 2.7	10
	Cu-1—s	43.8 ± 2.6	10
	Cu-1—Sh	3.9 ± 1.3	10
III	*k—st*	25.9 ± 5.2	10
IV	*a—f*	44.7 ± 1.4	10
	a—Svp-1	6.8 ± 3.2	4
V	*Ag-C—Es-1*	11.2 ± 5.3	5
	Ag-C—Es-2	5.6 ± 3.9	5
	Es-1—Es-2	8.9 ± 3.8	12
	Es-1—Es-3	8.9 ± 3.8	12
	Es-2—Es-4	9.6 ± 1.6	13
	Es-3—Es-4	1.5 ± 0.7	13
VI	*Gl-1—h*	9.7 ± 2.1	9
VII	*H-5—lx*	4.9 ± 1.5	6

Contributor: Roy Robinson

References

1. Brdička, R. 1968. Acta Univ. Carol. Med. 14:93-98.
2. DeWitt, C. E., and M. McCullough. 1975. Transplantation 19:310-317.
3. French, E. A., et al. 1971. Biochem. Genet. 5:395-404.
4. Gasser, D. L. 1972. Ibid. 6:61-63.
5. Gasser, D. L., et al. 1973. Ibid. 10:203-217.
6. Kren, V. 1975. Acta Univ. Carol. Monogr. 68.
7. Moutier, R., et al. 1973. Biochem. Genet. 8:321-328.
8. Moutier, R., et al. 1973. Ibid. 9:109-115.
9. Moutier, R., et al. 1973. Ibid. 10:395-398.
10. Robinson, R. 1972. Gene Mapping in Laboratory Mammals. Plenum, London. pt. B.
11. Robinson, R. 1978. In H. J. Baker, et al., ed. Biology of the Laboratory Rat. Academic, New York.
12. Womack, J. E. 1973. Biochem. Genet. 9:13-24.
13. Womack, J. E., and M. Sharp. 1976. Genetics 82:665-625.

70. RECOMBINANT STRAINS: RAT

The H-1 terminology shown in broken brackets ⟨ ⟩ is synonymous with the Ag-B terminology.

Strain or Haplotype Symbol	Origin	Phenotype				Generations of Inbreeding	Source	Reference
		Ag-B ⟨*H-1*⟩	*Ag-C*	*MLR*	*Ir*			
Naturally Occurring								
B3	Detected by typing with standard reagents and assay techniques	3 ⟨*n*⟩	2	4	GLT: h	F10	N, 1973	4
KGH	Detected by typing with standard reagents and assay techniques	7 ⟨*g*⟩	2	1	GLT: 1	F39 + F17	HVD, 1968	2,3
MNR	Detected by typing with standard reagents and assay techniques	4 ⟨*a*⟩	1	5	GLT: h	F54 + F4	N, 1976	6
Laboratory Derived								
ar1	(LEW.1A x LEW.1W)F_2	4 ⟨*a*⟩	..	*w*	*w*	....	....	5
1R[1/]	(DA x HO)F_1 x HO	4 ⟨*a*⟩	..	*c*	....	....	....	1

[1/] Also includes Ia-like antigens.

Contributors: Thomas J. Gill III and H. W. Kunz

References

1. Butcher, G. W., and J. C. Howard. 1977. Nature (London) 266:362-364.
2. Cramer, D. V., et al. 1974. J. Immunogenet. 1:421-428.
3. Kunz, H. W., and T. J. Gill, III. 1974. Ibid. 1:413-420.
4. Kunz, H. W., et al. 1977. Immunogenetics 5:271-283.
5. Štark, O., et al. 1977. Ibid. 5:183-187.
6. Štark, O., et al. 1978. J. Immunogenet. (in press).

71. CHROMOSOME POLYMORPHISMS: RAT

Combinations of polymorphic chromosome markers in 12 inbred strains of rats are given. **Chromosome No.**: st = subtelocentric; t = telocentric; + = large C-band; – = no, or very small, C-band.

Strain	Chromosome No.					
	3	4	5	7	9	X
ACI	st	–	–	–	+	t
ALB	st	–	–	–	+	t
BUF	st	+	–	+	+	t
F344	st	+	–	–	+	t
KYN	st	–	+	+	–	st
LEJ	st	–	+	+	–	t

Strain	Chromosome No.					
	3	4	5	7	9	X
NIG-III	t	+	–	+	–	t
SD[1/]	t	–	–	–	+	t
T[2/]	t	–	–	–	+	st
TO	st	–	–	+	–	t
W	st	+	–	+	–	st
WKA	st	+	–	+	–	st

[1/] Sprague-Dawley. [2/] Maintained by Dr. K. Takewaki, University of Tokyo.

Contributor: Motomichi Sasaki

Reference: Sasaki, M. 1978. Exp. Anim. (Tokyo) 27(1):123-130.

Data are for *Rattus norvegicus*.

Part I. Modern Cytogenetic Techniques (Conventional Staining)

The use of modern cytogenetic techniques (conventional staining) resulted in improved karyotype analysis and chromosome polymorphism observation in the rat. This progress in karyological studies was associated with three procedures that originated independently: (i) treatment of cells with hypotonic solutions for the purpose of swelling the cells to improve chromosome dispersion; (ii) use of colchicine, or its derivative demecolcine (colcemid), to accumulate mitoses and to clarify chromosome morphology; and (iii) development of cell culture techniques for chromosome analysis.

Karyotypes are those observed in normal somatic or germ cells, where 2n = 42. A = acrocentric; M = metacentric (or median metacentric); S = subtelocentric (or subterminal); SM = submetacentric (or submedian metacentric); T = telocentric (or terminal). Since the terms acrocentric and telocentric are inconsistently used in the literature, T, with respect to autosomes, may stand for either telocentric or acrocentric. **Remarks**: Chromosome Pair No. depends on the system of each author. Figures in heavy brackets are reference numbers.

Strain (Synonym)	Tissue or Cells	Technique	Karyotype	Remarks	Reference
ACI, BUF (Buffalo)	Bone marrow	Hypotonic sodium citrate; air-drying	8T; 5S; 7M; X:A; Y:A	Pair No. 3 subtelocentric	54
	Bone marrow & transplantable hepatomas	Hypotonic sodium citrate; air-drying	8T; 5S; 7M; X:A; Y:A	Among 35 hepatomas, 6 had normal chromosome no., but other 29 were aneuploid	29, 31
ALB (Albany), CW, NIG-IV	Bone marrow	Hypotonic sodium citrate; air-drying	8T; 5S; 7M; X:A; Y:A	Pair No. 3 subtelocentric	54
AO, AU, DA, HO	Lymph nodes	....	8T; 5S; 7M; X:A; Y:A	Pair No. 3 polymorphic in HO strain. *Ag-B* locus examined in pair No. 3.	14
August-Capushon	Bone marrow	Hypotonic sodium citrate; air-drying	8T; 5S; 7M; X:A; Y:A	Chromosomal polymorphism described for X & pair No. 3	48
BN	Bone marrow	Hypotonic sodium citrate; air-drying	8T; 5S; 7M; X:S; Y:A	....	15
	Blood, cultured leukocytes	Cultivation with phytohemagglutinin (PHA); hypotonic treatment of Moorhead, et al. [26]; air-drying	8T; 5S; 7M; X:A; Y:A	....	37
DONRYU	Bone marrow	Hypotonic sodium citrate; air-drying	9T; 4S; 7M; X:A; Y:A	Pair No. 3 acrocentric	54
	Bone marrow & cultured cells from tail tip	Hypotonic sodium citrate; air-drying	9T; 4S; 7M; X:A; Y:A	Pair No. 13 polymorphic	21
	Transplantable myeloblastic leukemias	Hypotonic sodium citrate; air-drying	9T; 4S; 7M; X:A; Y:A	Chromosomes analyzed from 4 sublines of myeloblastic leukemias induced by 1-butyl-1-nitrosourea	16
	Liver & tumors developing after administration of *N*-nitrosodiethylamine (DEN)	Hypotonic sodium citrate; air-drying	Corresponds to standard karyotype [9]	Cytogenetic changes (monosomy, trisomy, markers) occur during DEN administration	12
F344 (Fischer)	Bone marrow	Hypotonic sodium citrate; air-drying	8T; 5S; 7M; X:A; Y:A	Pair No. 3 subtelocentric	54
	Thyroid tumors	Method of Al-Saadi & Beierwaltes [2]	8T; 5S; 7M; X:A; Y:A	Dependent tumors had normal karyotype, but autonomous tumors showed aneuploidy. Pair No. 15 missing in many tumors.	3

continued

72. KARYOLOGY: RAT

Part I. Modern Cytogenetic Techniques (Conventional Staining)

Strain ⟨Synonym⟩	Tissue or Cells	Technique	Karyotype	Remarks	Reference
	Bone marrow	Hypotonic sodium citrate; air-drying	8T; 5S; 7M; X:A; Y:A	No correlation between karyotype & susceptibility to mammary cancer induction	36
	Transitional & autonomous thyroid tumors	Cultivation according to Al-Saadi, et al. [3]	8T; 5S; 7M; X:A; Y:A	In most tumors, one chromosome of the metacentric group is missing	49
	Normal thyroid & dependent, transitional, & autonomous thyroid tumors	Cultivation; hypotonic sodium citrate; air-drying according to Al-Saadi, et al. [3]	8T; 5S; 7M; X:A; Y:A	Dependent, transitional, & autonomous tumors show minor, moderate, & severe chromosome alterations, respectively	4
Holtzman	Bone marrow & cultured mammary cancer cells	Hypotonic treatment according to Tjio, et al. [47]	8T; 5S; 7M; X:A; Y:A	Sex chromosomes polymorphic. Absence of many subterminal chromosomes observed in cultured tumor cells.	35
King-Holtzman, hybrids	Fetal liver	Procedure developed by Moore [27]	8T; 2A; 3S; 7M; X:A; Y:A	Sex chromatin & idiogram from rats with abnormal reproductive organs examined	1
LEW ⟨Lewis⟩	Bone marrow	Hypotonic sodium citrate; air-drying	8T; 5S; 7M; X:A; Y:A	Polymorphic X may be acrocentric or subtelocentric	15
	Blood, cultured leukocytes	Cultivation with phytohemagglutinin ⟨PHA⟩; hypotonic treatment of Moorhead, et al. [26]; air-drying	8T; 5S; 7M; X:S; Y:T	Y larger than pair No. 13. Pair No. 3 seems to have satellite. Pair No. 19 polymorphic.	37
	Cultured lung fibroblasts (∿10-11 passages)	Drying method according to Moorhead, et al. [28]	8T; 5S; 7M; X:S	Arranged by Hungerford & Nowell system. Pair No. 3 subtelocentric. Late replicating X found in ♀ by [H^3]dThd pulse labeling.	8
LE ⟨Long-Evans⟩	Bone marrow	Hypotonic sodium citrate; air-drying	8T; 5S; 7M; X:A; Y:A	Pair No. 3 subtelocentric	54
	Bone marrow, spleen, & induced leukemias	Hypotonic sodium citrate; air-drying	8T; 5S; 7M; X:A; Y:A	C-1 trisomy found in many leukemias induced by 9,10-dimethyl-1,2-benzanthracene ⟨DMBA⟩	40
	Bone marrow, spleen, & induced leukemias	Hypotonic sodium citrate; air-drying	8T; 5S; 7M; X:A; Y:A	C-1 & A-6 trisomies found in many leukemias	18
	Bone marrow	Hypotonic sodium citrate; air-drying	8T; 5S; 7M; X:A; Y:A	Chromosome pairs No. 3 & No. 12 polymorphic. No correlation between karyotype & susceptibility to mammary cancer induction.	36
	Bone marrow, spleen, & induced leukemias	Hypotonic sodium citrate; air-drying	8T; 5S; 7M; X:A; Y:A	Leukemias with C-1 trisomy different in biological character from those with normal karyotype	41
	Induced leukemias	Hypotonic sodium citrate; air-drying	8T; 5S; 7M; X:A; Y:A	C-1 trisomy occurs by specific vulnerability to 9,10-dimethyl-1,2-benzanthracene ⟨DMBA⟩	38, 39
Marshall, OM ⟨Osborne-Mendel⟩	Bone marrow	Hypotonic sodium citrate; air-drying	8T; 5S; 7M; X:A; Y:A	No correlation between karyotype & susceptibility to mammary cancer induction	36
NIG-III	Bone marrow	Hypotonic sodium citrate; air-drying	9T; 4S; 7M; X:A; Y:A	Pair No. 3 acrocentric	54

continued

72. KARYOLOGY: RAT

Part I. Modern Cytogenetic Techniques (Conventional Staining)

Strain ⟨Synonym⟩	Tissue or Cells	Technique	Karyotype	Remarks	Reference
SD ⟨Sprague-Dawley⟩	Muscle cell lines	Hypotonic sodium citrate; air-drying	8T; 2A; 3S; 7M; X:A; Y:A	Diploid karyotype maintained for 380 generations	17
	Blood, cultured leukocytes	Cultivation with phytohemagglutinin ⟨PHA⟩; hypotonic treatment of Moorhead, et al. [26]; air-drying	8T; 5S; 7M; X:A; Y:A	....	37
	Spermatogonia, bone marrow, & cultured kidney cells	Hypotonic sodium citrate; air-drying; autoradiography	....	Late replication observed in Y chromosome & in one X chromosome of ♀	44
	Bone marrow	Hypotonic sodium citrate; air-drying	8T; 5S; 7M; X:A; Y:A	No correlation between karyotype & susceptibility to mammary cancer induction	36
Swiss albino	Bone marrow	Hypotonic sodium citrate; air-drying	9T; 4S; 7M; X:A; Y:A	Pairs No. 3, No. 12, & No. 13 polymorphic	6
Wayne-pink-eyed	Lymphoblasts	Water treatment & squash	8T; 5S; 7M; X:S; Y:A	One of pair No. 2 & pair No. 7 have secondary constriction	32
	Hepatic cells from newborn, original hepatomas, & transplantable tumors	Hypotonic Ringer's solution & squash	8T; 5S; 7M; X:A	Large V-shaped chromosomes observed in primary hepatomas induced by azo dye	53, 55
	Bone marrow	Hypotonic sodium citrate; air-drying	8T; 5S; 7M; X:A; Y:A	Pair No. 3 subtelocentric	54
WF ⟨Wistar-Furth⟩	Peripheral blood & bone marrow cells from normal & leukemic rats	Procedure according to Nowell, et al [30] for blood; procedure according to Hungerford & Nowell [15] for bone marrow	8T; 5S; 7M; X:A; Y:A	No consistent chromosome abnormalities found in leukemic rats. Marker chromosome found in one rat.	10, 25
	Normal bone marrow, cultured embryonic & adult tissues, sarcomas induced by Rous viruses	Procedure according to Tjio & Whang [46] for normal tissue; procedure according to Mitelman & Mark [24] for tumors	8T; 5S; 7M; X:A; Y:A	∿80% of primary sarcomas have normal diploid stemline. Sequential & predetermined karyotype changes occur in early tumor progression.	22
	Sarcomas induced by Rous virus	Procedure according to Mitelman & Mark [24]	8T; 5S; 7M; X:A; Y:A	Close karyotypic relation found between primary & metastatic tumors	23
Wistar	Male germ cells	Water treatment; Feulgen squash	Largest pair, No. 1, subtelocentric. Several pairs of autosomes median or submedian metacentrics.	Y heteropyknotic at metaphaes. Kinetochore of X located at proximal end.	33
	Male germ cells	Water treatment & squash	11A; 4S; 5M; X:A; Y:A	Pairs No. 2 & No. 7 carry nucleolus organizer	34
	Liver from newborn & ascites hepatoma 7974	YO-II solution according to Yosida & Ogawa [56]; squash	11T or S; 9SM or M; X:A; Y:A	48 chromosomes counted in hepatoma; includes dot-like & large V-shaped chromosomes	57

continued

Part I. Modern Cytogenetic Techniques (Conventional Staining)

Strain ⟨Synonym⟩	Tissue or Cells	Technique	Karyotype	Remarks	Reference
	Bone marrow, spleen, lymph nodes, & cells from chloroleukemic rats	Hypotonic sodium citrate; air-drying	8T; 5S; 7M; X:A; Y:A	No consistent or specific chromosome changes in tumors	30
	Bone marrow	Hypotonic sodium citrate; air-drying	9T; 4S; 7M; X:A; Y:A	Chromosome polymorphism described for X & pair No. 3	48
Wistar (inbred)	Bone marrow	Hypotonic sodium citrate; air-drying	8A; 4S; 8M; X: A; Y: A.	Chromosomes divided into 2 groups	11
Wistar (non-inbred)	Bone marrow	Hypotonic sodium citrate; air-drying	8T; 5S; 7M; X:A; Y:A	Polymorphic X may be acrocentric or subtelocentric	15
	Peripheral blood & bone marrow cells from normal & leukemic rats	Procedure according to Nowell, et al. [30] for blood; procedure according to Hungerford & Nowell [40] for bone marrow	8T; 5S; 7M; X:A; Y:A	No consistent chromosome abnormalities found in leukemic rats	10, 25
Wistar/Ms, Wistar/T	Bone marrow	Hypotonic sodium citrate; air-drying	9T; 4S; 7M; X:A; Y:A	Pair No. 3 acrocentric	54
WKA ⟨Wistar-King A⟩	Hepatic cells from newborn; original hepatomas, & transplantable tumors	Hypotonic Ringer's solution & squash	8T; 5S; 7M; X:A	Large V-shaped chromosomes observed in primary hepatomas induced by azo dye	53, 55
	Bone marrow	Hypotonic sodium citrate; air-drying	8T; 5S; 7M; X:A; Y:A	Pair No. 3 subtelocentric	54
	Cultured lung tissue	Water treatment; air-drying	8T; 5S; 7M; X:A; Y:A	Pattern of DNA synthesis examined using [H^3]dThd	42
	Hepatic cells from regenerating liver & primary hepatomas	Hypotonic KCl; air-drying	8T; 5S; 7M; X:A; Y:A	Among 19 hepatomas, 10 euploid, 1 tetroploid, 2 mosaic, and 6 aneuploid	13
YOS	Bone marrow	Hypotonic sodium citrate; air-drying	8T; 5S; 7M; X:A; Y:A	Pair No. 3 subtelocentric	54
(YOS x Wistar/Ms)F_1	Bone marrow	Hypotonic sodium citrate; air-drying	....	Pair No. 3 heteromorphic in telo- & subtelocentrics	54
Laboratory stock	Bone marrow	Hypotonic sodium citrate; air-drying	8T; 5S; 7M; X:A; Y:A	Compared with *R. rattus*. Pair No. 19 (probably No. 13 by other authors) polymorphic	5
Wild	Bone marrow	Hypotonic sodium citrate; air-drying	....	Pair No. 3 heteromorphic in telo- & subtelocentrics	54
Unspecified	Cultured embryonic tissues	Pretreatment with hypotonic Gey's solution & squash	12T; 2S; 5M; X:S; Y:S	X chromosome between 2nd & 3rd autosomes in length	19
	Male germ cells & 4 ascites tumors	Hypotonic Ringer's solution & squash	8T; 7S; 5M; X:S; Y:S	Large V- or J-shaped chromosomes in tumors caused by fusion of some autosomes	50-52
	Spermatogonia, embryonic tissues, regenerating liver, & Yosida sarcoma	Orcein squash	8T; 5S; 7M; X:A; Y:A	A great number of structural rearrangements included in tumor cells	45
	Spermatogonia & other embryonic & adult tissues	Water treatment & squash	5T; 9S; 6M; X:S; Y:M	Distribution of chromosome number from 32-84 in somatic cells	43

continued

Part I. Modern Cytogenetic Techniques (Conventional Staining)

Strain ⟨Synonym⟩	Tissue or Cells	Technique	Karyotype	Remarks	Reference
	Male germ cells & Yosida sarcoma	Water treatment & squash	8T; 5S; 7M; X:S; Y:A	Idiogram of 4 sublines from original Yosida stock tumor	20
	Bone marrow	Diluted phosphate solution; air-drying	8T; 5S; 7M; X:A; Y:A	Pair No. 13 polymorphic in telo- & subtelocentrics	7

Contributor: Toshide H. Yosida

References

1. Allison, J. E., et al. 1965. Anat. Rec. 153:85-92.
2. Al-Saadi, A., and W. H. Beierwaltes. 1966. Cancer Res. 26:676-688.
3. Al-Saadi, A., and W. H. Beierwaltes. 1967. Ibid. 27: 1831-1842.
4. Al-Saadi, A., and G. J. Mizejewski. 1972. Ibid. 32: 501-505.
5. Badr, F. M., and R. S. Badr. 1973. Proc. Egypt. Acad. Sci. 24:147-154.
6. Bhatnagar, V. S. 1976. Cytologia 41:671-677.
7. Bianchi, N. O., and O. Molina. 1966. J. Hered. 57:231-232.
8. Chang, T. H., et al. 1965. Can. J. Genet. Cytol. 7: 571-582.
9. Committee for a Standardized Karyotype of *Rattus norvegicus*. 1973. Cytogenet. Cell Genet. 12:199-205.
10. Dowd, G., et al. 1964. Blood 23:564-571.
11. Fitzgerald, P. H. 1961. Exp. Cell Res. 25:191-193.
12. Hitachi, M., et al. 1974. J. Natl. Cancer Inst. 53:507-516.
13. Hori, S. H., and M. Sasaki. 1969. Cancer Res. 29:880-891.
14. Howard, J. C. 1971. Transplantation 12:95-97.
15. Hungerford, D. A., and P. C. Nowell. 1963. J. Morphol. 113:275-285.
16. Ishidate, M., et al. 1974. J. Natl. Cancer Inst. 53:773-781.
17. Krooth, R. S., et al. 1964. Ibid. 32:1031-1044.
18. Kurita, Y., et al. 1968. Cancer Res. 28:1738-1752.
19. Makino, S., and T. C. Hsu. 1954. Cytologia 19:23-28.
20. Makino, S., and M. Sasaki. 1958. J. Natl. Cancer Inst. 20:465-487.
21. Masuji, H. 1970. Acta Med. Okayama 24:81-91.
22. Mitelman, F. 1971. Hereditas 69:155-186.
23. Mitelman, F. 1972. Ibid. 70:1-14.
24. Mitelman, F., and J. Mark. 1970. Ibid. 65:227-235.
25. Moloney, W., et al. 1965. Blood 26:341-353.
26. Moore, J. E. S., and G. Arnold. 1905. Proc. R. Soc. London B77:563-570.
27. Moore, K. L. 1955. Lancet 2:57-58.
28. Moorhead, P. S., and P. C. Nowell. 1964. Methods Med. Res. 10:310-322.
29. Nowell, P. C., and H. P. Morris. 1969. Cancer Res. 29: 969-970.
30. Nowell, P. C., et al. 1963. J. Natl. Cancer Inst. 30: 687-703.
31. Nowell, P. C., et al. 1967. Cancer Res. 27:1565-1579.
32. Ohno, S., and R. Kinoshita. 1955. Exp. Cell Res. 8: 558-562.
33. Ohno, S., et al. 1958. Cytologia 23:422-428.
34. Ohno, S., et al. 1959. Exp. Cell Res. 16:348-357.
35. Rees, E. D., and D. Mukerjee. 1964. Proc. Soc. Exp. Biol. Med. 117:869-871.
36. Rees, E. D., et al. 1968. Cancer Res. 28:823-830.
37. Rieke, W. O., and R. Schwarz. 1964. Anat. Rec. 150: 383-390.
38. Sugiyama, T. 1971. Proc. Natl. Acad. Sci. USA 68: 2761-2764.
39. Sugiyama, T. 1971. J. Natl. Cancer Inst. 47:1267-1275.
40. Sugiyama, T., et al. 1967. Science 158:1058-1059.
41. Sugiyama, T., et al. 1969. Cancer Res. 29:1117-1124.
42. Takagi, N., and S. Makino. 1966. Chromosoma 18: 359-370.
43. Tanaka, T., and K. Kano. 1957. Proc. Int. Genet. Symp. 1956, pp. 196-201.
44. Tiepolo, L., et al. 1967. Cytogenetics 6:51-66.
45. Tjio, J. H., and A. Levan. 1956. Hereditas 42:218-234.
46. Tjio, J. H., and J. Whang. 1962. Stain Technol. 37:17-20.
47. Tjio, J. H., et al. 1959. Proc. Natl. Acad. Sci. USA 45:1008-1016.
48. Udalova, L. D. 1968. Tsitologiya 10:733-742.
49. Wu, C., et al. 1971. Cancer Res. 31:577-582.
50. Yosida, T. H. 1955. Annu. Rep. Natl. Inst. Genet. Jpn. 5:17-18.
51. Yosida, T. H. 1955. Ibid. 5:66-67.
52. Yosida, T. H. 1955. Proc. Jpn. Acad. 31:237-242.
53. Yosida, T. H. 1957. Proc. Int. Genet. Symp. 1956, pp. 210-215.

continued

Part I. Modern Cytogenetic Techniques (Conventional Staining)

54. Yosida, T. H., and K. Amano. 1965. Chromosoma 16:658-667.
55. Yosida, T. H., and T. Ishihara. 1958. Senshokutai 37-38:1276-1281.
56. Yosida, T. H., and Y. Ogawa. 1956. Annu. Rep. Natl. Inst. Genet. Jpn. 6:20-21.
57. Yosida, T. H., et al. 1960. Jpn. J. Genet. 35:35-40.

Part II. Differential Staining Techniques

Cytogenetic studies of the rat were greatly facilitated by the advent of differential staining techniques by which the banding patterns of chromosomes can be distinguished (*see* Figures 1-4). These new techniques also made possible the staining of constitutive and facultative heterochromatin, and also of sister chromatids in the metaphase. **Technique**: BrdUrd = bromodeoxyuridine; EDTA = ethylenediaminetetraacetic acid; q. m. = quinacrine mustard dihydrochloride. **Remarks**: SCE = sister chromatid exchange.

Strain ⟨Synonym⟩	Tissue or Cells	Technique	Remarks	Reference
ACI	Cultured cells from tail tip	C-banding [27]	C-bands of 9 *Rattus* species compared; C-bands distinct in pairs No. 4, 6, 10, 12, & 14-16	32
AS, LIS ⟨Hooded Lister⟩	Cultured fibroblast cells from embryo	G-banding by trypsin	Banding pattern shows no difference between strains	2
Buf ⟨Buffalo⟩	Cultured cells from embryos & normal cultured hepatocytes from adults	Q-banding by q. m. (*see* Fig. 1 & 2). G-banding by trypsin-EDTA.	A single No. 3 has satellite. Banded chromosomes in malignant liver cells observed.	11
DONRYU	Epithelial cell lines	G-banding by trypsin	Karyotypes of diploid epithelial cell lines similar to normal diploid ones	11
	Cultured epithelial cell lines	G-banding by trypsin	Non-random chromosome aberrations found	9
F344 ⟨Fischer⟩	Cultured cells from embryos & normal cultured hepatocytes from adults	Q-banding by q. m. (*see* Fig. 1 & 2). G-banding by trypsin-EDTA.	No. 3 pair has satellite. Banded chromosomes in malignant liver cells observed.	11
	Normal somatic cells & leukemias induced by viruses & chemicals	Q-banding by q. m. G-banding [23] (*see* Fig. 3 & 4).	Q- & G- bands same as for WKA strain. Banding karyotypes in diploid tumors indistinguishable from normal somatic complex.	15
LE ⟨Long-Evans⟩	Cultured bone marrow cells	Cultured for 26 h in BrdUrd-containing medium, and differential staining of sister chromatids by Giemsa	Technique of differential staining of sister chromatids from bone marrow cells introduced	26
	Cultured embryonic lung cells	Differential staining of sister chromatids by BrdUrd observed by UV optics	Spontaneous SCE frequency, ∿8.5/cell in vitro	3
SD ⟨Sprague-Dawley⟩	Cultured lymphocytes & induced hepatomas	Q-banding by q. m. (*see* Fig. 1 & 2)	Pairs No. 3 & 12 subtelocentrics. Y appearance varies under fluorescent staining. Structural rearrangements in hepatomas observed in Q-bands.	31
	Fibroblast cultures	G-banding [21] (*see* Fig. 3 & 4)	Association of pairs No. 5 & 11 observed in metaphase	22

continued

72. KARYOLOGY: RAT

Part II. Differential Staining Techniques

Strain ⟨Synonym⟩	Tissue or Cells	Technique	Remarks	Reference
	Cultured cells from muscle, cartilage & Novikoff hepatoma	C-banding [25]. G-banding [23] (*see* Fig. 3 & 4).	Number of chromosome pairs do not show obvious bands. C-bands of hepatoma cells similar to those of normal cells. Y appears to be totally heterochromatic.	30
	Cultured cells from fetal tissues	G-banding by trypsin	Similarities of G-bandings in rat & mouse	17
	Cell lines & tumors	G-bandings [2]	Patterns of chromosome aberrations in tumors analyzed	18
WF ⟨Wistar-Furth⟩	Cultured embryonic; cells induced rat sarcomas	G-banding by trypsin + EDTA tetrasodium salt ⟨Versene⟩	Chromosomes grouped into 4 groups (A to D). Marker chromosomes in sarcomas analyzed. Model shown for G-band analysis in solid tumors.	4,5
	Sarcomas induced by 9,10-dimethyl-1,2-benzanthracene ⟨DMBA⟩	G-banding by trypsin + EDTA tetrasodium salt ⟨Versene⟩	Extra A2 chromosome included in 12 of 13 sarcomas induced by ⟨DMBA⟩.	7
	Sarcomas induced by 3-methylcholanthrene ⟨MC⟩ & benzo[*a*]pyrene ⟨BP⟩	G-banding by trypsin + EDTA tetrasodium salt ⟨Versene⟩	10 rat sarcomas induced by MC and 13 induced by BP have same markers	6
	Bone marrow, sarcomas induced by Rous sarcoma virus	G-banding [12]	Banding karyotype normal in sarcoma	13
Wistar	Cultured cells (?)	G-banding by trypsin	G-bands comparable to those found by other authors working on different strains. X chromosome polymorphic.	8
	Femur bone marrow	BrdUrd infused through tail vein for 24 h; bone marrow preparation made. Hypotonic KCl flame-dried and stained with 2-[2-(4-hydroxyphenyl)-6-benzimidazolyl]-6-(1-methyl-4-piperazyl)-benzimidazol-trihydrochloride ⟨33258 Hoechst⟩	Spontaneous SCE frequency in vivo, ∿0.75/cell/cell cycle	29
Wistar/Mk	Cultured lung cells & bone marrow	Q-banding by q. m. (*see* Fig. 1 & 2)	Same banding pattern as for WKA strain	14
WKA ⟨Wistar-King A⟩	Cultured lung cells & bone marrow	Q-banding by q. m. (*see* Fig. 1 & 2)	Same banding pattern as for Wistar/Mk strain	14
	Bone marrow & cultured fibroblasts	Q-banding [16]. G-banding [24] (*see* Fig. 3 & 4). C-banding [28].	Similar bands in 3 *Rattus* species	16
	Cultured cells from adult tail tip	G-banding by sodium dodecyl sulfate	Similarity of G-bands in 13 *Rattus* species	33
	Normal somatic cells & leukemias induced by viruses & chemicals	Q-banding by q. m. G-banding (*see* Fig. 3 & 4).	Q- and G-bands same as for F344 strain. Banding karyotypes in diploid tumors indistinguishable from normal somatic complex.	15
	Cultured fetal tissues & Yosida sarcoma	Q-banding by q. m. G-banding [23] (*see* Fig. 3 & 4).	Banding karyotypes of tumor cells deviated greatly from those of normal somatic cells	20

continued

Part II. Differential Staining Techniques

Strain ⟨Synonym⟩	Tissue or Cells	Technique	Remarks	Reference
Unspecified	Cultured cells	G-banding by trypsin & urea	G-bands in X of mammals similar	19
	Normal somatic cells	Q-banding, C-banding	Autosomes divided into 2 groups by banding patterns: acrocentrics (or subtelocentrics) and metacentrics (or submetacentrics)	1

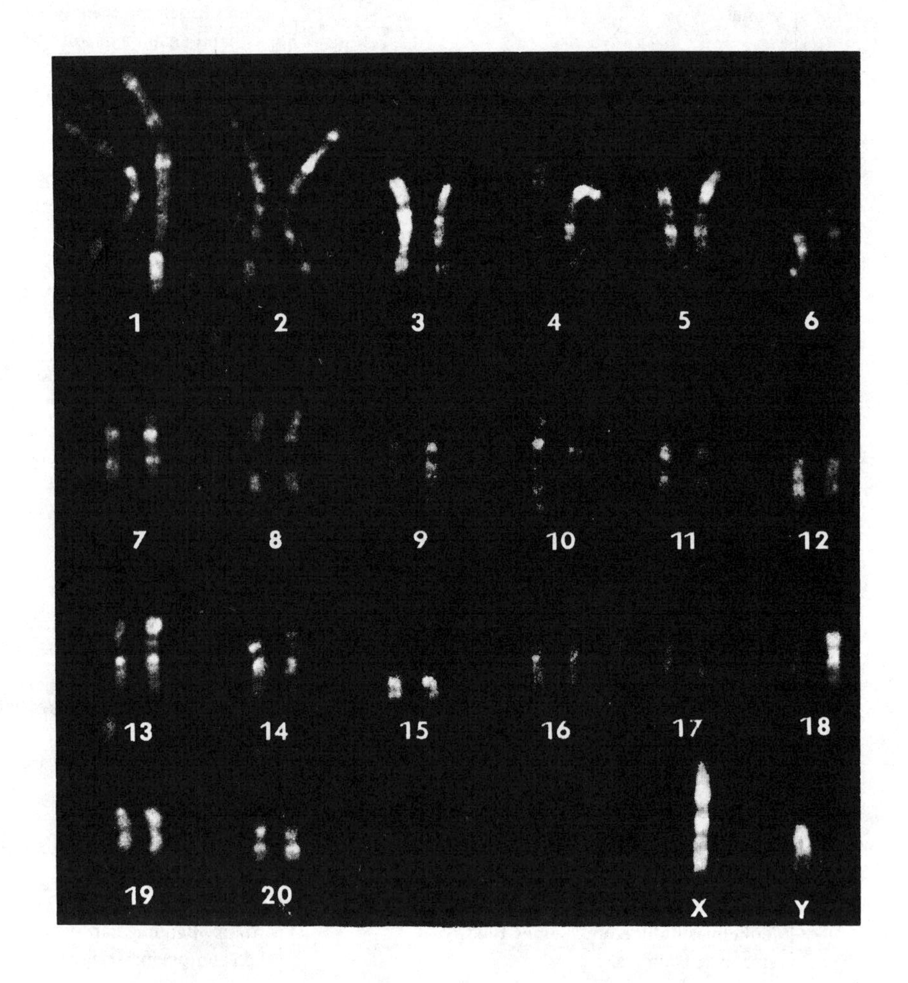

FIGURE 1. QUINACRINE FLUORESCENT BANDING KARYOTYPE OF THE MALE RAT [ref. 1]

continued

Part II. Differential Staining Techniques

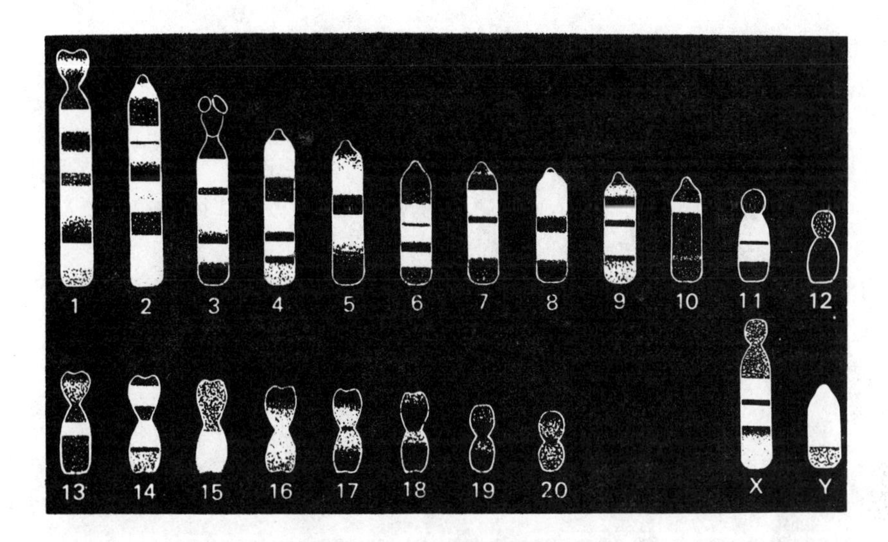

FIGURE 2. IDEOGRAM OF QUINACRINE FLUORESCENT BANDING KARYOTYPE OF THE MALE RAT [ref. 1]

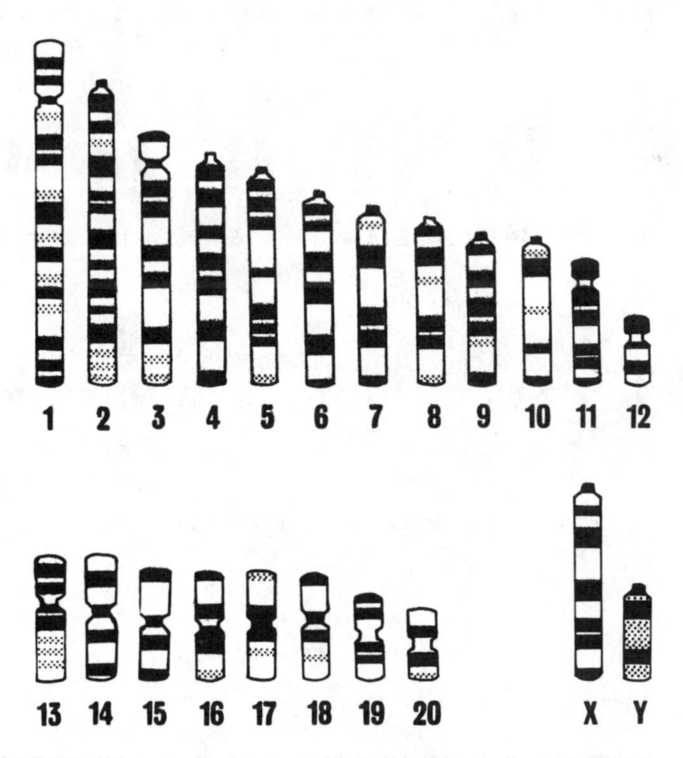

FIGURE 3. IDEOGRAM OF GIEMSA BANDING KARYOTYPE OF THE MALE RAT [ref. 1]

continued

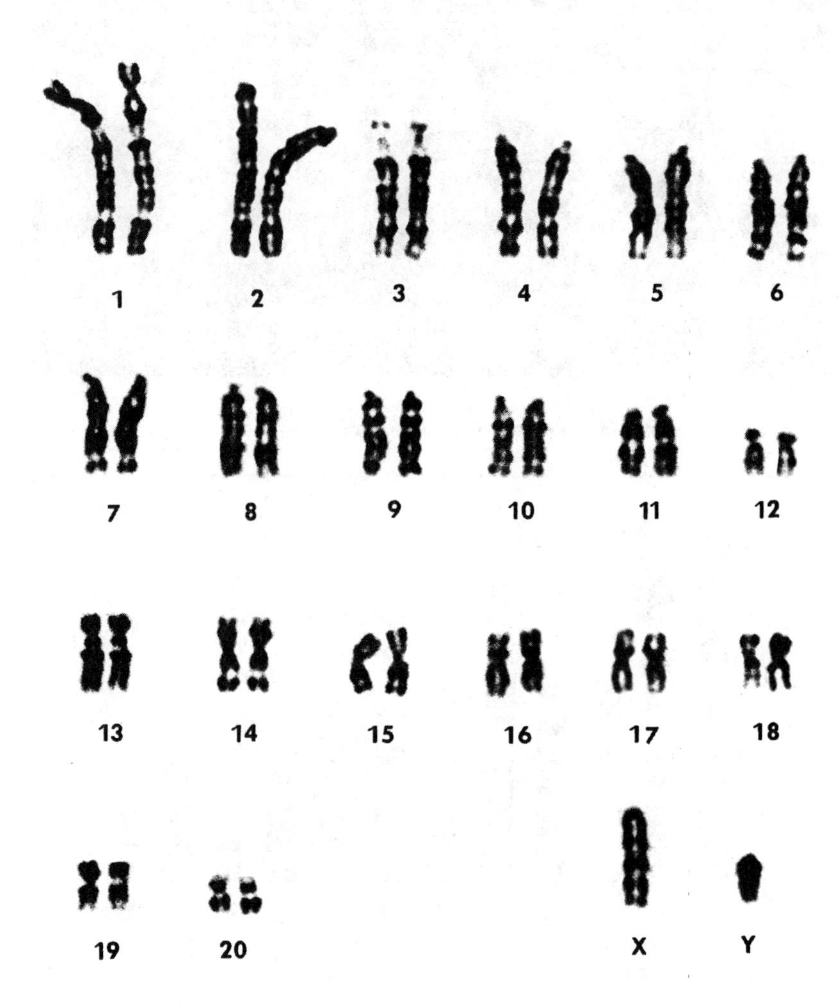

FIGURE 4. GIEMSA BANDING KARYOTYPE OF THE MALE RAT [ref. 1]

continued

72. KARYOLOGY: RAT

Part II. Differential Staining Techniques

Contributor: Toshide H. Yosida

References

1. Committee for a Standardized Karyotype of *Rattus norvegicus*. 1973. Cytogenet. Cell Genet. 12:199-205.
2. Gallimore, P. H., and C. R. Richardson. 1973. Chromosoma 41:259-263.
3. Kato, H. 1977. Int. Rev. Cytol. 49:55-97.
4. Levan, G. 1974. Hereditas 77:37-52.
5. Levan, G. 1974. Ibid. 78:273-290.
6. Levan, G., and A. Levan. 1975. Ibid. 79:161-198.
7. Levan, G., et al. 1974. Ibid. 77:263-280.
8. Manolache, M. 1975. Stud. Cercet. Biol. Ser. Biol. Anim. 27:307-311.
9. Masuji, H. 1974. Gann 65:429-438.
10. Masuji, H., et al. 1974. Acta Med. Okayama 28:281-293.
11. Miller, D. A., et al. 1972. Cancer Res. 32:2375-2382.
12. Mitelman, F., et al. 1974. Scand J. Haematol. 13:87-92.
13. Mitelman, F., et al. 1975. Hereditas 80:291-296.
14. Mori, M., and M. Sasaki. 1973. Chromosoma 40:173-182.
15. Mori, M., and M. Sasaki. 1974. J. Natl. Cancer Inst. 52:153-160.
16. Mori, M., et al. 1973. Jpn. J. Genet. 48:381-383.
17. Nesbitt, M. N. 1974. Chromosoma 46:217-224.
18. Olinici, C. D., and J. A. DiPaolo. 1974. J. Natl. Cancer Inst. 52:1627-1634.
19. Pathak, S., and A. D. Stock. 1974. Genetics 78:703-714.
20. Sasaki, M., et al. 1974. J. Natl. Cancer Inst. 52:1307-1315.
21. Schnedl, W. 1971. Nature (London) 233:93-94.
22. Schnedl, W., and M. Schnedl. 1972. Cytogenetics 11:188-196.
23. Seabright, M. 1971. Lancet 2:971-972.
24. Seabright, M. 1973. Ibid. 1:1249-1250.
25. Stefos, K., and F. E. Arrighi. 1971. Exp. Cell Res. 18:228-231.
26. Sugiyama, T., et al. 1976. Nature (London) 259:59-60.
27. Sugiyama, T., et al. 1969. Cancer Res. 29:1117-1124.
28. Sumner, A. T. 1972. Exp. Cell Res. 75:304-306.
29. Tice, R., et al. 1976. Ibid. 102:426-429.
30. Ünakul, W., and T. C. Hsu. 1972. J. Natl. Cancer Inst. 49:1425-1431.
31. Walman, S. R., et al. 1972. Science 175:1267-1269.
32. Yosida, T. H. 1975. Proc. Jpn. Acad. 51:659-663.
33. Yosida, T. H., and T. Sagai. 1973. Chromosoma 41:93-101.

Part III. Chromosome Numbering Systems: Conventional vs. Differential Staining Techniques

The numbering systems for rat chromosomes by various investigators are not similar because of the difficulty of chromosome pair identification in slides made by conventional staining techniques. With the advent of chromosome banding techniques, it became easy to identify each chromosome pair in the rat genome. (*See* Figures 1-4 in Part II of this table for karyotypes obtained using the quinacrine fluorescent and Giemsa banding techniques.) Almost all chromosome pairs in the rat can be distinguished with these two methods, although identification of pairs No. 16-18 is difficult.

To avoid the confusion of multiple numbering systems for rat chromosomes, a standard system based on G- and Q-banding patterns was proposed by the Committee for a Standardized Karyotype of *Rattus norvegicus* **[1]**. According to the Committee, chromosomes are numbered and arranged in order of decreasing length, with the exception of chromosomes 11 and 12 and the sex chromosomes. The short acrocentric chromosomes 11 and 12 are placed just after the longer acrocentric or telocentric chromosomes. The sex chromosomes are placed apart from the autosomes.

Differential Staining: **Chromosome Length** is expressed as percent of the haploid autosome complement. **Chromosome No. Assignment** is that proposed by the Committee for a Standardized Karyotype of *Rattus norvegicus* **[1]**. **Position of Centromere**: acrocentric includes telocentric (or terminal); metacentric includes median metacentrics; submetacentric includes submedian metacentrics; subtelocentric includes subterminals. Figures in heavy brackets are reference numbers.

Differential Staining		Chromosome No. Assignment by Conventional Staining According to						Position of Centromere
Chromosome No. Assignment	Chromosome Length	Tjio & Levan	Hungerford & Nowell	Krooth, et al.	Yosida & Amano	Kurita, et al.	Yosida	
1	10.62	1	1	1	1	B-1	1	Subtelocentric
2	9.46	2	2	2	2	C-1	2	Acrocentric
3	7.14	4	3	4	3	B-2	3	Subtelocentric (polymorphic)
4	6.98	3	4	3	4	C-2	4	Acrocentric
5	6.38	5		5	5	C-4	5	Acrocentric

continued

72. KARYOLOGY: RAT

Part III. Chromosome Numbering Systems: Conventional vs. Differential Staining Techniques

Differential Staining		Chromosome No. Assignment by Conventional Staining According to						Position of Centromere
Chromosome No. Assignment	Chromosome Length	Tjio & Levan	Hungerford & Nowell	Krooth, et al.	Yosida & Amano	Kurita, et al.	Yosida	
6	5.58	6			6	C-5	6	Acrocentric
7	5.41	7			7	C-6	7	Acrocentric
8	5.02	8			8	C-7	8	Acrocentric
9	4.76	9			9	C-8	10	Acrocentric
10	4.51	10	10	10	10	C-9	11	Acrocentric
11	3.81	14	12	12	14	B-4	12	Subtelocentric (polymorphic ?)
12	2.68	18	13	18	18	B-5	13	Subtelocentric (polymorphic)
13	4.16	12	11	11	12	B-3	9	Submetacentric
14	4.08	11	14	13	11	A-1	14	Metacentric
15	3.90	13	15		13	A-2	15	Metacentric
16	3.55	15	16		15	A-3	16	Metacentric
17	3.51	16	17		16	A-4	17	Metacentric
18	3.36	17	18	17	17	A-5	18	Metacentric
19	2.70	19	19	19	19	A-6	19	Metacentric
20	2.36	20	20	20	20	A-7	20	Metacentric
X	5.73	X	X	X	X	C-3(X)	X	Acrocentric (polymorphic)
Y	3.20	Y	Y	Y	Y	Y	Y	Acrocentric (polymorphic)
[1]	[1]	[5]	[2]	[3]	[7]	[4]	[6]	....

Contributor: Toshide H. Yosida

References

1. Committee for a Standardized Karyotype of *Rattus norvegicus*. 1973. Cytogenet. Cell Genet. 12:199-205.
2. Hungerford, D. A., and P. C. Nowell. 1963. J. Morphol. 113:275-285.
3. Krooth, R. S., et al. 1964. J. Natl. Cancer Inst. 32: 1031-1044.
4. Kurita, Y., et al. 1968. Cancer Res. 28:1738-1752.
5. Tjio, J. H., and A. Levan. 1956. Hereditas 42:218-234.
6. Yosida, T. H. 1973. Chromosoma 41:93-101.
7. Yosida, T. H., and K. Amano. 1965. Ibid. 16:658-667.

73. COAT AND EYE COLOR GENES: RAT

Gene Symbol	Gene Name	Bearing Strains	Reference
A	Agouti	AO, BD I, BD III, BD VIII, BD IX, CPBB, DA, IS, MNR[1/], TMB, WAG[2/]	1
a^m	Agouti-melanic	....	5
a	Non-agouti	ACH, ACI, ACP, AGA, AGUS, ALB, A35322, BDE, BD II, BD IV, BD V, BD VI, BD VII, BD X, BN, BP, BS, CAP, COP, F344, HS, INR, IR, LA/N, LEJ, LEP, LEW, MAXX, MNR[1/], MNRA, MR, MSUBL/Icgn, M520, NBR, NIG III, PETH, PVG, RCS, SEL, S5B, TMD, WA, WAG[2/], Z61	7
B	Non-brown	AGUS, AO, B, BD II, BD X, BUF, DA, F344, LEW, MNR, MR, M520, NIG III, WA, WAG, Z61	1,2
b	Brown	ALB, BN, LA/N	7
C	Non-albino	A990, BD I, BD III, BD IV, BD V, BD VI, BD VII, BD VIII, BD IX, DA, E3, IS, NIG-III TMB, TMD	1,2

[1/] Sublines differ and one colony is segregating at this locus. [2/] Depending on subline.

continued

73. COAT AND EYE COLOR GENES: RAT

Gene Symbol	Gene Name	Bearing Strains	Reference
c^d	Ruby-eyed dilute	IR (?)	7
c^h	Himalayan	....	6
c	Albino	AGUS, AO, AS, AS2, AVN, B, BD II, BD X, BIRMA, BIRMB, BROFO, BUF, CAR, CAS, DONRYU, F344, G/Cpb, GH, K, LEW, MNR, MNRA, MR, M14, M17, M520, NSD, NZR/Gd, OKA/Wsl, OM, PA, R, RHA/N, RLA/N, SD, SHR, S5B, TO, W, WA, WAB, WAG, WF, WKA, WKY/N, WM, WN, Z61	7
d	Dilute	ALB	7
e (?)	Yellow	....	3,8
f	Fawn	....	7
H^{re}	Restricted	....	4
H	Non-hooded	AGUS, BD I, BD VI, BD VII, BD IX, IS, LA/N, NIG-III, WAG[2]	1,2
h^i	Irish	ACI, ACP, BN, BP, TMB	7
h	Hooded	ACH, AO, AUG, A990, A7322, A28807, A35322, BDE, BD II, BD III, BD IV, BD V, BD VIII, BD X, CAP, COP, CPBB, E3, F344, GHA, HS, INR, IR, LEJ, LEP, LEW, MAXX, MNR, MNRA, MR, M520, PETH, PVG, RCS, WA, WAG[2], Z61	7
h^n	Notch	KYN	7
P	Non-pink-eyed dilute	AGUS, AO, BD II, BD IV, BD VI, BD VIII, BD IX, F344, LEW, MNR, MR, M520, WA, WAG, Z61	1,2
p	Pink-eyed dilution	AUG, A7322, BD I, BD III, BD V, BD VII, BD X, PETH, RCS	7
p^m	Ruby-eyed dilution	NIG-III	7
r	Red-eyed dilution	....	7
s	Silvering	....	7
sd	Sand	....	5
wb	White belly	....	8,9

[2] Depending on subline.

Contributors: Roy Robinson; Michael F. W. Festing

References

1. Festing, M. F. W. Unpublished. MRC Laboratory Animals Centre, Carshalton, Surrey, England, 1977.
2. Festing, M. F. W., and J. Staats. 1973. Transplantation 16:221-245.
3. Figala, J. 1963. Vestn. Cesk. Spol. Zool. 27:209-210.
4. Gumbreck, L. G., et al. 1972. J. Exp. Zool. 180:333-349.
5. Macy, R. M., and A. J. Stanley. 1973. J. Hered. 64:98.
6. Moutier, R., et al. 1973. Ibid. 64:303-304.
7. Robinson, R. 1965. Genetics of the Norway Rat. Pergamon, London.
8. Robinson, R. 1978. In H. J. Baker, et al., ed. Biology of the Rat. Academic, New York. pp.
9. Vasenius, L. 1972. Z. Versuchstierkd. 14:142-147.

74. LENGTH AND WEIGHT OF BODY STRUCTURES: RAT

Males and females of the following six strains [ref. 1, 2] were examined: AGUS/Lac, LH/Lac (not previously described), PVG/Lac, WA/Lac, WAG/Lac, and Porton/Lac (a commonly used outbred stock included for comparison). Animals were SPF category 4-star [ref. 4], fed PRD diet sterilized by γ-radiation at 2.5 Mrads. The animal room temperature was 20 ± 2°C, and the lighting cycle was 12 hours light and 12 hours darkness. All animals were weaned at 21 days. At a selected age each rat was killed with an overdose of CO_2 and examined post-mortem. Body weight, body length, and tail length were determined. The animal was then dissected and the various structures removed and weighed. For paired organs, except for lungs, only the left was taken (consult reference 3 for anatomical descriptions). A mean of 18.5 ± 6.7 rats of each sex of each strain was examined, ranging from a mean age of 17.2 ± 3.2 days to

continued

107.1 ± 11.8 days. A curve of the second degree polynomial ($\hat{Y} = a + bX + cX^2$) was fitted to each set of organ weights (for each sex of each strain), and the estimated weight of each organ at a particular age was determined; these weights are presented in the table below. All data were previously unpublished.

Specification	Male					Female				
AGUS/Lac										
Age, d	25	40	55	70	85	25	40	55	70	85
Body weight, g	58.8	135.0	199.3	251.7	292.3	64.1	112.3	150.3	178.2	195.8
Body length, cm	12.86	16.60	19.50	21.55	22.76	13.16	16.24	18.55	20.09	20.85
Tail length, cm	8.87	12.61	15.55	17.69	19.03	9.44	12.84	15.36	17.00	17.76
Tibia length, mm	20.49	26.05	30.62	34.20	36.79	21.02	26.03	30.00	32.93	34.82
Weight of structure, mg										
Brain	1508.0	1675.8	1803.8	1891.9	1940.3	1449.3	1589.5	1692.0	1756.8	1783.9
Eye	74.1	91.5	106.6	119.5	130.2	79.5	94.1	106.3	116.3	124.0
Salivary glands[1]	83.0	138.0	184.1	221.3	249.7	92.1	132.0	161.1	179.2	186.4
Stomach	440.8	813.1	1093.6	1282.2	1379.1	479.6	718.8	899.5	1021.8	1085.7
Liver	3100	7392	10861	13505	15326	3311	5908	7803	8995	9485
Cecum	292.1	546.6	730.6	844.2	887.4	288.4	501.3	652.0	740.2	766.2
Lungs	610.2	817.8	998.0	1150.9	1276.4	602.0	772.2	899.4	983.5	1024.5
Heart	307.3	539.2	730.8	882.2	993.3	306.7	466.9	592.5	683.6	740.1
Spleen	228.8	397.9	527.5	617.7	668.4	231.6	337.7	421.4	482.7	521.4
Lymph node[2]	2.4	9.5	14.8	18.4	20.1	3.9	7.5	10.8	14.0	16.9
Thymus	257.9	426.1	523.8	550.8	507.4	210.7	297.8	348.8	363.8	342.6
Kidney	353.7	667.5	926.5	1130.9	1280.5	377.2	580.0	725.3	812.9	842.9
Bladder	43.6	62.1	76.9	88.2	95.7	39.5	53.8	65.1	73.4	78.6
Testis	236.4	644.9	959.4	1180.0	1306.5	....	....	....	....	....
Testis fat	74.2	454.0	826.8	1192.7	1551.6	....	....	....	....	....
Ovary	....	....	....	....	....	16.5	31.3	41.7	47.8	49.6
Uterine horn	....	....	....	....	....	29.0	60.7	89.2	114.4	136.5
Uterine fat	....	....	....	....	....	49.2	329.7	575.5	786.6	963.0
Adrenal gland	8.6	14.7	19.5	23.1	25.4	7.6	15.4	21.7	26.8	30.5
Biceps femoris	437.7	1050.8	1594.7	2069.5	2475.0	460.0	820.7	1129.5	1386.3	1591.3
Tibia	101.5	213.4	315.4	407.5	489.7	112.4	198.3	271.1	330.8	377.4
LH/Lac										
Age, d	25	40	55	70	85	25	40	55	70	85
Body weight, g	68.7	148.0	216.9	275.4	323.5	60.4	114.7	157.7	189.2	209.3
Body length, cm	12.77	16.43	19.28	21.34	22.59	12.40	15.14	17.50	19.13	20.03
Tail length, cm	10.92	15.32	18.68	21.00	22.28	10.12	14.03	16.98	18.99	20.04
Tibia length, mm	20.42	26.56	31.53	35.34	37.98	19.13	25.20	29.92	33.28	35.30
Weight of structure, mg										
Brain	1513.3	1728.4	1892.1	2004.4	2065.3	1473.2	1642.4	1765.3	1841.8	1871.9
Eye	77.3	96.5	112.2	124.5	133.5	71.0	87.7	101.6	112.7	120.9
Salivary glands[1]	86.5	152.0	207.0	251.4	285.4	83.9	132.2	167.5	189.8	199.1
Stomach	532.3	942.1	1245.2	1441.7	1531.6	456.6	746.2	969.2	1125.8	1215.9
Liver	3875	8901	13115	16517	19107	3563	6619	8949	10554	11433
Cecum	400.4	694.8	901.6	1020.8	1052.3	306.7	510.9	655.0	738.9	762.7
Lungs	534.8	818.9	1026.3	1156.9	1210.8	442.8	645.5	785.8	863.7	879.1
Heart	312.7	560.4	766.4	930.8	1053.5	278.2	458.2	595.2	689.0	739.7
Spleen	282.7	474.5	614.1	701.5	736.7	217.0	328.8	412.3	467.4	494.2
Lymph node[2]	5.7	10.3	13.4	15.0	15.0[3]	4.1	7.8	11.2	14.3	17.0

[1] Submaxillary and major sublingual salivary glands. [2] Superficial cervical lymph node. [3] Indicates an apparent reduction in organ weight with age; these results are probably due to sampling variation.

continued

Specification	Male					Female				
Thymus	251.4	403.7	495.2	526.0	496.1	213.0	334.5	407.8	433.0	410.0
Kidney	399.9	731.1	994.7	1190.6	1318.9	352.0	576.8	744.8	856.0	910.3
Bladder	46.7	69.6	87.9	101.4	110.3	40.9	63.9	81.7	94.3	101.7
Testis	240.7	652.6	984.0	1235.0	1405.5	....	....	....	....	....
Testis fat	122.7	816.3	1516.9	2224.4	2938.9	....	....	....	....	....
Ovary	....	....	....	....	....	14.2	28.2	38.9	46.1	49.9
Uterine horn	....	....	....	....	....	29.5	90.3	135.3	164.3	177.3
Uterine fat	....	....	....	....	....	111.7	900.6	1667.8	2413.3	3136.9
Adrenal gland	8.1	14.7	20.0	23.9	26.6	9.1	16.8	23.3	28.4	32.3
Biceps femoris	358.4	863.0	1325.6	1746.3	2125.0	363.0	785.3	1113.8	1348.5	1489.5
Tibia	120.4	238.9	345.4	439.9	522.4	104.8	209.9	293.7	356.2	397.5
PVG/Lac										
Age, d	25	40	55	70	85	25	40	55	70	85
Body weight, g	65.3	130.9	185.1	227.7	259.0	58.2	100.5	133.5	157.1	171.5
Body length, cm	12.54	15.89	18.46	20.25	21.25	12.18	15.01	17.09	18.44	19.04
Tail length, cm	9.97	14.10	17.20	19.28	20.33	10.10	13.57	16.09	17.65	18.25
Tibia length, mm	21.24	25.26	29.24	33.18	37.08	20.19	25.60	29.79	32.76	34.50
Weight of structure, mg										
Brain	1460.6	1606.7	1715.7	1787.8	1822.9	1393.8	1530.9	1625.5	1677.7	1687.4
Eye	67.6	82.1	94.6	105.2	113.8	62.0	78.5	92.0	102.6	110.1
Salivary glands[1]	72.3	144.1	197.7	232.8	249.6	76.1	120.9	154.7	177.6	189.4
Stomach	411.6	639.0	809.7	923.6	980.7	376.1	544.3	666.8	743.6	774.8
Liver	3736	7811	10823	12772	13658	3140	5322	6903	7883	8263
Cecum	340.4	578.3	752.4	863.0	909.9	282.4	436.0	551.5	628.8	668.1
Lungs	652.7	909.5	1068.7	1130.5	1094.8[3]	524.4	702.3	824.3	890.3	900.4
Heart	286.4	463.3	600.3	697.6	755.1	259.5	386.1	476.8	531.6	550.5
Spleen	240.0	341.5	419.8	475.0	507.1	189.9	278.4	341.4	379.0	391.2
Lymph node[2]	5.9	12.6	18.1	22.3	25.3	3.9	9.3	14.3	17.2	18.7
Thymus	188.6	269.5	312.6	318.1	285.9	162.7	205.0	228.1	232.1	216.8
Kidney	392.4	653.0	853.0	992.3	1071.0	355.3	523.3	640.0	705.4	719.3
Bladder	38.7	53.5	66.1	76.4	84.4	36.6	46.1	53.9	60.0	64.4
Testis	307.1	755.1	1085.9	1299.6	1396.2	....	....	....	....	....
Testis fat	107.2	575.5	989.8	1350.3	1656.8	....	....	....	....	....
Ovary	....	....	....	....	....	14.7	26.7	35.5	41.0	43.2
Uterine horn	....	....	....	....	....	32.4	73.4	101.8	117.8	121.2
Uterine fat	....	....	....	....	....	122.1	560.4	895.6	1127.6	1256.6
Adrenal gland	9.9	17.1	22.5	26.0	27.7	8.8	18.5	26.1	31.5	34.8
Biceps femoris	395.1	962.2	1491.9	1984.2	2439.2	388.6	754.9	1053.0	1282.9	1444.6
Tibia	104.6	224.1	327.2	414.1	484.5	105.0	192.1	264.2	321.3	363.4
WA/Lac										
Age, d	25	40	55	70	85	25	40	55	70	85
Body weight, g	61.1	139.9	206.9	262.2	305.7	50.9	103.2	146.1	179.5	203.6
Body length, cm	12.27	15.91	18.80	20.92	22.29	11.74	15.08	17.60	19.29	20.16
Tail length, cm	9.45	14.15	17.76	20.27	21.67	9.10	13.41	16.55	18.51	19.30
Tibia length, mm	19.18	25.82	31.13	35.09	37.69	18.87	24.23	28.48	31.63	33.66
Weight of structure, mg										
Brain	1431.0	1614.0	1757.4	1861.2	1925.3	1351.5	1496.2	1610.0	1692.9	1744.8
Eye	67.1	86.3	102.7	116.1	126.7	65.2	83.5	99.0	111.7	121.7
Salivary glands[1]	72.4	130.6	175.8	207.9	226.8	69.5	118.9	155.4	179.2	190.1
Stomach	452.0	859.1	1162.4	1362.2	1458.3	401.8	696.1	910.4	1044.7	1098.9

[1] Submaxillary and major sublingual salivary glands. [2] Superficial cervical lymph node. [3] Indicates an apparent reduction in organ weight with age; these results are probably due to sampling variation.

continued

Specification	Male					Female				
Liver	3436	8414	12273	15013	16632	2911	6204	8434	9600	9704
Cecum	328.3	713.8	984.6	1140.5	1181.7	295.5	527.4	689.1	780.5	801.6
Lungs	469.7	805.4	1062.2	1240.1	1339.1	395.0	619.2	780.4	878.7	914.1
Heart	293.1	553.9	762.0	917.5	1020.4	251.2	432.2	568.0	658.7	704.2
Spleen	233.3	453.7	617.4	724.5	775.0	182.9	349.4	477.7	567.6	619.3
Lymph node[2]	4.0	11.2	16.7	20.5	22.6	3.1	6.8	11.2	16.1	21.7
Thymus	164.3	279.0	350.8	379.8	365.8	148.1	224.8	270.7	286.0	270.6
Kidney	359.5	756.5	1066.9	1290.4	1427.2	295.9	513.9	677.2	785.7	839.5
Bladder	36.3	51.7	63.4	71.4	75.7	32.0	42.8	50.9	56.2	58.8
Testis	297.8	756.8	1130.7	1401.4	1569.1	....	....	....	....	....
Testis fat	130.3	537.2	966.4	1417.7	1891.3	....	....	....	....	....
Ovary	....	....	....	....	....	12.0	30.6	44.3	53.1	56.9
Uterine horn	....	....	....	....	....	27.5	66.3	101.0	131.6	158.1
Uterine fat	....	....	....	....	....	16.3	124.6	431.6	937.3	1641.6
Adrenal gland	10.0	14.1	17.2	19.5	20.8	8.3	18.1	25.5	30.7	33.5
Biceps femoris	377.4	1049.4	1602.2	2035.8	2350.1	270.8	739.5	1127.6	1435.3	1662.5
Tibia	109.8	230.0	336.8	430.3	510.5	88.7	177.5	254.6	320.3	374.4
WAG/Lac										
Age, d	25	40	55	70	85	25	40	55	70	85
Body weight, g	44.8	109.3	167.7	219.9	266.0	47.9	92.9	129.1	156.5	175.2
Body length, cm	11.78	15.80	18.79	20.75	21.69	12.01	15.40	17.79	19.19	19.60
Tail length, cm	9.35	13.10	15.99	18.00	19.15	9.29	12.86	15.39	16.89	17.34
Tibia length, mm	19.13	25.68	30.76	34.37	36.51	19.21	24.93	29.16	31.92	33.18
Weight of structure, mg										
Brain	1481.6	1680.4	1823.8	1911.8	1944.6	1435.9	1595.8	1714.6	1791.2	1825.7
Eye	65.2	79.8	94.1	108.0	121.5	63.6	84.1	98.2	106.0	107.4
Salivary glands[1]	79.4	133.0	178.5	215.8	245.0	80.1	127.9	162.0	182.5	189.2
Stomach	449.5	836.9	1108.3	1263.7	1303.0	438.7	722.2	920.8	1034.5	1063.4
Liver	2475	7415	10857	12800	13244	2663	5337	7283	8500	8989
Cecum	262.5	602.8	827.7	937.2	931.5[3]	243.2	461.0	606.6	680.2	681.8
Lungs	412.8	767.5	1006.2	1129.1	1136.0	430.9	620.9	752.7	826.2	841.5
Heart	213.6	495.6	708.9	853.3	928.9	250.5	418.3	542.5	623.0	659.9
Spleen	158.4	305.6	410.2	472.2	491.6	140.9	247.1	326.4	378.6	406.8
Lymph node[2]	2.8	6.5	10.4	14.6	18.9	3.0	7.5	11.4	14.6	17.2
Thymus	155.6	305.5	389.8	408.5	361.6	163.6	268.2	330.7	350.9	329.0
Kidney	301.7	632.9	891.9	1078.6	1193.2	323.8	533.1	687.2	786.0	829.6
Bladder	31.0	37.3	43.9	50.9	58.3	25.8	33.1	38.5	42.2	44.1
Testis	180.7	635.0	1004.3	1288.4	1487.4	....	....	....	....	....
Testis fat	74.6	361.7	725.8	1167.1	1685.4	....	....	....	....	....
Ovary	....	....	....	....	....	11.3	23.4	32.2	38.0	40.6
Uterine horn	....	....	....	....	....	13.1	54.2	86.5	110.0	124.8
Uterine fat	....	....	....	....	....	40.7	311.2	613.2	946.7	1311.6
Adrenal gland	7.1	13.3	18.1	21.5	23.4	7.3	15.3	21.6	26.3	29.3
Biceps femoris	235.4	852.8	1406.4	1896.2	2322.1	275.6	669.7	998.6	1262.4	1461.1
Tibia	84.8	201.0	304.1	394.1	471.0	91.7	183.2	256.4	311.4	348.1
Porton/Lac										
Age, d	25	40	55	70	85	25	40	55	70	85
Body weight, g	63.4	160.4	238.4	297.2	336.9	61.2	125.0	176.8	216.6	244.2
Body length, cm	12.75	17.03	20.02	21.71	22.11	12.40	16.16	18.77	20.24	20.57
Tail length, cm	9.30	14.52	18.06	19.93	20.12	9.29	14.31	17.43	18.68	18.03[3]

[1] Submaxillary and major sublingual salivary glands. [2] Superficial cervical lymph node. [3] Indicates an apparent reduction in organ weight with age; these results are probably due to sampling variation.

continued

Specification	Male					Female				
Tibia length, mm	20.36	27.33	32.66	36.37	38.43	20.07	26.11	30.70	33.84	35.52
Weight of structure, mg										
Brain	1606.8	1834.4	1999.7	2102.6	2143.3	1544.4	1756.0	1919.5	2035.0	2102.3
Eye	74.0	93.3	109.3	122.0	131.5	71.2	89.0	104.3	116.9	126.9
Salivary glands[1]	90.5	170.6	229.6	267.5	284.4	87.9	142.9	184.4	212.3	226.7
Stomach	511.3	1065.8	1392.7	1492.1	1363.9[3]	532.0	876.1	1122.3	1270.5	1320.7
Liver	2771	9629	13985	15839	15190[3]	3106	6778	9491	11246	12044
Cecum	317.9	659.8	883.1	988.1	974.5[3]	272.2	475.3	645.8	783.9	889.4
Lungs	581.8	1072.6	1407.5	1586.5	1609.6	589.2	780.3	947.4	1090.7	1209.9
Heart	292.3	608.5	828.4	952.0	979.3	303.3	504.4	650.4	741.4	777.3
Spleen	335.2	661.6	817.0	801.4[3]	614.9[3]	321.5	541.4	648.8	643.6[3]	525.7[3]
Lymph node[2]	6.7	16.6	24.9	31.4	36.3	5.4	16.3	22.0	22.5	18.0[3]
Thymus	244.1	554.4	686.7	640.9	417.1	261.4	396.2	475.7	500.1	469.2
Kidney	344.8	766.3	1051.6	1201.0	1214.2	354.5	615.9	799.3	904.7	932.0
Bladder	37.9	55.4	68.6	77.6	82.3	35.6	43.1	52.3	63.1	75.7
Testis	243.2	816.2	1224.0	1466.4	1543.7	....	....	....	....	....
Testis fat	99.8	608.6	1120.2	1634.6	2151.8	....	....	....	....	....
Ovary	....	....	....	....	....	15.2	26.7	39.5	53.4	68.9
Uterine horn	....	....	....	....	....	22.6	67.9	109.7	147.9	182.5
Uterine fat	....	....	....	....	....	69.2	411.0	1003.8	1847.5	2942.1
Adrenal gland	9.5	18.2	23.9	26.5	26.1[3]	9.5	18.3	25.9	32.4	37.7
Biceps femoris	353.2	1148.9	1845.3	2442.5	2940.4	350.1	939.1	1419.7	1791.9	2055.6
Tibia	121.0	264.0	386.7	489.2	571.3	117.9	217.3	307.1	387.1	457.3

1/ Submaxillary and major sublingual salivary glands. 2/ Superficial cervical lymph node. 3/ Indicates an apparent reduction in organ weight with age; these results are probably due to sampling variation.

Contributor: Jon A. Turton

References

1. Festing, M. F. W., ed. 1975. International Index of Laboratory Animals. Laboratory Animals Centre, Carshalton, Surrey, England.
2. Festing, M. F. W., and J. Staats. 1973. Transplantation 16:221-245.
3. Greene, E. C. 1959. Anatomy of the Rat. Hafner, New York.
4. Medical Research Council. 1974. The Accreditation and Recognition Schemes for Suppliers of Laboratory Animals. Laboratory Animals Centre, Carshalton, Surrey, England.

75. AGING: RAT

Rearing or Holding System: BR = barrier reared; C = conventional; SPF = specific pathogen-free. Values in parentheses are ranges, estimate "c" (*see* Introduction).

Part I. Longevity

Data in brackets refer to the column heading in brackets.

Strain	Rearing or Holding System	Specification	Sex	Value [Maximum Value] d	Reference
Inbred					
ACI/N	C	Mean life-span	♂	791	14
			♀	756	14
BN/Bi	SPF	Mean life-span	♂	855 [1320]	12
			♀	915 [1620]	12

continued

75. AGING: RAT

Part I. Longevity

Strain	Rearing or Holding System	Specification	Sex	Value [Maximum Value] d	Reference
F344	SPF	50% survival time	♂[1,2]	825 (390-1080)	23
			♀[1]	780 (480-1140)	23
			♀[2]	810 (360-1100)	23
	BR	Mean life-span	♂	870 (600-1050)	4
F344[3]	C	Mean survival time	......	717 (330-960)	7,11
		50% survival time	♂	659	3
				840	9
SHR[4]	C	Mean life-span	......	540	16
WAG/Rij	C	Mean survival time	♀	930 (300-1290)	2
	SPF	50% survival time	♂	645 [990]	13
			♀	900 [1410]	13
Outbred					
Cataract	C	....	♂	636	13
			♀	834	13
Crl:CD(SD)BR	....	Mean survival time	......	729	20
		Mode	......	780	20
		50% survival time	......	820	21
Long Evans	C	50% survival time	♂	825 (249-990)	11
			♀	840 (480-990)	11
SD/Jcl	....	Mean survival time	♂	633	15
			♀	717	15
Sim:CD(SD)	....	Mean survival time (on a semipurified diet)	♂	706	17
			♀	756	17
Sprague-Dawley	Unspecified or C	50% survival time	......	600 [1095]	5
				504 [840]	1
		Mean survival time			
		With exercise	♂	605	19
			♀	665	19
		Without exercise	♂	474	19
			♀	476	19
Wistar	Unspecified or C	Mean survival time	♂	566	6
				690 [990]	10
			♀	607	6
				720 [1170]	10
		50% survival time	♂	522	13
				690 [1140]	8
				721	22
			♀	723	13
				793	22
				810 [1290]	8
Alderley Park strain 1	C	50% survival time	♂[1]	660 (30-870)	18
			♂[2]	660 (30-930)	18
			♀[1]	720 (150-1050)	18
			♀[2]	804 (90-1050)	18
	SPF	50% survival time	♂[2]	750 (420-1020)	18
			♀[1]	822 (300-1371)	18
			♀[2]	870 (450-1350)	18

1/ Breeders. 2/ Non-breeders. 3/ Synonym: Fischer 344. 4/ Wistar origin.

continued

75. AGING: RAT

Part I. Longevity

Contributors: Bennett J. Cohen and Nancy Hulett

References

1. Bilder, G. E. 1975. J. Gerontol. 30:641-646.
2. Boorman, G. A., and C. F. Hollander. 1973. Ibid. 28: 152-159.
3. Chesky, J., and M. Rockstein. 1976. Exp. Aging Res. 2:399-407.
4. Coleman, G. L., et al. 1977. J. Gerontol. 32:258-278.
5. Couser, W. G., and M. M. Stilmant. 1976. Ibid. 31: 31-22.
6. Dalderup, L. M., and W. Visser. 1969. Nature (London) 222:1050-1052.
7. Davey, F. R., and W. C. Moloney. 1970. Lab. Invest. 23:327-334.
8. Gsell, D. 1964. Int. Z. Vitaminforsch. Suppl. 9:114-125.
9. Heywood, R. 1973. Lab. Anim. 7:19-27.
10. Hirokawa, K. 1975. Mech. Aging Dev. 4:301-316.
11. Hoffman, H. J. 1978. U.S. Dep. HEW Publ. (NIH) 78-161:19-34.
12. Hollander, C. F. 1976. Lab. Anim. Sci. 26(2-2):320-328.
13. Léonard, A., et al. 1971. Exp. Gerontol. 6:293-296.
14. Maekawa, A., and S. Odashima. 1975. J. Natl. Cancer Inst. 55:1437-1445.
15. Muraoka, Y., et al. 1977. Exp. Anim. 26:1-13.
16. National Research Council, Committee on Care and Use of Spontaneously Hypertensive (SHR) Rats. 1976. ILAR News 19:G1-G20.
17. Nolen, G. A. 1972. J. Nutr. 102:1477-1494.
18. Paget, G. E., and P. G. Lemon. 1965. In W. E. Ribelin and J. B. McCoy, ed. Pathology of Laboratory Animals. Thomas, Springfield, IL. pp. 382-405.
19. Retzlaff, E., et al. 1966. Geriatrics 21:171-177.
20. Ross, M. H. 1959. Fed. Proc. Fed. Am. Soc. Exp. Biol. 18:1190-1207.
21. Ross, M. H. 1961. J. Nutr. 75:197-210.
22. Schlettwein-Gsell, D. 1970. Gerontology 16:111-115.
23. Wexler, B. C. 1976. J. Gerontol. 25:373-380.

Part II. Physiological and Biochemical Alterations

Age: A hyphen between ages indicates the range encompassed in obtaining an average quantitative value; "to" between ages indicates the range encompassed in determining a change (increase, decrease, or no change), with minimum age as time of initial observation and maximum age as time of final observation. Plus/minus (±) values are standard errors unless otherwise indicated.

Strain ⟨Synonym⟩	Rearing or Holding System	Specification	Sex	Age	Observation	Reference
Inbred						
BN/Bi	C	Urine production, ml/24 h	♂	3-4 mo	9 ± 2[1]	18
				34 mo	12 ± 4[1]	
		Urine osmolality, mosm/liter	♂	3-4 mo	1550 ± 300[1]	18
				34 mo	970 ± 100[1]	
		Blood urea nitrogen, mg/dl	♂	3-4 mo	59 ± 10[1]	18
				34 mo	51 ± 4[1]	
DONRYU	....	Tibialis anterior muscle				
		Weight	♂	6 to 24 mo	Decreases 47%	43
		White fiber size	♂	6 to 24 mo	Decreases	43
		White fiber:red fiber ratio	♂	6 to 24 mo	Decreases	43
F344	Unspecified	Right atrium (isolated tissue)				
		Duration of action potential plateau	♂	5-7 to 15-19 mo	Younger group 150-280% >older group	9
		Time to achieve 95% repolarization	♂	48-56 to 68-74 wk	Increases with age; younger rats ∿40% <older rats	9

[1] Plus/minus (±) value is standard deviation.

continued

Part II. Physiological and Biochemical Alterations

Strain (Synonym)	Rearing or Holding System	Specification	Sex	Age	Observation	Reference
	BR	Serum components				
		Potassium	♂	4 to >30 mo	No change	10
		Sodium	♂	4 to >30 mo	No change	10
		Cholesterol	♂	4 to >30 mo	Increases	10
		Total protein	♂	4 to >30 mo	Decreases	10
		Albumin	♂	4 to >30 mo	Decreases	10
		α_1-Globulin	♂	4 to >30 mo	Increases	10
		α_2-Globulin	♂	4 to >30 mo	No change	10
		β-Globulin	♂	4 to >30 mo	No change	10
		γ-Globulin	♂	4 to >30 mo	No change	10
		Creatinine	♂	4 to >30 mo	No change	10
	SPF	Urinary protein, ∿mean value, $mg^{-1}\cdot 100\ g^{-1}\cdot 24\ h^{-1}$				
		Total	♂	1-1.5 mo 17-20 mo	2.0 6.5	26
		Albumin	♂	1-1.5 mo 17-20 mo	0.5 3.9	26
		Albumin synthesis: [^{3}H]leucine incorporation, $counts\cdot min^{-1}\cdot mg\ albumin^{-1}$	♂	1-1.5 mo 17-20 mo	1.4×10^{-2} 2.4×10^{-2}	26
F344 (Fischer 344)	C	Heart rate (immobilized & unanesthetized)	♂	12 mo 28 mo	425 beats/min 404 beats/min	32
		O_2 consumption, ml/kg body $wt^{0.75}$	♂	200 d 700 d	763 840 (10% increase)	40
			♀	200 d 700 d	761 857 (13% increase)	
		Cerebral cortex: calcium levels	♂	1 to 28 mo	Decreases	30
		Heart				
		ADP-stimulated mitochondria	♂	8-9 to 24 mo	Decreases	31
		Enzymes				
		Amine oxidase (flavin-containing)[2]	♂	2 & 4 to 22 mo	No change	31
		Aromatic L-amino-acid decarboxylase[3]	♂	2 & 4 to 22 mo	Increases	31
		Catechol methyltransferase	♂	2 & 4 to 22 mo	No change	31
		Atria				
		Water, %	♂	6 mo 28 mo	82.47 83.26	14
		Extracellular, ml/g wet tissue	♂	6 mo 28 mo	0.31 0.27	14
		Potassium, meq/kg wet tissue	♂	6 mo 28 mo	69.08 77.55	14
		Sodium, meq/kg wet tissue	♂	6 mo 28 mo	59.78 43.58	14
		Norepinephrine content	♂	1 to 28 mo	Decreases	31

[2] Synonym: Monoamine oxidase. [3] Synonym: DOPA decarboxylase.

continued

Part II. Physiological and Biochemical Alterations

Strain ⟨Synonym⟩	Rearing or Holding System	Specification	Sex	Age	Observation	Reference
		Ventricles				
		Histamine content	♂	2, 4, & 22 mo	Increases	31
		Norepinephrine				
		Content	♂	1 to 28 mo	Decreases	31
		Uptake	♂	2, 4, & 22 mo	Decreases	
		Serotonin content	♂	2, 4, & 22 mo	No change	31
Outbred						
Carworth ⟨CFN⟩	C	Blood pressure	♂	2.5 to 16 mo	Increases	46
		Aorta				
		Total dry weight	♂	2.5 to 16 mo	Increases	46
		Diameter	♂	2.5 to 16 mo	Increases	46
		Medial area	♂	2.5 to 16 mo	Increases	46
		Wall				
		Thickness	♂	2.5 to 16 mo	Increases	46
		Connective tissue staining	♂	2.5 to 16 mo	Increases	46
		Tension per lamellar unit (calculated)	♂	2.5 to 16 mo	Increases	46
		Stress	♂	2.5 to 16 mo	No change	46
		Amount of elastin, collagen, & non-collagenous alkali-soluble proteins	♂	2.5 to 16 mo	Increases	46
		% of non-collagenous alkali-soluble proteins	♂	2.5 to 16 mo	Decreases	46
		% of collagen	♂	2.5 to 16 mo	Increases	46
		% of elastin	♂	2.5 to 16 mo	No change	46
		Lysine in elastin	♂	2.5 to 16 mo	Increases	46
Chbb:THOM	SPF	Testis: steroid dehydrogenase activity in interstitial cells	♂	10 wk to 2 yr	Increases	23
Crl:CD(SD)BR	....	Brain weight	♂	....	Increases with age	19
		Kidney weight	♂	....	Increases with age	7
		% of body weight	♂	<220 d	Decreases	7
				>220 d	Increases	
		Gastrointestinal enzymes				
		Amylase, g substrate hydrolyzed·min^{-1}·pancreas^{-1}	..	12 mo	25.4 ± 4.5	25
				25 mo	0 ± 0	
		Chymotrypsinogen, mmoles substrate hydrolyzed·min^{-1}·pancreas^{-1}	..	12 mo	12.9 ± 1.7	25
				25 mo	4.79 ± 0.58	
		Triacylglycerol lipase[4], g substrate hydrolyzed·min^{-1}·pancreas^{-1}	..	12 mo	3.06 ± 0.45	25
				25 mo	1.06 ± 0.15	
		Trypsinogen, mmoles substrate hydrolyzed·min^{-1}·pancreas^{-1}	..	12 mo	3.26 ± 0.37	25
				25 mo	1.17 ± 0.15	
		Hepatic enzyme activity (animals sacrificed at age indicated)				
		Adenosinetriphosphatase, units	..	21 d	∿5.3	33
				220 d	∿13.3	
				795 d	∿11.4	

[4] Synonym: Lipase.

continued

Part II. Physiological and Biochemical Alterations

Strain (Synonym)	Rearing or Holding System	Specification	Sex	Age	Observation	Reference
		Alkaline phosphatase, units	..	21 d 220 d 795 d	∿1.19 ∿0.73 ∿0.64	33
		Catalase, units	..	21 d 220 d 795 d	∿0.047 ∿0.084 ∿0.07	33
		Histidine ammonia-lyase[5/], units	..	21 d 220 d 795 d	∿1.1 ∿3.2 ∿2.8	33
Holtzman (SD)	C	Liver				
		Intralobular extrahepatocyte space, % of intralobular tissue	..	Young adult Retired breeder	18 13	36, 37
		Hepatocyte, centrolobular				
		Mean volume, % of intralobular tissue	..	Young adult to retired breeder	Increases 5%	36, 37
		Cytoplasmic volume, % of hepatocyte volume	..	Young adult Retired breeder	74 80	36, 37
		Mitochondrial volume, % of cytoplasmic volume	..	Young adult to retired breeder	No significant change with age	36, 37
		Lysosomal volume, % of cytoplasmic volume	..	Young adult Retired breeder	0.4 0.6	36, 37
		Smooth endoplasmic reticulum				
		Surface density, cm^2/cm^3 of parenchymal ground substance	..	Young adult to retired breeder	Young adult ∿14% >retired breeder	36, 37
		Volume density, % of parenchymal ground substance	..	Young adult Retired breeder	10 8	36, 37
		Rough endoplasmic reticulum				
		Surface density, cm^2/cm^3 of parenchymal ground substance	..	Young adult Retired breeder	2.5 ± 0.17 2.8 ± 0.16	36, 37
		Volume density, % of parenchymal ground substance	..	Young adult Retired breeder	4 4	36, 37
		Golgi				
		Surface density, cm^2/cm^3 parenchymal ground substance	..	Young adult Retired breeder	0.36 ± 0.05 0.26 ± 0.04	36, 37
		Volume density, % of parenchymal ground substance	..	Young adult Retired breeder	2.2 1.4	36, 37
		Nuclear volume, % of total hepatocyte volume	..	Young adult Retired breeder	8 7	36, 37
		Nuclear:cytoplasmic ratio	..	Young adult Retired breeder	0.112 0.093	36, 37

[5/] Synonym: Histidase.

continued

Part II. Physiological and Biochemical Alterations

Strain ⟨Synonym⟩	Rearing or Holding System	Specification	Sex	Age	Observation	Reference
LE ⟨Long Evans⟩	C	Reproductive status				
		Nearly normal	♀	>21 mo	18/108	35
		Prolonged estrus with cystic follicles	♀	>21 mo	17/108	35
		Prolonged diestrus with large corpora lutea	♀	>21 mo	10/108	35
		Anestrus with atrophic ovaries	♀	>21 mo	63/108	35
		Acyclic	♀	>32 mo	Nearly all	35
		Serum				
		Luteinizing hormone ⟨LH⟩ levels	♀	4-6 to 23-30 mo	Higher in young proestrus & old constant estrus rats than in young estrus, young diestrus, or old pseudopregnant rats	45
		After luteinizing hormone—releasing hormone ⟨LH-RH⟩ injection	♀	4-6 to 23-30 mo	Old rats have smaller increases	45
		Prolactin, ng/ml	♀	2.5 mo 20 mo	83 ± 14 126 ± 19	8
		Testosterone, ng/ml	♂	4 mo 22-26 mo	1-18 0.55	20
		Ovary				
		Glucose-6-phosphate dehydrogenase activity	♀	Young adult to 2 yr	50% decrease	20
		20α-Hydroxysteroid dehydrogenase activity, mU/mg protein[6]	♀	4 mo 18 mo	19.9 2.8	20
		3β-Hydroxy-Δ^5-steroid dehydrogenase ⟨Δ^5-3β-OHSD⟩ activity (to measure 4-androstene-3,17-dione[7]), $\mu g \cdot min^{-1} \cdot mg^{-1}$	♀	6 mo 24 mo	0.13 0.06	21
		Brain: serotonin levels	..	10 to 32 mo	Increases	30
OFA	SPF	Urine				
		Volume, ml/24 h	♂	6 & 13 wk 26 & 52 wk	11.4 ± 1.1 9.5 ± 1.1	29
		pH	♂	6 & 13 wk 26 & 52 wk	5.9 ± 0.1 6.0 ± 0.1	29
		Potassium, meq·100 kg body $wt^{-1} \cdot 24\ h^{-1}$	♂	6 & 13 wk 26 & 52 wk	254.0 ± 15.2 340.5 ± 53.4	29
		Sodium, meq·100 kg body $wt^{-1} \cdot 24\ h^{-1}$	♂	6 & 13 wk 26 & 52 wk	87.0 ± 4.7 44.8 ± 7.5	29
SD ⟨Sprague-Dawley⟩	Unspecified or C	Skin graft rejection time	♂	24 wk 108 wk	6 d, minimum 13 d, maximum	20
			♀	4 wk 89 wk	6 d, minimum 16.8 d, maximum	
		Heart weight	♂	....	Slightly increases with age	5

[6] mU = milliunits; one unit = NADP formation of 1 $\mu mole \cdot min^{-1} \cdot mg\ protein^{-1}$. [7] Synonym: Δ^4-Androstenedione.

continued

Part II. Physiological and Biochemical Alterations

Strain (Synonym)	Rearing or Holding System	Specification	Sex	Age	Observation	Reference
		% of body weight	♂	219 d	0.29 ± 0.01[1]	5
				903 d	0.47 ± 0.12[1]	
			♀	501 d	0.4 ± 0.05[1]	
				990 d	0.6 ± 0.09[1]	
		Rate of clearance by reticuloendothelial system, (V_{max}), $mg \cdot kg^{-1} \cdot min^{-1}$	♂	4 wk	1.13	6
				64 wk	8.0	
			♀	4 wk	15.2	
				64 wk	4.2	
		Incidence of hypertensiveness (⩾141 mm Hg), %	♂	>600 d	31	5
			♀	>600 d	19	
		Brain				
		Acid phosphatase activity	..	8 wk to 25 mo	Increases	31
		Adenylate cyclase activity				
		In caudate area & cerebellum	..	3 wk to 24 mo	Increases	30
		In cortex & hippocampus	..	3 wk to 24 mo	No change	30
		Catecholamines, intrastriatal spread	♀	<9 mo to 24-26 mo	Decreases ∿50%	31
		Spermidine	..	Young to 26 mo	Increases	30
		Spermine	..	Young to 26 mo	No change	30
		Lungs				
		Triglyceride, μmoles	♂	6 mo	23.8 ± 9.6	12
				22 mo	111.9 ± 46.1	
			♀	6 mo	11.2 ± 3.7	
				22 mo	83.6 ± 52.6	
		Phospholipid, mg	♂	6 mo	24.6 ± 6.1	12
				22 mo	59.9 ± 18.0	
			♀	6 mo, 22 mo	Both ages similar	
		Cholesterol	♂	6 mo, 22 mo	Both ages similar	12
			♀	6 mo, 22 mo	Both ages similar	
		Blood serum values				
		Triglyceride, μmoles/ml	♂	6 mo	1.18 ± 0.21	12
				22 mo	9.71 ± 2.20	
			♀	6 mo	0.51 ± 0.11	
				22 mo	7.19 ± 1.03	
		Phospholipid, mg/dl	♂	6 mo	145 ± 15	12
				22 mo	382 ± 92	
			♀	6 mo	139 ± 25	
				22 mo	214 ± 41	
		Cholesterol, mg/dl	♂	6 mo	107 ± 12	12
				22 mo	253 ± 47	
			♀	6 mo, 22 mo	Both ages similar	
		Heart				
		Microsomal enzymes				
		3-Hydroxybutyrate dehydrogenase	♂	16 to 24 mo	Small increase	16
		NADH dehydrogenase[8]	♂	16 to 24 mo	No change	16
		NADPH cytochrome reductase	♂	16 to 24 mo	No change	16

[1] Plus/minus (±) value is standard deviation. [8] Synonym: NADH cytochrome *c* reductase.

continued

Part II. Physiological and Biochemical Alterations

Strain ⟨Synonym⟩	Rearing or Holding System	Specification	Sex	Age	Observation	Reference
		Succinate cytochrome *c* reductase	♂	16 to 24 mo	No change	16
		Whole: microsomal phospholipid	♂	16 to 24 mo	No change	16
		Myocardium				
		Water				
		Total, g/kg dry fat-free tissue	♂	1.6 mo 35.4 mo	3682 3382	31
		Extracellular, g/kg dry fat-free tissue	♂	1.6 mo 35.4 mo	1522 1141	31
		Calcium, meq/kg dry fat-free tissue	♂	1.6 mo 35.4 mo	30.7 21.2	31
		Chloride, meq/kg dry fat-free tissue	♂	1.6 mo 35.4 mo	175 141	31
		Magnesium, meq/kg dry fat-free tissue	♂	1.6 mo 35.4 mo	82.7 47.4	31
		Phosphorus, meq/kg dry fat-free tissue	♂	1.6 mo 35.4 mo	344 339	31
		Potassium, meq/kg dry fat-free tissue	♂	1.6 mo 35.4 mo	396 355	31
		Sodium, meq/kg dry fat-free tissue	♂	1.6 mo 35.4 mo	2407 112	31
		Urinary protein, mg/d	.. .. ..	3 & 6 mo 12 mo 24 mo	<10 >20 140 ± 50	11
		Renal enzymes				
		Glucose-6-phosphatase	..	....	Mild decrease with age	39
		α-Glucosidase[9]	..	....	Mild decrease with age	24
		NADH dehydrogenase[8]	..	....	Mild decrease with age	16
		Phosphodiesterase	..	....	Mild decrease with age	24
		Succinate cytochrome *c* reductase	..	....	Mild decrease with age	16
		Ovary: 3β-Hydroxy-Δ^5-steroid dehydrogenase ⟨Δ^5-3β-OHSD⟩ activity (to measure Δ^5-androstenedione), $\mu g \cdot min^{-1} \cdot mg^{-1}$	♀	6 mo 24 mo	0.16 0.09	21
		Uterus: alkaline phosphatase, μg phenolphthalein$\cdot g^{-1} \cdot h^{-1}$	♀	6 mo 18 mo 24 mo	1.95 ± 0.16 1.08 ± 0.16 0.69 ± 0.09	20
SD (CFE)	SPF	Oxygen consumption, liter$\cdot kg^{-1} \cdot h^{-1}$	♂	43 d 160 d 579 d	0.910 0.800 0.755	42
Wistar	Unspecified or C	Transit time, h				
		Cecum	♂	19 d 17-20 mo	1.53 7.62	44
		Large intestine	♂	19 d 17-20 mo	11.02 18.30	44

[8] Synonym: NADH cytochrome *c* reductase. [9] Synonym: Maltase.

continued

Part II. Physiological and Biochemical Alterations

Strain (Synonym)	Rearing or Holding System	Specification	Sex	Age	Observation	Reference
		Heart				
		Weight: body weight ratio	♂	7 mo	∿2.8 × 10^{-3}	17
				24 mo	∿3.6 × 10^{-3}	
			♀	7 mo	∿3.1 × 10^{-3}	
				24 mo	∿4.5 × 10^{-3}	
		Baroreceptor activity	♂	12 to 24 mo	Decreases	31
		Ventricular trabeculae carneae stress-relaxation resting tensions	♂	11-13 to 25-29 mo	Greater in older group	28
		Heart rate (rats caged, previously cannulated, unanesthetized), beats/min	♂	12 mo	314	34
				24 mo	331	
		Cardiac output, ml/min	♂	12 mo	127	34
				24 mo	106	
		Aortic flow, ml·min⁻¹·kg⁻¹	♂	12 mo	277	34
				24 mo	237	
		Blood pressure, mm Hg	♂	12 mo	Systolic, 135; diastolic, 94	34
				24 mo	Systolic, 125; diastolic, 87	
		Mean arterial pressure, mm Hg	..	6 mo	93.73 ± 3.12	22
				12 mo	....	
				24 mo	104.07 ± 3.02	
		Left ventricular diastolic pressure, mm Hg	..	6 mo	1.83 ± 0.19	22
				12 mo	3.31 ± 0.036	
				24 mo	3.34 ± 0.35	
		Right atrial pressure, mm Hg	..	6 mo	2.67 ± 0.22	22
				12 mo	3.63 ± 0.29	
				24 mo	4.33 ± 0.24	
		Kidney				
		Glomeruli, μ				
		Short diameter	..	75-100 d	73.43 ± 1.093	1
			..	868-1170 d	96.45 ± 0.689	
		Long diameter	..	75-100 d	89.28 ± 0.999	1
			..	868-1170 d	117.76 ± 0.809	
		p-Aminohippurate tubular maxima ⟨Tm_{PAH}⟩, mg·min⁻¹·100 g	♀	12 mo	0.33 + 0.02	15
				20 mo	0.25 + 0.1 (25% decrease)	
		Inulin clearance	♀	12 to 20 mo	No significant change	15
		Psoas major muscle				
		$\sqrt{\text{Area}}$ of transverse section (as a measure of muscle volume), mm	♂	3 mo	6.28 ± 0.374[10]	43
				12 mo	8.6 ± 0.105[10]	
				24 mo	6.25 ± 0.263[10]	
		White fibers, μm	♂	3 mo	34.8 ± 1.378[10]	43
				12 mo	39.9 ± 1.212[10]	
				24 mo	30.5 ± 0.945[10]	
		no.	♂	3 mo	5781 ± 593.9[10]	43
				12 mo	12349 ± 914.6[10]	
				24 mo	8603 ± 704.3	
		Red fibers, no.	♂	3 mo	18960 ± 2149.6[10]	43
				24 mo	6591 ± 561.1[10]	

[10] Plus/minus (±) values are not labeled in the reference as to whether they are standard deviation or standard error.

continued

Part II. Physiological and Biochemical Alterations

Strain ⟨Synonym⟩	Rearing or Holding System	Specification	Sex	Age	Observation	Reference
		Tibialis anterior muscle				
		Weight	♂	12 to 24 mo	Decreases 53%	43
			♀	12 to 24 mo	Decreases 50%	
		White fibers, μm	♂	3 mo	33.4 ± 1.981[10]	43
				24 mo	24.0 ± 1.384[10]	
			♀	3 mo	25.8 ± 2.146[10]	
				24 mo	23.1 ± 0.691[10]	
		no.	♂	3 to 24 mo	Increases 54.0%	43
			♀	3 to 24 mo	No significant change	
		Red fibers, μm	♂	3 mo	23.2 ± 0.994[10]	43
				24 mo	24.8 ± 1.384[10]	
			♀	3 mo	20.9 ± 1.036[10]	
				24 mo	22.7 ± 0.729[10]	
		no.	♂	3 to 24 mo	Decreases 49.3%	43
			♀	3 to 24 mo	Decreases 55.3%	
		Total myelin lipids	..	20 d to 20 mo	Generally decrease	30
		Brain				
		Cholesterol	♂	5 to 20 mo	Increases	38
		Enzymes				
		Acetylcholinesterase				
		Prosencephalon	..	10 to 36 mo	No change	30
		Cerebellum	..	10 to 36 mo	Decreases	30
		Brain stem	..	10 to 36 mo	Increases	30
		Adenosinetriphosphatase, Na^+- & K^+-stimulated				
		Prosencephalon	..	10 to 36 mo	No change	30
		Cerebellum	..	10 to 36 mo	Decreases	30
		Brain stem	..	10 to 36 mo	Increases	30
		Heart enzymes				
		Lactate dehydrogenase activity	♀	1 d to 30 wk	Increases	41
				30 to 96 wk	Decreases	
		Cytoplasm: malate dehydrogenase	♀	22 & 96 wk	Decreases	31
		Mitochondria				
		Cytochrome *c* oxidase	..	12-14 to 24-27 mo	Decreases	31
		Malate dehydrogenase	♀	22 & 96 wk	Decreases	31
		Oxidative phosphorylation	..	12-14 to 24-27 mo	No change	31
		Blood urea nitrogen, mg/dl	♂	13 mo	20	17
				19 mo	60	
				24 mo	40	
			♀	13 mo	19	
				19 mo	22	
				24 mo	30	
		Serum				
		Cholesterol, mg/dl	♂	3 mo	75	17
				13 mo	250	
			♀	3 mo	75	
				13 mo	180	

[10] Plus/minus (±) values are not labeled in the reference as to whether they are standard deviation or standard error.

continued

Part II. Physiological and Biochemical Alterations

Strain (Synonym)	Rearing or Holding System	Specification	Sex	Age	Observation	Reference
		Creatinine, mg/dl	♂	13 mo 19 mo	1 2.1	17
		Renal enzymes				
		Acid phosphatase	♂	Young to 20 mo	Decreases	24
		Adenosinetriphosphatase, Na^+- & K^+-stimulated	♂	Young to 20 mo	Decreases	4
		Alkaline phosphatase	♂	Young to 20 mo	Decreases	2, 24
		Cathepsin	♂	Young to 20 mo	Increases	2
		Glucose-6-phosphatase	♂	Young to 20 mo	Decreases	2, 16
		Succinate dehydrogenase	♂	Young to 20 mo	Decreases	3
		Succinate dehydrogenase[11]	♂	Young to 20 mo	Decreases	24
		Renal inactivation and hydrolysis of parathyroid hormone[12]	♂	6 to 24 mo	Decreases 50%	13
		Urine: proteinuria (dipstick)	♂♀	>3 mo	Appeared after 3 mo of age	17
	Germfree	Hematocrit, %	♂♀	24 mo	45 (43-45)	27
		Leukocytes, no./μl	♂♀	24 mo	3400 (1800-7400)	27
Alderley Park strain 1	C	Serum prolactin (rats ether-killed), ng/ml	♂	4 mo 21 mo	83.1 ± 54 133 ± 16.1	8

[11] Synonym: Succinoxidase. [12] Synonym: Parathormone.

Contributors: Bennett J. Cohen and Nancy Hulett

References

1. Andrew, W., and D. Pruett. 1957. Am. J. Anat. 100: 51-79.
2. Barrows, C. H., and L. M. Roeder. 1961. J. Gerontol. 16:321-325.
3. Barrows, C. H., et al. 1958. Ibid. 13:351-355.
4. Beauchene, R. E., et al. 1965. Ibid. 20:306-310.
5. Berg, B. N., and C. R. Harmison. 1955. Ibid. 10:416-419.
6. Bilder, G. E. 1975. Ibid. 30:641-646.
7. Bras, G., and M. H. Ross. 1964. Toxicol. Appl. Pharmacol. 6:247-262.
8. Bruni, J. F., et al. 1976. IRCS Med. Sci. 4:256.
9. Cavoto, F. V., et al. 1974. Am. J. Physiol. 226:1293-1297.
10. Coleman, G. L., et al. 1977. J. Gerontol. 32:258-278.
11. Couser, W. G., and M. M. Stilmant. 1976. Ibid. 31:13-22.
12. Flodh, H., et al. 1974. Z. Versuchstierkd. 16:299-312.
13. Fujita, T., et al. 1971. J. Gerontol. 26:20-23.
14. Goldberg, P. B., et al. 1975. Fed. Proc. Fed. Am. Soc. Exp. Biol. 34:188-190.
15. Gregory, J. G., and C. H. Barrows. 1969. J. Gerontol. 24:321-323.
16. Grinna, L. S., and A. A. Barber. 1972. Biochim. Biophys. Acta 288:347-353.
17. Hirokawa, K. 1975. Mech. Aging Dev. 4:301-316.
18. Hollander, C. F. 1970. Exp. Gerontol. 5:313-321.
19. Koama, C., et al. 1969. Lab. Anim. Care 19:747-755.
20. Leathem, J. H. 1978. U.S. Dep. HEW Publ. (NIH) 78-161:141-154.
21. Leathem, J. H., and B. H. Shapiro. 1975. Proc. Soc. Exp. Biol. Med. 148:793-794.
22. Lee, J. C., et al. 1972. Am. J. Physiol. 222:432-438.
23. Lehmann, L. H., et al. 1975. Beitr. Pathol. Anat. Allg. Pathol. 156:109-116.
24. Lowenstein, L. M. 1978. U.S. Dep. HEW Publ. (NIH) 78-161:233-240.
25. Masoro, E. J. 1978. Ibid. 78-161:245-254.
26. Obenrader, M., et al. 1974. Exp. Gerontol. 9:173-180.
27. Pollard, M., and M. Kajima. 1970. Am. J. Pathol. 61: 25-32.
28. Retzlaff, E., et al. 1965. Geriatrics 21:171-177.
29. Richardson, B., and H. Luginbuhl. 1976. Virchows Arch. A370:13-19.

continued

Part II. Physiological and Biochemical Alterations

30. Roberts, J., and S. I. Baskin. 1978. U.S. Dep. HEW Publ. (NIH) 78-161:165-172, 191-202.
31. Roberts, J., and P. B. Goldberg. Ibid. 78-161:155-164, 207-232.
32. Roberts, J., and P. B. Goldberg. 1975. Adv. Exp. Med. Biol. 61:119-148.
33. Ross, M. H. 1959. Fed. Proc. Fed. Am. Soc. Exp. Biol. 18:1190-1207.
34. Rothbaum, D. A., et al. 1973. J. Gerontol. 28:287-292.
35. Schlettwein-Gsell, D. 1970. Gerontology 16:111-115.
36. Schmucker, D. L., and A. L. Jones. 1975. Proc. 10th Int. Congr. Gerontol. 1:114-116.
37. Schmucker, D. L., et al. 1974. J. Gerontol. 29:506-513.
38. Serougre-Gautheror, C., and F. Chevallier. 1973. Biochim. Biophys. Acta 316:244-250.
39. Shukla, S. D., and M. S. Kanungo. 1968. Exp. Gerontol. 3:31-34.
40. Simms, H. S., and B. N. Berg. 1957. J. Gerontol. 12: 244-252.
41. Singh, S. N., and M. S. Kanungo. 1968. J. Biol. Chem. 243:4526-4529.
42. Stupfel, M., et al. 1975. J. Gerontol. 30:154-156.
43. Tauchi, H., et al. 1971. Gerontology 17:219-227.
44. Varga, F. 1976. Digestion 14:319-324.
45. Watkins, B. E., et al. 1975. Endocrinology 97:543-548.
46. Wolinsky, H. 1972. Circ. Res. 30:301-309.

Part III. Lesions

Specification: IgG = immunoglobulin G; IgM = immunoglobulin M. **Age:** Fn[3] = footnote 3.

Strain ⟨Synonym⟩	Rearing or Holding System	Specification	Sex	Age	Incidence[1]	Reference
		Non-neoplastic				
Inbred						
BN/Bi	C	Heart: endocardial lesions	♂	31(21-39) mo	7%	4
		Clinical paralysis with associated skeletal muscle atrophy & disk degeneration	♂	....	2%	8
FW49/Biberach	SPF	Testis				
		Testicular tubular degeneration or atrophy	♂	720-778 d	24/90	21
		Interstitial cell hyperplasia or tumor-like lesions	♂	720-778 d	70/90	21
F344[2]	BR	Skin & adnexa	♂	6-12 mo 24-30 mo	15.4% 32.6%	9
		Central nervous system	♂	24-30 mo	23.4%	9
		Eye	♂	6-12 mo 24-30 mo	15.4% 15.2%	9
		Digestive system	♂	6-12 mo 24-30 mo	7.7% 9.5%	9
		Liver	♂	6-12 mo 24-30 mo	14.6% 100%	9
		Pancreas	♂	6-12 mo 24-30 mo	7.7% 17.0%	9
		Respiratory system	♂	6-12 mo 24-30 mo	76.9% 66.0%	9
		Heart	♂	6-12 mo 24-30 mo	84.6% 80.9%	9

[1] Unless otherwise specified. [2] Data includes tumors and non-neoplastic lesions.

continued

Part III. Lesions

Strain ⟨Synonym⟩	Rearing or Holding System	Specification	Sex	Age	Incidence[1]	Reference
		Spleen	♂	24-30 mo	51.1%	9
		Urinary bladder	♂	24 to 30 mo	Increases	9
		Testis	♂	6-12 mo	31.8%	9
				24-30 mo	100%	
		Accessory sex gland	♂	6-12 mo	38.5%	9
				24-30 mo	71.9%	
		Mammary gland	♂	6-12 mo	7.7%	9
				24-30 mo	10.6%	
		Adenohypophysis	♂	6-12 mo	15.4%	9
				24-30 mo	23.4%	
		Thyroid & parathyroid	♂	24-30 mo	9.5%	9
		Adrenal gland	♂	6-12 mo	23.1%	9
				24-30 mo	34.0%	
		Muscle	♂	24-30 mo	4.4%	9
		Bone	♂	24-30 mo	10.9%	9
F344 ⟨Fischer 344⟩	C	Severe pulmonary infection	♂♀	Fn[3]	100%	11
		Myocardial degeneration	♂	All ages[4]	Avg, 24%	28
			♀	All ages[5]	Avg, 12%	
		Nephrosis	♂	All ages[4]	Avg, 40%	28
			♀	All ages[5]	Avg, 17%	
OM/N	C	Retinal degeneration	..	⩾10 mo	High	34
SHR[6]	C	Cerebellar infarction & hemorrhage				23
		With blood pressure <200 mm Hg	♂♀	Fn[3]	8%	
		With blood pressure >200 mm Hg	♂♀	Fn[3]	12%	
		Myocardial infarction & fibrosis				23
		With blood pressure <200 mm Hg	♂♀	Fn[3]	19%	
		With blood pressure >200 mm Hg	♂♀	Fn[3]	79%	
		Angionecrosis				23
		With blood pressure <200 mm Hg	♂♀	Fn[3]	37%	
		With blood pressure >200 mm Hg	♂♀	Fn[3]	57%	
		Nephrosclerosis				
		Benign				23
		With blood pressure <200 mm Hg	♂♀	Fn[3]	5%	
		With blood pressure >200 mm Hg	♂♀	Fn[3]	22%	
		Malignant				23
		With blood pressure <200 mm Hg	♂♀	Fn[3]	12%	
		With blood pressure >200 mm Hg	♂♀	Fn[3]	18%	
WAG/Rij	C	Liver: hepatic cysts	♀	3 yr	30%	5
		Pancreas: acinar atrophy[7]	♀	....	25%; severity increases with age	5
		Heart: endocardial disease	..	32 mo[8]	4%	4
		Clinical paralysis with associated skeletal muscle atrophy & disk degeneration	..	15-42 mo	2%	8
Outbred						
August	C	Mesenteric arteries				
		1st series	♂♀	400 d	27%	24

1/ Unless otherwise specified. 3/ At death. 4/ Median age of onset, >14 mo. 5/ Median age of onset, 20-34 mo. 6/ Wistar origin. 7/ Severity increases with age. 8/ Mean age.

continued

Part III. Lesions

Strain ⟨Synonym⟩	Rearing or Holding System	Specification	Sex	Age	Incidence[1]	Reference
		2nd series	♂	400 d	13.2%	24
			♀	400 d	40.8%	
		3rd series	♂	400 d	45.4%	24
			♀	400 d	43.0%	
		Pancreas: microlesions including interstitial inflammation, fibrosis, & vascular lesions, as well as parenchymal & islet cell changes	..	14-16 mo	63/64	17
	CD-BR	Pancreas: microlesions including primarily changes in acinar cells	♂	3 mo	7/12	17
			♀	3 mo	5/12	
Civo	C	Heart: endocardial disease; rats sacrificed by 25 mo	..	23(16-25) mo	1%	4
Crl:CD(SD)BR	Unspecified	Spinal nerve root degeneration	♂♀	18-20 mo	66.7%	13
			♂♀	>20 mo	96%; increases with age	
		Retinal lesions	♂♀	1 yr	0/7	29
			♂♀	2 yr	5/75; overall incidence in ♀ is twice that in ♂	
		Otoconia in both saccule & utricle	♂	32 mo	Normal as shown by scanning electron microscopy	27
		Dental malalignment	♂	∼760 d[3]	4/63	20
		Lungs: pneumonia	♂	∼760 d[3]	3/63	20
		Circulatory system				
		Myocardial degeneration	♂	1250 d[3,9]	81%	30
		Periarteritis	♂	1250 d[3,9]	59%	30
		Septicemia secondary to skin infection	♂	∼760 d[3]	1/63	20
		Kidneys				
		Glomerulonephritis	♂	∼760 d[3]	15/63	20
		Progressive	♂	500-599 d[3]	0.7%	7
				1000-1099 d[3]	66.7%	
		Plus chronic nephrosis	♂	1250 d[3,9]	9.8%	30
		Interstitial infiltration	♂	Fn[3]	7.7%	7
		Acute tubular necrosis	♂	Fn[3]	1.0%	7
		Acute pyelonephritis	♂	Fn[3]	5.3%	7
		Muscular dystrophy	♂	∼760 d[3]	5/63	20
		Muscular degeneration	♂	1250 d[3,9]	100%	30
Holtzman (SD)	C	Retinal lesions	♂♀	6 mo	0/2	29
				1 yr	3/38	
				2 yr	31/235	
LE ⟨Long Evans⟩	C	Mesenteric arterial sclerosis	♂	>800 d	1.5%	24
		Chronic nephritis	..	24 mo	High	22
OM	C	Retinal degeneration	..	⩾10 mo	High	34
SD ⟨Sprague-Dawley⟩	Unspecified or C	Cerebral cortex membranous bodies	♂♀	3 mo	None	33
				12-15 mo	50%	
				28 mo	83%	
		Lens changes				
		Cortex, anterior	♂♀	10-12 wk	0	14
				20 mo	4.5%	2
				105-119 wk	1%	14

[1] Unless otherwise specified. [3] At death. [9] Maximum age.

continued

Part III. Lesions

Strain (Synonym)	Rearing or Holding System	Specification	Sex	Age	Incidence[1]	Reference
		posterior	♂♀	10-12 wk	1.0%	14
				20 mo	14.4%	2
				105-119 wk	8.4%	14
		Suture lines	♂♀	10-12 wk	28.2%	14
				105-119 wk	2.0%	
		Anterior Y suture line, heightening	♂♀	20 mo	28.8%	2
		Nucleus	♂♀	10-12 wk	12%	14
				105-119 wk	100%	
		Bilateral optical discontinuity	♂♀	20 mo	72.1%	2
		Hyaloids	♂♀	10-12 wk	22.5%	14
				105-119 wk	1.0%	
		Other	♂♀	10-12 wk	0.5%	14
				105-119 wk	0.5%	
		Lung: pulmonary foam cells	♂	6 mo	7/15	12
				22 mo	10/19	
			♀	6 mo	8/14	
				22 mo	14/20	
		Polyarteritis nodosa	♂♀	755 d	4/13	36
		In relation to no. of times bred				
		Virgin	♂	....	8%	35
			♀	....	9%	
		Bred 4 times	♀	....	6%	35
		Bred 6 times	♂	....	14%	35
		Bred 9 times	♀	....	90%	35
		Kidney: deposition of immunoglobulins, especially IgM	..	3 to 24 mo	Increases	10
SD (CFE)	SPF	Aortic medial calcinosis	♂	597 d; 700 d	Both ages similar	32
		Nephrosclerosis	♂	597 d	60.5%	32
				700 d	62.06%	
SD/Jcl	Unspecified	Fatty change in liver	♂	13-15 mo	2/5	22
				22-24 mo	10/19	
				31-36 mo	2/3	
			♀	13-15 mo	4/5	
				22-24 mo	10/16	
				31-36 mo	3/5	
		Pancreatic islet fibrosis	♂	13-15 mo	3/5	22
				22-24 mo	6/19	
				31-36 mo	0/3	
			♀	13-15 mo	0/5	
				22-24 mo	1/16	
				31-36 mo	0/5	
		Cardiopathy	♂	13-15 mo	2/5	22
				22-24 mo	17/19	
				31-36 mo	2/3	
			♀	13-15 mo	1/5	
				22-24 mo	3/16	
				31-36 mo	2/5	

[1] Unless otherwise specified.

continued

Part III. Lesions

Strain (Synonym)	Rearing or Holding System	Specification	Sex	Age	Incidence[1]	Reference
		Nephropathy	♂	13-15 mo	3/5	22
				22-24 mo	15/19	
				31-36 mo	3/3	
			♀	13-15 mo	1/5	
				22-24 mo	3/16	
				31-36 mo	2/5	
		Parathyroid hyperplasia	♂	13-15 mo	2/5	22
				22-24 mo	10/19	
				31-36 mo	0/3	
			♀	13-15 mo	0/5	
				22-24 mo	1/16	
				31-36 mo	1/5	
Sherman	C	Lens changes				
		Total	♂♀	4-5 mo	5.0%	3
				13-14 mo	40.7%	
				22-24 mo	86.2%	
		Cortex, anterior	♂♀	4-5 mo	2.5%	3
				13-14 mo	17.4%	
				22-24 mo	37.1%	
		posterior	♂♀	4-5 mo	0.8%	3
				13-14 mo	4.6%	
				22-24 mo	49.1%	
		Anterior Y suture	♂♀	4-5 mo	0	3
				13-14 mo	0	
				22-24 mo	18.1%	
		Nucleus, adult	♂♀	4-5 mo	0.8%	3
				13-14 mo	18.6%	
				22-24 mo	50.0%	
		fetal	♂♀	4-5 mo	0.8%	3
				13-14 mo	1.2%	
				22-24 mo	0.9%	
Wistar	Unspecified or C	Kidney				
		Disease	♂	Fn[3]	90%	15
			♀	Fn[3]	60%	
		Lesions	♂	Fn[3]	100%	15
			♀	Fn[3]	80%	
			♂	>13 mo	100%[10]	15
			♀	>24 mo	80%	
		Blood vessels				
		Basophilic deposits in wall	..	868-983 d	47.4%	1
				1000-1170 d	54.5%	
		Lymphocytic infiltration	..	868-983 d	100%	1
				1000-1170 d	100%	
		Glomeruli				
		Basement membrane thickness	♂♀	....	Increases with age	15
		Cytoplasmic fusion of podocytes	..	....	Increases with age	15
		Fibrosis of glomerular tufts				
		Slight	..	868-983 d	44.4%	1
			..	1000-1170 d	36.4%	

[1] Unless otherwise specified. [3] At death. [10] Glomerular lesions precede tubular and interstitial lesions; more severe in males.

continued

Part III. Lesions

Strain (Synonym)	Rearing or Holding System	Specification	Sex	Age	Incidence[1]	Reference
		Moderate to marked	..	868-983 d	10.5%	1
			..	1000-1170 d	0	
		Colloid	..	868-983 d	21.1%	1
			..	1000-1170 d	27.3%	
		IgG deposition	..	....	Increases with age	15
		Tubule				
		Aberrant cells	..	868-983 d	89.5%	1
			..	1000-1170 d	100%	
		Colloid	..	868-983 d	84.2%	1
			..	1000-1170 d	100%	
		Elution of kidney-bound immunoglobulin	♂	....	Increases with age	15
	SPF	Central nervous system				
		Anterior brain				
		White matter: mild cellularity	..	....	Most	31
		Ventricular system: dilated	..	....	5/24	31
		Gracile tract: extensive fiber loss	♂	....	25%	31
		Purkinje cells	..	....	Sometimes decreased in number compared to 1-yr-old rats	31
		Various areas				
		Eosinophilic bodies	..	....	100%	31
		Myelin sheath: distended	..	....	A few	31
		Brachial nerve[11]				
		Mild lesions	..	Fn[3]	22/24	31
		Moderate lesions	..	Fn[3]	2/24	31
		Sciatic nerve[11]				
		Mild lesions	..	Fn[3]	12/24	31
		Moderate lesions	..	Fn[3]	12/24	31
		Periarteritis nodosa	..	....	25%	31
		Nephrosis	..	....	80%	31
		Skeletal muscle changes				
		Front legs	..	....	16/24	31
		Hind legs	..	....	24/24	31
		Posterior paralysis	..	....	25%	31
	Germfree	Mesenteric arterial sclerosis	♂♀	Fn[3]	3.5%	24
		Spleen	♂♀	....	100%	26
		Kidney				
		1st study	♂♀	....	6/16	26
		2nd study	♂♀	....	0	26
Neoplastic						
Inbred ACI/N	C	Tumors[12]				
		Total	♂	Fn[3]	56.4%	21
			♀	Fn[3]	51.7%	
		Testis: interstitial cell	♂	Fn[3]	45.5%	21
		Uterus	♀	Fn[3]	12.0%	21
		Mammary gland	♀	Fn[3]	11.0%	21

[1] Unless otherwise specified. [3] At death. [11] Nerve lesions more severe in males. [12] Mean age for onset of all tumors, >73 wk. Main ones occurred at >100 wk.

continued

Part III. Lesions

Strain ⟨Synonym⟩	Rearing or Holding System	Specification	Sex	Age	Incidence[1]	Reference
		Pituitary	♀	Fn[3]	21.1%	21
		Adrenal	♂	Fn[3]	16.4%	21
BN/Bi	C	Malignant melanoma	♂	35 mo[8]	3%	16
			♀	29 mo[8]	3%	
		Pancreatic islet cell adenoma	♂	35 mo[8]	10%	16
			♀	40 mo[8]	1%	
		Lymphosarcoma	♂	25 mo[8]	13%	16
			♀	28 mo[8]	3%	
		Urinary bladder carcinoma	♂	30 mo[8]	23%	16
			♀	27 mo[8]	5%	
		Mammary gland				
		Fibroadenoma	♂	36 mo[8]	3%	16
			♀	31 mo[8]	11%	
		Carcinoma	♀	30 mo[8]	3%	16
		Pituitary adenoma	♂	21 mo[8]	8%	16
			♀	28 mo[8]	23%	
		Adrenal cortex				
		Adenoma	♂	27 mo[8]	13%	16
			♀	28 mo[8]	22%	
		Carcinoma	♀	32 mo[8]	5%	16
BN/BiRij	C	Tumors				
		Ureter	♂	7-18 mo	3/43	6
				19-36 mo	3/53	
				37-48 mo	0	
			♀	7-18 mo	21/107	
				19-36 mo	33/142	
				37-54 mo	0	
		Urinary bladder	♂	7-18 mo	12/43	6
				19-36 mo	17/53	
				37-48 mo	0	
			♀	7-18 mo	1/107	
				19-36 mo	1/142	
				37-54 mo	1/16	
F344	BR	Cutaneous squamous papilloma	♂	24-30 mo	2.2%	9
		Fibromas, subcutaneous & dermal	♂	24-30 mo	8.7%	9
		Subcutaneous angioma	♂	24-30 mo	2.1%	9
		Lipoma	♂	24-30 mo	2.1%	9
		Pancreatic islet cell lesions	♂	24-30 mo	2.2%	9
		Monocytic leukemia[13]	♂	24-30 mo	29.8%	9
		Generalized lymphosarcoma	♂	24-30 mo	2.1%	9
		Generalized reticulum cell sarcoma	♂	24-30 mo	2.1%	9
		Testis: interstitial cell lesions	♂	24-30 mo	97.6%	9
		Mammary gland lesions	♂	24-30 mo	4.3%	9
		Pituitary lesions	♂	6-12 mo	7.7%	9
				24-30 mo	17.0%	
		Thyroid adenoma	♂	24-30 mo	2.1%	9
		Thyroid medullary carcinoma	♂	24-30 mo	4.3%	9
		Adrenocortical lesions	♂	24-30 mo	4.3%	9
		Pheochromocytoma	♂	24-30 mo	4.3%	9

[1] Unless otherwise specified. [3] At death. [8] Mean age. [13] Synonym: Mononuclear cell leukemia.

continued

Part III. Lesions

Strain ⟨Synonym⟩	Rearing or Holding System	Specification	Sex	Age	Incidence[1]	Reference
F344[14] ⟨Fischer 344⟩	C	Subcutaneous tumors, variable type	♂♀	....	10.4%	11
		Leukemia & related lesions	♂	All ages[4]	Avg, 31%	28
			♀	All ages[4]	Avg, 21%	
		Monocytic[13]	♂♀	....	24.4%	11
		Testis: interstitial cell tumor	♂	....	65-100%	11
				All ages[4]	86%	28
		Mammary gland lesions	♂	All ages[5]	Avg, 23%	28
			♀	All ages[4]	Avg, 41%	
		Pituitary lesions	♂	All ages[4]	Avg, 24%	28
			♀	All ages[5]	Avg, 36%	
WAG/Rij	C	Tumors				
		Total[15]	♀	>3 yr[3]	95%	5
		Chromophobe adenoma	♀	>3 yr[3]	69%	5
		Thyroid medullary carcinoma	♀	>3 yr[3]	39%	5
		Adrenal cortical	♀	<2 yr[3]	14%	5
				>2 yr[3]	30%	
	SPF	Tumors				
		Lymphosarcoma	♀	31 mo[3,8]	0.5%	16
		Testis: mesothelioma	♂	22 mo[3,8]	5%	16
		Granulosa & theca cell	♀	36 mo[3,8]	0.3%	16
		Mammary gland				
		Fibroadenoma	♀	30 mo[3,8]	14%	16
		Carcinoma	♀	29 mo[3,8]	8%	16
		Pituitary adenoma	♂	21 mo[3,8]	96%	16
			♀	31 mo[3,8]	68%	
		Thyroid medullary carcinoma	♂	22 mo[3,8]	25%	16
			♀	31 mo[3,8]	40%	
		Adrenal cortical	♂	21 mo[3,8]	7%	16
			♀	32 mo[3,8]	31%	
Outbred						
Chbb:THOM	SPF	Testis: interstitial cells				
		Hyperplasia, string-like	♂	10 wk	No change	19
				2 yr	6/18	
		nodular	♂	10 wk	No change	19
				2 yr	12/18	
Crl:CD(SD)BR	BR	Pituitary adenoma	♂	1250 d[3,9]	20%	30
LE ⟨Long Evans⟩	C	Thyroid adenoma	..	2 yr	40%	18
Wistar	Unspecified or C	Tumors				
		Total	♂	Fn[3]	33%	15
			♀	Fn[3]	45%	
		Liver	♂	Fn[3]	2/6	15
		Lung	♀	Fn[3]	1/9	15
		Malignant lymphoma	♂	Fn[3]	2/6	15

1/ Unless otherwise specified. 3/ At death. 4/ Median age of onset, >14 mo. 5/ Median age of onset, 20-34 mo. 8/ Mean age. 9/ Maximum age. 13/ Synonym: Mononuclear cell leukemia. 14/ Other tumors: miscellaneous tumors of the skin and uterus [ref. 11, 28]; tumors of the epidermal appendages, liver, & thyroid [ref. 11]; tumors of the brain, salivary gland, pancreas, large intestine, bladder, urethra, prostate, vagina, parathyroid, & bone [ref. 28]; also mesothelioma & pheochromocytomas [ref. 28]. 15/ Includes lymphosarcoma, mammary fibroadenoma, & mammary carcinoma.

continued

75. AGING: RAT

Part III. Lesions

Strain (Synonym)	Rearing or Holding System	Specification	Sex	Age	Incidence[1]	Reference
		Mammary gland	♀	Fn[3]	2/9	15
		Pituitary	♂	Fn[3]	1/6	15
			♀	Fn[3]	6/9	
		Thyroid	♂	Fn[3]	1/6	15
	SPF	Tumors				
		Pituitary adenoma	..	Fn[3]	35%	31
		Mammary adenoma	♀	Fn[3]	25%	31
	Germfree	Tumors				
		Thymus, group 1	♂♀	....	8/16	26
		group 2	♂♀	....	3/33	
		Mammary gland, group 1	♂♀	....	6/16	26
		group 2	♂♀	....	2/33	
		Prostate gland adenocarcinoma	♂	22 mo	1	25
				32-40 mo	3/31	

[1] Unless otherwise specified. [3] At death.

Contributors: Bennett J. Cohen and Nancy Hulett

References

1. Andrew, W., and D. Pruett. 1957. Am. J. Anat. 100: 51-79.
2. Balazs, T., and L. Rubin. 1971. Lab. Anim. Care 21: 267-268.
3. Balazs, T., et al. 1970. Lab. Anim. Care 20:215-219.
4. Boorman, G. A. 1973. Arch. Pathol. 96:39-45.
5. Boorman, G. A., and C. F. Hollander. 1973. J. Gerontol. 28:152-159.
6. Boorman, G. A., and C. F. Hollander. 1974. J. Natl. Cancer Inst. 52:1005-1008.
7. Bras, G., and M. H. Ross. 1964. Toxicol. Appl. Pharmacol. 6:247-262.
8. Burek, J. D., et al. 1976. Vet. Pathol. 13:321-331.
9. Coleman, G. L., et al. 1977. J. Gerontol. 32:258-278.
10. Couser, W. G., and M. M. Stilmant. 1976. Ibid. 31: 13-22.
11. Davey, F. R., and W. C. Moloney. 1970. Lab. Invest. 23:327-334.
12. Flodh, H., et al. 1974. Z. Versuchstierkd. 16:299-312.
13. Gilmore, S. A. 1972. Anat. Rec. 174:251-258.
14. Heywood, R. 1973. Lab. Anim. 7:19-27.
15. Hirokawa, K. 1975. Mech. Aging Dev. 4:301-316.
16. Hollander, C. F. 1976. Lab. Anim. Sci. 26(2-2):320-328.
17. Kendry, G., and F. J. C. Roe. 1969. Lab. Anim. 3: 207-220.
18. Leathem, J. H. 1978. U.S. Dep. HEW Publ. (NIH) 78-161:141-154.
19. Lehmann, L. H., et al. 1975. Beitr. Pathol. Anat. Allg. Pathol. 156:109-116.
20. Lesser, G. T., et al. 1973. Am. J. Physiol. 225:1472-1478.
21. Lutzen, L., and H. Ueberberg. 1973. Beitr. Pathol. Anat. Allg. Pathol. 149:377-385.
22. Muraoka, Y., et al. 1977. Exp. Anim. 26:1-13.
23. National Research Council, Committee on Care and Use of Hypertensive (SHR) Rats. 1976. ILAR News 19(3):G1-G20.
24. Opie, E. L., et al. 1970. Arch. Pathol. 89:306-313.
25. Pollard, M. 1973. J. Natl. Cancer Inst. 51:1235-1241.
26. Pollard, M., and M. Kajima. 1970. Am. J. Pathol. 61: 25-32.
27. Ross, M. D. Unpublished. Univ. Michigan, Dep. Anatomy, Ann Arbor, 1977.
28. Sass, B., et al. 1975. J. Natl. Cancer Inst. 54:1449-1456.
29. Schardein, J. L., et al. 1975. Lab. Anim. Sci. 25:323-326.
30. Simms, H. S., and B. N. Berg. 1957. J. Gerontol. 12: 244-252.
31. Steenis, G. van, and R. Kroes. 1971. Vet. Pathol. 8: 320-332.
32. Stupfel, M., et al. 1975. J. Gerontol. 30:154-156.
33. Vaughn, D. W. 1976. J. Neuropathol. Exp. Neurol. 35:152-166.
34. von Sallmann, L., and P. Grimes. 1972. Arch. Ophthalmol. 88:404-411.
35. Wexler, B. C. 1976. J. Gerontol. 25:373-380.
36. Yang, Y. H. 1965. Lab. Invest. 14:81-88.

For more information on controlled breeding of inbred strains and the influence of genetic factors on the incidence of neoplasms, consult references 1, 15, 17, and 19 in Part I.

Part I. Spontaneous

For additional information on spontaneous tumors, consult references 6-8, 12-14, 23, 31, 34, 39-43, 47, 52, and 53.

Strain ⟨Synonym⟩	Organ	Tumor ⟨Synonym⟩	Sex & Age	Tumor Incidence	Reference
ACH	Lymph nodes, ileocecal	Lymphosarcoma	....	High	20
ACI ⟨AXC⟩	Anterior pituitary	Adenomas[1]	....	15%	20
	Lymph nodes	Lymphatic leukemias[2]	....	....	26
	Prostate	Adenocarcinomas	♂, >34 mo	17%	48
	Testis	Interstitial cell tumors	♂, old	25%	20
ACI/N	Adrenal medulla	Pheochromocytomas	>18 mo[3]	5-10%	24,33
	Anterior pituitary	Adenomas[1]	>18 mo[4]	15-40%	24,33
	Mammary gland	Fibroadenomas, adenomas, carcinomas	♀, >24 mo	11%[5]	33
	Testis	Interstitial cell tumors	♂, 12-18 mo	20%	24
			♀, >18 mo	85%	
	Uterus	Adenocarcinomas primarily[6]	♀, >18 mo	8-12%	24,33
ALB ⟨Albany⟩	Mammary gland	Adenocarcinomas	♀	....	18,56
BN/Bi	Adrenal cortex	Adenomas	♂, 33 mo	12%	25
			♀, 28 mo	22%	
		Carcinomas	♀, 32 mo	5%	25
	Anterior pituitary	Adenomas[1]	♂, 21 mo	8%	25
			♀, 28 mo	23%	
	Cervix & vagina	Sarcomas primarily[7]	♀, >24 mo	20%	9
	Hematopoietic system	Lymphosarcomas	♂, 25 mo	13%	25
			♀, 28 mo	3%	
	Mammary gland	Carcinomas	♀, 30 m0	3%	25
		Fibroadenomas	♀, 31 mo	11%	25
	Pancreatic islets of Langerhans	Adenomas	♂, 35 mo	10%	25
	Ureter	Transitional cell papillomas & carcinomas, & squamous cell carcinomas	♂	6%	3
			♀	20%	
	Urinary bladder	Transitional cell carcinomas	♂, 30 mo	23%	3,25
			♀, 27 mo	5%	
BUF	Adrenal cortex	Adenomas primarily[8]	Old	25%	20
	Anterior pituitary	Adenomas[1]	Old	30%	20
	Thymus	Thymomas	♂	54%	57
			♀	39%	
	Thyroid	Parafollicular cell ⟨C-cell, medullary⟩ tumors[9]	>24 mo	25%[10]	32
BUF/N	Adrenal cortex	Adenomas primarily[8]	>18 mo	30-70%	24
	Adrenal medulla	Pheochromocytomas	>18 mo	5-40%	24
	Anterior pituitary	Adenomas[1]	<18 mo	5-20%	24
			>18 mo	55-75%	
	Uterus	Endometrial stromal tumors	♀, >18 mo[11]	22%	24
COP ⟨Copenhagen⟩	Thymus	Thymomas	....	High	17,20
F344 ⟨Fischer⟩	Adrenal medulla	Pheochromocytomas	....	23%[12]	28
	Anterior pituitary	Adenomas[1]	♂	24%[10]	44
			♀	36%[10]	

[1] Primarily chromophobe adenomas, but chromophobe carcinomas are seen occasionally. [2] Transplantable. [3] Breeders. [4] More often found in females. [5] Two-thirds are benign. [6] Also endometrial polyps and smooth-muscle tumors. [7] Also some squamous cell carcinomas. [8] Also some carcinomas. [9] Benign and malignant. [10] Increases with age. [11] Virgins. [12] 27% bilateral.

continued

Part I. Spontaneous

Strain (Synonym)	Organ	Tumor (Synonym)	Sex & Age	Tumor Incidence	Reference
	Hematopoietic system	Leukemias	♂ ♀	31%[10] 21%[10]	35,44
	Mammary gland	Fibroadenomas	♂ ♀	23% 41%	44
	Testis	Interstitial cell (Leydig cell) tumors[13]	♂, >18 mo	60-90%	11,28
	Thyroid	Parafollicular cell (C-cell, medullary) tumors[9]	>24 mo	22%[10]	32
	Uterus	Endometrial stromal tumors[14]	♀, 22 mo	21%	28
F344/N	Adrenal cortex	Adenomas primarily[8]	>18 mo	≯10%	24
	Adrenal medulla	Pheochromocytomas	>18 mo	10-45%	24
	Anterior pituitary	Adenomas[1]	>18 mo	25%	24
	Testis	Interstitial cell tumors	♂, <18 mo ♂, >18 mo	30%[3]; 70%[15] 100%	24
	Uterus	Endometrial stromal tumors	♀, >18 mo	33%[3]; 75%[11]	24
LOU/C/Wsl	Lymph nodes, ileocecal	Plasmacytomas, malignant (Immunocytomas)[16]	♂ ♀	30% 16%	2,16
M520	Adrenal medulla	Pheochromocytomas	....	21-25%	20
	Uterus	Endometrial stromal tumors	♀, >21 mo	30%	50
M520/N	Adrenal cortex	Adenomas primarily[8]	>18 mo	20-45%	24
	Adrenal medulla	Pheochromocytomas	>18 mo	60-85%	24
	Anterior pituitary	Adenomas[1]	>18 mo	20-40%	24
	Testis	Interstitial cell tumors	♂, >18 mo[15]	35%	24
	Uterus	Endometrial stromal tumors	♀, >18 mo	12-50%	24
Nb	Adrenal gland	Carcinoma[17]	....	....	37
NZR/Gd	Heart	Atriocaval epithelial mesothelioma	....	20%	22
OM (Osborne-Mendel)	Adrenal cortex	Adenomas primarily[8]	....	70%	20
	Anterior pituitary	Adenomas[1]	♂ ♀	60%[10] 30%[10]	45
	Hematopoietic system	Reticulum cell sarcomas	>12 mo	15%	46
	Lymph nodes, pulmonary	Lymphomas, malignant	....	22%	36
	Mammary gland	Fibroadenomas	♀	26-52%	49
	Ovary	Ovarian tumors	♀	21-25%	20
	Thyroid	Parafollicular cell (C-cell, medullary) tumors[9]	>24 mo	33%[10]	32
OM/N	Adrenal cortex	Adenomas primarily[8]	<18 mo >18 mo	50-70% 75-95%	24
	Anterior pituitary	Adenomas[1]	>18 mo	15-20%	24
	Mammary gland	Fibroadenomas	♀	25-30%	24
	Ovary	Ovarian tumors	♀	20-25%	24
	Uterus	Endometrial stromal tumors	♀, >18 mo	10-15%	24
WAG/Rij	Adrenal cortex	Adenomas	♂, 21 mo ♀, 32 mo	7% 31%	25
	Anterior pituitary	Adenomas[1]	♂, 21 mo ♀, 31 mo	96% 68%	25
	Mammary gland	Adenocarcinomas	♀, 29 mo	8%	25
		Fibroadenomas	♀, 30 mo	14%	25

[1] Primarily chromophobe adenomas, but chromophobe carcinomas are seen occasionally. [3] Breeders. [8] Also some carcinomas. [9] Benign and malignant. [10] Increases with age. [11] Virgins. [13] Many are steroid-secreting; frequently bilateral. [14] Primarily benign. [15] Non-breeders. [16] Immunoglobin-secreting. [17] Hormone-dependent.

continued

76. TUMORS: RAT

Part I. Spontaneous

Strain (Synonym)	Organ	Tumor (Synonym)	Sex & Age	Tumor Incidence	Reference
	Thyroid	Parafollicular cell (C-cell, medullary) tumors[2,9,18]	♂, 22 mo ♀, 31 mo	25% 40%	4,5,25,32
	Tunica vaginalis testis	Mesotheliomas	♂, 22 mo	5%	25
W (Wistar)	Adrenal medulla	Pheochromocytomas	....	....	21
	Anterior pituitary	Adenomas[1,19]	♀	27%	54
	Mammary gland	Adenomas	♀, old	....	12,21,58
	Prostate	Adenocarcinomas	♂, >30 mo	13%	38
WF (Wistar Furth)	Anterior pituitary	Adenomas[1]	....	....	27,45,55
		Adenomas[1,2,20]	....	High	20
	Hematopoietic system	Leukemias[2]	....	18%	20,35
	Mammary gland	Adenocarcinoma, MT/W9[21]	....	....	10,29
		Fibroadenomas	♀, 20 mo	21%	30
Wistar-Af/Han	Brain	Meningiomas	....	....	51
WN	Anterior pituitary	Adenomas[1]	Old	21-25%	20
		Fibroadenomas	....	30-50%	20
WN/N	Adrenal cortex	Adenomas primarily[8]	>18 mo	8-12%	24
	Adrenal medulla	Pheochromocytomas	>18 mo	25-50%	24
	Anterior pituitary	Adenomas[1]	>18 mo	40-93%[10]	24
	Mammary gland	Fibroadenomas[9]	♀	30-50%	24

[1] Primarily chromophobe adenomas, but chromophobe carcinomas are seen occasionally. [2] Transplantable. [8] Also some carcinomas. [9] Benign and malignant. [10] Increases with age. [18] Calcitonin-secreting. [19] Hormone-secreting. [20] Mammotropic. [21] Prolactin (mammotropin) - dependent/independent variants.

Contributors: Dawn G. Goodman and Norman H. Altman; Hervé Bazin, Guy Burtonboy, and Andrée Beckers; R. W. Baldwin

References

1. Bazin, H., and A. Beckers. 1976. Nobel Symp. 33: 125-151.
2. Bazin, H., et al. 1973. J. Natl. Cancer Inst. 51:1359-1361.
3. Boorman, G. A., and C. F. Hollander. 1974. Ibid. 52: 1005-1008.
4. Boorman, G. A., et al. 1972. Arch. Pathol. 94:35-41.
5. Boorman, G. A., et al. 1974. J. Natl. Cancer Inst. 53: 1011-1015.
6. Bullock, F. D., and M. R. Curtis. 1931. Am. J. Cancer 15:67-121.
7. Bullock, F. D., and G. L. Rohdenburg. 1917. J. Cancer Res. 2:39-61.
8. Burek, J. D., and C. F. Hollander. 1977. J. Natl. Cancer Inst. 58:99-105.
9. Burek, J. D., et al. 1976. Ibid. 57:549-551.
10. Butler, T. P., and P. M. Gullino. 1975. Cancer Res. 35:512-516.
11. Cockrell, B. Y., and F. M. Garner. 1976. Comp. Pathol. Bull. 8:2-4.
12. Crain, R. C. 1958. Am. J. Pathol. 34:311-323.
13. Curtis, M. R., et al. 1931. Am. J. Cancer 15:67-121.
14. Davis, R. K., et al. 1956. Cancer Res. 16:194-197.
15. Dees, J. H., et al. 1976. J. Natl. Cancer Inst. 57:779-808.
16. Druckrey, H., et al. 1961. Naturwissenschaften 48: 722-723.
17. Dunning, W. F., and M. R. Curtis. 1946. Cancer Res. 6:61-81.
18. Durin, P. W., et al. 1966. Ibid. 26:400-411.
19. Eker, R., and J. Mossige. 1961. Nature (London) 189: 858-859.
20. Festing, M. F. W., and J. Staats. 1973. Transplantation 16:221-245.
21. Gilbert, M., and C. Gilman. 1958. S. Afr. J. Med. Sci. 23:257-272.
22. Goodall, C. M., et al. 1975. J. Pathol. 116:239-250.
23. Guerin, M. 1954. Tumeurs spontanées des animaux de laboratoire. Legrand, Paris. pp. 41-133.
24. Hansen, C. T., et al. 1974. U.S. Dep. HEW Publ. (NIH) 74-606.
25. Hollander, C. F. 1976. Lab. Anim. Sci. 26:320-328.
26. Iglesias, R., and E. Mardones. 1956. Proc. Am. Assoc. Cancer Res. 2:121.
27. Ito, A., et al. 1972. J. Natl. Cancer Inst. 49:701-711.
28. Jacobs, B. B., and R. A. Huseby. 1967. Ibid. 39:303-309.
29. Kim U., and M. J. Depowski. 1975. Cancer Res. 35: 2068-2077.
30. Kim, U., et al. 1960. J. Natl. Cancer Inst. 24:1031-1055.
31. Kinkel, V. H. J. 1971. Z. Versuchstierkd. 13:97-100.
32. Lindsay, S., et al. 1968. Arch. Pathol. 86:353-364.
33. Maekawa, A., and S. Odashima. 1975. J. Natl. Cancer Inst. 55:1437-1445.
34. Miyamoto, M., and S. Takizawa. 1975. Ibid. 55:1471-1472.
35. Moloney, W. C., et al. 1970. Cancer Res. 30:41-43.
36. Nelson, A. A., and H. J. Morris. 1941. Arch. Pathol. 3:578-584.

continued

Part I. Spontaneous

37. Noble, R. L., et al. 1975. Cancer Res. 35:766-780.
38. Pollard, M. 1973. J. Natl. Cancer Inst. 51:1235-1241.
39. Pollard, M., and B. A. Teah. 1963. Ibid. 31:457-465.
40. Prejean, J. D., et al. 1973. Cancer Res. 33:2768-2773.
41. Ratcliffe, H. L. 1940. Am. J. Pathol. 16(3):237-255.
42. Ratcliffe, H. L. 1963. In E. J. Farris and J. Q. Griffith, Jr., ed. The Rat in Laboratory Investigation. Ed. 2. Hafner, New York. pp. 521-524.
43. Ross, M. H., and G. Bras. 1965. J. Nutr. 87:245-260.
44. Saas, B., et al. 1975. J. Natl. Cancer Inst. 54:1449-1456.
45. Saxton, J. A., Jr., and J. B. Graham. 1944. Cancer Res. 4:168-175.
46. Saxton, J. A., Jr., et al. 1948. Acta Unio Int. Contra Cancrum 6:423-431.
47. Schardein, J. C., et al. 1968. Pathol. Vet. 5:238-252.
48. Shain, S. A., et al. 1975. J. Natl. Cancer Inst. 55:177-180.
49. Sher, S. P. 1972. Toxicol. Appl. Pharmacol. 22:562-588.
50. Snell, K. C. 1965. In W. E. Ribelin and J. R. McCoy, ed. The Pathology of Laboratory Animals. Thomas, Springfield, IL. pp. 241-302.
51. Sumi, N., et al. 1976. Arch. Toxicol. 35:1-13.
52. Thompson, S. W. 1961. J. Natl. Cancer Inst. 27:1037-1057.
53. Thompson, S. W., and R. D. Hunt. 1963. Ann. N.Y. Acad. Sci. 108:832-845.
54. Untae, K., et al. 1960. J. Natl. Cancer Inst. 24:1031-1055.
55. Wolfe, J. M., et al. 1938. Am. J. Cancer 34:352-372.
56. Wright, A. W., et al. 1940. Am. J. Pathol. 16:817-834.
57. Yamada, S., et al. 1973. Gann 64:287-291.
58. Young, S., and R. C. Hallowes. 1973. World Health Organ. Int. Agency Res. Cancer Sci. Publ. 5, 1(1):31-75.

Part II. Transplantable

Characteristics: CCL = Certified Cell Line. For additional information on transplantable tumors, consult reference 19. Information on a collection of frozen transplantable tumor-cell lines can be obtained by writing to: Frozen Tissue Tumor Bank, Mammalian Genetics & Animal Production Section, Division of Cancer Treatment, National Cancer Institute, Bldg. 201/Ft. Detrick, Frederick, Maryland 21701, USA.

Strain ⟨Synonym⟩	Organ	Tumor	Characteristics	Reference
COP ⟨Copenhagen⟩	Prostate	Adenocarcinoma, R3327	Androgen-sensitive/insensitive lines	20,21
Harlan Wistar	Mammary gland	Carcinoma, Walker rat	CCL 38	12
LOU/C/Wsl	Lymphoid tissue	Immunocytomas, malignant	Immunoglobulin-secreting	5,7
LOU/M/Wsl	Lymphoid tissue	Immunocytomas, malignant	Immunoglobulin-secreting	5,7
W/Not	Mammary gland	Adenocarcinomas, Sp 4	...	3
		Adenocarcinomas, Sp 15	...	3
	Skin	Squamous cell carcinomas, Sp1	Metastasizes	2
WF ⟨W/Fu⟩	...	Lymphomas, malignant	...	11
⟨Wistar Furth⟩	Pituitary	Epithelial-like tumor	CCL 82, growth-hormone-producing	10,13
Unspecified	Brain	Glial cell tumor	CCL 107, S-100 protein-secreting	1,4
		Schwannomas	...	9
	Liver	Hepatomas, Morris	CCL 144, albumin-secreting	18
		Yoshida	Ascites	22,23
	Unspecified	Leukemias	...	8
		Polyoma-virus-induced tumor	...	17
		Sarcomas, Jensen	CCL 45, asparagine-requiring	14,15
		Schmidt-Ruppin	CCL 47, virally induced	16
		SV40-induced tumor	...	6

Contributors: Hervé Bazin, Guy Burtonboy, and Andrée Beckers

References

1. Albright, A. L., et al. 1977. Cancer Res. 37(1):2512-2522.
2. Baldwin, R. W. 1966. Int. J. Cancer 1:257-264.
3. Baldwin, R. W., and M. J. Embleton. 1969. Ibid. 4:430-439.
4. Benda, P., et al. 1968. Science 161:370-371.
5. Burtonboy, G., et al. 1973. Eur. J. Cancer 9:259-262.
6. Chassouk, O. 1975. Int. J. Cancer 16:515-525.
7. Curtis, M. R., and W. F. Dunning. 1940. Am. J. Cancer 3:299-309.

continued

Part II. Transplantable

8. Dunning, W. F., and M. R. Curtis. 1957. J. Natl. Cancer Inst. 19:845-853.
9. Fields, K. L. 1975. Proc. N.Y. Acad. Sci. 72:1296-1300.
10. Furth, J., et al. 1953. Proc. Soc. Exp. Biol. Med. 84: 253-254.
11. Glaser, M., et al. 1974. Cancer Res. 34:2165-2171.
12. Hull, R. N., et al. 1956. Anat. Rec. 124:490.
13. Ito, A. 1976. Am. J. Pathol. 83:424-426.
14. Jensen, C. O. 1909. Z. Krebsforsch. 7:45-54.
15. McEuen, C. S. 1939. Am. J. Cancer 36:383-385.
16. Nichols, W. W. 1963. Hereditas 50:53-80.
17. Rhim, J. S. 1974. Proc. Soc. Exp. Biol. Med. 147: 730-735.
18. Richardson, U. I., et al. 1969. J. Cell Biol. 40:236-247.
19. Roberts, D. C., and B. Drobycz. 1977. Series of Cross References Bibliography. Imperial Cancer Research Fund, Mill Hill, London, England.
20. Voigt, W., and W. F. Dunning. 1974. Cancer Res. 34:1447-1450.
21. Voigt, W., et al. 1975. Ibid. 35:1840-1846.
22. Yoshida, T. 1957. Virchows Arch. 330:85-93.
23. Yoshida, T., et al. 1944. Proc. Imp. Acad. (Tokyo). 20:611.

Part III. Radiation-Induced

Strain ⟨Synonym⟩	Organ	Tumor	Inducer	Reference
Albino Wistar	Lung	Carcinomas	^{106}Ru	4
SD ⟨Sprague-Dawley⟩	Paranasal sinus	Epidermoid carcinomas	$[^{144}Ce]CeCl_4$	3
Wistar L	Kidneys	Carcinomas & adenomas 1/	X rays	5
Unspecified	Skeletal system	Osteosarcomas	^{32}P	2
			^{89}Sr	9
			^{144}Ce	9
			^{232}Pu	6
			^{238}U	9
	Skin	Carcinomas	Ultraviolet	7,8
		Basal cell	X rays	1

1/ Less frequently, sarcomas.

Contributors: Hervé Bazin and Guy Burtonboy

References

1. Albert, R. E., et al. 1969. Cancer Res. 29:658-668.
2. Cobb, L. M. 1970. Br. J. Cancer 24:294-299.
3. Jasmin, J. R., et al. 1977. J. Natl. Cancer Inst. 58(3): 423-429.
4. Kuschner, M., and S. Laskin. 1971. Am. J. Pathol. 64: 183-196.
5. Maldague, P., et al. 1958. Sang 29:751-763.
6. Siniakov, E. G. 1975. Radiobiologiya 15:937-940.
7. Stenbaeck, F. 1975. Oncology 31:209-225.
8. Wiskemann, A. 1975. Lichtkrebs Strahlenther. 150: 195-198.
9. Young, S., and R. C. Hallowes. 1973. World Health Organ. Int. Agency Res. Cancer Sci. Publ. 5, 1(1):31-75.

Part IV. Virally Induced

Strain ⟨Synonym⟩	Organ	Tumor	Inducer	Reference
BN	Unspecified	MST-1 sarcoma	Moloney sarcoma virus	4
Cb/Se (Cb)	Lymph nodes	Plasmacytomas	Murine sarcoma virus	10

continued

76. TUMORS: RAT

Part IV. Virally Induced

Strain ⟨Synonym⟩	Organ	Tumor	Inducer	Reference
LEW ⟨Lewis albino⟩	Unspecified	Sarcomas	Feline RNA virus	8
			Rous sarcoma virus, Schmidt-Ruppin strain	2,12
NBR Pl/Cr ⟨NZB[1]⟩	Skeletal system	Osteosarcomas	Murine sarcoma virus	9,11
OM ⟨Osborne-Mendel⟩	Lymph nodes	Plasmacytomas	Murine sarcoma virus	10
SD ⟨Sprague-Dawley⟩	Lymph nodes	Plasmacytomas	Murine sarcoma virus	10
WF ⟨W/Fu⟩	Unspecified	Lymphoma, (C58NT)D	Gross leukemia virus	3
WKA ⟨Wistar King A⟩	Lymph nodes	Lymphoma	Friend erythroleukemia virus	5,6
			Gross leukemia virus	7
			Rauscher erythroleukemia virus	7
Unspecified	Lymph nodes	Leukemias	Mouse leukemia passage A virus	1,13
	Thymus	Thymic tumors	Murine leukemia virus	13

1/ Young rats.

Contributors: Guy Burtonboy, Hervé Bazin, and Andrée Beckers; R. W. Baldwin

References

1. Gross, L., et al. 1961. Proc. Soc. Exp. Biol. Med. 106: 890-893.
2. Harris, R. J. C., and F. C. Chesterman. 1964. Natl. Cancer Inst. Symp. 17:321-335.
3. Herberman, R. B., and M. E. Oren. 1971. J. Natl. Cancer Inst. 46:391-396.
4. Jones, J. M., and J. D. Feldman. 1975. Ibid. 55:995-997.
5. Kobayashi, H., et al. 1969. Cancer Res. 29:1385-1392.
6. Kobayashi, H., et al. 1972. Gann Monogr. 12:73-87.
7. Kobayashi, H., et al. 1973. Cancer Res. 33:1598-1603.
8. Mawdesley-Thomas, L. E., and A. J. Newman. 1974. J. Pathol. 112:107-116.
9. Olson, H. M., and C. C. Capen. 1977. J. Natl. Cancer Inst. 58:433-438.
10. Ribacchi, R., and G. Giraldo. 1966. Lav. Ist. Anat. Istol. Patol. Univ. Studi Perugia 26:149-156.
11. Soehner, R. L., et al. 1969. Bibl. Haematol. (Basel) 36:593-599.
12. Svoboda, J. 1962. Folia Biol. (Prague) 8:215-219.
13. Young, S., and R. C. Hallowes. 1973. World Health Organ. Int. Agency Res. Cancer Sci. Publ. 5, 1(1):31-75.

Part V. Chemically Induced

Strain ⟨Synonym⟩	Organ	Tumor	Inducer	Reference
ACI/N	Brain	Unspecified	3-Acetyl-1-methyl-1-nitrosourea	14
	Colon	Adenocarcinoma, R2	*N*-Methyl-*N*′-nitro-*N*-nitrosoguanidine	12
BD IX	Colon	Adenocarcinoma, DHA	1,2-Dimethylhydrazine	17
	Intestine	Adenocarcinomas	4-Amino-2′,3-dimethylbiphenyl	33
			1,2-Dimethylhydrazine	6,10,18
	Unspecified	Leukemia, L5222	1-Ethyl-1-nitrosourea	35
BUF ⟨Buffalo⟩	Kidney	Adenocarcinoma, MK1	*N*-(4′-Fluoro-4-biphenyl)acetamide	20
	Liver	Hepatoma, 5123	*N*-2-Fluorenylphthalamic acid	19
		7777	*N*-2-Fluorenylphthalamic acid	19
DONRYU[1]	Hematopoietic system	Leukemias	1-Propyl-1-nitrosourea	25
		Erythroleukemias	1-Ethyl-1-nitrosourea	24
F344 ⟨Fischer 344⟩	Kidneys	Adenocarcinomas	*N*-(4′-Fluoro-4-biphenyl)acetamide	7
	Mammary gland	Adenocarcinoma, 13762	9,10-Dimethyl-1,2-benzanthracene[2]	5

1/ Female. 2/ Synonym: 7,12-Dimethylbenz[*a*]anthracene.

continued

Part V. Chemically Induced

Strain ⟨Synonym⟩	Organ	Tumor	Inducer	Reference
	Skin	Squamous cell carcinoma, Svoboda[3]	Lasiocarpine	29
	Thyroid	Thyroid tumors, 1-8	2-Thiouracil	16,32
HO ⟨Hooded⟩	Subcutaneous tissue	Sarcoma, HSN	Benzo[*a*]pyrene	8
LE ⟨Long-Evans⟩	Pancreas	Adenocarcinomas	9,10-Dimethyl-1,2-benzanthracene[2]	4
Nb	Cervix	Carcinoma[4]	Estrogen	21
SD ⟨Sprague-Dawley⟩	Hematopoietic system	Leukemias	1-Butyl-1-nitrosourea	22,23
	Mammary gland	Adenocarcinomas	2-Fluorenamine	33
W ⟨Wistar⟩	Hematopoietic system	Leukemias	Methylcholanthrene	27
	Liver	Hepatomas	*p*-Dimethylaminoazobenzene	11,15,26
	Lymph nodes	Lymphomas	Silicon dioxide[5]	31
W ⟨Wistar⟩[1]	Mammary gland	Adenocarcinomas	3-Methylcholanthrene	33
W/Not	Liver	Hepatoma, D23	*p*-Dimethylaminoazobenzene	1
	Mammary gland	Adenocarcinoma, AAF57	*N*-2-Fluorenylacetamide	2
	Subcutaneous tissue	Sarcoma, Mc 7	3-Methylcholanthrene	3
		Mc 57	3-Methylcholanthrene	3
WF ⟨W/Fu⟩	Colon	Adenocarcinoma, NG-W1	*N*-Methyl-*N'*-nitro-*N*-nitrosoguanidine	28
Unspecified	Esophagus	Squamous cell carcinomas	*N*-Methyl-*N*-nitrosoaniline	9,33
	Hematopoietic system	Leukemias	1-Butyl-3-dimethyl-1-nitrosourea	30
			Ethylnitrosourea	13
	Skin	Basal cell carcinomas	Anthramine	33,34
			3-Methylcholanthrene	33,34
		Squamous cell carcinomas	9,10-Dimethyl-1,2-benzanthracene[2]	33

[1] Female. [2] Synonym: 7,12-Dimethylbenz[a]anthracene. [3] Metastasizes (to the lung). [4] Hormone-dependent. [5] Synonym: Silica.

Contributors: R. W. Baldwin; Hervé Bazin, Guy Burtonboy, and Andrée Beckers

References

1. Baldwin, R. W., and C. R. Barker. 1967. Int. J. Cancer 2:355-364.
2. Baldwin, R. W., and M. J. Embleton. 1969. Ibid. 4: 47-53.
3. Baldwin, R. W., and M. V. Pimm. 1973. Br. J. Cancer 28:281-287.
4. Bockman, D. E., et al. 1976. J. Natl. Cancer Inst. 57: 931-936.
5. Bogden, A. E., et al. 1974. Cancer Res. 34:1627-1631.
6. Burdette, W. J. 1973. Comp. Pathol. Bull. 5:1-5.
7. Dees, J. H., et al. 1976. J. Natl. Cancer Inst. 57:779-808.
8. Denham, S., et al. 1975. Transplantation 19:102-114.
9. Druckrey, H., et al. 1961. Naturwissenschaften 48: 722-723.
10. Druckrey, H., et al. 1967. Ibid. 54:285-286.
11. Edmondson, H. A. 1958. U.S. Armed Forces Inst. Pathol. Fasc. 25.
12. Goto, K., et al. 1975. Gann 66:89-93.
13. Hadjiolov, D. 1972. Z. Krebsforsch. 77:98-100.
14. Maekawa, A., et al. 1976. Ibid. 86:195-207.
15. Maldague, P. 1966. Pathol. Eur. 1(4):321-409.
16. Mandato, E., et al. 1975. Cancer Res. 35:3089-3093.
17. Martin, F., et al. 1975. Int. J. Cancer 15:144-151.
18. Maruyama, K., et al. 1975. Bibl. Haematol. 40:93-95.
19. Morris, H. P., and B. P. Wagner. 1968. Methods Cancer Res. 3:125-152.
20. Morris, H. P., et al. 1970. Cancer Res. 30:1362-1369.
21. Noble, R. L., et al. 1975. Ibid. 35:766-780.
22. Odashima, S. 1972. Gann Monogr. 12:283-296.
23. Odashima, S., et al. 1975. Gann 66:615-621.
24. Ogiu, T., et al. 1974. Ibid. 65:377.
25. Ogiu, T., et al. 1975. J. Natl. Cancer Inst. 54:887-893.
26. Richardson, U. I., et al. 1969. J. Cell Biol. 40:236-247.
27. Shay, H., et al. 1951. Cancer Res. 11:29-34.
28. Steele, G., and H. O. Sjögren. 1974. Ibid. 34:1801-1807.
29. Svoboda, D. J., and J. K. Reddy. 1974. J. Natl. Cancer Inst. 53:1415-1418.
30. Takeuchi, M., et al. 1976. Ibid. 56:1177-1181.
31. Wagner, M., and J. C. Wagner. 1972. Ibid. 49:81-89.

continued

32. Wollman, S. H. 1961. Ibid. 26:473-488.
33. Young, S., and R. C. Hallowes. 1973. World Health Organ. Inst. Agency Res. Cancer Sci. Publ. 5, 1(1):31-75.
34. Zackheim, H. S., et al. 1959. J. Invest. Dermatol. 33:57-64.
35. Zeller, W. J., et al. 1975. Cancer Res. 35:1168-1174.

77. GENETICALLY CONTROLLED BIOCHEMICAL VARIANTS: RAT

Gene Symbol	Gene Name	Linkage Group	Reference
Alp	Alkaline phosphatase	....	8
Es-1	Esterase-1	V	1
Es-2	Esterase-2	V	7,13
Es-3	Esterase-3	V	14
Es-4	Esterase-4	V	15,16
Es-5	Esterase-5	V	10
Fum	Fumarate hydratase[1]	....	4
Gl-1	Plasma protein	VI	11
G-3-pdh[2]	Glyceraldehyde-phosphate dehydrogenase[3]	....	4
α-Gpdh[2]	Glycerol-3-phosphate dehydrogenase (NAD^+)[4]	....	4
Hbb	Hemoglobin β-chain	I	2,5
Pgd	Phosphogluconate dehydrogenase	....	3
Pgm	Phosphoglucomutase	....	9
Svp	Seminal vesicle protein	IV	6
Xdh	Xanthine dehydrogenase	....	4
No gene symbol	2,3-Diphosphoglycerate	....	12

[1] Synonym: Fumarase. [2] The Committee on Standardized Genetic Nomenclature of the Rat is currently considering rules for rat gene nomenclature. The Committee hopes to use extensively in its guidelines the established rules for mouse gene nomenclature. *G-3-pdh* and *α-Gpdh* do not conform to the rules of mouse gene nomenclature, and will probably be changed after the Committee issues its own set of guidelines. [3] Synonym: Glyceraldehyde-3-phosphate dehydrogenase. [4] Synonym: α-Glycerophosphate dehydrogenase.

Contributor: James E. Womack

References

1. Augustinsson, K. B., and B. Henricson. 1966. Biochim. Biophys. Acta 123:323-331.
2. Brdička, R. 1966. Folia Biol. (Prague) 12:305-306.
3. Carter, N. D., and C. W. Parr. 1969. Nature (London) 224:1214.
4. Eriksson, K., et al. 1976. Heredity 37:341-349.
5. French, E. A., et al. 1971. Biochem. Genet. 5:397-404.
6. Gasser, D. L. 1972. Ibid. 6:61-63.
7. Gasser, D. L., et al. 1973. Ibid. 10:207-217.
8. Jimenez-Marin, D. 1974. J. Hered. 65:235-237.
9. Koga, A., et al. 1972. Jpn. J. Genet. 46:335-338.
10. Moutier, R., et al. 1973. Biochem. Genet. 9:109-115.
11. Moutier, R., et al. 1973. Ibid. 10:395-398.
12. Noble, N. A., and G. J. Brewer. 1977. Genetics 85:669-679.
13. Womack, J. E. 1972. J. Hered. 63:41-42.
14. Womack, J. E. 1972. Experientia 28:1372.
15. Womack, J. E. 1973. Biochem. Genet. 9:13-24.
16. Womack, J. E., and M. Sharp. 1976. Genetics 82:665-675.

78. IMMUNOGLOBULIN POLYMORPHISMS: RAT

The biochemical polymorphism best known among the rat serum proteins is undoubtedly that of the kappa type of the immunoglobulin light chains (for a review of this subject, consult reference 8). Since the nomenclature has not

continued

been standardized, Part I lists the designations used by various authors and their corresponding relationship. Part II shows the distribution of known allotypes of rat immunoglobulins: Iκ(1) of the light chains of kappa type, Iα(1) of the alpha heavy chains [ref. 3], and Iγ2b(1) of the gamma 2b heavy chains [ref. 4].

Part I. Nomenclature of Kappa-Chain Allotypic Alleles

The equivalence of the designations from references 1, 5, 6, and 9 have been confirmed; the others are probable.

Equivalent designations of allotypic alleles	Iκ(1a)	Ra-1	Non RI-1	RI-1b	RL2	–	W-1
	Iκ(1b)	Non Ra-1	RI-1	RI-1a	RL1	+	SD-1
Reference	5	2	10	6	9	7	1

Contributor: Hervé Bazin

References

1. Armerding, D. 1971. Eur. J. Immunol. 1:39-45.
2. Barabas, A. Z., and A. S. Kelus. 1967. Nature (London) 215:155-156.
3. Bazin, H., et al. 1974. J. Immunol. 112:1035-1041.
4. Beckers, A., and H. Bazin. 1975. Immunochemistry 12:671-675.
5. Beckers, A., et al. 1974. Ibid. 11:605-609.
6. Gutman, G. A., and I. L. Weissman. 1971. J. Immunol. 107:1390-1393.
7. Humphrey, R. L., and G. S. Santos. 1971. Fed. Proc. Fed. Am. Soc. Exp. Biol. 30:248.
8. Nezlin, R. S., and O. V. Rokhlin. 1976. Contemp. Top. Mol. Immunol. 5:161-184.
9. Rokhlin, O. V., and R. S. Nezlin. 1974. Scand. J. Immunol. 3:209-214.
10. Wistar, R., Jr. 1969. Immunology 17:23-32.

Part II. Allotype Distribution

Iκ(1): *Iκ(1b)* is differentiated from *Iκ(1a)* by antigenic properties of C part [ref. 11], and in CNBr cleavage products [ref. 13]; C region of *Iκ(1b)* differs from C region of *Iκ(1a)* in at least 11 positions including possible antigenic regions—153-155 and 184-189 residues from N end of chain [ref. 8]. *Iκ(1)* is unlinked to MHL, three coat color genes [ref. 7], and *Iα(1)* and *Iγ2b(1)* markers.

Iα(1): *a* alleles are differentiated from non-*a* alleles by antigenic properties; *Iα(1)* is unlinked to *Iκ(1)* markers, but linked to *Iγ2b(1)* markers [ref. 3, 4].

Iγ2b(1): *a* and *b* alleles are differentiated from one another by antigenic properties; *Iγ2b(1)* is unlinked to *Iκ(1)* markers, but linked to *Iα(1)* markers [ref. 4].

Holder: For full name, *see* list of HOLDERS at front of book.

Abbreviations: a = allele *a*; b = allele *b*; not a = absence of allele *a*; nd = not determined.

Strain ⟨Synonym⟩	Iκ(1)	Iα(1)	Iγ2b(1)	Holder	Reference
ACI	b	not a	a	MAI, PL	1,4,6,7,9
ACJ	b	not a	a	MAX	4,6
ACP	a	not a	a	PL	4,6
AGUS	a	not a	a	LAC	3,4,6
ALB ⟨Albany⟩	b	a	a	PL	4,6
AO	a	nd	nd	OXF	10
AS	a	a	a	MAX	1,3,4,6
ASH/W ⟨Wistar outbred⟩	a	nd	nd	....	12
AS2	a	a	a	MAX	1,3,4,6
Atrichis	a	a	a	ORL	3,4,6

continued

Part II. Allotype Distribution

Strain ⟨Synonym⟩	Iκ(1)	Iα(1)	Iγ2b(1)	Holder	Reference
AUG ⟨August⟩	a	not a	a	ORL, PL, WSL	1,3,4,6,9,12
AVN	a	not a	a	CUB, MAX	1,4,6,12
AXC	a	not a	a	PL, WSL	3,4,6
BD	b	nd	nd	....	6,12
BD V	b	not a	a	MAX	4,6
BD V	a	nd	nd	CUB	12
BD IX	a	not a	a	PFD	4,5
BD X	a	not a	a	CUB	4,6
BDE	a	nd	nd	HAN	1
BH	a	nd	nd	....	9
BICR (Marshall)	a	nd	nd	....	1
BIRMA	a	nd	nd	....	12
BIRMB	a	nd	nd	....	12
Black and White Hooded	b	nd	nd	CBI	2
BN	a	a	a	CUB, MAI, PFD, PIT, PL, ORL	1,3,4,6,7,9,12
BP	a	nd	nd	CUB	1,12
BS	a	a	a	MAX	1,4,6
BUF ⟨Buffalo⟩	a	a	a	MAI, PL	1,4,6,7,9,12
CAP	a	a	a	CUB	1,4,6
COP ⟨Copenhagen⟩	b	not a	a	MAX, PL	4,6
DA	b	not a	a	MAI, MAX, OXF, WSL	1,3,4,6,7,9,10,14
E3	a	nd	nd	HAN	1
F-45	a	nd	nd	....	12
F344	b	a	a	MAI	1,4,6,9
⟨Fischer⟩	b	a	a	CRL, ICO, PL	3,4,6,7,12
Gowans (Albino)	a	nd	nd	....	2
HS	a	nd	nd	CUB	1
HW	a	nd	nd	CUB	1
Hypodactyly	a	a	a	ORL	4,6
Jaundice ⟨GUNN⟩	b	not a	a	PL	3,4,6
KGH	a	a	a	PIT, WSL	5
L/E	a	nd	nd	....	12
LE ⟨Long Evans⟩	a	a	a	ORL, PL	3,4,6
LEP	a	not a	nd	CUB	1,6
LEW	a	nd	nd	MAI	1,7
⟨Lewis⟩	a	a	a	....	1,3,4,6,9,12,14
LIS ⟨Hooded Lister⟩	a	a	a	CBI, WSL	5
LOU/C	a	a	a	WSL	3,4,6,12
LOU/C.IH(AUG)	a	not a	a	WSL	4,5
LOU/C.IH(AxC)	a	not a	a	WSL	5
LOU/C.IH(OKA)	a	not a	b	WSL	4,6
LOU/C.IK(OKA)	b	a	a	WSL	4,5
LOU/M	a	a	a	N, WSL	4,6
MAR ⟨MARSHALL⟩	a	a	a	PL	1,4,6,12
MAXX	a	nd	nd	....	9
MSU	b	a	a	....	4,6,12
NBR	nd	not a	nd	....	6
OFA	b	not a	a	ICO	3,4,6
OKA ⟨OKAMOTO⟩	b	not a	b	WSL	3,4,6
OM ⟨Osborn-Mendel⟩	nd	not a	nd	PL	6
PB	a	nd	nd	CUB	1
PVG/c	a	a	a	OXF, LAC	3-6,10
R	a	nd	nd	....	12

continued

78. IMMUNOGLOBULIN POLYMORPHISMS: RAT

Part II. Allotype Distribution

Strain ⟨Synonym⟩	Iκ(1)	Iα(1)	Iγ2b(1)	Holder	Reference
RA	a	nd	nd	....	12
SD ⟨Sprague-Dawley⟩	b	not a	a	PL	1,4,6,12
SEL ⟨Selfed⟩	a	a	a	PL	4,6
SH	b	not a	b	PL	4,6
Sherman	a	a	a	ORL	3,4,6
S5B	nd	not a	nd	PL	6
W ⟨Wistar⟩	a	nd	nd	....	1,2,6,12
WAG ⟨Wistar AG⟩	a	a	a	CUB, ORL	1,3,4,6,12
WF	a	nd	nd	MAI	1,9
⟨Wistar Fu or W/Fu⟩	a	a	a	MAI, PL	4,6,7
WFH	a	nd	nd	....	9
Wistar Af	a	nd	nd	HAN	1
Wistar BB	a	nd	nd	CBI	1
Wistar R	a	a	a	....	3,4,6
YOS ⟨Yoshida⟩	b	a	a	PL	4,6
Zimmerman	a	not a	a	PL	4,6

Contributors: Hervé Bazin; Roald Nezlin; Radim Brdička

References

1. Armerding, D. 1971. Eur. J. Immunol. 1:39-45.
2. Barabas, A. Z., and A. S. Kelus. 1967. Nature (London) 215:155-156.
3. Bazin, H., et al. 1974. J. Immunol. 112:1035-1041.
4. Beckers, A., and H. Bazin. 1975. Immunochemistry 12:671-675.
5. Beckers, A., and H. Bazin. Unpublished. Univ. Louvain, Brussels, Belgium, 1977.
6. Beckers, A., et al. 1974. Immunochemistry 11:605-609.
7. Gutman, G. A., and I. L. Weissman. 1971. J. Immunol. 107:1390-1393.
8. Gutman, G. A., et al. 1975. Proc. Natl. Acad. Sci. USA 72:5046-5050.
9. Humphrey, R. L., and G. S. Santos. 1971. Fed. Proc. Fed. Am. Soc. Exp. Biol. 30:248.
10. Hunt, S. V., and A. F. Williams. 1974. J. Exp. Med. 139:479-496.
11. Nezlin, R. S., et al. 1974. Immunochemistry 11:517-518.
12. Rokhlin, O. V., and R. S. Nezlin. 1974. Scand. J. Immunol. 3:209-214.
13. Vengerova, T. I., et al. 1972. Immunochemistry 9:1239-1246.
14. Wistar, R., Jr. 1969. Immunology 17:23-32.

79. BLOOD PROTEIN POLYMORPHISMS: RAT

Except for the immunoglobulins (Table 78), few experimental studies have been devoted to polymorphisms of blood proteins. The scant data available are listed below.

Protein ⟨Synonym⟩	Locus	Allele	Bearing Strain ⟨Synonym⟩	Reference
Ferroxidase ⟨Ceruloplasmin⟩	......	One component	BD VII, BD VIII	3
		Two components	AVN, BD II, BP, LEW	
Haptoglobin	......	Slow	BP, LEW ⟨Lewis⟩	3
		Fast	AVN, BD II, BD VII, BD VIII	
Hemoglobin	*Hbb*	Type A	AVN, BD II, BD V, BD VII, CDF, WAG, Y59	2,4,5
		Type B	AGA, AUG, BP, CAP, LEP, LEW	

continued

79. BLOOD PROTEIN POLYMORPHISMS: RAT

Protein ⟨Synonym⟩	Locus	Allele	Bearing Strain ⟨Synonym⟩	Reference
Plasma protein (undetermined)	*Gl-1* 1/	*Gl-1*^{a}	AUG ⟨August⟩, BN, LEW ⟨Lewis⟩, WAG	8,9
		Gl-1^{b}	LE ⟨Long Evans⟩	
Seminal vesicle protein	*Svp-1*	*Svp-1*^{a}	ACI, AS, AS2, AUG, AVN, A28807, BD V, BS, COP, DA, F344 ⟨Fischer⟩, WAG	1,6-8,10
		Svp-1^{b}	BN, LE ⟨Long Evans⟩, LEP, LEW ⟨Lewis⟩, PVG ⟨Black-hooded⟩, WF ⟨Wistar-Furth⟩	
Transferrin	*Tf*	Fast	BS	1
		Slow	ACI, AS, AS2, AVN, A28807, BD V, BD VII, BN, BUF, COP, DA, LEP, LEW, WAG	

1/ Linked to the *h* ⟨hooded⟩ locus.

Contributors: Hervé Bazin; Radim Brdička; Michael F. W. Festing; Roald Nezlin

References

1. Bender, K., and E. Günther. 1978. Biochem. Genet. 16:(in press).
2. Brdička, R. and K. Šulc. 1965. Folia Biol. (Prague) 11:328-329.
3. Dolezalova, V., and Z. Brada. 1968. Comp. Biochem. Physiol. 26:301-309.
4. Festing, M. F. W., and J. Staats. 1973. Transplantation 16:221-245.
5. French, E. A., et al. 1971. Biochem. Genet. 5:397-404.
6. Gasser, D. L. 1972. Ibid. 6:61-63.
7. Moutier, R. 1975. Détection électrophorétique des polymorphismes biochimiques chez les mammifères: Application à l'étude genetique des populations. Centre Nationale de la Recherche Scientifique, Paris.
8. Moutier, R., et al. 1973. Biochem. Genet. 8:321-328.
9. Moutier, R., et al. 1973. Ibid. 10:395-398.
10. Moutier, R., et al. 1971. Exp. Anim. 4:7-18.

80. ENZYME POLYMORPHISMS: RAT

Data are for enzymatic polymorphisms detected in individual rats or in different strains; studies dealing with tissue-specific isoenzymes have not been included. The nomenclature for the esterase loci is only provisional. Further work is needed to determine which ones may be identical. The esterase loci are tightly linked: There are approximately 9% recombinants between *Es-1* and each of *Es-2*, *Es-3*, *Es-4*, and *Es-5*; no recombinants were found between *Es-2*, *Es-3*, *Es-4*, and *Es-5* [ref. 32, 50]. Gasser gave evidence also for linkage between histocompatibility locus *Ag-C* and *Es-1* (11.24-5.3% recombination), and between *Ag-C* and *Es-2* (5.56-3.93% recombination) [ref. 16]. For additional information on esterases, consult references 21, 27, 28, 32, and 42; on brain esterases and acetylcholinesterases, references 3 and 47; on liver esterases, reference 20; on enzymes of catecholamine biosynthesis (in addition to dopamine β-monooxygenase) in adrenal gland of SHR ⟨spontaneously hypertensive rats⟩, reference 34; on lipogenic enzymes in Fatty ⟨genetically obese⟩ rats, reference 48; on intestinal enzymes in specific-pathogen-free rats, reference 18. For source of **Enzyme Commission No.**, *see* Introduction. **Characteristics**: SGE = starch gel electrophoresis. **Locus & Allele**: AD = autosomal dominant; CD = codominant. Data in brackets refer to the column heading in brackets.

Enzyme ⟨Synonym⟩ [Enzyme Commission No.]	Tissue	Characteristics	Locus & Allele	Strain ⟨Synonym⟩	Reference
Acid phosphatase [3.1.3.2]	Osteoclasts	....	*ia*	Incisorless 1/	29
Adenylate cyclase [4.6.1.1]	Renal medulla	....	*di*	Diabetes insipidus (hereditary)	12
Aldehyde dehydrogenase [1.2.1.3]	Liver	Enzyme activity increased by phenobarbital administration in LE ⟨Long-Evans⟩ but not in F344 ⟨Fischer⟩ rats	An AD gene, unlinked with coat color	LE ⟨Long-Evans⟩; F344 ⟨Fischer⟩	11

1/ Osteopetrotic mutant.

continued

Enzyme ⟨Synonym⟩ [Enzyme Commission No.]	Tissue	Characteristics	Locus & Allele	Strain ⟨Synonym⟩	Reference
Alkaline phosphatase [3.1.3.1]	Plasma	Two phenotypes—slow-moving ⟨S⟩ & fast-moving ⟨F⟩ bands in SGE	2 CD alleles at 1 autosomal locus. Frequency: $F = 0.58$; $S = 0.42$.	LE ⟨Long-Evans⟩	20
	Serum	....	*tl*	Toothless[1]	10
Dopamine β-monooxygenase ⟨Dopamine β-hydroxylase⟩ [1.14.17.1]	Brain, vascular system	Enzyme activity varies	....	SHR ⟨Spontaneously hypertensive rats⟩	35
	Serum	Elevated activity	*di*	Diabetes insipidus (hereditary)	53
Esterases					
Carboxylesterase [3.1.1.1]	Hepatic microsomes	Major & minor bands in disk electrophoresis; or near absence of bands	Presence of bands controlled by an AD gene	Sprague-Dawley outbred albino	56
⟨Serum prealbumin esterase⟩	Serum	Esterase activity in prealbumin zone in SGE	$Es\text{-}1^a$, an AD gene[2]	A35, F344, F344/N, MNR/N, M520/N, PETH/N, SHR/N, S3	15, 50
				AUG/Cub, AVN/Cub, BD VII/Cub, BD X/Cub, BN/Cub, BP/Cub, CAP/Cub, DA/Cub, LEW/Cub	5
				BN, F344, LEW ⟨Lewis⟩, WF ⟨Wistar-Furth⟩	16
				Some Wistar outbred & wild *Rattus norvegicus*	1,44
		Absence of prealbumin esterase	$Es\text{-}1^b$[2]	ACC, ACI, ACI/N, BUF/N, INR, MR/N, RHA/N, W/N	15, 50
				ACI, A28807 ⟨AUG-28807⟩, BUF, DA	16
				BD V/Cub, LEP/Cub, WAG/RijCub	5
				Some Wistar outbred & wild *R. norvegicus*	1,44
Arylesterase ⟨Serum albumin esterase⟩ [3.1.1.2]	Serum	Fast migrating fraction in albumin zone in SGE	$Es\text{-}2^a$[2]	ACI, ACI/N, A35, BUF/N, F344, F344/N, INR, MNR/N, MR/N, M520/N, PETH/N, RHA/N, SHR/N, S3, W/N	50
				BD X/Cub	5
				LEW/Orl	30
				Some Wistar outbred & wild rats	43, 44, 52
		Absence of fast fraction in albumin zone in SGE	$Es\text{-}2^b$[2]	ACC	50
				AUG/Cub, AVN/Cub, BD VII/Cub, BN/Cub, DA/Cub, LEP/Cub	5
				AUG/Orl ⟨August/Orl⟩, LE/Orl ⟨Long-Evans/Orl⟩, WAG/Orl	30
				Some Wistar outbred & wild rats	43, 44, 52
		Additional fast-moving form	$Es\text{-}2^c$[2]	BN	16
				LEW/Cub, WAG/RijCub	5
		Slow-moving form	$Es\text{-}2^d$[2]	LEW ⟨Lewis⟩	16
				CAP/Cub	5

[1] Osteopetrotic mutant. [2] *Es-1*, *Es-2*, & *Es-3* loci in close linkage.

continued

Enzyme ⟨Synonym⟩ [Enzyme Commission No.]	Tissue	Characteristics	Locus & Allele	Strain ⟨Synonym⟩	Reference
		....	$Es\text{-}2^l$	BD V/Cub	5
			$Es\text{-}2^t$	BP/Cub	5
Serum sex-influenced esterase	Serum	Present in SGE analysis of sera of sexually mature ♀	Controlled by a gene linked to the other esterase loci	ACI, AUG ⟨August⟩, BH, BUF ⟨Buffalo⟩, DA, HO	16
Other esterases	Digestive tract: small intestine	Presence of fast component in SGE	$Es\text{-}3^a$ [2]	ACI, ACI/N, BUF/N, F344, F344/N, INR, MR/N, M520/N, PETH/N, S3; also some wild *R. norvegicus*	50, 51
		Presence of slow component in SGE	$Es\text{-}3^b$ [2]	ACC, A35, MNR/N, RHA/N, SHR/N, W/N; also some wild *R. norvegicus*	50, 51
	Testis & lung	Slow band	$Es\text{-}3^a$ [2]	AUG/Orl ⟨August/Orl⟩, WAG/Orl	30-32
		Fast band	$Es\text{-}3^b$ [2]	LE/Orl ⟨Long-Evans/Orl⟩	30-32
			$Es\text{-}3^c$ [2]	BN/Orl, LEW/Orl	30-32
	Kidney	More anodal bands in SGE	$Es\text{-}4^a$	ACC, SHR/N	50
		More cathodal bands in SGE	$Es\text{-}4^b$	ACI, ACI/N, A35, BUF/N, F344, F344/N, INR, MNR/N, MR/N, M520/N, PETH/N, RHA/N, S3, W/N	50
	Liver; lung; testis, epididymis, prostate	Slow band	$Es\text{-}4^a$	ACI, AUG/Orl ⟨August/Orl⟩, AVN, BD V, BD VII, BS, BUF, COP, DA, LE/Orl ⟨Long-Evans/Orl⟩, LEP, WAG/Orl	2,30, 32
		Fast band	$Es\text{-}4^b$	AS, AS2, BN/Orl, LEW/Orl	2,30, 32
	Liver; epididymis	In more anodal zone: double-banded pattern	$Es\text{-}5^a$	AS, BN/Orl, LEP, LEW/Orl	2,30, 32
		Single-banded pattern of slightly greater mobility	$Es\text{-}5^b$	ACI, AS2, AUG/Orl ⟨August/Orl⟩, AVN, BD V, BD VII, BS, BUF, COP, DA, LE/Orl ⟨Long-Evans/Orl⟩, WAG/Orl	2,30, 32
	Heart	Slow-moving ⟨S⟩ & fast-moving ⟨F⟩ bands in SGE	2 CD alleles at 1 autosomal locus. Frequency: $F = 0.44$; $S = 0.56$	LE/Cal ⟨Long-Evans/Cal⟩	20
β-Fructofuranosidase ⟨Sucrase⟩ [3.2.1.26] activity	Jejunum	....	....	Diabetic rats	37
Fumarate hydratase ⟨Fumarase⟩ [4.2.1.2]	Liver	....	Fum^a	BN, LIS ⟨Hooded Lister⟩, LEW, PVG	7
			Fum^b	ACI, AUG, BD IX, BUF, DA, F344 ⟨Fischer⟩, Jaundice ⟨GUNN⟩, LE, WAG, WR	7,20
Glucose-6-phosphate dehydrogenase [1.1.1.49]	Erythrocytes	....	....	LE ⟨Long-Evans⟩, SD ⟨Sprague-Dawley⟩	21

[2] *Es-1, Es-2,* & *Es-3* loci in close linkage.

continued

Enzyme ⟨Synonym⟩ [Enzyme Commission No.]	Tissue	Characteristics	Locus & Allele	Strain ⟨Synonym⟩	Reference
Glycerol kinase ⟨Glycerokinase⟩ [2.7.1.30]	Liver; adipose tissue	....	....	BHE ⟨Carbohydrate-sensitive⟩, Fatty	14, 49
Histidine decarboxylase [4.1.1.22]	Aorta	....	....	SHR ⟨Spontaneously hypertensive rats⟩	19
Phosphoglucomutase [2.7.5.1]	Erythrocytes	Two phenotypes—slow-moving (a, b) & fast-moving (c, d, e) bands in SGE	Phenotype frequency: slow, 49.2%; fast, 18%	Japanese wild *R. norvegicus*, & laboratory rats	23
Phosphogluconate dehydrogenase [1.1.1.43]	Erythrocytes	Two components—slow-moving ⟨S⟩ & fast-moving ⟨F⟩	2 CD alleles at 1 autosomal locus: *Pgd*		
			Pgd^a (controls S)	Wistar outbred; wild *R. norvegicus*[4/]; 10 laboratory strains	78
			Pgd^b (controls F)	British wild *R. norvegicus*[5/]; Japanese wild *R. norvegicus*[6/]; *R. rattus*	8,23, 38, 45
			Pgd^s[7/]	ACI, AS, AS2, AVN, A28807, BD V, BD VII, BN, BS, BUF, COP, DA, F344 ⟨Fischer⟩, LEW, WAG	2,20
				Wistar outbred; wild rats	8,38, 45
			Pgd^f[8/]	LEP; wild rats	2,8, 38, 45
RNA nucleotidyltransferase ⟨RNA polymerase⟩ [2.7.7.6]	Liver	....	*fa*	Fatty ⟨Genetically obese rats⟩	13
Thymidine kinase [2.7.1.75]	Fat cells ⟨Adipocytes⟩	....	*fa*	Fatty ⟨Obese rats⟩	9
UDPglucuronosyltransferase ⟨Glucuronyl transferase⟩ [2.4.1.17]	Intestine; liver	....	*j*	Jaundice ⟨Gunn⟩[9/]	22, 39, 46

[4/] Frequency: Pgd^a, 0.061; Pgd^b, 0.939. Some *R. norvegicus* have mixed phenotype. [5/] Most animals have F; S only in 1.5%. [6/] Most animals have F (39.1%); S in 19.6%. [7/] Probably same as Pgd^a. [8/] Probably same as Pgd^b. [9/] For additional information on defect of UDPglucuronosyltransferase in Gunn rats, consult references 4, 6, 17, 24-26, 33, 36, 40, 41, 54, and 55.

Contributors: Roald Nezlin; Radim Brdička; R. Moutier; Michael F. W. Festing

References

1. Augustinsson, K. -B., and B. Henricson. 1966. Biochim. Biophys. Acta 124:323-331.
2. Bender, K., and E. Günther. 1978. Biochem. Genet. 16:(in press).
3. Bennett, E. L., et al. 1966. J. Neurochem. 13:563-572.
4. Berry, C. S., et al. 1972. Biochem. Biophys. Res. Commun. 49:1366-1375.
5. Brdička, R. 1974. Folia Biol. (Prague) 20:350-353.
6. Calvey, T. N., et al. 1970. Biochem. J. 119:659-663.
7. Carleer, J., and M. Ansay. 1976. Int. J. Biochem. 7: 565-566.
8. Carter, N. D., and C. W. Parr. 1969. Nature (London) 224:1214-1215.
9. Cleary, M. P., et al. 1975. Fed. Proc. Fed. Am. Soc. Exp. Biol. 34:908(Abstr. 3901).
10. Cotton, W. R., and J. F. Gaines. 1974. Proc. Soc. Exp. Biol. Med. 146:554-561.
11. Deitrich, R. A. 1971. Science 173:334-336.
12. Dousa, T. P., et al. 1975. Endocrinology 97:802-807.
13. Fillios, L. C., and O. Yokono. 1966. Metabolism 15: 279-285.
14. Gardner, L. B., and S. Reiser. 1976. Proc. Soc. Exp. Biol. Med. 153:158-160.

continued

15. Gasser, D. L., and J. Palm. 1972. Genetics 71:S19-(Abstr).
16. Gasser, D. L., et al. 1973. Biochem. Genet. 10:207-217.
17. Greenfield, S., and A. P. N. Majumdar. 1974. J. Neurol. Sci. 22:83-89.
18. Hietanen, E., and O. Hanninen. 1972. Ann. Zool. Fenn. 9:32-34.
19. Hollis, T. M., et al. 1972. Fed. Proc. Fed. Am. Soc. Exp. Biol. 31:292(Abstr. 447).
20. Jimenez-Marin, D. 1974. J. Hered. 65:235-237
21. Jimenez-Marin, D., and H. C. Dessauer. 1973. Comp. Biochem. Physiol. 46:487-490.
22. Kandall, S. R., et al. 1973. J. Pediatr. 82:1013-1019.
23. Koga, A., et al. 1972. Jpn. J. Genet. 47:335-338.
24. Majumdar, A. P. N. 1974. Neurobiology 4:425-431.
25. Majumdar, A. P. N., and S. Greenfield. 1974. Biochim. Biophys. Acta 335:260-263.
26. Majumdar, A. P. N., et al. 1973. Scand. J. Clin. Lab. Invest 31:219-223.
27. Manda, M., and Y. Oki. 1969. Jpn. J. Zootech. Sci. 40:363-369.
28. Manda, M., and S. Nishida. 1970. Ibid. 41:250-253.
29. Marks, S. C., Jr. 1973. Am. J. Anat. 138:165-178.
30. Moutier, R. 1975. Détection électrophorétique des polymorphismes biochimiques chez les mammifères: Application a l'étude génétique des populations. Centre Nationale de la Recherche Scientifique, Paris.
31. Moutier, R., et al. 1973. Biochem. Genet. 8:321-328.
32. Moutier, R., et al. 1973. Ibid. 9:109-115.
33. Mulder, G. J. 1972. Biochim. Biophys. Acta 289:284-292.
34. Nagatsu, I., et al. 1971. Experientia 27:1013-1014.
35. Nagatsu, T., et al. 1976. Science 191:290-291.
36. Nakata, D., et al. 1976. Proc. Natl. Acad. Sci. USA 73:289-292.
37. Olsen, W. A., and L. Rogers. 1971. J. Lab. Clin. Med. 77:838-842.
38. Paar, C. W. 1966. Nature (London) 210:487-489.
39. Puukka, R., et al. 1973. Biochem. Genet. 9:343-349.
40. Schmid, R., et al. 1958. J. Clin. Invest. 37:1123-1130.
41. Schutta, H. S., and L. Johnson. 1969. J. Pediatr. 75:1070-1079.
42. Serov, O. L. 1972. Isozyme Bull. 5:34.
43. Serov, O. L. 1973. Biochem. Genet. 9:117-131.
44. Serov, O. L. 1973. Genetika 9:49-56.
45. Serov, O. L., and G. P. Manchenko. 1974. Ibid. 10(3):40-43.
46. Strebel, L., and G. B. Odell. 1971. Pediatr. Res. 5:548-559.
47. Sviridov, S. M., et al. 1971. Biochem. Genet. 5:379-396.
48. Taketomi, S., et al. 1975. Horm. Metab. Res. 7:242-246.
49. Thenen, S. W., and J. Mayer. 1975. Proc. Soc. Exp. Biol. Med. 148:953-957.
50. Womack, J. E. 1973. Biochem. Genet. 9:13-24.
51. Womack, J. E. 1972. Experientia 28:1372.
52. Womack, J. E. 1972. J. Hered. 63:41-42.
53. Wooten, G., et al. 1975. J. Neural Transm. 36:107-112.
54. Yamada, N., et al. 1976. Proc. Jpn. Acad. 52:152.
55. Yeary, R. A., and R. H. Grothaus. 1971. Lab. Anim. Sci. 21:362-366.
56. Zemaitis, M. A., et al. 1974. Biochem. Genet. 12:295-308.

81. IMMUNOGENETIC CHARACTERIZATION: RAT

Generations Inbred: The first F is the generations at which the animals were received, and the second F is the number of generations since acquisition. Fx indicates no definitive information available, but probably inbred more than 20 generations. **Characteristics**: GLT = poly($Glu^{52}Lys^{33}Tyr^{15}$); (H,G)-A—L = poly(L-His,L-Glu)-poly(DL-Ala)-poly(L-Lys); (Phe,G)-A—L = poly(L-Phe,L-Glu)-poly(DL-Ala)-poly(L-Lys); (T,G)-A—L = poly(L-Tyr,L-Glu)-poly(DL-Ala)-poly(L-Lys); (T,G)-Pro—L = poly(L-Tyr,L-Glu)-poly(L-Pro)-poly(L-Lys); AIC = autoimmune complex nephritis; X-D1 indicates cross reactivity with D1 reagent, but not necessarily possession of the Ag-D1 allele; hr = high response; mr = moderate response; lr = low response. **Source** and **Maintained by**: For full name, *see* list of HOLDERS at front of book. Figures in heavy brackets are reference numbers.

Strain ⟨Synonym⟩	Source & Date of Acquisition	Maintained by	Generations Inbred	Characteristics
ACI ⟨AXC 9935, Irish⟩	MAI, 1971	MAX	Fx + F10	*H-1*[a] [9]; (H,G)-A—L [6]; hr to (T,G)-A—L [5]
	CR, 1971	PIT	Fx + F18	*H-1*[a], Ag-B4; Ag-C2; MLR-4; hr to GLT [11]
	HOK, 1974	HKM	F109 + F4	*H-1*[a], Ag-B4; Ag-C2 [15]

continued

Strain ⟨Synonym⟩	Source & Date of Acquisition	Maintained by	Generations Inbred	Characteristics
ACP 9935	CR, 1971	PIT	Fx + F11	*H-1*^{a}, Ag-B4; Ag-C1; MLR-4; hr to GLT [11]
ALB ⟨Albany⟩	CR, 1971	PIT	Fx + F11	*H-1*^{b}, Ag-B6; Ag-C2; MLR-6; lr to GLT [11]
	N, 1974	WIS	F62 + F6	Ag-B6; Ag-C2; Ag-A; Ag-D: not D1 [12,13]
AS	CSR, 1967	MAX	Fx + F22	*H-1*^{l} [14]; mr to (H,G)-A—L [6]; lr to (T,G)-A—L [4]; lr to (T,G)-Pro—L [7]; susceptible to AIC [8]
AS2	CSR, 1967	MAX	Fx + F23	*H-1*^{f} [16]; hr to (T,G)-Pro—L [7]; lr to (T,G)-A—L [4]
A28807 ⟨AUG 28807, August-28807⟩	CR, 1972	PIT	F60 + F11	*H-1*^{c}, Ag-B5; Ag-C1; MLR-5; hr to GLT [11]
		WIS	F57 + F13	Ag-B5; Ag-C1; Ag-A; Ag-D: X-D1 [12,13]
AVN	CUB, 1971	MAX	Fx + F23	*H-1*^{a} [17]; hr to (T,G)-A—L [6]; lr to (T,G)-Pro—L[7]
BD V	DR, 1971	MAX	Fx + F13	*H-1*^{d} [2]; lr to (T,G)-Pro—L [7]
BD VII	DR, 1976	MAX	Fx + F2	*H-1*^{e} [2]; hr to (H,G)-A—L [5]; hr to (Phe,G)-A—L [5]; lr to (T,G)-A—L [5]; lr to (T,G)-Pro—L [7]
BN	CUB, 1970	MAX	Fx + F15	*H-1*^{n} [2]; lr to (T,G)-A—L [4]; mr to (T,G)-Pro—L [7]; resistant to AIC [8]
	CR, 1971	PIT	Fx + F16	*H-1*^{n}, Ag-B3; Ag-C1; MLR-3; lr to GLT [11]
	SS, 1976	WIS	F57 + F3	Ag-B3; Ag-C1; Ag-A; Ag-D: not D1 [12,13]
BS	OSU, 1967	MAX	Fx + F19	*H-1*^{l} [14]; lr to (T,G)-A—L [4]
BUF ⟨Buffalo⟩	HOK, 1974	HKM	F64 + F7	*H-1*^{b}, Ag-B6; Ag-C1 [15]
	CR, 1971	PIT	Fx + F16	*H-1*^{b}, Ag-B6; Ag-C1; MLR-6; lr to GLT [11]
B3	PIT	PIT	F12	Ag-B3; MLR-4; hr to GLT [3]
COP ⟨COP-2331⟩	N, 1971	MAX	Fx + F9	Ag-B4 [12]
DA	ZBZ, 1967	MAX	Fx + F16	*H-1*^{a} [6,14]; lr to (H,G)-A—L [6]; hr to (T,G)-A—L [4]; lr to (T,G)-Pro—L [7]
	PM, 1972	PIT	F9 + F15	*H-1*^{a}, Ag-B4; Ag-C2; MLR-4; hr to GLT [11]
	DBW, 1967	WIS	F25 + F19	Ag-B4; Ag-C2; Ag-A; Ag-D: X-D1 [12,13]
F344 ⟨Fischer 344⟩	HOK, 1974	HKM	F107 + F6	*H-1*^{l}, Ag-B1; Ag-C1 [1,15]
	CR, 1971	PIT	(Fx + F18)1/ + F15	*H-1*^{l}, Ag-B1; Ag-C1; MLR-1; lr to GLT [11]
HO	OX, 1972	PIT	Fx + F11	*H-1*^{c}, Ag-B5; Ag-C2; MLR-5; hr to GLT [11]
KGH	HVD, 1968	PIT	F39 + F17	*H-1*^{g}, Ag-B7; Ag-C2; MLR-1; lr to GLT [11]
LEJ ⟨Long-Evans⟩	HOK, 1974	HKM	F22 + F5	*H-1*^{w}, Ag-B2; Ag-C2 [1,15]
LEP	CUB, 1973	MAX	Fx + F37	*H-1*^{w} [2]; hr to (H,G)-A—L [6]; lr to (Phe,G)-A—L [5]; lr to (T,G)-A—L [6]; lr to (T,G)-Pro—L [7]
LEW	CUB, 1969	MAX	Fx + F22	*H-1*^{l} [18]; mr to (H,G)-A—L [4]; hr to (Phe,G)-A—L [5]; lr to (T,G)-A—L [4]; mr to (T,G)-Pro—L [7]; susceptible to AIC [8]
	N, 1972	PIT	F36 + F7	*H-1*^{l}, Ag-B1; Ag-C1; MLR-1; lr to GLT [11]
LGE	CR, 1971	PIT	Fx + F15	*H-1*^{w}, Ag-B2; Ag-C2; MLR-2; mr to GLT [11]
MAXX	MAI, 1973	PIT	F20 + F10	*H-1*^{n}, Ag-B3; Ag-C1; MLR-3; lr to GLT [11]
MR	N, 1975	PIT	F59 + F5	*H-1*^{d}, Ag-B9; Ag-C1; hr to GLT [11]
M520	N, 1976	PIT	F123 + F4	*H-1*^{b}, Ag-B6; Ag-C1; MLR-6; lr to GLT [11]
NBR	CR, 1977	PIT	Fx + F2	Ag-B1; Ag-C2; MLR-1; lr to GLT [11]
OKA	WSL, 1975	PIT	Fx + F5	*H-1*^{k}, Ag-B8; Ag-C1; MLR-7; lr to GLT [11]
PVG	G, 1973	PIT	Fx + F10	*H-1*^{c}, Ag-B5; Ag-C2; MLR-5; hr to GLT [11]
SD ⟨Sprague-Dawley⟩	MK, 1976	HKM	F22 + F2	*H-1*^{w} (?), Ag-B2; Ag-C2 [1,15]
TO ⟨Tokyo⟩	MK, 1974	HKM	F47 + F5	Ag-B (?); Ag-C1 [1,15]
W ⟨Wistar/Mk⟩	MK, 1974	HKM	F70 + F7	*H-1*^{k}, Ag-B8; Ag-C1 [1,15]
WAG/Rij	CUB, 1970	MAX	Fx + F12	*H-1*^{w} [10]; hr to (H,G)-A—L [6]; lr to (T,G)-A—L [4]; lr to (T,G)-Pro—L [7]
WF	CR, 1971	PIT	Fx + F12	*H-1*^{w}, Ag-B2; Ag-C2; MLR-2; mr to GLT [11]
WKA ⟨Wistar King A⟩	HOK, 1974	HKM	F206 + F5	*H-1*^{k}, Ag-B8; Ag-C1 [1,15]
	HOK, 1973	PIT	F205 + F9	*H-1*^{k}, Ag-B8; Ag-C2; MLR-7; lr to GLT [11]
YO38366	CR, 1972	PIT	Fx + F14	*H-1*^{w}, Ag-B2; Ag-C2; MLR-2; mr to GLT [11]

1/ Fx + F18 when received.

continued

81. IMMUNOGENETIC CHARACTERIZATION: RAT

Contributors: Thomas J. Gill III and H. W. Kunz; Eberhard Günther; M. Aizawa; Dietrich Götze; Michael F. W. Festing

References

1. Festing, M., and J. Staats. 1973. Transplantation 16: 221-245.
2. Gill, T. J., III, et al. 1977. Immunogenetics 5:271-283.
3. Günther, E., et al. 1972. Eur. J. Immunol. 2:151-155.
4. Günther, E., et al. 1973. Transplant. Proc. 5:1467-1469.
5. Günther, E., et al. 1975. Ibid. 7(Suppl. 1):147-150.
6. Günther, E., et al. 1976. J. Immunol. 117:2047-2052.
7. Křen, V. 1974. Transplantation 17:148-152.
8. Kunz, H. W., et al. 1974. J. Immunogenet. 1:277-287.
9. Palm, J. 1971. Transplantation 11:175-183.
10. Palm, J., and G. Black. 1974. Ibid. 11:184-189.
11. Shoji, R. 1977. Proc. Jpn. Acad. 53:54-57.
12. Štark, O., and M. Hauptfeld. 1969. Folia Biol. (Prague) 15:35-40.
13. Štark, O., and V. Křen. 1967. Ibid. 13:93-99.
14. Štark, O., and V. Křen. 1967. Ibid. 13:312-316.
15. Štark, O., et al. 1968. Ibid. 14:169-175.
16. Štark, O., et al. 1968. Ibid. 14:425-432.
17. Štark, O., et al. 1969. Ibid. 15:259-262.
18. Stenglein, B., et al. 1975. J. Immunol. 115:895-897.

82. ALLOANTIGENIC SYSTEMS: RAT

Part I. Tissue Distribution of Alloantigenic Alleles

Data in brackets refer to the column heading in brackets.

Locus ⟨Synonym⟩ [Antiserum]	No. of Alleles	Linkage [Group]	Tissue Distribution	Reference
Ag-B ⟨*H-1*⟩	12[1/]	[VIII]	All cells	2,4,7,14,15,18,19
H-2	2	....	Erythrocytes, skin	9
H-3	2	[V]	Erythrocytes, skin	9
H-4	2	*c*-locus [I]	Skin	9
H-5	2	*lx*-locus [VII]	Skin	9
H-X	.....	X-chromosome	Skin	11
H-Y	.....	....	Skin	1,21
Ag-C	2	*Es-1*, *Es-2* loci [V]	Erythrocytes, liver, spleen	6,12,16,17
Ag-D	2	....	Erythrocytes	2,13,16
Ag-F	4	*c*-locus [I]	Lymphocytes	3
Ly-1 [AS anti-LEW]	2	....	Thymus, lymphocytes	5
[AS anti-LEW]	.....	....	Thymus, peripheral T-cells	20
[BH anti-LEW]	.....	....	Thymus, peripheral T-cells	10
[HO anti-AUG]	.....	....	Peripheral T-cells	8

[1/] 37 specificities.

Contributor: Eberhard Günther

References

1. Billingham, R. E., et al. 1962. Proc. Natl. Acad. Sci. USA 48:138-147.
2. Bogden, A. E., and P. M. Aptekman. 1960. Cancer Res. 20:1272-1282.
3. DeWitt, C. W., and M. McCullough. 1975. Transplantation 19:310-317.
4. Elkins, W. L., and J. Palm. 1966. Ann. N.Y. Acad. Sci. 129:573-580.
5. Fabre, J. W., and P. J. Morris. 1974. Tissue Antigens 4:238-246.
6. Gasser, D. L., et al. 1973. Biochem. Genet. 10:207-217.
7. Günther, E., and O. Štark. 1977. In D. Götze, ed. The Major Histocompatibility System in Man and Animals. Springer-Verlag, New York. pp. 207-253.
8. Howard, J. C., and D. C. Scott. 1974. Immunology 27:903-922.
9. Křen, V., et al. 1973. Transplant. Proc. 5:1463-1466.
10. Lubaroff, D. M. 1973. Ibid. 5:115-118.

continued

Part I. Tissue Distribution of Alloantigenic Alleles

11. Mullen, Y., and W. H. Hildemann. 1972. Transplantation 13:521-529.
12. Owen, R. D. 1962. Ann. N.Y. Acad. Sci. 97:37-42.
13. Palm, J. 1962. Ibid. 97:57-68.
14. Palm, J. 1964. Transplantation 2:603-612.
15. Palm, J. 1970. Ibid. 9:161-163.
16. Palm, J., and G. Black. 1971. Ibid. 11:184-189.
17. Poloskey, P. E., et al. 1975. J. Immunogenet. 2:179-187.
18. Štark, O., et al. 1967. In Institut Nationale de la Recherche Agronomique, ed. Polymorphismes biochemiques des animaux. Paris. pp. 501-506.
19. Štark, O., et al. 1967. Folia Biol. (Prague) 13:85-92.
20. Wonigeit, K., and R. Stumpenhorst. 1975. Z. Immunitaetsforsch. Allerg. Klin. Immunol. 150:257-258.
21. Zeiss, I. M., et al. 1962. Transplant. Bull. 30:161.

Part II. Strain Distribution of *Ag-B* ⟨*H-1*⟩ Haplotypes

Figures in heavy brackets are reference numbers.

Allele	Strain ⟨Synonym⟩
Ag-B1 ⟨$H\text{-}1^{l}$⟩	AGA [16], AGUS [2], AS [16], BD III [14], BH [19], BS [4,16], CAR [8], CAS [8], CDF [16], F344 [1,8,18], HCS [5], HS [14], LEW [13,16], NBR/1 [8], Paralyzed/N [9], S5B [8]
Ag-B2 ⟨$H\text{-}1^{w}$⟩	AO [5,8], A2 [8], BDE [14], BD II [16], BIRM 4B [5], BIRM 5A [5], BN.B2 [8], CAM [5], E3 [18], LE ⟨Long-Evans⟩ [9], LEP [15], LEW.1W [6], LGE [3], LOU [4,9], NBR/2 [9], NH [5], OM [8], PETH/N [9], R [5], RHA [5,9], SAL [5], Slonaker [8], VM [10], WA [5], WAG [17], WF [8], WR [16], YO38366 [9]
Ag-B3 ⟨$H\text{-}1^{n}$⟩	BN [8,15], BP.1N [6], LEW.1N [6], MAXX [9]
Ag-B4 ⟨$H\text{-}1^{a}$⟩	ACI [5,8], ACP [8], AVN [11], BN.B4 [8], COP [8], DA [4,8,16], LEW.1A [6], MNR/N [9]
Ag-B5 ⟨$H\text{-}1^{c}$⟩	AUG [2], A7322 [2], A28807 ⟨AUG 28807⟩ [8,18], CAP [15], CB [5], HO [5,8], LEW.1C [6], LIS [5], MNR/Brh [5], PD [18], PVG [5], Y59 [10]
Ag-B6 ⟨$H\text{-}1^{b}$⟩	ALB [8], BN.1B [6], BP [12], BUF [8], LA/N [9], LEW.1B [6], M520 [5,8], NSD [2], SD [8]
Ag-B7 ⟨$H\text{-}1^{g}$⟩	KGH [7]
Ag-B8 ⟨$H\text{-}1^{k}$⟩	OKA [3], SHR [3], WKA [7]
Ag-B9 ⟨$H\text{-}1^{d}$⟩	BD I [14], BD IV [14], BD V [15], BD VI [14], BD VIII [14], BD IX [14], BD X [15], LEW.1D [6], MR/Psy [5]
Ag-B10 ⟨$H\text{-}1^{f}$⟩	AS2 [10], LEW.1F [6]
$H\text{-}1^{e}$	BD VII [15]
$H\text{-}1^{h}$	HW [18], LEW.1H [6]

Contributors: Eberhard Günther; Michael F. W. Festing

References

1. Elkins, W. L., and J. Palm. 1966. Ann. N.Y. Acad. Sci. 129:573-580.
2. Festing, M. F. W., and J. Staats. 1973. Transplantation 16:221-245.
3. Gill, T. J., III. Unpublished. Univ. Pittsburgh, School Medicine, Dep. Pathology, Pittsburgh, 1977.
4. Günther, E., et al. 1975. Transplant. Proc. 7(Suppl. 1):147-150.
5. Křen, V. 1974. Transplantation 17:148-152.
6. Křen, V., et al. 1973. Transplant. Proc. 5:1463-1466.
7. Kunz, H. W., and T. J. Gill. 1974. J. Immunogenet. 1:413-420.
8. Palm, J. 1971. Transplantation 11:175-183.
9. Poloskey, P. E., et al. 1975. J. Immunogenet. 2:179-187.
10. Štark, O., and M. Hauptfeld. 1969. Folia Biol. (Prague) 15:35-40.
11. Štark, O., and V. Křen. 1967. Ibid. 13:93-99.
12. Štark, O., and V. Křen. 1967. Ibid. 13:299-305.
13. Štark, O., and V. Křen. 1967. Ibid. 13:312-316.
14. Štark, O., and I. M. Zeiss. 1970. Z. Versuchstierkd. 12:27-40.
15. Štark, O., et al. 1968. Folia Biol. (Prague) 14:169-175.
16. Štark, O., et al. 1968. Ibid. 14:425-432.
17. Štark, O., et al. 1969. Ibid. 15:259-262.
18. Štark, O., et al. 1971. Transplant. Proc. 3:165-168.
19. Wilson, D. B., and P. C. Nowell. 1971. J. Exp. Med. 133:442-453.

continued

82. ALLOANTIGENIC SYSTEMS: RAT

Part III. Strain Distribution of Other Alloantigenic Alleles

Locus[1]	Allele[1]	Strain ⟨Synonym⟩	Reference
Ag-C	*1*	ACP, AUG, A28807 ⟨AUG 28807⟩, BN, BUF, B3, CAR, F344, HW, LEW, MAXX, MNR/N, MR/N, M520, OM, Paralyzed/N, RHA/N, WKA, Z61	4-6
	2	ACI, ALB, AS2, BD V, BD VII, BN, CAS, COP, DA, HO, KGH, LA/N, LE ⟨Long-Evans⟩, LGE, LOU/C, NBR/1, NBR/2, NSD, OKA, PETH/N, PVG, RLA/N, SD, SHR/N, S5B, WAG, WF, YO38366	
Ag-F	*1*	ACI, AUG, A28807, DA, F344, M520, WF	2
	2	LEW	
	3	BN, MAXX	
	4	BUF	
MLR	*1*	F344, KGH, LEW, NBR	1
	2	LGE, WF, YO	
	3	BN, MAXX	
	4	ACI, ACP, DA	
	5	AUG, HO, PVG	
	6	ALB, BUF, M520	
	7	OKA, WKA	
"T-type"[2]	1[2]	AO, AUG, A28807	3
	2[2]	HO, LEW	
	3[2]	BH, DA, F344	

[1] Unless otherwise indicated. [2] HO anti-AUG antiserum reactivity.

Contributors: Eberhard Günther; Michael F. W. Festing; Thomas J. Gill III

References

1. Cramer, D. V., et al. 1977. Transplant. Proc. 9:559-562.
2. DeWitt, C. W., and M. McCullough. 1975. Transplantation 19:310-317.
3. Howard, J. C., and D. C. Scott. 1974. Immunology 27:903-922.
4. Kunz, H. W., and T. J. Gill III. 1978. J. Immunogenet. 5:(in press).
5. Palm, J., and G. Black. 1971. Transplantation 11:184-189.
6. Poloskey, P. E., et al. 1975. J. Immunogenet. 2:179-187.

83. IMMUNOLOGICAL RESPONSIVENESS: RAT

Response Determined: Ab = antibody; DHR = delayed-type hypersensitivity reaction; Ly = anamnestic lymph-cell response in vitro.

Stimulus ⟨Synonym⟩	Response Determined	Strain Response			Genetic Control	Reference
		High	Low	Intermediate		
Antigens						
Synthetic polypeptides						
Poly($Glu^{50}Ala^{50}$)	Ab, DHR, Ly	ACI, AUG, BUF, F344, LEW, M520, WF	BN	......	*H-1*	1,2
Poly($Glu^{60}Ala^{30}Tyr^{10}$)	Ab	......	......	......	......	1
	DHR	AUG, BUF, COP, F344, LEW, M520, WF	ACI, BN	......	......	1

continued

Stimulus 〈Synonym〉	Response Determined	Strain Response: High	Strain Response: Low	Strain Response: Intermediate	Genetic Control	Reference
Poly(Glu[52]Lys[33]Tyr[15])	Ab, DHR, Ly	ACI, ACP, AUG, COP, DA, PVG 〈HO〉	ALB, BN, BUF, CAR, CAS, F344, KGH, LEW, MAXX, M520, NBR/1, WKA	LE, NBR/2, OM, WF, YO38366	*H-1*, plus genetic background gene(s)	3,4,13, 21
Poly(Glu[50]Tyr[50])	Ab, DHR, Ly	ACI, AUG, BUF, COP	BN, F344, LEW, M520, WF	....	*H-1*	1,2
Poly(Tyr-Glu-Ala-Gly)	Ab	ACI, AUG, WF	BN, BUF, LEW, MAXX	F344	*H-1*, plus genetic background gene	15
(H,G)-A—L	Ab, Ly	AUG, BD VII, LEP, LEW.1W, LOU	ACI, ACP, DA, LEW.1A, LEW.1B, LEW.1D, LEW.1F, LEW.1N	AS, BS, LEW	*H-1*	6-8
(Phe,G)-A—L	Ab, Ly	AUG, BD VII, LEW, LEW.1A, LEW.1D, LEW.1F	LEP, LEW.1N, LEW.1W	....	*H-1*	6,8
(Phe,G)-Pro—L	Ab	LEW.1A, LEW.1F	....	LEW.1N, LEW.1W	*H-1*	9
Pro—L	Ab	LEW.1F	LEW.1A	LEW, LEW.1D, LEW.1W	*H-1*	9
T_4-A—L	Ab	LEW.1A	LEW.1D	....	*H-1*	20
T_6-A—L	Ab	LEW.1A	LEW.1D	....	*H-1*	20
(T,G)-A—L	Ab, Ly	ACI, AVN, LEW.1A	AS, AS2, BD VII, BN, BS, DA	AUG	*H-1*	5,8,11
(T,G)-Pro—L	Ab	LEW.1F	LEW.1A	....	....	9
Other antigens						
Bovine serum albumin 〈BSA〉						
1 μg	Ab	....	ACI, AUG, BN, BUF, M520	F344, LEW, WF	....	1
10 μg	Ab	BN, F344, WF	BUF	ACI, AUG, LEW, M520	....	1
Bovine gamma globulin 〈BGG〉	Ab	LE[1]	F344[2]	....	One gene, not *H-1*-linked	25
Dextran, ovomucoid	Anaphylactoid reaction	Wistar reactors	Wistar non-reactors	....	One autosomal, recessive gene	10,27
2,4-Dinitrophenyl-bovine gamma globulin 〈DNP-BGG〉	Anti-DNP Ab affinity	....	....	....	....	14
Encephalitogenic fragment of guinea pig basic protein	DHR, Ly	LEW	BN	....	....	16
Group A streptococcal vaccine	Ab, idiotypes	Selectively bred	Selectively bred, AUG, BN, BUF, COP, F344, M520, WF	....	Not *H-1*-linked	22-24
Lactate dehydrogenase (porcine), isoenzyme A_4 〈LDH-A_4〉	Ab, Ly	AS2, AUG, BD I, BD II, BD VII, BDE, BN, BUF, F344, LEW, LEW.1D, WF	ACI, ACP, COP, DA, LEW.1A	....	*H-1*	30,31

[1] Synonym: Long-Evans. [2] Synonym: Fischer 344.

continued

83. IMMUNOLOGICAL RESPONSIVENESS: RAT

Stimulus ⟨Synonym⟩	Response Determined	Strain Response			Genetic Control	Reference
		High	Low	Intermediate		
Phytohemagglutinin ⟨PHA⟩	Ab	AS	BN	....	....	12
Sheep erythrocytes	Ab, macrophage function	LE[1/]	F344[2/]	....	One gene, not *H-1*-linked	25,26
Mitogens						
Concanavalin A ⟨Con A⟩, Phytohemagglutinin ⟨PHA⟩	[^{3}H]thymidine uptake	AS, LEW	BN	....	One or few genes, not *H-1*-linked	17-19, 28,29

[1/] Synonym: Long-Evans. [2/] Synonym: Fischer 344.

Contributor: Eberhard Günther

References

1. Armerding, D., et al. 1974. Immunogenetics 1:329-339.
2. Armerding, D., et al. 1974. Ibid. 1:340-351.
3. Cramer, D. V., et al. 1977. Cell. Immunol. 28:167-173.
4. Gill, T. J., III, et al. 1970. J. Immunol. 105:14-28.
5. Günther, E., et al. 1972. Eur. J. Immunol. 2:151-155.
6. Günther, E., et al. 1973. Transplant. Proc. 5:1467-1469.
7. Günther, E., and E. Rüde. 1975. J. Immunol. 115:1387-1393.
8. Günther, E., and E. Rüde. 1976. Immunogenetics 3:261-269.
9. Günther, E., et al. 1976. J. Immunol. 117:2047-2052.
10. Harris, J. M., et al. 1963. Genet. Res. 4:346.
11. Koch, C. 1974. Immunogenetics 1:118-125.
12. Koch, C. 1976. Scand. J. Immunol. 5:1149-1153.
13. Kunz, H. W., et al. 1974. J. Immunogenet. 1:277-287.
14. Lamelin, J.-P., and W. E. Paul. 1971. Int. Arch. Allergy 40:351-360.
15. Luderer, A. A., et al. 1976. J. Immunol. 117:1079-1084.
16. McFarlin, D. E., et al. 1975. J. Exp. Med. 141:72-81.
17. Newlin, C. M., and D. L. Gasser. 1973. J. Immunol. 110:622-628.
18. Nielsen, H. E., and C. Koch. 1975. Scand. J. Immunol. 4:31-36.
19. Nielsen, H. E., and C. Koch. 1976. Ibid. 5:1139-1147.
20. Rüde, E., and E. Gunther. 1974. Prog. Immunol. II, 2:223-233.
21. Shonnard, J. W., et al. 1976. J. Immunogenet. 3:61-70.
22. Stankus, R. P., and C. A. Leslie. 1974. J. Immunol. 113:1859-1863.
23. Stankus, R. P., and C. A. Leslie. 1975. Immunogenetics 2:29-38.
24. Stankus, R. P., and C. A. Leslie. 1976. Ibid. 3:65-73.
25. Tada, N., et al. 1974. J. Immunogenet. 1:265-275.
26. Tada, N., et al. 1976. Ibid. 3:49-60.
27. West, G. B. 1974. Int. Arch. Allergy 47:296-305.
28. Williams, R. M., et al. 1973. J. Immunol. 111:1571-1578.
29. Williams, R. M., et al. 1973. Ibid. 111:1579-1584.
30. Würzburg, U. 1971. Eur. J. Immunol. 1:496-497.
31. Würzburg, U., et al. 1975. Ibid. 3:762-766.

84. CHEMISTRY OF ALLOANTIGENS: RAT

Mol Wt & Physical Characteristics: Molecular weight, given in daltons, is generally for the basic component. **Mol Size** = molecular size, in daltons, estimated by gel filtration. Data in brackets refer to the column heading in brackets.

Antigen	Gene Symbol [Allospecificity]	Strain	Source Material	Method of Isolation	Method of Identification	Mol Wt & Physical Characteristics [Mol Size]	Chemistry	Reference
Major histocompatibility	*Ag-B* [Ag-B1]	F344[1/]	Liver cell membrane	Papain digestion	....	37,000 & 11,000[2/] [59,000]; 25,000 [25,000]	....	10, 11

[1/] Synonym: Fischer 344. [2/] Identified as β_2-microglobulin.

continued

Antigen	Gene Symbol [Allospecificity]	Strain	Source Material	Method of Isolation	Method of Identification	Mol Wt & Physical Characteristics [Mol Size]	Chemistry	Reference
	Ag-B [Ag-B4]	ACI	Liver cell membrane	Papain digestion	....	37,000 & 11,000[2] [59,000]; ∿20,000 & 11,000[2] [35,000]; 25,000 [25,000]	....	10, 11
		DA	Lymphoid cell	3*M* KCl-gel filtration; 3*M* KCl-immunosorbent chromatography	Serological-radiolabel	30,000-35,000. Monomer; hydrophobic behavior.	Peptide or glycopeptide; no amino sugars; amino acid analysis in ref. 1	1,3, 15, 19, 20
Minor histocompatibility	*Ag-F*[3] [Ag-F]	F344[1]	Lymphoid cell	Non-ionic detergent; immune precipitation in polyethylene glycol	Serological-radiolabel	35,000-40,000. Monomer; hydrophobic behavior.	Peptide or glycopeptide; amino acid analysis in ref. 21; basic:acidic amino acids (2:1); *N*-terminal amino acid is aspartic acid	7,21
	Ag-G	....	....	3*M* KCl-immunosorbent chromatography	Serological-radiolabel	10,000-20,000. Monomer.	Peptide or glycopeptide	2,8
β_2-Microglobulin	....	....	....	Non-ionic detergent	Serological-radiolabel	11,600. Monomer; non-covalently bound to Ag-B.	Peptide; *N*-terminal amino acid is isoleucine; for partial sequence consult ref. 17 and 18	16-18
IgE-binding site	....	....	....	Non-ionic detergent; immune precipitation with anti-IgE	Binding to IgE	62,000. Possible polymer of 2×10^6.	Protein	4,5, 12
Acetylcholine receptor	....	....	....	Non-ionic detergent; sucrose gradient centrifugation; affinity chromatography	Binding to α-neurotoxin	250,000	....	13, 14

[1] Synonym: Fischer 344. [2] Identified as β_2-microglobulin. [3] May be confined to peripheral thymus-dependent lymphocytes [ref. 6 and 9].

Contributors: C. W. DeWitt; Nobuyuki Tanigaki and David Pressman

References

1. Callahan, G. N., and C. W. DeWitt. 1975. J. Immunol. 114:776-778.
2. Callahan, G. N., and C. W. DeWitt. 1975. Ibid. 114: 779-781.
3. Callahan, G. N., et al. 1974. Immunology 27:1141-1146.
4. Carson, D. A., et al. 1975. J. Immunol. 114:158-160.
5. Conrad, D. H., and A. Froese. 1976. Ibid. 116:319-326.
6. DeWitt, C. W. Unpublished. Univ. Utah, School Medicine, Salt Lake City, UT, 1977.
7. DeWitt, C. W., and M. McCullough. 1975. Transplantation 19:310-317.
8. DeWitt, C. W., and M. McCullough. 1977. Transplant. Proc. 9:625-627.
9. Howard, J. C., and D. W. Scott. 1974. Immunology 27:903-922.
10. Katagiri, M., et al. 1975. Transplantation 19:230-239.
11. Katagiri, M., et al. 1975. Ibid. 20:135-141.
12. König, W., and K. Ishizaka. 1974. J. Immunol. 113: 1237-1245.

continued

13. Lindstrom, J. M., et al. 1976. Ann. N.Y. Acad. Sci. 274:254-274.
14. Lindstrom, J. M., et al. 1976. J. Exp. Med. 144:726-738.
15. Palm, J. 1971. Transplantation 11:175-183.
16. Poulik, M. D. 1976. Prog. Clin. Biol. Res. 5:155-177.
17. Poulik, M. D. Unpublished. William Beaumont Hospital, Dep. Immunochemistry, Royal Oak, MI, 1977.
18. Poulik, M. D., et al. 1975. Fed. Proc. Fed. Am. Soc. Exp. Biol. 34:945.
19. Stroehmann, I., and C. W. DeWitt. 1972. Immunology 23:921-928.
20. Stroehmann, I., and C. W. DeWitt. 1972. Ibid. 23:929-935.
21. Williams, P. B. C., and C. W. DeWitt. 1976. J. Immunol. 117:33-39.

85. TRANSPLANT SURVIVAL TIMES: RAT

Data are for vascularized organs, skin, and pancreatic islets. **Major Histocompatibility Complex Phenotype**: Alternate nomenclatures *Ag-B* and *H-1* are shown; entries are grouped by donor incompatibility, and within groups, by recipient incompatibility. **Survival Time**: The end point for heart rejection is usually cessation of systole, while kidney end-point is death from uremia; hence, the latter time is approximately 2 days beyond a comparable rejection end-point. Mean survival time with kidney grafts is often misleading, since the distribution of times may be skewed with a proportion of long survivors. This phenomenon is a result of spontaneous reversal of rejection, and may be dependent, in part, on ischemic times during surgery [34]. Ranges without confidence limits are given, since in many cases results were pooled from different sources.

Major Histocompatibility Complex Phenotype		Strain Combination		Survival Time, d[1]	Reference
Ag-B	*H-1*	Donor	Recipient		
Skin					
1/1→1/1	*l/l→l/l*	AS	BS	13-17	1,46
		BS	AS	12	46
		AS	HS	13	46
		HS	AS	13	46
		AS	LEW	13	10
		LEW	AS	12	10
		BS	HS	12	46
		HS	BS	14	46
		F344	LEW	7-10	12,13,18,32,44,45
		LEW	F344	10-12	14,43,44
		(LEW x F344)F_1	LEW	11	23
1/1→2/2	*l/l→w/w*	AGUS	W[2]	8	3
4/1→4/4	*a/l→a/a*	(DA x LEW)F_1	DA	8	10
1/1→5/5	*l/l→c/c*	AGUS	PVG	9	3
1/1→6/6	*l/l→b/b*	F344	BUF	10	45
		LEW	BUF	10	18,45
1/1→10/10	*l/l→f/f*	AS	AS2	9-10	17,32
				10	33
2/2→1/1	*w/w→l/l*	W[2]	AGUS	9	3
		W[2]	AS	9	32
2/2→3/3	*w/w→n/n*	W[2]	BN	10	32
1/3→1/1	*l/n→l/l*	(LEW x BN)F_1	LEW	8-9	22,30
3/3→1/1	*n/n→l/l*	BN	LEW	10	32,45
4/1→1/1	*a/l→l/l*	(DA x LEW)F_1	LEW	9	8,10
1/5→1/1	*l/c→l/l*	(AS x AUG)F_1	AS	8	39
5/5→1/1	*c/c→l/l*	PVG	AGUS	8	3,4

[1] Unless otherwise indicated. [2] Animal was described as a "Wistar rat" without further characterization.

continued

Major Histocompatibility Complex Phenotype		Strain Combination		Survival Time, d	Reference
Ag-B	*H-1*	Donor	Recipient		
5/5→10/10	*c/c*→*f/f*	AUG	AS2	9	33
1/6→1/1	*l/b*→*l/l*	(LEW x BUF)F_1	LEW	7	18
6/6→1/1	*b/b*→*l/l*	BUF	LEW	7-11	18,27,32
10/10→1/1	*f/f*→*l/l*	AS2	AS	8-15	1,2,17,32,33
		Kidney			
1/1→1/1	*l/l*→*l/l*	AS	LEW	>300	10,36
		LEW	AS	>300	10,36
		F344	LEW	89->300	12,18,32,44
		LEW	F344	46->500	18,44
		(LEW x F344)F_1	LEW	90-180	23
1/1→3/3	*l/l*→*n/n*	LEW	BN	8-9	26
1/1→4/4	*l/l*→*a/a*	LEW	DA	7-21	9
4/1→4/4	*a/l*→*a/a*	(DA x LEW)F_1	DA	13->300	9,10
1/1→6/6	*l/l*→*b/b*	LEW	BUF	5-9	26
1/1→9/9	*l/l*→*d/d*	LEW	BD V	8	37
1/1→10/10	*l/l*→*f/f*	AS	AS2	9	32
				38-412	33
1/3→1/1	*l/n*→*l/l*	(LEW x BN)F_1	LEW	9->300	15,22,30
3/3→1/1	*n/n*→*l/l*	BN	LEW	7-9	15,32
3/3→2/2	*n/n*→*w/w*	BN	WAG	9-55	15,41,42
4/1→1/1	*a/l*→*l/l*	(DA x LEW)F_1	LEW	10	10
4/4→1/1	*a/a*→*l/l*	DA	LEW	9	9
1/5→1/1	*l/c*→*l/l*	(AS x AUG)F_1	AS	9-11	7,39
		(AS x HO)F_1	AS	10	19
1/6→1/1	*l/b*→*l/l*	(LEW x BUF)F_1	LEW	14->200	18,26
6/6→1/1	*b/b*→*l/l*	BUF	LEW	14-28	18,20,27,32
1/9→1/1	*l/d*→*l/l*	(LEW x BD V)F_1	LEW	15-146	35
9/9→1/1	*d/d*→*l/l*	BD V	LEW	10-16	37
		BD V	AS	9-10	37
10/10→1/1	*f/f*→*l/l*	AS2	AS	8-10	2,32,33
10/10→4/4	*f/f*→*a/a*	AS2	DA	7-36	9
4/10→4/4	*a/f*→*a/a*	(DA x AS2)F_1	DA	7->300	9
1/10→4/4	*l/f*→*a/a*	(LEW x AS2)F_1	DA	>200	9
		Heart			
1/1→1/1	*l/l*→*l/l*	F344	LEW	9-15	13,45
1/1→2/2	*l/l*→*w/w*	LEW	WF	11	14
1/3→2/3	*l/n*→*w/n*	(LEW x BN)F_1	(WF x BN)F_1	12	14
1/1→2/3	*l/l*→*w/n*	LEW	(WF x BN)F_1	9	14
1/1→3/3	*l/l*→*n/n*	LEW	BN	7	14
1/3→3/3	*l/n*→*n/n*	(LEW x BN)F_1	BN	12	14
2/2→1/1	*w/w*→*l/l*	W[2/]	AS	9	32
		WF	LEW	6	6
2/3→1/3	*w/n*→*l/n*	(WF x BN)F_1	(LEW x BN)F_1	8	14
2/3→3/3	*w/n*→*n/n*	(WF x BN)F_1	BN	10	14
		(WAG x BN)F_1	BN	10	24,25
2/2→3/3	*w/w*→*n/n*	WAG	BN	8-9	24,25
		W[2/]	BN	10	32
		WF	BN	7	5
2/2→5/5	*w/w*→*c/c*	AO	HO	7	38

continued

Major Histocompatibility Complex Phenotype		Strain Combination		Survival Time, d[1]	Reference
Ag-B	*H-1*	Donor	Recipient		
2/3→1/1	*w/n→l/l*	(WF x BN)F_1	LEW	7	14
2/3→1/6	*w/n→l/b*	(WF x BN)F_1	(LEW x BUF)F_1	8	14
2/3→6/6	*w/n→b/b*	(WF x BN)F_1	BUF	9	14
1/3→1/1	*l/n→l/l*	(AS x BN)F_1	AS	19	28
		(F344 x BN)F_1	F344	13	28
		(LEW x BN)F_1	LEW	7-12	14,16,30,40
3/3→1/1	*n/n→l/l*	BN	LEW	6-10	14,16,38,40,45
3/3→2/2	*n/n→w/w*	BN	WAG	9-11	42
1/3→1/6	*l/n→l/b*	(LEW x BN)F_1	(LEW x BUF)F_1	6	14
3/3→6/6	*n/n→b/b*	BN	BUF	7	14
4/4→1/1	*a/a→l/l*	ACI	F344	6-7	14,16
		ACI	LEW	6	14,16,21
4/4→1/3	*a/a→l/n*	ACI	(LEW x BN)F_1	6	14
4/4→3/3	*a/a→n/n*	ACI	BN	6	14
4/4→5/5	*a/a→c/c*	DA	HO	7	38
5/5→5/5	*c/c→c/c*	AUG	HO	>200	38
1/5→1/1	*l/c→l/l*	(AS x AUG)F_1	AS	8	39
5/5→1/1	*c/c→l/l*	AUG	AS	8	38
5/5→2/2	*c/c→w/w*	AUG	W[2]	8	38
6/6→1/1	*b/b→l/l*	BUF	LEW	7	29
1/6→1/1	*l/b→l/l*	(LEW x BUF)F_1	LEW	11	14
6/6→2/2	*b/b→w/w*	BUF	WF	6	14
1/6→2/2	*l/b→w/w*	(LEW x BUF)F_1	WF	6	14
1/6→2/3	*l/b→w/n*	(LEW x BUF)F_1	(WF x BN)F_1	8	14
1/6→3/3	*l/b→n/n*	(LEW x BUF)F_1	BN	8	14
1/6→1/3	*l/b→l/n*	(LEW x BUF)F_1	(LEW x BN)F_1	13	14
6/6→3/3	*b/b→n/n*	BUF	BN	9	14
10/10→1/1	*f/f→l/l*	AS2	AS	7-106	2
Spleen					
1/1→2/2	*l/l→w/w*	AGUS	W[2]	>3 mo	3
1/1→5/5	*l/l→c/c*	AGUS	PVG	>8 mo	3
2/2→1/1	*w/w→l/l*	W[2]	AGUS	>10 mo	3
5/1→1/1	*c/l→l/l*	(PVG x AGUS)F_1	AGUS	No rejection	40
5/5→1/1	*c/c→l/l*	PVG	AGUS	2-3 wk	3,4
Intestine					
1/3→1/1	*l/n→n/n*	(LEW x BN)F_1	LEW	10	30
Pancreas					
1/3→1/1	*l/n→n/n*	(LEW x BN)F_1	LEW	15	30
Pancreatic Islets[3]					
4/1→1/1	*a/l→l/l*	(DA x LEW)F_1	LEW	4	11
4/4→1/1	*a/a→l/l*	DA	LEW	4	11
Hindlimb					
1/5→1/1	*l/c→l/l*	(AS x AUG)F_1	AS	16	31

[1] Unless otherwise indicated. [2] Animal was described as a "Wistar rat" without further characterization. [3] By blood glucose.

continued

85. TRANSPLANT SURVIVAL TIMES: RAT

Contributor: Charles B. Carpenter

References

1. Allardyce, R. A., et al. 1977. Transplantation 24:159-161.
2. Bildsoe, P., et al. 1970. Transplant. Rev. 3:36-45.
3. Bitter-Suermann, H. 1974. Nature (London) 247:465-468.
4. Bitter-Suermann, H. 1974. Transplantation 18:515-519.
5. Björck, L., et al. 1977. Ibid. 23:383-386.
6. Dittmer, J., and M. Bennett. 1975. Ibid. 19:295-301.
7. Fabre, J. W., and J. R. Batchelor. 1975. Ibid. 20:473-479.
8. Fabre, J. W., and P. J. Morris. 1973. Ibid. 15:397-403.
9. Fabre, J. W., and P. J. Morris. 1974. Ibid. 18:429-435.
10. Fabre, J. W., and P. J. Morris. 1975. Ibid. 19:121-133.
11. Finch, D. R. A., and P. J. Morris. 1977. Ibid. 23:386-388.
12. Freeman, J. S., and D. Steinmuller. 1969. Ibid. 8:530-533.
13. Goodnight, J. E., et al. 1976. Ibid. 22:391-397.
14. Guttmann, R. D. 1974. Ibid. 17:383-386.
15. Guttmann, R. D., et al. 1967. Ibid. 5:668-681.
16. Guttmann, R. D., et al. 1974. Ibid. 18:93-98.
17. Heslop, B. F. 1968. Aust. J. Exp. Med. Sci. 46:479.
18. Hildemann, W. H., and Y. Mullen. 1973. Transplantation 15:231-237.
19. Hutchinson, I. V., et al. 1976. Ibid. 22:273-280.
20. Ippolito, R. J., et al. 1974. Ibid. 17:89-96.
21. Kuromachi, T., et al. 1975. Ibid. 20:167-170.
22. Lie, T. S., et al. 1976. Ibid. 21:103-109.
23. Mahabir, R. N., et al. 1969. Ibid. 8:369-378.
24. Marquet, R. L., et al. 1976. Ibid. 21:454-459.
25. Marquet, R. L., et al. 1977. Ibid. 23:199.
26. Mullen, Y., and W. H. Hildemann. 1971. Transplant. Proc. 3:668-672.
27. Mullen, Y., and W. H. Hildemann. 1975. Transplantation 20:281-290.
28. Nielsen, H. E., et al. 1975. Ibid. 19:360-362.
29. Pettirossi, O., et al. 1976. Ibid. 21:408-411.
30. Pettirossi, O., et al. 1976. Ibid. 21:403-407.
31. Poole, M., et al. 1976. Ibid. 22:108-111.
32. Sakai, A. 1969. Ibid. 8:882-889.
33. Salaman, J. R. 1971. Ibid. 11:63-70.
34. Schanzer, H., et al. 1974. Ibid. 18:417-420.
35. Thoenes, G. H. 1975. In W. H. Hildemann and A. A. Benedict, ed. Immunologic Phylogeny. Plenum, New York. pp. 457-466.
36. Thoenes, G. H., and E. White. 1973. Transplantation 15:308-311.
37. Thoenes, G. H., et al. 1974. Immunogenetics 3:239-253.
38. Tilney, N. L. 1974. Transplantation 17:561-567.
39. Tilney, N. L., and P. R. F. Bell. 1974. Ibid. 18:31-37.
40. Tilney, N. L., et al. 1975. Ibid. 20:323-330.
41. Tinbergen, W. J. 1968. Ibid. 6:203-207.
42. van Bekkum, D. W., et al. 1969. Ibid. 8:678-688.
43. Warren, R. P., et al. 1973. Transplant. Proc. 5:717-719.
44. White, E., and W. H. Hildemann. 1969. Transplantation 8:602-617.
45. Wildstein, A., et al. 1976. Ibid. 21:129-132.
46. Zeiss, I. M. 1967. Ibid. 5:1393-1399.

86. ALLOGRAFT SURVIVAL: RAT

Major Histocompatibility Complex Phenotype: Data are for alternate nomenclatures *Ag-B* and *H-1*; entries are grouped by donor incompatibility, and, within groups, by recipient incompatibility. **Graft Survival Time** is given as the range, standard error of the mean, or 95% confidence limits.

Major Histocompatibility Complex Phenotype		Strain Combination		Graft Survival Time, d	Reference
Ag-B	*H-1*	Donor	Recipient		
Skin					
1/1→1/1	*l/l*→*l/l*	HS[1]	HS[2]	61.8 ± 6.7	28
		F344	LEW	10.0	2
				10.2 ± 1.2	26
				10.0(9.3-10.7)	4
				10.3 ± 0.8	5

[1] Male donor. [2] Female recipient.

continued

Major Histocompatibility Complex Phenotype		Strain Combination		Graft Survival Time, d	Reference
Ag-B	*H-1*	Donor	Recipient		
		LEW	F344	13.8	2
		LEW[1]	LEW[2]	>100	21
1/1→3/3	*l/l→n/n*	LEW	BN	10.0(9.1-10.9)	3
1/1→6/6	*l/l→b/b*	F344	BUF	9.9 ± 1.1	27
2/2→5/5	*w/w→c/c*	A2/1	HO	8.0	1
3/3→3/3	*n/n→n/n*	BN[1]	BN[2]	13.6 ± 1.5[3]	21
3/3→1/1	*n/n→l/l*	BN	LEW	7.7 ± 0.4	5
				10.2 ± 0.8	23
				10.2(9.4-11.0)	3
Kidney[4]					
1/1→1/1	*l/l→l/l*	F344	LEW	>100	15,26
1/1→5/5	*l/l→c/c*	AGUS	PVG	8.7(8.1-9.3)	19
1/1→6/6	*l/l→b/b*	F344	BUF	13.5(6-23)	17
1/3→1/1	*l/n→l/l*	(LEW x BN)F_1	LEW	(7-16)	22
				12.0(7-19)	13
				12[5]; 41[6]	20
				16.1 ± 1.7	14
				17.3 ± 2.7	24
3/3→1/1	*n/n→l/l*	BN	LEW	(5-8)	29
2/3→2/2	*w/n→w/w*	(WF x BN)F_1	WF	8.7 ± 1.8	10
1/4→1/1	*l/a→l/l*	(LEW x ACI)F_1	LEW	(8-12)	22
1/5→1/1	*l/c→l/l*	(AS x HO)F_1	AS	10.2(10-11)	11
1/6→1/1	*l/b→l/l*	(LEW x BUF)F_1	LEW	(8->90)	22
6/6→1/1	*b/b→l/l*	BUF	F344	68(40-104)	17
Heart					
1/1→1/1	*l/l→l/l*	F344	LEW	8.8	2
				12.8 ± 1.3	6
				13.2 ± 1.1	4
				13.4 ± 0.9	5
				14.7 ± 1.0	27
		LEW	F344	>30	2,25
1/1→2/2	*l/l→w/w*	LEW	WF	11.3 ± 1.4	7
1/3→2/3	*l/n→w/n*	(LEW x BN)F_1	(WF x BN)F_1	12.1 ± 2.1	7
1/1→2/3	*l/l→w/n*	LEW	(WF x BN)F_1	8.8 ± 1.2	7
1/1→3/3	*l/l→n/n*	LEW	BN	7.3 ± 0.2	7
1/3→3/3	*l/n→n/n*	(LEW x BN)F_1	BN	11.8 ± 0.6	7
1/4→4/4	*l/a→a/a*	(LEW x ACI)F_1	ACI	7.5(6.7-8.4)	9
1/6→6/6	*l/b→b/b*	(LEW x BUF)F_1	BUF	5.5(5.1-5.9)	9
1/6→2/2	*l/b→w/w*	(LEW x BUF)F_1	WF	6.3 ± 0.2	7
1/6→2/3	*l/b→w/n*	(LEW x BUF)F_1	(WF x BN)F_1	8.5 ± 0.7	7
1/6→3/3	*l/b→n/n*	(LEW x BUF)F_1	BN	7.7 ± 0.4	7
2/2→5/5	*w/w→c/c*	A2/1	HO	15.0 ± 7.0	12
2/3→1/3	*w/n→l/n*	(WF x BN)F_1	(LEW x BN)F_1	8.1 ± 0.6	7
2/3→1/1	*w/n→l/l*	(WF x BN)F_1	LEW	7.0 ± 0.5	7
2/3→3/3	*w/n→n/n*	(WF x BN)F_1	BN	9.8 ± 0.4	7
		(WAG x BN)F_1	BN	(7-11)	16
2/3→1/6	*w/n→l/b*	(WF x BN)F_1	(LEW x BUF)F_1	8.0 ± 0.6	7

[1] Male donor. [2] Female recipient. [3] Most large grafts >2-4 cm^2 survive >100 d. Data are for grafts <2.0 cm^2. [4] Survival time refers to animal survival, not graft survival. Death due to complications of uremia. Technical factors and uremia may influence rate of rejection, thus model is variable. [5] Short ischemic time. [6] Long ischemic time.

continued

Major Histocompatibility Complex Phenotype		Strain Combination		Graft Survival Time, d	Reference
Ag-B	*H-1*	Donor	Recipient		
2/3→6/6	*w/n*→*b/b*	(WF x BN)F_1	BUF	9.0 ± 1.2	7
3/3→1/1	*n/n*→*l/l*	BN	LEW	6.6 ± 1.1	27
				8.1 ± 0.4	5
				10.4 ± 2.3	7
1/3→1/1	*l/n*→*l/l*	(LEW x BN)F_1	LEW	7.3 ± 0.3	7
2/3→2/2	*w/n*→*w/w*	(WF x BN)F_1	WF	7.7(6.3-9.4)	9
4/3→4/4	*a/n*→*a/a*	(ACI x BN)F_1	ACI	6.9(5.4-8.8)	9
3/3→6/6	*n/n*→*b/b*	BN	BUF	7.1 ± 0.5	7
1/3→1/6	*l/n*→*l/b*	(LEW x BN)F_1	(LEW x BUF)F_1	6.0 ± 0.4	7
4/4→1/1	*a/a*→*l/l*	ACI	F344	6.7 ± 0.2	7
		ACI	LEW	6.1 ± 0.1	7
1/4→1/1	*l/a*→*l/l*	(LEW x ACI)F_1	LEW	6.6(5.5-7.9)	9
4/4→1/3	*a/a*→*l/n*	ACI	(LEW x BN)F_1	6.1 ± 0.1	7
4/4→3/3	*a/a*→*n/n*	ACI	BN	6.5 ± 0.2	7
4/3→3/3	*a/n*→*n/n*	(ACI x BN)F_1	BN	5.5(4.7-6.5)	9
5/5→3/3	*c/c*→*n/n*	BROFO	BN	(7-8)	16
6/6→1/1	*b/b*→*l/l*	BUF	LEW	7.0 ± 0.1	18
1/6→1/1	*l/b*→*l/l*	(LEW x BUF)F_1	LEW	10.2(8.6-12.2)	9
				10.7 ± 0.4	7
6/6→2/2	*b/b*→*w/w*	BUF	WF	6.2 ± 0.2	7
6/6→3/3	*b/b*→*n/n*	BUF	BN	9.3 ± 0.6	7
1/6→1/3	*l/b*→*l/n*	(LEW x BUF)F_1	(LEW x BN)F_1	13.1 ± 3.2	7
			Heart [7]		
1/1→2/2	*l/l*→*w/w*	LEW	WF	1.4(1.1-1.9)	8
1/3→2/3	*l/n*→*w/n*	(LEW x BN)F_1	(WF x BN)F_1	2.8(1.9-4.1)	8
1/1→2/3	*l/l*→*w/n*	LEW	(WF x BN)F_1	4.0(2.5-6.4)	8
1/1→3/3	*l/l*→*n/n*	LEW	BN	0.5(0.1-1.6)	8
1/3→3/3	*l/n*→*n/n*	(LEW x BN)F_1	BN	1.5(1.0-2.2)	9
1/1→4/4	*l/l*→*a/a*	LEW	ACI	3.3(2.4-4.6)	8
1/4→4/4	*l/a*→*a/a*	(LEW x ACI)F_1	ACI	5.6(4.8-6.5)	9
1/6→2/2	*l/b*→*w/w*	(LEW x BUF)F_1	WF	0.013(0.010-0.016)	8
1/6→2/3	*l/b*→*w/n*	(LEW x BUF)F_1	(WF x BN)F_1	0.4(0.2-0.6)	8
1/6→3/3	*l/b*→*n/n*	(LEW x BUF)F_1	BN	0.03(0.02-0.05)	8
1/6→6/6	*l/b*→*b/b*	(LEW x BUF)F_1	BUF	0.6(0.3-1.3)	9
2/3→1/3	*w/n*→*l/n*	(WF x BN)F_1	(LEW x BN)F_1	2.1(1.9-2.3)	8
2/2→3/3	*w/w*→*n/n*	WF	BN	0.1(0.06-0.16)	8
2/3→3/3	*w/n*→*n/n*	(WF x BN)F_1	BN	2.1(1.1-4.0)	9
2/3→1/1	*w/n*→*l/l*	(WF x BN)F_1	LEW	2.5(2.2-2.9)	8
2/3→1/6	*w/n*→*l/b*	(WF x BN)F_1	(LEW x BUF)F_1	2.0(2.0-2.0)	8
2/3→6/6	*w/n*→*b/b*	(WF x BN)F_1	BUF	1.7(1.5-1.9)	8
3/3→1/1	*n/n*→*l/l*	BN	LEW	2.0(0.8-5.2)	8
1/3→1/1	*l/n*→*l/l*	(LEW x BN)F_1	LEW	4.0(3.5-4.6)	9
2/3→2/2	*w/n*→*w/w*	(WF x BN)F_1	WF	2.0(0.7-5.6)	9
4/3→4/4	*a/n*→*a/a*	(ACI x BN)F_1	ACI	3.7(3.3-4.1)	9
3/3→6/6	*n/n*→*b/b*	BN	BUF	2.6(2.1-3.1)	8
1/3→1/6	*l/n*→*l/b*	(LEW x BN)F_1	(LEW x BUF)F_1	3.3(2.5-4.3)	8
4/4→1/1	*a/a*→*l/l*	ACI	F344	0.02(0.01-0.04)	8
		ACI	LEW	0.06(0.02-0.24)	8
1/4→1/1	*l/a*→*l/l*	(LEW x ACI)F_1	LEW	1.6(1.2-2.1)	9
4/4→1/3	*a/a*→*l/n*	ACI	(LEW x BN)F_1	0.2(0.1-0.3)	8
4/4→3/3	*a/a*→*n/n*	ACI	BN	0.13(0.08-0.20)	8

[7] Allograft survival after donor skin graft presensitization. Three donor strain skin grafts. Median survival time is given.

continued

86. ALLOGRAFT SURVIVAL: RAT

Major Histocompatibility Complex Phenotype		Strain Combination		Graft Survival Time, d	Reference
Ag-B	*H-1*	Donor	Recipient		
4/3→3/3	*a/n*→*n/n*	(ACI x BN)F_1	BN	0.3(0.1-0.6)	9
6/6→1/1	*b/b*→*l/l*	BUF	LEW	2.7(2.1-3.5)	8
1/6→1/1	*l/b*→*l/l*	(LEW x BUF)F_1	LEW	4.0(3.5-4.5)	9
6/6→2/2	*b/b*→*w/w*	BUF	WF	0.8(0.6-1.0)	8
6/6→2/3	*b/b*→*w/n*	BUF	(WF x BN)F_1	1.5(0.8-2.9)	8
6/6→3/3	*b/b*→*n/n*	BUF	BN	0.5(0.2-1.3)	8
1/6→1/3	*l/b*→*l/n*	(LEW x BUF)F_1	(LEW x BN)F_1	2.7(2.2-3.3)	8

Contributor: Ronald D. Guttmann

References

1. Anderson, N. F., et al. 1967. Lancet 1:1126.
2. Barker, C. F., and R. E. Billingham. 1971. Transplant. Proc. 3:172-175.
3. Billingham, R. E., et al. 1962. J. Natl. Cancer Inst. 28: 365-435.
4. Freeman, J. S., and D. Steinmuller. 1969. Transplantation 8:530-532.
5. Freeman, J. S., et al. 1970. Circulation 41(Suppl. 2): 86-89.
6. Freeman, J. S., et al. 1971. Transplant. Proc. 3:580-582.
7. Guttmann, R. D. 1974. Transplantation 17:383-386.
8. Guttmann, R. D. 1976. Ibid. 22:583-588.
9. Guttmann, R. D. 1977. Ibid. 23:153-157.
10. Heron, I. 1972. Acta Pathol. Microbiol. Scand. 80:9-16.
11. Hutchinson, I. V., et al. 1976. Transplantation 22: 273-280.
12. Jenkins, A. M., and M. F. A. Woodruff. 1971. Ibid. 12:57-60.
13. Kawabe, K., et al. 1972. Ibid. 13:21-26.
14. Lie, T. S., et al. 1976. Ibid. 21:103-109.
15. Mahabir, R. N., et al. 1969. Ibid. 8:369-377.
16. Marquet, R. L., et al. 1976. Ibid. 21:454-459.
17. Murray, D. E. 1969. Nature (London)223:87-88.
18. Pettirossi, O., et al. 1976. Transplantation 21:408-411.
19. Salaman, J. R. 1972. Ibid. 14:74-78.
20. Schanzer, H., et al. 1974. Ibid. 18:417-420.
21. Silvers, W. K., et al. 1977. Immunogenetics 4:85-100.
22. Soulillou, J. P., et al. 1976. J. Exp. Med. 143:405-421.
23. Steinmuller, D., and L. J. Weiner. 1963. Transplantation 8:369-377.
24. Stuart, F. P., et al. 1968. Surgery 64:17-24.
25. Warren, R. P., et al. 1973. Transplant. Proc. 5:717-719.
26. White, E., and W. H. Hildemann. 1969. Science 162: 1293-1294.
27. Wildstein, A., et al. 1976. Transplantation 21:129-132.
28. Zeiss, I. M. 1966. Ibid. 4:48-55.
29. Zimmerman, C. E. 1973. Ibid. 15:519-521.

87. TISSUE AND ORGAN TRANSPLANTATION: RAT

Part I. Graft-versus-Host Reaction

Data give incidence of deaths due to runt syndrome after intravenous administration of spleen cells into neonatal rats. **Cell Dose**: Numbers are given in millions of cells, i.e., each number should be multiplied by 10^6.

Incompatibility	Strain Combination		Cell Dose	No. Dead/Total No.	Mortality, %
	Donor	Recipient			
H-1	LEW	LEW.1A	10	12/16	75
	LEW.1A	LEW	10	10/11	91
	LEW	LEW.1B	5	21/21	100
			20	7/8	87

continued

Part I. Graft-versus-Host Reaction

Incompatibility	Strain Combination		Cell Dose	No. Dead/Total No.	Mortality, %
	Donor	Recipient			
	LEW.1B	LEW	5	10/12	83
			20	10/10	100
	LEW	LEW.1N	10	2/6	33
			30	7/7	100
	LEW.1N	LEW	10	5/5	100
			30	10/10	100
	LEW	LEW.1W	15	5/9	55
			30	5/5	100
	LEW.1W	LEW	10	10/10	100
Non-H-1	AVN	LEW.1A	20	0/5	0
	BN	LEW.1N	10	0/7	0
			30	0/5	0
	BP	LEW.1B	10	1/17	6
			30	6/15	40
	WP	LEW.1W	30	0/6	0
H-1 plus non-H-1	AVN	LEW	10	10/25	40
	BN	LEW	30	2/12	17
	BP	LEW	5	9/10	90
			20	8/8	100
	WP	LEW	20	0/9	0

Contributor: Eberhard Günther

Reference: Kren, V., et al. 1970. Folia Biol (Prague) 16:305-313.

Part II. Graft Survival Time

Strain Combination: In hybrids, the female parent is always listed first. Plus/minus (±) values are standard deviations unless otherwise indicated. Values in parentheses are ranges, estimate "c" unless otherwise indicated (*see* Introduction). Data in brackets refer to the column heading in brackets.

Incompatibility [Phenotype]	Strain Combination		Graft Median Survival Time[1], d	Reference
	Donor	Recipient		
		Skin[2]		
H-1 plus non-H-1	ACI	LEW	6.9 ± 0.3 (6-8)	5
	AS	AS2	9.6	19
			10.1(9.6-10.6)[b]	20
			10.2 ± 0.1; 6.6 ± 0.2[3]	13,30
	AS2	AS	8.3(7.6-9.0)[b]	20
			9.6 ± 0.2; 6.1 ± 0.1[3]	13,30
			10.2	19
	(AS x AS2)F_1	AS	9.8 ± 0.26	19
	(AS x AS2)F_1	AS2	9.6 ± 0.19	19

[1] Unless otherwise indicated. [2] Determination of survival end point for skin transplants in rats is usually more difficult to achieve than in mice. Variations for the same combination from different references may also be due to different graft sizes. Furthermore, the grafting technique (e.g., body localization) may influence the survival data. [3] Second set of grafts.

continued

Part II. Graft Survival Time

Incompatibility [Phenotype]	Strain Combination		Graft Median Survival Time[1], d	Reference
	Donor	Recipient		
	AS2	BS	10.2 ± 0.1; 6.9 ± 0.2[3]	13,30
	BS	AS2	9.5 ± 0.1; 6.7 ± 0.2[3]	13,30
	AS2	HS	9.5 ± 0.2; 6.7 ± 0.2[3]	13,30
	HS	AS2	10.3 ± 0.2; 6.8 ± 0.1[3]	13,30
	AUG	AS2	8.6(8.0-9.3)[b]	20
	BN	LEW	7.0 ± 0.2	21
			10.1	19
	LEW	BN	8.1 ± 0.4	21
			9.0 ± 0.29; 6.1 ± 0.15[3]	1
	BUF	LEW	6.8(6.2-7.5)[b]	16,17
			11	19
	LEW	BUF	9.7(9.1-10.4)[b]	16,17
	(LEW x BUF)F_1	LEW	7.3(6.7-8.0)[b]	16,17
	DA	LEW	7.1 ± 0.17	21
	WIS[4]	BN	10.4	19
	WN	AS	9.4	19
Non-H-1	AS	BS	12.6 ± 0.2; 9.4 ± 0.2[3]	13,30
	BS	AS	11.5 ± 0.2; 9.1 ± 0.2[3]	13,30
	AS	HS	13.3 ± 0.2; 7.7 ± 0.3[3]	13,30
	HS	AS	12.9 ± 0.2; 8.2 ± 0.2[3]	13,30
	BS	HS	11.5 ± 0.2; 9.1 ± 0.2[3]	13,30
	HS	BS	13.9 ± 0.3; 9.8 ± 0.1[3]	13,30
	F344	LEW	6.6(5.9-7.5)[b]	16,17
			7.6 ± 0.4	21
			9.0 ± 0.3 (8-11)	5
			10.0 ± 0.8	3
			10.3(9.1-11.3)[b] (8-12); 6.2(5.9-6.5)[b] (6-9)[3]	29
			10.5 ± 0.5	19
	LEW	F344	10.4(9.6-11.3)[b]	16,17
			11.6 ± 0.3	21
			13.0 ± 1.5	3
Sex-associated				
♂→♂				
$[H\text{-}X^f \rightarrow H\text{-}X^l]$	F344	(LEW x F344)F_1	259(171-390)[b] (110->390)	17
$[H\text{-}X^l \rightarrow H\text{-}X^f]$	LEW	(F344 x LEW)F_1	335(177-637)[b] (76->696)	17
$[H\text{-}Y^f \rightarrow H\text{-}Y^l]$	F344	(F344 x LEW)F_1	>>360 (221-700)	17
$[H\text{-}Y^l \rightarrow H\text{-}Y^f]$	LEW	(LEW x F344)F_1	>>360	17
♂→♀	WN[4,5]	WN[4,5]	103 ± 10.5	19
$[H\text{-}Y^b]$	BUF	BUF	143(119-171)[b] (115-250)	17
$[H\text{-}Y^f]$	F344	F344	99(87-112)[b] (76-138)	17
$[H\text{-}Y^l]$	LEW	LEW	110(98-124)[b] (91-174)	17
Xenograft	Rabbit	Rat	(5-11)[6]	2
			(13-23)[7]	2
		Kidney[8]		
H-1 (all congenic lines)	LEW	LEW.1D	7.4(6.5-9.1)[b] (6-13)	26

[1] Unless otherwise indicated. [3] Second set of grafts. [4] Described as Wistar rats without further characterization. [5] Inbred strain. [6] Small grafts; 20/24 grafts destroyed at 9th day. [7] Large grafts; 25/27 grafts destroyed at 17th day. [8] The broad ranges of survival in semiallogenic, congenic, or non-H-1-different combinations very likely should be regarded as a consequence of an "autoenhancement" counteracting the rejection process in face of alloantigens which may be qualitatively, and/or quantitatively, less potent in turning on the rejection.

continued

Part II. Graft Survival Time

Incompatibility [Phenotype]	Strain Combination		Graft Median Survival Time[1/], d	Reference
	Donor	Recipient		
	LEW.1D	LEW	9.8(9.1-10.6)[b] (8-13)	26
	(LEW x LEW.1D)F_1	LEW	(75-293)	23
	(LEW x LEW.1D)F_1	LEW.1D	(9->244)	23
	LEW	LEW.1F	(22-184)	18
	LEW.1F	LEW	(5-105)	18
	LEW	LEW.1N	(9-87)	23
	LEW.1D	LEW.1N	(7-11)	23
	LEW.1N	LEW.1D	(7-24)[9/]	23
	LEW.1N	LEW.1F	(6->158)	18
	LEW.1W	LEW	(8-172)	23
H-1 plus non-H-1	AS	AS2	125(38-412)[b]	20
			(7-186)	18
	AS2	AS	8.0(7-9.2)[b]	20
			(10-12)	4
	(AS x AS2)F_1	AS	11 ± 0.6	19
	(AS x AS2)F_1	AS2	21.4 ± 0.7	19
	AS	BD V	8.0(8-9)	26
	BD V	AS	9.4(9-10)	26
	(AS x BD V)F_1	AS	(46-158)	26
	AS	LEW.1D	7.2(6.6-7.9)[b] (6-9)	26
	LEW.1D	AS	8.5(8-10)	26
	(AS x LEW.1D)F_1	AS	(187-336)	23
	(LEW.1D x AS)F_1	LEW.1D	(7-254)	23
	AS	LEW.1F	(8-171)	18
	AS2	DA	(7-36)	6
	AUG	AS	8.5	9
	(AUG x AS)F_1	AS	8.8 ± 1.0	27
			9(8-12)	10
	BD V	BN	(7-8)	24
	BN	BD V	(9-11)	24
	BD V	LEW	11.0(8.9-13.5)[b] (10-16)	26
	LEW	BD V	8.0	26
	(LEW x BD V)F_1	LEW	(15-146)	23
	BN	LEW	6.8 ± 1.2	15
			(8-11)	23
	LEW	BN	(8-9)	23
	(LEW x BN)F_1	LEW	9.6 ± 1.2 & 1.6 ± 1.0[10/]	15
			17.3 ± 2.7	22
	BN	WAG	14(9-55)	28
	BUF	LEW	18.5(15.2-22.5)[b] (8-59)	16
			18.9 & 294[10/]	14
	LEW	BUF	7.7	14
			7.7(7.4-8.0)[b] (5-9)	16
	(LEW x BUF)F_1	LEW	12.5(9.3-16.8)[b] (7-201)	16
	DA	LEW	(8-10)	6
	LEW	DA	(7-21)	6
	(DA x LEW)F_1	DA	(13->300)	8
	(DA x LEW)F_1	LEW	(9-10)	8
	(LEW x AS2)F_1	DA	>200	7
	LEW	AS2	(8-205)	18

[1/] Unless otherwise indicated. [9/] One graft survived 137 days. [10/] 2 modes of survival times.

continued

Part II. Graft Survival Time

Incompatibility [Phenotype]	Strain Combination		Graft Median Survival Time[1/], d	Reference
	Donor	Recipient		
	(LEW.1D x BD V)F_1	AS	(8-10)	26
	(LEW.1D x BD V)F_1	LEW	7.4(7-10)	26
	(LEW.1N x BN)F_1	LEW	(8-11)	23
Non-H-1	AS	LEW	>300[11/]	25
	LEW	AS	>300[11/]	25
	BD V	LEW.1D	>480[11/]	23
	LEW.1D	BD V	>300[11/]	23
	BN	LEW.1N	(31-409)	23
	LEW.1N	BN	(>79->310)	23
	F344	LEW	121(92-158)[b] (76-197)	16
	LEW	F344	>500[11/] (455->664)[b]	16
	LEW.1A	DA	>100[11/]	11
Heart				
H-1 & non-H-1[12/]				
Haplotype matched[13/]	....	....	9.8[14/] ± 4.3 ± 0.6[15/]	12
Haplotype mismatched[16/]	....	....	7.9[14/] ± 2.8 ± 0.3[15/]	12

1/ Unless otherwise indicated. 11/ Average survival time. 12/ Summary of data from 25 strain combinations. 13/ 56 animals. 14/ Mean survival time. 15/ Standard error. 16/ 124 animals.

Contributors: Gunther H. Thoenes and Eberhard Günther

References

1. Ballantyne, D. L., and P. Nathan. 1968. Transplantation 6:342-350.
2. Ballantyne, D. L., et al. 1969. Ibid. 7:274-280.
2. Barker, C. F., and R. E. Billingham. 1971. Transplant. Proc. 3:172-175.
4. Bildsøe, P., et al. 1970. Transplant. Rev. 3:36-45.
5. Cho, S. I., et al. 1972. Transplantation 13:486-492.
6. Fabre, J. W., and P. J. Morris. 1974. Ibid. 18:429-435.
7. Fabre, J. W., and P. J. Morris. 1974. Ibid. 18:436-442.
8. Fabre, J. W., and P. J. Morris. 1975. Ibid. 19:121-133.
9. French, M. E., and J. R. Batchelor. 1969. Lancet 2: 1103-1106.
10. French, M. E., and J. R. Batchelor. 1972. Transplant. Rev. 13:115-141.
11. Günther, E., and Wagner. Unpublished. Max Planck Inst. für Immunobiologie, Freiburg, Germany, 1978.
12. Guttmann, R. D. 1974. Transplantation 17:383-386.
13. Heslop, B. F. 1969. Aust. J. Exp. Biol. Med. Sci. 46: 479-491.
14. Ippolito, R. J., et al. 1972. Transplantation 14:183-190.
15. Lucas, Z. J., et al. 1970. Fed. Proc. Fed. Am. Soc. Exp. Biol. 29:2041-2047.
16. Mullen, Y., and W. H. Hildemann. 1971. Transplant. Proc. 3:669-672.
17. Mullen, Y., and W. H. Hildemann. 1972. Transplantation 13:521-529.
18. Paris, A., et al. 1978. Ibid. 25:252-254.
19. Sakai, A. 1969. Ibid. 8:882-889.
20. Salaman, J. R. 1971. Ibid. 11:63-70.
21. Silvers, W. K., and R. E. Billingham. 1969. Ibid. 8: 167-178.
22. Stuart, F. P., et al. 1968. Science 160:1463-1465.
23. Thoenes, G. H. 1975. In W. H. Hildemann and A. A. Benedict, ed. Immunologic Phylogeny. Plenum, New York. pp. 457-466.
24. Thoenes, G. H. 1978. Eur. Surg. Res. 10:294.
25. Thoenes, G. H., and E. White. 1973. Transplantation 15:308-311.
26. Thoenes, G. H., et al. 1974. Immunogenetics 3:239-253.
27. Tilney, N. L., and P. R. F. Bell. 1974. Transplantation 18:31-37.
28. Tinbergen, W. J. 1968. Ibid. 6:203-207.
29. White, E., and W. H. Hildemann. 1969. Transplant. Proc. 1:395-399.
30. Zeiss, J. M. 1967. Transplantation 5:1393-1399.

continued

Part III. Enhancement

Induction of immunological unresponsiveness in adults, by active and/or passive donor-specific immunization (enhancement) before transplantation, is most successful for organs but much less so for skin. Many variables of different experimental protocols (timing of presensitization, route of injection, dosage, mode of solubilization, etc.) affect enhancement. Third party sensitization is effective in some cases, possibly due to shared alloantigenic specificities or non-H-1 antigens. Also, so-called autoenhancement (without active pre-immunization) may participate in non-H-1-different-H-1-identical combinations, especially in many of the semiallogenic combinations investigated in the past. The graft can survive as long as the individual, but survival does not necessarily imply a healthy and functioning graft. **Type of Enhancement**: A = active and donor specific; B = blood; L = lymphoid cells; P = passive- and donor-specific; S = solubilized cellular material; 3 = third-party-specific; 0 = no treatment. **Fraction Surviving**: number surviving longer than controls/total number, or number surviving indefinitely/total number. Plus/minus (±) values are standard errors.

Incompatibility	Strain Combination		Type of Enhancement	Graft Median Survival Time, d	Fraction Surviving	Reference
	Donor	Recipient				
			Kidney			
H-1	LEW	LEW.1D	A, S	>130	4/4	11
	LEW	LEW.1N	A, B	>45[1]	12/12	12
			A, 3, B	(9->21)[1]	0/6	
H-1 plus non-H-1	AS	LEW.1D	A, S	(>31->235)	9/9	11
	AS2	DA	0	(7-36)	.....	4
			P	>200	4/4	
			P, 3	>200	7/8	
	LEW	BN	A, B	(14-41)	6/6	12
			A, 3, B	(8-9)	0/6	
Semiallogenic	(AS x AUG)F_1	AS	0	(9-11)	.....	2
			P	(216->250)	5/5	
	(DA x LEW)F_1	DA	0	(13->300)	.....	4
			P	>200	5/5	
	(DA x LEW)F_1	LEW	0	(9-13)	.....	3
			P	(22-29)[2]	5/5	
			P	(15-36)	5/5	1
			P + ALG[3]	>110	5/6	
	(LEW x AS2)F_1	DA	0	>200	.....	4
			P	>200	.....	
	(LEW x BN)F_1	LEW	A, 3, S	10.6 ± 2.7 (5-33)	.....	8
			A, S	44.7 ± 13.2 (12->100)	4/6	
			0	7.3 ± 0.5 (6-8)	.....	9
			A, 3	8.7 ± 1.7 (6-18)	.....	
			A, S	43.3 ± 11.8 (14->160)	6/9	
			Heart			
H-1 plus non-H-1	WAG	BN	0	10	.....	10
			A, L	>200	15/15	
Semiallogenic	(LEW x BN)F_1	LEW	0	7.3(6-9)	.....	7
			A	14.9(8-21)	9/10	
			Pancreatic Islets			
H-1 plus non-H-1	DA	LEW	0	(2-5)	.....	5
			P	(5-11)	4/5	
	LEW	DA	0	4	.....	5
			P	(6-8)	5/5	

[1] Compare figure. [2] One graft survived 127 days. [3] Synonym: Anti-lymphocyte-globulin.

continued

Part III. Enhancement

Incompatibility	Strain Combination		Type of Enhancement	Graft Median Survival Time, d	Fraction Surviving	Reference
	Donor	Recipient				
Semiallogenic	(DA x LEW)F_1	DA	0	(4-6)	.....	5
			P	(22->36)	5/5	
	(DA x LEW)F_1	LEW	0	(5-8)	.....	5
			P	(12->36)	5/5	
Muscle						
H-1 plus non-H-1	LEW	AVN	P	>30	0/4	6
Non-H-1	LEW.1A	AVN	P	>30	5/10	6

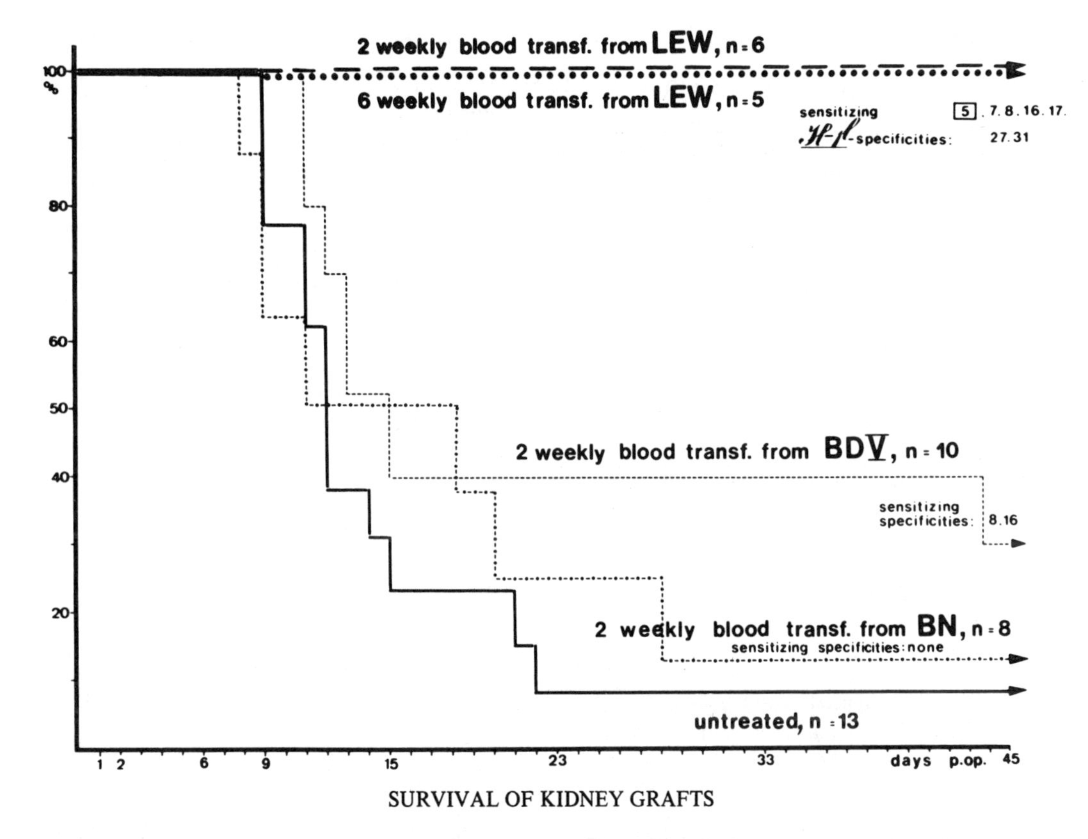

SURVIVAL OF KIDNEY GRAFTS

KIDNEYS FROM LEW $H\text{-}1^l \rightarrow$ LEW.1N $H\text{-}1^n$ (Congenic Resistant) AFTER PRETREATMENT WITH BLOOD FROM: THE KIDNEY DONOR (LEW); A PARTLY RELATED THIRD-PARTY STRAIN (BD V); or AN UNRELATED STRAIN (BN), matched for the recipient at *H-1*, different for the recipient at non-*H-1* antigens

Contributor: Gunther H. Thoenes

References

1. Batchelor, J. R., et al. 1972. Transplantation 13:610-613.

2. Fabre, J. W., and J. R. Batchelor. 1975. Ibid. 20:269-271.

continued

87. TISSUE AND ORGAN TRANSPLANTATION: RAT

Part III. Enhancement

3. Fabre, J. W., and P. J. Morris. 1972. Ibid. 13:604-609.
4. Fabre, J. W., and P. J. Morris. 1974. Ibid. 18:436-442.
5. Finch, D. R. A., and P. J. Morris. 1976. Ibid. 22:508-512.
6. Gutman, E., et al. 1976. Ibid. 21:220-224.
7. Guttmann, R. D., and R. E. Falk. 1974. Ibid. 17:228-231.
8. Haefen, U. von, et al. 1973. Ibid. 16:295-303.
9. Kim, J. P., et al. 1972. Ibid. 13:322-329.
10. Marquet, R. L., et al. 1971. Transplant. Proc. 3:708-710.
11. Thoenes, G. H. Unpublished. Medizinische Klinik Innenstadt, Univ. München, Germany, 1978.
12. Thoenes, G. H. 1978. Eur. Surg. Res. 10:294.

88. IMMUNOLOGICALLY MEDIATED DISEASES: RAT

Susceptibility: S = susceptible; I = intermediate; R = resistant. **Genetics of Disease**: *H-1* ⟨Ag-B⟩ = major histocompatibility system of the rat.

Disease ⟨Synonym⟩	Induction of Disease	Assay of Disease	Rat		Genetics of Disease	Reference
			Susceptibility	Strain ⟨Synonym⟩		
Adjuvant-induced arthritis ⟨Adjuvant disease; AA⟩	Cell-wall preparations—of *Mycobacterium, Streptomyces, Corynebacterium, Nocardia,* & *Staphylococcus aureus*—in mineral oil	Joint swelling, mucocutaneous lesions	S	LE ⟨Long-Evans⟩; LEW ⟨Lewis⟩; SD ⟨Sprague-Dawley⟩; Wistar, dextran-R	Polygenic; susceptibility dominant	9,23,35, 38,39,52
			I	LE ⟨Long-Evans⟩		
			R	AVN; Wistar, dextran-NR		
Experimental allergic encephalomyelitis ⟨EAE⟩	Brain & spinal cord homogenate, plus adjuvant	Clinical signs	S	LEW	Susceptibility recessive	39
			R	AVN; LE ⟨Long-Evans⟩; (AVN x LEW)F_1		
	Brain & spinal cord homogenate, basal encephalitogenic protein, plus adjuvant	Clinical signs, histology	S	AGUS; AS; AUG; BH; BN[1/]; CAR; CAS; DA; F344 ⟨Fischer⟩; LEW ⟨Lewis⟩; NBR; (BN x LEW)F_1; (DA x LEW)F_1	Polygenic; *H-1* controlled; susceptibility dominant	11,12,15, 18,24,28, 31,34,49, 55
			R	BN; BN.B4; LH		
Experimental allergic neuritis ⟨EAN⟩	Peripheral nerve homogenate, plus adjuvant	Clinical signs, histology	S	F344; SD ⟨Sprague-Dawley⟩	....	4,29
Experimental autoimmune myasthenia gravis ⟨EAMG⟩	Acetylcholine-receptor protein, plus adjuvants	Antibody by radioimmunoassay	S	LEW ⟨Lewis⟩	....	26
Experimental allergic myositis ⟨EAM⟩	Muscle homogenate or extract, plus adjuvant	Histology, creatine kinase ⟨CPK⟩ activity	S	PVG/c	....	7,32
Experimental autoallergic sialadenitis	Submandibular gland homogenate, plus adjuvant	Histology	S	LEW ⟨Lewis⟩	....	54

1/ Susceptible when using special immunization procedure [28].

continued

Disease ⟨Synonym⟩	Induction of Disease	Assay of Disease	Rat: Susceptibility	Rat: Strain ⟨Synonym⟩	Genetics of Disease	Reference
Spontaneous autoimmune thyroiditis ⟨SAT⟩	None	Histology, antibody	S	BUF ⟨Buffalo⟩	Resistance dominant	16,33,43,45
			R	F344; LE ⟨Long-Evans⟩; LEW; RAH; SD ⟨Sprague-Dawley⟩; Wistar		
Experimental autoimmune thyroiditis ⟨EAT⟩	Thyroid extract, thyroglobulin, plus adjuvant	Histology, antibody	S	AO; AUG; BUF; DA; LIS ⟨HL⟩; LEW; LH; SD; Wistar	....	20,36,41,46,50
			R	AS; CAM; PVG/c ⟨HO⟩; WAG		
Thyroiditis	T-cell depletion	Histology, antibody	S	AUG; BUF; PVG/c; Wistar	....	37,44
			I	LIS; WAG		
			R	CAM		
	3-Methylcholanthrene	Histology, antibody	S	BUF	....	14,43-45
			R	ACI; F344; LEW; Marshall; OM; (BUF x LEW)F_1		
	9,10-Dimethyl-1,2-benzanthracene[2], trypan blue, carbon tetrachloride	Histology	S	BUF	....	40
Experimental autoimmune adrenalitis ⟨EAA⟩	Adrenal homogenate or extract, plus adjuvant	Histology	S	LEW ⟨Lewis⟩	....	1,19,30
Nephrotoxic serum nephritis ⟨Masugi's nephritis; NTN⟩	Heterologous anti-kidney antibody	Proteinuria, histology	S	SD	....	51
Autologous immune-complex nephritis ⟨AIC⟩	Kidney homogenate, renal tubular epithelial antigen, plus adjuvant	Proteinuria, histology	S	AS; F344; LEW ⟨Lewis⟩; LEW.1D; SD ⟨Sprague-Dawley⟩	*H-1* controlled	6,13,17,47,51,53
			R	BN; BUF; LEW.BN; LEW.1N		
	Kidney homogenate, plus adjuvant	Proteinuria, histology	S	BUF; SD	....	22
Serum sickness type of glomerulonephritis	Antigen-antibody complexes	Proteinuria, histology	S	Carworth W; Holtzman	....	3,8
Experimental membranous glomerulonephritis	Renal glycopeptide, plus adjuvant	Proteinuria, histology	S	Wistar	....	42
Experimental tubulointerstitial disease ⟨Experimental anti-tubular basement membrane nephritis; tubulointerstitial nephritis; ⟨TIN⟩	Kidney homogenate, tubular basement membrane, plus adjuvant	Proteinuria, histology	S	BN; SD; (LEW x BN)F_1	....	22,25,48

[2] Synonym: 7,12-Dimethylbenz[*a*]anthracene.

continued

Disease ⟨Synonym⟩	Induction of Disease	Assay of Disease	Rat		Genetics of Disease	Reference
			Susceptibility	Strain ⟨Synonym⟩		
$HgCl_2$-induced anti-glomerular basement membrane antibody production	$HgCl_2$ injections	Proteinuria, histology	S	BN	2-3 genes, 1 gene *H-1* linked	5
			R	AUG; LEW ⟨Lewis⟩; LEW.1N; PVG/c; Wistar		
$HgCl_2$-induced nephritis	$HgCl_2$ injections	Urine, histology	S	LEW; Wistar	....	2,21
Experimental autoimmune orchitis ⟨Experimental aspermatogenesis; EAO⟩	Testis homogenate, plus adjuvant	Histology	S	LEW ⟨Lewis⟩; Sherman; Wistar	....	10,27

Contributors: Larry D. Bacon and Noel R. Rose; Eberhard Günther

References

1. Andrada, J. A., et al. 1968. Lab. Invest. 19:460-465.
2. Bariety, J., et al. 1976. Am. J. Pathol. 65:293-302.
3. Benacerraf, B., et al. 1960. J. Exp. Med. 111:195-200.
4. Crawford, C. L., et al. 1974. Nature (London) 251: 223-225.
5. Druet, E., et al. 1977. Eur. J. Immunol. 7:348-351.
6. Edgington, T. S., et al. 1976. Science 155:1432-1434.
7. Esiri, M. M., and I. C. M. MacLennan. 1974. Clin. Exp. Immunol. 17:139-150.
8. Fennell, R. H., and V. M. Pardo. 1967. Lab. Invest. 17:481-488.
9. Freeman, P. C., and G. B. West. 1972. Br. J. Pharmacol. 44:327.
10. Freund, J., et al. 1954. Proc. Soc. Exp. Biol. Med. 87:408-411.
11. Gasser, D. L., et al. 1973. Science 181:872-873.
12. Gasser, D. L., et al. 1975. J. Immunol. 115:431-433.
13. Glassock, R. J., et al. 1968. J. Exp. Med. 127:573-588.
14. Glover, E. L., et al. 1969. Arch. Environ. Health 18: 901.
15. Gonatas, N. K., and J. C. Howard. 1974. Science 186: 839-841.
16. Hajdu, A., and G. Rona. 1969. Experientia 25:1325-1327.
17. Heyman, W., et al. 1959. Proc. Soc. Exp. Biol. Med. 100:660-664.
18. Hughes, R. A. C., and J. Stedronska. 1973. Immunology 24:879-884.
19. Irino, T., and A. Grollman. 1968. Metabolism 17: 1365-1375.
20. Jones, H. E., and I. M. Roitt. 1961. Br. J. Exp. Pathol. 42:546-557.
21. Kelchner, J., et al. 1976. Experientia 32:1204-1208.
22. Klassen, J., et al. 1971. Lab. Invest. 25:577-585.
23. Koga, T., et al. 1973. Proc. Soc. Exp. Biol. Med. 143: 824-827.
24. Kornblum, J. 1968. J. Immunol. 101:702.
25. Lehmann, D. H., et al. 1974. Kidney Int. 5:187-195.
26. Lennon, V. A., et al. 1975. J. Exp. Med. 141:1365-1375.
27. Levine, S., and R. Sowinski. 1970. Am. J. Pathol. 59: 437-451.
28. Levine, S., and R. Sowinski. 1975. J. Immunol. 114: 597-601.
29. Levine, S., and E. J. Wenk. 1963. Proc. Soc. Exp. Biol. Med. 113:898-900.
30. Levine, S., and E. J. Wenk. 1968. Am. J. Pathol. 52: 41-53.
31. Martenson, R. E., et al. 1975. J. Immunol. 114:592-596.
32. Morgan, G., et al. 1971. Arthritis Rheum. 14:599.
33. Noble, B., et al. 1976. J. Immunol. 117:1447-1455.
34. Paterson, P. Y. 1968. In P. A. Miescher and H. J. Müller-Eberhard, ed. Textbook of Immunopathology. Grune and Stratton, New York. v. 1, pp. 132-149.
35. Pearson, C. M., et al. 1961. J. Exp. Med. 113:485-509.
36. Penhale, W. J., et al. 1975. Clin. Exp. Immunol. 19: 179-191.
37. Penhale, W. J., et al. 1975. Ibid. 21:362-375.
38. Perlik, F., and Z. Zidek. 1973. Ann. Rheum. Dis. 32: 72-74.
39. Perlik, F., and Z. Zidek. 1974. Z. Immunitaetsforsch. Allerg. Klin. Immunol. 147:191-193.
40. Reuber, M. D., and E. L. Glover. 1969. Experientia 25:753.
41. Rose, N. R. 1975. Cell. Immunol. 18:360-364.
42. Shibata, S., et al. 1972. Lab. Invest. 27:457-465.
43. Silverman, D. A., and N. R. Rose. 1971. Proc. Soc. Exp. Biol. Med. 138:579-584.
44. Silverman, D. A., and N. R. Rose. 1974. Science 184: 162-163.
45. Silverman, D. A., and N. R. Rose. 1975. J. Immunol. 114:145-147.

continued

46. Silverman, D. A., and N. R. Rose. 1975. Ibid. 114:148-150.
47. Stenglein, B., et al. 1975. Ibid. 115:895-897.
48. Sugusaki, T., et al. 1973. Lab. Invest. 28:658-671.
49. Swierkosz, J. E., and R. E. Swanborg. 1975. J. Immunol. 115:631-633.
50. Twarog, F. J., et al. 1970. Proc. Soc. Exp. Biol. Med. 133:185-189.
51. Unanue, E. R., and F. J. Dixon. 1967. Adv. Immunol. 6:1-90.
52. Van Arman, C. G. 1976. Fed. Proc. Fed. Am. Soc. Exp. Biol. 35:2442-2446.
53. Watson, J. I., and F. J. Dixon. 1966. Proc. Soc. Exp. Biol. Med. 121:216-223.
54. White, S. C., and G. W. Casarett. 1974. J. Immunol. 112:178-185.
55. Williams, R. M., and M. J. Moore. 1973. J. Exp. Med. 138:775-783.

89. SUSCEPTIBILITY TO INFECTIOUS DISEASES: RAT

Source: For full name, *see* list of HOLDERS at front of book; local = inbred strains developed in contributor's laboratory. **Response**: WBC = leukocytes. Data in brackets refer to the column heading in brackets.

Organism ⟨Synonym⟩	Strain or Stock ⟨Synonym⟩	Source	Infection: Response	Infection: Phenotypic Marker	Reference
			Mycoplasmas		
Probable *Mycoplasma* species	Albino Norway	Local	Young, 33%; adult, 69%	Incidence of otitis media	9
	SD ⟨Sprague-Dawley⟩	Local	3 wk, 11%; 12 wk, 10%; 52 wk, 25%	Incidence of otitis media	4
	Wistar	Wistar Institute	3 wk, 13%; 12 wk, 21%; 52 wk, 60%		
	Wild type	Local	2%	Incidence of otitis media	9
			Bacteria		
Escherichia coli	Blu:(LE) ⟨Long-Evans (Blue Spruce)⟩	BLU	90%	Renal infection after bladder inoculation	12
	Caw:CFE (SD) ⟨Sprague-Dawley⟩	CAW	11%		
Haemophilus influenzae b	SD ⟨Sprague-Dawley⟩	HLA	High responders (>10-fold); intermediate responders (2-10-fold); non-responders (<2-fold)	Immune response after experimental bacteremia & meningitis—increase in antibodies	8
Mycobacterium tuberculosis	LE ⟨Long-Evans⟩	....	Mean survival, 255 d	Survival after intravenous inoculation	13
	SD ⟨Sprague-Dawley⟩	....	Mean survival, 404 d		
	Wistar	....	Mean survival, 361 d		
Salmonella enteritidis	D	Local	100%	Mortality after intraperitoneal inoculation[1]	5
	Ra	Local	84.7%		
	Wistar B	Local	80.6%		
	Wistar S	Local	93.5%		
	Wild type	Local	95.7%		
Streptococcus pneumoniae	Wistar	Local	45%	Nasopharyngeal carriage[2]	7
	Wild type	Local	0		
	F_1 hybrid	Local	2%		
	Blu:(LE) ⟨Long-Evans (Blue Spruce)⟩	BLU	♂, 90%; ♀, 64%	Mortality 14 d after intratracheal inoculation	12
	Caw:CFE (SD) ⟨Sprague-Dawley⟩	CAW	♂, 100%; ♀, 93%		

[1] Breeding for resistance was possible, appeared to be polygenic, and was generally dominant. [2] Resistance to carriage dominant.

continued

89. SUSCEPTIBILITY TO INFECTIOUS DISEASES: RAT

Organism (Synonym)	Strain or Stock (Synonym)	Source	Infection: Response	Infection: Phenotypic Marker	Reference
Yersinia pestis	SD (Sprague-Dawley)	....	Deaths, 55%; survived with immune response, 22%; survived without immune response, 22%	Mortality & immune response to subcutaneous inoculation	2
Protozoa					
Eimeria miyairii	Hi	Local	190.8 ± 7.8	No. of oocysts excreted after feeding viable oocysts	1
	Lambert	Local	34.1 ± 3.9		
	Lo	Local	81.8 ± 5.5		
	Wistar	Local	78.1 ± 5.8		
	Wistar A	Local	162.8 ± 7.2		
E. nieschulzi	CD[3]	CRL	2500 oocysts	Optimal size of inoculum necessary to produce total resistance (immunity) to experimental challenge with oocysts	6
	CDF[4]	CRL	3500 oocysts		
Cestodes					
Cysticercus fasciolaris	AUG (August)	Local	37%	Incidence of hepatic sarcomas after feeding ova of *Taenia taeniaeformis*[5]	3
	AXC	Local	67%		
	Copenhagen	Local	63%		
	F344 (Fischer 344)	Local	60%		
		Local	67%		
	Marshall	Local	37%		
	Zimmerman	Local	58%		
Taenia taeniaeformis	F344 (Fischer)	....	No., 4.8 ± 0.65; diameter, 3.6 ± 0.13 mm	Number & size of hepatic cysts	10
	Holtzman	....	No., 8.6 ± 0.85; diameter, 2.6 ± 0.24 mm		
	OM (Osborne-Mendel)	....	No., 6.8 ± 0.81; diameter, 1.3 ± 0.12 mm		
	SD (Sprague-Dawley)	....	Similar to Holtzman strain		
Other					
None	AUG (August)	Local	WBC, 14.6 ± 0.38 x $10^3/\mu l$; longevity, 14.03 ± 0.04 mo	Mean blood leukocyte count & longevity	11
	AxC	Local	WBC, 17.5 ± 0.33 x $10^3/\mu l$; longevity, 21.6 ± 0.13 mo		
	Copenhagen	Local	WBC, 25.1 ± 0.5 x $10^3/\mu l$; longevity, 19.3 ± 0.13 mo		
	F344 (Fischer 344)	Local	WBC, 14.0 ± 0.10 x $10^3/\mu l$; longevity, 9.37 ± 0.14 mo		
	Marshall	Local	WBC, 17.1 ± 0.24 x $10^3/\mu l$; longevity, 13.53 ± 0.09 mo		
	Zimmerman	Local	WBC, 15.5 ± 0.23 x $10^3/\mu l$; longevity, 11.86 ± 0.17 mo		

[3] Outbred. [4] Inbred. [5] Resistance was dominant.

Contributor: Richard L. Myerowitz

References

1. Becker, E. R., and P. R. Hall. 1933. Parasitology 25: 397-401.
2. Chen, T. H., and K. F. Meyer. 1974. J. Infect. Dis. 129:S62-S71.

continued

3. Curtis, M. R., et al. 1933. Am. J. Cancer 17:894-923.
4. Freudenberger, C. B. 1932. Anat. Rec. 54:179-184.
5. Irwin, M. R. 1933. J. Immunol. 24:297-348.
6. Liburd, E. M. 1973. Can. J. Zool. 51:273-279.
7. Mirick, G. S., et al. 1950. Am. J. Hyg. 52:48-53.
8. Myerowitz, R. L., and C. W. Norden. 1977. Infect. Immun. 15:1200-1208.
9. Nelson, J. B., and J. W. Gowen. 1930. J. Infect. Dis. 46:53-63.
10. Olivier, L. 1962. J. Parasitol. 48:373-378.
11. Reich, C., and W. F. Dunning. 1941. Science 93:429-430.
12. Stratman, S. L., and M. Conjeros. 1969. Lab. Anim. Care 19:742-745.
13. Tobach, E., and H. Bloch. 1955. Adv. Tuberc. Res. 6:62-89.

90. HORMONE POLYMORPHISMS: RAT

Hormonal abnormalities detected in genetically different animals or in different strains are presented. For further information on diabetes insipidus ⟨Brattleboro⟩ rats, consult references 3, 9, 10, 20, 22, and 28. Data in brackets refer to the column heading in brackets.

Hormone ⟨Synonym⟩ or Hormone Source	Strain ⟨Synonym⟩ or Condition	Reference
Androgens	Testicular feminization ⟨Tfm⟩	21
Insulin	BHE ⟨Carbohydrate-sensitive⟩	4
	Fatty ⟨Genetically obese Zucker rat⟩	25,30,31
Neurophysin	Diabetes insipidus (hereditary hypothalamic) ⟨Brattleboro⟩	7,24
Pituitary hormone	Pseudohermaphroditism in genetically ♂ rat	11
Pituitary-thyroid axis	SHR ⟨Spontaneously hypertensive rats⟩	14
	SPR ⟨Stroke-prone rats⟩	
Prostaglandins	SHR	1,23
Renin [3.4.99.19][1]	SHR, SPR	5,8,12,16,18,26
Thyroid hormone	Fatty ⟨Genetically obese rats⟩	6
	SHR	15,17
Thyrotropin ⟨TSH⟩	SHR	13,27
	Fatty ⟨Genetically obese rats⟩	29
Thyroxine	Jaundice ⟨Gunn⟩	2
Vasopressin ⟨Antidiuretic hormone; ADH⟩	Diabetes insipidus (hypothalamic)	19

[1] Enzyme Commission No. For source, *see* Introduction.

Contributors: R. Moutier; Radim Brdička

References

1. Armstrong, J. M., et al. 1976. Nature (London) 260:582-586.
2. Bastomsky, C. H. 1973. Endocrinology 92:35-40.
3. Bauman, J. W., et al. 1969. Acta Endocrinol. 61:720-728.
4. Berdanier, C. D. 1974. Diabetologia 10:691-695.
5. Bianchi, G., et al. 1975. Circ. Res. 26(6, Suppl. 1):153-161.
6. Bray, G. A., and D. A. York. 1971. Endocrinology 88:1095-1099.
7. Cheng, K. W., et al. 1972. Ibid. 90:1055-1063.
8. de Jong, W., et al. 1972. Proc. Soc. Exp. Biol. Med. 139:1213-1216.
9. Dlouhá, H., et al. 1976. Experientia 32:59.
10. Dyball, R. E. J. 1974. J. Endocrinol. 60:135-143.
11. Goldman, A. S., et al. 1975. Ibid. 64:249-255.
12. Gresson, C. R. 1972. Proc. Univ. Otago Med. Sch. 50:51-52.
13. Kojima, A., et al. 1975. Proc. Soc. Exp. Biol. Med. 149:661-663.
14. Kojima, A., et al. 1975. Ibid. 150:571-573.
15. Kojima, A., et al. 1976. Endocrinology 98:1109-1115.
16. Lee, D. R. 1973. Proc. Univ. Otago Med. Sch. 51:34-35.
17. Manger, W. M., and S. C. Werner. 1973. Fed. Proc. Fed. Am. Soc. Exp. Biol. 32:749(Abstr. 3013).
18. Matsunaga, M., et al. 1975. Jpn. Circ. J. 39:1305-1311.

continued

19. Möhring, B., and J. Möhring. 1975. Life Sci. 17:1307-1314.
20. Möhring, J., et al. 1972. Ibid. 11:65-72.
21. Naess, O., et al. 1976. Endocrinology 99:1295-1303.
22. Norstrom, A. 1973. Z. Zellforsch. Mikrosk. Anat. 140:413-424.
23. Sirois, P., and D. J. Gagnon. 1974. Experientia 30: 1418-1419.
24. Sokol, H. W., et al. 1976. Endocrinology 98:1176-1188.
25. Stern, J., et al. 1972. Proc. Soc. Exp. Biol. Med. 139: 66-69.
26. Vincent, M., et al. 1975. J. Physiol. (Paris) 71: 152A.
27. Werner, S. C., et al. 1975. Proc. Soc. Exp. Biol. Med. 148:1013-1017.
28. Wimersma Greidanus, Tj. B. van, et al. 1974. Experientia 30:1217-1218.
29. York, D. A., et al. 1972. Endocrinology 90:67-72.
30. Zucker, L. M., and H. N. Antoniades. 1970. Fed. Proc. Fed. Am. Soc. Exp. Biol. 29:379(Abstr. 758).
31. Zucker, L. M., and H. N. Antoniades. 1972. Endocrinology 90:1320-1330.

91. REPRODUCTIVE ENDOCRINOLOGY: RAT

Hormone Measured: FSH = follicle-stimulating hormone; LH = luteinizing hormone; LH-RH = luteinizing hormone—releasing hormone. **Value**: Gonadotropin serum concentrations are expressed as ng/ml of the appropriate NIAMDD rat reference preparation. Where necessary, concentrations have been calculated on the basis of the following potency estimates—LH-RP-1 = 0.03 x NIH-LH-S1; FSH-RP-I = 2.1 x NIH-FSH-S1; and Prolactin RP-1 = 11 IU/mg. Gonadotropin concentrations in the anterior pituitary gland are expressed as μg/mg (wet weight), or as μg/pituitary gland relative to the appropriate NIAMDD rat reference preparation. Hypothalamic hormone ⟨LH-RH or GN-RH⟩ concentrations are expressed in terms of the mass of the synthetic decapeptide. **Change**: ↑ = increase(d); ↓ = decrease(d). Data in brackets refer to the column headings in brackets.

Stock or Strain ⟨Synonym⟩	Age, d [Body Wt, g]	Sampling Period [Interval Between Samples]	Hormone	Measurement [Change]	Reference
			Puberty: Male		
Crl:(WI) ⟨Wistar (Charles River)⟩	1-68, plus fetus	75 d [4 d]	Gonadotropins		9
			Serum		
			FSH	300-800 ng/ml [↑ on days 40-50]	
			LH	35-50 ng/ml	
			Anterior pituitary		
			FSH	0.1-35 μg/mg [↑ on day 40]	
			LH	0.15-90 μg/mg [Progressive ↑]	
			Hypothalamic, brain: LH-RH	100-12,000 pg/hypothalamus [Progressive ↑]	9
Holtzman	1-12, plus adult	11 d [1-2 d]	Gonadotropins, serum		23
			FSH	200-500 ng/ml	
			LH	1-3 ng/ml	
			Prolactin	3-6 ng/ml	
	5-40, plus adult	35 d [2-5 d]	Gonadotropins, serum		32
			FSH	300-800 ng/ml [↑ on days 20-35]	
			LH	40-110 ng/ml [Peak at day 20]	
			Steroids, serum: testosterone	0.5-3 ng/ml [Peak at day 20]	32
	14-79	65 d [3-6 d]	Gonadotropins, serum		38
			FSH	450-800 ng/ml	
			LH	8-60 ng/ml [↑ on day 25]	
			Prolactin	5-50 ng/ml [↑ on days 35-45]	
			Steroids, serum: testosterone, plus other androgens	0.5-3 ng/ml [↑ on days 58-79]	38

continued

Stock or Strain ⟨Synonym⟩	Age, d [Body Wt, g]	Sampling Period [Interval Between Samples]	Hormone	Measurement [Change]	Reference
Holtzman (SD) ⟨Sprague-Dawley (Holtzman)⟩	15-60, plus adult	45 d [5-10 d]	Gonadotropins		12
			Serum: FSH	250-650 ng/ml [↑ on day 35]	
			Anterior pituitary		
			FSH	10-60 μg/mg [↑ on day 35]	
			LH	15-45 μg/mg	
Hooded Liverpool	21-81, plus adult	60 d [2-10 d]	Gonadotropins		15
			Serum		
			FSH	400-1400 ng/ml [↑ on days 35-55]	
			LH	1-60 ng/ml [↑ on days 50-55]	
			Anterior pituitary		
			FSH	10-160 μg/mg [↑ on days 50-60]	
			LH	40-350 μg/pituitary [↑ on days 50-60]	
			Steroids, serum: testosterone	0.3-5 ng/ml [↑ on days 59-81]	15
Ivanovac	16-90	75 d [1-10 d]	Steroids, serum		25
			Testosterone	0.5-2.5 ng/ml [↑ on days 30-90]	
			Stanolone[1/]	0.05-0.2 ng/ml [↑ on day 38]	
SD ⟨Sprague-Dawley⟩	1-47, plus adult	46 d [1-2 d]	Gonadotropins, serum		16
			FSH	200-500 ng/ml [↑ on days 41-47]	
			LH	20-50 ng/ml	
			Prolactin	20-125 ng/ml [↑ on days 31-45]	
			Steroids, serum		16
			Estradiol	20-330 pg/ml [↑ on days 1-2 & 11-19]	
			Progesterone	2-5 ng/ml [↑ on days 27-40]	
			Testosterone	1-2 ng/ml [↑ on days 20-40]	
	10-90	80 d [5-10 d]	Steroids, serum		33
			5α-Androstane-3α,17β-diol	1.2-2.2 ng/ml [↑ on day 20]	
			Testosterone, plus stanolone[1/]	0.2-5.2 ng/ml [↑ on day 60]	
	25	24 h [4 h]	Steroids, serum		34
			5α-Androstane-3α,17β-diol	0.4-2.6 ng/ml [↑ at 0800 & 1600]	
			Testosterone, plus stanolone[1/]	0.2-1.4 ng/ml [↑ at 0800]	
Wistar	10-90	80 d [5-30 d]	Steroids, serum		37
			Testosterone	0.5-4 ng/ml [Peak at day 20]	
			Stanolone[1/]	0.1-0.8 ng/ml	
	21-91	70 d [14 d]	Gonadotropins, serum		48
			FSH	450-1200 ng/ml [↑ on day 35]	
			LH	35-50 ng/ml	
Wistar-Holtzman mix	5-45	40 d [5 d]	Gonadotropins, serum		37
			FSH	450-900 ng/ml	
			LH	15-40 ng/ml	
Puberty: Female					
Crl:(WI) ⟨Wistar (Charles River)⟩	1-68, plus fetus	75 d [4 d]	Gonadotropins		9
			Serum		
			FSH	200-1000 ng/ml [↑ on days 8-20]	
			LH	30-90 ng/ml	
			Anterior pituitary		
			FSH	0.1-30 μg/mg [↑ on day 22]	

1/ Synonym: Dihydrotestosterone.

continued

Stock or Strain ⟨Synonym⟩	Age, d [Body Wt, g]	Sampling Period [Interval Between Samples]	Hormone	Measurement [Change]	Reference
			LH	0.15-45 µg/mg [Progressive ↑ on days 1-20]	
			Hypothalamic, brain: LH-RH	100-6000 pg/hypothalamus [Progressive ↑ to day 40]	9
Holtzman	1-12, plus adult	11 d [1-2 d]	Gonadotropins, serum		23
			LH	1-4 ng/ml	
			Prolactin	3-6 ng/ml	
Holtzman (SD) ⟨Sprague-Dawley (Holtzman)⟩	15-60, plus adult	45 d [5 d]	Gonadotropins		13
			Serum: FSH	200-650 ng/ml [↑ on day 15]	
			Anterior pituitary		
			FSH	3-25 µg/mg [↑ on day 15]	
			LH	7.5-30 µg/mg [↑ on day 15]	
SD ⟨Sprague-Dawley⟩	1-47, plus adult	46 d [1-2 d]	Gonadotropins, serum		16
			FSH	100-900 ng/ml [↑ on days 9-17]	
			LH	20-500 ng/ml [Variable on days 11-23]	
			Prolactin	10-200 ng/ml [↑ after day 25]	
			Steroids, serum		16
			Estradiol	25-300 pg/ml [↑ on days 9-21]	
			Progesterone	2-10 ng/ml	
			Testosterone	0.2-0.4 ng/ml	
Wistar-Holtzman mix	5-45	40 d [5 d]	Gonadotropins, serum		37
			FSH	300-1600 ng/ml [↑ on days 12-15]	
			LH	12-42 ng/ml [↑ on day 10]	
Intact Adult: Male					
Charles River	[360-500]	8 h [30 min]; 8 d [24 h]	Steroids, serum: testosterone	1-15 ng/ml [Rapid fluctuations]	2
Crl:CD ⟨Charles River (CD)⟩	[180-200]	1 d [3 h]	Gonadotropins, serum		27
			FSH	600-850 ng/ml [↑ at 1900]	
			LH	10-35 ng/ml	
			Steroids, serum		27
			Progesterone	0.05-0.7 ng/ml [↑ at 1900]	
			Testosterone	0.5-3.5 ng/ml [↑ at 1600]	
			Stanolone[1]	0.02-0.12 ng/ml	
			Hypothalamic: LH-RH		27
			Serum	1.5-3 pg/ml	
			Brain	2-3 ng/hypothalamus	
SD ⟨Sprague-Dawley⟩	[320-420]	1400-1600 [Single sample]	Gonadotropins		5
			Serum		
			FSH	376 ng/ml	
			LH	17 ng/ml	
			Prolactin	57 ng/ml	
			Anterior pituitary		
			LH	22.5 µg/mg	
			Prolactin	0.91 µg/mg	
			Steroids, serum: testosterone	3.07 ng/ml	5
Wistar	Adult	1 d [2 h]	Gonadotropins, serum		31
			FSH	600-1300 ng/ml [↑ at 1300]	
			LH	60-100 ng/ml	
			Prolactin	25-55 ng/ml	

[1] Synonym: Dihydrotestosterone.

continued

Stock or Strain ⟨Synonym⟩	Age, d [Body Wt, g]	Sampling Period [Interval Between Samples]	Hormone	Measurement [Change]	Reference
		1 d [3-5 h]	Gonadotropins		8
			Serum		
			FSH	300-400 ng/ml	
			LH	18-50 ng/ml	
			Anterior pituitary		
			FSH	9-15 μg/mg	
			LH	18-31 μg/mg	
			Hypothalamic, brain: LH-RH	8.3-14.5 ng/hypothalamus	8
Intact Adult: Female, 4-Day Cycle					
Crl:CD ⟨Charles River (CD)⟩	[180-220]	4-d cycle [1-6 h]	Gonadotropins, serum: LH	15-1000 ng/ml [↑ on day of estrus]	26
			Steroids, serum		26
			Estradiol	10-60 pg/ml	
			Progesterone	2-21 ng/ml [↑ on day of diestrus & day of proestrus]	
			Hypothalamic: LH-RH		26
			Serum	5-14 pg/ml	
			Brain	2.2-3 ng/hypothalamus	
Crl:CD (SD) ⟨Sprague-Dawley (C. River, CD)⟩	Adult	4-d cycle [3-5 h]	Gonadotropins, serum: LH	20-1300 ng/ml [Peak on day of proestrus]	28
			Steroids, serum		28
			Estradiol	6-40 pg/ml	
			Progesterone	4-30 ng/ml	
Crl:(SD) ⟨Sprague-Dawley (Charles River)⟩	....	4-d cycle [2-24 h]	Gonadotropins, serum		42
			FSH	50-400 ng/ml	
			LH	100-6000 ng/ml	
			Steroids, serum: progesterone	7-23 ng/ml [↑ on day of proestrus]	42
SD ⟨Sprague-Dawley⟩	[150-175]	4-d cycle [3 h]	Gonadotropins, serum		7
			FSH	125-300 ng/ml [Proestrous ↑ to 600 ng/ml]	
			LH	10-40 ng/ml [Proestrous ↑ to 3400 ng/ml]	
			Prolactin	10-30 ng/ml [Proestrous ↑to 450 ng/ml]	
			Steroids, serum		7
			Estradiol	20-28 pg/ml [Progressive ↑ to proestrus]	
			Progesterone	5-20 ng/ml [↑ to 60 ng/ml on day of proestrus]	
	[200-250]	4-d cycle [2 h]	Gonadotropins, serum		46
			FSH	50-150 ng/ml [500 ng/ml on day of proestrus & on day of estrus]	
			LH	15-45 ng/ml [2000 ng/ml on day of proestrus]	
			Prolactin	20-280 ng/ml	

continued

Stock or Strain (Synonym)	Age, d [Body Wt, g]	Sampling Period [Interval Between Samples]	Hormone	Measurement [Change]	Reference
			Steroids, serum		46
			Estradiol	5-45 pg/ml [Progressive ↑ to proestrus]	
			Progesterone	5-25 ng/ml [50 ng/ml on day of proestrus]	
	[280-300]	Proestrus (afternoon) [5-60 min]	Gonadotropins, serum: LH	20-2000 ng/ml [↑ at 1600-1800]	4
Intact Adult: Female, 5-Day Cycle					
Holtzman	120	Proestrus & estrus [1-6 h]	Gonadotropins, serum		11
			FSH	120-600 ng/ml	
			LH	100-1000 ng/ml	
	[225-275]	5-d cycle [3-12 h]	Gonadotropins, serum: LH	20-1100 ng/ml	24
			Steroids, serum		24
			Estradiol	5-70 pg/ml [Progressive ↑ to proestrus]	
			Progesterone	8-45 ng/ml [↑ at proestrus]	
	Retired breeders	5-d cycle [2 h]	Gonadotropins, serum		22
			FSH	100-500 ng/ml [↑ on day of proestrus & on day of estrus]	
			LH	5-40 ng/ml [↑ to 900 ng/ml on day of proestrus]	
Long Evans	Adult	34 h [2-27 h]	Gonadotropins, serum		43
			FSH	80-570 ng/ml	
			LH	20-1400 ng/ml	
			Prolactin	5-460 ng/ml[2]	
Wistar substrain	Adult	Diestrus & proestrus [4 h]	Gonadotropins, serum: LH	40-450 ng/ml[3] 40-1740 ng/ml[4]	41
			Steroids, serum: progesterone	3-25 ng/ml[3] 3-18 ng/ml[4]	41
Post-Copulation: Male					
SD (Sprague-Dawley)	110	0-60 min post-copulation [4-30 min]	Gonadotropins, serum		29
			FSH	150 ng/ml [None]	
			LH	24-36 ng/ml	
			Prolactin	50-210 ng/ml	
			Steroids, serum: testosterone	1.5-2.9 ng/ml	29
Post-Copulation: Female					
Charles River	Adult	1430-2130 on day of proestrus [30-120 min]	Gonadotropins, serum		40
			FSH	600-700 ng/ml [None]	
			LH	300-1230 ng/ml [None]	
Holtzman	Mature	0-16 h post-copulation [4 h]	Gonadotropins, serum		30
			FSH	300 ng/ml [None]	

[2] Compared with 4-day cycle in reference. [3] 4-day cycle. [4] 5-day cycle.

continued

Stock or Strain (Synonym)	Age, d [Body Wt, g]	Sampling Period [Interval Between Samples]	Hormone	Measurement [Change]	Reference
			LH	15 ng/ml [↑ at 4 h]	
			Prolactin	25-110 ng/ml [↑ at 4-16 h]	
SD (Sprague-Dawley)	[150-175]	48 h [3 h]	Gonadotropins, serum		7
			FSH	50-550 ng/ml	
			LH	10-3600 ng/ml	
			Prolactin	20-300 ng/ml	
			Steroids, serum		7
			Estradiol	20-85 pg/ml	
			Progesterone	3-55 ng/ml	
Pseudopregnancy					
SD (Sprague-Dawley)	24	4 d [2-4 h]	Gonadotropins, serum: prolactin	15-75 ng/ml [Nocturnal surge only]	45
	[200-250]	3 d [2 h]	Gonadotropins, serum		46
			FSH	100-300 ng/ml	
			LH	10-20 ng/ml	
			Prolactin	20-320 ng/ml [Surges at 0500 & 1800]	
			Steroids, serum		46
			Estradiol	4-15 pg/ml	
			Progesterone	5-45 ng/ml [Progressive ↑]	
	[200-275][5]	11 d [2 h]	Gonadotropins, serum: prolactin	20-400 ng/ml [Nocturnal surge only, at 0300-0500]	17
Wistar Inbred	[180-210]	17 d [1-3 d]	Gonadotropins, serum		14
			FSH	120-170 ng/ml	
			LH	20-35 ng/ml	
			Prolactin	15-140 ng/ml [Surges twice daily]	
			Steroids, serum: progesterone	30-80 ng/ml [↑ on days 5-8]	14
Substrain	[170-220]	12 d [1-2 d]	Gonadotropins, serum		49
			FSH	20-150 ng/ml	
			LH	5-30 ng/ml	
			Steroids, serum		49
			Estradiol	2-15 pg/ml	
			Progesterone	25-90 ng/ml [↑ on day 8]	
Pregnancy					
Holtzman	Adult	23 d [1 d]	Gonadotropins, serum		30
			FSH	50-150 ng/ml [↑ on day 18]	
			LH	3-26 ng/ml [↑ on day 23]	
			Prolactin	12-36 ng/ml [No peaks]	
Holtzman (SD) (Sprague-Dawley (Holtzman))	....	23 d, plus 1 d postpartum [12 h]	Gonadotropins, serum		35
			LH	5-50 ng/ml	
			Prolactin	8-400 ng/ml [↑ on day 23]	
			Steroids, serum: progesterone	5-130 ng/ml [↑ on days 15-19]	35

[5] Catheterized animals.

continued

Stock or Strain ⟨Synonym⟩	Age, d [Body Wt, g]	Sampling Period [Interval Between Samples]	Hormone	Measurement [Change]	Reference
SD ⟨Sprague-Dawley⟩	[150-175]	Days 1-4 of pregnancy [3 h]	Gonadotropins, serum		7
			FSH	125-250 ng/ml	
			LH	10-50 ng/ml	
			Prolactin	25-450 ng/ml [2 peaks/d]	
			Steroids, serum		7
			Estradiol	15-30 pg/ml	
			Progesterone	10-80 ng/ml [Progressive ↑]	
Wistar	Adult	16 d [1-4 d]	Gonadotropins: LH		6
			Serum	9-40 ng/ml [↓ on days 12-16]	
			Anterior pituitary	8-34 µg/mg [↑ on day 16]	
			Steroids, serum: estradiol	3-8 pg/ml	6
Lactation					
SD ⟨Sprague-Dawley⟩	Adult	Days 2-20 postpartum [5 d]	Gonadotropins		44
			Serum		
			FSH	58-94 ng/ml	
			LH	1.6-6.8 ng/ml [↑ on day 20]	
			Anterior pituitary		
			FSH	1.0-3.8 µg/mg [↑ on day 20]	
			LH	3.2-10.2 µg/mg	
			Steroids, serum		44
			Estradiol	2.6-17.5 pg/ml [Progressive ↑]	
			Progesterone	13-69 ng/ml [↑ on day 10]	
Castration: Male					
Holtzman	25, plus adult	1-11 d post-castration [0.5-2 d]	Gonadotropins: LH		50
			Serum	120-510 ng/ml [Progressive ↑]	
			Anterior pituitary	9-48 µg/mg	
	50	2-24 h post-castration [2-12 h]	Gonadotropins, serum		19
			FSH	400-1400 ng/ml [↑ lasts from 8 h post-castration to 24 h post-castration]	
			LH	7-200 ng/ml [↑ lasts from 8 h post-castration to 24 h post-castration]	
	75	5 min-2 h post-castration [5-15 min]	Steroids, serum		10
			Testosterone	5.5 ng/ml [↓ to 1 ng/ml in 30 min]	
			Stanolone[1]	0.35 ng/ml [↓ to 0.13 ng/ml in 15 min]	
	Adult	1-61 d post-castration [1-18 d]	Gonadotropins: LH		20
			Serum	300-750 ng/ml	
			Anterior pituitary	35-175 µg/mg [Progressive ↑]	
	Adult[6]	1-2 h [3 min]	Gonadotropins, serum: LH	500-1500 ng/ml [Episodic ↑, 20-30 min]	21
SD ⟨Sprague-Dawley⟩	55-60	1-60 d post-castration [1-30 d]	Gonadotropins		1
			Serum		
			FSH	1000-1400 ng/ml	
			LH	400-1400 ng/ml	

[1] Synonym: Dihydrotestosterone. [6] Long-term castrates.

continued

Stock or Strain (Synonym)	Age, d [Body Wt, g]	Sampling Period [Interval Between Samples]	Hormone	Measurement [Change]	Reference
			Anterior pituitary		
			FSH	90-100 μg/mg protein	
			LH	250-700 μg/mg protein	
			Hypothalamic: LH-RH		1
			Serum	30-42 pg/ml [None]	
			Brain	1.8-5 ng/hypothalamus [↓ on day 7]	
Wistar	21	1-70 d post-castration [14 d]	Gonadotropins, serum		48
			FSH	900-2500 ng/ml	
			LH	35-400 ng/ml	
Castration: Female					
Holtzman	25, plus adult	1-8 d post-castration [0.5-6 d]	Gonadotropins: LH		50
			Serum	30-225 ng/ml	
			Anterior pituitary	7.5-18 μg/mg	
	Adult	1-35 d post-castration [3-7 d]	Gonadotropins, serum: LH	50-500 ng/ml [Progressive ↑]	20
	[220-250]	1-16 d post-castration [2 d]	Gonadotropins, serum		18
			FSH	100-2000 ng/ml [Progressive ↑]	
			LH	20-530 ng/ml [Progressive ↑]	
SD (Sprague-Dawley)	80	1-6 h post-castration on day of proestrus [1-2 h]	Gonadotropins, serum		36
			FSH	230-1300 ng/ml	
			LH	60-900 ng/ml	
			Steroids, serum		36
			Estradiol	40-50 pg/ml	
			Progesterone	5-15 ng/ml	
	[200-220][6]	1200-1800 [10 min]	Gonadotropins, serum: LH	500-1000 ng/ml [Episodic variations]	3
Wistar	[130-180][6]	2.5 h, light:dark [5 min]	Gonadotropins, serum: LH	1000-2000 ng/ml [Episodic variations]	47

[6] Long-term castrates.

Contributor: Vernon L. Gay

References

1. Badger, T. M., et al. 1978. Endocrinology 102:136-141.
2. Bartke, A., et al. 1973. Ibid. 92:1223-1228.
3. Blake, C. A. 1974. Ibid. 95:813-817.
4. Blake, C. A. 1976. Ibid. 98:445-450.
5. Blake, C. A., et al. 1973. Ibid. 92:1419-1425.
6. Brown-Grant, K., et al. 1972. J. Endocrinol. 53:31-35.
7. Butcher, R. L., et al. 1975. Endocrinology 96:576-586.
8. Chiappa, S. A., and G. Fink. 1977. J. Endocrinol. 72:195-210.
9. Chiappa, S. A., and G. Fink. 1977. Ibid. 72:211-224.
10. Coyotupa, J., et al. 1973. Endocrinology 92:1579-1581.
11. Daane, T. A., and A. F. Parlow. 1971. Ibid. 88:653-663.
12. Debeljuk, L., et al. 1972. Ibid. 90:585-588.
13. Debeljuk, L., et al. 1972. Ibid. 90:1499-1502.
14. de Greef, W. J., et al. 1977. Ibid. 101:1054-1063.
15. de Jong, F. H., and R. M. Sharpe. 1977. J. Endocrinol. 75:197-207.
16. Dohler, K. D., and W. Wuttke. 1975. Endocrinology 97:898-907.

continued

17. Freeman, M. E., and J. D. Neill. 1972. Ibid. 90:1292-1294.
18. Gay, V. L., and R. L. Hauger. 1977. Biol. Reprod. 16: 527-535.
19. Gay, V. L., and J. T. Kerlan. 1978. Arch. Androl. 1: 257-266.
20. Gay, V. L., and A. R. Midgley, Jr. 1969. Endocrinology 84:1359-1364.
21. Gay, V. L., and N. A. Sheth. 1972. Ibid. 90:158-162.
22. Gay, V. L., et al. 1970. Fed. Proc. Fed. Am. Soc. Exp. Biol. 29:1880-1887.
23. Goldman, B. D., et al. 1971. Endocrinology 88:771-776.
24. Goodman, R. L. 1978. Ibid. 102:142-150.
25. Gupta, D., et al. 1975. Steroids 25:33-42.
26. Kalra, P. S., and S. P. Kalra. 1977. Acta Endocrinol. 85:449-455.
27. Kalra, P. S., and S. P. Kalra. 1977. Endocrinology 101:1821-1827.
28. Kalra, S. P., and P. S. Kalra. 1974. Ibid. 95:1711-1718.
29. Kamel, F., et al. 1977. Ibid. 101:421-429.
30. Linkie, D. M., and G. D. Niswender. 1972. Ibid. 90: 632-637.
31. McLean, B. K., et al. 1977. Neuroendocrinology 23: 23-30.
32. Miyachi, Y., et al. 1973. Endocrinology 92:1-6.
33. Moger, W. H. 1977. Ibid. 100:1027-1032.
34. Moger, W. H., and P. R. Murphy. 1977. J. Endocrinol. 75:177-178.
35. Morishige, W. K., et al. 1973. Endocrinology 92:1527-1530.
36. Nequin, L. G., et al. 1975. Ibid. 97:718-724.
37. Ojeda, S. R., and V. D. Ramirez. 1972. Ibid. 90:466-472.
38. Piacsek, B. E., et al. 1978. Am. J. Physiol. 234:E262-E266.
39. Podesta, E. J., and M. A. Rivarola. 1974. Endocrinology 95:455-461.
40. Rodgers, C. H., and N. B. Schwartz. 1973. Ibid. 92: 1475-1479.
41. Schoot, P. van der, and W. J. de Greef. 1976. J. Endocrinol. 70:61-68.
42. Shaikh, A. A., and S. A. Shaikh. 1975. Endocrinology 96:37-44.
43. Smith, E. R., et al. 1973. Ibid. 93:756-758.
44. Smith, M. S., and J. D. Neill. 1977. Biol. Reprod. 17: 255-261.
45. Smith, M. S., and J. A. Ramaley. 1978. Endocrinology 102:351-357.
46. Smith, M. S., et al. 1975. Ibid. 96:219-226.
47. Soper, B. D., and R. F. Weick. 1977. Neuroendocrinology 23:306-311.
48. Swerdloff, R. S., et al. 1971. Endocrinology 88:120-128.
49. Welschen, R., et al. 1975. J. Endocrinol. 64:37-47.
50. Yamamoto, M., et al. 1970. Endocrinology 86:1102-1111.

92. HYPERTENSION: RAT

Part I. Strains and Origins of Hypertensive Rats

Holder: For full name, *see* list of HOLDERS at front of book.

Strain ⟨Synonym⟩	Origin	Generations Inbred	Establishment: Year of Publication	Establishment: Year of First Report	Characteristics	Holder	Reference
ALR ⟨Arteriolipidosis-prone rat⟩	SHR	10	1977	1976	Reactive hypercholesterolemia, acute arterial fat deposition	JSP	8,9
GH ⟨Genetically hypertensive rat⟩	Wistar-Otago	>30	1958	....	Hypertension	OMR	7
Milan-SHR	Wistar	>20	1974	....	Hypertension	CMML	1
Obese-SHR	SHR x Sprague-Dawley	...	1972	1971	Obesity, hypertension, endocrine dysfunction, hyperlipemia	N, CWRP	3
Salt-S ⟨Salt-sensitive rat⟩	Sprague-Dawley	...	1963	....	Salt-sensitive hypertension	BNL	2
SHR ⟨Spontaneously hypertensive rat⟩	Wistar-Kyoto	42	1963	1962	Hypertension, hypertensive cardiovascular lesions	N, JSP, OKJ	4,5

continued

Part I. Strains and Origins of Hypertensive Rats

Strain (Synonym)	Origin	Generations Inbred	Establishment		Characteristics	Holder	Reference
			Year of Publication	Year of First Report			
SHRSP (Stroke-prone spontaneously hypertensive rat)	SHR	17	1974	1973	Severe hypertension, cerebrovascular lesions (hemorrhage and/or infarction)	N, JSP, OKJ	6,10

Contributor: Yukio Yamori

References

1. Bianchi, G., et al. 1974. Life Sci. 14:339-347.
2. Dahl, L. K., et al. 1962. J. Exp. Med. 115:1173-1190.
3. Koletsky, S. 1972. In K. Okamoto, ed. Spontaneous Hypertension. Igaku-Shoin, Tokyo. pp. 194-197.
4. Okamoto, K. 1972. Int. Rev. Exp. Pathol. 7:227-270.
5. Okamoto, K., and K. Aoki. 1963. Jpn. Circ. J. 27: 282-293.
6. Okamoto, K., et al. 1974. Circ. Res. 34-35(Suppl. 1): 143-153.
7. Smirk, F. H., and W. H. Hall. 1958. Nature (London) 182:727-728.
8. Yamori, Y. 1977. Jpn. Heart J. 18:602-603.
9. Yamori, Y. 1977. In E. Inoue and H. Nishimura, ed. Gene-Environment Interaction in Common Diseases. Univ. Kyoto, Kyoto, Japan.
10. Yamori, Y., et al. 1974. Jpn. Circ. J. 38:1095-1100.

Part II. Blood Pressure in Stroke-Prone and Stroke-Resistant Spontaneously Hypertensive Rats

Data are for spontaneously hypertensive rats (SHR) that have been inbred for 30 to 32 generations. Each point represents the mean plus/minus the standard error for 20 to 140 cases.

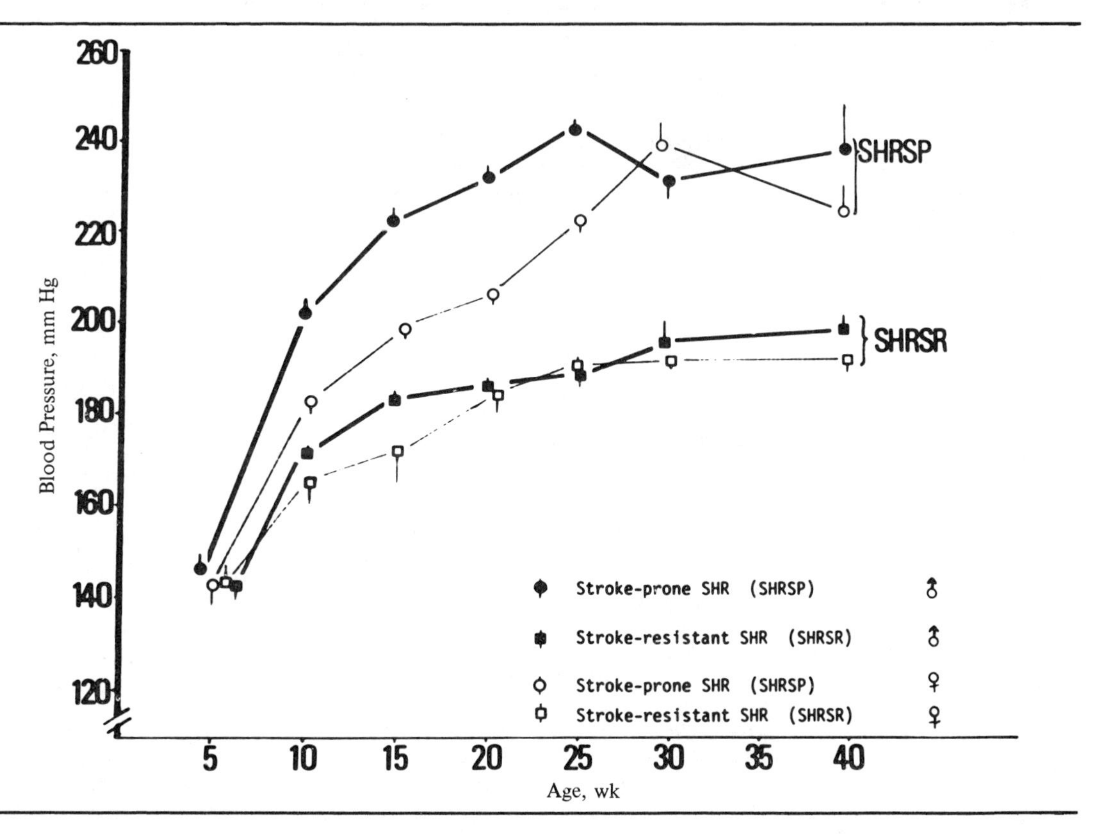

Contributor: Yukio Yamori

Reference: Yamori, Y., et al. 1976. Stroke 7(1):46-53.

continued

92. HYPERTENSION: RAT

Part III. Incidence of Complications in Autopsied Spontaneously Hypertensive Rats

Data are for 177 males and 158 females.

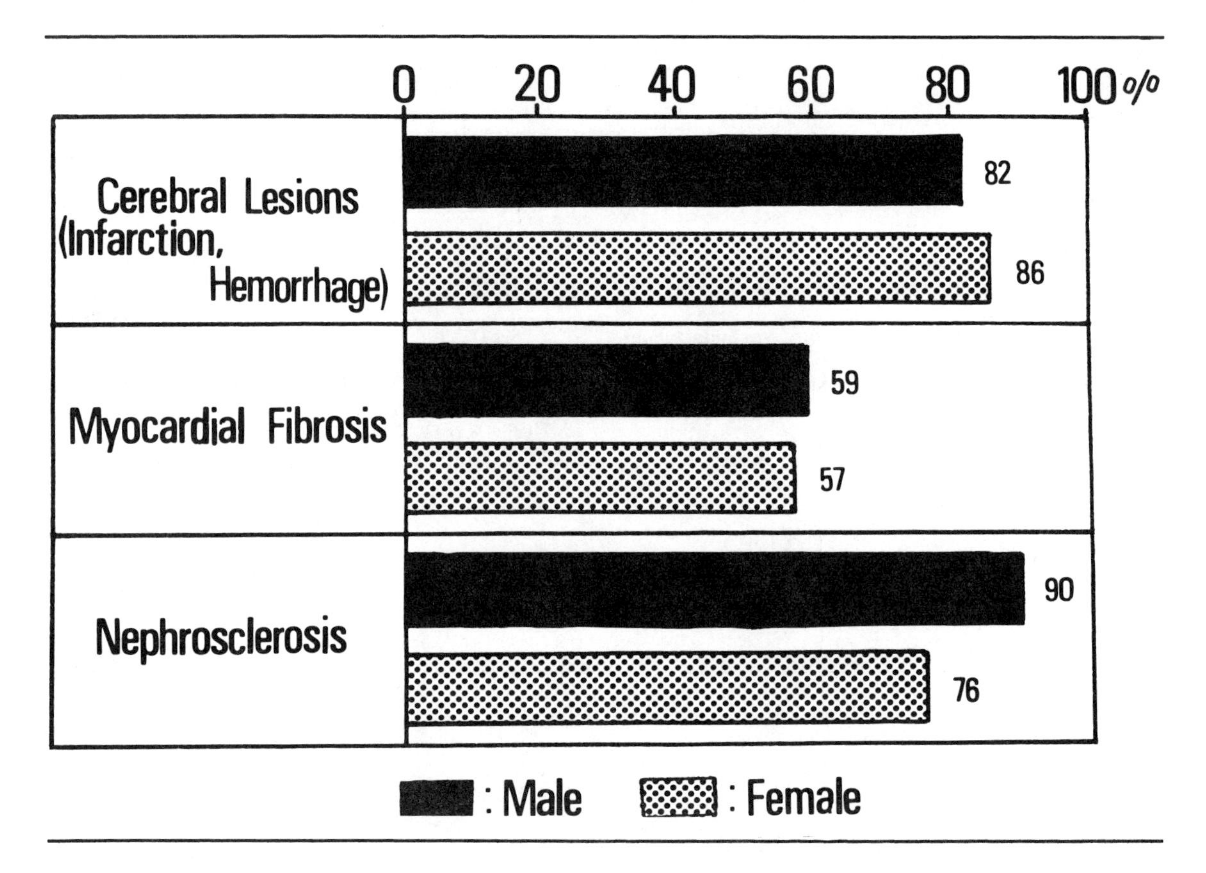

Contributor: Yukio Yamori

Reference: Yamori, Y., et al. Unpublished. Dep. Pathology, Kyoto Univ., Kyoto, Japan, 1978.

93. BEHAVIOR GENETICS: RAT

Genetic definition of the rat in the behavioral literature is frequently inadequate or in error. The essence of genetic definition is the existence of a meaningful and accurate genealogy which only follows from an unbroken history of the application of a systematic breeding method and the continuous maintenance of accurate breeding records. A variety of breeding plans have been devised for specific purposes. The data presented in this table are limited to those lines which can readily be replicated in the sense of reproducing the same gene distribution to obtain the same data. This criterion eliminates three rather substantial bodies of data which the behavioral investigator might expect to see represented. Certain commercial stocks carrying a vendor's name are frequently assumed—contrary to fact—to be genetically defined. Some of the best-known lines in behavioral research have been the product of two-way selection studies. It does not necessarily follow that genetic definition of the difference between two lines defines the lines themselves. The well-known Tryon stocks are no longer widely used, and are not supported by a systematic intra-line breeding plan or records. The Maudsley lines are widely used and are systematically bred, with a recognized breeding method supported by adequate records. However, the breeding method was chosen to maximize the information about the differential effects of selection, not to create defined lines for other uses.

Standard rules for nomenclature define an inbred strain, or simply strain, as animals that have been brother-sister mated for a minimum of 20 generations [20]. The rules define a subline as one which has been separated after eight or more generations. Most laboratory workers interpret this to mean

continued

that a strain is a single line from the original stock through at least the 8th generation. It is this identity which facilitates replication. The Committee on Standardized Genetic Nomenclature for Mice assumed but did not state this when they drew up the rules, although Falconer makes it explicit in his paper on breeding methods [ref. 18]. The Maudsley rats were inbred from the first generation, but were not limited to a single line of descent within each of the two selection groups. Thus, within a literal interpretation of the international rules, they constitute two strains, while under Falconer's definition and Committee intent they constitute a substantially larger number of strains. Each of these lines is a replication of the selective breeding experiment. However, labeling the original dozen lines of each stock as a single strain does not permit ready replication of the genotypes, as one may not know which of the inbred lines was used in a previous experiment without tracing the specific genealogies of the specific litters used. Data below are only for those Maudsley lines which can be uniquely identified and for which genealogies have been verified.

Part I. Strains and Other Stocks Used in Behavioral Research

A strain is here defined as a single line with a number of sublines, all of which trace back to points of origin subsequent to the attainment of a high degree of inbreeding [ref. 18]. Strain designations follow standardized nomenclature [ref. 20]. Stock designations are those most used in the literature. Breeding method terms are defined in reference 18.

Designation	Remarks	Reference
Strains		
ACI	Synonym: AXC 9935	11,16,51-59,68-70,93,108,110
A990	Synonym: I	11,52-59,63
A35322	...	11,51-59,110
B	...	63
BN	...	108
BUF	...	93,108
F344	Synonyms: J; Fischer	4,16,52-59,63,68-70,75,87,93,94,108-110
INR	Synonym: Iowa Nonreactive. Descendants of Hall Nonreactive.	48-50,52-59,61,62
IR	Synonyms: Walker; Iowa Reactive. Descendants of Maier animals.	52-59,61,80,81
LEW	...	16,75,108
MNR/Har	Line of MNR stock[1]. MNR/N is a related subline	19,54-60,82,88-90,92,97-103,112,114
MNRA	Synonym: MNR-a. Line of MNR stock[1].	52-59,111
MR/Har	Line of MR stock[1]. Relationship to MR/N unknown.	19,52-60,82,88-90,92,97-103,111
MR/N	Line of MR stock[1]. Relationship to MR/Har unknown.	112
OM	...	78
PVG	...	11,51
R	...	77
SHR	...	17,78
TS1	Descendants of Tryon Maze Bright	52-59
TS3	Descendants of Tryon Maze Dull	52-59
WAG	...	11,51-59,96
WF	...	16,93
Other Stocks		
LRS; WRS	Synonyms: Loser Runway Stock; Winner Runway Stock. Differentially selected for losing and winning in a runway competition task. Closed stocks.	25,83

[1] For a more complete description, *see* the MNR, MR entry under "Other Stocks" (*see* center heading).

continued

93. BEHAVIOR GENETICS: RAT

Part I. Strains and Other Stocks Used in Behavioral Research

Designation	Remarks	Reference
MNR; MR	Synonyms: Maudsley Nonreactive; Maudsley Reactive. Designates a number of different strains from two-way selection for low defecation and high defecation in an open field. Inbred parallel-line stocks. With some exceptions[2], the different lines are not individually named and cannot be distinctively identified other than by tracing the entire pedigree litter by litter.	2,3,5-14,21-23,26,27,29,30-47,64-67,71-74, 76,79,84-86,95,96,106,115-121
RLA; RHA; RCA	Synonyms: Roman Low Avoidance; Roman High Avoidance; Roman Control Avoidance. Differentially selected for low and high avoidance conditioning. Closed stocks—RLA & RHA are outcrossed, RCA is random-bred from original RLA-RHA gene pool.	1,6,7,15,24,27,28,34,67,96-105
TMB; TMD	Synonyms: Tryon Maze Bright or Berkeley S1; Tryon Maze Dull or Berkeley S3. Differentially selected for high and low maze performance. Closed stocks. No further breeding method or records. Considerable genetic drift.	91,96,107,113

[2] Individual names are indicated under "Strains" (*see* center heading).

Contributor: Gordon M. Harrington

References

1. Bignami, G. 1965. Anim. Behav. 13:221-227.
2. Blizard, D. A. 1970. Psychonom. Sci. 19:145-146.
3. Blizard, D. A. 1971. J. Comp. Physiol. Psychol. 76: 282-289.
4. Bonnet, K. A., and K. E. Petersen. 1975. Pharmacol. Biochem. Behav. 3:47-55.
5. Brewster, D. J. 1968. J. Hered. 59:283-285.
6. Brewster, D. J. 1969. J. Genet. Psychol. 115:217-227.
7. Brewster, D. J. 1972. Ann. N.Y. Acad. Sci. 197:49-53.
8. Broadhurst, P. L. 1957. Br. J. Psychol. 48:1-12.
9. Broadhurst, P. L. 1958. Ibid. 49:12-20.
10. Broadhurst, P. L. 1958. Bull. Br. Psychol. Soc. 34:2A.
11. Broadhurst, P. L. 1960. In H. J. Eysenck, ed. Experiments in Personality. Routledge and Kegan Paul, London. pp. 1-102.
12. Broadhurst, P. L. 1962. Psychol. Rep. 10:65-66.
13. Broadhurst, P. L., and H. Wallgren. 1964. Q. J. Stud. Alcohol. 25:476-489.
14. Broadhurst, P. L., and R. H. J. Watson. 1964. Anim. Behav. 12:42-51.
15. Coyle, J. T., Jr., et al. 1973. Psychopharmacologia 31:25-34.
16. Dewsbury, D. A. 1975. J. Comp. Physiol. Psychol. 88: 713-722.
17. Eichelman, B., et al. 1973. Physiol. Behav. 10:301-304.
18. Falconer, D. S. 1957. In A. N. Worden and W. Lane-Petter, ed. The UFAW Handbook on the Care and Management of Laboratory Animals. Ed. 2. Universities Federation for Animal Welfare, London. pp. 85-107.
19. Ferraro, D. P., and K. M. York. 1968. Psychonom. Sci. 10:177-178.
20. Festing, M., and J. Staats. 1973. Transplantation 16: 221-245.
21. Feuer, G. 1963. J. Physiol. (London) 169:43P-44P.
22. Feuer, G. 1963. In J. K. Grant, ed. The Control of Lipid Metabolism. Academic, London. pp. 147-152.
23. Feuer, G. 1969. In E. Bajusz, ed. Physiology and Pathology of Adaptation Mechanisms. Pergamon, Oxford. pp. 213-233.
24. Fleming, J. C., and P. L. Broadhurst. 1975. Psychopharmacologia 42:147-152.
25. Fukusawa, T., et al. 1975. Behav. Biol. 15:333-342.
26. Garg, M. 1969. Psychopharmacologia 14:150-156.
27. Garg, M. 1969. Ibid. 14:432-483.
28. Garg, M. 1969. Ibid. 15:408-414.
29. Garg, M. 1970. Ibid. 18:172-179.
30. Garg, M., and H. C. Holland. 1967. Life Sci. 6:1987-1997.
31. Garg, M., and H. C. Holland. 1968. Int. J. Neuropharmacol. 7:55-59.
32. Garg, M., and H. C. Holland. 1968. Psychopharmacologia 12:96-103.
33. Garg, M., and H. C. Holland. 1968. Ibid. 12:127-132.
34. Garg, M., and H. C. Holland. 1969. Ibid. 14:426-431.
35. Gray, J. A. 1965. Can. J. Psychol. 19:74-82.
36. Gray, J. A., and S. Levine. 1964. Nature (London) 201:1198-1200.
37. Gray, J. A., et al. 1965. Anim. Behav. 13:33-45.
38. Gregory, K. 1967. Life Sci. 6:1253-1256.
39. Gregory, K. 1967. Act. Nerv. Super. 9:140-144.
40. Gregory, K. 1968. Psychopharmacologia 13:29-34.
41. Gregory, K., and E. Liebelt. 1967. Act. Nerv. Super. 9:137-139.
42. Gregory, K., et al. 1967. Life Sci. 6:981-988.

continued

Part I. Strains and Other Stocks Used in Behavioral Research

43. Gupta, B. D., and K. Gregory. 1967. Psychopharmacologia 11:365-371.
44. Gupta, B. D., and H. C. Holland. 1969. Ibid. 14:95-105.
45. Gupta, B. D., and H. C. Holland. 1969. Int. J. Neuropharmacol. 8:227-234.
46. Gupta, B. D., and H. C. Holland. 1972. Neuropharmacology 11:23-30.
47. Gupta, B. D., and H. C. Holland. 1972. Ibid. 11:31-38.
48. Hall, C. S. 1934. J. Comp. Psychol. 18:385-403.
49. Hall, C. S. 1938. Sigma Xi Q. 26:17-27.
50. Hall, C. S. 1951. In S. S. Stevens, ed. Handbook of Experimental Psychology. Wiley, New York. pp. 304-329.
51. Harrington, G. M. 1966. Psychonom. Sci. 6:103-104.
52. Harrington, G. M. 1968. Dev. Psychobiol. 1:211-218.
53. Harrington, G. M. 1968. Ibid. 1:245-253.
54. Harrington, G. M. 1971. Psychonom. Sci. 23:348-349.
55. Harrington, G. M. 1971. Ibid. 23:363-364.
56. Harrington, G. M. 1972. Ibid. 27:51-53.
57. Harrington, G. M. 1979. Bull. Psychonom. Soc. 13:(in press).
58. Harrington, G. M. 1979. Ibid. 14:(in press).
59. Harrington, G. M. Unpublished. Univ. Northern Iowa, Dep. Psychology, Cedar Falls, 1978.
60. Harrington, G. M., and J. R. Hanlon. 1966. Psychonom. Sci. 6:425-426.
61. Harrington, G. M., and G. R. Kohler. 1966. Ibid. 6:419-420.
62. Harrington, G. M., and G. R. Kohler. 1966. Psychol. Rep. 18:803-808.
63. Heinze, W. J. 1974. Behav. Genet. 4:125-131.
64. Holland, H. C., and B. D. Gupta. 1966. Psychopharmacologia 9:419-425.
65. Holland, H. C., and B. D. Gupta. 1967. Act. Nerv. Super. 9:134-136.
66. Imada, H. 1970. Annu. Rev. Anim. Psychol. 20:1-6.
67. Imada, H. 1972. J. Comp. Physiol. Psychol. 79:474-480.
68. Jakubczak, L. F. 1973. Bull. Psychonom. Soc. 1:395-396.
69. Jakubczak, L. F. 1973. Proc. Annu. Conv. Am. Psychol. Assoc. 81st, pp. 761-762.
70. Jakubczak, L. F. 1976. Physiol. Behav. 17:251-258.
71. Joffe, J. M. 1964. Psychonom. Sci. 1:185-186.
72. Joffe, J. M. 1965. Nature (London) 208:815-816.
73. Joffe, J. M. 1965. Science 150:1844-1845.
74. Joffe, J. M. 1969. Prenatal Determinants of Behaviour. Pergamon, Oxford.
75. Katzev, R. D., and S. K. Mills. 1974. J. Comp. Physiol. Psychol. 87:661-671.
76. Keehn, J. D. 1972. Psychonom. Sci. 29:20-22.
77. Knoll, B. 1972. Act. Nerv. Super. 14:293.
78. Lamprecht, F., et al. 1974. Psychosom. Med. 36:298-303.
79. Lonowski, D. J., et al. 1975. Bull. Psychonom. Soc. 6:629-631.
80. Maier, N. R. F. 1943. J. Comp. Psychol. 35:327-335.
81. Maier, N. R. F., and N. M. Glaser. 1940. Ibid. 30:413-418.
82. Martin, L. K., and B. J. Powell. 1970. Psychonom. Sci. 13:44-45.
83. Masur, J., and M. A. C. Benedito. 1974. Nature (London) 249:284.
84. Mikhail, A. A. 1969. Psychonom. Sci. 15:245-247.
85. Mikhail, A. A. 1972. J. Psychosom. Res. 16:115-122.
86. Mikhail, A. A., and P. L. Broadhurst. 1965. Ibid. 8:477-479.
87. Mills, S. K., and R. D. Katzev. 1974. J. Comp. Physiol. Psychol. 87:672-680.
88. Powell, B. J. 1967. Proc. Annu. Conv. Am. Psychol. Assoc. 75th, pp. 69-70.
89. Powell, B. J. 1970. Ibid. 78th, pp. 825-826.
90. Powell, B. J., and D. J. Hopper. 1971. Psychonom. Sci. 22:167-168.
91. Powell, B. J., and M. Leach. 1967. Ibid. 9:287-288.
92. Powell, B. J., and M. North-Jones. 1974. Dev. Psychobiol. 7:145-148.
93. Ray, O. S., and R. J. Barrett. 1975. Behav. Biol. 15:391-417.
94. Ray, O. S., and R. J. Barrett. 1973. In C. Eisdorfer and W. E. Fann, ed. Psychopharmacology and Aging. Plenum, New York. pp. 17-39.
95. Rick, J. T., et al. 1967. Comp. Biochem. Physiol. 20:1009-1012.
96. Rick, J. T., et al. 1971. Brain Res. 32:234-238.
97. Satinder, K. P. 1971. J. Comp. Physiol. Psychol. 76:359-364.
98. Satinder, K. P. 1972. Ibid. 80:422-434.
99. Satinder, K. P. 1972. Psychonom. Sci. 29:291-293.
100. Satinder, K. P. 1975. J. Stud. Alcohol 36:1493-1507.
101. Satinder, K. P. 1976. Psychopharmacology 48:235-237.
102. Satinder, K. P. 1976. Anim. Learn. Behav. 4:172-176.
103. Satinder, K. P. 1976. J. Comp. Physiol. Psychol. 90:946-957.
104. Satinder, K. P. 1977. Ibid. 91:1326-1336.
105. Satinder, K. P., and K. D. Hill. 1974. Ibid. 86:363-374.
106. Savage, R. D. 1965. Behav. Res. Ther. 2:281-283.
107. Searle, L. V. 1949. Genet. Psychol. Monogr. 39:279-325.
108. Segal, D. S., et al. 1972. Behav. Biol. 7:75-81.
109. Shurman, A. J., and R. D. Katzev. 1975. J. Comp. Physiol. Psychol. 88:548-553.
110. Sines, J. O. 1962. J. Genet. Psychol. 101:209-216.

continued

Part I. Strains and Other Stocks Used in Behavioral Research

111. Slater, J., et al. 1977. Pharmacol. Biochem. Behav. 6(5):511-520.
112. Sudak, H. S., and J. W. Maas. 1964. Science 146:418-420.
113. Tryon, R. C. 1940. 39th Yearb. Natl. Soc. Study Educ. (1):111-119.
114. Wachs, T. D. 1974. Dev. Psychobiol. 5:385-392.
115. Weldon, E. 1977. Br. J. Psychol. 58:253-259.
116. Weldon, E. 1968. Psychonom. Sci. 10:239-240.
117. Weldon, E. 1968. Ibid. 12:83-84.
118. Wilcock, J. 1966. Bull. Br. Psychol. Soc. 19:A28.
119. Wilcock, J. 1968. Anim. Behav. 16:294-297.
120. Wilcock, J., and P. L. Broadhurst. 1967. J. Comp. Physiol. Psychol. 63:335-338.
121. Wraight, K. B., et al. 1967. Anim. Behav. 15:287-290.

Part II. Characteristics of Genetically Defined Strains

Characteristics of genetically defined lines are classified (by purpose of supporting research) as systematic description, discovery of existing genetic differences, or analysis of effects of selection. Figures in heavy brackets are reference numbers. Plus/minus (±) values are standard deviations.

Category & Characteristics	Strain	No. & Sex	Value
Systematic Description			
Anatomy: Organ weights, mg/kg body wt [10-12]			
Brain	ACI	>20♂	7193 ± 456
		>20♀	9944 ± 487
	A990	>20♂	5303 ± 506
		>20♀	8362 ± 576
	A35322	>20♂	6682 ± 757
		>20♀	9473 ± 534
	F344	>20♂	5899 ± 753
		>20♀	9034 ± 547
	INR	>20♂	5258 ± 847
		>20♀	8076 ± 664
	IR	>20♂	6432 ± 613
		>20♀	9002 ± 606
	MNR/Har	>20♂	6744 ± 651
		>20♀	10230 ± 583
	MNRA	>20♂	6975 ± 536
		>20♀	10953 ± 551
	MR/Har	>20♂	5986 ± 457
		>20♀	9603 ± 632
	TS1	>20♂	4775 ± 515
		>20♀	7596 ± 533
	TS3	>20♂	5462 ± 608
		>20♀	7809 ± 621
	WAG	>20♂	5803 ± 1029
		>20♀	9232 ± 663
Spleen	ACI	>20♂	1541 ± 85
		>20♀	1819 ± 139
	A990	>20♂	1721 ± 133
		>20♀	2242 ± 213
	A35322	>20♂	2366 ± 178
		>20♀	2705 ± 181
	F344	>20♂	1585 ± 102
		>20♀	2099 ± 140
	INR	>20♂	1816 ± 238
		>20♀	2056 ± 161
	IR	>20♂	1247 ± 124
		>20♀	1544 ± 123
	MNR/Har	>20♂	1539 ± 125
		>20♀	1883 ± 113
	MNRA	>20♂	1781 ± 243
		>20♀	2203 ± 226
	MR/Har	>20♂	2214 ± 187
		>20♀	2701 ± 296
	TS1	>20♂	1628 ± 122
		>20♀	2109 ± 160
	TS3	>20♂	1864 ± 245
		>20♀	2358 ± 145
	WAG	>20♂	1344 ± 183
		>20♀	1644 ± 102
Thymus	ACI	>20♂	537 ± 81
		>20♀	976 ± 153
	A990	>20♂	430 ± 126
		>20♀	779 ± 294
	A35322	>20♂	514 ± 199
		>20♀	725 ± 192
	F344	>20♂	451 ± 147
		>20♀	743 ± 210
	INR	>20♂	681 ± 209
		>20♀	986 ± 206
	IR	>20♂	610 ± 222
		>20♀	1027 ± 252
	MNR/Har	>20♂	630 ± 240
		>20♀	1109 ± 264
	MNRA	>20♂	512 ± 127
		>20♀	852 ± 143
	MR/Har	>20♂	720 ± 145
		>20♀	1094 ± 240
	TS1	>20♂	506 ± 131

continued

Part II. Characteristics of Genetically Defined Strains

Category & Characteristics	Strain	No. & Sex	Value
		>20♀	662 ± 152
	TS3	>20♂	366 ± 130
		>20♀	623 ± 192
	WAG	>20♂	674 ± 269
		>20♀	1158 ± 231
Pituitary	ACI	>20♂	33.2 ± 3.5
		>20♀	58.1 ± 9.3
	A990	>20♂	25.9 ± 5.6
		>20♀	55.6 ± 12.4
	A35322	>20♂	31.0 ± 3.2
		>20♀	48.8 ± 6.9
	F344	>20♂	31.7 ± 5.8
		>20♀	60.4 ± 13.4
	INR	>20♂	26.2 ± 5.8
		>20♀	51.8 ± 11.6
	IR	>20♂	34.3 ± 3.1
		>20♀	57.8 ± 5.6
	MNR/Har	>20♂	30.7 ± 3.3
		>20♀	65.3 ± 8.1
	MNRA	>20♂	29.4 ± 6.2
		>20♀	53.0 ± 9.0
	MR/Har	>20♂	31.4 ± 4.6
		>20♀	48.8 ± 10.4
	TS1	>20♂	34.4 ± 5.0
		>20♀	54.3 ± 9.5
	TS3	>20♂	32.1 ± 6.2
		>20♀	49.0 ± 7.5
	WAG	>20♂	31.5 ± 6.0
		>20♀	53.7 ± 8.9
Thyroid	ACI	>20♂	61.4 ± 12.9
		>20♀	82.5 ± 15.0
	A990	>20♂	67.0 ± 19.0
		>20♀	90.3 ± 23.2
	A35322	>20♂	58.2 ± 16.5
		>20♀	84.2 ± 15.9
	F344	>20♂	63.7 ± 15.2
		>20♀	84.0 ± 22.8
	INR	>20♂	66.0 ± 12.5
		>20♀	82.5 ± 11.6
	IR	>20♂	53.4 ± 10.4
		>20♀	68.9 ± 17.4
	MNR/Har	>20♂	59.1 ± 14.2
		>20♀	77.4 ± 21.5
	MNRA	>20♂	66.2 ± 13.5
		>20♀	86.4 ± 19.5
	MR/Har	>20♂	48.0 ± 15.0
		>20♀	76.0 ± 21.7
	TS1	>20♂	70.4 ± 12.3
		>20♀	93.8 ± 13.0
	TS3	>20♂	53.4 ± 10.0
		>20♀	79.0 ± 17.6

Category & Characteristics	Strain	No. & Sex	Value
	WAG	>20♂	57.4 ± 18.2
		>20♀	79.0 ± 13.0
Adrenals			
Right	ACI	>20♂	50.3 ± 5.8
		>20♀	97.0 ± 13.6
	A990	>20♂	45.7 ± 7.2
		>20♀	97.2 ± 16.8
	A35322	>20♂	42.2 ± 5.3
		>20♀	73.6 ± 10.5
	F344	>20♂	39.3 ± 9.4
		>20♀	75.4 ± 13.3
	INR	>20♂	34.4 ± 6.2
		>20♀	72.5 ± 11.5
	IR	>20♂	29.9 ± 7.0
		>20♀	64.3 ± 11.8
	MNR/Har	>20♂	42.4 ± 10.0
		>20♀	96.3 ± 17.3
	MNRA	>20♂	41.5 ± 6.7
		>20♀	94.2 ± 20.7
	MR/Har	>20♂	45.0 ± 9.4
		>20♀	97.8 ± 13.2
	TS1	>20♂	38.5 ± 7.9
		>20♀	81.6 ± 12.5
	TS3	>20♂	37.0 ± 5.4
		>20♀	65.2 ± 12.8
	WAG	>20♂	33.8 ± 7.2
		>20♀	60.1 ± 13.4
Left	ACI	>20♂	57.4 ± 9.3
		>20♀	110.0 ± 9.0
	A990	>20♂	47.3 ± 9.4
		>20♀	107.7 ± 20.6
	A35322	>20♂	47.9 ± 6.6
		>20♀	90.3 ± 9.2
	F344	>20♂	41.0 ± 8.2
		>20♀	79.2 ± 17.3
	INR	>20♂	38.3 ± 7.1
		>20♀	79.5 ± 12.6
	IR	>20♂	36.0 ± 7.6
		>20♀	71.6 ± 15.4
	MNR/Har	>20♂	46.7 ± 7.3
		>20♀	105.2 ± 13.0
	MNRA	>20♂	45.1 ± 8.9
		>20♀	104.0 ± 22.6
	MR/Har	>20♂	49.4 ± 8.1
		>20♀	104.7 ± 25.0
	TS1	>20♂	44.6 ± 11.0
		>20♀	91.7 ± 16.3
	TS3	>20♂	42.5 ± 6.7
		>20♀	75.0 ± 14.5
	WAG	>20♂	40.1 ± 8.1
		>20♀	62.3 ± 9.9

continued

Part II. Characteristics of Genetically Defined Strains

Category & Characteristics	Strain	No. & Sex	Value
Physiology: O_2 consumption1/, kcal·h^{-1}·m^{-2} [10-12]	ACI	>20♂	37.4 ± 8.2
		>20♀	43.9 ± 12.5
	A990	>20♂	32.8 ± 6.6
		>20♀	40.8 ± 9.0
	A35322	>20♂	32.6 ± 9.3
		>20♀	42.4 ± 10.0
	F344	>20♂	34.6 ± 6.1
		>20♀	37.1 ± 10.4
	INR	>20♂	40.1 ± 10.3
		>20♀	41.6 ± 8.3
	IR	>20♂	31.9 ± 5.5
		>20♀	33.7 ± 6.9
	MNR/Har	>20♂	31.7 ± 5.2
		>20♀	36.7 ± 7.8
	MNRA	>20♂	37.3 ± 8.2
		>20♀	45.9 ± 12.4
	MR/Har	>20♂	37.0 ± 6.8
		>20♀	38.8 ± 7.5
	TS1	>20♂	34.2 ± 9.1
		>20♀	39.2 ± 10.9
	TS3	>20♂	42.3 ± 10.1
		>20♀	44.8 ± 10.4
	WAG	>20♂	32.7 ± 6.0
		>20♀	40.0 ± 9.4
Activity			
Home cage movement, relative measure in uncalibrated units proportional to kinetic energy [10-12]	ACI	>20♂	3.2 ± 0.88
		>20♀	2.9 ± 0.81
	A990	>20♂	4.5 ± 2.6
		>20♀	4.1 ± 3.1
	A35322	>20♂	5.0 ± 1.8
		>20♀	5.6 ± 2.3
	F344	>20♂	3.2 ± 1.2
		>20♀	3.0 ± 1.3
	INR	>20♂	7.1 ± 2.3
		>20♀	6.1 ± 2.1
	IR	>20♂	3.4 ± 1.5
		>20♀	3.0 ± 1.3
	MNR/Har	>20♂	2.6 ± 0.43
		>20♀	2.2 ± 0.46
	MNRA	>20♂	4.4 ± 1.9
		>20♀	3.8 ± 1.6
	MR/Har	>20♂	5.6 ± 1.1
		>20♀	4.8 ± 1.1
	TS1	>20♂	3.6 ± 1.3
		>20♀	3.6 ± 1.4
	TS3	>20♂	7.6 ± 1.8
		>20♀	7.7 ± 1.9
	WAG	>20♂	4.2 ± 1.5
		>20♀	3.6 ± 1.4
Lever press operant rate, five 1000-s sessions [10-12]	ACI	>20♂	7.1 ± 5.6
		>20♀	13.0 ± 10.9
	A990	>20♂	11.4 ± 12.6
		>20♀	11.2 ± 12.8
	A35322	>20♂	13.7 ± 7.6
		>20♀	17.4 ± 10.7
	F344	>20♂	12.3 ± 11.6
		>20♀	10.2 ± 8.8
	INR	>20♂	32.1 ± 29.1
		>20♀	39.6 ± 38.2
	IR	>20♂	2.3 ± 2.5
		>20♀	3.4 ± 5.1
	MNR/Har	>20♂	3.4 ± 4.4
		>20♀	2.0 ± 3.9
	MNRA	>20♂	11.3 ± 10.5
		>20♀	11.0 ± 12.4
	MR/Har	>20♂	10.9 ± 16.0
		>20♀	5.8 ± 4.7
	TS1	>20♂	3.2 ± 3.9
		>20♀	2.1 ± 1.9
	TS3	>20♂	15.7 ± 13.5
		>20♀	16.5 ± 13.6
	WAG	>20♂	10.9 ± 8.0
		>20♀	5.1 ± 4.5
Open field ambulation, 4.63 X m for 4 d, tested 4 min/d [9-12]	ACI	>20♂	166 ± 97
		>20♀	223 ± 73
	A990	>20♂	159 ± 58
		>20♀	188 ± 60
	A35322	>20♂	181 ± 80
		>20♀	261 ± 77
	F344	>20♂	189 ± 74
		>20♀	226 ± 92
	INR	>20♂	360 ± 90
		>20♀	392 ± 57
	IR	>20♂	235 ± 74
		>20♀	316 ± 89
	MNR/Har	>20♂	194 ± 60
		>20♀	205 ± 56
	MNRA	>20♂	296 ± 99
		>20♀	289 ± 68
	MR/Har	>20♂	196 ± 55
		>20♀	224 ± 73
	TS1	>20♂	109 ± 79
		>20♀	165 ± 78
	TS3	>20♂	251 ± 64
		>20♀	307 ± 82
	WAG	>20♂	337 ± 75
		>20♀	342 ± 52
Rotating cage, revolutions for 4 d, tested 1000 s/d [8,12]	ACI	>20♂	380 ± 228
		>20♀	770 ± 283
	A990	>20♂	244 ± 100
		>20♀	446 ± 156
	A35322	>20♂	509 ± 188
		>20♀	923 ± 331
	F344	>20♂	377 ± 325

1/ As measure of basal metabolic rate.

continued

Category & Characteristics	Strain	No. & Sex	Value
		>20♀	480 ± 199
	INR	>20♂	432 ± 113
		>20♀	529 ± 158
	IR	>20♂	331 ± 185
		>20♀	521 ± 322
	MNR/Har	>20♂	742 ± 238
		>20♀	1301 ± 337
	MNRA	>20♂	536 ± 215
		>20♀	901 ± 240
	MR/Har	>20♂	345 ± 226
		>20♀	592 ± 315
	TS1	>20♂	342 ± 165
		>20♀	719 ± 241
	TS3	>20♂	417 ± 171
		>20♀	663 ± 285
	WAG	>20♂	543 ± 271
		>20♀	1005 ± 314
Stabilimeter (photo cell), crossings for 5 d, tested 1000 s/d [10-12]	ACI	>20♂	318 ± 91
		>20♀	387 ± 101
	A990	>20♂	468 ± 102
		>20♀	498 ± 96
	A35322	>20♂	620 ± 133
		>20♀	756 ± 204
	F344	>20♂	399 ± 192
		>20♀	482 ± 108
	INR	>20♂	381 ± 82
		>20♀	403 ± 98
	IR	>20♂	326 ± 81
		>20♀	390 ± 140
	MNR/Har	>20♂	415 ± 69
		>20♀	439 ± 67
	MNRA	>20♂	523 ± 102
		>20♀	524 ± 142
	MR/Har	>20♂	434 ± 76
		>20♀	533 ± 116
	TS1	>20♂	343 ± 71
		>20♀	395 ± 77
	TS3	>20♂	421 ± 89
		>20♀	507 ± 98
	WAG	>20♂	365 ± 92
		>20♀	396 ± 96
Curiosity & neophobia, latencies, s [7,12]			
Cage emergence			
Familiar environment	ACI	>20♂	217 ± 102
		>20♀	182 ± 113
	A990	>20♂	159 ± 106
		>20♀	135 ± 94
	A35322	>20♂	107 ± 85
		>20♀	85 ± 76
	F344	>20♂	207 ± 103
		>20♀	162 ± 117
	INR	>20♂	68 ± 75
		>20♀	47 ± 48
	IR	>20♂	244 ± 84
		>20♀	194 ± 116
	MNR/Har	>20♂	174 ± 115
		>20♀	202 ± 100
	MNRA	>20♂	164 ± 108
		>20♀	171 ± 98
	MR/Har	>20♂	201 ± 103
		>20♀	233 ± 89
	TS1	>20♂	203 ± 104
		>20♀	144 ± 102
	TS3	>20♂	111 ± 79
		>20♀	128 ± 114
	WAG	>20♂	116 ± 95
		>20♀	126 ± 110
Unfamiliar environment	ACI	>20♂	230 ± 77
		>20♀	189 ± 93
	A990	>20♂	163 ± 97
		>20♀	135 ± 92
	A35322	>20♂	142 ± 104
		>20♀	103 ± 85
	F344	>20♂	160 ± 98
		>20♀	142 ± 91
	INR	>20♂	47 ± 40
		>20♀	29 ± 30
	IR	>20♂	186 ± 114
		>20♀	114 ± 90
	MNR/Har	>20♂	203 ± 123
		>20♀	215 ± 107
	MNRA	>20♂	166 ± 93
		>20♀	165 ± 102
	MR/Har	>20♂	215 ± 107
		>20♀	157 ± 89
	TS1	>20♂	156 ± 103
		>20♀	117 ± 86
	TS3	>20♂	68 ± 60
		>20♀	63 ± 55
	WAG	>20♂	62 ± 71
		>20♀	93 ± 94
Exploring strange object	ACI	>20♂	173 ± 121
		>20♀	79 ± 82
	A990	>20♂	89 ± 110
		>20♀	68 ± 84
	A35322	>20♂	103 ± 93
		>20♀	53 ± 37
	F344	>20♂	140 ± 111
		>20♀	166 ± 110
	INR	>20♂	37 ± 37
		>20♀	47 ± 77
	IR	>20♂	200 ± 129
		>20♀	165 ± 116
	MNR/Har	>20♂	118 ± 103
		>20♀	118 ± 118
	MNRA	>20♂	156 ± 110

continued

Part II. Characteristics of Genetically Defined Strains

Category & Characteristics	Strain	No. & Sex	Value
		>20♀	151 ± 99
	MR/Har	>20♂	204 ± 94
		>20♀	137 ± 91
	TS1	>20♂	137 ± 98
		>20♀	117 ± 88
	TS3	>20♂	93 ± 60
		>20♀	97 ± 91
	WAG	>20♂	50 ± 78
		>20♀	67 ± 97
Approaching human hand	ACI	>20♂	135 ± 120
		>20♀	64 ± 81
	A990	>20♂	68 ± 102
		>20♀	75 ± 113
	A35322	>20♂	111 ± 116
		>20♀	51 ± 83
	F344	>20♂	169 ± 138
		>20♀	126 ± 128
	INR	>20♂	20 ± 40
		>20♀	16 ± 24
	IR	>20♂	185 ± 131
		>20♀	127 ± 130
	MNR/Har	>20♂	178 ± 130
		>20♀	157 ± 117
	MNRA	>20♂	72 ± 107
		>20♀	54 ± 78
	MR/Har	>20♂	178 ± 131
		>20♀	223 ± 115
	TS1	>20♂	68 ± 101
		>20♀	60 ± 87
	TS3	>20♂	17 ± 34
		>20♀	39 ± 94
	WAG	>20♂	21 ± 21
		>20♀	48 ± 100
Reactivity: Open-field defecation, fecal boli for 4 d, tested 4 min/d [9-12]	ACI	>20♂	16.7 ± 9.9
		>20♀	12.0 ± 5.9
	A990	>20♂	19.8 ± 8.9
		>20♀	16.0 ± 9.6
	A35322	>20♂	18.9 ± 7.4
		>20♀	20.8 ± 6.9
	F344	>20♂	12.2 ± 5.2
		>20♀	14.0 ± 4.5
	INR	>20♂	13.3 ± 5.8
		>20♀	8.9 ± 6.0
	IR	>20♂	14.0 ± 6.4
		>20♀	12.9 ± 7.8
	MNR/Har	>20♂	10.8 ± 10.9
		>20♀	10.5 ± 8.4
	MNRA	>20♂	10.5 ± 8.8
		>20♀	9.1 ± 8.9
	MR/Har	>20♂	22.4 ± 6.8

Category & Characteristics	Strain	No. & Sex	Value
		>20♀	19.8 ± 5.5
	TS1	>20♂	13.0 ± 5.5
		>20♀	12.3 ± 5.0
	TS3	>20♂	12.2 ± 5.7
		>20♀	10.0 ± 7.7
	WAG	>20♂	12.4 ± 6.2
		>20♀	13.0 ± 10.0
Problem solving			
Hebb-Williams maze [5,6]			
Mean errors	ACI	6♂	263.5 ± 55.5
		6♀	182.8 ± 37.0
	A990	6♂	160.0 ± 36.1
		6♀	147.3 ± 18.1
	A35322	6♂	182.0 ± 38.2
		6♀	200.8 ± 31.9
	F344	6♂	172.2 ± 15.0
		6♀	144 ± 26.7
	INR	6♂	187.5 ± 15.8
		6♀	222.7 ± 62.1
	IR	6♂	211.8 ± 45.6
		6♀	218.0 ± 37.5
	MNRA	6♂	136.3 ± 26.8
		6♀	138.5 ± 40.6
	MR/Har	6♂	148.8 ± 28.9
		6♀	156.3 ± 39.7
	TS1	6♂	169.3 ± 46.9
		6♀	215.3 ± 34.7
	TS3	6♂	154.0 ± 38.7
		6♀	147.8 ± 42.4
	WAG	6♂	126.0 ± 15.1
		6♀	147 ± 23.3
Mean goal latency, log s	ACI	>20♂	25.1 ± 1.1
		>20♀	23.2 ± 1.1
	A990	>20♂	22.3 ± 1.6
		>20♀	22.2 ± 0.7
	A35322	>20♂	24.5 ± 1.5
		>20♀	23.0 ± 1.2
	F344	>20♂	24.5 ± 1.1
		>20♀	22.9 ± 1.1
	INR	>20♂	23.5 ± 0.6
		>20♀	23.5 ± 0.8
	IR	>20♂	23.1 ± 1.1
		>20♀	22.2 ± 1.8
	MNRA	>20♂	21.5 ± 1.9
		>20♀	21.0 ± 1.2
	MR/Har	>20♂	21.8 ± 1.5
		>20♀	21.4 ± 0.7
	TS1	>20♂	24.2 ± 1.2
		>20♀	24.2 ± 0.6
	TS3	>20♂	23.6 ± 1.4

continued

Category & Characteristics	Strain	No. & Sex	Value
		>20♀	20.7 ± 0.5
	WAG	>20♂	22.0 ± 1.0
		>20♀	21.1 ± 0.9
Passive avoidance: Latency to enter shock compartment, s (maximum = 90 s) [10-12]	ACI	>20♂	90.0 ± 0
		>20♀	85.4 ± 19.0
	A990	>20♂	61.8 ± 38.8
		>20♀	56.9 ± 34.5
	A35322	>20♂	71.9 ± 29.7
		>20♀	84.9 ± 13.8
	F344	>20♂	44.7 ± 36.0
		>20♀	57.0 ± 37.1
	INR	>20♂	60.6 ± 33.0
		>20♀	45.9 ± 32.8
	IR	>20♂	18.5 ± 29.9
		>20♀	17.0 ± 30.4
	MNR/Har	>20♂	49.4 ± 34.8
		>20♀	60.0 ± 34.5
	MNRA	>20♂	64.8 ± 35.0
		>20♀	67.0 ± 34.8
	MR/Har	>20♂	28.6 ± 29.1
		>20♀	16.3 ± 22.9
	TS1	>20♂	80.2 ± 25.6
		>20♀	83.9 ± 19.4
	TS3	>20♂	63.8 ± 36.3
		>20♀	72.8 ± 32.6
	WAG	>20♂	74.0 ± 29.4
		>20♀	69.2 ± 32.8
Shuttle-conditioned avoidance: Responses/50 min [10-12]	ACI	>20♂	5.5 ± 9.5
		>20♀	6.4 ± 11.9
	A990	>20♂	10.4 ± 12.4
		>20♀	7.3 ± 11.3
	A35322	>20♂	13.8 ± 11.1
		>20♀	15.0 ± 9.1
	F344	>20♂	13.1 ± 15.9
		>20♀	14.4 ± 16.8
	INR	>20♂	18.8 ± 12.9
		>20♀	12.4 ± 13.3
	IR	>20♂	3.1 ± 1.6
		>20♀	7.0 ± 9.7
	MNR/Har	>20♂	9.0 ± 7.3
		>20♀	4.5 ± 5.1
	MNRA	>20♂	8.8 ± 10.8
		>20♀	6.5 ± 11.9
	MR/Har	>20♂	15.2 ± 13.4
		>20♀	6.8 ± 6.8
	TS1	>20♂	10.2 ± 9.6
		>20♀	12.0 ± 12.9
	TS3	>20♂	15.2 ± 9.8
		>20♀	15.9 ± 12.3
	WAG	>20♂	11.9 ± 13.2
		>20♀	4.7 ± 7.4
Operant conditioning acquisition			
Food contingent bar presses to auditory cue stimulus, ten 100-s sessions [10-12]	ACI	>20♂	27.8 ± 38.2
		>20♀	75.4 ± 56.3
	A990	>20♂	79.5 ± 56.6
		>20♀	55.4 ± 56.7
	A35322	>20♂	81.4 ± 40.1
		>20♀	80.9 ± 45.8
	F344	>20♂	48.0 ± 55.2
		>20♀	54.9 ± 57.7
	INR	>20♂	65.3 ± 29.4
		>20♀	73.1 ± 33.3
	IR	>20♂	17.4 ± 31.6
		>20♀	27.1 ± 40.1
	MNR/Har	>20♂	91.3 ± 56.4
		>20♀	57.5 ± 51.5
	MNRA	>20♂	80.8 ± 36.2
		>20♀	77.4 ± 39.6
	MR/Har	>20♂	25.0 ± 31.2
		>20♀	49.2 ± 44.6
	TS1	>20♂	19.2 ± 42.7
		>20♀	20.1 ± 30.5
	TS3	>20♂	73.8 ± 48.7
		>20♀	76.9 ± 42.3
	WAG	>20♂	86.1 ± 47.9
		>20♀	71.0 ± 34.8
Light contingent bar presses, five 1000-s sessions [10-12]	ACI	>20♂	52.8 ± 62.6
		>20♀	89.3 ± 84.0
	A990	>20♂	127.0 ± 98.6
		>20♀	95.7 ± 93.4
	A35322	>20♂	96.5 ± 81.2
		>20♀	115.5 ± 89.8
	F344	>20♂	54.1 ± 49.9
		>20♀	65.6 ± 57.2
	INR	>20♂	165.6 ± 68.2
		>20♀	195.3 ± 73.9
	IR	>20♂	6.1 ± 8.8
		>20♀	7.2 ± 14.7
	MNR/Har	>20♂	81.8 ± 57.7
		>20♀	54.8 ± 54.8
	MNRA	>20♂	100.0 ± 44.3
		>20♀	81.4 ± 53.8
	MR/Har	>20♂	51.3 ± 41.9
		>20♀	56.3 ± 33.9
	TS1	>20♂	16.7 ± 27.5
		>20♀	15.4 ± 29.2
	TS3	>20♂	107.7 ± 93.7
		>20♀	112.5 ± 107.2
	WAG	>20♂	101.9 ± 68.2
		>20♀	109.9 ± 64.6

continued

Part II. Characteristics of Genetically Defined Strains

Category & Characteristics	Strain	No. & Sex	Value
Runway acquisition: Mean time per trial, s, tested 10 trials/d for 10 d [10-12]	ACI	>20♂	145 ± 79
		>20♀	86 ± 71
	A990	>20♂	144 ± 82
		>20♀	164 ± 103
	A35322	>20♂	98 ± 64
		>20♀	92 ± 86
	F344	>20♂	209 ± 84
		>20♀	164 ± 76
	INR	>20♂	60 ± 37
		>20♀	27 ± 9
	IR	>20♂	143 ± 78
		>20♀	103 ± 53
	MNR/Har	>20♂	77 ± 80
		>20♀	45 ± 72
	MNRA	>20♂	48 ± 63
		>20♀	31 ± 52
	MR/Har	>20♂	90 ± 74
		>20♀	37 ± 29
	TS1	>20♂	138 ± 82
		>20♀	52 ± 31
	TS3	>20♂	59 ± 44
		>20♀	34 ± 28
	WAG	>20♂	62 ± 57
		>20♀	33 ± 27
Existing Genetic Differences			
Avoidance learning			
Shuttle avoidances, mean no. [17]			
Day 1	ACI	....	12
	BUF	....	19
	F344	....	24
	WF	....	10
Day 3	ACI	....	23
	BUF	....	21
	F344	....	25
	WF	....	20
Day 5	ACI	....	23
	BUF	....	24
	F344	....	25
	WF	....	22
Shuttle escapes, ratio of conditioned responses/unit time to inter-trial responses/unit time, day 2 [13]	F344	24♀	3.01
	LEW	24♀	0.63
Reproductive behavior: % copulating in four 15-min intervals [3]	ACI	18♂	83
	F344	27♂	78
	LEW	33♂	48
	WF	35♂	54

Category & Characteristics	Strain	No. & Sex	Value
Shock-induced fighting responses [14]			
Blood pressure, mm Hg			
Pre-fighting	OM	8♂, 8♀	133[2]
	SHR	8♂, 8♀	167[3]
Post-fighting	OM	8♂, 8♀	111[2]
	SHR	8♂, 8♀	163[3]
Serum dopamine β-monooxygenase[4], units	OM	8♂, 8♀	8.7 ± 5.6
	SHR	8♂, 8♀	3.0 ± 2.0
Stress response [23]			
Ambulation, 1-foot squares entered/2 min	ACI	11♂	28.36
	A35322	14♂	20.28
	F344	14♂	26.93
Rearings, no./2 min	ACI	11♂	10.91
	A35322	14♂	12.0
	F344	14♂	8.93
Defecation, fecal boli/2 min	ACI	11♂	2.82
	A35322	14♂	1.21
	F344	14♂	0.93
Stomach lesions, %	ACI	11♂	64
	A35322	14♂	21
	F344	14♂	57
Activity & brain enzymes [22]			
Spontaneous activity: Ambulation	ACI	7♂	92 ± 77
	BN	7♂	59 ± 40
	BUF	7♂	137 ± 77
	F344	7♂	46 ± 11
	LEW	7♂	114 ± 63
Tyrosine 3-monooxygenase[5], mean net counts/min ^3H_2O released·20 min^{-1}·mg protein^{-1}			
In corpus striatum	ACI	10♂	15572 ± 2852
	BN	5♂	15079 ± 2196
	BUF	10♂	12485 ± 1758
	F344	15♂	17944 ± 1619
	LEW	15♂	14867 ± 1026
In midbrain	ACI	5♂	1707 ± 237
	BN	10♂	2432 ± 224
	BUF	5♂	1596 ± 311
	F344	5♂	2613 ± 148
	LEW	15♂	1831 ± 379
Selection Effects			
Dominance in competition [4,15]			
Weight at 80 d, g	LRS	8♂	153 ± 15
		15♀	120 ± 10
	WRS	8♂	164 ± 18
		15♀	135 ± 15
Defecation, fecal boli/3 min	LRS	8♂	0.5
		15♀	1.0

2/ Standard deviation of difference between pre- & post-fighting values = 10. 3/ Standard deviation of difference between pre- & post-fighting values = 18. 4/ Synonym: Dopamine β-hydroxylase. 5/ Synonym: Tyrosine hydoxylase.

continued

Part II. Characteristics of Genetically Defined Strains

Category & Characteristics	Strain	No. & Sex	Value
	WRS	8♂	4.5
		15♀	3.0
Runway competition, % winning 90% of contests	LRS	8♂	0
		15♀	20
	WRS	8♂	100
		15♀	80
Foodhole competition, time licking, s/300 s	LRS	8♂	144
		15♀	149
	WRS	8♂	144
		15♀	150
Avoidance conditioning			
Avoidance responses, mean no. day 4 [21]			
Two-way avoidance	RCA	48♂♀	4.0
	RHA	48♂♀	7.8
	RLA	48♂♀	0.2
One-way avoidance	RCA	48♂♀	5.4
	RHA	48♂♀	9.2
	RLA	48♂♀	3.6
Unconditioned escape response threshold, volts			
[21]	RCA	24♂	150
		24♀	103
	RHA	24♂	193
		24♀	145
	RLA	24♂	217
		24♀	230
[20]	MNR/Har	12♂	130
		12♀	100
	MR/Har	12♂	140
		12♀	105
	RCA	12♂	165
		12♀	125
	RHA	12♂	165
		12♀	130
	RLA	12♂	260
		12♀	220
Reversal learning, mean no. of errors [16]	MNR/Har	24♂	2.53 ± 0.31
	MR/Har	24♂	2.29 ± 0.31
Response to cat, rabbit, & box			
General activity, 22.5 X 22.5 cm sections crossed/300 s [19]			
Cat	MNR/Har	10	30
	MR/Har	10	12
	RCA	10	27
	RHA	10	12
	RLA	10	11
Rabbit	MNR/Har	10	47
	MR/Har	10	20
	RCA	10	27
	RHA	10	17
	RLA	10	9
Box	MNR/Har	10	53
	MR/Har	10	29

Category & Characteristics	Strain	No. & Sex	Value
	RCA	5♂, 5♀	62
	RHA	5♂, 5♀	11
	RLA	5♂, 5♀	15
Approach behavior, sum scores (approach = 1; touch = 2) [19]			
Cat	MNR/Har	5♂, 5♀	5
	MR/Har	5♂, 5♀	9
	RCA	5♂, 5♀	3
	RHA	5♂, 5♀	4
	RLA	5♂, 5♀	2
Rabbit	MNR/Har	5♂, 5♀	14
	MR/Har	5♂, 5♀	9
	RCA	5♂, 5♀	8
	RHA	5♂, 5♀	4
	RLA	5♂, 5♀	4
Box	MNR/Har	5♂, 5♀	14
	MR/Har	5♂, 5♀	10
	RCA	5♂, 5♀	14
	RHA	5♂, 5♀	7
	RLA	5♂, 5♀	4
Alcohol consumption			
Preference: Intake ratio—$\sin^{-1}$ (alcohol solution/total fluid) [1]			
Solution concentration, 0.1%	RCA	10♂	48 ± 14
		10♀	49 ± 6
	RHA	10♂	44 ± 10
		10♀	43 ± 10
	RLA	10♂	45 ± 14
		10♀	47 ± 7
Solution concentration, 1.0%	RCA	10♂	53 ± 23
		10♀	57 ± 8
	RHA	10♂	43 ± 10
		10♀	45 ± 16
	RLA	10♂	37 ± 14
		10♀	48 ± 6
Solution concentration, 10%	RCA	10♂	23 ± 20
		10♀	27 ± 15
	RHA	10♂	28 ± 12
		10♀	49 ± 12
	RLA	10♂	18 ± 4
		10♀	29 ± 15
Selection: Consumption of 10% alcohol solution [18]			
Self selection			
Ratio solution to total fluid intake	MNR/Har	6♂, 6♀	0.057
	MR/Har	6♂, 6♀	0.161
	RCA	6♂, 6♀	0.124
	RHA	6♂, 6♀	0.189
	RLA	6♂, 6♀	0.119
ml alcohol ingested/kg body wt	MNR/Har	6♂, 6♀	1.4
	MR/Har	6♂, 6♀	3.4
	RCA	6♂, 6♀	2.0
	RHA	6♂, 6♀	3.0

continued

Part II. Characteristics of Genetically Defined Strains

Category & Characteristics	Strain	No. & Sex	Value
	RLA	6♂, 6♀	2.2
Forced selection ml alcohol ingested/kg body wt	MNR/Har	6♂, 6♀	11.5
	MR/Har	6♂, 6♀	12.0
	RCA	6♂, 6♀	6.0
	RHA	6♂, 6♀	9.0
	RLA	6♂, 6♀	8.0
Total fluid intake for 3 d alcohol & 2 d water, ml/kg	MNR/Har	6♂, 6♀	256
	MR/Har	6♂, 6♀	204
	RCA	6♂, 6♀	151
	RHA	6♂, 6♀	166
	RLA	6♂, 6♀	187
Neurochemical correlates			
Serotonin, mean limbic, μg/g [25]	MNR/Har	12♂	0.50
		6♀	0.36
	MR/N	12♂	0.67
		6♀	0.37
Tyrosine metabolism, measured by [^{3}H]L-tyrosine tracer, % of total radioactivity [2]			
Free tyrosine	RHA	8♂	47
	RLA	8♂	51
Dopamine	RHA	8♂	2.1
	RLA	8♂	3.4
Norepinephrine	RHA	8♂	0.8

Category & Characteristics	Strain	No. & Sex	Value
	RLA	8♂	1.0
Non-catecholamines	RHA	8♂	1.9
	RLA	8♂	2.3
Norepinephrine concentration [24]			
Telencephalon, ng/g	MNRA	13	368 ± 40
	MR/Har	13	376 ± 41
Hypothalamus, ng/g	MNRA	13	2127 ± 218
	MR/Har	13	1770 ± 214
Brainstem, ng/g	MNRA	13	473 ± 54
	MR/Har	13	475 ± 59
Heart, ng/g	MNRA	13	1546 ± 217
	MR/Har	13	1123 ± 172
Spleen, ng/g	MNRA	6	3757 ± 578
	MR/Har	6	1938 ± 490
Adrenal, mg/pair	MNRA	6	20.61 ± 0.99
	MR/Har	6	16.32 ± 0.64
Adrenal enzyme activity, nmoles·adrenal^{-1}·h^{-1} [2]			
Tyrosine-3-monooxygenase[5]	RHA	8♂	20 ± 3
	RLA	8♂	15 ± 3
Dopamine β-monooxygenase[4]	RHA	8♂	146 ± 40
	RLA	8♂	83 ± 32
Noradrenalin *N*-methyltransferase[6]	RHA	8♂	3.0 ± 0.08
	RLA	8♂	2.8 ± 0.3

[4] Synonym: Dopamine β-hydroxylase. [5] Synonym: Tyrosine hydroxylase. [6] Synonym: Phenylethanolamine-*N*-methyltransferase.

Contributor: Gordon M. Harrington

References

1. Brewster, D. J. 1969. J. Genet. Psychol. 115:217-227.
2. Coyle, J. T., Jr., et al. 1973. Psychopharmacologia 31:25-34.
3. Dewsbury, D. A. 1975. J. Comp. Physiol. Psychol. 88:713-722.
4. Fukusawa, T., et al. 1975. Behav. Biol. 15:333-342.
5. Harrington, G. M. 1968. Dev. Psychobiol. 1:211-218.
6. Harrington, G. M. 1968. Ibid. 1:245-253.
7. Harrington, G. M. 1971. Psychonom. Sci. 23:348-349.
8. Harrington, G. M. 1971. Ibid. 23:363-364.
9. Harrington, G. M. 1972. Ibid. 27:51-53.
10. Harrington, G. M. 1979. Bull. Psychonom. Soc. 13:(in press).
11. Harrington, G. M. 1979. Ibid. 14:(in press).
12. Harrington, G. M. Unpublished. Univ. Northern Iowa, Dep. Psychology, Cedar Falls, 1978.
13. Katzev, R. D., and S. K. Mills. 1974. J. Comp. Physiol. Psychol. 87:661-671.
14. Lamprecht, F., et al. 1974. Psychosom. Med. 36:298-303.
15. Masur, J., and M. A. C. Benedito. 1974. Nature (London) 249:284.
16. Powell, B. J. 1970. Proc. Annu. Conv. Am. Psychol. Assoc. 78th, pp. 825-826.
17. Ray, O. S., and R. J. Barrett. 1975. Behav. Biol. 15:391-417.
18. Satinder, K. P. 1972. J. Comp. Physiol. Psychol. 80:422-434.
19. Satinder, K. P. 1976. Anim. Learn. Behav. 4:172-176.
20. Satinder, K. P. 1976. J. Comp. Physiol. Psychol. 90:946-957.
21. Satinder, K. P. 1977. Ibid. 91:1326-1336.
22. Segal, D. S., et al. 1972. Behav. Biol. 7:75-81.
23. Sines, J. O. 1962. J. Genet. Psychol. 101:209-216.
24. Slater, J., et al. 1977. Pharmacol. Biochem. Behav. 6:511-520.
25. Sudak, H. S., and J. W. Maas. 1964. Science 146:418-420.

INDEX

Strains

To facilitate location of related strains and substrains, strain designations are presented in the following order:

Major strain
- Substrains
- Congenics
- Hybrids
- Strains derived by egg transfer or egg transplant from major strain
- Strains derived by foster nursing of major strain

Major strain designations are listed alphabetically within the index. Their substrains and derived strains are presented alphabetically within the above categories.

Strain designations containing numbers are listed numerically after the last entry bearing the same initial letter or letters (e.g., AS2 follows ASW/Sn; A2 follows Azoxymethane).

Strain designations commencing with numbers can be found at the end of the index, after the listings for the letter "Z."

Genes

Gene symbols are given in italics, and are accompanied, where available, by the pertinent gene name in brackets. Dominant and recessive gene symbols of the same locus appear together. Gene symbols within a group are listed in alphabetical order. Gene names are listed alphabetically within the index.

* indicates diagram or graph
fn indicates footnote material
hn indicates headnote material

A (strain): mouse
- adrenal cortical tumors, 178
- alloantigens, 114, 116-118
- characteristics, 21
- congenic-inbred line progenitor, 128
- genes, 21
- hemangioendothelioma, 173
- histocompatibility, 120
- Ia specificities, 146
- Ig heavy chain linkage groups, 111
- leukemia antigens, 168
- neoplasms, 169
- ovarian tumors, 194-195, 198-200, 202-203
- pituitary tumors, 182-183, 189
- pulmonary tumors, 162-163
- recombinant haplotype, 126
- subline genealogy, 18*
- testicular tumors, 206 hn, 206-207

A/Bcr, 18*
A/Bi, 18*
A/BiMtz, 18*
A/Br, 18*
A/BrA, 18*, 84, 89, 112
A/BrA$_f$, 18*
A/Cam, 18*
A/Fa, 18*, 92
A/H, 18*
A/He
- biochemical variation, 84, 90-91
- characteristics, 21
- complement factors, 103
- genealogy, 16*, 18*
- genes, 21
- histocompatibility, 121
- proteins, 104
- pulmonary tumors, 161
- testicular teratomas, 208 fn

A/HeHa, 18*, 88-89, 124
A/HeJ, 45, 47, 81-92, 106, 112, 167
A/HeJax, 18*
A/HeN, 106
A/J
- biochemical variation, 81-83, 85-92
- body weight, 45
- cellular alloantigens, 114
- characteristics, 21
- complement factors, 103
- genealogy, 16*
- genes, 21, 30
- histocompatibility, 120
- *If-1* allele, 112
- IgC_H alleles, 109
- leukemia antigens, 167-168
- proteins, 104, 106
- reproduction, 45

A/Jax, 18*
A/JBoy, 115, 120
A/Ki, 18*
A/Kl, 18*
A/Lac, 18*, 92
A/Li, 18*
A/Man, 18*
A/Ms, 18*
A/Mtz, 18*
A/Mv, 114
A/N, 18*
A/Orl, 92
A/Ph, 44
A/Rl, 18*
A/Sf, 18*
A/Sn, 18*, 21, 103, 114, 121
A/SnA, 84
A/SnEg, 130
A/SnSf, 127-128
A/St, 18*, 86-87
A/StHa, 18*
A/Wy, 18*
A/WySn, 16*, 124-125, 128, 169
A/WySnSg, 124
A congenics, 103, 115, 123-128, 130, 146-147

A hybrids, 161-163, 165, 179, 182-183, 185
A_f hybrids, 159-160, 179
A or *a* (gene): mouse
A [agouti], 17*, 30, 36, 77 fn
a [non-agouti]
chromosome aberrations, 42
chromosome position, 38*
histocompatibility linkage, 122, 123*
inbred strain genealogy, 17*
mutations, 30, 71
pleiotropic behavior, 155
strain distribution, 21-28, 30, 36
a^{da} [non-agouti dark belly], 71
a^e [extreme non-agouti], 30, 36, 71
A^i [intermediate agouti], 30, 71
A^{iy} [intermediate yellow], 30, 71
a^l [non-agouti lethal], 63, 71
a^m [mottled agouti], 71
A^s [agouti-suppressor], 36, 71
A^{sy} [sienna yellow], 30, 71
a^t [black and tan], 17*, 30, 36, 71
a^{td} [tanoid], 30, 71
a^u [agouti umbrous], 71
A^{vy} [viable yellow]
hepatomas, 164
immune system, 68
mammary tumorigenesis, 161
mutation, 71
pleiotropic behavior, 155
strain distribution, 29, 36
A^w [white-bellied agouti], 23, 25, 28, 30, 36, 71
a^x [lethal non-agouti], 63, 71
A^y [yellow]
hepatomas, 164
immune system, 68
inbred strain genealogy, 16*
lethality, 63
mammary tumorigenesis, 161
mutations, 71
phenotypic characteristics, 101
pleiotropic behavior, 155
strain distribution, 25, 30, 36
A or *a* (gene): rat
A [agouti], 234, 239-241, 243, 245, 247, 252, 272
a [non-agouti], 234, 238-253, 259, 272
a^m [agouti-melanic], 272
AA (strain): rat, 237
AAF57 adenocarcinoma: rat, 304
Aal [active avoidance learning] (gene): mouse, 22-23, 154
Aap [active avoidance performance] (gene): mouse, 154
AB (strain): mouse, 87
ab [asebia] (gene): mouse, 30, 39*, 60, 74
ABC (strain): mouse, 177
Abdomen irradiation: mouse, 189
Abelson MuLV (virus): mouse, 166
abn [abnormal] (gene): mouse, 57
Abnormal (gene): mouse (see *abn*)
Abnormal skin & hair (gene): mouse (see spf^{ash})
ABP (strain): mouse
ABP/J, 86
ABP/Le, 82-83, 86
Absent corpus callosum (gene): mouse (see *ac*)
ac [absent corpus callosum] (gene): mouse, 58, 149
ACC (strain): rat, 310-311
Accessory cusp: mouse, 214, 219
Accessory sex gland: rat, 290
"A" cell tumor: mouse, 180
2-Acetylaminofluorene: rat, 248, 253
Acetylcholine receptor: rat, 320
Acetylcholine-receptor protein: rat, 334
Acetylcholinesterase(s): rat, 287, 309 hn
β-D-*N*-Acetylhexosaminidase (gene): mouse (see *Hex-1*)
3-Acetyl-1-methyl-1-nitrosourea: rat, 303
Acf-1 [albumin] (gene): mouse, 21-28, 104
ACH (strain): rat, 234, 238, 272-273, 298
Achondroplasia (gene): mouse (see *cn*)
ACI (strain): rat
alloantigen chemistry, 320
alloantigenic alleles, 316-317
allograft survival, 325-326
behavior, 356-360
blood protein polymorphism, 309
characteristics, 238-239
chromosome polymorphism, 260
coat color, 238-239, 272-273
enzyme polymorphisms, 310-312
immunogenetic characterization, 313
immunoglobulin polymorphism, 306
immunologically mediated disease, 335
immunological responsiveness, 317-318
karyology, 261, 266
O_2 consumption, 356
organ weights, 354-355
origin, 234, 238-239
research use, 236, 351
transplant donor, 323, 328
tumors, 298
ACI/N, 277, 294-295, 298, 303, 310-311
(ACI x BN)F_1, 326-327
Acid phosphatase: rat, 284, 288, 309
Acid phosphatase [ACP-1] (protein): mouse, 97*, 99
Acid phosphatase—kidney (gene): mouse (see *Apk*)
Acid phosphatase—liver (gene): mouse (see *Apl*)
Acinar cells: rat, 290-291
ACJ (strain): rat, 306
ACP (strain): rat, 239, 272-273, 306, 316-318
ACP 9935 (strain): rat, 314
ACP 9935 Irish piebald (strain): rat, 239 (*see also* ACP)
Acp-1: mouse (see *Apl*)
Acp-2: mouse (see *Apk*)
ACTH (chemical): mouse, 101 (*see also* Adrenocorticotropic hormone)
ACTH (chemical): rat, 247
Action potential: rat, 279
Active avoidance learning (gene): mouse (see *Aal*)
Active avoidance performance (gene): mouse (see *Aap*)
Ad [adult obesity & diabetes] (gene): mouse, 38*, 155
Adenine phosphoribosyltransferase [APRT] (protein): mouse, 96*, 98
Adenoacanthoma: mouse, 26
Adenocarcinomas: mouse, 22, 178-179, 195, 201-202
Adenocarcinomas: rat, 239, 243, 297-301, 303-304
Adenohypophysis: rat, 290 (*see also* Anterior pituitary: rat)
Adenomas: mouse
adrenal, 26, 185
adrenal cortical, 178-179
chromophobe, 181-184, 186, 188, 190
induced, 162-163, 178-179, 181-186, 188-190, 197 hn, 198-204

ovarian, 193 hn, 194-196, 197 hn, 198-204
pituitary, 181-186, 188-190
pulmonary, 27, 161-163
spontaneous, 161-162, 194 hn, 194-196
Adenomas: rat
adrenal cortical, 242, 252, 295, 298-299
chromophobe, 252, 296
induced, 302
kidney, 302
mammary, 297, 300
pancreatic islet, 242, 252, 295, 298
pituitary, 242, 244, 252, 295-300
spontaneous, 298-300
thyroid, 295-296
Adenosine deaminase: mouse, 226-227
Adenosinetriphosphatase: rat, 281, 287-288
Adenylate cyclase: rat, 284, 309
Adenylate kinase [AK] (proteins): mouse, 96*, 98
Adh-1 [alcohol dehydrogenase] (gene): mouse, 77, 80, 220, 225-226
Adipocytes: rat, 312
Adipose (gene): mouse (see db^{ad})
Adipose tissue: rat, 312
Adjuvant disease: rat, 334
Adjuvant-induced arthritis: rat, 246, 334
Adnexa: rat, 289
Adrenal, duplicated ectopic: mouse, 60 fn
Adrenal adenomas: mouse, 26, 185
Adrenal cortex carcinoma: rat, 295
Adrenal cortical adenomas: mouse, 178-179
Adrenal cortical adenomas: rat, 242, 252, 295, 298-299
Adrenal cortical tumors: mouse, 28, 178-180
Adrenal cortical tumors: rat, 243, 248-249, 253, 296, 298-300
Adrenal dysplasia: mouse, 183
Adrenalectomy: mouse, 180, 190, 192
Adrenal enzyme activity: rat, 362
Adrenal glands: mouse, 21, 23-25, 80, 101
Adrenal glands: rat, 247, 274-277, 290, 309 hn, 355, 362
Adrenalitis, autoimmune: rat, 335
Adrenal medullary tumors: rat, 248, 253, 298-300
Adrenal pheochromocytoma: rat, 242
Adrenal regeneration hypertension: rat, 252
Adrenal tumors: rat, 239-240, 295 (*see also* specific adrenal tumor)
Adrenal X-zone: mouse, 27, 101
Adrenocortical carcinoma: mouse, 23, 183
Adrenocortical lesions: rat, 295
Adrenocortical lipid depletion (gene): mouse (see *ald*)
Adrenocortical lipid depletion-1 (gene): mouse (see *ald-1*)
Adrenocortical lipid depletion-2 (gene): mouse (see *ald-2*)
Adrenocortical tumors: mouse, 28 (*see also* Adrenal cortical tumors: mouse)
Adrenocorticotropic hormone: mouse, 178 (*see also* ACTH: mouse)
Adult obesity and diabetes (gene): mouse (see *Ad*)
AE (strain): mouse, 88
AE/Wf, 81-82, 86, 89
AEJ/Gn (strain): mouse, 30
AET (chemical): mouse, 204
A_f-MTV (virus): mouse, 160
African pygmy mice, 212
ag [agitans] (gene): mouse, 39*, 149
AG (strain): mouse
AG/Cam, 36
AG/JNt, 81-82, 89
AGA (strain): rat, 239, 272, 308, 316
Ag-A [agglutinogen-A] (gene): rat, 257, 314
Ag-B (gene): rat (see also *H-1*: rat)
coat color alleles, 241 fn
histocompatibility antigens, 319-320
immunologically mediated diseases, 334 hn, 334-336
strain distribution, 246, 255-256, 260, 313-314, 316
tissue distribution, 315
transplant survival, 321 hn, 321-323, 324 hn, 324-327
Ag-C [agglutinogen-C] (gene): rat, 256-257, 259 hn, 259, 313-315, 317
Ag-D [agglutinogen-D] (gene): rat, 257, 314-315
Ag-E [histocompatibility] (gene): rat, 257
Ag-F [lymphocytotoxin] (gene): rat, 257, 259, 315, 317, 320
Ag-G (gene): rat, 320
Agglutinogen-A (gene): rat (see *Ag-A*)
Agglutinogen-C (gene): rat (see *Ag-C*)
Agglutinogen-D (gene): rat (see *Ag-D*)
Aggressiveness: mouse, 22
Aggressiveness: rat, 252
Aging: rat, 277-297
Aging, red cell: mouse, 85 fn
Aging factor: mouse, 193 hn
Agitans (gene): mouse (see *ag*)
Agouti (gene): mouse (see *A*: mouse)
Agouti (gene): rat (see *A*: rat)
Agouti-melanic (gene): rat (see a^m: rat)
Agouti-suppressor (gene): mouse (see A^s)
Agouti umbrous (gene): mouse (see a^u)
Ags [α-galactosidase] (gene): mouse, 77, 80, 97*, 99, 226
AGUS (strain): rat
alloantigenic allele, 316
allograft survival, 325
characteristics, 239
coat color, 239, 272-273
eye color, 273
immunoglobulin polymorphism, 306
immunologically mediated disease, 334
origin, 239
transplant survival, 321, 323
AGUS/Lac, 274
Ah [aromatic hydrocarbon responsiveness] (gene): mouse, 21-28, 77, 81
ak [aphakia] (gene): mouse, 30, 39*, 58, 60 fn
$AKD2F_1$ (strain): mouse, 211
AKR (strain): mouse
alloantigens, 114, 116-118
biochemical variation, 81
centromeric heterochromatin, 44
characteristics, 21
cholesterol, 101
complement factors, 103
congenic-inbred line progenitor, 111, 125
genes, 21
haplotypes, 124
hemangioma, 173

IgC_H alleles, 109
Ig heavy chain linkage groups, 111
leukemia antigens, 167-168
leukemia incidence, 168
leukemia virus origin, 165
lymphoma, 168
ovarian tumors, 194, 201
proteins, 104
AKR/Cum, 17*, 83-84, 86-92, 106
AKR/Fu, 84
AKR/FuA, 17*, 21, 83-84, 86-92, 106
AKR/FuRdA, 83-84, 86-92, 106, 112
AKR/J
biochemical variation, 81-92
body weight, 45
congenic-inbred line progenitor, 128
genealogy, 17*
genes, 31, 33, 69 fn
If-1 allele, 112
IgC_H alleles, 109
ovarian tumors, 194, 202
recombinant-inbred line progenitor, 37
reproduction, 45, 47
urinary proteins, 106
vaginal septa, 47
viral antigens, 167
AKR/JBoy, 125
AKR/Lw, 17*, 83-84, 86-91
AKR/LwN, 106
AKR/N, 37
AKR/Orl, 92
AKR/Sm, 88
AKR/Sn, 89, 121
AKR.B6 congenics, 124
AKR.M, 126, 128, 146
AKR hybrids, 170, 211
AKR (virus): mouse, 227 hn
AKR-LI (virus): mouse, 165, 230
AKR type C virus: mouse, 98
Akv-1 [AKR leukemia virus inducer-1] (gene): mouse, 38*, 96*, 98, 165, 167
Akv-2 (gene): mouse, 167
AKXD (strain): mouse, 37
AKXL (strain): mouse, 37
AL (strain): mouse, 16*, 47, 92, 103-104, 109, 111
AL/N
biochemical variation, 82, 88-90
characteristics, 21
congenic-inbred line progenitor, 111, 115
genes, 21
leukemia antigens, 167
urinary proteins, 106
Al [alopecia] (gene): mouse, 39*, 74
Ala-1 (gene): mouse, 116
Alanine aminotransferase: mouse, 78, 99
ALB (strain): rat
alloantigenic alleles, 316-317
characteristics, 239
chromosome polymorphism, 260
coat color, 239, 272-273
congenic-inbred line progenitor, 256
immunogenetic characterization, 314
immunoglobulin polymorphism, 306
immunological responsiveness, 318
karyology, 261
origin, 239
tumors, 298
Alb-1 [albumin] (gene): mouse
biochemical locus, 38*, 96*
phenotypic expression, 81
polymorphisms, 220, 225-226
proteins, 77, 98, 104
strain distribution, 81
Albany (strain): rat (*see also* ALB)
characteristics, 239
congenic-inbred line progenitor, 256
immunogenetic characterization, 314
immunoglobulin polymorphism, 306
karyology, 261
tumors, 298
Albino (gene): mouse (see *c*: mouse)
Albino (gene): rat (see *c*: rat)
Albino (stock): mouse, 16*-17*, 208
Albino (strain): rat, 234
Albino Norway (strain): rat, 337
Albino Wistar (strain): rat, 302
Albumin: mouse, 98
Albumin: rat, 280
Albumin (genes): mouse (see *Acf-1; Alb-1*)
Albumin esterase, serum: rat, 310-311
ALB-1C (strain): mouse, 104
Alcohol: rat, 237, 251, 361-362
Alcohol dehydrogenase (gene): mouse (see *Adh-1*)
ald [adrenocortical lipid depletion] (gene): mouse, 38*, 155
ald-1 [adrenocortical lipid depletion-1] (gene): mouse, 101
ald-2 [adrenocortical lipid depletion-2] (gene): mouse, 101
Aldehyde dehydrogenase: rat, 309
Aldehyde oxidase (gene): mouse (see *Aox*)
Aldolase—liver: mouse, 220 hn
Alkaline phosphatase: mouse, 80, 92, 101
Alkaline phosphatase (gene): rat (see *Alp*)
Alkaline phosphatase (protein): rat, 282, 285, 288, 310
Alleles: mouse
alloantigens, 113-114, 116-118
endocrine functions, 101-102
genes, 105, 108 hn, 108-109
hepatomas, 164
histocompatibility loci, 119-122
H-2 gene complex, 138-140
If-1 locus, 112
leukemia antigens, 167-168
nomenclature rules, 12-14
proteins, 103-104
strain distribution, 80-93
Alleles: rat, 306 hn, 306-312, 315-317
Alleles, excluded: mouse, 21-28
Alleles—inbreeding coefficient: laboratory animals, 4
Alleles, mutant: laboratory animals, 4
Allergic encephalitis: mouse, 22
Allergic encephalomyelitis: rat, 239-240, 242, 246, 249, 252, 334
Allergic myositis: rat, 334
Allergic neuritis: rat, 334
Alloantigenic alleles: rat, 315-317
Alloantigens: mouse, 113-118, 136, 141, 143
Alloantigens: rat, 319-320
Allogeneic skin grafts: mouse, 69
Allogeneic target antigens: mouse, 136

Allograft rejection: mouse, 68-69
Allograft survival: rat, 324-327
Allospecificities: rat, 319-320
Allotypes, immunoglobulin: mouse, 107-111
Alloxan diabetes: rat, 250
Alopecia (gene): mouse (see *Al*)
Alopecia periodica (gene): mouse (see *ap*)
Alopécie (gene): mouse (see *alp*)
Alp [alkaline phosphatase] (gene): rat, 257, 305
alp [alopécie] (gene): mouse, 74
Alpha heavy chains, immunoglobulin: rat, 306 hn, 306-308
ALR (strain): rat, 348
Altitude: mouse tumors, 184, 200
AM (strain): rat, 239
am [amputated] (gene): mouse, 38*, 56
am [anemia-2] (gene): rat, 257
Ambulation: rat, 240, 245, 247, 251-252, 360
AMDIL (strain): rat, 239
Ames dominant spotting (gene): mouse (see W^a)
Ames dwarf (gene): mouse (see *df*)
Ames waltzer (gene): mouse (see *av*)
Amine oxidase (flavin-containing): rat, 280
L-Amino-acid decarboxylase, aromatic: rat, 280
Amino acid metabolism, aromatic: mouse, 151
Amino acids: rat, 320
Amino acid sequence, *N*-terminal: mouse, 147
o-Aminoazotoluene: mouse, 173-175
γ-Aminobutyric acid: rat, 238 hn, 251-252
4-Amino-2′,3-dimethylbiphenyl: rat, 303
2-Aminoethylisothiuronium bromide hydrobromide: mouse, 204
p-Aminohippurate tubular maxima: rat, 286
δ-Aminolevulinate dehydratase (gene): mouse (see *Lv*)
Aminopeptidase: mouse, 98
Amphotropic viruses: mouse, 227 hn, 228, 229 hn, 229-230
Amphotropic virus replication [RAM-1]: mouse, 96*, 98
Amputated (gene): mouse (see *am*: mouse)
A-MTV (virus): mouse, 159
Amy-1 [salivary α-amylase] (gene): mouse
  biochemical locus, 39*, 97*
  phenotypic expression, 81
  protein control, 77, 99
  strain distribution, 21, 23, 25, 27, 81
Amy-2 [pancreatic α-amylase] (gene): mouse
  biochemical locus, 39*, 97*
  electrophoretic polymorphism, 225
  phenotypic expression, 81
  protein control, 77, 99
  strain distribution, 25, 27, 81
Amylase: rat, 281
α-Amylase: mouse, 77, 99
n-Amylhydrazine hydrochloride: mouse, 176, 201
Amyloidosis: mouse, 22, 24, 26-28
an [anemia] (gene): mouse, 30, 38*, 65
an [anemia-1] (gene): rat, 257, 259
ANA (strain): rat, 237
Anamnestic lymph-cell response: rat, 317 hn, 317-318
Anaphylactoid reaction: rat, 318
Anaphylactoid reaction (gene): rat (see *dx*)
Anaplastic sarcoma: mouse, 166
Androgen hormone metabolism-1 (gene): mouse (see *Hom-1*)
Androgen receptor protein: mouse, 80
Androgen receptors: mouse, 102
Androgens: mouse, 80 fn (*see also* specific androgen)
Androgens: rat, 339 (*see also* specific androgen)
5α-Androstane-3α,17β-diol: rat, 341
Δ^4-Androstenedione: rat, 283 fn
4-Androstene-3,17-dione: rat, 283
Anemia: mouse, 24, 26, 28, 30, 65-66
Anemia (gene): mouse (see *an*: mouse)
Anemia-1 (gene): rat (see *an*: rat)
Anemia-2 (gene): rat (see *am*: rat)
Anencephaly: mouse, 149
Anestrus: rat, 283
Aneuploidy: rat, 261
Angioma: mouse, 173-174, 176-177, 200
Angioma: rat, 295
Angionecrosis: rat, 290
Angiopericytoma: mouse, 176
Angiosarcomas: mouse, 174, 176-177
Angora (gene): mouse (see *go*)
Aniline metabolism: rat, 239, 243-244, 248
Anophthalmia: mouse, 149-150
Anophthalmia: rat, 247
Anterior cusp: mouse, 214, 217, 219
Anterior pituitary: mouse, 102
Anterior pituitary: rat, 340 hn, 340-343, 346-347 (*see also* Adenohypophysis)
Anterior pituitary adenomas: rat, 298-300
Anterior pituitary tumors: rat, 243, 248, 253
Anteroexternal cusp: mouse, 217, 219
Anthramine: rat, 304
Antibodies: mouse, 26, 69
Antibodies: rat, 317 hn, 317-319, 335-337
Antibody-forming cells: mouse, 116-117
Antidiuretic hormone: rat, 339
Anti-DNA autoantibody: mouse, 69
Antigenic determinants: mouse, 110-111, 129 hn
Antigenic serum substance (gene): mouse (see *Sas-1*)
Antigenic specificities: mouse, 137 hn, 138-139
Antigens: mouse, 112-113, 135*, 136, 142-144, 147-148, 167-168
Antigens: rat, 257, 317-320, 335
Antiglomerular basement membrane antibodies: rat, 336
Anti-hypertensive drugs: rat, 251
Anti-kidney antibody: rat, 335
Antinuclear antibodies: mouse, 26, 69
Antisera: mouse, 145
Antisera: rat, 315
Antithymocyte autoantibody: mouse, 69
Antitubular basement membrane nephritis: rat, 335
AO (strain): rat
  alloantigenic alleles, 316-317
  coat color, 239, 272-273
  eye color, 273
  immunoglobulin polymorphism, 306
  immunologically mediated disease, 335
  karyology, 261
  origin, 239
  transplant donor, 322
ao [apampischo] (gene): mouse, 74
Aorta: rat, 281, 312
Aorta weight: rat, 281
Aortic blood flow: rat, 286
Aortic medial calcinosis: rat, 292
Aox [aldehyde oxidase] (gene): mouse, 38*, 77, 81, 96*, 98

ap [alopecia periodica] (gene): mouse, 74
Ap-1 (gene): mouse, 220
Apampischo (gene): mouse (see *ao*)
Aperistalsis: rat, 241
Aphakia (gene): mouse (see *ak*)
Apk [acid phosphatase—kidney] (gene): mouse, 77, 81, 97*, 99, 226
Apl [acid phosphatase—liver] (gene): mouse, 39*, 77, 81, 97*, 99
Aplasia: mouse, 24
Appendicular skeletal malformations: mouse, 56-57
Approach behavior: rat, 358, 361
AQR (strain): mouse, 126, 147
Aqueductal stenosis, secondary: mouse, 151
arl (haplotype): rat, 260
A region, *H-2* gene complex: mouse, 126-127
Aromatic L-amino-acid decarboxylase: rat, 280
Aromatic amino acid metabolism: mouse, 151
Aromatic hydrocarbon responsiveness (gene): mouse (see *Ah*)
Arterial fat deposition: rat, 348
Arterial pressure, mean: rat, 286
Arterial sclerosis, mesenteric: rat, 291, 294
Arteries: rat, 249, 290-291
Arteriolipidosis-prone rat, 348
Arteriosclerosis: mouse, 27
Arthritis, adjuvant-induced: rat, 246, 334
Arylesterase: rat, 310
Arylsulfatase B (gene): mouse (see *As-1*)
Arylsulfatase B regulation (gene): mouse (see *Asr-1*)
AS (strain): rat
alloantigenic alleles, 316
blood protein polymorphism, 309
characteristics, 239-240
coat color, 239, 273
congenic-inbred line progenitor, 256
enzyme polymorphism, 311-312
eye color, 273
graft survival, 325, 328-331
immunogenetic characterization, 314
immunoglobulin polymorphism, 306
immunologically mediated diseases, 334-335
immunological responsiveness, 318-319
karyology, 266
origin, 239-240
transplant survival 321-323, 332
AS.AS2, 256
AS hybrids, 321-323, 325, 328, 330, 332
As-1 [arylsulfatase B] (gene): mouse, 21-24, 27, 77, 81
AS anti-LEW antiserum: rat, 315
Ascites hepatomas: rat, 244, 263
Ascites sarcoma, rat, 243
Ascites tumors: rat, 248-249, 264
Asebia (gene): mouse (see *ab*)
ASH/W (strain): rat, 306 (*see also* Wistar outbred)
ash [ashen] (gene): mouse, 71
Ashen (gene): mouse (see *ash*)
asp [audiogenic seizure prone] (gene): mouse, 38*, 149, 154
Aspartate aminotransferase: mouse, 99, 226
Aspermatogenesis: rat, 336
ASR-1 (strain): mouse, 147
Asr-1 [arylsulfatase B regulation] (gene): mouse, 23, 77, 81
Astrocytes, abnormal: mouse, 149
ASW/Sn (strain): mouse, 194, 202-203
AS2 (strain): rat
alloantigenic alleles, 316-317
blood protein polymorphism, 309
coat color, 240, 273
congenic-inbred line progenitor, 256
enzyme polymorphism, 311-312
eye color, 273
graft survival, 328
immunogenetic characterization, 314
immunoglobulin polymorphism, 306
immunological responsiveness, 318
origin, 240
transplant survival, 321-323, 332
AS2.AS, 256
at [atrichosis] (gene): mouse, 38*, 59, 74
at [atrichosis] (gene): rat, 257
Ataxia: mouse, 66-67, 149-153
Ataxia (gene): mouse (see *ax*)
Ataxic gait: mouse, 152-153
Athymism: mouse, 27
Atomic bomb: mouse tumors, 179, 200-201
Atresia, vaginal: rat, 241
Atria: rat, 279-280
Atrial pressure: rat, 286
Atrial tumors: rat, 249
Atrichis (strain): rat, 306
Atrichosis (gene): mouse (see *at*: mouse)
Atrichosis (gene): rat (see *at*: rat)
Atriocaval epithelial mesotheliomas: rat, 249, 299
AU (strain): mouse, 21, 103-104, 124
AU/SsJ, 81-92, 105-106
AU (strain): rat, 261
Audiogenic seizure(s): mouse, 24, 27, 149
Audiogenic seizure(s): rat, 243
Audiogenic seizure prone (gene): mouse (see *asp*)
AUG (strain): rat
alloantigenic alleles, 316-317
blood protein polymorphisms, 308-309
characteristics, 240
coat color, 240, 273
congenic-inbred line progenitor, 255
enzyme polymorphism, 311
eye color, 273
immunoglobulin polymorphism, 307
immunologically mediated diseases, 334-336
immunological responsiveness, 317-318
infectious diseases, 338
origin, 234, 240
transplant donor, 322-323, 329-330
AUG/Cub, 310
AUG/Orl, 310-311
AUG congenics, 256
(AUG x AS)F_1, 330
August (strain): rat, 290-291 (*see also* AUG)
August/Orl, 310-311
August-Capushon (strain): rat, 261
August990 (strain): rat (*see* A990)
August7322 (strain): rat (*see* A7322)
August28807 (strain): rat (*see* A28807)
AUG28807 (strain): rat (*see* A28807)
Aurothioglucose: mouse, 163
Autoallergic sialadenitis: rat, 334
Autoantibodies: mouse, 69
Autoimmune adrenalitis: rat, 335

Autoimmune complex nephritis: rat, 313 hn, 314
Autoimmune hemolytic anemia: mouse, 26
Autoimmune myasthenia gravis: rat, 334
Autoimmune myocarditis: rat, 246
Autoimmune orchitis: rat, 336
Autoimmune thyroiditis: rat, 240, 243, 249, 252, 335
Autoimmunity: mouse, 68
Autologous immune-complex nephritis: rat, 335
Autosomal striping (gene): mouse (see *Sta*)
Autosomes: mouse, 40*, 44 hn, 219
Autosomes: rat, 264, 268
av [Ames waltzer] (gene): mouse, 38*, 149, 155
AVN (strain): rat
alloantigenic alleles, 316
blood protein polymorphisms, 308-309
characteristics, 240
coat color, 240, 273
congenic-inbred line progenitor, 255
enzyme polymorphisms, 311-312
eye color, 273
immunogenetic characterization, 314
immunoglobulin polymorphism, 307
immunologically mediated diseases, 334
immunological responsiveness, 318
origin, 240
transplant donor, 328, 333
AVN/Cub, 310
(AVN x LEW)F_1, 334
Avoidance behavior: mouse, 155 hn, 155-156
Avoidance behavior: rat, 359
Avoidance conditioning: rat, 352, 361
Avoidance learning: rat, 237, 360 (see also *Aal* [active avoidance learning])
ax [ataxia] (gene): mouse, 30, 39*, 58, 149, 155
AXC (strain): rat, 298, 307, 338 (*see also* ACI)
AXC9935 (strain): rat, 351 (*see also* ACI)
AXC9935 Irish (strain): rat, 238-239, 313 (*see also* ACI)
AXC9935 Irish piebald (strain): rat, 238 (*see also* ACP)
Axial skeletal malformations: mouse, 55-57
Axon abnormalities: mouse, 66-67
AY (strain): mouse, 84, 87, 89
AY/Nga, 84, 88
AZF_1 (strain): mouse, 179
Azoxymethane: mouse, 209
A2 (strain): rat, 316
A2/1, 325
A2G (strain): mouse
biochemical variation, 81, 83-84, 86-92
body weight, 45
genealogy, 16*
mutant gene, 36
protein, 104
reproduction, 45
A2G/A, 86, 106
A2G/GA, 112
A2G/Lac, 85
A35 (strain): rat, 310-311
A990 (strain): rat, 234, 240, 272-273, 351, 354-360
A7322 (strain): rat, 234, 240, 273, 316
A28807 (strain): rat
alloantigenic alleles, 316-317
blood protein polymorphism, 309
characteristics, 240
coat color, 240, 273
congenic-inbred line progenitor, 256
enzyme polymorphisms, 310, 312
immunogenetic characterization, 314
origin, 234, 240
A35322 (strain): rat, 234, 240, 272-273, 351, 354-360

B (strain): rat, 241, 272-273, 351
B or *b* (gene): mouse
b [brown]
chromosome aberration, 42
chromosome position, 38*
enzyme activity, 77 hn
histocompatibility linkage, 122, 123*
inbred strain genealogy, 17*
mutation, 71
protein control, 77 fn
strain distribution, 21-28, 30
urinary proteins, 106 hn
b^c [cordovan], 30, 71
B^{lt} [light], 25, 30, 71
B^w [white-based brown], 71
B or *b* (gene): rat
B [non-brown], 243, 248, 272
b [brown], 234, 239, 242, 246, 259, 272
BA (strain): mouse, 47, 88
BA/Nmg, 81-82, 89
ba [bare] (gene): mouse, 60
BAB-14 (strain): mouse, 111
Backcrosses; backcrossing: laboratory animals, 4*, 4-5
Backcrosses: mouse, 11, 29-36, 157 hn-158 hn, 162-164
Backcrosses, introgressive: mouse, 111 hn, 111
Bacteremia: rat, 337
Bacterial infections: rat, 337-338
Bacterial lipopolysaccharides: mouse, 23, 68 hn, 69, 113
Bacteria-resistant stock: mouse, 17*
Bacteria-susceptible stock: mouse, 17*
Bacteriophage T-4: mouse, 69
Bagg (strain): mouse, 16* (*see also* BALB/c)
Bagg albino (strain): mouse, 178 (*see also* BALB/c)
bal [balding] (gene): mouse, 60
Balance defect: mouse, 151
BALB (strain): mouse
BALB-A, 197-198
BALB/c
adrenal cortical tumors, 178
alloantigens, 114, 116-118
biochemical variation, 83, 87-88, 90
body weight, 45
characteristics, 22
complement factors, 103
congenic-inbred line progenitor, 6*, 111, 124
genealogy, 17*
genes, 22, 36, 105
haplotypes, 124-125
hemangiomas, 173
IgC_H alleles, 108-109
Ig heavy chain linkage groups, 110
intestinal tumors, 208, 209 hn, 209
leukemia antigens, 168
leukemia virus origin, 165-166
mammary tumors, 22, 157 hn-158 hn, 159-160
neoplasms, 170
ovarian tumors, 194, 196-198, 200-204
pituitary tumors, 182, 184
proteins, 104
recombinant-inbred line progenitor, 7*

reproduction, 45
subline genealogy, 18*
testicular tumors, 206 hn, 207
BALB/cA, 18*
BALB/cAnDe, 161, 170
BALB/cAnN, 106
BALB/cBoy, 120
BALB/cBy
biochemical variation, 85, 88, 90
congenic-inbred line progenitor, 124, 128
genealogy, 16*, 18*
histocompatibility, 120
If-1 allele, 112
mutant gene, 101
mutant line progenitor, 129
recombinant-inbred line progenitor, 37
urinary proteins, 106
BALB/cByA, 84
BALB/cByJ, 82, 86
BALB/cCbSe, 202
BALB/cCd, 22
BALB/cCr, 173, 194, 210
BALB/cCrgl, 18*, 84, 91
BALB/cCrglA, 112
BALB/cCrN, 18*
BALB/cDe, 83, 114
BALB/cDeHe, 161
BALB/c_fRIII, 169, 194
BALB/cGif, 88
BALB/cGn, 18*
BALB/cGr, 32, 85
BALB/cGw, 18*
BALB/cHeA, 84, 112
BALB/cHeAn, 18*
BALB/cHf, 18*, 90, 106
BALB/cHu, 18*, 114
BALB/cJ
biochemical variation, 81-93
body weight, 45
centromeric heterochromatin, 44
congenic-inbred line progenitor, 124-125, 128
genealogy, 16*
histocompatibility, 121
If-1 allele, 112
reproduction, 45, 47
BALB/cJax, 18*
BALB/cJPas, 37
BALB/cKa, 18*
BALB/cKh, 129-130, 210
BALB/cLac, 18*
BALB/cN, 18*, 93, 167
BALB/cOrl, 92
BALB/cOs, 173, 194, 201, 203
BALB/cPi, 173, 194, 211
BALB/cRl, 18*
BALB/cSc, 18*
BALB/cSn, 18*
BALB/cWt, 18*, 47
BALB/c π, 16*
BALB congenics, 111, 120, 124-125, 128, 170
(BALB x C57BL/6)F_1, 211
BALB/c hybrids, 182, 184, 211
BALB-MTV (virus): mouse, 160
BALB viruses: mouse, 165, 230
BALB 3T3 cell line: mouse, 227 hn
Bald (gene): mouse (see *hr^ba^*)
Balding (gene): mouse (see *bal*)
Ballantyne's spotting (gene): mouse (see W^b)
Banding karyotype: rat, 266, 268*-270*
Banding patterns, chromosome: mouse, 40*
Bare (gene): mouse (see *ba*)
Bare patches (gene): mouse (see *Bpa*)
Bar presses: rat, 359
Basal cell carcinomas: rat, 302, 304
Basal metabolic rate: rat, 356 fn
Basement membrane disorders: rat, 293, 335-336
Basophilic adenomas: mouse, 183
B(black hooded) strain: rat, 249 (*see also* PVG)
BBT/Ha (strain): mouse, 86
BC (strain): mouse, 181-182
BC x C57, 182
bc [bouncy] (gene): mouse, 39*, 149
B-cell lymphoma: mouse, 166
B-cell mitogens: mouse, 69
B cells: mouse, 69, 116-117
BC-8 (strain): mouse, 111
BD (strains): rat, 307
BD I, 241, 272-273, 316, 318
BD II, 241, 272-273, 308, 316, 318
BD III, 241, 272-273, 316
BD IV, 241, 272-273, 316
BD V
alloantigenic alleles, 316-317
blood protein polymorphism, 308-309
characteristics, 241
coat color, 241, 272-273
congenic-inbred line progenitor, 256
enzyme polymorphisms, 311-312
eye color, 273
immunogenetic characterization, 314
immunoglobulin polymorphism, 307
origin, 241
transplant survival, 322, 330-331, 333*
BD V/Cub, 310-311
BD VI, 241, 272-273, 316
BD VII, 241, 272-273, 308-309, 311-312, 314, 316-318
BD VII/Cub, 310
BD VIII, 241, 272-273, 308
BD IX, 241, 272-273, 303, 307, 311, 316
BD X, 241, 272-273, 307, 316
BD X/Cub, 310
bd [bradypneic] (gene): mouse, 30
BDE (strain): rat, 241, 272-273, 307, 316, 318
BDH-SPF (strain): mouse, 187
BDL (strain): mouse, 36
BDP (strain): mouse
alloantigens, 114, 116-118
biochemical variation, 88
characteristics, 22
complement factors, 103
genes, 22
H-2 haplotype & markers, 124
Ig heavy chain linkage groups, 110
leukemia antigens, 168
proteins, 104
BDP/J, 16*, 45, 81-92, 106, 121
Bedding: mouse tumors, 165
Behavior: mouse, 68, 151, 154-155
Behavior: rat, 236, 239-242, 244-249, 251-252, 356-362

Behavior genetics: rat, 350-362
Beige (gene): mouse (see *bg*)
Bell-flash ovulation (gene): mouse (see *Bfo*)
Belly spot and tail (gene): mouse (see *Bst*)
Belted (gene): mouse (see *bt*)
Belted-2 (gene): mouse (see *bt-2*)
Bence-Jones proteins: rat, 247
Bent-tail (gene): mouse (see *Bn*)
Benzo[*a*]pyrene: mouse, 201
Benzo[*a*]pyrene: rat, 267-304
3,4-Benzopyrene: mouse, 201
Benzylpenicilloyl$_{25}$-bovine gamma globulin: mouse, 143
Benzylpenicilloyl$_{4}$-bovine pancreatic ribonuclease: mouse, 143
Berkeley S1 (strain): rat (*see* TMB)
Berkeley S3 (strain): rat (*see* TMD)
bf [buff] (gene): mouse, 30, 38*, 71
Bfo [bell-flash ovulation] (gene): mouse, 101, 154
bg [beige] (gene): mouse, 30, 39*, 68, 71
Bge [β-galactosidase—electrophoresis] (gene): mouse, 77, 81-82, 96*, 99
Bgs [β-galactosidase activity] (gene): mouse, 21-28, 38*, 77, 82, 96*, 98
Bgt [β-galactosidase, temporal] (gene): mouse, 77, 82
BH (strain): rat, 307, 311, 316-317, 334
bh [brain hernia] (gene): mouse, 58, 60 fn, 149
BH anti-LEW antiserum: rat, 315
BHE (strain): rat, 312, 339
Biceps femoris: rat, 274-277
BICR (Marshall) strain: rat, 307
BIMA (strain): mouse, 159-160
    BIMA/A, 112
BIR/A (strain): mouse, 112
BIRMA (strain): rat, 241, 273, 307
BIRMB (strain): rat, 241, 273, 307
Birmingham A (strain): rat, 241 (*see also* BIRMA)
Birmingham B (strain): rat, 241 (*see also* BIRMB)
BIRM 4B (strain): rat, 316
BIRM 5A (strain): rat, 316
Birth weight: mouse, 25
Bittner virus: mouse, 159
BL (strain): mouse, 109, 111, 159-160
    BL/De, 90, 106
BL (gene): mouse, 115 (see also *Ea-4*)
bl [blebbed] (gene): mouse, 38*, 57
Black agouti: mouse, 196 fn
Black and tan (gene): mouse (see a^t)
Black and White Hooded (strain): rat, 307
Black-eyed white (gene): mouse (see mi^{bw})
Black-hooded (strain): rat, 309 (*see also* PVG)
Black Praha (strain): rat, 242 (*see also* BP)
Bladder, urinary: rat, 242, 290, 295
Bladder calculi: rat, 243
Bladder exstrophy: mouse, 60 fn
Bladder tumors: rat, 295, 296 fn
Bladder weight: rat, 274-277
Bld [blind] (gene): mouse, 39*, 58
Blebbed (gene): mouse (see *bl*)
Blebs: mouse, 60, 151
Blebs (gene): mouse (see *my*)
Bleeding, umbilical: mouse, 65
Blind (gene): mouse (see *Bld*)
Blindness: mouse, 152
Blind sterile (gene): mouse (see *bs*)
BL-MTV (virus): mouse, 160
BLN/Nmg (strain): mouse, 47
Blood: rat, 261-263
Blood catalase activity: mouse, 21
Blood flow, aortic: rat, 286
Blood pressure: mouse, 22-23, 26-28
Blood pressure: rat, 250, 281, 286, 290, 349*, 360
Blood pressure, systolic: rat, 239, 248-249, 253
Blood protein polymorphisms: rat, 308-309
Blood urea nitrogen: rat, 279, 287
Blood vessels, kidney: rat, 293
Blood vessel tumor: mouse, 176
Blotchy (gene): mouse (see Mo^{blo})
Blu:(LE) strain: rat, 337
bm [brachymorphic] (gene): mouse, 30, 39*
BN (strain): rat
        alloantigenic alleles, 316-317
        allograft survival, 325-327
        blood protein polymorphism, 309
        brain enzymes, 360
        characteristics, 242
        coat color, 242, 272-273
        congenic-inbred line progenitor, 255-256
        enzyme polymorphisms, 310-312
        immunogenetic characterization, 314
        immunoglobulin polymorphism, 307
        immunologically mediated diseases, 334-336
        immunological responsiveness, 317-319
        karyology, 261
        origin, 234, 242
        research use, 236, 351
        transplant donor, 328
        transplant survival, 321-323, 328-332, 333*
        virally induced tumor, 302
    BN/Bi, 277, 279, 289, 295, 298
    BN/BiRij, 295
    BN/Cub, 310
    BN/Orl, 311
    BN congenics, 255-256, 316, 334
    (BN x LEW)F_1, 334
Bn [bent-tail] (gene): mouse, 30, 39*, 56
BNXAKN (strain): mouse, 37
Body fat: mouse, 28
Body fat: rat, 245
Body length: rat, 273-276
Body + head length: mouse, 212, 214, 217, 219
Body size: mouse, 48-49, 102
Body size: rat, 241, 244, 257-258
Body weight: mouse, 45-49, 219
Body weight: rat, 236, 245, 250, 273-276, 360
Bone lesions: rat, 290
Bone marrow: mouse, 114
Bone marrow cells: rat, 261-267
Bones, parietal: mouse, 214
Bone tumors: rat, 296 fn
Bony labyrinth defects: mouse, 149-150
Bouncy (gene): mouse (see *bc*)
Bovine gamma globulin: rat, 318
Bovine serum albumin: rat, 318
BP (strain): rat, 242, 255, 272-273, 307-308, 316, 328
    BP/Cub, 310-311
    BP.1N, 255, 316
BP/Ww (strain): mouse, 83-84, 86-89, 91
bp [brachypodism] (gene): mouse, 30, 38*, 42, 56
Bpa [bare patches] (gene): mouse, 39*, 44, 74
Br [brachyrrhine] (gene): mouse, 59, 60 fn

Brachial nerve: rat, 294
Brachymorphic (gene): mouse (see *bm*)
Brachyphalangy (gene): mouse (see Xt^{bph})
Brachypodism (gene): mouse (see *bp*)
Brachyrrhine (gene): mouse (see *Br*)
Brachyury (strain): mouse, 124
Brachyury (gene): mouse (see *T*)
Bradypneic (gene): mouse (see *bd*)
Brain: mouse, 77-78, 114-117, 150-151
Brain: rat
aging, 281, 283-284, 287, 294
enzymes, 309 hn, 310, 360
reproductive hormones, 340, 342, 347
tumors, 296 fn, 300-301, 303
weight, 274-277, 281, 354
Braincase: mouse, 217
Brain hernia (gene): mouse (see *bh*)
Brainstem: mouse, 67, 149
Brainstem: rat, 287, 362
Brattleboro (strain): rat, 237, 339 hn, 339
Breakpoints, chromosome: mouse, 44
Breast: mouse, 193 hn (*see also* Mammary entries)
Breast fibroadenoma: rat, 252 (*see also* Mammary entries)
Breeders: mouse, 21-24, 26, 45-48, 177 fn, 181 fn-182 fn
Breeding: rat, 236, 337 fn
Breeding performance: rat, 239, 242, 244-245, 247-249, 253 (*see also* Reproduction: rat)
B region, *H-2* gene complex: mouse, 126-127
Brindled (gene): mouse (see Mo^{br})
BROFO (strain): rat, 242, 273, 326
Bromodeoxyuridine: rat, 266 hn, 266-267
Bronchiogenic carcinoma: rat, 240
Brown (gene): mouse (see *b*: mouse)
Brown (gene): rat (see *b*: rat)
BRS (strain): mouse
BRS/Se, 88
BRS/St, 89
BRSUNT (strain): mouse, 103, 110
BRSUNT/N, 81-82, 88-90, 106
BRSUNT/St, 81-82
Brucella abortus: mouse, 69
Brucella antigen: mouse, 68
BRVR (strain): mouse, 22, 88, 103
BRVR/Pl, 81-82, 87, 89
BRVR/SrCr, 45
BRVS (strain): mouse, 17*
BRVS/Pl, 81-82, 87, 89
BS (strain): mouse, 91
BS (strain): rat
alloantigenic alleles, 316
blood protein polymorphism, 309
characteristics, 242
coat color, 242, 272
enzyme polymorphisms, 311-312
immunogenetic characterization, 314
immunoglobulin polymorphism, 307
immunological responsiveness, 318
origin, 242
transplant survival, 321, 329
bs [blind-sterile] (gene): mouse, 58
BSL (strain): mouse, 110
Bst [belly spot and tail] (gene): mouse, 56, 71
BSVS (strain): mouse
biochemical variation, 88
characteristics, 22
complement factors, 103
congenic-inbred line progenitor, 128
genealogy, 17*
genes, 22
Ia specificities, 146
Ig heavy chain linkage groups, 111
recombinant haplotype, 126
BSVS/SrCr, 45
bt [belted] (gene): mouse, 30, 36, 39*, 71
bt-2 [belted-2] (gene): mouse, 71
BTOs (strain): mouse, 194, 201, 203
BU (strain): mouse, 17*
BUA (strain): mouse, 17*, 47, 88
BUA/Hn, 81-82, 87, 89
BUB (strain): mouse, 17*, 47, 103-104, 124
BUB/Bn, 106 hn, 114, 116-118
BUB/BnJ, 81-91, 114
BUB/J, 90
BUF (strain): rat
alloantigenic alleles, 316-317
avoidance learning, 360
behavioral research, 351
blood protein polymorphisms, 309
characteristics, 242-243
chromosome polymorphism, 260
coat color, 242, 272-273
congenic-inbred line progenitor, 256
enzymes, 310-312, 360
eye color, 273
immunogenetic characterization, 314
immunoglobulin polymorphism, 307
immunologically mediated diseases, 335
immunological responsiveness, 317-318
karyology, 261, 266
origin, 242-243
transplant survival, 321-323, 325-327, 329-330
tumors, 298, 303
BUF/N, 298, 310-311
(BUF x LEW)F_1, 335
Buff (gene): mouse (see *bf*)
Buffalo (strain): rat (*see also* BUF)
characteristics, 242-243
coat color, 242
enzyme polymorphism, 311
immunogenetic characterization, 314
immunoglobulin polymorphism, 307
immunologically mediated disease, 335
karyology, 261, 266
origin, 242-243
tumors, 303
Busulfan: mouse, 201
1-Butyl-3-dimethyl-1-nitrosourea: rat, 304
Butylnitrosourea: mouse, 209
1-Butyl-1-nitrosourea: rat, 261, 304
BXD (strain): mouse, 37
BXH (strain): mouse, 37, 106
BXHN (strain): mouse, 37
B3 (strain): rat, 243, 256, 260, 314, 317
B6 (strain): mouse, 125 (*see also* C57BL/6)
B6 congenics, 88, 115, 119-120, 124-129
B6 hybrids, 195, 211
B6-*H-2* mutants, 129
B6.C-*H-2* mutants, 129

B10 (strain): mouse, 104, 119, 124, 140 (*see also* C57BL/10)
- B10-H-1 congenics, 159-160
- B10-H-2 congenics, 129, 159-160
- B10-H-3^b, 159-160
- B10-H-8^b, 159-160
- B10-H-9^b, 159
- B10.A, 44, 103, 119, 124, 126-128
- B10.A(1R), 126, 128, 146
- B10.A(2R), 92, 103, 126, 128, 146
- B10.A(3R), 126, 128, 146
- B10.A(4R), 92, 103, 126, 128, 146
- B10.A(5R), 92, 103, 126, 128, 146
- B10.A(15R), 126, 128, 146
- B10.A(18R), 126, 128, 146
- B10.AK-H-1, 119
- B10.AKM, 92, 103, 126, 128
- B10.AKM/Sn, 170
- B10.AKR, 125
- B10.AL, 126-128
- B10.AM, 126, 128, 146
- B10.AQR, 128
- B10.ASW, 125
- B10.BDR congenics, 126, 128, 146
- B10.BR congenics, 92, 103, 124-125, 128, 167
- B10.BSVS, 126, 128
- B10.BUA congenics, 130, 132-133
- B10.BYR, 126, 128
- B10.C, 119
- B10.CA congenics, 130, 132-133
- B10.CBA, 125
- B10.CE(62NX), 119
- B10.CHR51, 130, 132-133
- B10.CNB, 125
- B10.C3H, 119
- B10.DA, 126
- B10.DA(80NS), 128
- B10.D1/Ph, 125
- B10.D1/Y, 125
- B10.D2
  - biochemical variation, 92
  - centromeric heterochromatin, 44
  - complement factors, 103
  - congenic-inbred line progenitor, 111, 128
  - histocompatibility, 119
  - Ia specificities, 146
  - *Ss* allele, 140
- B10.D2/n, 106, 124
- B10.D2/nSn, 30-31, 82-83, 86, 88, 112, 124
- B10.D2/o, 106, 124
- B10.D2/oSn, 88, 112
- B10.D2b.Ig, 111
- B10.D2-H-2^{dm1}, 129
- B10.D2(M504), 92, 128
- B10.D2(R101), 126, 128, 146
- B10.D2(R103), 92, 126, 128, 146
- B10.D2(R106), 126, 128, 146
- B10.D2(R107), 126, 128, 146
- B10.D2(58N)/Sn, 171
- B10.F congenics, 115, 125, 128
- B10.G, 124-125, 140, 146
- B10.GAA20, 130-131
- B10.GD, 126, 128
- B10.HTG, 126, 146
- B10.HTT, 92, 103, 126, 128, 146
- B10.K, 119, 125, 140, 146
- B10.KEA congenics, 130-133
- B10.KPA42, 130-131
- B10.KPB congenics, 130-133
- B10.KR, 119
- B10.L, 115
- B10.LG, 126, 128
- B10.LIB18, 132-133
- B10.LP, 92, 119
- B10.M, 103, 124, 127, 130, 146
- B10.M/Sn, 124
- B10.M-H-2^{fm2}, 130
- B10.M(11R), 126-127, 147
- B10.M(17R), 126-127, 147
- B10.MOL1, 132-133
- B10.NB, 125
- B10.OH, 126, 128
- B10.OL, 126, 128
- B10.P congenics, 119, 124-125, 128, 140, 146
- B10.PL congenics, 92, 124-125
- B10.Q, 125
- B10.Q/Sg, 124
- B10.QAR, 126, 128
- B10.QSR2, 127-128
- B10.RB7, 130-131
- B10.RIII congenics, 115, 124-125, 146
- B10.S congenics, 103, 124-126, 128, 140, 146-147
- B10.SAA48, 130-131
- B10.SM congenics, 124-125
- B10.SNA70, 127-128, 132-133
- B10.STA10, 132-133
- B10.STC congenics, 132-133
- B10.T(6R), 126, 128, 147
- B10.TL, 126, 128
- B10.VW, 119
- B10.W congenics, 130-133
- B10.WB congenics, 92, 119, 124-125
- B10.WOA1, 130-131
- B10.WR7, 103, 127-128
- B10.Y, 125
- B10.129 congenics, 115, 119

C (strain): mouse, 207 (*see also* BALB/c)
- C/St: mouse, 16*
- C-H-2^{dm2}, 129
- C.B6-H-2^{dm4}, 130
- C hybrids, 207, 211

C or *c* (gene): mouse
- *C* [colored], 36
- *c* [albino]
  - biochemical locus, 38*, 96*
  - chromosome aberrations, 42
  - histocompatibility linkage, 122, 123*
  - inbred strain genealogy, 17*
  - mutations, 30, 36, 71
  - phenotypic expression, 82
  - pleiotropic behavior, 155
  - protein control, 77, 98
  - strain distribution, 21-22, 25-26, 28, 30, 36, 82
- c^{ch} [chinchilla], 16*, 24-25, 27-28, 30, 71
- c^{e} [extreme dilution], 23-24, 30, 36, 71, 196
- c^{h} [Himalayan], 30, 71
- c^{i} [intense chinchilla], 71
- c^{m} [chinchilla-mottled], 71
- c^{p} [platinum], 30, 71

C or *c* (gene): rat
C [non-albino], 240-241, 243-244, 246 fn, 248, 272
c [albino], 239-253, 273, 315
c^d [ruby-eyed dilute], 245, 273
c^h [Himalayan], 273
Ca [caracul] (gene): mouse, 30, 39*, 74
Ca^d [directional caracul], 74
Ca [cataract] (gene): rat, 257
CAF_1 (strain): mouse, 211
Calcareous heart deposits: mouse, 24
Calcinosis, aortic medial: rat, 292
Calcitonin-secreting tumors: rat, 300 fn
Calcium, serum: mouse, 101
Calcium, tissue: rat, 280, 285
Calculi, bladder: rat, 243
Calf type I collagen: mouse, 143
Calf type I procollagen peptide: mouse, 143
Calvino (gene): mouse (see *cv*)
CAL-9 (strain): mouse, 111
CAL-20 (strain): mouse, 111
CAM (strain): rat, 316, 335
cAMP levels, liver: mouse, 136
can [cartilage anomaly] (gene): mouse, 56
Cancer, mammary: mouse, 21, 26, 157, 181
Cancer cell karyology, mammary: rat, 262
Cancer research: laboratory animals, 1
Candida albicans: mouse, 68
CAP (strain): rat, 243, 272-273, 307-308, 316
CAP/Cub, 310
Capillary hemangiomas: mouse, 173-174
CAR (strain): rat, 236, 243, 273, 316-318, 334
Car-1 [carbonic anhydrase-1] (gene): mouse, 21-28, 38*, 77, 82, 96*, 98
Car-2 [carbonic anhydrase-2] (gene): mouse, 21-28, 38*, 77, 83, 96*, 98
Caracul (strain): mouse, 124
Caracul (gene): mouse (see *Ca*: mouse)
Carbamic acid esters: mouse, 175
1-Carbamyl-2-phenylhydrazine: mouse, 177
Carbohydrate-sensitive (strain): rat, 312, 339
Carbonate dehydratase: mouse, 77, 98
Carbon clearance, intravascular: mouse, 68
Carbonic anhydrase (genes): mouse (see *Car-1*; *Car-2*)
Carbon tetrachloride: rat, 335
Carboxylesterase: rat, 310
Carcinogenesis, mammary: mouse, 158 hn
Carcinogens: mouse, 28, 193 hn, 201-204, 209-210
Carcinoma, Walker rat, 301
Carcinoma 5: rat, 249
Carcinoma No. 10 Lewis: rat, 246
Carcinoma 338: rat, 248
Carcinoma 343: rat, 248
Carcinomas: mouse
adrenal cortical 23, 178-180
colon, 209 hn
induced, 183, 202, 204, 208 hn-209 hn, 209-210
intestinal, 208 hn, 208-210
mammary, 27
ovarian, 194-195, 202, 204
pituitary, 183
spontaneous, 28, 194-195, 208
squamous cell, 28
Carcinomas: rat
adrenal cortex, 295, 298
adrenal gland, 299
basal cell, 302, 304
bronchiogenic, 240
cervix, 304
chromophobe, 298 fn-300 fn
colon, 252
epidermoid, 302
Harderian gland, 248
induced, 302, 304
kidney, 302
lung, 302
mammary, 244, 248, 295-296, 296 fn, 298
spontaneous, 249, 298-299
squamous cell, 242, 298, 298 fn, 301, 304
thyroid, 242-244, 249, 252, 295-296
transitional cell, 298
transplantable, 240, 301
ureter, 242
urinary bladder, 242
Carcinosarcoma, Walker 256: rat, 244, 250
Cardiac hypertrophy: rat, 245
Cardiac output: rat, 286
Cardiopathy: rat, 292
Cardiovascular disease: rat, 250
Cardiovascular lesions, hypertensive: rat, 348
Caries, dental: rat, 243
Caroli CI (virus): mouse, 231
Cartilage anomaly (gene): mouse (see *can*)
Cartilage cells: rat, 267
Carworth (strain): rat, 281
Carworth W (strain): rat, 335
CAS (strain): rat, 236, 243, 273, 316-318, 334
CAS-IU-1 (virus): mouse, 230
Castration: mouse, 178-179, 184, 197
Castration: rat, 252, 346-347
CAST-8155 (virus): mouse, 231
cat [cataract lens] (gene): rat, 257
Catabolism, renal prostaglandin: rat, 245
Catalase (gene): mouse (see *Cs-1*)
Catalase activity: mouse, 21, 99
Catalase activity: rat, 282
Catalase degradation (gene): mouse (see *Ce-1*)
Cataract (strain): rat, 278
Cataract (gene): rat (see *Ca*: rat)
Cataract lens (gene): rat (see *cat*)
Cataractous (gene): rat (see *ct*: rat)
Cataracts: rat, 249-250
Catecholamines: rat, 251, 284, 309 hn
Catechol methyltransferase: rat, 280
Cathepsin: rat, 288
Cattanach's translocation: mouse, 42 fn
Caudal rachischisis: mouse, 152
Caudal spina bifida: mouse, 153
Caudate area, brain: rat, 284
Cavernous hemangiomas: mouse, 173-174
Caw:CFE (SD) strain: rat, 337
CB (strain): rat, 316
cb [cerebral degeneration] (gene): mouse, 58, 149
cb [chubby] (gene): rat, 257
Cb/Sc (CB) strain: rat, 302
CBA (strain): mouse
adrenal cortical adenoma, 178
biochemical variation, 83
body weight, 45
complement factors, 103

haplotypes, 124
hemoglobin gene, 105
Ig heavy chain linkage groups, 110
leukemia antigens, 168
mammary tumors, 159
ovarian tumors, 194, 198
pituitary tumors, 182-183, 187-188, 191
proteins, 104
reproduction, 45, 47
subline genealogy, 16*, 19*
testicular tumors, 206 hn, 207
tumorigenesis inhibition, 158 hn
CBA/A, 19*
CBA/An, 19*
CBA/Bln, 19*
CBA/Br, 19*
CBA/BrA, 16*, 88-89, 91-92, 112
CBA/Ca, 16*, 19*, 22, 36, 114, 121
CBA/CaGnJ, 90
CBA/CaH, 36
CBA/CaJ, 81-83, 85-91
CBA/CaJax, 19*
CBA/CaLac, 36
CBA/CaLacSto, 130
CBA/Cam, 19*, 36, 85
CBA/Cbi, 19*
CBA/Cl, 19*, 36
CBA/Fa, 19*, 92
CBA/Gn, 19*
CBA/Gr, 19*
CBA/H
biochemical variation, 87
cellular alloantigen, 114
centromeric heterochromatin, 44
genealogy, 19*
intestinal tumors, 209
life-span, 47
lymphoma, 170
CBA/H-T6
biochemical variation, 82-83, 86, 90, 92
characteristics, 22
genes, 22, 36
intestinal tumors, 209
lymphocyte alloantigens, 116-118
rat chromosome marker analogue, 249
CBA/J
alloantigens, 113-114, 116-118
biochemical variation, 81-92
body weight, 45
centromeric heterochromatin, 44
characteristics, 22
congenic-inbred line progenitor, 125
fibrosarcoma, 211
genealogy, 16*
genes, 22
histocompatibility, 121
leukemia antigens, 167-168
lymphoma, 170
reproduction, 45, 47
vaginal septa, 47
CBA/Jax, 19*
CBA/Ki, 19*
CBA/Lac, 19*, 92
CBA/Man, 19*
CBA/Ms, 19*
CBA/N, 113
CBA/Orl, 92
CBA/Ph, 125
CBA/St, 16*, 19*
CBA/StKi, 19*
CBA/H-*nu*/*nu*, 170
CBA-$H\text{-}2^{km1}$, 130
CBA hybrids, 182, 184-187, 195
CBA_f/Lw, 90, 106
CB-20 (strain): mouse, 111
C-cell tumor: rat, 298-300
CD (strain): rat, 338
Cd [crooked] (gene): mouse, 38*, 57, 149
Cd-1 (gene): mouse, 23
CDF (strain): rat, 308, 316, 338
cdm [cadmium resistance] (gene): mouse, 39*
CE (strain): mouse
carcinoma, 178
characteristics, 23
complement factors, 103
congenic-inbred line progenitor, 111
genes, 23
H-2 haplotype & markers, 124
IgC_H alleles, 109
Ig heavy chain linkage groups, 111
lymphocyte alloantigens, 116-118
mammary tumors, 159
ovarian tumors, 193 hn, 195
pituitary tumors, 182
proteins, 104
CE/J
biochemical variation, 81-83, 85-92
body weight, 47
cellular alloantigens, 114
complement factors, 103
genealogy, 17*
histocompatibility, 121
If-1 allele, 112
IgC_H alleles, 109
leukemia antigens, 168
protein, 104
reproduction, 47
urinary proteins, 106
CE/Lac, 85, 92
CE/Wy, 196
CE hybrids, 179-180, 182-183, 193 hn, 196
Ce-1 [catalase degradation] (gene): mouse, 22-24, 28, 83
Ce-2 [kidney-catalase] (gene): mouse, 77, 83, 97*, 99
^{144}Ce: rat, 302
[^{144}Ce]$CeCl_4$: rat, 302
Cecum: rat, 274-277, 285
Cedar shavings: mouse tumors, 165
Cell(s) (*see* specific cell)
Cell body atrophy: mouse, 66
Cell-mediated lympholysis allogeneic target antigens: mouse, 136 hn, 136
Cell membrane molecules: mouse, 135*
Cell surface antigen [H-Y]: mouse, 97*, 99
Cellular alloantigens: mouse, 113-114, 136
Central nervous system: mouse (*see* CNS entries)
Central nervous system: rat, 289, 294
Centrolobular hepatocyte: rat, 282
Centromere: mouse, 13, 134*
Centromere: rat, 271-272

Centromeric heterochromatin: mouse, 13, 44
Cerebellar degeneration: mouse, 151
Cerebellar hemorrhage: rat, 290
Cerebellar infarction: rat, 290
Cerebellum: mouse, 152-153
Cerebellum: rat, 284, 287
Cerebral cortex: mouse, 152
Cerebral cortex: rat, 280, 284, 291
Cerebral degeneration (gene): mouse (see *cb*: mouse)
Cerebral hernia: mouse, 149
Cerebral lesions: rat, 350*
Cerebrovascular lesions: rat, 349
Ceruloplasmin: rat, 308
Cerv C viruses: mouse, 231
Cervical spinal cord: mouse, 67
Cervical tumors: rat, 242, 298, 304
Cestode infections: rat, 338
CFCW/R1 (strain): mouse, 81-82, 89-90, 106
CF-LP (strain): mouse, 187
CFN (strain): rat, 281
CFW (strain): mouse, 173
    CFW/N, 106, 209
CF1 (strain): mouse, 166, 174, 194, 200-203, 209
ch [congenital hydrocephalus] (gene): mouse, 39*, 57, 149
Charles River (strain): rat, 342, 344
    Charles River (CD), 342-343
Chbb:THOM (strain): rat, 281, 296
Chediak-Higashi syndrome: mouse, 68
Chemical carcinogens: mouse, 193 hn (*see also* Carcinogens; specific carcinogen)
Chemotaxis, granulocyte: mouse, 68
CHI (strain): mouse, 16*, 103, 124
    CHI/St, 103
Chinchilla (gene): mouse (see c^{ch})
Chinchilla-mottled (gene): mouse (see c^{m})
Chloride: rat, 285
Chlormadinone acetate: mouse, 187
Chloroform fumes: mouse, 23-24
Chlorotrianisene: mouse, 206 hn, 206
Chlorpromazine avoidance (gene): mouse (see *Cpz*)
cho [chondrodysplasia] (gene): mouse, 57
Cholangioma: mouse, 68
Cholesterol: mouse, 101, 206 hn
Cholesterol: rat, 280, 284, 287
Cholesterol-supplemented diet: mouse, 188
Chondrodysplasia (gene): mouse (see *cho*)
Choriomeningitis, lymphocytic: mouse, 227 hn
Chromatids, sister: rat, 266 hn, 266-267
Chromophobe adenomas: mouse, 181-184, 186, 188, 190
Chromophobe adenomas: rat, 252, 296, 298 fn-300 fn
Chromophobe carcinomas: rat, 298 fn-300 fn
Chromosome marker analogue: rat, 249
Chromosome number, diploid: rat, 233
Chromosomes: mouse (*see also* X & Y chromosomes: mouse)
        aberrations, 41-44
        anemia development, 65-66
        banding patterns, 40*
        behavioral genes, 154-156
        biochemical loci, 96*-97*, 98-99
        biochemical variation, 77-80
        coat color, 71-73
        complement factors, 103
        congenital malformations, 55-60
        endocrine disorders, 101-102
        erythrocyte production, 65-66
        genetically transmitted viruses, 167
        hemoglobin genes, 105
        histocompatibility systems, 121, 123*
        immune system development, 68-69
        lethal & semilethal genes, 63-64
        leukemogenesis, 168
        linkage maps, 38*-39*, 96*-97*, 123*
        mammary tumors, 158, 161
        muscle development, 66-67
        neurologic disorders, 149-153
        proteins, 98-99, 103-104
        skin, 74-76
Chromosomes: rat, 260, 261 hn, 261-268, 268*-270*, 271-272 (*see also* X & Y chromosomes: rat)
Chubby (gene): rat (see *cb*: rat)
Chymotrypsinogen: rat, 281
C-I (strain): mouse, 198
Ci (gene): mouse, 136, 141
Circling: mouse, 149-153, 155 hn, 155-156
Circulatory system: rat, 291
Civo (strain): rat, 291
cl [clubfoot] (gene): mouse, 56
Cleft lip: mouse, 21, 60 fn
Cleft palate: mouse, 21, 60 fn
Cleft palate: rat, 251
Clones, virus: mouse, 227 hn, 228-229
Cloud-gray (gene): mouse (see Sl^{cg})
Clubfoot (gene): mouse (see *cl*)
Clubfoot, congenital: rat, 253
Cm [coloboma] (gene): mouse, 30, 38*, 58, 149
CN (strain): mouse, 208
cn [achondroplasia] (gene): mouse, 30, 38*, 56
CNS congenital malformations: mouse, 58
CNS reduplication: mouse, 150
CNS sudanophilic leukodystrophy: mouse, 150
co [cocked] (gene): mouse, 39*
CO_{13} (strain): mouse, 106 hn
^{60}Co: mouse, 200
Coat color: mouse
        Asian species, 214, 217, 219
        chromosome aberrations, 42 hn, 42
        genes affecting, 70 hn, 71-73
        histocompatibility linkage, 122, 123*
        inbred strain genealogy, 16*
Coat color: rat, 238-253, 272-273
Coefficient of inbreeding: laboratory animals, 2, 4, 4*
Coelomys, 212, 214
Coh [coumarin hydroxylase] (gene): mouse, 38*, 77, 83, 96*, 98
Coiled (gene): mouse, 57
Coisogenic strains: laboratory animals, 6
Coisogenic strains: mouse, 11 (*see also* specific strain)
Collagen: rat, 281
Coloboma (gene): mouse (see *Cm*)
Colon: mouse, 151-152, 209 hn
Colon adenocarcinoma: rat, 303-304
Colon carcinoma: rat, 252
Colored (gene): mouse (see *C*: mouse)
Committee on Standardized Genetic Nomenclature for Mice, 9
Competition dominance: rat, 360
Complement components: mouse, 103, 135*, 136
Complement receptor lymphocyte ontogeny: mouse, 136 hn, 136

Complement receptor lymphocytes (gene): mouse (see *Crl-1*)
Concanavalin A: mouse, 68 hn, 68
Concanavalin A: rat, 319
Congenic lines: laboratory animals, 5-6, 6*
Congenic strains: mouse (*see also* specific strain)
anemia, 65-66
antigenic specificities, 130-133
biochemical variation, 81-83, 86-89, 92
cellular alloantigens, 114-115
centromeric heterochromatin, 44
complement components, 103
fibrosarcoma, 211
haplotypes, 124-133
hepatomas, 163
histocompatibility, 119-121, 121 fn-122 fn
Ia specificities, 145-147
If-1 alleles, 112
immunoglobulin allotypes, 111
leukemia, 167, 169
lymphoma, 169
mammary tumors, 159-160
mutations, 29-35
neoplasms, 169 hn, 170-172
nomenclature rules, 10-11
ovarian tumors, 194, 201
progenitors, 111, 115, 124-125, 127-130
proteins, 103-107
Ss alleles, 140
testicular teratomas, 207-208
uses, 6
Congenic strains: rat, 255-256
Congenital cleft palate: rat, 251
Congenital clubfoot: rat, 253
Congenital cystic kidney: mouse, 26
Congenital cystic ovaries: mouse, 24
Congenital hydrocephalus (gene): mouse (see *ch*)
Congenital hydronephrosis: rat, 236, 242
Congenital malformations: mouse, 55-60 (*see also* specific malformation)
Congenital malformations: rat, 238 (*see also* specific malformation)
Congenital testicular teratomas: mouse, 28
Congenital urogenital abnormalities: rat, 236
Contrasted (gene): mouse (see Sl^{con})
Convulsions: mouse, 150
COP (strain): rat (*see also* Copenhagen)
alloantigenic alleles, 316-317
blood protein polymorphism, 309
characteristics, 243
coat color, 243, 272-273
enzyme polymorphisms, 311-312
immunogenetic characterization, 314
immunoglobulin polymorphism, 307
immunological responsiveness, 318
origin, 234, 243
tumors, 298, 301
Copenhagen (strain): rat, 298, 301, 307, 338 (*see also* COP)
Copenhagen 2331 (strain): rat, 243 (*see also* COP)
Copulation: rat, 344-345, 360
COP-2331 (strain): rat, 243, 314 (*see also* COP)
Cordovan (gene): mouse (see b^c)
Corn oil supplemented diet: mouse, 188
Corpora lutea: rat, 283
Corpulent (gene): rat (see *cp*)
Corpus striatum enzyme: rat, 360
Corticosterone: mouse, 101
Corynebacterium: rat, 334
p-Coumarate 3-monooxygenase: mouse, 77, 98
Coumarin hydroxylase (gene): mouse (see *Coh*)
Cowlick (gene): rat (see *cw*: rat)
cp [corpulent] (gene): rat, 257
CPB (strain): mouse
CPB-H, 47
CPB-Mo/Cpb, 81-82, 89
CPB-R/Cpb, 81-82, 87
CPB (strain): rat
CPB-B, 243 (*see also* CPBB)
CPB-G, 245 (*see also* G/Cpb)
CPB-WE, 252
CPBB (strain): rat, 243, 272-273
Cpl-1 [plasma corticosterone level-1] (gene): mouse, 22-23, 101
Cpl-2 [plasma corticosterone level-2] (gene): mouse, 22, 101
Cpz [chlorpromazine avoidance] (gene): mouse, 154
cr [crinkled] (gene): mouse, 39*, 60, 74
cr [crawler] (gene): rat, 257
Crab-like walk: mouse, 152
Cranioschisis: mouse, 153
Cranioschisis (gene): mouse (see *crn*)
Cranium size: mouse, 217
Crawler (gene): rat (see *cr*: rat)
Creatinine: rat, 280, 288
C region, *H-2* gene complex: mouse, 126-127
cri [cribriform degeneration] (gene): mouse, 30, 38*, 60, 149
Cribriform degeneration (gene): mouse (see *cri*)
Crinkled (gene): mouse (see *cr*: mouse)
Crinkly-tail (gene): mouse (see *cy*)
Crl-1 [complement receptor lymphocytes] (gene): mouse, 103
Crl:CD (strain): rat, 342-343
Crl:CD(SD) strain: rat, 343
Crl:CD(SD)BR (strain): rat, 278, 281-282, 291, 296
Crl:(SD) strain: rat, 343
Crl:(WI) strain: rat, 340-342
crn [cranioschisis] (gene): mouse, 58
Crooked (gene): mouse (see *Cd*)
Crosses, reciprocal: mouse, 164
CS (strain): mouse, 47, 84, 88-89, 91
Cs [catalase] (gene): mouse, 38*
Cs-1 [catalase] (gene): mouse, 22-23, 77, 83, 96*, 98
ct [curly tail] (gene): mouse, 57, 149
ct [cataractous] (gene): rat, 257
CTM (strain): mouse, 174
Cu-1 [curly-1] (gene): rat, 257, 259
Cu-2 [curly-2] (gene): rat, 257
Curly-1 (gene): rat (see *Cu-1*)
Curly-2 (gene): rat (see *Cu-2*)
Curly coat: rat, 234
Curly tail (gene): mouse (see *ct*: mouse)
Curly whiskers (gene): mouse (see *cw*: mouse)
Curtailed (gene): mouse (see T^c)
Cusps: mouse, 212, 214, 217, 219
Cutaneous squamous papilloma: rat, 295
cv [calvino] (gene): mouse, 74
CW (strain): rat, 261
cw [curly whiskers] (gene): mouse, 38*, 74
cw [cowlick] (gene): rat, 257

CXB (strain): mouse, 37
CXBD (strain): mouse, 87-88, 90, 106, 114
  CXBD/By, 112
CXBE (strain): mouse, 87-88, 90, 106, 114
  CXBE/By, 112
CXBG (strain): mouse, 87-88, 90, 106, 114
  CXBG/By, 112
CXBH (strain): mouse, 87-88, 90, 106, 114
  CXBH/By, 112
CXBI (strain): mouse, 87-88, 90, 106, 114
  CXBI/By, 112
CXBJ (strain): mouse, 87-88, 90, 106, 114
  CXBJ/By, 112
CXBK (strain): mouse, 87-88, 90, 106, 114
  CXBK/By, 112
CXD (strain): mouse, 37
cy [crinkly-tail] (gene): mouse, 57
Cyclic adenosine 3′,5′-monophosphate (*see* cAMP)
Cystadenomas: mouse, 196, 199, 200, 203-204
Cysticercus fasciolaris: rat, 338
Cysticercus infections: rat, 240, 243-244, 248, 253
Cystic kidney: mouse, 26
Cystic kidney: rat, 238
Cystic ovaries: mouse, 24
Cystine diet: mouse, 163
Cysts, hepatic: rat, 290, 338
Cytochrome *c* oxidase: mouse, 79
Cytogenetic techniques: rat, 261-265
Cytologic marker translocation: mouse, 22
Cytomegalovirus: mouse, 227 hn
Cytoplasm: rat, 282, 287
Cytosol: mouse, 102
Cytotoxicity, cell-mediated: mouse, 136 hn, 136
Cytotoxic targets: mouse, 135*
C3fv-1 (gene): mouse, 167
C3H (strain): mouse
    biochemical variation, 80-81, 83, 90
    cellular alloantigens, 115
    congenic-inbred line progenitor, 125, 128
    fibrosarcomas, 211
    hemangiomas, 174
    hemoglobin genes, 105
    Ig heavy chain linkage groups, 110
    immune response, 113
    intestinal tumors, 208-210
    leukemia virus origin, 165
    ovarian tumors, 194, 198, 201-202
    pituitary tumors, 182-184, 187
    proteins, 103-104
    reproduction, 45, 47
    subline genealogy, 19*
    testicular tumors, 205 hn, 206-207
  C3H/An, 16*, 19*, 114, 116-118, 168
  C3H/AnCum, 105
  C3H/Bcr, 19*, 178
  C3H/Bi, 16*, 19*, 115, 121, 124, 179
  C3H/BiIcrf, 19*
  C3H/Ca, 19*
  C3H/Crgl, 19*
  C3H/De, 19*
  C3H/Di, 16*, 44
  C3H/DiA, 84
  C3H/DiSn, 19*, 124, 128
  C3H/Ep, 19*
  C3H/Fg, 167, 169
C3H/Fs, 19*
C3H/Gif, 19*
C3H/Gs, 19*
C3H/H, 19*, 198
C3H/Ha, 19*
C3H/HaN, 209
C3H/He
  alloantigens, 113-114, 116-118
  biochemical variation, 83-84, 87-89, 91-92
  body weight, 45
  characteristics, 23
  complement factors, 103
  genealogy, 16*, 19*
  genes, 23, 36
  hepatomas, 163
  intestinal tumors, 210
  leukemia antigens, 167-168
  proteins, 104
  pulmonary tumors, 162
  reproduction, 45
C3H/HeA, 19*, 86, 90
C3H/HeA$_f$, 19*
C3H/HeCrgl, 179
C3H/HeDiSn, 114, 124-125, 128, 170
C3H/He$_f$, 36
C3H/He$_f$A, 112
C3H/He$_f$Lac, 36
C3H/HeIcrf, 19*
C3H/HeJ
  biochemical variation, 80-92
  body weight, 45
  centromeric heterochromatin, 44
  If-1 allele, 112
  immune response, 113
  mutant genes, 30-33, 35
  recombinant-inbred line progenitor, 37
  reproduction, 45
  urinary proteins, 106
C3H/HeLac, 85, 88
C3H/HeN, 37
C3H/HeNWe, 91
C3H/HeOrl, 86, 89, 92
C3H/HeOs, 174, 194, 201, 203
C3H/J, 45, 112, 127-128
C3H/Jax, 19*
C3H/JSf, 115, 124-125, 128
C3H/Ki, 19*
C3H/Lac, 19*, 112
C3H/Lee, 19*
C3H/Lw$_f$, 106
C3H/Mai, 19*
C3H/Ms, 19*
C3H/N, 19*, 196
C3H/Rl, 19*
C3H/Sf, 19*
C3H/Sn, 16*, 119
C3H/St
  biochemical variation, 81-82, 87, 89, 93
  cellular alloantigen, 114
  genealogy, 16*, 19*
C3H/StHa, 88-89
C3H/StKi, 19*
C3H/Tw, 179, 194, 200
C3H-A^{vy} congenics, 163, 165, 194, 201
C3H.A/Ha, 124

C3H.B, 105
C3H.B10, 103, 124
C3H.HTG(82NS), 128
C3H.JK, 124-125
C3H.K, 119, 211
C3H.K/Sn, 170
C3H.NB, 124-125
C3H.OH, 92, 103, 126, 128, 146
C3H.OH/Sf, 128
C3H.OH/Sn, 128
C3H.OL, 92, 103, 126, 128, 146
C3H.Q, 124-125
C3H.SW, 103, 123-124
C3H.W3lp, 103
C3H x A, 206
(C3H x A)F_1, 179, 183, 195, 202
C3H x BC, 182
(C3H x C3H-A^{vy})F_1 x C3H, 164
C3H x C57BL, 187
(C3H x C57BL/6)F_1, 83
(C3H x ICRC)F_1, 195
(C3H x Y)F_1, 164
(C3H x YBR)F_1, 164
(C3H x 129)F_1, 195, 198
C3H/He x C57BL, 196
C3Hb, 194, 201-203 (*see also* C3H_f)
C3H_e/A, 194, 200
C3H_eB/De, 19*, 162-163, 170, 193 hn, 195
C3H_eB/Fe, 19*, 104, 193 hn, 195
C3H_eB/FeJ
biochemical variation, 81-91
body weight, 45
life-span, 45
mutations, 29-30, 32, 34
reproduction, 45, 47
urinary proteins, 106
vaginal septa, 47
C3H_eB/FeJax, 19*
C3H_eB/FeJOrl, 92
C3H_eB/Os, 203
C3H_eB x C57BL, 187
C3H_eB x SWR x C57L x 129/Rr, 181
C3H_f
hemangiomas, 174
mammary tumors, 157 hn-158 hn, 158-160
neoplasms, 170
ovarian tumors, 194, 201-203
pituitary tumors, 182-183, 187, 191
C3H_f/An, 115, 121
C3H_f/Bi, 114, 116-118, 124-125, 168
C3H_f/De, 90, 106
C3H_f/Dp, 194, 203
C3H_f/He, 45, 83, 194
C3H_f/HeHa, 124
C3H_f/Lac, 36
C3H_f/Rl, 85
C3H_f/Umc, 47
C3H_fB, 193 hn, 195, 198
C3H_fB/He, 19*, 162-163
C3H_fB/HeDe, 19*
C3H_fB/HeHa, 19*
C3H_fB/HeN, 19*
C3H_fC/Crgl, 19*
C3H_f hybrids, 159-160, 164, 170, 179
C3H-MTV (virus): mouse, 159
C3H_f-MTV (virus): mouse, 160
C57 (strain): mouse, 181-183, 188, 198, 208, 209 hn
C57 hybrids, 181, 183, 185
C57$_e$/Ha, 81-82, 86
C57BL (strain): mouse
biochemical variation, 80, 83, 90, 92
body weight, 45
cellular alloantigens, 113-114
congenic-inbred line progenitor, 111
genes, 31, 33, 105
hemangiomas, 174
Ig heavy chain linkage groups, 110
intestinal tumors, 208
leukemia virus origin, 166
mammary carcinogenesis, 158 hn
mammary tumors, 159-160
nodular hyperplasia, 179
ovarian tumors, 194, 198, 201-202
pituitary tumors, 182-184, 186-188, 190-191
reproduction, 45, 47
subline genealogy, 20*
testicular tumors, 206 hn, 207
tumorigenesis inhibition, 158 hn
C57BL/A, 20*
C57BL/An, 20*, 83, 90, 106
C57BL/Bcr, 20*
C57BL/Fa, 20*
C57BL/FnLn, 91
C57BL/Fr, 20*
C57BL/Ge, 20*
C57BL/Go, 20*, 36
C57BL/GoH, 36
C57BL/Gr, 20*, 85
C57BL/H, 20*
C57BL/Ha, 83, 209
C57BL/HaAo, 125
C57BL/He
biochemical variation, 83, 90
cellular alloantigen, 114
genealogy, 20*
neoplasms, 170
ovarian tumors, 194
pituitary tumors, 190
urinary proteins, 106
C57BL/HeDe, 170
C57BL/Hf, 20*
C57BL/How, 20*
C57BL/Icrf, 20*, 194, 211
C57BL/J, 93
C57BL/Ka, 20*, 106, 111
C57BL/KaLw, 87
C57BL/Ks
biochemical variation, 87-88
cellular alloantigen, 114
characteristics, 23
genealogy, 16*, 20*
genes, 23, 30-33
H-2 haplotype & markers, 124
histocompatibility, 121
C57BL/KsJ, 81-83, 85-86, 88, 90, 106
C57BL/Lac, 88, 112
C57BL/LacH, 20*
C57BL/LiA, 89, 112
C57BL/Mcl, 20*
C57BL/MHeA, 112

C57BL/Rij, 20*, 106
C57BL/Rl, 20*
C57BL/St, 20*
C57BL/1Umc, 20*
C57BL/6 (*see also* B6)
alloantigens, 114-116, 118
biochemical variation, 83, 88-89
body weight, 45
characteristics, 23
complement factors, 103
congenic-inbred line progenitor, 6*, 111, 128
genealogy, 17*
genes, 23, 33-34
haplotypes, 123, 125
histocompatibility, 120-121
IgC_H alleles, 108-109
Ig heavy chain linkage groups, 110
intestinal tumors, 209-210
leukemia antigens, 167-168
proteins, 104
recombinant-inbred line progenitor, 7*
reproduction, 45, 47
C57BL/6Bg, 93
C57BL/6Bom, 20*
C57BL/6By
anemia, 65-66
biochemical variation, 82-83, 85
congenic-inbred line progenitor, 124, 128
genealogy, 20*
histocompatibility, 119-120
If-1 allele, 112
mutant genes, 30-35
mutant line progenitor, 129
recombinant-inbred line progenitor, 37
C57BL/6ByA, 84, 112
C57BL/6ByJ, 81
C57BL/6Ha, 31 fn, 81
C57BL/6Hb, 20*
C57BL/6He, 20*
C57BL/6Hf, 90, 106
C57BL/6Icr, 20*
C57BL/6J
anemia, 65-66
biochemical variation, 81-92
body weight, 45
congenic-inbred line progenitor, 115, 125, 128
genealogy, 16*
histocompatibility, 119, 121
If-1 allele, 112
intestinal tumors, 209
lymphocytic neoplasm, 170
mutant genes, 29-31, 31 fn, 32-36
ovarian tumors, 194, 202
recombinant-inbred line progenitor, 37
reproduction, 45, 48
urinary proteins, 106
C57BL/6Jax, 20*
C57BL/6JBoy, 115, 120, 124-125
C57BL/6JPas, 37
C57BL/6Kh, 129-130
C57BL/6Ml, 20*
C57BL/6N, 20*, 37, 106, 208
C57BL/6NIcr, 211
C57BL/6Orl, 92
C57BL/6Os, 174, 194, 201, 203
C57BL/6Y, 20*
C57BL/6YEg, 129
C57BL/10 (*see also* B10)
biochemical variation, 83, 87, 89-90, 92
cellular alloantigens, 114-115
characteristics, 23
complement factors, 103
genes, 23
Ia specificities, 146
Ig heavy chain linkage groups, 110
leukemia antigens, 167-168
mammary tumors, 159
proteins, 104
C57BL/10Bg, 93
C57BL/10Ch, 20*
C57BL/10Fo, 20*
C57BL/10Gn, 20*, 33, 92
C57BL/10J
biochemical variation, 81-83, 85-90, 92-93
body weight, 45
centromeric heterochromatin, 44
congenic-inbred line progenitor, 115, 125
genealogy, 16*
If-1 allele, 112
mutation, 36
reproduction, 45
urinary proteins, 106
C57BL/10Jax, 20*
C57BL/10N, 90, 106
C57BL/10Ph, 20*
C57BL/10Sc, 20*, 106
C57BL/10ScSn
biochemical variation, 85
body weight, 45
cellular alloantigens, 114
fibrosarcoma, 211
H-2 haplotype & markers, 123
life-span, 45
lymphoma, 170
mutant gene, 34
C57BL/10ScSnA, 81, 112
C57BL/10ScSnEg, 129
C57BL/10ScSnPh, 44
C57BL/10Sf, 20*
C57BL/10Sn
biochemical variation, 82-83, 85-86, 88-89
congenic-inbred line progenitor, 115, 124-125, 128
genealogy, 16*, 20*
histocompatibility, 119
If-1 allele, 112
mutant line progenitor, 130
C57BL/10SnEg, 115, 125, 128-129
C57BL/10SnKlj, 128
C57BL/10SnPh, 125, 128
C57BL/10SnSf, 127-128
C57BL/10SnSg, 125, 127-128
C57BL/10Y, 20*, 125
C57BL/Icrf-a^t, 194, 211
C57BL/6-*Tla*, 114
C57BL/6.Ig, 111
C57BL/6J-bg^J/bg^J, 81-82, 89
C57BL/6J-*le rd*, 81-82
C57BL/6J-*ot*/+, 89
C57BL/6J-*Pt*/+, 106

C57BL/10Gn-*lu*, 92
C57BL x A, 206
C57BL x A_f, 160
(C57BL x A/He)F_1, 208
(C57BL x A/J)F_1, 198
C57BL x BALB/c, 159
C57BL x CBA, 184-187, 190, 206
(C57BL x CBA) x CBA, 187
(C57BL x CE)F_1, 179
C57BL x C3H, 177, 184-187, 195, 204
C57BL x $C3H_f$, 159
(C57BL x C3H/An_f)F_1, 170
C57BL x C57BR, 181
(C57BL x DBA)F_1, 177, 180, 183
C57BL x DBA_f, 159, 182, 188, 191
(C57BL x DBA/2_f)F_1, 182
C57BL x I, 159
(C57BL x NH)F_1, 179
C57BL x RIII, 160, 182
(C57BL/He x $C3H_f$/He)F_1, 170
(C57BL/6 x A)F_1, 211
(C57BL/6 x C3H)F_1, 188-189, 195, 199-200
C57BL/6 x C3H/He, 196
(C57BL/6 x C57BR)F_1, 211
(C57BL/6 x DBA/2)F_1, 211
(C57BL/6J x WC/Re)F_1, 170
C57BR (strain): mouse, 103, 110, 116-118, 124, 168, 174
C57BR/a, 16*
C57BR/cd
alloantigens, 114, 116-118
biochemical variation, 87
characteristics, 23
complement factors, 103
congenic-inbred line progenitor, 125
genealogy, 16*
genes, 23
pituitary tumors, 181
proteins, 104
C57BR/cdJ, 44-45, 81-92, 106, 113, 121, 167-168
C57BR/cdLac, 85
C57BR/cdOrl, 92
C57BR/J, 81
C57L (strain): mouse
alloantigens, 114, 116-118
biochemical variation, 86-87, 89, 91-92
characteristics, 24
complement factors, 103
genes, 24
H-2 haplotype & markers, 123
Ig heavy chain linkage groups, 110
leukemia antigens, 167-168
pituitary tumors, 181, 187, 189
proteins, 104
C57L/He, 90, 106, 162
C57L/HeDe, 171
C57L/J
biochemical variation, 81-83, 85-91
body weight, 45
centromeric heterochromatin, 44
congenic-inbred line progenitor, 115
genealogy, 16*
histocompatibility, 121
ovarian tumors, 194, 202
recombinant-inbred line progenitor, 37
reproduction, 45, 48
urinary proteins, 106
C57L/Lac, 85, 92
(C57L x A)F_1: mouse
adrenal cortical tumors, 179-180
fibrosarcoma, 211
neoplasms, 171
ovarian tumors, 195, 198, 200, 204
pituitary tumors, 181, 183, 187-191
(C57L x A/He)F_1, 182, 189, 200-201
(C57L x A/He)F_1J, 195, 204
C57L x RIII, 184
(C57L/J x A/HeJ)F_1, 189-190, 192
C58 (strain): mouse
alloantigens, 114, 116-118
biochemical variation, 87
characteristics, 24
complement factors, 103
genes, 24
hemangiomas, 174
H-2 haplotype & markers, 124
Ig heavy chain linkage groups, 110
leukemia antigens, 168
leukemia incidence, 169
lymphoma, 169
proteins, 104
C58/J
antigens, 167
biochemical variation, 81-84, 86-92
body weight, 45
genealogy, 16*
histocompatibility, 121
recombinant-inbred line progenitor, 37
reproduction, 45, 48
urinary proteins, 106
C58/Lw, 90, 106
(C58 x $C3H_f$)F_1, 171
C121 (strain): mouse, 183

D (strain): rat, 337
d [dilute] (gene): mouse
chromosome position, 38*
enzyme activity, 83
inbred strain genealogy, 17*
mutation, 71
pleiotropic behavior, 155
protein control, 77
strain distribution, 22, 24-26, 31, 83
d^l [dilute lethal], 31, 71, 101, 149
d^s [slight dilution], 71
d^+ [dense], 31
d^{15} [dilute-15], 71, 149
d [dilute] (gene): rat, 239, 273
DA (strain): mouse, 103-104, 111, 116-118, 126
DA/Hu, 17*, 121
DA/HuSn, 114, 128
DA (strain): rat
alloantigen chemistry, 320
alloantigenic alleles, 316-317
blood protein polymorphism, 309
characteristics, 243
coat color, 243, 272
congenic-inbred line progenitor, 256
enzyme polymorphisms, 310-312
eye color, 272
graft donor, 322-323, 329-330, 332

- immunogenetic characterization, 314
- immunoglobulin polymorphism, 307
- immunologically mediated diseases, 334-335
- immunological responsiveness, 318
- karyology, 261
- origin, 243
- transplant survival, 321-322, 330-333

DA/Cub, 310
DA.B3, 256
DA hybrids, 260, 321-323, 330, 332-334
da [dark] (gene): mouse, 38*, 71
Dancer (gene): mouse (see *Dc*)
Danforth's short tail (gene): mouse (see *Sd*)
Dao (gene): mouse, 22-23, 26
Dappled (gene): mouse (see Mo^{dp})
Dark (gene): mouse (see *da*)
Dark foot pads (gene): mouse (see *Dfp*)
Dark pink-eye (gene): mouse (see p^d)
db [diabetes] (gene): mouse, 31, 38*, 101, 155
db^{ad} [adipose], 101
DBA (strain): mouse
- adrenal cortical tumors, 179
- biochemical variation, 81, 83
- hemangiomas, 174-175, 177
- inbred strain genealogy, 17*
- intestinal tumors, 208-210
- ovarian tumors, 194-195
- pituitary tumor, 183
- subline genealogy, 20*

DBA/Ha, 86
DBA/HeA, 84, 89, 112
DBA/He$_f$A, 112
DBA/LiA, 20*, 84, 112
DBA/LiA$_f$, 20*, 112
DBA/LiHa, 81-82, 86
DBA/Ms, 20*
DBA/1
- alloantigens, 114, 116-118
- biochemical variation, 85, 87, 92
- characteristics, 24
- complement factors, 103
- congenic-inbred line progenitor, 125
- genes, 24
- Ig heavy chain linkage groups, 111
- leukemia antigens, 168
- proteins, 104

DBA/1Bg, 93
DBA/1Cbi, 20*
DBA/1FeHu, 20*
DBA/1J
- biochemical variation, 81-92
- body weight, 46
- centromeric heterochromatin, 44
- congenic-inbred line progenitor, 124
- genealogy, 16*
- *H-2* haplotype & markers, 124
- *If-1* allele, 112
- leukemia antigens, 168
- life-span, 46
- reproduction, 46
- urinary proteins, 106

DBA/1Jax, 20*
DBA/1Lac, 20*
DBA/1Sf, 20*
DBA/1Sn, 121
DBA/1Y, 20*
DBA/2
- alloantigens, 114, 116-118
- biochemical variation, 83-84, 91-92
- body weight, 46
- characteristics, 24
- complement factors, 103
- congenic-inbred line progenitor, 111, 124, 128
- genes, 24, 101, 105
- IgC_H alleles, 109
- Ig heavy chain linkage groups, 111
- intestinal tumors, 209
- leukemia antigens, 168
- pituitary tumor, 184
- proteins, 104
- reproduction, 46
- testicular tumors, 206 hn, 207

DBA/2Cbi, 20*
DBA/2Crgl, 20*
DBA/2De, 20*
DBA/2DeJ, 82-83, 86, 90
DBA/2DeJax, 20*
DBA/2$_e$BDe, 171
DBA/2$_e$De, 90, 106
DBA/2$_f$, 203
DBA/2Fs, 20*
DBA/2Gif, 20*
DBA/2Ha, 20*, 89
DBA/2He, 20*
DBA/2HeA, 20*
DBA/2Hf, 20*
DBA/2Hu, 20*
DBA/2Icrf, 20*
DBA/2J
- biochemical variation, 81-92
- body weight, 46
- cellular alloantigen, 113
- centromeric heterochromatin, 44
- congenic-inbred line progenitor, 127-128
- fibrosarcoma, 211
- genealogy, 16*
- *H-2* haplotype & markers, 124
- histocompatibility, 121
- *If-1* allele, 112
- life-span, 46
- lymphomas, 171
- mutant genes, 30-31
- ovarian tumors, 194, 202
- recombinant-inbred line progenitor, 37
- reproduction, 46, 48
- testicular teratomas, 208 fn

DBA/2Jax, 20*
DBA/2JPas, 37
DBA/2Ki, 20*
DBA/2Lac, 20*
DBA/2Lw, 20*
DBA/2Lw$_f$, 106
DBA/2Mai, 20*
DBA/2N, 20*, 106, 167
DBA/2Orl, 92
DBA/2Rl, 20*
DBA/2Sf, 20*
DBA/2Umc, 20*
DBA/2We, 20*
DBA/2Wy, 196

DBA/2Y, 20*
DBA congenics (*see* D2.GD)
DBA hybrids, 177, 180, 182-184, 193 hn, 196, 198
DBA_f, 158 hn, 159
DBA-MTV (virus): mouse, 159
Dc [dancer] (gene): mouse, 31, 39*, 59, 149, 155
DD (strain): mouse, 88, 104, 109-110
DD/He, 82, 87, 89-90, 106
DD/HeA, 81, 83-84, 86-89, 91-92
DD/I, 210
DD/N, 175
DD_f, 160
ddbbaa (stock): mouse, 16*
DDD (strain): mouse, 175
DDI (strain): mouse, 48, 87, 175
DDK (strain): mouse, 48, 84, 87-89, 91-92
DD-MTV (virus): mouse, 160
ddO (strain): mouse, 198, 204
DDT (chemical): mouse, 201
ddY/F (strain): mouse, 194, 200
DE (strain): mouse, 24, 104, 111
DE/Cv, 81, 88
DE/J, 17*, 81-92, 103
(DE x DBA)F_1, 180, 183
de [droopy-ear] (gene): mouse, 39*, 57
Deaf (gene): mouse (see v^{df})
Deafness: mouse, 149-153
Deafness (gene): mouse (see *dn*)
Death: mouse, 48, 53-54, 194-204 (*see also* Lethality; Mortality: mouse)
Death: rat, 327-328 (*see also* Mortality: rat)
Defecation: rat
open-field, 240, 244-245, 247, 249, 251-252, 352, 358
stress response, 360-361
Degeneration, motor neuron: mouse, 67
Delayed-splotch (gene): mouse (see Sp^d)
Delestrogen: mouse, 186
Delmadinone acetate: mouse, 187
Den [denuded] (gene): mouse, 31, 74
Denervation, progressive motor: mouse, 67
Dense (gene): mouse (see d^+)
Dental caries: rat, 236, 243
Dental malalignment: rat, 291
Denuded (gene): mouse (see *Den*)
Denver-type nomenclature: mouse, 14
dep [depilated] (gene): mouse, 38*, 74
Depilated (gene): mouse (see *dep*)
Dermal fibroma: rat, 295
Dervish (gene): mouse (see *dv*)
DES (chemical): mouse, 183-185, 187, 190-192, 206-207
Developmental stages, prenatal: laboratory mammals, 50-53
Dextran: mouse, 112
Dextran: rat, 252, 318
Dey [Dickie's small eye] (gene): mouse, 58
df [Ames dwarf] (gene): mouse, 39*, 68, 101
Dfp [dark foot pads] (gene): mouse, 71
Dh [dominant hemimelia] (gene): mouse, 38*, 57, 60 fn, 68
DHA adenocarcinoma: rat, 303
DHS/Mk (strain): mouse, 82, 87, 89
di [duplicate incisors] (gene): mouse, 60
di [diabetes insipidus] (gene): rat, 237, 257, 309-310
Diabetes (gene): mouse (see *db*)
Diabetes: rat, 311
Diabetes, alloxan: rat, 250
Diabetes, hereditary hypothalamic: rat, 339
Diabetes insipidus (strain): rat, 309-310, 339 hn, 339
Diabetes insipidus (gene): rat (see *di*: rat)
Diabetes insipidus, nephrogenic: mouse, 27
Diabetes mellitus: mouse, 25
Diabetes mutation: mouse, 23
Diaphragm abnormalities: mouse, 60 fn
Diastolic pressure, ventricular: rat, 286
Dibenz[*a,h*]anthracene: mouse, 174
1,2:5,6-Dibenzanthracene: mouse, 177, 201, 209
Dickie's small eye (gene): mouse (see *Dey*)
Diestrus: rat, 283, 343
Diet: mouse tumor, 162-165, 188
Diet, fat: rat, 246, 249
Diethyl ether: mouse, 175
Diethylstilbestrol: mouse, 183-185, 187, 206 hn, 206-207 (*see also* DES)
Diethylstilbestrol: rat, 239, 243-244
Differential locus: congenic lines, 6
Digestive system: mouse, 68
Digestive system: rat, 289
Digestive tract: rat, 311
Dihydrotestosterone: rat, 341 fn-342 fn, 346 fn
Dilute (gene): mouse (see *d*: mouse)
Dilute (gene): rat (see *d*: rat)
Dilute-lethal (gene): mouse (see d^l)
Dilute-15 (gene): mouse (see d^{15})
Dilution-Peru (gene): mouse (see *dp*)
p-Dimethylaminoazobenzene: rat, 304
p-Dimethylaminobenzene-1-azo-2-naphthalene: mouse, 174
1-(4-Dimethylaminobenzylidene)indene: mouse, 201
4-(*p*-Dimethylaminostyryl)quinoline: mouse, 201
7,12-Dimethylbenzanthracene: mouse, 179 fn
7,12-Dimethylbenz[*a*]anthracene: rat, 303 fn-304 fn, 335 fn
9,10-Dimethyl-1,2-benzanthracene: mouse, 173-174, 176, 179, 193 hn, 201-202
9,10-Dimethyl-1,2-benzanthracene: rat, 262, 267, 303-304, 335
Dimethylhydrazine: mouse, 177, 209
1,2-Dimethylhydrazine: rat, 303
1,2-Dimethylhydrazine dihydrochloride: mouse, 176
Dimethylnitrosamine: mouse, 174, 176, 203
Diminutive (gene): mouse (see *dm*)
2,3-Dinitrophenol: rat, 238 hn, 247
Dinitrophenyl-bovine gamma globulin: mouse, 143
2,4-Dinitrophenyl-bovine gamma globulin: rat, 318
Dinitrophenyl bovine insulin: mouse response, 143
Dinitrophenyl-ovomucoid: mouse, 143
Dip-1 [dipeptidase] (gene): mouse
biochemical locus, 96*
chromosome position, 38*
phenotypic variation, 83
polymorphism, 221
protein control, 77, 98
strain distribution, 21-28, 83
Dipeptidase: mouse, 77, 226-227
Dipeptidase (gene): mouse (see *Dip-1*)
2,3-Diphosphoglycerate: rat, 305
Diplococcus pneumoniae: mouse, 68 fn
Diploid chromosome number: rat, 233
Directional caracul (gene): mouse (see Ca^d)
Disease (*see* specific disease)

Disease-free stocks: rat, 237
Disk degeneration: rat, 289-290
Disk electrophoresis: rat, 310
Disorganization (gene): mouse (see *Ds*)
Disoriented (gene): mouse (see *Do*)
dl [downless] (gene): mouse, 38*, 74
DLS (strain): mouse, 48
    DLS/Le, 31, 33
DM (strain): mouse, 88
dm [diminutive] (gene): mouse, 31, 38*, 55, 65
DMBA (chemical): mouse, 201-202, 204
DMC (strain): mouse, 48
dn [deafness] (gene): mouse, 150
DNA, denatured: mouse, 113
DNP hapten: rat, 247
Do [disoriented] (gene): mouse, 150, 155
Dog chow diet: mouse, 163
Dominant hemimelia (gene): mouse (see *Dh*)
Dominant mutation: mouse nomenclature rules, 12
Dominant reduced ear (gene): mouse (see *Dre*)
Dominant spotting (gene): mouse (see *W*)
Dominant spotting-x (gene): mouse (see W^x)
Dominant syndactylism (gene): mouse (see *Dsy*)
DONRYU (strain): rat, 244, 261, 266, 273, 279, 303 (*see also* YOS)
DOPA decarboxylase: rat, 280 fn
Dopamine: rat, 362
Dopamine β-hydroxylase: rat, 251, 310, 360 fn, 362 fn
Dopamine β-monooxygenase: rat, 251, 310, 360, 362
Dorsal kyphosis: mouse, 122
Dorsal (spinal) roots: mouse, 67, 150
Double toe (gene): mouse, 58
Downless (gene): mouse (see *dl*)
dp [dilution-Peru] (gene): mouse, 39*, 71
dr [dreher] (gene): mouse, 38*, 58, 150
Dre [dominant reduced ear] (gene): mouse, 38*, 59
D region, *H-2* gene complex: mouse
        associated function, 135*
        gene products, 147 hn
        genetic fine structure map, 134*
        genetic traits, 136
        H-2 haplotype specificities, 137 hn, 138-140
        mutant haplotypes, 129 hn, 129
        recombinant haplotypes, 126-127, 135*
        serological markers, 123 hn, 123-124
Dreher (gene): mouse (see *dr*)
Droopy-ear (gene): mouse (see *de*)
Ds [disorganization] (gene): mouse, 39*, 57
Dsy [dominant syndactylism] (gene): mouse, 56
dt [dystonia musculorum] (gene): mouse, 31, 38*, 66, 150
DU/Ei (strain): mouse, 87
du [ducky] (gene): mouse, 31, 38*, 101, 150, 155
    du^{td} [torpid], 101, 150
Ducky (gene): mouse (see *du*)
Dunning hepatoma: rat, 244
Dunning leukemia: rat, 244
Duplicated ectopic adrenal: mouse, 60 fn
Duplicate incisors (gene): mouse (see *di*: mouse)
Dusty (gene): mouse (see sl^{du})
dv [dervish] (gene): mouse, 155
DW (strain): mouse, 24, 88, 104
    DW/J, 31-32, 82-91, 171
    DW/Orl, 92
dw [dwarf] (gene): mouse
        chromosome position, 39*
        endocrine effects, 101
        immune system development, 68
        mammary tumorigenesis, 161
        strain distribution, 24, 31, 36
dw-1 [dwarf-1] (gene): rat, 257
dw-2 [dwarf-2] (gene): rat, 257
Dwarf (gene): mouse (see *dw*)
Dwarf-1 (gene): rat (see *dw-1*)
Dwarf-2 (gene): rat (see *dw-2*)
Dwarfism: mouse, 24, 60, 65
dx [anaphylactoid reaction] (gene): rat, 252, 257
dy [dystrophia muscularis] (gene): mouse, 31, 36, 38*, 67, 150
Dysplasia, adrenal: mouse, 183
Dysrhaphic disorders: mouse, 149-151
Dystonia musculorum (gene): mouse (see *dt*)
Dystrophia muscularis (gene): mouse (see *dy*)
D1 (strain): mouse (*see* DBA/1)
    D1 congenics, 124
D2 (strain): mouse (*see* DBA/2)
    D2.GD, 126, 128, 146
D23 hepatoma: rat, 304

E/Gw (strain): mouse, 92
E or *e* (gene): mouse
    e [recessive yellow], 31, 38*, 72
    E^{so} [sombre], 31, 42, 72
    E^{tob} [tobacco darkening], 72
e [yellow] (gene): rat, 273
Ea-1 [erythrocyte antigen-1] (gene): mouse, 22-24, 28, 38*, 113
Ea-2 [erythrocyte antigen-2] (gene): mouse, 21-28, 114-115
Ea-3 [erythrocyte antigen-3] (gene): mouse, 22-23, 114-115
Ea-4 [erythrocyte antigen-4] (gene): mouse, 21-28, 114-115
Ea-5 [erythrocyte antigen-5] (gene): mouse, 21, 23-28, 114
Ea-6 [erythrocyte antigen-6] (gene): mouse, 21-26, 28, 38*, 114-115
Ea-7 [erythrocyte antigen-7] (gene): mouse, 21-28, 114-115
Ear defects: mouse, 59, 150-151
Ear duct tumor: rat, 239
Earlier X-zone degeneration (gene): mouse (see *Ex*)
Ear malformed (gene): mouse (see *Em*)
Ear size: mouse, 214
E-B+ (strain): mouse, 106
eb [eye-blebs] (gene): mouse, 38*, 57, 150
EBT/Ha (strain): mouse, 86
ecl [epistatic circling of C57L/J] (gene): mouse, 155
Ecotropic viruses: mouse, 227 hn, 228-229, 229 hn, 230-231
Ecotropic virus replication [REC-1]: mouse, 96*, 98
ecs [epistatic circling of SWR/J] (gene): mouse, 155
Ectromelia virus: mouse, 21, 23
Edema: mouse, 60
Eg [endoplasmic β-glucuronidase] (gene): mouse, 38*, 77, 83
Egasyn: mouse, 77

Egg-sperm contact time: laboratory mammals, 52 hn
Egg-supplemented diet: mouse, 188
Egg transfer: mouse nomenclature, 11
Eh [hairy ears] (gene): mouse, 31, 39*, 74, 122, 123*
Ehrlich ascites tumor: mouse, 165
EI (strain): mouse, 181
Eimeria miyairii: rat, 338
E. nieschulzi: rat, 338
El [epilepsy] (gene): mouse, 150
Elastin: rat, 281
Electrophoresis, starch gel: rat, 309 hn, 310-312
Electrophoretic polymorphisms: mouse, 220-225
Electrophoretic variation: rat, 257-258
Em [ear malformed] (gene): mouse, 59
Ema [electrophoretic mobility & agglutinability] (gene): mouse, 21-24, 26-28
Embryo developmental stages: laboratory mammals, 50-53
Embryo mortality: mouse, 48
Embryo mortality: rat, 238
Embryonic cell surface antigen [*T*] (protein): mouse, 97*, 99
Embryonic ear malformations: mouse, 151
Embryonic tissue karyology: rat, 263-265, 266-267
Embryo preservation by freezing: mouse nomenclature, 12
Emotionality: rat, 236
Encephalitic viruses: mouse, 22
Encephalitis, experimental allergic: mouse, 22
Encephalomyelitis, experimental allergic: rat, 239-240, 242, 246, 249, 252, 334
Endocardial disease: rat, 252
Endocardial lesions: rat, 289-291
Endocrine dysfunction: rat, 348
Endocrine effects: mouse, 101-102
Endocrine system prenatal development: mouse, 50-52
Endocrinology, reproductive: rat, 340-347
Endometrial polyps: rat, 298 fn
Endometrial stromal tumors: rat, 298-299
Endoplasmic β-glucuronidase (gene): mouse (see *Eg*)
Endoplasmic reticulum, liver: rat, 282
Endotoxemia: mouse, 69
Enolase [ENO-1] (protein): mouse, 96*, 98, 226
Enovid: mouse, 198 fn
Entamoeba histolytica: rat, 239, 249
Environment exploration: rat, 357-358
Enzyme activity: mouse nomenclature rules, 14
Enzyme polymorphisms: rat, 309-312
Enzymes: mouse, 77-93, 96*-97*, 98-99
Enzymes: rat, 280-282, 284-288, 309 hn, 360, 362
Eo [eye opacity] (gene): mouse, 58
Eosinophilic bodies: rat, 294
ep [pale ears] (gene): mouse, 31, 39*, 72, 122, 123*
Epidermal cells: mouse, 114
Epidermoid carcinomas: rat, 302
Epididymis: rat, 311
Epilepsy (gene): mouse (see *El*)
Epinephrine: mouse, 101
Epistatic circling of C57L/J (gene): mouse (see *ecl*)
Epistatic circling of SWR/J (gene): mouse (see *ecs*)
Epithelial cell karyology: rat, 266
Epithelial mesotheliomas, atriocaval: rat, 249
Epithelial tumors: rat, 242, 301
Er [repeated epilation] (gene): mouse, 31, 63, 74
E region, *H-2* gene complex: mouse, 126-127, 134*
Erp-1 [erythrocytic protein-1] (gene): mouse, 23, 27, 77, 83
Erythrocyte alloantigen 1: mouse, 113
Erythrocyte alloantigen 2.1: mouse, 143
Erythrocyte count: mouse, 23-24, 27
Erythrocyte membrane molecules: mouse, 135*
Erythrocytes: mouse, 65-66, 77-79, 113-115, 136, 139 hn, 147
Erythrocytes: rat, 311-312, 315
Erythrocytes, sheep: mouse response, 68-69
Erythrocytes, sheep: rat response, 243-244
Erythrocytic protein-1 (gene): mouse (see *Erp-1*)
Erythroleukemia: mouse, 166
Erythroleukemia: rat, 303
Erythroleukemia virus, Friend: rat, 303
Erythroleukemia virus, Rauscher: rat, 303
Erythropoiesis, extramedullary: mouse, 26
Erythropoietic stem cells: mouse, 66
Esc [esterase concentration] (gene): rat, 257
Escape response threshold: rat, 361
Escherichia coli: rat, 337
Esophagus: rat, 304
Esterase: mouse, 80, 93, 226
Esterase (genes): mouse
- *Es-1* [esterase-1]
  - biochemical locus, 38*, 96*
  - biochemical variation, 83-84, 104
  - electrophoretic polymorphism, 221
  - histocompatibility linkage, 122, 123*
  - phenotypic expression, 83-84
  - protein control, 78, 98
  - strain distribution, 21-28, 83-84, 104
- *Es-2* [esterase-2]
  - biochemical locus, 38*, 96*
  - biochemical variation, 104
  - electrophoretic polymorphism, 221, 225-226
  - phenotypic expression, 84
  - protein control, 78, 98
  - strain distribution, 21-28, 84, 104
- *Es-3* [esterase-3]
  - biochemical locus, 39*, 97*
  - electrophoretic polymorphism, 221
  - phenotypic expression, 84
  - protein control, 78, 99
  - strain distribution, 21-28, 84
- *Es-5* [esterase-5]
  - biochemical locus, 38*, 96*
  - electrophoretic polymorphism, 221-222
  - phenotypic expression, 84
  - protein control, 78, 98
  - strain distribution, 21-27, 84
- *Es-6* [esterase-6], 21-28, 38*, 78, 84, 96*, 98
- *Es-7* [esterase-7], 38*, 78, 84, 96*, 98, 225
- *Es-8* [esterase-8]
  - biochemical locus, 38*, 96*
  - electrophoretic polymorphism, 225
  - phenotypic expression, 84
  - protein control, 78, 98
  - strain distribution, 21-28, 84
- *Es-9* [esterase-9], 27, 78, 85, 96*, 98
- *Es-10* [esterase-10]
  - biochemical locus, 39*, 97*
  - electrophoretic polymorphism, 225-226
  - phenotypic expression, 85
  - protein control, 78, 99
  - strain distribution, 21-28, 85
- *Es-11* [esterase-11], 78, 85, 96*, 98

Es-12 [esterase-12], 78, 85
Es-13 [esterase-13], 96*, 98
Esterase (genes): rat
Es-1 [esterase-1]
alloantigenic systems, 315
effect, 257
linkage group, 259 hn, 259, 305
polymorphism, 310, 310 fn-311 fn
Es-2 [esterase-2], 257, 259, 305, 310-311, 310 fn-311 fn, 315
Es-3 [esterase-3], 257, 259 hn, 259, 305, 310 fn-311 fn, 311
Es-4 [esterase-4], 257, 259 hn, 259, 305, 311
Es-5 [esterase-5], 257, 259 hn, 305, 311
Es-12 [esterase-12], 257
Esterase, sex-dependent: rat, 259 hn
Esterase concentration (gene): rat (see *Esc*)
Esterase polymorphism: rat, 309 hn, 310-311
Estradiol: mouse, 187
Estradiol: rat, 341-347
Estradiol benzoate: mouse, 185-186, 206 hn, 206
Estradiol dipropionate: mouse, 186
Estradiol 17-valerate: mouse, 186
Estrogen(s): mouse, 28, 183-187, 190, 198, 206-207
Estrogen(s): rat, 240, 253, 304
Estrone: mouse, 186-187, 191
Estrus: mouse, 47-49
Estrus: rat, 283, 343-344
Ethanol: mouse, 23-24
Ethylenediaminetetraacetic acid: rat, 266 hn, 266
Ethylenediaminetetraacetic acid tetrasodium salt: rat, 267
Ethylene glycol: mouse, 175, 204
Ethylene oxide products: mouse, 27
Ethylhydrazine hydrochloride: mouse, 176, 202
Ethylmorphine metabolism: rat, 242-244, 246, 248
Ethylnitrosoguanidine: mouse, 209-210
Ethylnitrosourea: rat, 303-304
1-Ethyl-1-nitrosourea: mouse, 202
Ethynodiol diacetate: mouse, 187
Ethynylestradiol: mouse, 187
Ex [earlier X-zone degeneration] (gene): mouse, 38*, 101
Exa [exploratory activity] (gene): mouse, 22-23, 154
Exencephaly (gene): mouse (see *xn*)
Exploratory activity (gene): mouse (see *Exa*)
Extent of zona glomerulosa (gene): mouse (see *Ezg*)
Extracellular water: rat, 280, 285
Extrahepatocyte space, intralobular: rat, 282
Extra-toes (gene): mouse (see *Xt*)
Extreme dilution (gene): mouse (see c^e)
Extreme non-agouti (gene): mouse (see a^e)
Exudate cells, peritoneal: mouse, 114, 117
ey-1 [eyeless-1] (gene): mouse, 59, 150
ey-2 [eyeless-2] (gene): mouse, 59
Eye: rat, 257-258, 274-277, 289
Eyeball muscle: mouse, 60 fn
Eye-blebs (gene): mouse (see *eb*)
Eye color: mouse, 70 hn
Eye color: rat, 273
Eye defects: mouse, 58-59, 149-152
Eye-ear reduction (gene): mouse (see *Ie*)
Eyeless (genes): mouse (see *ey-1; ey-2*)
Eyeless white (gene): mouse (see mi^{ew})
Eyelids-open-at-birth (gene): mouse (see *o*)
Eye opacity (gene): mouse (see *Eo*)
Eye size: mouse, 212, 214
Ezg [extent of zona glomerulosa] (gene): mouse, 101
E3 (strain): rat, 244, 272-273, 307, 316

F (strain): mouse, 24, 90, 106
F/St
biochemical variation, 81-82, 88-89
cellular alloantigen, 114
complement factors, 103
congenic-inbred line progenitor, 115, 125, 128
genealogy, 16*
leukemia antigens, 168
F/StAo, 125
F/StLac, 85
f [flexed-tail] (gene): mouse, 31, 39*, 42, 57, 65, 72
f [fawn] (gene): rat, 259, 273
fa [fatty] (gene): rat, 237, 257, 312
Falling: mouse, 152
Fat [fat] (gene): mouse, 101
Fat, body: mouse, 28
Fat, body: rat, 245
Fat cells: rat, 312
Fat deposition, acute arterial: rat, 348
Fat diet: rat, 246, 249
Fatty (strain): rat, 309 hn, 312, 339
Fatty (gene): rat (see *fa*)
Fatty liver: rat, 244
Fawn (gene): rat (see *f*: rat)
F.B congenics: rat, 256
FBJ (virus): mouse, 166 fn
fd [fur deficient] (gene): mouse, 31, 38*, 74
Fecal boli: rat, 358
Feces: mouse, 114
FEDEX-AVN sarcoma: rat, 240
Feeding behavior: mouse, 155 hn, 155-156
Feline RNA virus: rat, 303
Femur bone marrow karyology: rat, 267
Feral mouse, 85 (*see also* Wild mouse)
Ferridextron: rat, 240
Ferroxidase: rat, 308
Fertility: mouse, 21, 23, 28, 47-49
Fertility: rat, 239, 241-242, 246
Fetal anemia: mouse, 24
Fetal hematoma (gene): mouse (see *fh*)
Fetal liver: mouse, 65
Fetal resorption: rat, 239
Fetal tissue karyology: rat, 262, 267
fh [fetal hematoma] (gene): mouse, 57
fi [fidget] (gene): mouse
chromosomal aberration, 42
chromosome position, 38*
congenital malformations, 57, 60 fn
neurologic mutation, 150
pleiotropic behavior, 155
strain distribution, 31
Fiber denervation, muscle: mouse, 67
Fibers, muscle: rat, 279, 286-287
Fibroadenoma: rat
mammary, 239, 242, 252, 295-296, 296 fn, 298-300
subcutaneous, 244
Fibroangioma: mouse, 176
Fibroblast karyology: rat, 262, 266-267
Fibromas: mouse, 211 fn
Fibromas: rat, 244, 295
Fibrosarcoma: mouse, 166, 188, 195, 210-211

Fibrosarcomas: rat, 244, 246
Fibrosis: rat, 290-294, 350*
Fidget (gene): mouse (see *fi*)
Fighting, shock-induced: rat, 360
Fischer (strain): rat (*see* F344)
Fischer 344 (strain): rat (*see* F344)
  Fischer 344/Mai, 256
Fk [fleck] (gene): mouse, 63, 72
Fkl [freckled] (gene): mouse, 39*, 72
FL (strain): mouse, 90, 106, 124
  FL/1Re
    anemia, 24, 65-66
    biochemical variation, 81-83, 85-88
    characteristics, 24
    genes, 24, 31, 34
    reproduction, 48
  FL/2, 103
  FL/2Re, 83, 85-87, 114, 116-118, 121
  FL/4Re, 82-83, 85
  FL/4ReII, 86
fl [flipper-arm] (gene): mouse, 56
Flaky tail (gene): mouse (see *ft*)
fld [forelimb deformity] (gene): mouse, 56
Fleck (gene): mouse (see *Fk*)
Flexed-tail (gene): mouse (see *f*: mouse)
Flipper-arm (gene): mouse (see *fl*)
2-Fluorenamine: rat, 304
N-2-Fluorenylacetamide: rat, 248, 253, 304
N-2-Fluorenylphthalamic acid: rat, 303
N-(4′-Fluoro-4-biphenyl)acetamide: rat, 303
Fn [funny tail] (gene): mouse, 55
Follicles, cystic: rat, 283
Follicles, polyovular: mouse, 24, 48
Follicle-stimulating hormone: mouse, 101
Follicle-stimulating hormone: rat, 340 hn, 340-347
Follicular structures, ovarian: mouse, 102
Food contingent bar presses: rat, 359
Foodhole competition: rat, 361
Foot pads: mouse, 212
For-1 [formamidase-1] (gene): mouse, 78, 85
For-5 [formamidase] (gene): mouse, 97*, 99
Foramina, incisive: mouse, 212, 214, 217-219
Foramina, posterior palatine: mouse, 217
Forelimb deformity (gene): mouse (see *fld*)
Forequarter motor denervation: mouse, 67
Formamidase (gene): mouse (see *For-5*)
Formamidase-1 (gene): mouse (see *For-1*)
Foster nursing: mouse, 11, 158 hn
fr [frizzy] (gene): mouse, 38*, 74
Freckled (gene): mouse (see *Fkl*)
Freezing, preservation by: mouse nomenclature rules, 12
Friend erythroleukemia virus: rat, 303
Friend lymphatic leukemia viruses: mouse, 165-166
Friend virus leukemogenesis: mouse, 168 fn
Frizzy (gene): mouse (see *fr*)
Frl [fur-loss] (gene): mouse, 75
Frowzy (gene): mouse (see fz^{fy})
β-Fructofuranosidase: rat, 311
Fructose-bisphosphate aldolase—liver: mouse, 220 hn
FS/Ei (strain): mouse, 82-83, 86
fs [furless] (gene): mouse, 31, 39*, 75
FSB/Gn (strain): mouse, 87
  FSB/GnEi, 31
ft [flaky tail] (gene): mouse, 39*, 59, 75
Fu [fused] (gene): mouse, 31, 39*, 57, 122, 123*, 150
  Fu^{ki} [kinky], 57, 150
Fum [fumarate hydratase] (gene): rat, 305, 311
Fumarase: mouse, 220 hn
Fumarase (gene): rat (see *Fum*)
Fumarate hydratase: mouse, 220 hn
Fumarate hydratase (gene): rat (see *Fum*)
Funny tail (gene): mouse (see *Fn*)
Fur: mouse, 212, 214, 217, 219
Fur deficient (gene): mouse (see *fd*)
Furless (gene): mouse (see *fs*)
Fur-loss (gene): mouse (see *Frl*)
Fused (gene): mouse (see *Fu*)
Fused phalanges (gene): mouse (see sy^{fp})
Fuzzy (gene): mouse (see *fz*: mouse)
Fuzzy (gene): rat (see *fz*: rat)
Fuzzy tail (gene): mouse (see *fzt*)
Fv-1 [Friend virus susceptibility-1] (gene): mouse, 21-28, 38*, 96*, 98, 167, 227 hn
Fv-2 [Friend virus susceptibility-2] (gene): mouse, 21-28, 38*, 96*, 98, 168
FW49/Biberach (strain): rat, 289
FZ (strain): mouse, 111
fz [fuzzy] (gene): mouse, 31, 38*, 75
  fz^{fy} [frowzy], 75
fz [fuzzy] (gene): rat, 257, 259
fzt [fuzzy tail] (gene): mouse, 75
F_1 hybrids: laboratory animals, 2, 3
F_1 hybrids: mouse, 11, 143 fn, 193 hn, 206 hn (*see also* specific hybrid strain)
F_1 hybrids: rat, 337 (*see also* specific hybrid strain)
F9 antigen: mouse, 148
F-45 (strain): rat, 307
F344 (strain): rat
    aging, 280-281, 289-290, 295-296
    alloantigenic alleles, 316-317
    alloantigens, 319-320
    behavioral research, 351
    blood protein polymorphism, 309
    characteristics, 244, 354-360
    chromosome polymorphism, 260
    coat color, 244, 272-273
    congenic-inbred line progenitor, 256
    enzyme polymorphism, 309-312
    eye color, 244
    immunogenetic characterization, 314
    immunoglobulin polymorphism, 307
    immunologically mediated diseases, 334-335
    immunological responsiveness, 317-319
    infectious disease susceptibility, 338
    karyology, 261-262, 266
    longevity, 278
    origin, 234, 244
    transplant survival, 321-326, 329, 331
    tumors, 296, 298-299, 303-304
  F344/Mai, 256
  F344/N, 299, 310-311
  F344 congenics (*see* F.B congenics)
  F344 hybrids, 323, 329

G/Cpb (strain): rat, 245, 273
G (gene): mouse, 78, 86-87 (see also *Gus*: mouse)
Ga [graying with age] (gene): mouse, 72
Gait, abnormal: mouse, 149-150, 152-153
Galactokinase (gene): mouse (see *Glk*)
Galactokinase (protein): mouse, 78, 99

α-Galactosidase (gene): mouse (see *Ags*)
β-Galactosidase (protein): mouse, 77, 98-99
β-Galactosidase activity (gene): mouse (see *Bgs*)
β-Galactosidase—electrophoresis (gene): mouse (see *Bge*)
β-Galactosidase, temporal (gene): mouse (see *Bgt*)
Ganglia: mouse, 67, 150-152
Gaping lids (gene): mouse (see *gp*)
Gamma *2b* heavy chains, immunoglobulin: rat, 306-308
γ-paraproteinemia: mouse, 27
γ rays: mouse, 189, 200-201
Gastrointestinal enzymes: rat, 281
Gastrointestinal tract neurons: mouse, 152
Gastrointestinal tract prenatal development: mouse, 50-52
Gastroschisis: mouse, 60 fn
Gdc-1 [NAD-α-glycerol phosphate dehydrogenase] (gene): mouse
  biochemical locus, 39*, 97*
  electrophoretic polymorphism, 222, 225
  phenotypic expression, 85
  protein control, 78, 99
  strain distribution, 21-27, 85
Gdr-1 [glucose-6-phosphate dehydrogenase regulator] (gene): mouse, 78, 85
Gdr-2 [glucose-6-phosphate dehydrogenase regulator] (gene): mouse, 21-24, 26-28, 78, 85
Gene (*see* specific gene)
Genealogy: mouse, 16*-20*
Gene nomenclature rules: mouse, 12-14
Generations inbred: mouse, 21-37
Generations inbred: mouse nomenclature rules, 10-12
Generations inbred: rat, 238 hn, 238-253, 313-314
Gene symbols: mouse nomenclature rules, 9, 12, 14
Genetically hypertensive rat, 348 (*see also* GH)
Genetically obese Zucker rat, 339
Genetic fine structure map: mouse, 134*
Genetic fixation: laboratory animals, 5*
Genetic hypertension (strain): rat, 245 (*see also* GH)
Genetic variants affecting enzyme activity: mouse nomenclature rules, 14
Genome fixation: laboratory animals, 2
Genome purity: laboratory animals, 5, 5*
Germ cell karyology: rat, 261 hn, 263-265
Germ cells: mouse, 102
Germfree strains: mouse, 211 fn
Germfree strains: rat, 237, 242
Germinal proviruses: mouse, 157 hn
Gestational death: mouse, 49, 53-55
GH (strain): rat, 236, 245, 273, 348
GHA (strain): rat, 245, 273
Giemsa banding karyotype: rat, 269*-270*
Gix (gene): mouse, 116
G_{IX} antigen: mouse, 98, 168 fn
gl [gray-lethal] (gene): mouse, 38*, 72, 101
Gl-1 [plasma protein] (gene): rat, 257, 259, 305, 309
Gland (*see* specific gland)
Glia, abnormal: mouse, 149
Glial cell tumor: rat, 301
Gliosis, spinal: mouse, 228 hn
Glk [galactokinase] (gene): mouse, 39*, 78, 85, 97*, 99
Glo-1 (gene): mouse, 97*, 99
α-Globulin: mouse, 105 hn
α-Globulins: rat, 280
β-Globulin: rat, 280
γ-Globulin: rat, 280
Glomerular lesions: mouse, 26
Glomeruli: rat, 248, 286, 293 fn, 293-294
Glomerulonephritis: mouse, 69
Glomerulonephritis: rat, 291, 335
Glomerulosclerosis: mouse, 26
Glucose-6-phosphatase: rat, 285, 288
Glucose-6-phosphate dehydrogenase: rat, 283, 311
Glucose-6-phosphate dehydrogenase [G6PD] (protein): mouse, 78, 97*, 99, 226-227
Glucose-6-phosphate dehydrogenase, autosomal (gene): mouse (see *Gpd-1*)
Glucose-6-phosphate dehydrogenase, sex-linked (gene): mouse (see *Gpd-2*)
Glucose-6-phosphate dehydrogenase regulator (genes): mouse (see *Gdr-1; Gdr-2*)
Glucose phosphate isomerase (gene): mouse (see *Gpi-1*)
Glucosephosphate isomerase [GPI] (protein): mouse, 78, 227
α-Glucosidase: rat, 285
β-Glucuronidase (gene): mouse (see *Gus*)
β-Glucuronidase: mouse, 21, 28, 47, 78-79
β-Glucuronidase, temporal (gene): mouse (see *Gut*)
β-Glucuronidase regulation (gene): mouse (see *Gur*)
Glucuronyl transferase: rat, 312
Glutamate oxaloacetate transaminase: mouse, 226
Glutamate oxaloacetate transaminase (mitochondrial)-2 (gene): mouse (see *Got-2*)
Glutamate oxaloacetate transaminase (soluble)-1 (gene): mouse (see *Got-1*)
Glutamate-pyruvate transaminase: mouse, 78, 99
Glutamate pyruvate transaminase (gene): mouse (see *Gpt-1*)
Glutathione reductase (gene): mouse (see *Gr-1*)
Glutathione reductase (NAD(P)H): mouse, 78
Glyceraldehyde-phosphate dehydrogenase (gene): rat (see *G-3-pdh*)
Glycerol kinase: rat, 312
Glycerol-3-phosphate dehydrogenase (NAD^+): mouse, 78, 99, 227
Glycerol-3-phosphate dehydrogenase (NAD^+) gene: rat (see *α-Gpdh*)
Glyceryl monostearate: mouse, 188
Glycopeptides: rat, 320, 335
Glycoprotein: mouse, 147 hn, 147-148
Glyoxalase: mouse, 99
gm [gunmetal] (gene): mouse, 31, 39*, 72
Gnotobiotic stock: rat, 237
go [angora] (gene): mouse, 31, 38*, 75, 122, 123*
Golgi: rat, 282
Gonadectomy: mouse
  adrenal adenomas, 26
  adrenal cortical tumors, 179-180
  adrenocortical carcinoma, 23
  ovarian tumors, 193 hn, 198, 204
  pituitary tumors, 182-183, 190-191
Gonadotropins: mouse, 101-102
Gonadotropins: rat, 340 hn, 340-347
Gonad prenatal development: man & laboratory animals, 53
Gonad prenatal development: mouse, 51, 53
Got-1 [glutamate oxaloacetate transaminase (soluble)-1] (gene): mouse, 39*, 78, 85, 97*, 99, 225
Got-2 [glutamate oxaloacetate transaminase (mitochondrial)-2] (gene): mouse, 21-28, 38*, 78, 85, 96*, 98
Gowans albino (strain): rat, 307
gp [gaping lids] (gene): mouse, 59

Gpd-1 [glucose-6-phosphate dehydrogenase, autosomal] (gene): mouse
biochemical locus, 38*, 96*
electrophoretic polymorphism, 222, 225
histocompatibility linkage, 122, 123*
phenotypic expression, 86
protein control, 78, 98
strain distribution, 21-28, 86
Gpd-2 [glucose-6-phosphate dehydrogenase, sex-linked] (gene): mouse, 222
α-Gpdh [glycerol-3-phosphate dehydrogenase (NAD^+)] (gene): rat, 305
G-3-pdh [glyceraldehyde-phosphate dehydrogenase] (gene): rat, 305
Gpi-1 [glucose phosphate isomerase] (gene): mouse
biochemical locus, 38*, 96*
electrophoretic polymorphism, 222
histocompatibility linkage, 122, 123*
phenotypic expression, 86
protein control, 78, 98
strain distribution, 21-28, 86
Gpt-1 [glutamate pyruvate transaminase] (gene): mouse
biochemical locus, 39*, 97*
electrophoretic polymorphism, 225-226
phenotypic expression, 86
protein control, 78, 99
strain distribution, 21-28, 86
GR (strain): mouse, 84, 157 hn, 158
GR/A, 81-82, 89, 116-118
GR/N, 81-82, 88-89
GR/Rl, 82, 87-88
gr [grizzled] (gene): mouse, 31, 38*, 43-44, 57, 60 fn, 72
Gr-1 [glutathione reductase] (gene): mouse, 21-28, 38*, 78, 86, 96*, 98
Gracile tract: rat, 294
Graft rejection: mouse, 68-69, 129 hn, 129-130
Graft rejection: rat, 283
Grafts: mouse, 181-182, 190-191, 193 hn, 204 (*see also* Transplants)
Graft-versus-host reaction: mouse, 68 hn, 68-69, 129 hn, 129-130, 135*
Graft-versus-host reaction: rat, 238 hn, 242
Graft-versus-host reaction—stimulating antigens: mouse, 136 hn, 136, 141 hn, 141
Granule cells, cerebellar: mouse, 152-153
Granules, lysosomal: mouse, 68
Granulocyte chemotaxis: mouse, 68
Granulocytes: mouse, 21
Granulosa cells: rat, 296
Granulosa cell tumors: mouse, 193 hn-194 hn, 194-196, 197 hn, 198, 199 hn, 199-204
Graying with age (gene): mouse (see *Ga*)
Gray-lethal (gene): mouse (see *gl*)
Gray-lethal linkage testing stock: mouse, 125, 128
Greasy (gene): mouse (see *Gs*)
G region, *H-2* gene complex: mouse
associated function, 135*
genetic fine structure map, 134*
genetic trait, 136
recombinant haplotypes, 126-127, 135*
serological markers, 123 hn, 124
specificities by haplotype, 137 hn, 138 fn
Grizzle-belly (gene): mouse (see Sl^{gb})
Grizzled (gene): mouse (see *gr*)
GR-MTV (virus): mouse, 160
Gross leukemia virus: rat, 303
Gross murine leukemia virus: mouse, 68 hn, 68
Gross passage A virus: mouse, 165
Gross virus, resistance to (gene): mouse (see *Rgv-1*)
Gross virus leukemogenesis: mouse, 168 fn
Growth hormone: mouse, 101 hn, 101-102
Growth hormone: rat, 248, 252
Growth period: rat, 236
Growth rate, reduced: mouse, 101
GRS (strain): mouse, 24, 87, 91
GRS/A, 81, 83-84, 86, 88-92, 106, 112
Grüneberg's lethal (gene): rat (see *lg*: rat)
Gs [greasy] (gene): mouse, 39*, 75
gt [gyre-tail] (gene): mouse, 55
Guinea pig basic protein: rat response, 246, 318
Gunmetal (gene): mouse (see *gm*)
GUNN (strain): rat, 237, 307, 311-312, 312 fn, 339
Gur [β-glucuronidase regulation] (gene): mouse
biochemical locus, 38*, 96*
phenotypic expression, 86
protein control, 78, 79 fn, 98
strain distribution, 21-27, 86
Gus [β-glucuronidase] (gene): mouse
biochemical locus, 38*, 96*
electrophoretic polymorphism, 225
phenotypic expression, 86-87
protein control, 78, 98
strain distribution, 21-28, 86-87
Gut [β-glucuronidase, temporal] (gene): mouse, 78, 87
Gv-1 [Gross virus antigen-1] (gene): mouse, 167-168
Gv-2 [Gross virus antigen-2] (gene): mouse, 38*, 96*, 98, 167-168
Gy [gyro] (gene): mouse, 39*, 56, 150
Gyre-tail (gene): mouse (see *gt*)
Gyro (gene): mouse (see *Gy*)

H [histocompatibility] (genes): mouse
H-1 [histocompatibility-1], 21-28, 38*, 119, 121-122, 123*
H-2 [histocompatibility-2] (see also *H-2* gene complex)
biochemical locus, 97*, 99
chromosome aberration, 42
chromosome position, 123*
congenic lines, 124-125, 127-128
mammary tumorigenesis, 161
strain distribution, 21-28
H-2A, 136, 141
H-2C, 136, 141
H-2D [H-2 antigen D], 39*, 134*-135*, 136-139, 147
H-2G [H-2 antigen G], 39*, 134*-135*, 136-139
H-2K [H-2 antigen K], 39*, 134*-135*, 136-140, 147
H-2L, 136-138
H-3 & *H-4* [histocompatibility-3 & 4], 21-28, 38*, 119, 121-122, 123*
H-5, 114 (see also *Ea-5*)
H-6, 114-115 (see also *Ea-6*)
H-7 [histocompatibility-7], 21-28, 119, 121-122, 123*
H-8 [histocompatibility-8], 21-28, 119, 121-122
H-9 [histocompatibility-9], 21-24, 26-28, 119, 121-122
H-10 & *H-11* [histocompatibility-10 & 11], 119
H-12 [histocompatibility-12], 21-28, 119, 121-122
H-13 [histocompatibility-13], 21-28, 38*, 119, 121-122, 123*
H-14, 114-115 (see also *Ea-2*)

H-15 & *H-16* [histocompatibility-15 & 16], 22-23, 38*, 119, 123*
H-17 [histocompatibility-17], 22-23, 119
H-18 & *H-19* [histocompatibility-18 & 19], 22-23, 38*, 119, 123*
H-20 & *H-21* [histocompatibility-20 & 21], 22-23, 119
H-22 [histocompatibility-22], 22-23, 38*, 119, 123*
H-23 to *H-30* [histocompatibility-23 to 30], 22-23, 119-120
H-31 & *H-32* [histocompatibility-31 & 32], 21, 23, 120, 123*, 136
H-33 [histocompatibility-33], 22, 120, 123*, 136
H-34 to *H-38* [histocompatibility-34 to 38], 22-23, 120
H-X, 120, 123*
H-Y, 114, 120, 123*
New loci, 38*-39*, 121, 123*
H [histocompatibility] (genes): rat
H-1 [histocompatibility-1]
immunologically mediated diseases, 334 hn, 334-336
immunological responsiveness, 317-319
linkage, 259 hn
main effect, 257
strain distribution, 255-256, 260, 316
tissue distribution, 315
transplant survival, 321 hn, 321-323
H-2 to *H-4* [histocompatibility-2 to 4], 257, 315
H-5 [histocompatibility-5], 257, 259, 315
H-X, 315
H-Y, 257, 315
H or *h* (gene): rat
H [non-hooded], 239, 241, 246, 246 fn, 248, 252, 273
h [hooded], 239-241, 243-250, 252-253, 259, 273, 309 fn
h^i [Irish], 238-239, 242, 251, 273
h^n [notch], 246, 273
H^{re} [restricted], 273
HA/Icr (strain): mouse, 88-89
ha [hemolytic anemia] (gene): mouse, 31, 65
Haemophilus influenzae b: rat, 337
Hair, genes affecting: mouse, 31, 68, 74-76, 122, 123*
Hair anomaly: rat, 257
Hair color: mouse, 24-25 (*see also* Coat color: mouse)
Hair interior defect (gene): mouse (see *hid*)
Hairless (gene): mouse (see *hr*: mouse)
Hairless (gene): rat (see *hr*: rat)
Hairlessness: mouse, 25
Hair-loss (gene): mouse (see *hl*)
Hairpin-tail (gene): mouse (see T^{hp})
Hairy ears (gene): mouse (see *Eh*)
Hall (strain): mouse, 175
Hammer-toe (gene): mouse (see *Hm*)
H-2 antigen D (gene): mouse (see *H-2D*)
H-2 antigen G (gene): mouse (see *H-2G*)
H-2 antigenic specificities: mouse, 130-133
H-2 antigen K (gene): mouse (see *H-2K*)
Hao-1 [α-hydroxyacid oxidase] (gene): mouse
biochemical locus, 38*, 96*
phenotypic expression, 87
protein control, 78, 98
strain distribution, 22-23, 26, 87
Haplotypes: mouse
H-1, 121-122
H-2
antigenic specificities, 137 hn, 138-139
congenic lines, 124-125, 127-128, 130-133
Ia specificities, 145-147
immune response, 142-144
inbred strains, 16 hn, 16*-17*
independent, 123-124, 142-144
mutant, 129-130
recombinant, 125-128, 135, 142-144
serological markers, 123-124
Ss alleles, 140
wild-derived, 130-133
H-3 & *H-4*, 121-122
H-7 to *H-9*, 121-122
H-12 & *H-13*, 121-122
Haplotypes: rat, 241 fn, 260, 313-314, 316
Hapten-directed responses: mouse, 143
Haptoglobin: rat, 308
Harderian gland carcinoma 2226: rat, 248
Hardy-Weinberg formula: mouse heterozygosity, 220 hn
Hare tail (gene): mouse (see *Hi*)
Harlan Wistar (strain): rat, 301
Harvard caries susceptible (strain): rat, 245 (*see also* HCS)
Harvey sarcoma virus: mouse, 166
Haze (gene): mouse (see $ru\text{-}2^{hz}$)
Hba [hemoglobin α-chain] (gene): mouse
biochemical locus, 39*, 97*
electrophoretic polymorphism, 222
phenotypic expression, 87
protein control, 78, 99
strain distribution, 21-28, 31, 87, 105
Hbb [hemoglobin β-chain] (gene): mouse
biochemical locus, 38*, 96*
electrophoretic polymorphism, 222
histocompatibility linkage, 122, 123*
phenotypic expression, 87
protein control, 78, 98
strain distribution, 21-28, 31, 87, 105
Hbb [hemoglobin β-chain] (gene): rat, 257, 259, 305, 308
hbd [hemoglobin deficit] (gene): mouse, 65
Hbx [hemoglobin x-chain] (gene): mouse, 105
Hby [hemoglobin y-chain] (gene): mouse, 79, 87, 105
Hbz [hemoglobin z-chain] (gene): mouse, 105
Hc [hemolytic complement] (gene): mouse, 21-28, 38*, 103
HCS (strain): rat, 245, 316
Hct [single constriction] (gene): mouse, 75
Hd [hypodactyly] (gene): mouse, 38*, 57
hd [hypodactyly] (gene): rat, 257
he [hematoma] (gene): rat, 257, 259
Head abnormalities: mouse, 60 fn
Head-bleb (gene): mouse (see *heb*)
Head + body size: mouse, 212, 214, 217, 219
Head irradiation: mouse, 189
Head-shaking: mouse, 149-153
Heart: mouse, 21-22, 24, 78-79, 114
Heart: rat (*see also* Cardiac & Cardiovascular entries)
aging, 280-281, 284-285, 289-291
enzymes, 280, 284-285, 287, 311
lesions, 289-291
norepinephrine concentration, 362

transplant survival, 321 hn, 322-323, 325-327, 332
tumor, 299
weight, 245, 274-277, 283-284, 286
Heart rate: mouse, 27-28
Heart rate: rat, 245, 280, 286
heb [head-bleb] (gene): mouse, 57
Hebb-Williams maze: rat, 358-359
Helper factors, soluble: mouse, 141
Hemangiomas: mouse, 173-177
Hematocrit: mouse, 24
Hematocrit: rat, 288
Hematoma (gene): rat (see *he*)
Hematopoiesis: rat, 257
Hematopoietic histocompatibility: mouse, 136
Hematopoietic system tumors: rat, 298-300, 303-304
Hemimelic extra toes (gene): mouse (see *Hx*)
Hemoglobin: mouse, 24, 105 hn
Hemoglobin: rat, 308
Hemoglobin α-chain: mouse, 78, 99
Hemoglobin α-chain (gene): mouse (see *Hba*)
Hemoglobin β-chain: mouse, 78, 98
Hemoglobin β-chain (gene): mouse (see *Hbb*: mouse)
Hemoglobin β-chain (gene): rat (see *Hbb*: rat)
Hemoglobin β-chain [HBB] (protein): mouse, 226-227
Hemoglobin deficit (gene): mouse (see *hbd*)
Hemoglobin x-chain (gene): mouse (see *Hbx*)
Hemoglobin y-chain: mouse, 79
Hemoglobin y-chain (gene): mouse (see *Hby*)
Hemoglobin z-chain (gene): mouse (see *Hbz*)
Hemolytic anemia (gene): mouse (see *ha*)
Hemolytic anemia, autoimmune: mouse, 26
Hemolytic complement (gene): mouse (see *Hc*)
Hemolytic disease: mouse, 65
Hemopoietic function: mouse, 65
Hemorrhage: rat, 290, 349, 350*
Hemorrhagic ovaries: mouse, 22
Hen albumen lysozyme: mouse, 113, 143
Hepatic cells: rat, 263-264 (*see also* Hepatocytes)
Hepatic cysts: rat, 290, 338
Hepatic enzyme activity: rat, 281-282
Hepatic fusion (gene): mouse (see *hf*)
Hepatic metabolism: rat, 239, 242-244, 246, 248
Hepatic microsomes: rat, 310
Hepatic sarcoma: rat, 338
Hepatocytes: rat, 266, 282 (*see also* Hepatic cells)
Hepatomas: mouse, 22-23, 28, 68, 163-165
Hepatomas: rat, 243-244, 248, 261, 266-267, 301, 303-304
Hepatosplenic amyloidosis: mouse, 27
Hereditary hypothalamic diabetes: rat, 339
Hermaphrodites: mouse, 47
Hernia: mouse, 60 fn, 149
Heterochromatin variations, centromeric: mouse, 13, 44
Heteropyknotic Y chromosome: rat, 252
Heterozygosis, forced: mouse, 11, 29 hn, 30-35
Heterozygosity: mouse, 220-225
Heterozygote phenotype symbols: mouse nomenclature rules, 14
Heterozygotes: laboratory animals, 4
Heterozygous phenotype: mouse, 53-55
Hex-1 [β-D-*N*-acetylhexosaminidase] (gene): mouse, 79, 88
Hexokinase [HK-1] (protein): mouse, 97*, 99
Hexose 6-phosphate dehydrogenase: mouse, 78, 98
hf [hepatic fusion] (gene): mouse, 38*, 60
$HgCl_2$-induced antiglomerular basement membrane antibodies: rat, 336
$HgCl_2$-induced nephritis: rat, 336
H-2 gene complex: mouse, 123-127, 129-133, 134*-135*, 136-148 (see also *H* [histocompatibility]: mouse)
Hh (gene): mouse, 136
Hi (strain): rat, 338
Hi [hare tail] (gene): mouse, 55
hid [hair interior defect] (gene): mouse, 75
High-tail (gene): mouse (see *Ht*)
Himalayan (gene): mouse (see c^h: mouse)
Himalayan (gene): rat (see c^h: rat)
Hindbrain hydrocephalus: mouse, 150
Hindbrain myeloschisis: mouse, 152
Hindfoot: mouse, 217
Hindlimb: mouse, 60 fn, 67-68, 150-152
Hindlimb: rat, 323
Hindquarters, swaying: mouse, 151, 153
Hippocampus: rat, 284
His [histidase] (gene): mouse, 79, 88
Histamine: rat, 281
Histidase: mouse, 79
Histidase: rat, 282 fn
Histidase (gene): mouse (see *His*)
Histidine ammonia-lyase: mouse, 79
Histidine ammonia-lyase: rat, 282
Histidine decarboxylase: rat, 312
Histiocytes: mouse, 169 hn
Histocompatibility, hematopoietic: mouse, 136
Histocompatibility antigens: mouse, 141, 147
Histocompatibility antigens: rat, 319-320
Histocompatibility (genes): mouse (see *H* [histocompatibility]: mouse)
Histocompatibility (genes): rat (see *H* [histocompatibility]: rat; *Ag-E* [histocompatibility])
Histocompatibility linkage map: mouse, 122, 123*
Histocompatibility systems (except *H-2*): mouse, 118-122
Hk [hook] (gene): mouse, 31, 38*, 57
HL (strain): rat, 335
hl [hair-loss] (gene): mouse, 39*, 75
Hm [hammer-toe] (gene): mouse, 31, 38*, 56
Hnl [hypothalamic norepinephrine level] (gene): mouse, 22-23, 101
HO (strain): rat (*see also* PVG)
alloantigenic alleles, 316-317
characteristics, 245, 249
enzyme polymorphism, 311
immunogenetic characterization, 314
immunologically mediated disease, 335
karyology, 261
sarcoma, 304
transplant survival, 322-323, 325
HO anti-AUG antiserum: rat, 315, 317 fn
ho [hotfoot] (gene): mouse, 31, 150
Holtzman (strain): rat, 262, 335, 338, 340, 342, 344-347
Holtzman (SD) strain: rat, 282, 291, 341-342, 345
Hom-1 [androgen hormone metabolism-1] (gene): mouse, 101, 136
Homeostasis: mouse, 193 hn
Homozygosity: laboratory animals, 5, 5*, 6
Homozygotes: mouse, 63 hn, 64 fn, 150-152
Homozygous lethality: mouse, 53 hn, 53-54, 63 hn-64 hn, 65-66
Hooded (strain): rat, 304 (*see also* HO)
Hooded (gene): rat (see *h*)

Hooded Lister (strain): rat, 234 (*see also* LIS: rat)
Hooded Liverpool (strain): rat, 341
Hook (gene): mouse (see *Hk*)
hop [hop-sterile] (gene): mouse, 56, 150
Hopping: mouse, 150, 152
Hop-sterile (gene): mouse (see *hop*)
Hormones: mouse, 68, 182-187, 191-192, 193 hn, 197-198 (*see also* specific hormone)
Hormones: rat, 288, 339, 340 hn, 340-347 (*see also* specific hormone)
Hotfoot (gene): mouse (see *ho*)
House mice: principal species & subspecies, 212
hpy [hydrocephalic polydactyl] (gene): mouse, 38*, 57, 150
Hq [harlequin] (gene): mouse, 39*
HR (strain): mouse, 110, 129 hn, 129-130
  HR/DE
    biochemical variation, 90
    complement factors, 103
    hemangioendotheliomas, 175
    neoplasms, 171
    ovarian tumors, 195, 204
    urinary proteins, 107
hr (gene): mouse
  hr [hairless]
    chromosome position, 39*
    immune system development, 68
    leukemogenesis, 168
    strain distribution, 25, 31, 36, 75
  hr^{ba} [bald], 75
  hr^{rh} [rhino], 31, 75
hr [hairless] (gene): rat, 257
HRS (strain): mouse, 104, 114, 116-118, 169
  HRS/A, 87
  HRS/J, 25, 31, 82-91, 106
HS (strain): rat
    alloantigenic alleles, 316
    characteristics, 245
    coat color, 272-273
    immunoglobulin polymorphism, 307
    transplant survival, 321, 324, 329
HSFS/N (Swiss) strain: mouse, 116-118
HSN sarcoma: rat, 304
Hst-1 [hybrid sterility-1] (gene): mouse, 39*
Ht [high-tail] (gene): mouse, 39*, 55
HTG (strain): mouse, 114, 126, 128, 168
HTH (strain): mouse, 126, 146
HTI (strain): mouse, 114, 126, 146, 168
H-2-Tla gene complex: mouse, 136
HTT-non-inbred (strain): mouse, 128
Humoral response: mouse, 129 hn, 129-130
Hunt's caries resistant (strain): rat, 243 (*see also* CAR)
Hunt's caries susceptible (strain): rat, 243 (*see also* CAS)
HW (strain): rat, 307, 316-317
HW81/By (strain): mouse, 112
HW94/By (strain): mouse, 112
HW97/By (strain): mouse, 112
Hx [hemimelic extra toes] (gene): mouse, 31, 38*, 56
hy [hypotrichosis] (gene): rat, 257
hy-1 [hydrocephalus-1] (gene): mouse, 58
hy-2 [hydrocephalus-2] (gene): mouse, 58
hy-3 [hydrocephalus-3] (gene): mouse, 31, 38*, 58, 150
Hyaloid artery: rat, 249
Hyaloids: rat, 292
Hybrids: mouse, 161-164 (*see also* F_1 hybrids; specific hybrid strain)
Hybrids: rat, 262 (*see also* specific hybrid strain)
Hydrocephalic polydactyl (gene): mouse (see *hpy*)
Hydrocephalus: mouse, 149-152
Hydrocephalus: rat, 242, 245
Hydrocephalus (genes): mouse (see *hy-1; hy-2; hy-3*)
Hydrocephalus-like (gene): mouse (see *hy-like*)
Hydronephrosis: rat, 236, 242
Hydronephrosis (gene): rat (see *Ne*)
α-Hydroxyacid oxidase (gene): mouse (see *Hao-1*)
L-2-Hydroxyacid oxidase: mouse, 78, 98
3-Hydroxybutyrate dehydrogenase: rat, 284
3β-Hydroxy-Δ^5-steroid dehydrogenase activity: rat, 283, 285
20α-Hydroxysteroid dehydrogenase activity: rat, 283
hy-like [hydrocephalus-like] (gene): mouse, 58
Hyp [hypophosphatemia] (gene): mouse, 31, 39*, 56
Hyp-1 [hypertension-1] (gene): rat, 257
Hypercholesterolemia, reactive: rat, 348
Hyperglycemia: mouse, 101
Hyperimmunoglobulinemia: mouse, 69
Hyperinsulinemia: mouse, 69, 101
Hyperlipemia: rat, 348
Hyperplasia: mouse, 26, 193 hn
Hyperplasia: rat, 244, 253, 289, 293, 296
Hypersensitivity reaction: rat, 317 hn, 317-318
Hypertension: rat, 236, 239, 244-245, 250, 252, 348-349, 349*-350*
Hypertension-1 (gene): rat (see *Hyp-1*)
Hypertensive cardiovascular lesions: rat, 348
Hypertensiveness: rat, 284
Hypertonia: mouse, 149, 151-152
Hypochromic anemia, microcytic: mouse, 65
Hypodactyly (strain): rat, 307
Hypodactyly (gene): mouse (see *Hd*)
Hypodactyly (gene): rat (see *hd*)
Hypoimmunoglobulinemia: mouse, 69
Hypophosphatemia (gene): mouse (see *Hyp*)
Hypophyseal isografts: mouse, 181-182, 190-191
Hypophysectomy: mouse, 182
Hypoplastic kidney: rat, 238
Hypothalamic diabetes, hereditary: rat, 339
Hypothalamic hormones: rat, 340 hn, 340, 342-343, 347
Hypothalamic norepinephrine level (gene): mouse (see *Hnl*)
Hypothalamus: mouse, 101
Hypothalamus: rat, 250, 362
Hypotrichosis: rat, 257-258
Hypotrichosis (gene): rat (see *hy*)
Hypoxanthine phosphoribosyltransferase [HPRT] (protein): mouse, 97*, 99, 226-227

I (strain): mouse
    alloantigens, 114, 116-118
    characteristics, 25
    genes, 25
    Ig heavy chain linkage groups, 111
    leukemia antigens, 168
    pituitary tumors, 184
    testicular tumors, 206 hn, 207
  I/An, 90, 106
  I/AnN, 88-89
  I/Ao, 103

I/FnLn, 82
I/Ln, 106
I/LnJ, 81, 83-89, 91
I/LnRe, 87, 90
I/LnReJ, 82-83, 85-86
I/Sr, 168
I/St, 16*, 82, 88-89, 124-125
I/StDa, 114
I/StN, 81-82, 88-89
I hybrids, 184
I (strain): rat, 351 (*see also* A990)
Ia [immune response-associated antigen] (genes): mouse, 134, 134*-135*, 136, 141, 147
ia [incisorless] (gene): rat, 309 (see also *in*)
Ia lymphocyte alloantigens: mouse, 141
Ia specificities: mouse, 145-147
Iα (gene): rat, 256
IC (strain): mouse, 195, 200
IC/Le, 32
ic [ichthyosis] (gene): mouse, 32, 38*, 75
ic [infantile ichthyosis] (gene): rat, 246 fn
ICFW/Lac (strain): mouse, 85
Ichthyosis (gene): mouse (see *ic*: mouse)
ICI-SPF (strain): mouse, 195
ICR (strain): mouse, 176
ICR/Ha, 88-89, 209 fn, 209-210
ICR/Jcl, 195, 203-204
ICR x C57BL, 209
(ICRC x C3H)F_1 (strain): mouse, 195
Id-1 [isocitrate dehydrogenase] (gene): mouse
biochemical locus, 38*, 96*
electrophoretic polymorphisms, 223, 225-226
phenotypic expression, 88
protein control, 79, 98
strain distribution, 21-28, 88
Id-2 [isocitrate dehydrogenase, mitochondrial] (gene): mouse, 223
IDT (strain): mouse, 87
Ie [eye-ear reduction] (gene): mouse, 59
IF (strain): mouse, 25, 48, 159, 201-203
IF/BrA, 84, 87, 92, 112
IF/Or, 202
IF hybrids, 202-203
If-1 [interferon-1] (gene): mouse, 21-24, 26-27, 79, 88
If-1 alleles: mouse, 112
If-2 [interferon-2] (gene): mouse, 22-23, 79, 88
IFS (strain): mouse, 206 hn, 207
Ig [immunoglobulin] (genes): mouse
IgA, 107, 109-111 (see also *Ig-2*)
IgD, 109-111
IgF, 107, 109-111 (see also *Ig-4*)
IgG, 107-111 (see also *Ig-1*)
IgH, 107, 109-111 (see also *Ig-3*)
IgM, 109-111
Ig-1, 21-28, 107-109 (see also *IgG*)
Ig-2, 22-28, 107, 109
Ig-3, 21-28, 107, 109 (see also *IgH*)
Ig-4, 22-28, 107, 109 (see also *IgF*)
Ig-5, 107, 109
Ig-6, 107, 109
Ig allotype loci: mouse, 112-113
IgA myeloma protein: mouse, 143
IgA synthesis: rat, 247
IgC_H loci: mouse, 108-111
IgD: mouse, 107
IgE: mouse, 107
IgE antibody response: rat, 247
IgE-binding site: rat, 320
IgE synthesis: rat, 247
IgG, glomerular: rat, 294
IgG antibody response: mouse, 69
IgG chains: mouse, 107
IgG_1 synthesis: rat, 247
IgG_{2a} myeloma protein: mouse, 143
IgG_{2a} serum level: mouse, 69
Ig heavy chain loci: mouse, 107
Ig heavy chain loci: rat, 256
Ig light chain locus: rat, 256
Ig light chains: mouse, 107
IgM: mouse, 107
IgM: rat, 292
IgM antibody-forming cells: mouse, 116
IgM antibody response: mouse, 69
IgM serum levels: mouse, 22, 68
IgV_H: mouse, 111
Iγ2b (gene): rat, 256
IITES (strain): mouse, 84, 87-88
Iκ (gene): rat, 256
Ileocecal lymph node tumors: rat, 247, 298-299
Ileocecal mesentery: rat, 238
Immune complex nephritis: mouse, 69
Immune complex nephritis, autologous: rat, 335
Immune response: mouse, 112-113
Immune response: rat, 313-314
Immune response (genes): mouse (see *Ir*: mouse)
Immune response (genes): rat (see *Ir*: rat)
Immune response complex: rat, 259 hn
Immune suppression: mouse, 136, 141
Immune system development: mouse, 67-69
Immunocytomas: rat, 247, 301
Immunogenetic characterization: rat, 313-314
Immunoglobulin markers, lymphocyte membrane: mouse, 110 hn, 110-111
Immunoglobulins: mouse (see *Ig*)
Immunoglobulins: rat, 247, 257-258, 292, 294, 305-308
Immunoglobulin-secreting tumors: rat, 299 fn
Immunologically mediated diseases: rat, 334-336
Immunological responsiveness: rat, 317-319
Imperforate vagina: mouse, 26, 48
in [incisorless] (gene): rat, 257, 259 (see also *ia*)
Inbred strains (*see also* specific strain)
definition, 3-7, 10
historical development, 1-2, 9
inbreeding coefficient, 2, 4, 4*
mouse nomenclature rules, 9-15
research uses, 1-3
Incisive foramina: mouse, 212, 214, 217-219
Incisorless (strain): rat, 309
Incisorless (gene): rat (see *in*)
Incisors: mouse, 214, 217
Indophenol oxidase-1 (gene): mouse (see *Ipo-1*)
Infantile ichthyosis (gene): rat (see *ic*: rat)
Infarction: rat, 290, 349, 350*
Infectious diseases: mouse (*see* specific disease; specific organism)
Infectious diseases: rat, 337-338 (*see also* specific disease; specific organism)
Inferior vena cava tumor: rat, 249
Inner ear changes, degenerative: mouse, 122, 150-151

INR (strain): rat, 245, 272-273, 310-311, 351, 354-360
Insulin: mouse, 23, 101
Insulin: rat, 244, 246, 339
Intense chinchilla (gene): mouse (see c^i)
Interdigital foot pads: mouse, 212
Interferon: mouse, 79, 112 hn
Interferon (genes): mouse (see *If-1; If-2*)
Intermediate agouti (gene): mouse (see A^i)
Intermediate yellow (gene): mouse (see A^{iy})
Interorbitum: mouse, 212, 214, 217
Interpterygoid space: mouse, 212, 214, 217
Interstitial cells, testicular: rat, 244, 248, 281, 289, 294-296, 298-299
Interstitial cell testicular tumors: mouse, 198 fn, 206-207
Interstitial lesions, kidney: rat, 293 fn
Interstitial tissue: mouse, 48
Intestinal proteins: mouse, 78-80
Intestinal tumors: mouse, 208-210
Intestines: rat, 285, 303, 309 hn, 311-312, 323
Intralobular extrahepatocyte space: rat, 282
Inulin clearance: rat, 286
Inversions, chromosome: mouse, 13, 44
Iodine, thyroid: mouse, 102
Iodine-131: mouse, 187-188, 190-191
Iodine-131 metabolism: rat, 251
Iowa Non-reactive (strain): rat (*see* INR)
Iowa Reactive (strain): rat (*see* IR)
Ipo-1 [indophenol oxidase-1] (gene): mouse, 79, 88, 223
IR (strain): rat, 245, 272-273, 351, 354-360
Ir [immune response] (genes): mouse
Ir-A-CHO, 113
Ir-GAC, 113
Ir-1, 39*, 135*, 136, 141
Ir-1A, 134*-135*
Ir-1B, 134*-135*
Ir-2, 38*, 113
Ir-3, 113
Ir-4, 113
Ir-5, 22-23, 39*, 113
Ir [immune response] (genes): rat, 257
I regions, H-2 gene complex: mouse, 123 hn, 123-124, 134*-135*, 136, 141-144, 145 hn
Irish (gene): rat (see h^i)
Iron deficiency: rat, 252
Iron utilization, defective: mouse, 65
Irregular teeth (gene): mouse (see *It*)
IS (strain): mouse, 25
IS/Cam, 81-89, 91, 104
IS/CamEi, 86-87
IS (strain): rat, 245, 272-273
Is (gene): mouse, 136, 141
Ischemic time: rat transplants, 321 hn, 325 fn
Islets of Langerhans: mouse, 101
Islets of Langerhans: rat (*see* Pancreatic islets)
Isocitrate dehydrogenase (genes): mouse (see *Id-1; Id-2*)
Isocitrate dehydrogenase [ID-1] (protein): mouse, 226-227
Isoenzymes: mouse nomenclature rules, 14
Isografts: mouse, 21, 181-182
Isoleucine: rat, 320
Isoniazid: mouse, 202
It [irregular teeth] (gene): mouse, 60
ITES (strain): mouse, 84, 87-89, 92
iv [situs inversus viscerum] (gene): mouse, 60
Ivanovac (strain): rat, 341
IVCE (strain): mouse, 46, 48
IVCS (strain): mouse, 46, 48
IXBL (strain): mouse, 84, 86-87, 89

J (strain): rat, 351
j [jaundice] (gene): rat, 237, 258, 312
j [jaw-lethal] (gene): mouse, 59
ja [jaundiced] (gene): mouse, 32, 63, 65
Jackson circler (gene): mouse (see *jc*)
Jackson shaker (gene): mouse (see *js*)
Jackson waltzer (gene): mouse (see *jv*)
Jagged-tail (gene): mouse (see *jg*)
Japanese ruby (gene): mouse (see p^r)
Jaundice (strain): rat, 307, 311-312, 339
Jaundice (gene): rat (see *j*: rat)
Jaundiced (gene): mouse (see *ja*)
Jaw-lethal (gene): mouse (see *j*: mouse)
Jay's dominant spotting (gene): mouse (see W^j)
JBT (strain): mouse
JBT/Jd, 81, 86-88
JBT/Jn, 82
jc [Jackson circler] (gene): mouse, 32, 38*, 150, 155
JE (strain): mouse, 86
je [jerker] (gene): mouse, 32, 38*, 150, 155
Jejunum: rat, 311
Jensen sarcoma: rat, 248, 253, 301
Jerker (gene): mouse (see *je*)
jg [jagged-tail] (gene): mouse, 38*, 57
ji [jittery] (gene): mouse, 32, 38*, 150, 155
Jimpy (gene): mouse (see *jp*)
Jittery (gene): mouse (see *ji*)
JK (strain): mouse
alloantigens, 114, 116, 118
biochemical variation, 93
congenic-inbred line progenitor, 125
genealogy, 16*
pituitary tumors, 183
testicular tumors, 206 hn, 206
JK/St, 103, 124
jo [jolting] (gene): mouse, 150
Joined toes (gene): mouse (see *jt*)
Joints: mouse, 67
Jolting (gene): mouse (see *jo*)
jp [jimpy] (gene): mouse, 39*, 150, 155
jp^{msd} [myelin synthesis deficiency], 150
J region, *H-2* gene complex: mouse, 126-127
js [Jackson shaker] (gene): mouse, 32, 39*, 122, 123*, 150, 155
jt [joined toes] (gene): mouse, 56
JU (strain): mouse, 25, 86, 90
JU/Fa, 36, 48, 87, 92
JU/FaCt, 36
jv [Jackson waltzer] (gene): mouse, 32, 59, 151, 155

K (strain): mouse, 88
K/Gh, 81-82, 87
K (strain): rat, 245, 273
k [kinky] (gene): rat, 258-259
Kappa chain, immunoglobulin: mouse, 107
Kappa light chains, immunoglobulin: rat, 305 hn-306 hn, 306-308
Karyology: rat, 261-268, 268*-270*, 271-272
Kb [knobbly] (gene): mouse, 55
kd [kidney disease] (gene): mouse, 36, 38*
KE (strain): mouse, 25, 48, 88

KE/Kw, 81-83, 89
KGH (strain): rat, 246, 256, 260, 307, 314, 316-318
KH-1 (strain): mouse, 111
KH-2 (strain): mouse, 111
KI (strain): mouse, 48
Kidney, kidney tissue: mouse (*see also* Renal entries: mouse)
alloantigens, 114, 117
congenital disorders, 21-22, 24-27
grafts, 182, 190
protein distribution, 77-80
tumors, 182, 190, 198
Kidney, kidney tissue: rat (*see also* Renal entries: rat)
adenocarcinoma, 303
aging, 281, 286, 291-294
enzyme polymorphisms, 311
karyology, 263
lesions, 291-294
malformations, 238, 248
sarcoma, 246
transplant survival, 321 hn, 322, 325, 332, 333*
tumors, radiation-induced, 302
weight, 274-277, 281
Kidney-catalase (gene): mouse (see *Ce-2*)
Kidney disease (gene): mouse (see *kd*)
Kimbo-tail (gene): mouse (see *kt*)
King-Holtzman (strain): rat, 262
Kinked tail: mouse, 122
Kinky (gene): mouse (see Fu^{ki})
Kinky (gene): rat (see *k*)
Kinky-waltzer (gene): mouse (see *Kw*)
Kirsten sarcoma virus: mouse, 166
Kitty hawk wild mouse, 111
KK (strain): mouse, 25, 48, 84, 86-89, 91-92
KL (strain): mouse, 104
KL/Fa, 92
KM (stock): mouse, 198
Knobbly (gene): mouse (see *Kb*)
Kon's lethal (gene): rat (see *lk*)
KP (strain): mouse, 25, 48, 88
KP/Kw, 89
KR (strain): mouse, 48, 84, 86-89, 91-92
KR/Nga, 81-82, 87, 89
kr [kreisler] (gene): mouse, 38*, 59, 151
K region, *H-2* gene complex: mouse, 123-124, 126-127, 134*-135*, 136-140, 147 hn
Kreisler (gene): mouse (see *kr*)
KSA (strain): mouse, 87, 91-92
KSB (strain): mouse, 48, 84, 86-89, 91-92
KSB/Nga, 82, 86-87, 89
kt [kimbo-tail] (gene): mouse, 55
Kw [kinky-waltzer] (gene): mouse, 55, 151, 156
KX (strain): rat, 246 (*see also* Slonaker)
KY (strain): mouse, 111
ky [kyphoscoliosis] (gene): mouse, 36
KYN (strain): rat, 246, 260, 273
Kyoto notched (strain): rat, 246 (*see also* KYN)
Kyphoscoliosis (gene): mouse (see *ky*)
Kyphoscoliosis: rat, 245
Kyushu wild mouse, 109, 111
K9 & K19 calorie-restricted diets: mouse, 163

L (strain): mouse (*see* C57L)
L/St, 88-89, 103
L/E (strain): rat, 307
LA/N (strain): rat, 246, 272-273, 316-317
la [leaner] (gene): mouse, 58, 151 (see also tg^{la})
Lachrymal gland abnormalities: mouse, 60 fn
Lactate dehydrogenase: mouse, 79
Lactate dehydrogenase: rat, 287
Lactate dehydrogenase-A: mouse, 98
Lactate dehydrogenase (genes): mouse (see *Ld-A; Ld-B*)
Lactate dehydrogenase (porcine), isoenzyme A_4: rat, 318
Lactate dehydrogenase regulator (gene): mouse (see *Ldr-1*)
Lactate dehydrogenase regulator (protein): mouse, 98
Lactation: rat, 346
Lactoyl-glutathione lyase: mouse, 99
Lad [lymphocyte-activating determinant] (gene): mouse, 136, 141
LAF_1 (strain): mouse (*see also* (C57L x A)F_1)
adrenal cortical tumors, 179-180
fibrosarcomas, 211
ovarian tumors, 195, 198 fn, 200, 204
pituitary tumors, 183, 189-191
LAF_1J, 195, 204
Lambda chains, immunoglobulin: mouse, 107
Lambert (strain): rat, 338
Lap-1 [leucine arylaminopeptidase] (gene): mouse, 38*, 79*, 88, 96*, 98
Large intestine: rat, 285, 296 fn (*see also* Intestines)
LAS (strain): mouse, 92
Lasiocarpine: rat, 304
Lateral septal nucleus: mouse, 25
Lc [lurcher] (gene): mouse, 38*, 42, 58, 63, 151, 156
ld [limb-deformity] (gene): mouse, 32, 38*, 57
Ld-A [lactate dehydrogenase, A form] (gene): mouse, 223
Ld-B [lactate dehydrogenase, B form] (gene): mouse, 223
Ldh-1 (gene): mouse, 96*, 98
Ldr-1 [lactic dehydrogenase regulator] (gene): mouse
biochemical locus, 38*, 96*
electrophoretic polymorphism, 223
phenotypic expression, 89
protein control, 79, 98
strain distribution, 21-28, 89
LE strain: rat (*see also* Long Evans)
aging, 283, 291, 296
alloantigenic alleles, 316-317
blood protein polymorphism, 309
coat color, 246
enzyme polymorphisms, 309-311
immunoglobulin polymorphism, 307
immunologically mediated diseases, 334-335
immunological responsiveness, 318-319
infectious diseases, 337
karyology, 262, 266
lesions, 291, 296
tumors, 296, 304
LE/Cal, 311
LE/Orl, 310-311
le [light ear] (gene): mouse, 32, 38*, 72
Leaden (gene): mouse (see *ln*)
Leaner (gene): mouse (see *la*; tg^{la})
Learning, avoidance: rat, 360
Lecithin-supplemented diet: mouse, 188
Leg muscles: rat, 294
Leiomyoma: mouse, 208
Leiomyoma: rat, 242
LEJ (strain): rat, 260, 272-273 (*see also* Long Evans)
Lens changes: rat, 291-293
Lens reflection: rat, 249

LEP (strain): rat
Ag-B haplotype, 316
blood protein polymorphism, 308-309
coat color, 246, 272-273
enzyme polymorphisms, 311-312
immunogenetic characterization, 314
immunoglobulin polymorphisms, 307
immunological responsiveness, 318
origin, 246
LEP/Cub, 310
Lesions: rat, 289-297 (*see also* specific lesion)
Lesions, angiomatous: mouse, 173
Lethality: mouse, 63-66, 152 (*see also* Mortality)
Lethal milk (gene): mouse (see *lm*)
Lethal non-agouti (gene): mouse (see a^x)
Lethal spotting (gene): mouse (see *ls*)
Lethal yellow (gene): mouse, 161 (see also A^y)
Lethargic (gene): mouse (see *lh*)
Leucine aminopeptidase: mouse, 79, 98
Leucine arylaminopeptidase (gene): mouse (see *Lap-1*)
Leukemia: mouse, 22, 24-27, 68, 148, 168-169 (*see also* specific leukemia)
Leukemia: rat
aging, 295-296
chemically induced, 266-267, 303-304
karyology, 262, 266-267
spontaneous, 298-300
strain distribution, 244, 249, 252-253, 298-301, 303-304
Leukemia antigens: mouse, 148, 167-168
Leukemia-associated transplantation antigen: mouse, 143
Leukemia passage A virus, mouse: rat response, 303
Leukemia virus, Gross: rat, 303
Leukemia virus, murine: rat response, 303
Leukemia viruses: mouse, 68 hn, 68, 116 hn, 116, 165-168, 227-231 (*see also* specific leukemia virus)
Leukemogenesis: mouse, 167-168
Leukencephalosis (gene): mouse, 58
Leukocyte count: mouse, 23
Leukocyte count: rat, 337 hn, 338
Leukocyte reaction stimulation: mouse, 135*, 136 hn, 136, 141
Leukocytes: rat, 261-263, 288
Leukocytosis: mouse, 68
Leukodystrophy, CNS sudanophilic: mouse, 150
Lever press operant rate: rat, 356
LEW (strain): rat
alloantigenic alleles, 316-317
allograft survival, 324-327
behavioral research, 351
blood protein polymorphisms, 308-309
characteristics, 238 hn, 246, 360
coat color, 246, 272-273
congenic-inbred line progenitor, 255-256
enzyme polymorphisms, 310-312
eye color, 272-273
immunogenetic characterization, 314
immunoglobulin polymorphism, 307
immunologically mediated diseases, 334-336
immunological responsiveness, 317-319
karyology, 262
origin, 246
transplant survival, 321-323, 327-333, 333*
virally induced tumors, 303
LEW/Cub, 310
LEW/Mai, 256
LEW/Orl, 310-311
LEW congenics, 255-256, 316, 318, 327-333, 335-336
LEW hybrids, 260, 321-323, 325-327, 329-332, 335
Lewis (strain): rat, 246, 262, 307-310, 334-336 (*see also* LEW)
Lewis albino (strain): rat, 303 (*see also* LEW)
Leydig cell: rat, 299 (*see also* Interstitial cell, testicular: rat)
LG (strain): mouse, 25, 48, 104, 126
LG/Ckc, 120, 128, 147
LG/J, 81-92, 106
lg [Grüneberg's lethal] (gene): rat, 258-259
lg [lid gap] (gene): mouse, 59
lg^{ga} [ophthalmatrophy], 59
lg^{stn} [slit-lid], 59
LGE (strain): rat, 246, 314, 316-317
LH (strain): rat, 334-335
LH/Lac, 274-275
lh [lethargic] (gene): mouse, 38*, 68, 101, 151
Lid gap (gene): mouse (see *lg*: mouse)
Lids open (gene): mouse (see *lo*)
Life-span: mouse, 45-46, 48-49, 67, 69, 210 hn, 210-211 (*see also* Survival; Mortality: mouse)
Life-span: rat, 240-244, 252, 277-278 (*see also* Longevity)
Light (gene): mouse (see B^{lt})
Light contingent bar presses: rat, 359
Light ear (gene): mouse (see *le*)
Light-head (gene): mouse (see *te*)
Limb ataxia: mouse, 152
Limb-deformity (gene): mouse (see *ld*)
Limb muscles, abnormal: mouse, 60 fn
Linkage groups: mouse, 13 fn, 65-66, 116-118, 122 hn, 123, 157 hn, 162
Linkage groups: rat, 259, 305, 315
Linkage map, histocompatibility: mouse, 122, 123*
Lip, cleft: mouse, 21, 60 fn
Lipase: rat, 281 fn
Lipids: mouse, 21, 23-25
Lipids: rat, 287
Lipogenic enzymes: rat, 309 hn
Lipoma: rat, 253, 295
Lipopolysaccharides, bacterial: mouse, 23, 113
Liquid-supplemented diet: mouse, 188
LIS (strain): mouse, 159-160
LIS/A, 81-84, 86-92, 107
LIS (strain): rat, 266, 307, 311, 316, 335
lit [little] (gene): mouse, 32, 38*, 101
Litter size: mouse, 25, 27, 45-49
Litter size: rat, 236, 238-239, 242-244, 247-250, 253
Little (gene): mouse (see *lit*)
Liver; liver tissue: mouse (*see also* Hepatic entries: mouse)
alloantigens, 114, 117
anemia, 65
cAMP levels, 136
proteins, 77-80, 106 hn
tumors, 175 fn-176 fn
Liver; liver tissue: rat (*see also* Hepatic entries: rat)
aging, 282, 289-290, 292, 296, 296 fn
alloantigenic alleles, 315
cell membrane alloantigens, 319-320
enzyme polymorphisms, 309, 311-312
karyology, 261-264
lesions, 289-290, 292

nodular hyperplasia, 244
tumors, 296, 301, 303-304
weight, 274-277
Liver, fatty: rat, 244
Liver-specific F antigen: mouse, 143
Lizard (gene): mouse (see *lz*)
lk [Kon's lethal] (gene): rat, 258
lm [lethal milk] (gene): mouse, 32, 38*
ln [leaden] (gene): mouse, 16*, 25, 32, 38*, 72, 122, 123*
Lo (strain): rat, 338
lo [lids open] (gene): mouse, 59
Locomotor disorders: mouse, 151-153
Locus (*see* specific gene)
Locus symbol: mouse nomenclature rules, 12-15
Long-Evans (strain): rat, 234 hn, 234, 278, 314, 337, 344 (*see also* LE)
Long-Evans (Blue Spruce), 337
Long-Evans/Cal, 311
Long-Evans/Orl, 310-311
Long-Evans Praha, 246 (*see also* LEP)
Longevity: mouse (*see* Life-span: mouse)
Longevity: rat, 277-278, 338 (*see also* Life-span: rat)
Loop-tail (gene): mouse (see *Lp*)
Loser Runway (stock): rat, 351 (*see also* LRS)
LOU (strain): rat, 236, 316, 318
LOU/C, 246-247, 256, 307, 317
LOU/C/Wsl, 299, 301
LOU/M, 247, 307
LOU/M/Wsl, 301
LOU/C congenics, 256, 307
Louvain (strain): rat, 246-247 (*see also* LOU/C; LOU/M)
LP (strain): mouse
alloantigens, 114, 116-118
characteristics, 25
complement factors, 103
genes, 25
H-2 haplotype & markers, 123
Ig heavy chain linkage groups, 110
leukemia antigens, 168
proteins, 104
LP/J
biochemical variation, 81-92
body weight, 46
complement factors, 103
congenic-inbred line progenitor, 124-125
fibrosarcoma, 211
genealogy, 16*
histocompatibility, 121
lymphoma, 171
reproduction, 46
urinary proteins, 106
LP.RIII, 124-125
Lp [loop-tail] (gene): mouse, 32, 38*, 57, 60 fn, 151
lpr [lymphoproliferation] (gene): mouse, 69
Lps [Lps response] (gene): mouse, 69
Lps response (gene): mouse (see *Lps*)
LRS (stock): rat, 351, 360, 361
LS/Le (strain): mouse, 25, 30, 32, 48, 87
ls [lethal spotting] (gene): mouse, 25, 32, 38*, 42, 72, 151
lst [Strong's luxoid] (gene): mouse, 38*, 57, 65
LT (strain): mouse, 25, 193 hn, 196
LT/ChRe, 48
LT/ChReSv, 65, 82-83, 86, 88
LT/Re, 87, 91, 107
LT/Sv, 37
lt [lustrous] (gene): mouse, 32, 39*, 75
LTS (strain): mouse, 88
LTS/A, 81-84, 86-92, 106
LTS_f/A, 112
LTXB (strain): mouse, 37
lu [luxoid] (gene): mouse, 32, 38*, 56
Lumbarless (gene): mouse, 57
Lumbar spinal cord: mouse, 152
Lumbar vertebrae abnormality: rat, 245
Lumbosacral myeloschisis: mouse, 152
Lung; lung tissue: mouse, 22, 26-28, 51, 68, 77, 114
Lung; lung tissue: rat
aging, 284, 291-292, 296
enzyme polymorphisms, 311
karyology, 262, 264, 266-267
lesions, 291-292
tumors, 296, 302
weight, 274-277
Lupus nephritis: mouse, 26
Lurcher (gene): mouse (see *Lc*)
Lustrous (gene): mouse (see *lt*)
Luteinizing hormone: mouse, 101
Luteinizing hormone: rat, 283, 340 hn, 340-347
Luteinizing hormone—releasing hormone: rat, 283, 340 hn, 340, 342-343, 347
Luteoma: mouse, 194-202, 203 fn, 204
Luxate (gene): mouse (see *lx*: mouse)
Luxoid (gene): mouse (see *lu*)
Lv [δ-aminolevulinate dehydratase] (gene): mouse, 21-28, 38*, 79, 89, 96*, 98
lx [luxate] (gene): mouse, 32, 36, 38*, 57
lx [polydactyly-luxate] (gene): rat, 258-259, 315
LXB (strain): mouse, 37
Ly [lymphocyte antigen] (genes): mouse, 21-28, 38*-39*, 96*-97*, 98-99, 115-117
Ly-1 (gene): rat, 315
Lyb-2 [B-lymphocyte alloantigen] (gene): mouse, 116
Lymphatic leukemia: mouse, 21, 24
Lymphatic leukemia: rat, 298
Lymphatic system prenatal development: mouse, 50-52
Lymph cell response, anamnestic: rat, 317 hn, 317-318
Lymph nodes: mouse, 28, 68-69, 101, 114
Lymph nodes: rat, 247, 261, 264, 274-277, 298-299, 302-304
Lymphoblast: rat, 263
Lymphocyte antigens: mouse, 98-99, 115-118, 136, 141
Lymphocyte membrane immunoglobulin markers: mouse, 110 hn
Lymphocytes: mouse, 69, 114, 129 hn, 129-130, 141, 169 hn
Lymphocytes: rat, 266, 315
Lymphocytic choriomeningitis virus: mouse, 227 hn, 228
Lymphocytic neoplasms: mouse, 169-172
Lymphocytotoxin (gene): rat, (see *Ag-F*)
Lymphoid cells: mouse, 147
Lymphoid cells: rat, 320
Lymphoid tissue: rat, 301
Lymphoid tissues: mouse, 69
Lympholysis, cell-mediated: mouse, 129 hn, 129-130
Lymphomas: mouse, 166, 168-172, 228 hn, 228-229
Lymphomas: rat, 246, 249, 253, 296, 299, 301, 303-304
Lymphopenia: mouse, 68-69
Lymphoproliferation (gene): mouse (see *lpr*)
Lymphoreticular sarcomas: rat, 242

Lymphosarcomas: rat, 238, 244, 295-296, 298
Lynestrenol: mouse, 187
Lysine: rat, 281
Lysosomal granules: mouse, 68
Lysosomes: rat, 282
Lyt (genes): mouse, 116-117 (see also *Ly*)
lz [lizard] (gene): mouse, 39*, 67, 151

m [misty] (gene): mouse, 32, 38*, 42, 72
MA (strain): mouse, 25, 103-104, 110, 124, 159
MA/J
biochemical variation, 81-92
body weight, 46
cellular alloantigen, 114
complement factor, 103
genealogy, 16*
leukemia antigens, 168
reproduction, 46
urinary proteins, 106
MA/My, 106
MA/MyJ, 82-88, 116-118
MA_f/Sp, 198
ma [matted] (gene): mouse, 32, 39*, 75
Macrocytic anemia: mouse, 65-66
Macrophage factors: mouse, 141
Macrophage function: rat, 319
Macrophages: mouse, 69, 117, 147
Macula absence: mouse, 149
Magnesium: rat, 285
Mahoganoid (gene): mouse (see *md*)
Mahogany (gene): mouse (see *mg*)
Major histocompatibility antigen: rat, 319-320
Major histocompatibility complex (gene): mouse (see *H-2*)
Major histocompatibility complex phenotype: rat, 321-327
Major urinary protein: mouse, 79, 98
Major urinary protein-1 (gene): mouse (see *Mup-1*)
Malate dehydrogenase: rat, 287
Malate dehydrogenase, mitochondrial (gene): mouse (see *Mor-1*)
Malate dehydrogenases: mouse, 227
Male-specific histocompatibility antigen: mouse, 143
Malformations, congenital: mouse, 55-60
Malformed vertebra (gene): mouse (see *Mv*)
Malic enzyme: mouse, 227
Malic enzyme (genes): mouse (see *Mod-1; Mod-2*)
Malignant tumors: mouse, 174 fn, 177 fn, 179 (*see also* specific tumor)
Malignant tumors: rat (*see* specific tumor)
Maltase: rat, 285 fn
Mammae: mouse, 214
Mammary cancer: mouse, 21, 26, 157, 181-182, 190
Mammary cancer cell karyology: rat, 262
Mammary development, male: mouse, 60 fn
Mammary glands: mouse, 183
Mammary tumor agent: mouse, 21, 24-28
Mammary tumorigenesis: mouse, 161
Mammary tumors: mouse, 21-28, 68, 157-161, 181
Mammary tumors: rat, 239-240, 242-244, 248-249, 253, 294-301, 303-304
Mammary tumor virus inducer (genes): mouse (see *Mtv-1; Mtv-2*)
Mammectomy: mouse, 179
Mannosephosphate isomerase: mouse, 79, 98
Mannose phosphate isomerase (gene): mouse (see *Mpi-1*)
α-Mannosidase processing: mouse, 79, 98-99
α-Mannosidase processing (genes): mouse (see *Map-1; Map-2*)
Map-1 [α-mannosidase processing-1] (gene): mouse, 79, 89, 96*, 98
Map-2 [α-mannosidase processing-2] (gene): mouse, 79, 89, 97*, 99
MAR (strain): rat, 307
Marcel (gene): mouse (see *mc*)
Marker locus, *H-2* gene complex: mouse, 129 hn, 129-130, 134*-135*
Markers, immunoglobulin: mouse, 110 hn
Markers, serological: mouse, 123 hn, 123-124
Marker translocation, cytologic: mouse, 22
Maroon (gene): mouse (see $ru\text{-}2^{mr}$)
Marshall (strain): rat, 262, 307, 335, 338
Marshall 520, 248 (*see also* M520)
MAS (strain): mouse, 159-160
MAS/A, 81-84, 86-87, 89-92, 107, 112
MAS/J, 88
Masked (gene): rat (see *mk*: rat)
Masseteric knob: mouse, 217, 219
Masugi's nephritis: rat, 335
Maternal instincts: rat, 241
Mating preference: mouse, 136
Matted (gene): mouse (see *ma*)
Maudsley (strains): rat, 350 hn-351 hn (*see also* MR)
Maudsley nonreactive, 247, 352 (*see also* MNR)
Maudsley reactive, 247, 352 (*see also* MR)
MAXX (strain): rat, 247, 272-273, 307, 314, 316-318
Maze learning: rat, 236
Maze performance: rat, 352
mc [marcel] (gene): mouse, 38*, 75
MCF (viruses): mouse, 165, 230
md [mahoganoid] (gene): mouse, 32, 39*, 72
mdf [muscle deficient] (gene): mouse, 32
mdg [muscular dysgenesis] (gene): mouse, 60, 60 fn, 67
Mdh-1 (gene): mouse, 79, 89 (see also *Mod-1*)
me [motheaten] (gene): mouse, 32, 38*, 69, 75
Mean arterial pressure: rat, 286
Measles virus: mouse, 21-22
Mechlorethamine: mouse, 202
Mechlorethamine oxide: mouse, 175
med [motor end-plate disease] (gene): mouse, 39*, 67, 156
Megacolon: mouse, 25-26, 151-152
Megaesophagus: rat, 241
Melanin biosynthesis: mouse, 70 hn
Melanoblast survival: mouse, 71 hn
Melanoma: rat, 295
Melanoma S91: mouse, 24
Membrane immunoglobulin markers, lymphocyte: mouse, 110 hn
Membrane labyrinth defects: mouse, 149-150
Membrane molecules: mouse, 135*
Membranous glomerulonephritis: rat, 335
Meningiomas: rat, 300
Meningitis: rat, 337
Mesenchymal tissue: mouse, 152
Mesenchymal tumors: mouse, 174 fn, 177 fn
Mesenteric arterial sclerosis: rat, 291, 294
Mesenteric arteries: rat, 290-291
Mesentery, ileocecal: rat, 238
Mesotheliomas: rat, 249, 296, 299-300
Mestranol: mouse, 187, 198
Metabolic rate, basal: rat, 356 fn

Metaplasia: rat, 253
Metastases: rat, 242, 304 fn
Methylazoxyoctane: mouse, 177
Methylbis(β-chloroethyl)amine: mouse, 202
3-Methylcholanthrene: mouse, 163, 173-175, 177, 193 hn, 202-204, 209
3-Methylcholanthrene: rat, 243, 252, 267, 304, 335
20-Methylcholanthrene: mouse, 173-175, 177, 209 fn
N-Methyl-*N*′-nitro-*N*-nitrosoguanidine: mouse, 173
N-Methyl-*N*′-nitro-*N*-nitrosoguanidine: rat, 303-304
N-Methyl-*N*-nitrosoaniline: rat, 304
Methylnitrosoguanidine: mouse, 210
Methylnitrosourea: mouse, 210
Methylthiouracil: mouse, 191
mg [mahogany] (gene): mouse, 32, 36, 38*, 72
mh [mocha] (gene): mouse, 38*, 72
Mi or *mi* (gene): mouse
mi [microphthalmia], 32, 36, 38*, 59, 72, 151
Mi^{b} [microphthalmia-brownish], 72
mi^{bw} [black-eyed white], 72, 196 fn
mi^{ew} [eyeless white], 59
Mi^{or} [microphthalmia-Oak Ridge], 59, 72
mi^{rw} [red-eyed white], 32, 59, 72
mi^{sp} [microphthalma-spotted], 32, 72
Mi^{wh} [white], 32, 59, 72
mi^{ws} [white-spot], 72
Mi [microphthalmia] (gene): rat, 258
Michaelis constant: mouse, 80 hn, 81
Microcytic anemia (gene): mouse (see *mk*: mouse)
β_2-Microglobulin: mouse, 147 fn-148 fn
β_2-Microglobulin: rat, 320
Microphthalmia: mouse, 23
Microphthalmia: rat, 241, 247, 250
Microphthalmia (gene): mouse (see *mi*)
Microphthalmia (gene): rat (see *Mi*: rat)
Microphthalmia-brownish (gene): mouse (see Mi^{b})
Microphthalmia-Oak Ridge (gene): mouse (see Mi^{or})
Microphthalmia-spotted (gene): mouse (see mi^{sp})
Micropinna-microphthalmia (gene): mouse (see *Mp*)
Microsomal β-glucuronidase: mouse, 28
Microsomes: rat, 284-285, 310
Midbrain: rat, 360
Milan-SHR (strain): rat, 348
Milk agent or factor: mouse, 157 hn
Mimics, mutant: mouse nomenclature rules, 12
Miniature (gene): mouse (see *mn*)
Minor histocompatibility antigen: rat, 320
Misty (gene): mouse (see *m*)
Mitochondria: rat, 280, 282, 287
Mitochondrial aspartate aminotransferase: mouse, 78, 98
Mitochondrial glutamate-oxaloacetate transaminase: mouse, 78, 98
Mitochondrial isocitrate dehydrogenase [IDH-2] (protein): mouse, 96*, 98
Mitochondrial malate dehydrogenase: mouse, 79, 98
Mitochondrial malate dehydrogenase (decarboxylating): mouse, 79, 98, 227
Mitochondrial malic enzyme: mouse, 79, 98
Mitochondrial proline dehydrogenase: mouse, 80
Mitochondrial pyrroline-5-carboxylate reductase: mouse, 80
Mitochondrial pyruvate kinase [PK-3]: mouse, 97*, 99
Mitogens: rat, 319
Mitogens, B-cell: mouse, 69
Mixed leukocyte reaction: mouse, 135*, 136, 141
Mixed lymphocyte reaction: mouse, 129 hn, 129-130
MK (strain): mouse, 26
MK/Re, 32, 46, 48, 65, 81-83, 85-88
mk [masked] (gene): rat, 258
mk [microcytic anemia] (gene): mouse, 32, 39*, 65
MK1 adenocarcinoma: rat, 303
MLR (gene): rat, 260, 313-314, 317
Mls [mouse minor MLC-stimulating] (gene): mouse, 21-24, 26-27, 117
mn [miniature] (gene): mouse, 39*, 60
MNR (strain): rat, 236, 247, 260, 272-273, 351-352
MNR/Brh, 316
MNR/Har, 351, 354-362
MNR/N, 310-311, 316-317, 351
MNRA (strain): rat, 247, 272-273, 351, 354-360, 362
Mo [mottled] (gene): mouse, 39*, 63, 72, 75
Mo^{blo} [blotchy], 32, 72, 75
Mo^{br} [brindled], 32, 72, 75, 151
Mo^{dp} [dappled], 56, 72, 75
Mo^{vbr} [viable-brindled], 72, 75
Mocha (gene): mouse (see *mh*)
Mod-1 [malic enzyme, supernatant] (gene): mouse
biochemical locus, 38*, 96*
electrophoretic polymorphism, 223-224, 226
phenotypic expression, 89
protein control, 79, 98
strain distribution, 21-28, 89
Mod-2 [malic enzyme, mitochondrial] (gene): mouse
biochemical locus, 38*, 96*
electrophoretic polymorphism, 224
phenotypic expression, 89-90
protein control, 79, 98
strain distribution, 21-28, 89-90
Mol-A (strain): mouse, 91
Molars: mouse, 22, 212, 214, 217, 219
Molecular weight, alloantigen: rat, 319-320
Molecular weight, H-2 antigen: mouse, 137 hn, 147-148
MOL-NIH (virus): mouse, 231
Mol-O (strain): mouse, 91
Moloney sarcoma virus: rat, 302
Moloney virus: mouse, 166
MOL-8155 (virus): mouse, 231
Monoamine oxidase: rat, 280 fn
Monoclonal immunoglobulins: rat, 247
Monocytic leukemia: rat, 295-296
Mononuclear cell leukemia: rat, 244, 295 fn-296 fn
Monophenol monooxygenase: mouse, 77, 98
Monosomy: mouse nomenclature rules, 13
Monosomy: rat karyology, 261
MOPC 173 (protein): mouse, 143
MOPC 467 (protein): mouse, 143
MOR/Cv (strain): mouse, 81, 90
Mor-1 [malate dehydrogenase, mitochondrial] (gene): mouse, 38*, 79, 90, 96*, 98, 224-226
Mor-2 [NAD malate dehydrogenase, cytoplasmic] (gene): mouse, 224
Morris hepatoma: rat, 243, 301
Mortality: mouse, 23-25, 27, 48, 68 (*see also* Lethality; Life-span: mouse)
Mortality: rat, 238, 242, 327-328, 337-338 (*see also* Life-span; Longevity: rat)
Motheaten (gene): mouse (see *me*)
Motor coordination: mouse, 150, 155 hn, 155-156
Motor denervation, progressive: mouse, 67
Motor end-plate disease (gene): mouse (see *med*)
Motor neurons: mouse, 67, 153

Mottled (gene): mouse (see *Mo*)
Mottled agouti (gene): mouse (see a^m: mouse)
Mouse (*see* specific condition, gene, strain; *Mus*; Wild mouse)
Mouse leukemia passage A virus: rat response, 303
Mouse nomenclature rules, 9-15
Mouse serum antigen, 80
Mp [micropinna-microphthalmia] (gene): mouse, 32, 57, 151
Mph-1 [macrophage antigen-1] (gene): mouse, 21-28, 38*, 114, 117
Mpi-1 [mannose phosphate isomerase] (gene): mouse, 21-27, 38*, 79, 90, 96*, 98
MR (strain): rat, 236, 247, 272-273, 314, 351-352
  MR/Har, 351, 354-362
  MR/N, 310-311, 317, 351, 362
  MR/Psy, 316
MST-1 sarcoma: rat, 302
MSU (strain): rat, 307
MSUBL/Icgn (strain): rat, 247-248, 272
MSV (virus): mouse, 173, 177
MT (strain): mouse, 88
  MT/Mk, 89
MTV (viruses): mouse, 158-160
Mtv-1 [mammary tumor virus inducer 1] (gene): mouse, 157 hn-158 hn, 158
Mtv-2 [mammary tumor virus inducer 2] (gene): mouse, 157 hn-158 hn, 158
mu [muted] (gene): mouse, 39*, 59, 72
Mud [murine serum antigen] (gene): mouse, 21-28, 104
Mühlbock virus: mouse, 160
Mup [major urinary protein] (gene): mouse, 38*
Mup-a [major urinary protein] (gene): mouse, 106 hn, 106-107 (see also *Mup-1*)
Mup-1 [major urinary protein-1] (gene): mouse, 21-28, 79, 90, 96*, 98, 106 hn, 106
Murine leukemia virus: rat, 303
Murine leukemia viruses: mouse, 68 hn, 68, 116 hn, 116, 165-166, 227 hn-228 hn, 228-229
Murine sarcoma virus: rat, 302-303
Murine sarcoma viruses: mouse, 166, 173, 177
Murine type C viruses: mouse, 98, 165-168, 227-231
Mus spp.: classification & identification, 212-220
M. booduga, 212, 217, 219
M. caroli, 212, 217, 226-227, 229 hn, 230-231
M. cervicolor, 212, 217, 229 hn, 230-231
M. cervicolor cervicolor, 212, 227
M. cookii, 212, 217
M. cookii nagarum, 212, 217
M. cookii palnica, 212, 217
M. crociduroides, 212, 214
M. dunni, 212, 219, 227
M. famulus, 212, 214
M. fernandoni, 212, 214
M. fulvidiventris, 217-218
M. lepidoides, 217-218
M. mayori, 212, 214
M. musculus, 21 hn, 105 hn, 217, 220-226, 230-231
M. musculus bactrianus, 212, 219
M. musculus castaneus, 81-86, 88, 90, 212, 219, 225, 229 hn, 231
M. musculus domesticus, 80-81, 88, 212, 219, 225, 229 hn
M. musculus gansuensis, 212, 219
M. musculus homourus, 212, 219
M. musculus manchu, 212, 219
M. musculus molossinus
  biochemical variation, 80-86, 88, 90
  centromeric heterochromatin, 44 hn
  classification & identification, 212, 219
  electrophoretic polymorphism, 226
  heavy chain linkage groups, 111
  type C RNA viruses, 229 hn, 231
M. musculus mongolium, 212, 219
M. musculus musculus
  biochemical variation, 80, 85, 88, 90
  centromeric heterochromatin, 44 hn
  classification & identification, 212, 219
  electrophoretic polymorphism, 226
  type C RNA viruses, 229 hn, 230-231
M. musculus tantillus, 212, 219
M. musculus tytleri, 212, 219
M. musculus yamashinai, 212, 219
M. pahari, 212, 214
M. phillipsi, 212, 214
M. platythrix, 212, 214
M. poschiavinus, 41 hn, 41
M. saxicola, 212, 214
M. saxicola gurkha, 212
M. shortridgei, 212, 214, 227
M. terricolor, 219
M. vulcani, 212, 217
Muscle; muscle tissue: mouse, 66-67, 79, 114, 150 (*see also* specific muscle)
Muscle; muscle tissue: rat, 279, 286-287, 289-291, 333 (*see also* specific muscle)
Muscle cell karyology: rat, 263, 267
Muscle deficient (gene): mouse (see *mdf*)
Muscular degeneration & dystrophy: rat, 291
Muscular dysgenesis (gene): mouse (see *mdg*)
Mutant genes: mouse, 55-76 (*see also* specific gene)
Mutant genes: rat, 257-258 (*see also* specific gene)
Mutants: rat, 234, 236-237 (*see also* Mutant genes; specific gene: rat)
Mutations: inbred strains, 4
Mutations: mouse, 29-36, 149-153, 193 hn, 196 (*see also* Mutant genes; specific gene: mouse)
Mutations: mouse nomenclature rules, 12, 14
Muted (gene): mouse (see *mu*)
Mv [malformed vertebra] (gene): mouse, 55
MW (strain): rat, 248
MWT (strain): mouse, 66, 87
  MWT/Le, 30, 32, 34-35, 85
  MWT/WeGn, 90
  MWT/Ww, 82-84, 86-91, 107
my [blebs] (gene): mouse, 38*, 57, 60 fn, 151
Myasthenia gravis, autoimmune: rat, 334
Mycobacterium: rat, 334
M. tuberculosis: rat, 337
Mycoplasma: rat, 337
MYD/Le (strain): mouse, 32, 48
myd [myodystrophy] (gene): mouse, 32, 38*, 67
Myelencephalic blebs: mouse, 151
Myelin: mouse, 149, 151
Myelination, defective: mouse, 67
Myelin lipids: rat, 287
Myelin sheath: rat, 294
Myelin synthesis deficiency (gene): mouse (see jp^{msd})
Myeloblastic leukemias, transplantable: rat, 261
Myelomas: mouse, 117
Myeloschisis: mouse, 152

Myenteric plexus neurons: mouse, 152
Myleran: mouse, 201
Myoblast differentiation: mouse, 67
Myocardial degeneration: rat, 290-291
Myocardial fibrosis: rat, 290, 350*
Myocardial infarction: rat, 290
Myocarditis, autoimmune: rat, 246
Myocardium: rat, 285
Myodystrophy (gene): mouse (see *myd*)
Myopathy, progressive: mouse, 67
Myositis, allergic: rat, 334
Myotubes, striated: mouse, 67
M14 (strain): rat, 248, 273
M17 (strain): rat, 248, 273
M520 (strain): rat, 234, 248, 272-273, 299, 314, 316-318
  M520/N, 299, 310-311

N (strain): mouse, 16*, 106, 183
  N/St, 81-82, 88-89, 124
N [naked] (gene): mouse, 32, 39*, 60, 75
n [naked] (gene): rat, 258
NAD-α-glycerol phosphate dehydrogenase: mouse, 78, 227
NAD-α-glycerol phosphate dehydrogenase (gene): mouse (see *Gdc-1*)
NADH cytochrome *c* reductase: rat, 284 fn-285 fn
NADH dehydrogenase: rat, 284-285
NAD malate dehydrogenase, cytoplasmic (gene): mouse (see *Mor-2*)
NADPH cytochrome reductase: rat, 244, 284
Naked (gene): mouse (see *N*)
Naked (gene): rat (see *n*)
Nasals: mouse, 217
Nasopharyngeal carriage: rat, 337
National Cancer Institute diet: mouse, 165
NB (strain): mouse, 17*, 125
Nb (strain): rat, 299, 304
nb [normoblastic anemia] (gene): mouse, 32, 65
NBC (strain): mouse, 91
NBL (strain): mouse, 110
  NBL/N, 90, 106
NBR (strain): mouse, 48, 84, 87-89
NBR (strain): rat, 248, 272, 307, 314, 317, 334
  NBR Pl/Cr, 303
  NBR/1, 316-318
  NBR/2, 316-318
NC (strain): mouse, 48, 84, 87-89, 91-92
  NC/Nga, 89
nc [non-agouti curly] (gene): mouse, 72, 75
Ne [hydronephrosis] (gene): rat, 258
Necrosis, seminiferous tubule: mouse, 48
Necrosis, tubular: rat, 291
NEDH (strain): rat, 246 (*see also* Slonaker)
Neophobia: rat, 357
Neoplasm: mouse, 68, 169-172, 201
Neoplastic lesions: rat, 242, 294-297
Nephritis: mouse, 26, 69
Nephritis: rat, 244, 248, 291, 335-336
Nephrogenic diabetes insipidus: mouse, 27
Nephropathy: rat, 293
Nephrosclerosis: rat, 290, 292, 350*
Nephrosis: rat, 252, 290-291, 294
Nephrotoxic serum nephritis: rat, 335
Nerve disorders: mouse, 67, 150 (*see also* specific disorder)
Nerve disorders: rat, 291, 294 fn (*see also* specific disorder)
Nerves (*see* specific nerve)
Nervous (gene): mouse (see *nr*)
Nervous system development: mouse, 50-52
Neural tube defects: mouse, 150-151
Neuritis, allergic: rat, 334
Neuroepithelia, inner ear: mouse, 151
Neurologic mutants: mouse, 149-153
Neurons: mouse, 66-67, 152-153
Neurophysin: rat, 339
Neutrons: mouse, 180, 189-190, 201
Newcastle disease virus: mouse, 77 hn, 79, 112 hn
New Zealand black (strain): mouse, 17* (*see also* NZB)
New Zealand chocolate (strain): mouse, 17* (*see also* NZC)
New Zealand white (strain): mouse, 17* (*see also* NZW)
NFS/N (strain): mouse, 167
NGP/N (strain): mouse, 17*, 82, 88-89
NG-W1 adenocarcinoma: rat, 304
NH (strain): mouse, 16*, 26, 48, 111, 159, 179
  NH/Lw, 90, 106
  NH/LwN, 88-89
  NH/N, 81-82
  NH hybrids, 159, 180
NH (strain): rat, 316
NIG-III (strain): rat, 248, 260, 262, 272-273
NIG-IV (strain): rat, 261
NIH (strain): mouse, 167
  NIH Swiss, 209, 228 hn
NIH black (strain): rat, 248 (*see also* NBR: rat)
NIH 3T3 cell line: mouse, 227 hn
Nijmegen waltzer (gene): mouse (see *nv*)
Nil [neonatal intestinal lipidosis] (gene): mouse, 38*
Nipples: mouse, 47, 49
Nitrogen, blood urea: rat, 279, 287
Nitrogen mustard: mouse, 202
Nitrogen mustard *N*-oxide: mouse, 175
N-Nitrosodiethylamine: mouse, 176
N-Nitrosodiethylamine: rat, 261
N-Nitrosodimethylamine: mouse, 173-176, 203
N-Nitroso-*N*-methylurea: mouse, 203
NLC (strain): mouse, 88
  NLC/Rd, 81-82, 89
NMRI (strain): mouse, 84, 209 hn, 209
  NMRI/Lac, 36, 85
Nocardia: rat, 334
Nodular hyperplasia: mouse, 26, 179
Nodular hyperplasia: rat, 244, 296
Nodule-inducing virus: mouse, 160
Nomenclature: mouse, 3, 9-15
Non-agouti (gene): mouse (see *a*: mouse)
Non-agouti (gene): rat (see *a*: rat)
Non-agouti curly (gene): mouse (see *nc*)
Non-agouti dark belly (gene): mouse (see a^{da})
Non-agouti lethal (gene): mouse (see a^{l})
Non-albino (gene): rat (see *C*: rat)
Non-brown (gene): rat (see *B*: rat)
Non-erupted teeth (gene): mouse (see *tl*: mouse)
Non-hooded (gene): rat (see *H*: rat)
Non-neoplastic lesions: rat, 289-294
Non-pink-eyed dilute (gene): rat (see *P*)
Noradrenaline: rat, 251 (*see also* Norepinephrine: rat)
Noradrenalin *N*-methyltransferase: mouse, 80, 93
Noradrenalin *N*-methyltransferase: rat, 362

Norepinephrine: mouse, 101
Norepinephrine: rat, 251, 280-281, 362
Norethindrone: mouse, 198
Norethindrone acetate: mouse, 187
Norethisterone acetate: mouse, 187
Norethynodrel: mouse, 187, 198
Normoblastic anemia (gene): mouse (see *nb*)
Normochromic macrocytic anemia: mouse, 65
19-Norprogesterone: mouse, 198
Notch (gene): rat (see h^n)
Novikoff hepatoma: rat, 244, 267
Np-1 [nucleoside phosphorylase-1] (gene): mouse, 39*, 79, 90, 97*, 99, 225-226
nr [nervous] (gene): mouse, 32, 38*, 151, 156
NSD (strain): rat, 248, 273, 316-317
nu [nude] (gene): mouse
biochemical locus, 39*, 97*, 99
congenital defects, 60, 60 fn, 75
endocrine effects, 101
immune system development, 69
strain distribution, 27, 36
nu^{str} [nude-streaker], 33, 69, 75, 101
Nucleolus organizers: mouse nomenclature rules, 13
Nucleoside phosphorylase: mouse, 99, 227
Nucleoside phosphorylase-1 (gene): mouse (see *Np-1*)
Nucleus, lateral septal: mouse, 25
Nucleus, lens: rat, 292-293
Nude (gene): mouse (see *nu*)
Nudeness: mouse, 27
Nude stock: mouse, 176
Nude-streaker (gene): mouse (see nu^{str})
Nutrition research: rat, 237
nv [Nijmegen waltzer] (gene): mouse, 38*, 151, 156
NX8 (strain): mouse, 37
NZB (strain): mouse
antigens, 114, 116-118, 167-168
characteristics, 26
complement factors, 103-104
congenic-inbred line progenitor, 111
genealogy, 17*
genes, 26
H-2 haplotype & markers, 124
Ig heavy chain linkage groups, 111
leukemia virus origin, 166
reproduction, 48
NZB/Bl, 171, 195
NZB/BlN, 87
NZB/BlNJ, 81-82, 85-87, 90-91, 171
NZB/BlNRe, 85, 88
NZB/BlNReJ, 83, 86
NZB/BlUmc, 171
NZB/Hz, 84, 86, 88-91, 106
NZB/Icr, 37
NZB/J, 86
NZB/Lac, 36
NZB.Ig, 111
NZB (strain): rat, 303
NZB-IU (viruses): mouse, 166, 230
NZC (strain): mouse, 17*, 26, 87-88
NZC/Bl, 81-82, 87, 89, 193 hn, 196
NZO (strain): mouse, 26, 48, 88
NZO/Bl, 171, 195, 211
NZO/L, 81-82
NZR/Gd (strain): rat, 249, 273, 299
NZW (strain): mouse, 17*, 26, 36, 111, 114
NZW/BlN, 88-89
NZW/Cr, 46
NZW/N, 81-82
NZX (strain): mouse, 26, 48
NZY (strain): mouse, 26, 88, 181
NZY/Bcr, 202
NZY/Bl, 81-82, 87, 195

O [operator] (gene): mouse, 78 fn, 79, 90
o [eyelids-open-at-birth] (gene): mouse, 59
ob [obese] (gene): mouse
chromosome position, 38*
endocrine effects, 101
immune system development, 69
mammary tumorigenesis, 161
pleiotropic behavior, 156
strain distribution, 33, 36
Obese (strain): rats, 312
Obese-SHR, 348
Obese (gene): mouse (see *ob*)
Obesity: mouse, 48, 68-69, 101 (see also *ob*)
Obesity: rat, 246, 249, 251, 348
Obstructive hydrocephalus (gene): mouse (see *oh*)
oc [osteosclerotic] (gene): mouse, 39*, 56, 151
Och [ochre] (gene): mouse, 38*, 72
Ochre (gene): mouse (see *Och*)
Ocular retardation (gene): mouse (see *or*)
oe [open eyelids] (gene): mouse, 39*, 59
oel [open-eyelids with cleft palate] (gene): mouse, 59
OFA (strain): rat, 283, 307
oh [obstructive hydrocephalus] (gene): mouse, 58, 151
OIR (strain): mouse, 159-160
OKA (strain): rat, 249 fn, 256, 307, 314, 316-317
OKA/Wsl, 249, 273
Okamoto (strain): rat, 249, 307 (*see also* OKA)
ol [oligodactyly] (gene): mouse, 38*, 57, 60 fn
Oligodactyly (gene): mouse (see *ol*)
Oligosyndactylism (gene): mouse (see *Os*)
Oligotriche (gene): mouse (see *olt*)
olt [oligotriche] (gene): mouse, 75
OM (strain): rat
aging, 291
alloantigenic alleles, 316-317
behavioral research, 351
cestode infections, 338
characteristics, 249, 360
coat color, 249, 273
eye color, 273
immunoglobulin polymorphism, 307
immunologically mediated disease, 335
immunological responsiveness, 318
karyology, 262
origin, 249
tumors, 299, 303
OM/N, 290, 299
om [ovum mutant] (gene): mouse, 63
Omnivorousness: rat, 237
Oncogenic virus susceptibility: mouse, 136, 141
Ontogeny, complement receptor lymphocyte: mouse, 136 hn, 136
Oocysts: rat, 338
op [osteopetrosis] (gene): mouse, 39*, 60
op [osteopetrosis] (gene): rat, 258

Open eyelid, congenital: mouse, 60 fn
Open eyelids (gene): mouse (see *oe*)
Open-eyelids with cleft palate (gene): mouse (see *oel*)
Open-field ambulation: rat, 240, 245, 247, 251-252, 356
Open-field defecation: rat, 240, 244-245, 247, 249, 252, 358
Operator (gene): mouse (see *O*)
Ophthalmatrophy (gene): mouse (see lg^{ga})
Opisthotonus: mouse, 149, 151
Opisthotonus (gene): mouse (see *opt*)
Opossum (gene): mouse (see Ra^{op})
opt [opisthotonus] (gene): mouse, 38*, 151
Optic (*see also* Eye entries)
Optic cup development: mouse, 151
Optic nerve defects: mouse, 150-151
or [ocular retardation] (gene): mouse, 33, 59, 151
Orchiectomy: mouse, 179
Orchitis, autoimmune: rat, 336
Organ of Corti degeneration: mouse, 150
Organ weights: mouse, 136
Organ weights: rat, 274-277, 354-355
Ornithine carbamoyltransferase: mouse, 80
Ornithine transcarbamylase: mouse, 80
Orofacial defects: mouse, 60, 60 fn
Os [oligosyndactylism] (gene): mouse, 33, 38*, 57, 60 fn
Osborne-Mendel (strain): rat (*see* OM)
Oscillator (gene): mouse (see *ot*)
Osteoclasts: rat, 309
Osteogenic sarcoma 344: rat, 248
Osteopetrosis: mouse, 151
Osteopetrosis (gene): mouse (see *op*: mouse)
Osteopetrosis (gene): rat (see *op*: rat)
Osteosarcoma: mouse, 166
Osteosarcomas: rat, 302-303
Osteosclerotic (gene): mouse (see *oc*)
ot [oscillator] (gene): mouse, 151
Otitis media: rat, 337
Otocephaly: mouse, 60 fn
Otoconia: rat, 291
Otoliths: mouse, 151, 153
OUCW strain: mouse, 46
Ova: mouse, 48
Ovalbumin: mouse, 143
Ovalbumin: rat, 247
Ovarian follicular structures: mouse, 102
Ovarian tumors: mouse, 23, 25-26, 181, 193-204
Ovarian tumors: rat, 299
Ovariectomy: mouse, 178-179, 182, 192, 198, 204
Ovaries: mouse, 22, 24, 48
Ovaries: rat, 274-277, 283, 285
Ovary transplant: mouse nomenclature rules, 11
Ovary transplants: mouse, 180, 193 hn, 198, 204
Ova transfer: mouse, 11, 198
Ovomucoid: mouse, 143
Ovomucoid: rat, 318
Ovulation: mouse, 101, 154
Ovum mutant (gene): mouse (see *om*)
Ox-1 (gene): mouse, 23-24
Ox-2 (gene): mouse, 23-24
Oxazolone: mouse, 69
Oxygen consumption: mouse, 27
Oxygen consumption: rat, 280, 285, 356
O20 (strain): mouse
    biochemical variation, 83-84, 86-90, 92
    characteristics, 26
    genes, 26
    tumorigenesis inhibition, 158 hn
    tumors, 159-160, 183, 188, 191
    urinary proteins, 107
  O20/A, 81-82, 84, 86-87, 112
  O20 hybrids, 159, 181, 183, 188, 191

P (strain): mouse, 17*, 103-104
  P/A, 112
  P/J
    biochemical variation, 81-92
    body weight, 46
    characteristics, 26
    complement factors, 103
    congenic-inbred line progenitor, 125, 128
    genealogy, 16*
    genes, 26
    H-2 haplotype & markers, 124
    Ig heavy chain linkage groups, 110
    reproduction, 46
    urinary proteins, 107
p [pink-eyed dilution] (gene): mouse
    chromosome aberrations, 42
    chromosome position, 38*
    endocrine effects, 102
    histocompatibility linkage, 122, 123*
    inbred strain genealogy, 17*
    mutation, 72
    protein control, 77 fn
    strain distribution, 22, 25-26, 28, 33, 36
  p^{bs} [*p*-black-eyed sterile], 72
  p^{cp} [*p*-cleft palate], 60, 72
  p^{d} [dark pink-eye], 72
  p^{dn} [*p*-darkening], 72
  p^{m} [pink-eyed mottled], 72
  p^{r} [Japanese ruby], 72
  p^{s} [pink-eyed sterile], 72, 151
  p^{un} [pink-eyed unstable], 33, 72
  p^{x} [*p*-extra dark], 72
P or *p* (gene): rat
  P [non-pink-eyed dilute], 241, 273
  p [pink-eyed dilution], 234, 240-241, 243, 249-250, 259, 273
  p^{m} [ruby-eyed dilution], 248, 273
PA (strain): rat, 249, 273
pa [pallid] (gene): mouse
    chromosome aberrations, 42
    chromosome position, 38*
    ear defect, 59
    histocompatibility linkage, 122, 123*
    mutations, 72, 151
    strain distribution, 33
pad [paddle] (gene): mouse, 57
Paddle (gene): mouse (see *pad*)
P.A. King albino (strain): rat, 249
Palate, cleft: mouse, 21, 60 fn
Palate, cleft: rat, 251
Palatine foramina, posterior: mouse, 217
Pale ears (gene): mouse (see *ep*)
Pallid (gene): mouse (see *pa*)
Palpebra operta [*sic*] (gene): mouse (see *po*)
Pancreas: mouse, 77, 80, 198
Pancreas: rat, 248, 289-291, 296 fn, 304, 323

Pancreatic α-amlyase (gene): mouse (see *Amy-2*)
Pancreatic islets: rat
adenomas, 242, 252, 295, 298
aging, 291-292, 295
transplant survival, 321 hn, 323, 332-333
Panda-white (gene): mouse (see W^{pw})
Papillary adenoma: mouse, 195, 201-202
Papillary cystadenomas: mouse, 193 hn, 194, 200
Papilliferous cystadenoma: mouse, 196
Papillomas: rat, 243, 295, 298
Parabiosis: mouse, 179
Parafollicular cell tumors: rat, 298-300
Paralysis: mouse, 60 fn, 67, 149, 151, 153, 228 hn, 228-229
Paralysis: rat, 289-290, 294
Paralyzed/N (strain): rat, 316-317
Paranasal sinus: rat, 302
Parathormone: rat, 288 fn
Parathyroid: rat, 290, 293, 296 fn
Parathyroid hormone: rat, 288
Parenchymal cells, pancreatic: rat, 291
Paresis: mouse, 152
Parietal bones: mouse, 214
Pars intermedia: mouse, 181, 183, 186, 188
Passenger genes: mouse, 6
Passive performance (gene): mouse (see *Pp*)
Patch (gene): mouse (see *Ph*)
Patch-extended (gene): mouse (see Ph^e)
Pathogen-free stocks: rat, 237
PB (strain): rat, 307
PBB/Ld (strain): mouse, 48
p-black-eyed sterile (gene): mouse (see p^{bs})
PBR (strain): mouse, 106
pc [phocomelic] (gene): mouse, 57
Pca-1 (gene): mouse, 21-28, 117
p-cleft palate (gene): mouse (see p^{cp})
PD (strain): rat, 316
Pd [pyrimidine degradation] (gene): mouse, 22-28, 79, 90
p-darkening (gene): mouse (see p^{dn})
pe [pearl] (gene): mouse, 33, 39*, 72
Pearl (gene): mouse (see *pe*)
Pelage: mouse, 214
Pentobarbital sodium: rat, 239, 242, 244
Pep-c (gene): mouse, 77, 83 (see also *Dip-1*)
Peptidase [PEP] (proteins): mouse, 96*-97*, 98-99
Peptides: rat, 320
Periarteritis: rat, 291, 294
Pericardium, distended: mouse, 60 fn
Pericentric inversion: mouse nomenclature rules, 13
Peripheral nerve defects: mouse, 67, 150
Peripheral nerve homogenate: rat, 334
Peripheral T-cells: rat, 315
Peritoneal exudate cells: mouse, 114, 117
PERU (strain): mouse, 82-83, 85, 88, 101
PERU-Atteck (strain): mouse, 86
Peru-Emaus (strain): mouse, 85
PETH (strain): rat, 249, 250 fn, 272-273
PETH/N, 310-311, 316-317
p-extra dark (gene): mouse (see p^x)
Peyer's patches: mouse, 68-69
pf [pupoid-fetus] (gene): mouse, 38*, 57, 75
pg [pygmy] (gene): mouse, 38*, 59, 102
Pgd [6-phosphogluconate dehydrogenase] (gene): mouse, 38*, 79, 90, 96*, 98, 224-226
Pgd [phosphogluconate dehydrogenase] (gene): rat, 258, 305, 312
Pgk-1 [phosphoglycerate kinase-1] (gene): mouse, 79, 90, 97*, 99, 226
Pgk-2 [phosphoglycerate kinase-2] (gene): mouse, 39*, 79, 90, 97*, 99
Pgm [phosphoglucomutase] (gene): rat, 305
Pgm-1 [phosphoglucomutase-1] (gene): mouse
biochemical locus, 38*, 96
electrophoretic polymorphism, 224
phenotypic expression, 91
protein control, 79, 98
strain distribution, 21-28, 91
Pgm-2 [phosphoglucomutase-2] (gene): mouse
biochemical locus, 38*, 96*
electrophoretic polymorphism, 224
phenotypic expression, 91
protein control, 79, 98
strain distribution, 21-28, 91
pH, enzyme: mouse, 92
pH, urine: rat, 283
PH (strain): mouse, 48
PH/Re, 33, 81-91, 104
PH/ReJ, 84, 90
Ph [patch] (gene): mouse, 33, 38*, 58, 60 fn, 72
Ph^e [patch-extended], 58, 60 fn, 73
Phenicarbazide: mouse, 177
Phenobarbital: rat, 309
Phenotype: mouse, 53-55, 68-69, 80-93, 101-102, 142-144
Phenotype: rat, 260, 321-323, 324 hn, 324-327
Phenylalanine hydroxylase: mouse, 77
Phenylalanine 4-monooxygenase: mouse, 77
Phenylethanolamine *N*-methyltransferase: mouse, 80
Phenylethanolamine-*N*-methyltransferase: rat, 362 fn
β-Phenylhydrazine sulfate: mouse, 177
Pheochromocytomas: rat, 242, 252, 295, 298-300
Pheromonal response (gene): mouse (see *Phr*)
PHH (strain): mouse, 46, 103-104
Phk [phosphorylase kinase] (gene): mouse
biochemical locus, 39*, 97*
phenotypic expression, 91
protein control, 79, 99
strain distribution, 22-23, 25, 91
PHL (strain): mouse, 46, 88, 103-104
PHL/We, 81-82, 86-87, 89
Phocomelic (gene): mouse (see *pc*)
Phosphodiesterase: rat, 285
Phosphoglucomutase (gene): rat (see *Pgm*)
Phosphoglucomutase (genes): mouse (see *Pgm-1; Pgm-2*)
Phosphoglucomutase (protein): mouse, 79, 98
Phosphoglucomutase (protein): rat, 312
Phosphogluconate dehydrogenase (gene): rat (see *Pgd*: rat)
Phosphogluconate dehydrogenase (protein): mouse, 79, 98, 227
Phosphogluconate dehydrogenase (protein): rat, 312
6-Phosphogluconate dehydrogenase (gene): mouse (see *Pgd*: mouse)
Phosphoglycerate kinase (genes): mouse (see *Pgk-1; Pgk-2*)
Phosphoglycerate kinase (protein): mouse, 227
Phospholipid: rat, 284-285
Phospholipid deposits: mouse, 152
Phosphorus, myocardial: rat, 285

Phosphorus, serum: mouse, 101
Phosphorus-32: rat, 302
Phosphorylase kinase (gene): mouse (see *Phk*)
Phosphorylase kinase (protein): mouse, 79, 99
Phosphorylation, oxidative: rat, 287
Photoreceptor cells: mouse, 152
Phr [pheromonal response] (gene): mouse, 23, 27
Phytohemagglutinin: mouse, 68 hn, 68
Phytohemagglutinin: rat, 319
pi [pirouette] (gene): mouse, 33, 38*, 122, 123*, 151, 156
Pia-arachnoid membranes: mouse, 150
Picryl chloride: mouse, 68
Piebald (stock): mouse, 16*
Piebald (gene): mouse (see *s*: mouse)
Piebald-lethal (gene): mouse (see s^l)
Piebald mutant: rat, 234
Pigmentation, coat: mouse, 28 (*see also* Coat color)
Pigtail (gene): mouse, 57
Pine bedding: mouse, 165
Pink-eyed (stock): mouse, 16*
Pink-eyed dilution (gene): mouse (see *p*: mouse)
Pink-eyed dilution (gene): rat (see *p*: rat)
Pink-eyed mottled (gene): mouse (see p^m: mouse)
Pink-eyed sterile (gene): mouse (see p^s)
Pink-eyed unstable (gene): mouse (see p^{un})
Pintail (gene): mouse (see *Pt*)
Pirouette (gene): mouse (see *pi*)
Pituitary: mouse, 25-26
Pituitary: rat, 243-244, 355
Pituitary hormones: mouse, 68, 101
Pituitary hormones: rat, 339, 340 hn, 340-343, 346-347
Pituitary-thyroid axis: rat, 339
Pituitary tumors: mouse, 24, 181-192
Pituitary tumors: rat, 239, 242-244, 248-249, 252-253, 295-301
Pivoter (gene): mouse (see *Pv*)
pk [plucked] (gene): mouse, 33, 75
PL (strain): mouse, 26, 104, 110, 114, 116-118
PL/J
biochemical variation, 81, 83-92
complement factors, 103
congenic-inbred line progenitor, 115, 125
genealogy, 17*
H-2 haplotype & markers, 124
histocompatibility typing, 122
leukemia antigens, 168
lymphoma, 169
reproduction, 46
viral antigens, 167
Plantar foot pads: mouse, 212
Plasma: mouse, 101, 106 hn, 148, 220 hn
Plasma: rat, 248, 251, 310
Plasma corticosterone level (genes): mouse (see *Cpl-1; Cpl-2*)
Plasmacytomas: rat, 236, 246, 299, 302-303
Plasma protein (gene): rat (see *Gl-1*)
Plasma serotonin level (gene): mouse (see *Spl*)
Platinum (gene): mouse (see c^p)
Pleiotropic behavior: mouse, 155-156
Plucked (gene): mouse (see *pk*)
Plutonium-232: rat, 302
pn [pugnose] (gene): mouse, 39*
Pneumococcal polysaccharide, type III: mouse, 22, 69
Pneumonia: rat, 291
Pneumonitis, spontaneous: mouse, 68
Po [postaxial polydactyly] (gene): mouse, 56
po [palpebra operta [*sic*]] (gene): mouse, 59
Podocytes: rat, 293
POLY (strain): mouse, 90
POLY/St, 88-89
POLY-2, 107
Polyarteritis nodosa: rat, 292
Polycystic kidneys: mouse, 22, 26
Polycystic nephrosis: rat, 252
Polydactyly: rat, 253
Polydactyly (gene): mouse (see *py*)
Polydactyly-luxate (gene): rat (see *lx*: rat)
Polydipsia: mouse, 24-25, 27, 155
Polymorphisms: mouse, 220-227
Polymorphisms: rat, 261 hn, 261-265, 267, 305-312, 339
Polyoma virus: mouse, 176, 227 hn
Polyoma-virus-induced tumor: rat, 301
Polyovular follicles: mouse, 24, 48
Polypeptide antigens, synthetic: mouse, 142
Polypeptides: rat, 313-314, 317-318
Polyps, intestinal: mouse, 208 hn, 208, 209 fn
Polysaccharide, type III pneumococcal: mouse, 22, 68 hn, 69
Polyspermia: rat, 249, 252
Polysyndactyly (gene): mouse (see *Ps*)
Polyuria: mouse, 24-25
PON (strain): mouse, 91
Porcine lactic dehydrogenases: mouse, 144
Porcine tail (gene): mouse (see *pr*)
Porphobilinogen synthase: mouse, 79, 98
Porton/Lac (strain): rat, 276-277
Postaxial hemimelia (gene): mouse (see *px*)
Postaxial polydactyly (gene): mouse (see *Po*)
Posterior palatine foramina: mouse, 217
Potassium: rat, 280, 283, 285
Pp [passive performance] (gene): mouse, 23-24, 154
pr [porcine tail] (gene): mouse, 55, 151
Pre-1 [prealbumin-1 protein] (gene): mouse, 21-28, 79, 91, 104, 224
Pre-2 [prealbumin-2 protein] (gene): mouse, 79, 91, 104 (see also *Pre-4*)
Pre-4 (gene): mouse, 21-25, 91 (see also *Pre-2*)
Prealbumin protein (genes): mouse (see *Pre-1; Pre-2*)
Prealbumin proteins: mouse, 79, 220 hn
Prealbumins, urinary: mouse, 105 hn
Pregnancy: mouse, 47-49, 203-204
Pregnancy: rat, 345-346
Pregnancy-dependent mammary tumors: mouse, 157 hn
Prenatal mortality: mouse, 48
Prenatal mortality: rat, 238, 242
Prenatal stages: man & laboratory mammals, 52-53
Preservation by freezing: mouse nomenclature rules, 12
Pressure (*see* specific pressure)
Primitive streak: mouse, 50, 151
Private specificities, *H-2* gene complex: mouse, 137 hn, 138-140, 139 hn-140 hn
PRO (strain): mouse
PRO/J, 86
PRO/Re, 33, 83, 85, 87-88, 91, 107
Pro-1 [proline oxidase] (gene): mouse, 23-24, 28, 33, 80, 91
Problem solving: rat, 358
Proestrus: rat, 343-344
Progesterone: mouse, 190, 198
Progesterone: rat, 341-347

Progestins: mouse, 187, 198
Prolactin: mouse, 101 hn, 101
Prolactin: rat, 283, 288, 340-346
Prolactin-dependent tumor: rat, 300 fn
Prolapse, vaginal: rat, 240
Proline oxidase (gene): mouse (see *Pro-1*)
1-Propyl nitrosourea: rat, 303
Propylthiouracil: mouse, 188
Prosencephalon: rat, 287
Prostaglandins: rat, 245, 339
Prostate: rat
enzyme polymorphism, 311
tumors, 239, 243, 296 fn, 297-298, 300-301
Protein(s): mouse, 77-80, 96*-97*, 98-99, 103-107, 135*, 148 (*see also* specific protein)
Protein(s): rat, 280-281, 285, 320 (*see also* specific protein)
Protein, guinea pig basic: rat response, 246, 318
Proteinase (genes): mouse (see *Prt-1; Prt-2; Prt-3*)
Proteinase (protein): mouse, 80, 98
Proteins, Bence-Jones: rat, 247
Proteinuria: rat, 288
Protons: mouse, 188, 199
Protozoa infections: rat, 338
Proviruses: mouse, 157 hn
Prt-1 [proteinase-1] (gene): mouse, 21-24, 80, 91
Prt-2 [proteinase-2] (gene): mouse, 21-24, 38*, 80, 91, 96*, 98
Prt-3 [proteinase-3] (gene): mouse, 80, 91
Ps [polysyndactyly] (gene): mouse, 33, 38*, 56
Pseudoencephaly: mouse, 58, 149
Pseudohermaphroditism: rat, 339
Pseudopregnancy: mouse, 25, 48, 201-203
Pseudopregnancy: rat, 345
Psoas major muscle: rat, 286
p-sterile (gene): mouse (see p^s)
Pt [pintail] (gene): mouse, 33, 38*, 57, 106 hn
pu [pudgy] (gene): mouse, 38*, 56
Puberty: rat, 340-342
Public specificities, *H-2* gene complex: mouse, 137 hn, 140 hn, 140
Pudgy (gene): mouse (see *pu*)
Pulmonary adenomas: mouse, 27, 161-163
Pulmonary foam cells: rat, 292
Pulmonary infections: rat, 244, 290
Pulmonary lymph node malignant lymphomas: rat, 299
Pupoid-fetus (gene): mouse (see *pf*)
Purina chow: mouse, 165
Purine nucleoside phosphorylase: mouse, 79
Purkinje cells: mouse, 149, 151
Purkinje cells: rat, 294
Pv [pivoter] (gene): mouse, 59, 151, 156
PVG (strain): rat (*see also* HO)
alloantigenic alleles, 316-317
behavioral research, 351
blood protein polymorphism, 309
characteristics, 249
coat color, 249, 272-273
congenic-inbred line progenitor, 256
enzyme polymorphism, 311
immunogenetic characterization, 314
immunological responsiveness, 318
origin, 234, 249
transplant survival, 321, 323, 325
PVG/c, 307, 334-336
PVG/Lac, 275
PVG congenics, 256
(PVG x AGUS)F_1, 323
px [postaxial hemimelia] (gene): mouse, 38*, 57-58
py [polydactyly] (gene): mouse, 38*, 56
Pyelonephritis: rat, 291
Pygmy (gene): mouse (see *pg*)
Pygmy mouse, African, 212
Pyknosis: mouse, 48
Pyrimidine: mouse, 79
Pyrimidine degradation (gene): mouse (see *Pd*)
Pyromys, 212, 214
Pyrophosphate [PP] (protein): mouse, 97*, 99
P1534 leukemia: mouse, 24

Q [quinky] (gene): mouse, 38*, 57, 151
Qa-1 [lymphoid cell differentiation antigen-1] (gene): mouse, 117, 136
Qa-2 [lymphoid cell differentiation antigen-2] (gene): mouse, 117, 136
QF-5604 (strain): mouse, 48
QF-5612 (strain): mouse, 48
qk [quaking] (gene): mouse, 33, 39*, 53-54, 151, 156
Quaking (gene): mouse (see *qk*)
Quinacrine fluorescent banding karyotype: rat, 268*-269*
Quinacrine mustard dihydrochloride: rat, 266 hn, 266-267
Quinky (gene): mouse (see *Q*)
Quivering (gene): mouse (see *qv*)
qv [quivering] (gene): mouse, 38*, 151, 156

R (stock): mouse, 195, 200
R (strain): rat, 250, 273, 307, 316, 351
R (gene): mouse, 115 (see also *Ea-2* gene)
r [red-eyed dilution] (gene): rat, 259, 273
r [rodless retina] (gene): mouse, 152
Ra (strain): rat, 337
Ra [ragged] (gene): mouse, 33, 38*, 60, 75
Ra^{op} [opossum], 60, 75
Ra [renal adenoma] (gene): rat, 237
Rabbit skin graft: rat response, 329
Rachischisis, caudal: mouse, 152
Rachiterata (gene): mouse (see *rh*)
Radiation: mouse (*see also* X rays: mouse)
hemoglobin genes, 105 hn
leukemia virus, 166
mutations, 71 hn, 71-73, 82 fn
resistant strains, 22-23, 26-28
T/t mutants, 53-54
tumors, 179-180, 188-192, 193 hn, 199-201
Radiation: rat, 302 (*see also* X rays: rat)
Radiomimetic compound: mouse, 201
Radiothyroidectomy: mouse, 190-191
RAD LV (virus): mouse, 166
Ragged (gene): mouse (see *Ra*: mouse)
Ragweed pollen extract: mouse, 144
RAH (strain): rat, 335
Random-bred: laboratory animals, 2-3
Random-bred: mouse, 105 hn, 208 (*see also* specific strain)
Rat (*see* specific condition, gene, strain; *Rattus;* Wild rat)
Rat prenatal stages, 52-53
Rattus, 233
R. norvegicus, 233-234, 310-312
R. rattus, 233-234, 312
Rauscher erythroleukemia virus: rat, 303
Rauscher LLV (virus): mouse, 166

RBC esterases: mouse, 220 hn
rc [rough coat] (gene): mouse, 33, 75
RCA (strain): rat, 237, 352, 361-362
RCS (strain): rat, 249 fn, 250, 272-273
rd [retinal degeneration] (gene): mouse, 24, 26, 28, 33, 38*, 152, 156
rdy [retinal dystrophy] (gene): rat, 237, 249, 258
Re [rex] (gene): mouse, 27, 33, 39*, 75
    Re^wc [wavy coat], 42-43, 75
Re [rex] (gene): rat, 258
Reactive hypercholesterolemia: rat, 348
Rearings, no. of: rat, 360
Receptor protein: mouse, 77
Receptors, sensory: mouse, 67
Recessive mutations: mouse nomenclature rules, 12
Recessive spotting (gene): mouse (see *rs*)
Recessive *t*-alleles: mouse, 54 hn, 54-55
Recessive yellow (gene): mouse (see *e*: mouse)
Reciprocal crosses: mouse, 164
Reciprocal translocations: mouse, 13, 42-43
Recombinant haplotypes: mouse, 125-128, 135*, 146-147
Recombinant-inbred strains: laboratory animals, 6-7
Recombinant-inbred strains: mouse, 37 (*see also* specific strain)
Recombinant-inbred strains: rat, 246, 260 (*see also* specific strain)
Rectal squamous cell carcinoma: mouse, 208
Red-cell aging: mouse, 85 fn
Red-eyed dilution (gene): rat (see *r*: rat)
Red-eyed white (gene): mouse (see mi^{rw})
Red fibers: rat, 279, 286-287
Reduced pinna (gene): mouse (see *Rp*)
Regenerating liver cell karyology: rat, 264
Renal (*see also* Kidney; Tubular entries)
Renal abnormality: rat, 248
Renal adenoma (gene): rat (see *Ra*: rat)
Renal artery, ectopic: mouse, 60 fn
Renal capsule: mouse, 198
Renal disorders: mouse, 21, 26
Renal glycopeptide: rat, 335
Renal inactivation: rat, 288
Renal infection: rat, 337
Renal medulla: rat, 309
Renal prostaglandin catabolism: rat, 245
Renal tubular epithelial antigen: rat, 335
Renal tumor (gene): rat (see *Tu*)
Renin activity: mouse, 80, 93
Renin polymorphism: rat, 339
Repeated epilation (gene): mouse (see *Er*)
Reproduction: mouse, 45-49
Reproduction: rat, 238-240, 242-243, 245, 247-249, 253, 283
Reproductive behavior: rat, 360
Reproductive endocrinology: rat, 340-347
Reproductive organs: mouse, 65
Respiratory infections: rat, 237
Respiratory system: mouse, 50-52
Respiratory system: rat, 289
Restricted (gene): rat (see H^{re})
Reticular cell tumors: mouse, 24
Reticuloendothelial organs: mouse, 22
Reticulum, liver endoplasmic: rat, 282
Reticulum cell sarcomas: mouse, 27, 169-172
Reticulum cell sarcomas: rat, 295, 299
Retinal degeneration: rat, 249, 290-291
Retinal degeneration (gene): mouse (see *rd*)
Retinal dystrophy (gene): rat (see *rdy*)
Retinal layer, abnormal: mouse, 151
Retinal lesions: rat, 291
Reversal learning: rat, 361
Reversions to wild-type: mouse nomenclature rules, 12-13
Revolving (gene): mouse (see *Rv*)
Rex (gene): mouse (see *Re*: mouse)
Rex (gene): rat (see *Re*: rat)
RF (strain): mouse
        characteristics, 26
        complement factors, 103
        congenic-inbred line progenitor, 115
        genes, 26
        H-2 haplotype & markers, 124
        Ig heavy chain linkage groups, 111
        lymphocyte alloantigens, 116-118
        proteins, 104
        tumors, 176, 195, 199-203
    RF/J
        antigens, 114, 167-168
        biochemical variation, 81-92
        body weight, 46
        complement factors, 103
        genealogy, 17*
        histocompatibility typing, 122
        lymphoma, 169
        reproduction, 46
        urinary proteins, 106
    RF/Lw, 90, 106
    RF/Un, 171, 181, 188, 195, 199-201
    RF/Up, 195, 200, 202-203
    (RF x AK)F_1, 204
Rf [rib fusion] (gene): mouse, 33, 57
RFM (strain): mouse, 17*, 114, 124
    RFM/Un, 81-82, 84, 87, 89
    RFM_f/Un, 200-201
Rfv-1 (gene): mouse, 168
rg [rotating] (gene): mouse, 56, 156
Rgv-1 [resistance to Gross virus-1] (gene): mouse, 97*, 99, 136, 141, 168
rh [rachiterata] (gene): mouse, 38*, 56
RHA (strain): rat, 237, 316, 352, 361-362
    RHA/N, 250, 273, 310-311, 317
Rhabdomyosarcoma: mouse, 166
Rhino (gene): mouse (see hr^{rh})
RI-1 [immunoglobulin-1] (gene): rat, 258
Rib fusion (gene): mouse (see *Rf*)
RIET (strain): mouse, 159, 183
Right atrium: rat, 249, 279, 286
RIII (strain): mouse
        biochemical variation, 88
        cellular alloantigens, 114
        characteristics, 27
        congenic-inbred line progenitor, 111
        genealogy, 17*
        genes, 27
        IgC_H allotypic determinants, 109
        Ig heavy chain linkage groups, 111
        proteins, 104
        tumors, 193 hn, 206 hn, 207
    RIII/An, mouse, 46, 90, 103, 107
    RIII/Fa, 92
    RIII/SeA, 27, 81, 84, 87-88, 90, 92, 106
    RIII/Se_fA, 112

RIII/WyJ, 115, 125
RIII/2J, 81-91, 103-104, 107, 109, 116-118, 124
RIII$_f$, 160
RIII hybrids, 182, 184
RIII-MTV (virus): mouse, 160
Rjv-1 (gene): mouse, 167
rl [reeler] (gene): mouse, 36, 38*, 58, 152, 156
RLA (strain): rat, 237, 352, 361-362
RLA/N, 250, 273, 317
Rn [roan] (gene): mouse, 33, 39*, 73
RNA, synthetic: mouse, 69
RNA nucleotidyltransferase: rat, 312
RNA polymerase: rat, 312
RNA virus, feline: rat response, 303
RNA viruses, type C: mouse, 227-231
RNC (strain): mouse, 27
ro [rough] (gene): mouse, 38*, 75
Roan (gene): mouse (see *Rn*)
Robertsonian translocations: mouse, 13, 41
Rodless retina (gene): mouse (see *r*: mouse)
Roman Control Avoidance (strain): rat, 352 (*see also* RCA)
Roman High Avoidance (strain): rat, 352 (*see also* RHA)
Roman Low Avoidance (strain): rat, 352 (*see also* RLA)
ROP (strain): mouse, 104
ROP/Gn, 82-91
ROP/GnJ, 90
ROP/GnLe, 33
ROP/J, 84, 90
Rosette (gene): mouse (see *rst*)
Rostrum: mouse, 214, 217
Rotating (gene): mouse (see *rg*)
Rough (gene): mouse (see *ro*)
Rough coat (gene): mouse (see *rc*)
Rous sarcoma virus: rat, 303
Rous virus sarcoma karyology: rat, 263, 267
Rp [reduced pinna] (gene): mouse, 59
rs [recessive spotting] (gene): mouse, 33, 73
rst [rosette] (gene): mouse, 75
RSV/Le (strain): mouse, 33-34, 86
rt [runt] (gene): rat, 258
ru [ruby-eye] (gene): mouse, 33, 39*, 42, 73
ru-2 [ruby-eye-2] (gene): mouse, 33, 38*, 73
ru-2^{hz} [haze] (gene): mouse, 33, 73
ru-2^{mr} [maroon] (gene): mouse, 73
Ruby-eye (gene): mouse (see *ru*)
Ruby-eye-2 (gene): mouse (see *ru-2*)
Ruby-eyed dilute (gene): rat (see c^d)
Ruby-eyed dilution (gene): rat (see p^m: rat)
Rump-white (gene): mouse (see *Rw*: mouse)
Runt (gene): rat (see *rt*)
Runway activity: rat, 360-361
RUS (strain): mouse, 88
RUS/Rl, 89
Ruthenium-106: rat, 302
Rv [revolving] (gene): mouse, 156
Rv-1 [Rauscher leukemia virus susceptibility-1] (gene): mouse, 22-24
Rv-2 [Rauscher leukemia virus susceptibility-2] (gene): mouse, 22-24
Rw [rump-white] (gene): mouse, 38*, 63, 73
Rw [warfarin resistance] (gene): rat, 258-259
R2 adenocarcinoma: rat, 303
S (strain): mouse, 92, 195, 200
S/Gh, 82, 86, 89
S/Gn, 88
S/Gw, 92
s [piebald] (gene): mouse, 24, 26, 33, 39*, 43, 73, 152
s^l [piebald-lethal], 33, 73, 152
s [silvering] (gene): rat, 259, 273
sa [satin] (gene): mouse, 39*, 75
Sac [saccharin preference] (gene): mouse, 155
Saccharin preference (gene): mouse (see *Sac*)
Saccular otoconia: rat, 291
SAL (strain): rat, 316
sal [satin-like] (gene): mouse, 75
Saliva: mouse, 77
Salivary α-amylase (gene): mouse (see *Amy-1*)
Salivary glands: rat, 274-277, 296 fn
Salmonella: mouse, 22
S. enteritidis: rat, 337
Salt hypertension: rat, 244
Salt-sensitive hypertension: rat, 348
Salt-S or salt-sensitive (strain): rat, 348
Sand (gene): rat (see *sd*)
Sarcoma cell karyology: rat, 263-265, 267
Sarcomas: mouse, 27, 166, 169-173, 176
Sarcomas: rat, 240, 242-244, 246, 248, 251, 295, 298-299, 301-304, 338
Sas-1 [antigenic serum substance] (gene): mouse, 21-28, 80, 92, 104
Satin (gene): mouse (see *sa*)
Satin-like (gene): mouse (see *sal*)
SB/Le (strain): mouse, 87
sb [stub] (gene): mouse, 56
sb [stubby] (gene): rat, 258
sc [screw-tail] (gene): mouse, 57
SC-1 cell line: mouse, 227 hn
Scaly (gene): mouse (see *Sk*)
Scant hair (gene): mouse (see *sch*)
sch [scant hair] (gene): mouse, 33, 38*, 75
Schedule induced polydipsia (gene): mouse (see *Sip*)
Schmidt-Ruppin sarcoma: rat, 301, 303
Schneider (strain): mouse, 85
Schwannomas: rat, 301
Sciatic nerve: rat, 294
Sclerosis, mesenteric arterial: rat, 291, 294
Sco [scopolamine modification of exploratory activity] (gene): mouse, 22-23, 155
Scopolamine modification of exploratory activity (gene): mouse (see *Sco*)
Screw-tail (gene): mouse (see *sc*)
Scurfy (gene): mouse (see *sf*)
SD (strain): mouse, 48
SD (strain): rat (*see also* Sprague-Dawley)
aging, 283-285
alloantigenic alleles, 316-317
characteristics, 250
chromosome polymorphisms, 260
coat color, 250, 273
enzyme polymorphism, 311
eye color, 273
immunogenetic characterization, 314
immunoglobulin polymorphism, 308
immunologically mediated diseases, 334-335
infectious diseases, 337-338

karyology, 263, 266-267
origin, 250
reproductive hormones, 341-347
tumors, 302-304
SD (CFE), 285, 292
SD/Jcl, 278, 292-293
SD/N, 248 (*see also* NSD)
Sd [Danforth's short tail] (gene): mouse, 33, 38*, 42, 44, 57
sd [sand] (gene): rat, 273
se [short-ear] (gene): mouse
chromosome aberrations, 42
chromosome position, 38*
congenital malformations, 58, 60 fn
inbred strain genealogy, 17*
strain distribution, 26-27, 33, 36
se^{sv} [short-eared waltzer], 156
SEA (strain): mouse, 104, 109-110, 124
SEA/Gn, 81-89, 91, 106
SEA/GnJ, 17*, 33, 86-88, 90-91, 116-118
SEA/J, 109
sea [sepia] (gene): mouse, 73
Sea-level radiation: mouse, 200
SEC (strain): mouse, 103-105, 110, 124
SEC/1Gn, 17*, 27
SEC/1GnLe, 30, 33
SEC/1Re
anemia, 65
biochemical variation, 87-88, 91
cellular alloantigens, 114
mutations, 30-32, 34-35
reproduction, 48
SEC/1ReJ, 81-91, 106, 113, 116-118
Seizures: mouse, 24, 27, 149-152, 154
Seizures, audiogenic: rat, 243
SEL (strain): rat, 250, 272, 308
Selfed (strain): rat (*see* SEL)
Selfed 36670 (*see* SEL)
Semicircular canal abnormality: mouse, 152-153
Seminal vesicle protein: mouse, 80, 98
Seminal vesicle protein (genes): mouse (see *Svp-1; Svp-2*: mouse)
Seminal vesicle protein (genes): rat (see *Svp; Svp-1*: rat)
Seminal vesicles: mouse, 80, 101
Seminal vesicle transplants: mouse, 178-179
Seminiferous tubules: mouse, 48
Sensory ataxia: mouse, 66, 153
Sensory axon defects: mouse, 152
Sensory ganglion cells: mouse, 67
Sensory neuron atrophy: mouse, 66
Sensory organ development: mouse, 50-52
Sensory receptors: mouse, 67
Sep-1 (gene): mouse, 38*, 96*, 98
Sepia (gene): mouse (see *sea*)
Septa, vaginal: mouse, 23, 47
Septal nucleus, lateral: mouse, 25
Septicemia: rat, 291
Serological markers, *H-2* gene complex: mouse, 123-124, 126 hn, 126-127
Serotonin: mouse, 102
Serotonin: rat, 281, 283, 362
Serotypes, *H-2L*: mouse, 138
Serum: mouse, 22, 68-69, 77-80, 101, 135*
Serum: rat
aging, 280, 283-284, 287-288
enzyme polymorphisms, 310-311
hormones, 239, 244, 246, 252, 283, 340-347
shock-induced fighting, 360
Serum antigen: mouse, 80, 104
Serum antigen (gene): mouse (see *Ss*)
Serum esterase-1 (gene): mouse (see *Es-1*)
Serum nephritis, nephrotoxic: rat, 335
Serum substance (gene): mouse (see *Ss*)
Sex chromosomes (*see* X chromosome; Y chromosome)
Sex differentiation, prenatal: laboratory mammals, 51, 53
Sex gland, accessory: rat, 290
Sex-influenced esterase, serum: rat, 311
Sex-limited serum protein: mouse, 80, 140, 143, 148
Sex-limited serum protein (gene): mouse (see *Slp*)
Sex-linked anemia (gene): mouse (see *sla*)
Sex ratio: mouse, 25, 45-47
Sex-reversal (gene): mouse (see *Sxr*)
Sex tumor incidence: mouse, 161-165
Sexual maturity: mouse, 48-49
Sey [small eye] (gene): mouse, 36, 59, 152
SF (strain): mouse
SF/Cam, 27, 81-89, 104
SF/CamEi, 86
SF-5613, 49
SF-5621, 49
sf [scurfy] (gene): mouse, 39*, 75
sg [staggerer] (gene): mouse, 33, 38*, 58, 152, 156
SH (strain): rat, 308
Sh [shaggy] (gene): rat, 258-259
sh-1 [shaker-1] (gene): mouse, 33, 38*, 42, 58, 152, 156
sh-2 [shaker-2] (gene): mouse, 34, 39*, 59, 152, 156
Sh-3 [shaker-3] (gene): mouse, 57
Sha [shaven] (gene): mouse, 39*, 60, 75
Shaggy (gene): rat (see *Sh*)
Shaker (gene): rat (see *sr*: rat)
Shaker (genes): mouse (see *sh-1; sh-2; Sh-3*)
Shaker-short (gene): mouse (see *st*: mouse)
Shaker-with-syndactylism (gene): mouse (see *sy*)
Shambling (gene): mouse (see *shm*)
Shaven (gene): mouse (see *Sha*)
Sheep erythrocytes: mouse response, 68 hn, 68-69
Sheep erythrocytes: rat response, 243-244, 319
Sherman (strain): rat, 293, 308, 336
SHM/2Gn (strain): mouse, 49
shm [shambling] (gene): mouse, 34, 39*, 152, 156
SHN (strain): mouse, 49
sho [shorthead] (gene): mouse, 58
Shock avoidance: rat, 247, 251-252, 359
Shock-induced fighting: rat, 360
Short-ear (gene): mouse (see *se*)
Short-eared waltzer (gene): mouse (see se^{sv})
Shorthead (gene): mouse (see *sho*)
Short snout: mouse, 60 fn
Short tail: mouse, 53-54
SHR (strain): rat
aging, 290
alloantigenic alleles, 316
catecholamine synthesis, 309 hn
characteristics, 250-251
coat color, 250, 273
enzyme polymorphisms, 310, 312
eye color, 273
hormone polymorphism, 339
hypertension, 348-349, 349*
life-span, 278

origin, 250-251
research use, 236, 351
shock-induced fighting, 360
skin grafts, 249 fn
SHR/N, 310-311, 317
SHR x Sprague-Dawley, 348
SHR/GnEi (strain): mouse, 33-34
Shrew-like mice, 212
SHRSP (strain): rat, 349, 349*
SHRSR (strain): rat, 349*
Shuttle avoidance: rat, 359-360
si [silver] (gene): mouse, 38*, 73
Sialadenitis, autoallergic: rat, 334
Siderocytic anemia, transitory: mouse, 65
Sienna yellow (gene): mouse (see A^{sy})
Sig [sightless] (gene): mouse, 34, 38*, 57, 152
Sightless (gene): mouse (see *Sig*)
SIIT (strain): mouse, 84, 87-89, 92
Silica: rat, 304 fn
Silicon dioxide: rat, 304
Silver (gene): mouse (see *si*)
Silvering (gene): rat (see *s*: rat)
Silver mutant: rat, 234
Sim:CD(SD) strain: rat, 278
Single constriction (gene): mouse (see *Hct*)
Sinus, paranasal: rat, 302
Sip [scheduled induced polydipsia] (gene): mouse, 23-24, 155
Siren (gene): mouse (see *srn*)
Situs inversus viscerum (gene): mouse (see *iv*)
SJA (strain): mouse, 111
SJL (strain): mouse
alloantigens, 114, 116-118
characteristics, 27
congenic-inbred line progenitor, 111
genes, 27
H-2 haplotype & markers, 124
Ig heavy chain linkage groups, 110
leukemia antigens, 168
proteins, 103-104
SJL/J
biochemical variation, 81-92
body weight, 46
genealogy, 17*
histocompatibility typing, 122
leukemia, 168-169
lymphoma, 169
reproduction, 46, 49
reticulum cell neoplasms, 171
urinary proteins, 106
SJL/JDg, 46
SJL/Wn, 172
SJL/Wt, 49
SK (strain): mouse, 27
SK/Cam, 49, 81-82, 84-87, 89, 91, 104
SK/CamEi, 86
Sk [scaly] (gene): mouse, 34, 63, 75, 114
sk [skinny] (gene): rat, 258
Skeletal malformations, congenital: mouse, 55-58
Skeletal muscle: mouse, 67
Skeletal muscle: rat, 289-290, 294
Skeletal system: rat, 302-303
Skin: mouse, 26, 60, 74-77, 117-118, 147
Skin: rat, 289, 291, 315
Skin grafts: inbred lines, 3
Skin grafts: mouse, 69
Skin grafts: rat, 249 fn, 283, 321-322, 324-325
Skinny (gene): rat (see *sk*)
Skin sensory receptors: mouse, 67
Skin tumors: rat, 239, 296 fn, 301-302, 304
Skull abnormalities: mouse, 60 fn, 149
Skull size: mouse, 214, 217, 219
SL (strain): mouse, 169
SL/Ms, 172
Sl or *sl* (gene): mouse
Sl [steel]
anemia, 65
chromosome aberrations, 42-43
chromosome position, 38*
endocrine effects, 102
lethality, 63
leukemia antigens, 168
mutation, 34, 73
strain distribution, 34, 65
Sl^{cg} [cloud-gray], 59, 73
Sl^{con} [contrasted], 65, 73
Sl^{d} [steel-Dickie], 34, 65, 73
sl^{du} [dusty], 73
Sl^{gb} [grizzle-belly], 65, 73
Sl^{m} [steel-Miller], 73
Sl^{so} [sooty], 65, 73
sla [sex-linked anemia] (gene): mouse, 39*, 60, 65
Slaty (gene): mouse (see *slt*)
Sleek (gene): mouse (see *Slk*)
Slight dilution (gene): mouse (see d^{s})
Slit-lid (gene): mouse (see lg^{stn})
Slk [sleek] (gene): mouse, 75
SLN (strain): mouse, 49
Slonaker (strain): rat, 246, 316
Slp [sex-limited serum protein] (gene): mouse
complement components, 136
immune response, 143
phenotypic expression, 92
properties, 148
proteins, 80, 103
strain distribution, 21-26, 28, 92
slt [slaty] (gene): mouse, 34, 39*, 73
Slye (stock): mouse, 179, 181
SM (strain): mouse, 27, 49, 104, 106, 110
SM/J
alloantigens, 114, 116-118
biochemical variation, 81-92
body weight, 46
complement factors, 103
congenic-inbred line progenitor, 125
genealogy, 17*
H-2 haplotype & markers, 124
reproduction, 46
sm [syndactylism] (gene): mouse, 34, 56
Small eye (gene): mouse (see *Sey*)
Small intestine: mouse, 78
Small intestine: rat, 311
Smooth-muscle tumors: rat, 298 fn
sn [spinner] (gene): rat, 258
Snell's waltzer (gene): mouse (see *sv*)
sno [snubnose] (gene): mouse, 38*, 56
Snout, short: mouse, 60 fn
Snubnose (gene): mouse (see *sno*)
soc [soft coat] (gene): mouse, 39*, 75
Sodium: rat, 280, 283, 285

Soft coat (gene): mouse (see *soc*)
Soluble aspartate aminotransferase: mouse, 78, 99
Soluble glutamate oxaloacetate transaminase: mouse, 78, 99
Soluble isocitrate dehydrogenase: mouse, 79, 98
Soluble malate dehydrogenase: mouse, 79, 98
Soluble malic enzyme: mouse, 79, 98
Somatic cell karyology: rat, 261 hn, 266-268
Somatic proviruses: mouse, 157 hn
Sombre (gene): mouse (see E^{so})
Somites: mouse, 50, 51-53
Somites: rat, 53
Sooty (gene): mouse (see Sl^{so})
Sp [splotch] (gene): mouse, 34, 38*, 44, 57, 73, 152
  Sp^d [delayed-splotch], 34, 58, 73, 152
sp [spastic] (gene): rat, 258
spa [spastic] (gene): mouse, 34, 39*, 152, 156
Sparse coat (gene): mouse (see *spc*)
Sparse-fur (gene): mouse (see *spf*)
Spastic (gene): mouse (see *spa*)
Spastic (gene): rat (see *sp*)
Spaying: mouse, 179
Spaying: rat, 252
spc [sparse coat] (gene): mouse, 39*, 75
Specific gravity, urine: mouse, 24
Specificities, alloantigenic alleles: rat, 315 fn
Specificities, *H-2* gene complex: mouse, 137-140
Sperm; spermatozoa: mouse, 25, 47-48, 114, 147-148
Spermatocytes: mouse, 48
Spermatogonia: mouse, 48
Spermatogonia karyology: rat, 263-264
Sperm-egg contact time: rat, 52 hn
Spermidine: rat, 284
Spermine: rat, 284
spf [sparse fur] (gene): mouse, 39*, 75, 80, 92
  spf^{ash} [abnormal skin & hair], 75
sph [spherocytic anemia] (gene): mouse, 34, 63, 65
Spherocytic anemia (gene): mouse (see *sph*)
spi [spiral tail] (gene): mouse, 56
Spina bifida: mouse, 149, 153
Spinal cord: mouse, 67, 149-153
Spinal cord: rat, 245, 334
Spinal nerve root degeneration: rat, 291
Spinal roots, defective: mouse, 67
Spinner (gene): mouse (see *sr*: mouse)
Spinner (gene): rat (see *sn*)
Spino-cerebellar tract degeneration: mouse, 153
Spiny mice, 212
Spiral ganglion degeneration: mouse, 150
Spiral tail (gene): mouse (see *spi*)
Spl [plasma serotonin level] (gene): mouse, 23, 102, 156
Spleen; spleen tissue: mouse
    amyloidosis, 22
    cellular alloantigens, 114
    congenital malformation, 60 fn
    lymphoma, 168 hn, 169
    ovary transplants, 193 hn, 198
    protein distribution, 77-78
Spleen; spleen tissue: rat
    aging, 290, 294
    alloantigenic alleles, 315
    karyology, 262, 264
    mononuclear cell leukemia, 244
    norepinephrine concentration, 362
    transplant survival, 323
    weight, 274-277, 354
Spleen cell reactivity: mouse, 68-69
Splenectomy: mouse, 173-174, 201
Splenomegaly: mouse, 69
Splotch (gene): mouse (see *Sp*)
Spongiosis: mouse, 228 hn
Spontaneously hypertensive rats, 250, 348 (*see also* SHR: rat)
Spontaneous pneumonitis: mouse, 68
Spontaneous tumors: mouse, 28, 193 hn, 194-196, 207-208, 210-211
Spontaneous tumors: rat, 238-240, 243-246, 249, 253, 298-300
SPR (strain): rat, 339
Sprague-Dawley (strain): rat, 234 hn, 234, 278 (*see also* SD: rat)
  Sprague-Dawley (Charles River), 343
  Sprague-Dawley (C. River, CD), 343
  Sprague-Dawley (Holtzman), 341-342, 345
  Sprague-Dawley outbred albino, 310
Sprawling (gene): mouse (see *Swl*)
sq [squint] (gene): mouse, 59, 60 fn
Squamous cell carcinomas: mouse, 28, 208
Squamous cell carcinomas: rat, 242, 298, 298 fn, 301, 304
Squamous cell hyperplasia: rat, 253
Squamous cell metaplasia: rat, 253
Squamous papilloma, cutaneous: rat, 295
Squint (gene): mouse (see *sq*)
sr [shaker] (gene): rat, 258
sr [spinner] (gene): mouse, 152, 156
S region, *H-2* gene complex: mouse, 123 hn, 123-124, 126-127, 134*-135*, 147 hn
srn [siren] (gene): mouse, 58
Ss (gene): mouse
    alleles, 140
    chromosome position, 39*
    complement components, 103, 136
    enzyme activity, 92
    H-2 gene complex maps, 134*-135*
    properties, 140, 147 hn, 148
    protein control, 80
    strain distribution, 21-28, 92
ST (strain): mouse, 17*, 104, 110, 124, 162
  ST/a, 17*, 84, 86, 91-92
  ST/aEh, 81
  ST/b, 27, 92, 103-104
  ST/bJ, 17*, 46, 81-92, 106, 114, 116-118
st [shaker-short] (gene): mouse, 57
st [stub] (gene): rat, 258-259
Sta [autosomal striping] (gene): mouse, 63
Stabilimeter crossings: rat, 357
Staggerer (gene): mouse (see *sg*)
Staining techniques, karyology: rat, 261-268, 268*-270*, 271-272
Stanolone: rat, 341-342, 346
Staphylococcal nuclease: mouse, 144
Staphylococcus aureus: mouse, 68
S. aureus: rat, 334
STAR/N (strain): mouse, 88-89
Starch gel electrophoresis: rat, 309 hn, 310-312
stb [stubby] (gene): mouse, 34, 38*, 56, 60 fn
Steel (gene): mouse (see *Sl*)
Steel-Dickie (gene): mouse (see Sl^d)

Steel-Miller (gene): mouse (see Sl^m)
Stenosis, secondary aqueductal: mouse, 151
Sterility: mouse, 25, 28, 42 hn, 42-43, 48, 65, 102, 193 hn
Steroid dehydrogenase: rat, 281
Steroids, serum: rat, 340-347
Steroid-secreting tumors: rat, 299 fn
Stilbestrol: mouse, 183-815, 187 (*see also* DES; Diethylstilbestrol)
stm [stumpy] (gene): mouse, 56
STOLI (strain): mouse, 16*, 111
  STOLI/Lw, 92, 125
Stomach: mouse, 78
Stomach: rat, 274-277, 360
STR (strain): mouse
  STR/Cr, 46
  STR/N, 81-82, 88-90, 106, 110
  STR/1, 110
  STR/1N, 46, 81-82, 90, 106
Str [striated] (gene): mouse, 39*, 63, 75
Strain (*see* specific strain)
Strain symbols: mouse nomenclature rules, 10-12
Streaker (gene): mouse (see nu^{str})
Streptococcal Group A polysaccharide: mouse, 113
Streptococcal vaccine, Group A: rat, 318
Streptococcus pneumoniae: mouse, 68
S. pneumoniae: rat, 337
Streptomyces: rat, 334
Stress response: rat, 360
Striated (gene): mouse (see *Str*)
Striated muscle abnormalities: mouse, 60 fn
Striated myotubes: mouse, 67
Stria vascularis degeneration: mouse, 150
Stroke-prone rats (*see* SPR)
Stroke-prone SHR (strain): rat, 349*
Stroke-resistant SHR (strain): rat, 349*
Stromal tumors: mouse, 195, 203
Strong's dominant spotting (gene): mouse (see W^s)
Strong's luxoid (gene): mouse (see *lst*)
Strontium nitrate, radioactive: mouse, 188
Strontium-89: rat, 302
STS (strain): mouse, 159-160
  STS/A, 81-84, 86-92, 107, 112, 116-118
Stub (gene): mouse (see *sb*: mouse)
Stub (gene): rat (see *st*: rat)
Stubby (gene): mouse (see *stb*)
Stubby (gene): rat (see *sb*: rat)
Stumpy (gene): mouse (see *stm*)
STX/Le (strain): mouse, 86
su [surdescens] (gene): mouse, 152
Subcutaneous transplants: mouse, 181-182
Subcutaneous tumors: rat, 244, 295-296, 304
Submandibular gland homogenate: rat, 334
Submaxillary gland: mouse, 80
Subscripts: mouse nomenclature rules, 9, 11
Substrains: mouse, 18*-20* (*see also* specific substrain)
Substrains: mouse nomenclature rules, 10-12
Succinate cytochrome *c* reductase: rat, 285
Succinate dehydrogenase: rat, 288
Succinoxidase: rat, 288 fn
Sucrase: rat, 311
Superscripts: mouse nomenclature rules, 9, 13-14
Suppressor factors, soluble: mouse, 141
Supraorbital ridge: mouse, 212
Surdescens (gene): mouse (see *su*)
Survival: mouse, 47-48
sv [Snell's waltzer] (gene): mouse, 34, 38*, 152, 156
Svoboda carcinoma: rat, 304
Svp [seminal vesicle protein] (gene): rat, 305
Svp-1 [seminal vesicle protein] (gene): rat, 258-259, 309
Svp-1 [seminal vesicle protein-1] (gene): mouse
    biochemical locus, 38*, 96
    biochemical variation, 104
    phenotypic expression, 92
    protein control, 80, 98
    strain distribution, 21-24, 28, 92
Svp-2 [seminal vesicle protein-2] (gene): mouse, 21-24, 80, 92, 104
SV40-induced tumor: rat, 301
sw [swaying] (gene): mouse, 39*, 58, 152, 156
Swaying (gene): mouse (see *sw*)
Swiss (strain): mouse
    congenic-inbred line progenitor, 124-125
    genealogy, 17*
    leukemia virus, 165-166
    tumors, 163, 176, 187, 195, 201-204, 209
  Swiss albino, 177
  Swiss NIH, 177
  Swiss random, 187
Swiss albino (strain): rat, 263
Swl [sprawling] (gene): mouse, 67, 152
SWM/Ms (strain): mouse, 89
SWR (strain): mouse
    alloantigens, 114, 116-118
    biochemical variation, 83, 92
    characteristics, 27
    complement factors, 103
    genes, 27
    H-2 haplotype & markers, 124
    Ig heavy chain linkage groups, 111
    leukemia antigens, 168
    proteins, 104
    tumors, 162, 208
  SWR/De, 90, 106
  SWR/J
    biochemical variation, 81-93
    body weight, 46
    genealogy, 17*
    histocompatibility typing, 122
    leukemia antigens, 167-168
    neoplasms, 172
    recombinant-inbred line progenitor, 37
    reproduction, 46
    tumors, 195, 202
SWV (strain): mouse, 27, 49
SWXL (strain): mouse, 37
Sxr [sex reversal] (gene): mouse, 59
sy [shaker-with-syndactylism] (gene): mouse, 39*, 56, 152
  sy^{fp} [fused phalanges], 34, 56
Symbols: mouse nomenclature rules, 9-14
Syndactylism (gene): mouse (see *sm*)
Systole cessation: rat, 321 hn
Systolic blood pressure: rat, 239, 248-249, 253
S1 (strain): rat (*see* TMB)
S3 (strain): rat, 310-311 (*see also* TMD)
S5B (strain): rat, 248, 251, 272-273, 308, 316-317

T (strain): rat, 260
T or *t* (gene): mouse
  T [brachyury]
    biochemical locus, 39*, 97*

congenital malformation, 56
genetic fine structure map, 134*
H-2 antigenic specificities, 132-133
mutation, 53
protein control, 99
strain distribution, 34, 114
tissue distribution, 114
t; t alleles
congenital malformations, 58, 60 fn
H-2 antigenic specificities, 132-133
mutations, 54 hn, 54-55
lethal haplotypes, 64
properties, 147 hn, 148
T^c [curtailed], 53, 57, 60 fn
T^h [T-Harwell], 56
T^{hp} [hairpin-tail], 53, 56
T^{Or} [T-Oak Ridge], 54, 56
T^{Orl} [T-Orleans], 54, 56
Ta [tabby] (gene): mouse, 39*, 42, 57, 75
Tabby (gene): mouse (see *Ta*)
Taenia taeniaeformis: rat, 338
Tag [temporal α-galactosidase] (gene): mouse, 77, 81
Tail: mouse, 26, 53-55, 65-66, 217, 219
Tail cell karyology: rat, 261, 266-267
Tail hair depletion (gene): mouse (see *thd*)
Tail-kinks (gene): mouse (see *tk*)
Tail length: rat, 274-276
Tailless: rat, 239
Tail short (gene): mouse (see *Ts*)
Tail-zigzagged (gene): mouse (see *Tz*)
Tanoid (gene): mouse (see a^{td})
Taupe (gene): mouse (see *tp*)
Taxonomy: mouse, 212
Taxonomy: rat, 233
TB/Orl (strain): mouse, 92
tb [tumbler] (gene): mouse, 38*, 152, 156
TBR2 (strain): mouse, 147
tc [truncate] (gene): mouse, 38*, 57
T-cell depletion: rat, 335
T-cells: mouse, 69, 116-118
te [light-head] (gene): mouse, 73
Teetering (gene): mouse (see *tn*)
Teeth: mouse, 60 fn, 217 (*see also* Incisors; Molars)
Teflon disk: mouse, 174
Telencephalon: rat, 362
Temporal α-galactosidase (gene): mouse (see *Tag*)
Tennant LLV (virus): mouse, 166
TER/Sv (strain): mouse, 168 fn
Teratocarcinoma: mouse, 148, 193 hn, 198
Teratology: rat, 236
Teratomas: mouse, 28, 195-196, 198, 203, 207-208
Testes; testicular tissue: mouse
abnormalities, 25, 48
cellular alloantigens, 114
proteins, 79-80
tumors, 28, 198, 206-208
Testes; testicular tissue: rat
aging, 281, 289-290, 296
atrophy, 238
enzyme polymorphism, 311
fat, 274-277
tumors, 239, 244, 294-296, 298-299
weight, 274-277
Testicular feminization (gene): mouse (see *Tfm*: mouse)
Testicular feminization (gene): rat (see *Tfm*: rat)
Testis homogenate: rat, 336
Testosterone: mouse, 21, 106 hn, 136
Testosterone: rat, 283, 340-342, 344, 346
Tetanus toxoid: mouse, 68
2,3,7,8-Tetrachlorodibenzo-*p*-dioxin: mouse, 80 hn, 81
Tetramethylhydrazine hydrochloride: mouse, 176
TF (strain): mouse
TF/Gn, 86-87
TF/GnLe, 34
Tf (gene): rat, 309
tf [tufted] (gene): mouse, 34, 39*, 76, 122, 123*, 134, 134*
Tfm [testicular feminization] (gene): mouse, 39*, 59, 60 fn, 80, 92, 102
Tfm [testicular feminization] (gene): rat, 258, 339
tg [tottering] (gene): mouse, 34, 38*, 152, 156
tg^{la} [leaner], 34, 152 (see also *la*)
th [tilted head] (gene): mouse, 38*
T-Harwell (gene): mouse (see T^h)
thd [tail hair depletion] (gene): mouse, 76
Theca cell tumors: rat, 296
Theta alloantigens: mouse, 113, 143
thf [thin fur] (gene): mouse, 39*, 76, 134*
Thin fur (gene): mouse (see *thf*)
2-Thiouracil: rat, 304
Thoracic eventration: mouse, 60 fn
Thoracolumbar vertebrae malformations: rat, 245
Thoracoschisis: mouse, 60 fn
Thorium dioxide: mouse, 174
Thy-1 [thymus cell antigen-1] (gene): mouse, 21-28, 38*, 96*, 98, 115, 117
Thymectomy: mouse, 169, 198
Thymectomy: rat, 243
Thymic cortical atrophy: mouse, 68
Thymic leukemia: mouse, 168 fn
Thymidine kinase: rat, 312
Thymidine kinase [TK] (protein): mouse, 97*, 99
Thymidine uptake: rat, 319
Thymocyte alloantigen: mouse, 115
Thymocytes: mouse, 114, 148
Thymocytotoxic autoantibodies: mouse, 69
Thymomas: rat, 252, 298
Thymus; thymus tissue: mouse
alloantigens, 114, 116
leukemia, 165, 168-169
lymphocytic neoplasm, 170 fn
lymphoma, 168-169
mutant genes, 60 fn, 68-69, 101
Thymus; thymus tissue: rat, 243, 274-277, 297-298, 303, 315, 354-355
Thymus cell antigen: mouse (see *Thy-1*)
Thymus-independent antigens: mouse, 69, 113
Thymus leukemia alloantigens: mouse, 136
Thymus leukemia antigen: mouse (see *Tla*)
Thyroglobulin: mouse, 143
Thyroglobulin: rat, 335
Thyroid; thyroid tissue: mouse, 25, 102, 187-188, 191
Thyroid; thyroid tissue: rat
aging, 290, 295-297
hyperplasia, 253
karyology, 261-262
metaplasia, 253
mononuclear cell infiltration, 243
tumors, 242-244, 249, 252, 261-262, 295-300, 304

weight, 251, 355
Thyroid-blocking agent: mouse, 188
Thyroidectomy: mouse, 187-188, 190-191
Thyroid hormone: rat, 339
Thyroiditis: rat, 240, 243, 249, 252, 335
Thyrotropin: mouse, 101
Thyrotropin: rat, 251, 339
Thyroxine: mouse, 101
Thyroxine: rat, 239, 246, 339
ti [tipsy] (gene): mouse, 39*, 152
Tibia: rat, 274-277
Tibialis anterior muscle: rat, 279, 287
Tight skin (gene): mouse (see *Tsk*)
Tipsy (gene): mouse (see *ti*)
tk [tail kinks] (gene): mouse, 34, 36, 38*, 56
tl [non-erupted teeth] (gene): mouse, 56
tl [toothless] (gene): rat, 258, 310
Tla [thymus leukemia antigen] (gene): mouse
antigens, 115, 136, 148
biochemical locus, 39*, 97*
H-2 gene complex map, 134 hn, 134*-135*
histocompatibility linkage, 122, 123*
leukemia antigens, 168
protein control, 99
strain distribution, 21-28, 115, 168
Tla serological markers: mouse, 123 hn, 124, 124 fn, 126 hn, 126-127
TL region, *H-2* gene complex: mouse, 123 hn, 124, 134*-135*, 136
TM (strain): mouse, 28, 183, 186, 188
tm [tremulous] (gene): mouse, 152, 156
TMB (strain): rat, 236, 251, 272-273, 352
TMD (strain): rat, 236, 251, 272, 352
tn [teetering] (gene): mouse, 34, 39*, 152, 156
TO (strain): mouse, 85
TO (strain): rat, 251, 260, 273, 314
To [tortoiseshell] (gene): mouse, 34, 56, 73
T-Oak Ridge (gene): mouse (see T^{Or})
Tobacco darkening (gene): mouse (see E^{tob})
Toe malformations: mouse, 23
Toe-ulnar (gene): mouse (see *tu*)
Tokyo (strain): rat (*see* TO: rat)
Tol-1 [tolerance to BGG] (gene): mouse, 22, 24
4-*o*-Tolylazo-*o*-toluidine: mouse, 175
Toothless (strain): rat, 310
Toothless (gene): rat (see *tl*: rat)
T-Orleans (gene): mouse (see T^{Orl})
Torpid (gene): mouse (see du^{td})
Tortoiseshell (gene): mouse (see *To*)
Tottering (gene): mouse (see *tg*)
TP (strain): mouse, 49
tp [taupe] (gene): mouse, 34, 38*, 73
TR (strain): mouse, 49
TR/DiEi, 34
Tr [trembler] (gene): mouse, 39*, 152, 156
tr [trembler] (gene): rat, 258
Traits, *H-2* gene complex controlled: mouse, 136
Transferrin: rat, 309
Transferrin (gene): mouse (see *Trf*)
Transitional cell carcinomas: rat, 298
Transitional cell papillomas: rat, 298
Transitory siderocytic anemia: mouse, 65
Translocations: mouse, 12-13, 22, 41-43
Transplantable tumors: rat, 238-240, 243, 245, 247, 250, 261, 263
Transplantation antigens: mouse, 135*, 136
Transplants, tumor: mouse, 178-179, 181-182, 193 hn, 198
Transplant survival: rat, 321-323 (*see also* Allograft survival; Graft rejection: rat)
Transposition, chromosome: mouse nomenclature rules, 13
Trembler (gene): mouse (see *Tr*)
Trembler (gene): rat (see *tr*)
Trembling: mouse, 151
Tremor: mouse, 67, 149-153
Tremulous (gene): mouse (see *tm*)
Trf [transferrin] (gene): mouse
biochemical locus, 38*, 96*
electrophoretic polymorphism, 224
phenotypic expression, 92
protein control, 80, 98
strain distribution, 21-28, 92, 104
Triacylglycerol lipase: rat, 281
Tri-*p*-anisylchloroethylene: mouse, 206 hn
Trichonosis: rat, 257-258
Triethylenemelamine: mouse, 203
Triglyceride: rat, 284
Trimethylbenzanthracene: mouse, 174, 177, 203
Trimethylhydrazine hydrochloride: mouse, 176
Trinitrophenyl-$H\text{-}2D^d$ alloantigen: mouse, 144
2,4,6-Trinitrophenyl-mouse serum albumin: mouse, 144
Triosephosphate isomerase [TPI] (protein): mouse, 96*, 98
Tripeptidase: mouse, 227
Tripeptidase-1 (gene): mouse (see *Trp-1*)
1,1,2-Triphenylethylene: mouse, 206 hn, 206-207
Trisomy: mouse nomenclature rules, 13
Trisomy: rat, 261-262
Tropism: leukemia viruses, 165-166
Trp-1 [tripeptidase-1] (gene): mouse, 225
Truncal ataxia: mouse, 151-152
Truncate (gene): mouse (see *tc*)
Tryon Maze Bright (strain): rat (*see* TMB)
Tryon Maze Dull (strain): rat (*see* TMD)
Trypan blue: mouse, 174, 176
Trypan blue: rat, 335
Trypsin: rat, 266-268
Trypsinogen: rat, 281
Tryptophan hydroxylase: mouse, 80, 93
Tryptophan 5-monooxygenase: mouse, 80, 93
Ts [tail short] (gene): mouse, 34, 58, 66
TSI (strain): mouse, 159
TSI/A, 81-82, 88-89, 112
TSJ/Le (strain): mouse, 34
Tsk [tight skin] (gene): mouse, 34, 38*, 63, 76
TS1 (strain): rat, 351, 354-360 (*see also* TMB)
TS3 (strain): rat, 351, 354-360 (*see also* TMD)
T/t mutant stocks: mouse, 53-55, 132-133
Tu [renal tumor] (gene): rat, 258
tu [toe-ulnar] (gene): mouse, 56
Tubular adenomas: mouse, 193 hn-194 hn, 194-196, 197 hn, 198, 199 hn, 199-204
Tubular basement membrane: rat, 335
Tubular lesions: mouse, 26
Tubular lesions: rat, 293 fn
Tubular necrosis: rat, 291
Tubules, kidney: rat, 294
Tubules, seminiferous: mouse, 48
Tubulo-interstitial disease: rat, 335
Tufted (gene): mouse (see *tf*)

Tumbler (gene): mouse (see *tb*)
Tumorigenesis, mammary: mouse, 161
Tumors: mouse, 21-28, 68, 157-211 (*see also* specific tumor)
Tumors: rat, 238-253, 261-264, 267, 289 fn, 294-304
Tumor viruses, mammary: mouse, 157-161
Tunica vaginalis testis tumor: rat, 300
Turpentine: mouse, 23
TW (strain): mouse, 87
Tw [twirler] (gene): mouse, 34, 39*, 59, 152, 156
Twirler (gene): mouse (see *Tw*)
Tyrosinase: mouse, 77, 98
Tyrosine hydroxylase: rat, 360 fn, 362 fn
Tyrosine metabolism: rat, 362
Tyrosine-3-monooxygenase: rat, 360, 362
Tz [tail-zigzagged] (gene): mouse, 56
T6 (strain): mouse, 81
T138 (strain): mouse, 128

U (strain): rat, 251
U [umbrous] (gene): mouse, 73
ub [unbalanced] (gene): mouse, 59, 153
UDPglucuronosyltransferase: rat, 312, 312 fn
Ul [ulnaless] (gene): mouse, 56
Ulnaless (gene): mouse (see *Ul*)
Ultraviolet light: mouse, 176
Ultraviolet radiation: rat, 302
Umbilical abnormalities: mouse, 60 fn, 65
Umbrous (gene): mouse (see *U*)
Un or *un* (gene): mouse
    un [undulated], 34, 38*, 56, 122, 123*
    Un^s [undulated-short-tail], 56
Unbalanced (gene): mouse (see *ub*)
Underwhite (gene): mouse (see *uw*)
Undulated (gene): mouse (see *un*)
Undulated-short-tail (gene): mouse (see Un^s)
ur [urogenital] (gene): mouse, 57
Uranium-238: rat, 302
Urea: rat, 268
Urea nitrogen, blood: rat, 279, 287
Uremia: rat, 321 hn, 325 fn
Ureter tumors: rat, 242, 295, 298
Urethan: mouse, 174-177, 203-204
Urethral tumors, 296 fn
Urinary bladder lesion: rat, 290
Urinary bladder tumors: rat, 242, 295, 298
Urinary calculi: mouse, 28
Urinary proteins: mouse, 105-107
Urinary proteins: rat, 280, 285
Urine: mouse, 24, 79, 105 hn
Urine: rat, 279, 283, 288
Urogenital (gene): mouse (see *ur*)
Urogenital abnormalities: mouse, 59, 68
Urogenital abnormalities: rat, 236, 246
Urogenital syndrome (gene): mouse (see *us*)
Urogenital system development: mouse, 50-52
Uromucoid: mouse, 105 hn
us [urogenital syndrome] (gene): mouse, 56
Uterine horn: rat, 238, 274-277
Uterine tumors: rat, 239, 244, 248, 294, 296 fn, 298-299
Uterus: mouse, 114, 193 hn
Uterus: rat, 274-277, 285
Utricle: mouse, 149
UW (strain): mouse, 87
uw [underwhite] (gene): mouse, 34, 39*, 43, 73

V/Le (strain): mouse, 86
v [waltzer] (gene): mouse, 34, 38*, 43, 153, 156
    v^{df} [deaf] (gene): mouse, 59, 153
Va [varitint-waddler] (gene): mouse, 34, 39*, 73, 153, 156
Vaccine, Group A streptococcal: rat, 318
Vacillans (gene): mouse (see *vc*)
Vacuolated lens (gene): mouse (see *vl*)
Vagina, imperforate: mouse, 26, 48
Vaginal atresia: rat, 241
Vaginal prolapse: rat, 240
Vaginal septa: mouse, 23, 47-49
Vaginal tumors: rat, 242, 296 fn, 298
Variants: mouse nomenclature rules, 12-13
Varitint-waddler (gene): mouse (see *Va*)
Vascular disease: rat, 245
Vascular lesions: rat, 291
Vascular system: rat, 257, 310
Vascular system development: mouse, 50-52
Vascular tumor: mouse, 173
Vasopressin: mouse, 27
Vasopressin: rat, 339
vb [vibrator] (gene): mouse, 34, 39*, 153, 156
vb [vibrissaeless] (gene): rat, 258
vc [vacillans] (gene): mouse, 38*, 153, 156
Ve [velvet coat] (gene): mouse, 34, 39*, 64, 76
Velvet coat (gene): mouse (see *Ve*)
Vena cava tumor: rat, 249
Ventral spinal roots: mouse, 67
Ventricles: rat, 281
Ventricular diastolic pressure: rat, 286
Ventricular dilation: mouse, 151
Ventricular trabeculae carneae: rat, 286
Vermis, cerebellum: mouse, 152
Versene: rat, 267
Vertebrae, presacral: mouse, 21
Vertebral column, abnormal: mouse, 149
Vertebral malformation, thoracolumbar: rat, 245
Vestibulo-spinal tract: mouse, 153
Vestigial tail (gene): mouse (see *vt*)
V_H-*DEX* (gene): mouse, 112
vi [visceral inversion] (gene): mouse, 58
Viable-brindled (gene): mouse (see Mo^{vbr})
Viable dominant spotting (gene): mouse (see W^v)
Viable yellow (gene): mouse (see A^{vy})
Vibrator (gene): mouse (see *vb*: mouse)
Vibrissaeless (gene): rat (see *vb*: rat)
Vinyl chloride: mouse, 176
Viral antigens: mouse, 167
Virally induced tumors: rat, 301-303
Virgin: mouse
    life-span, 45-46
    mammary cancer, 21, 23
    tumors, 175 fn, 177 fn, 182 fn, 190 fn, 201-203
Virgin: rat, 248, 292
Viruses: mouse (*see also* specific virus)
    assay, 230-231
    genes affecting transmittal, 167-168
    origin, 165-166, 228-229
    resistance, 28, 157 hn-158 hn, 159-160
    susceptibility, 17*, 136, 141
    tumor incidence, 157 hn, 158-160, 173, 176-177
Visceral inversion (gene): mouse (see *vi*)
Visceral reduplication: mouse, 150
Visual tract, abnormal: mouse, 150

Vitamin K deficiency: mouse, 24, 27
vl [vacuolated lens] (gene): mouse, 38*, 58, 153
VM (strain): rat, 316
vt [vestigial tail] (gene): mouse, 39*, 56

W (strain): rat (*see also* Wistar)
characteristics, 251
chromosome polymorphisms, 260
coat color, 251, 273
eye color, 273
immunogenetic characterization, 314
immunoglobulin polymorphism, 308
origin, 251
transplant survival, 321-323
tumors, 300, 304
W/Fu, 301, 303-304, 308
W/N, 310-311
W/Not, 301, 304
W [dominant spotting] (gene): mouse
anemia, 66
chromosome position, 38*
leukemia antigens, 168
mutation, 73
phenotypic characteristics, 102
strain distribution, 28, 34-35, 66
W^a [Ames dominant spotting], 66, 73
W^b [Ballantyne's spotting], 66, 73
W^f [W-fertile], 73
W^j [Jay's dominant spotting], 66 fn, 73
W^{pw} [panda-white], 66, 73
W^s [Strong's dominant spotting], 66
W^v [viable dominant spotting], 35, 42, 44, 66, 73
W^x [dominant spotting-x], 35, 66 fn, 73
w [waltzing] (gene): rat, 258-259
WA (strain): rat, 252, 272-273, 316
WA/Lac, 275-276
wa-1 [waved-1] (gene): mouse, 38*, 59, 76
wa-2 [waved-2] (gene): mouse, 35, 39*, 59, 76
WAB (strain): rat, 252, 273
Wabbler-lethal (gene): mouse (see *wl*)
Waddler (gene): mouse (see *wd*)
WAG (strain): rat
alloantigenic alleles, 316-317
behavior, 356-360
blood protein polymorphism, 308-309
characteristics, 252
coat color, 252, 272-273
enzyme polymorphism, 311-312
eye color, 272-273
immunoglobulin polymorphism, 308
immunologically mediated diseases, 335
O_2 consumption, 356
organ weights, 354-355
research use, 351
transplant donor, 332
transplant survival, 322-323, 330
WAG/Lac, 276
WAG/Orl, 310-311
WAG/Rij, 252, 278, 290, 296, 299-300, 314
WAG/RijCub, 310
(WAG x BN)F_1, 322, 325
Walker (strain): rat, 351
Walker rat carcinoma, 301
Walker 256 tumor: rat, 244, 250
Waltzer (gene): mouse (see *v*)
Waltzer-type (gene): mouse (see *Wt*)
Waltzing (gene): rat (see *w*)
War [warfarin resistance] (gene): mouse, 38*
Warfarin resistance (gene): rat (see *Rw*: rat)
Water, tissue: rat, 280, 285
Waved (genes): mouse (see *wa-1; wa-2*)
Waved coat (gene): mouse (see *Wc*)
Wavy coat (gene): mouse (see Re^{wc})
Wavy fur (gene): mouse (see *Wf*)
Wayne-pink-eyed (strain): rat, 263
WB (strain): mouse, 28, 49, 90, 104, 106, 116-118
WB/J, 17*, 82-83, 85
WB/Re
anemia, 65-66
biochemical variation, 82, 86-87, 89-92
cellular alloantigens, 114
complement factors, 103
congenic-inbred line progenitor, 125
histocompatibility typing, 122
mutant genes, 30-32, 34-35
WB/ReCz, 87, 89
WB/ReJ, 81-82
wb [white belly] (gene): rat, 273
WC (strain): mouse, 28, 49, 90, 104, 106
WC/Re
anemia, 65-66
biochemical variation, 82-83, 85-87, 89, 92
genealogy, 17*
histocompatibility typing, 122
mutant gene, 34
Wc [waved coat] (gene): mouse, 39*, 64, 76
wd [waddler] (gene): mouse, 38*, 153, 156
WE/Cpb (strain): rat, 252
we [wellhaarig] (gene): mouse, 35, 38*, 42, 76, 122, 123*
Weakness: mouse, 67, 149
Weaver (gene): mouse (see *wv*)
Weight (*see* Body weight; Organ weights)
Wellhaarig (gene): mouse (see *we*)
WF (strain): rat
alloantigenic alleles, 316-317
avoidance learning, 360
behavioral research, 351
blood protein polymorphism, 309
characteristics, 252-253
coat color, 252, 273
congenic-inbred line progenitor, 256
enzyme polymorphism, 310
eye color, 273
immunogenetic characterization, 314
immunoglobulin polymorphism, 308
immunological responsiveness, 317-318
karyology, 263, 267
origin, 252
reproductive behavior, 360
transplant survival, 322-323, 325-327
tumors, 300-301, 303-304
(WF x BN)F_1, 322-323, 325-327
Wf [wavy fur] (gene): mouse, 76
W-fertile (gene): mouse (see W^f)
WFH (strain): rat, 308
WH (strain): mouse, 17*, 28, 49
WH/Re, 34, 66, 82, 87, 89, 106
wh [writher] (gene): mouse, 153, 156
Wheel activity: rat, 240, 244-245, 247, 251-252
Whirler (gene): mouse (see *wi*)

White (gene): mouse (see Mi^{wh})
White-based brown (gene): mouse (see B^w)
White-bellied agouti (gene): mouse (see A^w)
White belly (gene): rat (see *wb*)
White blood cells (*see* Leukocyte entries)
White fibers: rat, 279, 286-287
White matter, spinal cord: mouse, 149, 153
White matter lesions, brain: mouse, 150
White matter lesions, brain: rat, 294
White-spot (gene): mouse (see mi^{ws})
WHT (strain): mouse, 28
wi [whirler] (gene): mouse, 35, 38*, 59, 153, 156
WIF (strain): rat, 256 (*see also* WF)
Wild mouse (see also *Mus*)
biochemical variation, 81, 83-85
cellular alloantigens, 113
centromeric heterochromatin, 44 hn
classification & identification, 212-220
congenic-inbred line progenitor, 128
enzyme polymorphism, 220 hn, 220-227
heavy chain linkage group, 111
IgC_H allotypic determinants, 109
inbred strain genealogy, 17*
nomenclature rules, 12-13
protein polymorphism, 220 hn, 222, 224-227
Robertsonian translocations, 41
tumors, 160, 172, 177, 195
type C RNA viruses, 227-231
urinary proteins, 105 hn
"WILD"-MTV (virus): mouse, 160
Wild populations: research use, 3
Wild rat, 234, 310, 312, 337
Winner Runway Stock: rat (*see* WRS)
WIS (strain): rat, 329
Wistar (strain): rat (*see also* W)
aging, 285-288, 290 fn, 293-294
hypertensive strain progenitor, 348
immunoglobulin polymorphism, 308
immunologically mediated diseases, 335-336
immunological responsiveness, 318
infectious diseases, 337-338
karyology, 263-264
longevity, 278
origin, 234 hn, 234
reproductive hormones, 341-343, 345-347
transplant survival, 321 fn, 323 fn
tumors, 296-297, 300, 304
Wistar/Mk, 251, 267, 314
Wistar/Ms, 264
Wistar/T, 264
Wistar substrain, 344-345
Wistar A (strain): rat, 338
Wistar Af (strain): rat, 308
Wistar-Af/Han, 300
Wistar AG (strain): rat, 308 (*see also* WAG)
Wistar albino Boots (strain): rat (*see* WAB)
Wistar albino Glaxo (strain): rat (*see* WAG)
Wistar Alderley Park (strain): rat, 278, 288
Wistar B (strain): rat, 337
Wistar BB (strain): rat, 308
Wistar (Charles River) strain: rat, 340-342
Wistar dextran-NR (strain): rat, 318, 334
Wistar dextran-R (strain): rat, 318, 334
Wistar Fu (strain): rat (*see* WF)
Wistar Furth (strain): rat (*see* WF)
Wistar-Holtzman mix (strain): rat, 341-342
Wistar-Kyoto (strain): rat, 348
Wistar L (strain): rat, 302
Wistar-Otago (strain): rat, 348
Wistar outbred (strain): rat, 306, 310, 312
Wistar R (strain): rat, 308
Wistar S (strain): rat, 337
WK (strain): mouse, 28, 49
WK/Re
anemia, 66
biochemical variation, 82-83, 85-87
genealogy, 17*
histocompatibility typing, 122
mutant gene, 34
urinary proteins, 106
WKA (strain): rat
alloantigenic alleles, 316-317
characteristics, 253
chromosome polymorphisms, 260
coat color, 253, 273
congenic-inbred line progenitor, 256
eye color, 273
immunogenetic characterization, 314
immunological responsiveness, 318
karyology, 264, 267
origin, 249 fn, 253
tumors, 303
WKY/N (strain): rat, 253, 273
wl [wabbler-lethal] (gene): mouse, 35, 39*, 153, 156
WLHR/J (strain): mouse, 87
WLHR/Le (strain): mouse, 31, 35, 49
WLL (strain): mouse, 88, 206 hn, 207
WLL/Br, 92
WLL/BrA, 81-84, 86-90, 92, 106, 112
WLL/Br_eBA, 112
WLL_f, 159
WLL_f hybrids, 159
WLL-MTV (virus): mouse, 159
WM (strain): rat, 253, 273
WN (strain): mouse, 49, 83-84, 86-88
WN (strain): rat, 253, 273, 300, 329
WN/N, 300
WN1802 (viruses): mouse, 166
wo [wobbly] (gene): rat, 258
Wobbler (gene): mouse (see *wr*)
Wobbly (gene): rat (see *wo*)
Wound infection resistance: rat, 237
WP (strain): rat, 255, 328
WR (strain): rat, 253, 311, 316
wr [wobbler] (gene): mouse, 67, 153, 156
Writher (gene): mouse (see *wh*)
WRS (strain): rat, 351, 360, 361
Wt [waltzer-type] (gene): mouse, 59, 153, 156
wv [weaver] (gene): mouse, 39*, 58, 153, 156

X/Gf (strain): mouse, 28, 49, 88-89
X-1 (gene): mouse, 118
Xanthine dehydrogenase: mouse, 80, 93, 220 hn
Xanthine dehydrogenase (gene): rat (see *Xdh*)
Xanthine oxidase: mouse, 80, 93
Xce [X-chromosomal controlling element] (gene): mouse, 39*
X chromosome: mouse
banding pattern, 40*
centromeric heterochromatin, 44 hn, 44

gene inheritance, 91 fn
gene loci, 39*
immune response genes, 113
Mus classification & identification, 219
skin & hair mutations, 74-75
translocations, 13
X chromosome: rat, 260-265, 267-268, 268*-270*, 272, 315
XC test, virus: mouse, 228 hn, 228-229
Xdh [xanthine dehydrogenase] (gene): rat, 305
Xenogeneic skin grafts: mouse, 69
Xenotropic viruses: mouse, 227 hn, 230 hn, 230-231
Xid [X-linked immunodeficiency] (gene): mouse, 69
XLII (strain): mouse, 28, 195, 200
XLII/Orl, 84, 86-89, 92
X-linked defect: mouse, 22
X-linked immunodeficiency (gene): mouse (see *Xid*)
xn [exencephaly] (gene): mouse, 58
X rays; X-radiation: mouse
adrenal cortical tumors, 180
ovarian tumors, 199-200, 204
pituitary tumors, 188-189, 192
resistance, 23, 25-28
X rays: rat, 252, 302
Xt [extra-toes] (gene): mouse, 35, 39*, 58, 60 fn, 153
Xt^{bph} [brachyphalangy], 58
XVII (strain): mouse, 28, 88
XVII/Rd, 81-82, 89
X zone, adrenal: mouse, 27

Y (strain): mouse, 159, 206 hn, 207
YBR (strain): mouse, 16*, 28, 92, 111, 113-114, 124
YBR/Cv, 81, 83, 88-90, 92
YBR/HaCv, 81-82
YBR/He, 83, 90, 103, 107
YBR/HeWiHa, 92
YBR/WiHa, 82
YBR congenics, 107
YBR hybrids, 164, 172
Y chromosome: mouse, 13, 40*, 44 hn, 48, 114
Y chromosome: rat, 252, 261-267, 268*-270*, 272
Yellow (gene): mouse (see A^y)
Yellow (gene): rat (see *e*: rat)
Yellow mottling (gene): mouse (see *Ym*)
Yersinia pestis: rat, 338
Ym [yellow mottling] (gene): mouse, 64, 73
YO (strain): rat, 253, 317
Yolk sac blood islands: mouse, 66
YOS (strain): rat, 264, 308 (*see also* DONRYU)
(YOS x Wistar/Ms)F_1, 264
Yoshida (strain): rat (*see* YOS)
Yoshida tumors: rat, 243-244, 248, 251, 264-265, 267, 301
YO38366 (strain): rat, 256, 314, 316-318
YS (strain): mouse, 88
YS/Wf, 81-82, 87, 89
Y59 (strain): rat, 253, 308, 316

Z (strain): mouse, 178-179 (*see also* C3H/Bi)
ZAF_1 (strain): mouse, 179 (*see also* (C3H x A)F_1)
Zb (strain): mouse, 182
ZBC (strain): mouse, 179
Zimmerman (strain): rat, 308, 338
Zona glomerulosa: mouse, 27
Zucker (strain): rat, 237, 339
Zygomatic plate: mouse, 212, 214, 217, 219
Z61 (strain): rat, 234, 253, 272-273, 317

27HB/Rl (strain): mouse, 105
86 HB/Rl (strain): mouse, 105
101 (strain): mouse
antigens, 104, 168
biochemical variation, 90
characteristics, 28
genealogy, 16*
genes, 28
H-2 haplotype & markers, 124
ovarian tumors, 195
101/H, 49, 86
101/Rl, 81-82, 89
(101 x C3H)F_1, 196, 201
129 (strain): mouse
antigens, 114, 116-118, 167-168
centromeric heterochromatin, 44
characteristics, 28
complement factors, 103-104
genes, 28, 30
H-2 haplotype & markers, 123
Ig heavy chain linkage groups, 110
proteins, 103-104
129/J
biochemical variation, 81-92
body weight, 46
complement factors, 103
congenic-inbred line progenitor, 115
genealogy, 16*
leukemia antigens, 167-168
reproduction, 46
tumors, 207, 211
urinary proteins, 106
129/Re, 207
129/ReJ, 16*, 31, 82-83, 85-86, 91
129/ReWl, 91
129/Rr, 31, 88, 207
129/RrJ, 16*, 89, 172
129/RrLac, 85
129/Sn, 122
129/Sv
anemia, 65-66
biochemical variation, 86, 90
characteristics, 28
leukemia antigens, 168
mutant genes, 30, 33-34
testicular teratomas, 207
129/SvPas, 37
129/Sv congenics, 207-208
129XB (strain): mouse, 37
352HB/R1 (strain): mouse, 105

7407797
3 1378 00740 7797